**Student Solutions Manual**

# Mathematical Applications for the Management, Life, and Social Sciences

**ELEVENTH EDITION**

**Ronald J. Harshbarger**
University of South Carolina, Beaufort

**James J. Reynolds**
Clarion University of Pennsylvania



Prepared by

**Scott E. Barnett**
Henry Ford College

CENGAGE Learning

Australia • Brazil • Mexico • Singapore • United Kingdom • United States

CENGAGE
Learning



ISBN: 978-1-305-10806-6

**Cengage Learning**
20 Channel Center Street
Boston, MA 02210
USA

Cengage Learning is a leading provider of customized learning solutions with office locations around the globe, including Singapore, the United Kingdom, Australia, Mexico, Brazil, and Japan. Locate your local office at: **www.cengage.com/global**.

Cengage Learning products are represented in Canada by Nelson Education, Ltd.

To learn more about Cengage Learning Solutions, visit **www.cengage.com**.

Purchase any of our products at your local college store or at our preferred online store **www.cengagebrain.com**.

Printed in the United States of America
Print Number: 01 Print Year: 2014

# Table of Contents

# Chapter 0: Algebraic Concepts

## *Exercises 0.1*

**1.** $12 \in \{1, 2, 3, 4, \ldots\}$

**3.** $6 \notin \{1, 2, 3, 4, 5\}$

**5.** $\{1, 2, 3, 4, 5, 6, 7\}$

**7.** $\{x$: $x$ is a natural number greater than 2 and less than 8$\}$

**9.** $\varnothing \subseteq A$ since $\varnothing$ is a subset of every set. $A \subseteq B$ since every element of $A$ is an element of $B$. $B \subseteq B$ since a set is always a subset of itself.

**11.** No. $c \in A$ but $c \notin B$.

**13.** $D \subseteq C$ since every element of $D$ is an element of $C$.

**15.** $A \subseteq B$ and $B \subseteq A$. (Also $A = B$.)

**17.** Yes. $A \subseteq B$ and $B \subseteq A$. Thus, $A = B$.

**19.** No. $D \neq E$ because $4 \in E$ and $4 \notin D$.

**21.** $A$ and $B$ are disjoint since they have no elements in common. $B$ and $D$ are disjoint since they have no elements in common. $C$ and $D$ are disjoint.

**23.** $A \cap B = \{4, 6\}$ since 4 and 6 are elements of each set.

**25.** $A \cap B = \varnothing$ since they have no common elements.

**27.** $A \cup B = \{1, 2, 3, 4, 5\}$

**29.** $A \cup B = \{1, 2, 3, 4\}$ or $A \cup B = B$.

**For problems 31 - 41, we have $U = \{1, 2, 3, \ldots, 9, 10\}$.**

**31.** $A' = \{4, 6, 9, 10\}$ since these are the only elements in $U$ that are not elements of $A$.

**33.** $B' = \{1, 2, 5, 6, 7, 9\}$
$A \cap B' = \{1, 2, 5, 7\}$

**35.** $A \cup B = \{1, 2, 3, 4, 5, 7, 8, 10\}$
$(A \cup B)' = \{6, 9\}$

**37.** $A' = \{4, 6, 9, 10\}$
$B' = \{1, 2, 5, 6, 7, 9\}$
$A' \cup B' = \{1, 2, 4, 5, 6, 7, 9, 10\}$

**39.** $B' = \{1, 2, 5, 6, 7, 9\}$
$C' = \{1, 3, 5, 7, 9\}$
$A \cap B' = \{1, 2, 3, 5, 7, 8\} \cap \{1, 2, 5, 6, 7, 9\}$
$= \{1, 2, 5, 7\}$
$(A \cap B') \cup C' = \{1, 2, 3, 5, 7, 9\}$

**41.** $B' = \{1, 2, 5, 6, 7, 9\}$
$A \cap B' = \{1, 2, 3, 5, 7, 8\} \cap \{1, 2, 5, 6, 7, 9\}$
$= \{1, 2, 5, 7\}$
$(A \cap B')' \cap C = \{3, 4, 6, 8, 9, 10\} \cap \{2, 4, 6, 8, 10\}$
$= \{4, 6, 8, 10\}$

**For problems 43 and 45, we have $U = \{1, 2, 3, \ldots, 8, 9\}$.**

**43.** $A - B = \{1, 3, 7, 9\} - \{3, 5, 8, 9\} = \{1, 7\}$

**45.** $A - B = \{2, 1, 5\} - \{1, 2, 3, 4, 5, 6\} = \varnothing$ or $\{\ \}$

**47.** **a.** $L = \{2000, 2001, 2004, 2005, 2006, 2007, 2010, 2011, 2012\}$
$H = \{2000, 2001, 2006, 2007, 2008, 2010, 2011, 2012\}$
$C = \{2001, 2002, 2003, 2008, 2009\}$

**b.** no

**c.** $C'$ is the set of all years when the percentage change from low to high was 35% or less.

**d.** $H' = \{2002, 2003, 2004, 2005, 2009\}$
$C' = \{2000, 2004, 2005, 2006, 2007, 2010, 2011, 2012\}$
$H' \cup C' = \{2000, 2002, 2003, 2004, 2005, 2006, 2007, 2009, 2010, 2011, 2012\}$. $H' \cup C'$ is the set of years when the high was less than or equal to 11,000 or the percent change was less than or equal to 35%.

e. $L' = \{2002, 2003, 2008, 2009\}$
$L' \cap C = \{2002, 2003, 2008, 2009\}$.
$L' \cap C$ is the set of years when the low was less than or equal to 8,000 and the percent change was more than 35%.

**49.** **a.** From the table, there are 100 white Republicans and 30 non-white Republicans who favor national health care, for a total of 130.
**b.** From the table, there are 350 + 40 Republicans, and 250 + 200 Democrats who favor national health care, for a total of 840.
**c.** From the table, there are 350 white Republicans, and 150 white Democrats and 20 non-whites who oppose national health care, for a total of 520.

**51.** **a.** The key to solving this problem is to work from "the inside out". There are 40 aides in $E \cap F$. This leaves 65 – 40 = 25 aides who speak English but do not speak French. Also we have 60 – 40 = 20 aides who speak French but do not speak English. Thus there are 40 + 25 + 20 = 85 aides who speak English or French. This means there are 15 aides who do not speak English or French.

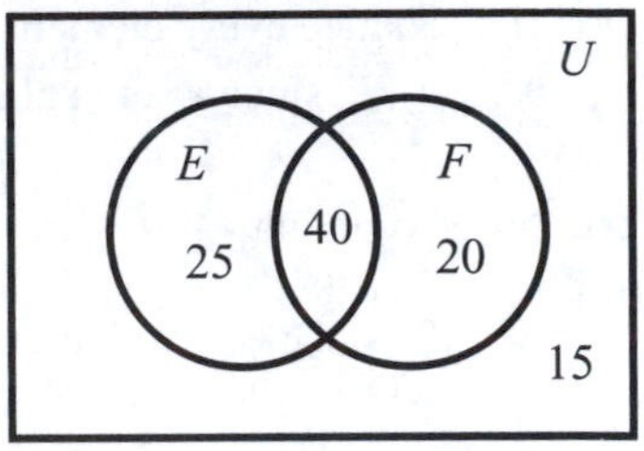


**b.** From the Venn diagram $E \cap F$ has 40 aides.
**c.** From the Venn diagram $E \cup F$ has 85 aides.
**d.** From the Venn diagram $E \cap F'$ has 25 aides.

**53.** Since 12 students take $M$ and $E$ but not $FA$, and 15 take $M$ and $E$, 3 take all three classes. Since 9 students take $M$ and $FA$ and we have already counted 3, there are 6 taking $M$ and $FA$ which are not taking $E$. Since 4 students take $E$ and $FA$ and we have already counted 3, there is only 1 taking $E$ and $FA$ but not taking $M$ also. Since 20 students take $E$ and we already have 16 enrolled in $E$, this leaves 4 taking only $E$. Since 42 students take $FA$ and we already have 10 enrolled in $FA$, this leaves 32 taking only $FA$. Since 38 students take $M$ and we already have 21 enrolled in $M$, this leaves 17 taking only $M$.

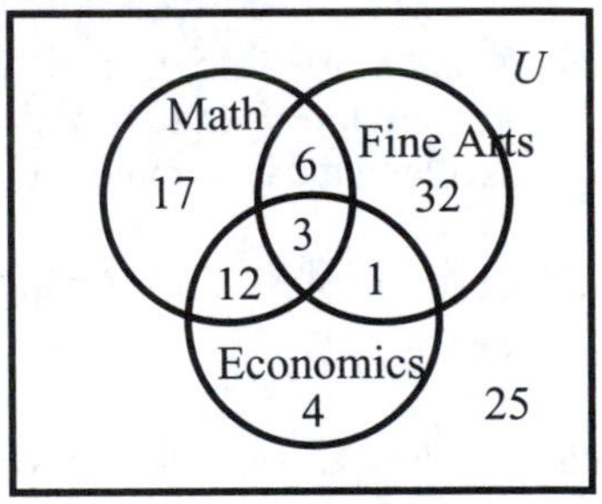


**a.** In the union of the 3 courses we have 17 + 12 + 3 + 6 + 32 + 1 + 4 = 75 students enrolled. Thus, there are 100 – 75 = 25 students who are not enrolled in any of these courses.
**b.** In $M \cup E$ we have 17 + 12 + 3 + 6 + 1 + 4 = 43 enrolled.
**c.** We have 17 + 32 + 4 = 53 students enrolled in exactly one of the courses.

**55.** **a.** and **b.**

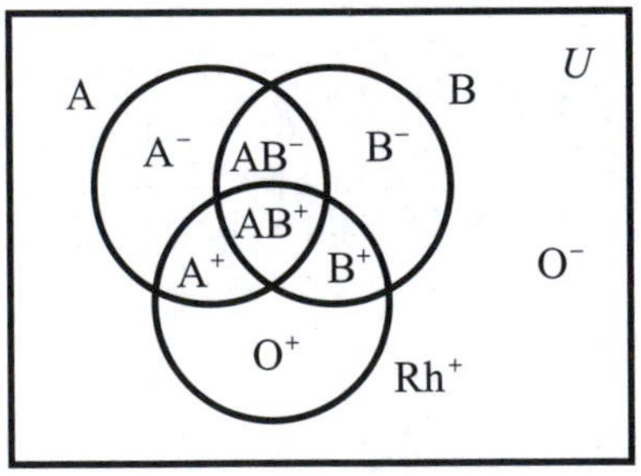


**c.** $A^+$ : 34%; $B^+$ : 9%; $O^+$ : 38%; $AB^+$ : 3%; $O^-$ : 7%; $A^-$ : 6%; $B^-$ : 2%; $AB^-$ : 1%

## *Exercises 0.2*

**1.** **a.** Note that $-\frac{\pi}{10} = \pi \cdot \left(-\frac{1}{10}\right)$, where $\pi$ is irrational and $-\frac{1}{10}$ is rational. The product of a rational number other than 0 and an irrational number is an irrational number.
**b.** –9 is rational and an integer.
**c.** $\frac{9}{3} = \frac{3}{1} = 3$. This is a natural number, an integer, and a rational number.
**d.** Division by zero is meaningless.

**3.** **a.** Commutative
**b.** Distributive
**c.** Associative
**d.** Additive Identity

**5.** $-6 < 0$

**7.** $-14 < -3$

**9.** $0.333 < \frac{1}{3}\left(\frac{1}{3} = 0.3333\cdots\right)$

**11.** $|-3| + |5| > |-3+5|$

**13.** $-3^2 + 10 \cdot 2 = -3^2 + 20 = -9 + 20 = 11$

**15.** $\frac{4+2^2}{2} = \frac{4+4}{2} = \frac{8}{2} = 4$

**17.** $\frac{16-(-4)}{8-(-2)} = \frac{16+4}{8+2} = \frac{20}{10} = 2$

**19.** $\frac{|5-2|-|-7|}{|5-2|} = \frac{|3|-|-7|}{|3|} = \frac{3-7}{3} = -\frac{4}{3}$

**21.** $\frac{(-3)^2 - 2\cdot 3 + 6}{4 - 2^2 + 3} = \frac{9-6+6}{4-4+3} = \frac{9}{3} = 3$

**23.** $\frac{-4^2 + 5 - 2\cdot 3}{5 - 4^2} = \frac{-16+5-6}{5-16} = \frac{-17}{-11} = \frac{17}{11}$

**25.** The entire line

**27.** (1, 3]; half-open interval

**29.** (2, 10); open interval

**31.** $-3 \le x < 5$

**33.** $x > 4$

**35.** $(-\infty, 4) \cap (-3, \infty) = (-3, 4)$

**37.** $x > 4$ and $x \ge 0$ = $(4, \infty)$

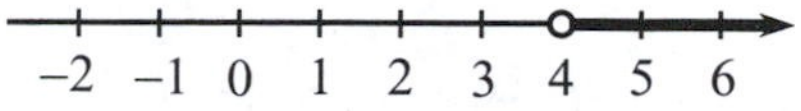


**39.** $[0, \infty) \cup [-1, 5] = [-1, \infty)$

**41.** $(-\infty, 0) \cup (7, \infty)$

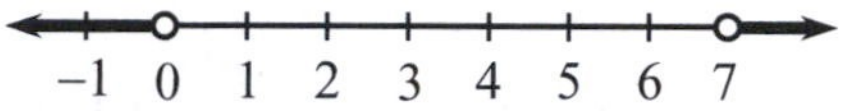


**43.** –0.000038585

**45.** 9122.387471

**47.** $\frac{2500}{[(1.1^6)-1]} = \frac{2500}{0.771561} = 3240.184509$

Equation (2) gives
$y = 0.00454(13)^2 + 0.126(13) + 0.271$
$= 2.68$ billion

**49.** **a.** $\$300.00 + \$788.91 = \$1088.91$
**b.** Federal withholding
$= 0.25(1088.91 - 54.45) = \$258.62$
**c.** Retirement: $0.05(1088.91) = \$54.45$
State tax = Retirement = \$54.45
Local tax = $0.01(1088.91) = \$10.89$
Federal tax = \$258.62 (from **b.** above)
Social Security and Medicare tax
$= 0.0765(1088.91) = \$83.30$
Total Withholding = \$461.71
Take-home = $1088.91 - 461.71 = \$627.20$

**51.** **a.** Equation (1) is more accurate.
Equation (1) gives
$y = 0.207(13) - 0.000370$
$= 2.69$ billion
Equation (2) gives
$y = 0.00454(13)^2 + 0.126(13) + 0.271$
$= 2.68$ billion

**b.** For 2018, Equation (1) gives
$y = 0.207(18) - 0.000370$
$= 3.73$ billion

For 2018, Equation (2) gives
$y = 0.00454(18)^2 + 0.126(18) + 0.271$
$= 4.01$ billion

**53. a.** $\$82,401 \le I \le 171,850;$
$\$171,851 \le I \le \$373,650;$
$I > \$373,650$

**b.** $T = \$4681.25$ for $I = \$34,000$
$T = \$16,781.25$ for $I = \$82,400$

**c.** $[4681.25,\ 16,781.25]$

## *Exercises 0.3*

**1.** $(-4)^4 = (-4)(-4)(-4)(-4) = 256$

**3.** $-2^6 = -1\cdot 2\cdot 2\cdot 2\cdot 2\cdot 2\cdot 2 = -64$

**5.** $3^{-2} = \dfrac{1}{3^2} = \dfrac{1}{9}$

**7.** $-\left(\dfrac{3}{2}\right)^2 = (-1)\left(\dfrac{3}{2}\right)\left(\dfrac{3}{2}\right) = -\dfrac{9}{4}$

**9.** 1.2 $\boxed{y^x}$ 4 $\boxed{=}$ 2.0736

**11.** 1.5 $\boxed{y^x}$ −5 $\boxed{=}$ 0.1316872428

**13.** $6^5 \cdot 6^3 = 6^{5+3} = 6^8$

**15.** $\dfrac{10^8}{10^9} = 10^{8-9} = 10^{-1} = \dfrac{1}{10}$

**17.** $\dfrac{9^4 \cdot 9^{-7}}{9^{-3}} = \dfrac{9^{4+(-7)}}{9^{-3}} = \dfrac{9^{-3}}{9^{-3}} = 9^{-3-(-3)} = 9^0 = 1$

**19.** $\left(3^3\right)^3 = 3^{3\cdot 3} = 3^9$

**21.** $\left(\dfrac{2}{3}\right)^{-2} = \left(\dfrac{3}{2}\right)^2 = \dfrac{9}{4}$

**23.** $-x^{-3} = -1\cdot x^{-3} = -1\cdot \dfrac{1}{x^3} = -\dfrac{1}{x^3}$

**25.** $xy^{-2}z^0 = x\cdot \dfrac{1}{y^2}\cdot 1 = \dfrac{x}{y^2}$

**27.** $x^3 \cdot x^4 = x^{3+4} = x^7$

**29.** $x^{-5}\cdot x^3 = x^{-5+3} = x^{-2} = \dfrac{1}{x^2}$

**31.** $\dfrac{x^8}{x^4} = x^{8-4} = x^4$

**33.** $\dfrac{y^5}{y^{-7}} = y^{5-(-7)} = y^{12}$

**35.** $\left(x^4\right)^3 = x^{3\cdot 4} = x^{12}$

**37.** $(xy)^2 = x^2y^2$

**39.** $\left(\dfrac{2}{x^5}\right)^4 = \dfrac{2^4}{\left(x^5\right)^4} = \dfrac{16}{x^{5\cdot 4}} = \dfrac{16}{x^{20}}$

**41.** $\left(2x^{-2}y\right)^{-4} = 2^{-4}x^8y^{-4} = \dfrac{x^8}{16y^4}$

**43.** $\left(-8a^{-3}b^2\right)\left(2a^5b^{-4}\right) = -16a^{-3+5}b^{2-4}$
$= -16a^2b^{-2}$
$= -\dfrac{16a^2}{b^2}$

**45.** $\left(2x^{-2}\right)\div\left(x^{-1}y^2\right) = \dfrac{2}{x^2}\div\dfrac{y^2}{x} = \dfrac{2}{x^2}\cdot\dfrac{x}{y^2} = \dfrac{2}{xy^2}$

**47.** $\left(\dfrac{x^3}{y^{-2}}\right)^{-3} = \dfrac{x^{-9}}{y^6} = \dfrac{1}{x^9}\cdot\dfrac{1}{y^6} = \dfrac{1}{x^9y^6}$

**49.** $\left(\dfrac{a^{-2}b^{-1}c^{-4}}{a^4b^{-3}c^0}\right)^{-3} = \left(\dfrac{b^2}{a^6c^4}\right)^{-3} = \left(\dfrac{a^6c^4}{b^2}\right)^3 = \dfrac{a^{18}c^{12}}{b^6}$

**51. a.** $\frac{2x^{-2}}{(2x)^2} = 2\cdot\frac{1}{x^2}\cdot\frac{1}{(2x)^2} = 2\cdot\frac{1}{x^2}\cdot\frac{1}{4x^2} = \frac{1}{2x^4}$

**b.** $\frac{(2x)^{-2}}{(2x)^2} = \frac{1}{(2x)^2}\cdot\frac{1}{(2x)^2} = \frac{1}{4x^2}\cdot\frac{1}{4x^2} = \frac{1}{16x^4}$

**c.** $\frac{2x^{-2}}{2x^2} = 2\cdot\frac{1}{x^2}\cdot\frac{1}{2x^2} = \frac{1}{x^4}$

**d.** $\frac{2x^{-2}}{(2x)^{-2}} = 2\cdot\frac{1}{x^2}\cdot(2x)^2 = 2\cdot\frac{1}{x^2}\cdot 4x^2 = 8$

**53.** $\frac{1}{x} = x^{-1}$

**55.** $(2x)^3 = 2^3x^3 = 8x^3$

**57.** $\frac{1}{(4x^2)} = \frac{1}{4}\cdot\frac{1}{x^2} = \frac{1}{4}x^{-2}$

**59.** $\left(\frac{-x}{2}\right)^3 = \frac{-x^3}{2^3} = -\frac{1}{8}x^3$

**61.** $P = 1200,\ i = 0.12,\ n = 5$

$S = P(1+i)^n$
$= 1200(1+0.12)^5$
$= 1200(1.12)^5$
$= \$2114.81$

$I = S - P = 2114.81 - 1200 = \$914.81$

**63.** $P = 5000,\ i = 0.115,\ n = 6$

$S = P(1+i)^n$
$= 5000(1+0.115)^6$
$= 5000(1.115)^6$
$= \$9607.70$

$I = S - P = 9607.70 - 5000 = \$4607.70$

**65.** $S = 15{,}000,\ n = 6,\ i = 0.115$

$P = S(1+i)^{-n}$
$= 15{,}000(1+0.115)^{-6}$
$= 15{,}000(1.115)^{-6}$
$= \$7806.24$

**67.** $I = 492.4(1.070)^t$

| | Year | 1980 | 2000 | 2008 |
|---|---|---|---|---|
| **a.** | $t$-value | 20 | 40 | 48 |
| **b.** | Income (in billions) | \$1905 | \$7373 | \$12,669 |

**c.** $I = 492.4(1.070)^{58} \approx \$24{,}922$ billion

**69.** $y = \frac{1095}{1+10.12(1.212)^{-t}}$

**a.**

| Year | 1990 | 2003 | 2012 |
|---|---|---|---|
| $t$ – value | 10 | 23 | 32 |
| Predicted number of endangered species | 442 | 976 | 1072 |

**b.** Year 2020: $t = 40$; $y = \frac{1095}{1+10.12(1.212)^{-40}} \approx 1090$

Increase between 2007 and 2020 is $1090 - 1037 = 53$ species

**c.** Two possibilities might be more environmental protections and the fact that there are only a limited number of species.

**d.** There are only a limited number of species. Also, below some threshold level the ecological balance might be lost, perhaps resulting in an environmental catastrophe (which the equation could not predict). To find the upper limit, which is 1095, compute $y$ for large $t$-values:

| Year | 2040 | 2100 | 2200 |
|---|---|---|---|
| $t-$value | 60 | 120 | 220 |
| Predicted number of endangered species | 1094.9 | 1095 | 1095 |

**71.** $H = 738.1(1.065)^t$

**a.** $t = 10$ corresponds to 2000.

**b.** $H = 738.1(1.065)^{10} \approx \$1385.5$ billion

**c.** $H = 738.1(1.065)^{20} \approx \$2600.8$ billion

**d.** $H = 738.1(1.065)^{28} \approx \$4304.3$ billion

## *Exercises 0.4*

**1. a.** Since $\left(\frac{16}{3}\right)^2 = \frac{256}{9}$ we have

$$\sqrt{\frac{256}{9}} = \frac{16}{3} \approx 5.33$$

**b.** $\sqrt{1.44} = 1.2$

**3. a.** $16^{3/4} = \left(\sqrt[4]{16}\right)^3 = 2^3 = 8$

**b.** $(-16)^{-3/2} = \left(\sqrt{-16}\right)^{-3}$ The square root of a negative number is not real.

**5.** $\left(\frac{8}{27}\right)^{-2/3} = \left(\frac{27}{8}\right)^{2/3} = \left(\sqrt[3]{\frac{27}{8}}\right)^2 = \left(\frac{3}{2}\right)^2 = \frac{9}{4}$

**7. a.** $64^{2/3} = \left(\sqrt[3]{64}\right)^2 = 4^2 = 16$

**b.** $(-64)^{-2/3} = \frac{1}{(-64)^{2/3}} = \frac{1}{\left(\sqrt[3]{-64}\right)^2}$

$$= \frac{1}{(-4)^2} = \frac{1}{16}$$

**9.** $\sqrt[9]{(6.12)^4} = (6.12)^{4/9} \approx 2.237$

**11.** $\sqrt{m^3} = m^{3/2}$

**13.** $\sqrt[4]{m^2n^5} = \left(m^2n^5\right)^{1/4} = m^{2/4}n^{5/4} = m^{1/2}n^{5/4}$

**15.** $2x^{\frac{1}{2}} = 2\sqrt{x}$

**17.** $x^{7/6} = \sqrt[6]{x^7}$

**19.** $-\left(\frac{1}{4}\right)x^{-5/4} = -\frac{1}{4}\cdot\frac{1}{x^{5/4}} = \frac{-1}{4\sqrt[4]{x^5}}$

**21.** $y^{1/4}\cdot y^{1/2} = y^{(1/4)+(1/2)} = y^{3/4}$

**23.** $z^{3/4}\cdot z^4 = z^{(3/4)+(16/4)} = z^{19/4}$

**25.** $y^{-3/2}\cdot y^{-1} = y^{(-3/2)-(2/2)} = y^{-5/2} = \frac{1}{y^{5/2}}$

**27.** $\frac{x^{\frac{1}{3}}}{x^{\frac{-2}{3}}} = x^{\left(\frac{1}{3}\right)-\left(\frac{-2}{3}\right)} = x^{\frac{3}{3}} = x$

**29.**

$$\frac{y^{-5/2}}{y^{-2/5}} = y^{(-5/2)-(-2/5)} = y^{(-25/10)+(4/10)}$$

$$= y^{-21/10} = \frac{1}{y^{21/10}}$$

**31.** $(x^{2/3})^{3/4} = x^{(2/3)(3/4)} = x^{2/4} = x^{1/2}$

**33.** $(x^{-1/2})^2 = x^{-1} = \dfrac{1}{x}$

**35.** $\sqrt{64x^4} = 8x^2$

**37.** $\sqrt{128x^4y^5} = \sqrt{64x^4y^4 \cdot 2y}$
$= \sqrt{64} \cdot \sqrt{x^4} \cdot \sqrt{y^4} \cdot \sqrt{2y} = 8x^2y^2\sqrt{2y}$

**39.** $\sqrt[3]{40x^8y^5} = \sqrt[3]{8x^6y^3 \cdot 5x^2y^2}$
$= \sqrt[3]{8} \cdot \sqrt[3]{x^6} \cdot \sqrt[3]{y^3} \cdot \sqrt[3]{5x^2y^2}$
$= 2x^2y\sqrt[3]{5x^2y^2}$

**41.** $\sqrt{12x^3y} \cdot \sqrt{3x^2y} = \sqrt{36x^5y^2} = \sqrt{36} \cdot \sqrt{x^5} \cdot \sqrt{y^2}$
$= 6x^2y\sqrt{x}$

**43.** $\sqrt{63x^5y^3} \cdot \sqrt{28x^2y} = \sqrt{9x^4y^2 \cdot 7xy} \cdot \sqrt{4x^2 \cdot 7y}$
$= 3x^2y\sqrt{7xy} \cdot 2x\sqrt{7y}$
$= 42x^3y^2\sqrt{x}$

**45.** $\dfrac{\sqrt{12x^3y^{12}}}{\sqrt{27xy^2}} = \sqrt{\dfrac{4x^2y^{10}}{9}} = \dfrac{2xy^5}{3}$

**47.** $\dfrac{\sqrt[4]{32a^9b^5}}{\sqrt[4]{162a^{17}}} = \sqrt[4]{\dfrac{16b^4}{81a^8} \cdot \dfrac{b}{1}} = \dfrac{2b}{3a^2}\sqrt[4]{b}$

**49.** $(A^9)^x = A^{9x}$
$A^{9x} = A^1$
$9x = 1$
$x = \dfrac{1}{9}$

**51.** $\left(\sqrt[7]{R}\right)^x = R^{x/7}$
$R^{x/7} = R^1$
$\dfrac{x}{7} = 1$
$x = 7$

**53.** $\sqrt{\dfrac{2}{3}} \cdot \dfrac{\sqrt{3}}{\sqrt{3}} = \dfrac{\sqrt{2} \cdot \sqrt{3}}{\sqrt{3} \cdot \sqrt{3}} = \dfrac{\sqrt{6}}{3}$

**55.** $\dfrac{\sqrt{m^2x}}{\sqrt{mx^2}} = \dfrac{\sqrt{m}}{\sqrt{x}} = \dfrac{\sqrt{m} \cdot \sqrt{x}}{\sqrt{x} \cdot \sqrt{x}} = \dfrac{\sqrt{mx}}{x}$

**57.**
$\dfrac{\sqrt[3]{m^2x}}{\sqrt[3]{mx^5}} = \dfrac{\sqrt[3]{m}}{\sqrt[3]{x^4}} = \dfrac{\sqrt[3]{m}}{\sqrt[3]{x^3} \cdot \sqrt[3]{x}} \cdot \dfrac{\sqrt[3]{x^2}}{\sqrt[3]{x^2}} = \dfrac{\sqrt[3]{mx^2}}{x\sqrt[3]{x^3}}$
$= \dfrac{\sqrt[3]{mx^2}}{x^2}$

**59.** $\dfrac{-2}{3\sqrt[3]{x^2}} = \dfrac{-2}{3} \cdot \dfrac{1}{x^{2/3}} = -\dfrac{2}{3}x^{-2/3}$

**61.** $3x\sqrt{x} = 3x \cdot x^{1/2} = 3x^{3/2}$

**63.** $\dfrac{3}{2}x^{1/2} = \dfrac{3}{2}\sqrt{x}$

**65.** $\dfrac{1}{2}x^{-1/2} = \dfrac{1}{2} \cdot \dfrac{1}{x^{1/2}} = \dfrac{1}{2\sqrt{x}}$

**67. a.** $R = 8.5 = \dfrac{17}{2}$ $\quad I = 10^{17/2} = \sqrt{10^{17}}$

**b.** $I = 10^{9.0} = 1{,}000{,}000{,}000$

**c.** $\dfrac{I_{2011}}{I_{1989}} = \dfrac{10^{9.0}}{10^{6.9}} = 10^{2.1} \approx 125.9$

**69. a.** $S = 1000\sqrt{\left(1 + \dfrac{r}{100}\right)^5}$

**b.** $S = 1000\sqrt{\left(1 + \dfrac{6.6}{100}\right)^5} \approx \$1173.26$

**71. a.** $P = 0.924t^{13/100} = 0.924\sqrt[100]{t^{13}}$

**b.**

| Year | $t$ | Population |
|---|---|---|
| 2005 | 5 | 1.1390 |
| 2010 | 10 | 1.2464 |
| 2045 | 45 | 1.5156 |
| 2050 | 50 | 1.5365 |

Change from 2005 to 2010: 0.1074 billion

Change from 2045 to 2050: 0.0209 billion

By 2045 and 2050 the population is much larger than earlier in the 21st century, and there is a limited number of people that any land can support—in terms of both space and food.

**73.** $k = 25, t = 10,\ q_0 = 98$

$q = q_0(2^{-t/k})$

$= 98(2^{-10/25})$

$= 98(2^{-2/5}) \approx 74$ kg

**75.** $P = P_0(2.5)^{ht} = 30,000(2.5)^{0.03(10)}$

$= 30,000(2.5)^{0.3} \approx 39,491$

**77. a.** $N = 500(0.02)^{(0.7)^t}$; at $t = 0$ we have $(0.7)^0 = 1$. Thus, $N = 500(0.02)^1 = 10$.

**b.** $N = 500(0.02)^{(0.7)^5}$

$= 500(0.02)^{0.16807}$

$= 259$

## *Exercises 0.5*

**1.** $10 - 3x - x^2$

**a.** The largest exponent is 2. The degree of the polynomial is 2.

**b.** The coefficient of $x^2$ is $-1$.

**c.** The constant term is 10.

**d.** It is a polynomial of one variable $x$.

**3.** $7x^2y - 14xy^3z$

**a.** The sum of the exponents in each term is 3 and 5, respectively. The degree of the polynomial is 5.

**b.** The coefficient of $xy^3z$ is $-14$.

**c.** The constant term is zero.

**d.** It is a polynomial of several (three) variables: $x$, $y$, and $z$.

**5.** $2x^5 - 3x^2 - 5$

**a.** $a_n x^n$ means $2 = a_5$.

**b.** $a_3 = 0$ (Term is $0x^3$)

**c.** $-3 = a_2$

**d.** $a_0 = -5$, the constant term.

**7.** $4x - x^2$

When $x = -2$,

$4x - x^2 = 4(-2) - (-2)^2$

$= -8 - 4$

$= -12.$

**9.** $10xy - 4(x - y)^2$

When $x = 5$ and $y = -2$,

$10xy - 4(x - y)^2 = 10(5)(-2) - 4(5 - (-2))^2$

$= -100 - 196$

$= -296.$

**11.** $\dfrac{2x - y}{x^2 - 2y}$

When $x = -5$ and $y = -3$,

$\dfrac{2x - y}{x^2 - 2y} = \dfrac{2(-5) - (-3)}{(-5)^2 - 2(-3)} = \dfrac{-10 + 3}{25 + 6} = -\dfrac{7}{31}.$

**13.** $1.98T-1.09(1-H)(T-58)-56.8$
$=1.98(74.7)-1.09(1-0.80)(74.7-58)-56.8$
$=147.906-3.6406-56.8=87.4654$

**15.** $(16pq-7p^2)+(5pq+5p^2)=21pq-2p^2$

**17.** $(4m^2-3n^2+5)-(3m^2+4n^2+8)$
$=4m^2-3n^2+5-3m^2-4n^2-8$
$=(4m^2-3m^2)-(3n^2+4n^2)+5-8$
$=m^2-7n^2-3$

**19.** $-[8-4(q+5)+q]=-[8-4q-20+q]$
$=-[-12-3q]$
$=12+3q$

**21.** $x^2-[x-(x^2-1)+1-(1-x^2)]+x$
$=x^2-[x-x^2+1+1-1+x^2]+x$
$=x^2-(x+1)+x$
$=x^2-x-1+x$
$=x^2-1$

**23.** $(5x^3)(7x^2)=35x^{3+2}=35x^5$

**25.** $(39r^3s^2)\div(13r^2s)=3r^{3-2}s^{2-1}=3rs$

**27.** $ax^2(2x^2+ax+ab)=2ax^4+a^2x^3+a^2bx^2$

**29.** $(3y+4)(2y-3)=6y^2-9y+8y-12$
$=6y^2-y-12$

**31.** $6(1-2x^2)(2-x^2)$
$=6(2-x^2-4x^2+2x^4)$
$=6(2-5x^2+2x^4)$
$=12x^4-30x^2+12$

**33.** $(4x+3)^2=16x^2+2(4x)(3)+9=16x^2+24x+9$

**35.** $(0.1-4x)(0.1+4x)=(0.1)^2-(4x)^2$
$=0.01-16x^2$

**37.** $9(2x+1)(2x-1)=9[(2x)^2-1^2]=9[4x^2-1]$
$=36x^2-9$

**39.** $\left(x^2-\frac{1}{2}\right)^2=x^4+2(x^2)\left(-\frac{1}{2}\right)+\left(-\frac{1}{2}\right)^2$
$=x^4-x^2+\frac{1}{4}$

**41.** $(0.1x-2)(x+0.05)=0.1x^2+0.005x-2x-0.10$
$=0.1x^2-1.995x-0.10$

**43.**
$$\begin{array}{r} x^2+2x+4 \\ \underline{x-2} \\ -2x^2-4x-8 \\ \underline{x^3+2x^2+4x\phantom{-8}} \\ x^3\phantom{+2x^2+4x}-8 \end{array}$$

**45.**
$$\begin{array}{r} x^5-2x^3+5 \\ \underline{x^3+5x} \\ 5x^6-10x^4\phantom{+5x^3}+25x \\ \underline{x^8-2x^6\phantom{-10x^4}+5x^3\phantom{+25x}} \\ x^8+3x^6-10x^4+5x^3+25x \end{array}$$

**47.** $\dfrac{18m^2n+6m^3n+12m^4n^2}{6m^2n}$
$=\dfrac{18m^2n}{6m^2n}+\dfrac{6m^3n}{6m^2n}+\dfrac{12m^4n^2}{6m^2n}=3+m+2m^2n$

**49.** $\dfrac{24x^8y^4+15x^5y-6x^7y}{9x^5y^2}$
$=\dfrac{24x^8y^4}{9x^5y^2}+\dfrac{15x^5y}{9x^5y^2}-\dfrac{6x^7y}{9x^5y^2}=\dfrac{8x^3y^2}{3}+\dfrac{5}{3y}-\dfrac{2x^2}{3y}$

**51.** $(x+1)^3=x^3+3(x^2)(1)+3(x)(1)^2+1^3$
$=x^3+3x^2+3x+1$

**53.** $(2x-3)^3=(2x)^3-3(2x)^2(3)+3(2x)(3)^2-3^3$
$=8x^3-36x^2+54x-27$

**55.**
$$\begin{array}{r} x^2-2x+5 \\ x+2\overline{)x^3 \qquad +x-1} \\ \underline{x^3+2x^2} \qquad\qquad \\ -2x^2+\ x-1 \\ \underline{-2x^2-4x} \quad \\ 5x-\ 1 \\ \underline{5x+10} \\ -11 \end{array}$$

Quotient: $x^2-2x+5-\dfrac{11}{x+2}$

**57.**
$$\begin{array}{r} x^2+3x-1 \\ x^2+1\overline{)x^4+3x^3 \qquad -x+1} \\ \underline{x^4 \qquad +x^2} \qquad\qquad \\ 3x^3-x^2-x+1 \\ \underline{3x^3 \qquad +3x} \quad \\ -x^2-4x+1 \\ \underline{-x^2 \qquad -1} \\ -4x+2 \end{array}$$

Quotient: $x^2+3x-1+\dfrac{-4x+2}{x^2+1}$

**59. a.** $(3x-2)^2-3x-2(3x-2)+5$
$=9x^2-12x+4-3x-6x+4+5$
$=9x^2-21x+13$

**b.** $(3x-2)^2-(3x-2)(3x-2)+5$
$=(3x-2)^2-(3x-2)^2+5$
$=5$

**61.** $x^{1/2}\left(x^{1/2}+2x^{3/2}\right)=x^{2/2}+2x^{4/2}=x+2x^2$

**63.** $\left(x^{1/2}+1\right)\left(x^{1/2}-2\right)=x-2x^{1/2}+x^{1/2}-2$
$=x-x^{1/2}-2$

**65.** $\left(\sqrt{x}+3\right)\left(\sqrt{x}-3\right)=\left(\sqrt{x}\right)^2-(3)^2=x-9$

**67.** $(2x+1)^{1/2}\left[(2x+1)^{3/2}-(2x+1)^{-1/2}\right]=(2x+1)^2-(2x+1)^0=4x^2+4x+1-1=4x^2+4x$

**69.** $R=55x$

**71. a.** $C=49.95+0.49x$
**b.** $C=49.95+0.49(132)=\$114.63$

**73. a.** $4000-x$
**b.** $0.10x$
**c.** $0.08(4000-x)$
**d.** $0.10x+0.08(4000-x)$ or $320+0.02x$

**75.** $V=x(15-2x)(10-2x)$

## *Exercises 0.6*

**1.** $9ab-12a^2b+18b^2=3b\left(3a-4a^2+6b\right)$

**3.** $4x^2+8xy^2+2xy^3=2x\left(2x+4y^2+y^3\right)$

**5.** $\left(7x^3-14x^2\right)+(2x-4)=7x^2(x-2)+2(x-2)$
$=(x-2)\left(7x^2+2\right)$

**7.** $6x-6m+xy-my=(6x-6m)+(xy-my)$
$=6(x-m)+y(x-m)$
$=(x-m)(6+y)$

**9.** $x^2+8x+12=(x+6)(x+2)$

**11.** $x^2-15x-16=(x-16)(x+1)$

**13.** $7x^2-10x-8$
$7x^2\cdot 8=56x^2$
The factors $-14x$ and $+4x$ give a sum of $-10x$.
$7x^2-10x-8=7x^2-14x+4x-8$
$=7x(x-2)+4(x-2)$
$=(x-2)(7x+4)$

**15.** $x^2-10x+25=x^2-2\cdot 5x+5^2=(x-5)^2$

**17.** $49a^2 - 144b^2 = (7a)^2 - (12b)^2$
$= (7a + 12b)(7a - 12b)$

**19. a.** $9x^2 + 21x - 8$
$9x^2(-8) = -72x^2$
The factors $24x$ and $-3x$ give a sum of $21x$.
$9x^2 + 21x - 8 = 9x^2 + 24x - 3x - 8$
$= 3x(3x + 8) - 1(3x + 8)$
$= (3x + 8)(3x - 1)$

**b.** $9x^2 + 22x + 8$
$9x^2 \cdot 8 = 72x^2$
The factors $18x$ and $4x$ give a sum of $22x$.
$9x^2 + 22x + 8 = 9x^2 + 18x + 4x + 8$
$= 9x(x + 2) + 4(x + 2)$
$= (x + 2)(9x + 4)$

**21.** $4x^2 - x = x(4x - 1)$

**23.** $x^3 + 4x^2 - 5x - 20 = x^2(x + 4) - 5(x + 4)$
$= (x + 4)(x^2 - 5)$

**25.** $x^2 - x - 6 = (x - 3)(x + 2)$
Note that two numbers whose product is $-6$ and whose sum is $-1$ are $-3$ and 2.

**27.** $2x^2 - 8x - 42 = 2(x^2 - 4x - 21) = 2(x - 7)(x + 3)$

**29.** $2x^3 - 8x^2 + 8x = 2x(x^2 - 4x + 4)$
$= 2x(x^2 - 2 \cdot 2x + 2^2)$
$= 2x(x - 2)^2$

**31.** $2x^2 + x - 6$
$2x^2 \cdot (-6) = -12x^2$
The factors $4x$ and $-3x$ give a sum of $x$.
$2x^2 + x - 6 = 2x^2 + 4x - 3x - 6$
$= 2x(x + 2) - 3(x + 2)$
$= (2x - 3)(x + 2)$

**33.** $3x^2 + 3x - 36 = 3(x^2 + x - 12) = 3(x + 4)(x - 3)$

**35.** $2x^3 - 8x = 2x(x^2 - 4) = 2x(x + 2)(x - 2)$

**37.** $10x^2 + 19x + 6$
$10x^2 \cdot 6 = 60x^2$
The factors $4x$ and $15x$ give a sum of $19x$.
$10x^2 + 19x + 6 = 10x^2 + 4x + 15x + 6$
$= 2x(5x + 2) + 3(5x + 2)$
$= (5x + 2)(2x + 3)$

**39.** $9 - 47x + 10x^2$
$9 \cdot 10x^2 = 90x^2$
The factors $-45x$ and $-2x$ give a sum of $-47x$.
$9 - 47x + 10x^2 = 9 - 45x - 2x + 10x^2$
$= 9(1 - 5x) - 2x(1 - 5x)$
$= (1 - 5x)(9 - 2x)$
or $(5x - 1)(2x - 9)$

**41.** $y^4 - 16x^4 = (y^2)^2 - (4x^2)^2$
$= (y^2 - 4x^2)(y^2 + 4x^2)$
$= (y - 2x)(y + 2x)(y^2 + 4x^2)$

**43.** $x^4 - 8x^2 + 16 = (x^2)^2 - 2 \cdot 4x^2 + 4^2 = (x^2 - 4)^2$
$= [(x - 2)(x + 2)]^2$
$= (x - 2)^2(x + 2)^2$

**45.** $4x^4 - 5x^2 + 1 = (4x^2 - 1)(x^2 - 1)$
$= (2x + 1)(2x - 1)(x + 1)(x - 1)$

**47.** $x^{3/2} + x^{1/2} = x^{1/2}(x^{2/2} + 1)$
$= x^{1/2}(x + 1)$
$? = x + 1$

**49.** $x^{-3} + x^{-2} = x^{-3}(1 + x^1)$
$= x^{-3}(1 + x)$
$? = 1 + x$

**51.** $x^3 + 3x^2 + 3x + 1 = (x + 1)^3$

**53.** $x^3 - 12x^2 + 48x - 64 = x^3 - 3(4x^2) + 3(16x) - 4^3$
$= x^3 - 3x^2(4) + 3x(4)^2 - 4^3$
$= (x - 4)^3$

**55.** $x^3 - 64 = x^3 - 4^3 = (x - 4)(x^2 + 4x + 16)$

**57.** $27+8x^3=3^3+(2x)^3=(3+2x)(9-6x+4x^2)$

**59.** $P+Prt=P(1+rt)$

**61.** $S=cm-m^2=m(c-m)$

**63. a.** In the form $px$ we have $p(10{,}000-100p)$.
$x = 10{,}000 - 100p$

**b.** If $p=38$, then $x=10{,}000-100\cdot 38=6200$.

**65. a.** $R=x(300-x)$

**b.** $P=300-x$

## *Exercises 0.7*

**1.** $\dfrac{18x^3y^3}{9x^3z}=\dfrac{2x^3y^3}{x^3z}=\dfrac{2y^3}{z}$

**3.** $\dfrac{x-3y}{3x-9y}=\dfrac{1(x-3y)}{3(x-3y)}=\dfrac{1}{3}$

**5.** $\dfrac{x^2-2x+1}{x^2-4x+3}=\dfrac{(x-1)(x-1)}{(x-3)(x-1)}=\dfrac{x-1}{x-3}$

**7.** $\dfrac{6x^3}{8y^3}\cdot\dfrac{16x}{9y^2}\cdot\dfrac{15y^4}{x^3}=\dfrac{6}{y^3}\cdot\dfrac{2x}{9y^2}\cdot\dfrac{15y^4}{1}=\dfrac{2}{1}\cdot\dfrac{2x}{3y^2}\cdot\dfrac{15y}{1}$

$=\dfrac{2}{1}\cdot\dfrac{2x}{y}\cdot\dfrac{5}{1}$

$=\dfrac{20x}{y}$

**9.** $\dfrac{8x-16}{x-3}\cdot\dfrac{4x-12}{3x-6}=\dfrac{8(x-2)}{x-3}\cdot\dfrac{4(x-3)}{3(x-2)}=\dfrac{8\cdot 4}{3}=\dfrac{32}{3}$

**11.** $\dfrac{x^2+7x+12}{3x^2+13x+4}\cdot\dfrac{9x+3}{1}$

$=\dfrac{(x+4)(x+3)}{(3x+1)(x+4)}\cdot\dfrac{3(3x+1)}{1}$

$=3(x+3)$

$=3x+9$

**13.** $\dfrac{x^2-x-2}{2x^2-8}\cdot\dfrac{18-2x^2}{x^2-5x+4}\cdot\dfrac{x^2-2x-8}{x^2-6x+9}$

$=\dfrac{(x-2)(x+1)}{2(x^2-4)}\cdot\dfrac{-2(x^2-9)}{(x-4)(x-1)}\cdot\dfrac{(x-4)(x+2)}{(x-3)(x-3)}$

$=\dfrac{(x-2)(x+1)}{2(x-2)(x+2)}\cdot\dfrac{-2(x-3)(x+3)}{(x-1)}\cdot\dfrac{(x+2)}{(x-3)(x-3)}$

$=-\dfrac{(x+1)(x+3)}{(x-1)(x-3)}$

**15.** $\dfrac{15ac^2}{7bd}\div\dfrac{4a}{14b^2d}=\dfrac{15ac^2}{7bd}\cdot\dfrac{14b^2d}{4a}$

$=\dfrac{15c^2}{1}\cdot\dfrac{2b}{4}$

$=\dfrac{15bc^2}{2}$

**17.** $\dfrac{y^2-2y+1}{7y^2-7y}\div\dfrac{y^2-4y+3}{35y^2}$

$=\dfrac{y^2-2y+1}{7y(y-1)}\cdot\dfrac{35y^2}{y^2-4y+3}$

$=\dfrac{(y-1)(y-1)}{7y(y-1)}\cdot\dfrac{35y^2}{(y-3)(y-1)}$

$=\dfrac{5y}{y-3}$

**19.** $\dfrac{x^2-x-6}{1}\div\dfrac{9-x^2}{x^2-3x}$

$=\dfrac{x^2-x-6}{1}\cdot\dfrac{x^2-3x}{-1(x^2-9)}$

$=\dfrac{(x-3)(x+2)}{1}\cdot\dfrac{x(x-3)}{-1(x-3)(x+3)}$

$=\dfrac{-x(x-3)(x+2)}{x+3}$

**21.** $\dfrac{2x}{x^2-x-2}-\dfrac{x+2}{x^2-x-2}=\dfrac{2x-x-2}{(x-2)(x+1)}$

$=\dfrac{x-2}{(x-2)(x+1)}$

$=\dfrac{1}{x+1}$

**23.** $$\frac{a}{a-2}-\frac{a-2}{a}=\frac{a}{a-2}\cdot\frac{a}{a}-\frac{a-2}{a}\cdot\frac{a-2}{a-2}$$
$$=\frac{a^2-(a^2-4a+4)}{a(a-2)}$$
$$=\frac{4a-4}{a(a-2)}$$
$$=\frac{4(a-1)}{a(a-2)}$$

**25.** $$\frac{x}{x+1}-x+1=\frac{x}{x+1}-\frac{x}{1}\cdot\frac{x+1}{x+1}+\frac{1}{1}\cdot\frac{x+1}{x+1}$$
$$=\frac{x-x^2-x+x+1}{x+1}$$
$$=\frac{-x^2+x+1}{x+1}$$

**27.** $$\frac{4a}{3x+6}+\frac{5a^2}{4x+8}=\frac{4a}{3(x+2)}+\frac{5a^2}{4(x+2)}=\frac{4a}{3(x+2)}\cdot\frac{4}{4}+\frac{5a^2}{4(x+2)}\cdot\frac{3}{3}=\frac{16a+15a^2}{12(x+2)}$$

**29.** $$\frac{3x-1}{2x-4}+\frac{4x}{3x-6}-\frac{x-4}{5x-10}=\frac{3x-1}{2(x-2)}+\frac{4x}{3(x-2)}-\frac{x-4}{5(x-2)}$$

$$=\frac{3x-1}{2(x-2)}\cdot\frac{3\cdot 5}{3\cdot 5}+\frac{4x}{3(x-2)}\cdot\frac{2\cdot 5}{2\cdot 5}-\frac{(x-4)}{5(x-2)}\cdot\frac{3\cdot 2}{3\cdot 2}$$
$$=\frac{(45x-15)+40x-6x+24}{30(x-2)}$$
$$=\frac{79x+9}{30(x-2)}$$

**31.** $$\frac{x}{x^2-4}+\frac{4}{x^2-x-2}-\frac{x-2}{x^2+3x+2}=\frac{x}{(x+2)(x-2)}+\frac{4}{(x-2)(x+1)}-\frac{x-2}{(x+2)(x+1)}$$
$$=\frac{x}{(x+2)(x-2)}\cdot\frac{x+1}{x+1}+\frac{4}{(x-2)(x+1)}\cdot\frac{x+2}{x+2}-\frac{x-2}{(x+2)(x+1)}\cdot\frac{x-2}{x-2}$$

$$=\frac{(x^2+x)+(4x+8)(x^2-4x+4)}{(x+2)(x+1)(x-2)}=\frac{9x+4}{(x+2)(x+1)(x-2)}$$

**33.** $\dfrac{-x^3+x}{\sqrt{3-x^2}}+\dfrac{2x\sqrt{3-x^2}}{1}$

$=\dfrac{-x^3+x}{\sqrt{3-x^2}}+\dfrac{2x\sqrt{3-x^2}}{1}\cdot\dfrac{\sqrt{3-x^2}}{\sqrt{3-x^2}}$

$=\dfrac{-x^3+x+2x\left(3-x^2\right)}{\sqrt{3-x^2}}$

$=\dfrac{-x^3+x+6x-2x^3}{\sqrt{3-x^2}}$

$=\dfrac{7x-3x^3}{\sqrt{3-x^2}}$

**35.** $\dfrac{\frac{3}{1}-\frac{2}{3}}{\frac{14}{1}}\cdot\dfrac{3}{3}=\dfrac{9-2}{14(3)}=\dfrac{7}{14(3)}=\dfrac{1}{6}$

**37.** $\dfrac{x+y}{\frac{1}{x}+\frac{1}{y}}=\dfrac{(x+y)}{\frac{1}{x}+\frac{1}{y}}\cdot\dfrac{xy}{xy}=\dfrac{xy(x+y)}{y+x}=xy$

**39.** $\dfrac{2-\frac{1}{x}}{2x-\frac{3x}{x+1}}=\dfrac{\frac{2}{1}-\frac{1}{x}}{\frac{2x}{1}-\frac{3x}{x+1}}\cdot\dfrac{x(x+1)}{x(x+1)}$

$=\dfrac{2x(x+1)-1(x+1)}{2x^2(x+1)-3x(x)}$

$=\dfrac{2x^2+x-1}{2x^3-x^2}$

$=\dfrac{(2x-1)(x+1)}{x^2(2x-1)}=\dfrac{x+1}{x^2}$

**41.** $\dfrac{\sqrt{a}-\frac{b}{\sqrt{a}}}{a-b}=\dfrac{\frac{\sqrt{a}}{1}-\frac{b}{\sqrt{a}}}{\frac{a-b}{1}}\cdot\dfrac{\sqrt{a}}{\sqrt{a}}=\dfrac{a-b}{\sqrt{a}(a-b)}$

$=\dfrac{1}{\sqrt{a}}$ or $\dfrac{\sqrt{a}}{a}$

**43.** **a.** $(2^{-2}-3^{-1})^{-1}=\left(\dfrac{1}{2^2}-\dfrac{1}{3}\right)^{-1}=\left(-\dfrac{1}{12}\right)^{-1}=-12$

**b.** $(2^{-1}+3^{-1})^2=\left(\dfrac{1}{2}+\dfrac{1}{3}\right)^2=\left(\dfrac{5}{6}\right)^2=\dfrac{25}{36}$

Hint: Work inside ( ) first when adding or subtracting is involved.

**45.** $\dfrac{2a^{-1}-b^{-2}}{(ab^2)^{-1}}=\dfrac{\frac{2}{a}-\frac{1}{b^2}}{\frac{1}{ab^2}}\cdot\dfrac{ab^2}{ab^2}=\dfrac{2b^2-a}{1}$ or $2b^2-a$

**47.** $\dfrac{1-\sqrt{x}}{1+\sqrt{x}}=\dfrac{1-\sqrt{x}}{1+\sqrt{x}}\cdot\dfrac{1-\sqrt{x}}{1-\sqrt{x}}=\dfrac{1-2\sqrt{x}+x}{1-x}$

**49.** $\dfrac{\sqrt{x+h}-\sqrt{x}}{h}=\dfrac{\sqrt{x+h}-\sqrt{x}}{h}\cdot\dfrac{\sqrt{x+h}+\sqrt{x}}{\sqrt{x+h}+\sqrt{x}}$

$=\dfrac{(x+h)-(x)}{h\left(\sqrt{x+h}+\sqrt{x}\right)}=\dfrac{h}{h\left(\sqrt{x+h}+\sqrt{x}\right)}$

$=\dfrac{1}{\sqrt{x+h}+\sqrt{x}}$

**51.**

$\dfrac{1}{a}+\dfrac{1}{b}+\dfrac{1}{c}=\dfrac{1}{a}\cdot\dfrac{bc}{bc}+\dfrac{1}{b}\cdot\dfrac{ac}{ac}+\dfrac{1}{c}\cdot\dfrac{ab}{ab}=\dfrac{bc+ac+ab}{abc}$

**53.** **a.** Avg. cost $=\dfrac{4000}{x}+\dfrac{55}{1}+\dfrac{0.1x}{1}$

$=\dfrac{4000+55x+0.1x^2}{x}$

**b.** Total cost = (Avg. cost)(number of units)

$=4000+55x+0.1x^2$

**55.** $SV=1+\dfrac{3}{t+3}-\dfrac{18}{(t+3)^2}=\dfrac{(t+3)^2+3(t+3)-18}{(t+3)^2}=\dfrac{t^2+6t+9+3t+9-18}{(t+3)^2}=\dfrac{t^2+9t}{(t+3)^2}$

# Chapter 0: Algebraic Concepts

## *Chapter 0 Review Exercises*

**1.** Yes. $B = \{1, 2, 3, 4, 5, 6, 7, 8\}$. Since every element of $A$ is also an element of $B$, $A$ is a subset of $B$.

**2.** No. $3 \notin \{x : x > 3\}$

**3.** No. $A$ and $B$ are not disjoint since each set contains the element 1.

**4.** $A = \{1, 2, 3, 9\}$ $B' = \{2, 4, 9\}$.
$A \cup B' = \{1, 2, 3, 4, 9\}$

**5.** $\{4, 5, 6, 7, 8, 10\} \cap \{1, 3, 5, 6, 7, 8, 10\}$
$= \{5, 6, 7, 8, 10\}$

**6.** $A = \{1, 2, 3, 9\}$ $B = \{1, 3, 5, 6, 7, 8, 10\}$
$A' = \{4, 5, 6, 7, 8, 10\}$
$A' \cap B = \{5, 6, 7, 8, 10\}$
$(A' \cap B)' = \{1, 2, 3, 4, 9\}$

**7.** $\{4, 5, 6, 7, 8, 10\} \cup \{2, 4, 9\}$
$= \{2, 4, 5, 6, 7, 8, 9, 10\}$
$(A' \cup B')' = \{2, 4, 5, 6, 7, 8, 9, 10\}' = \{1, 3\}$
$A \cap B = \{1, 2, 3, 9\} \cap \{1, 3, 5, 6, 7, 8, 10\}$
$= \{1, 3\}$ Yes.

**8.** **a.** $6 + \frac{1}{3} = \frac{1}{3} + 6$ illustrates the Commutative Property of Addition.
**b.** $2(3 \cdot 4) = (2 \cdot 3)4$ illustrates the Associative Property of Multiplication.
**c.** $\frac{1}{3}(6 + 9) = 2 + 3$ illustrates the Distributive Law.

**9.** **a.** irrational
**b.** rational, integer
**c.** undefined

**10.** **a.** $\pi > 3.14$
**b.** $-100 < 0.1$
**c.** $-3 > -12$

**11.** $|5 - 11| = |-6| = -(-6) = 6$

**12.** $44 \div 2 \cdot 11 - 10^2 = 22 \cdot 11 - 100 = 242 - 100 = 142$

**13.** $(-3)^2 - (-1)^3 = 9 - (-1) = 10$

**14.** $\frac{(3)(2)(15) - (5)(8)}{(4)(10)} = \frac{90 - 40}{40} = \frac{50}{40} = \frac{5}{4}$

**15.**
$$\begin{aligned} 2 - [3 - (2 - |-3|)] + 11 &= 2 - [3 - (2 - 3)] + 11 \\ &= 2 - [3 - (-1)] + 11 \\ &= 2 - [3 + 1] + 11 \\ &= 2 - 4 + 11 \\ &= 9 \end{aligned}$$

**16.** $-4^2 - (-4)^2 + 3 = -16 - 16 + 3 = -32 + 3 = -29$

**17.** $\frac{4 + 3^2}{4} = \frac{4 + 9}{4} = \frac{13}{4}$

**18.** $\frac{(-2.91)^5}{\sqrt{3.29^5}} \approx \frac{-208.6724}{19.6331} \approx -10.62857888$

**19.** **a.** [0, 5], closed

**b.** [−3, 7), half open

**c.** (−4, 0) open

**20.** **a.** $(-1, 16)$
$-1 < x < 16$
**b.** $[-12, 8]$
$-12 \le x \le 8$
**c.** $x < -1$

**21.** **a.** $\left(\frac{3}{8}\right)^0 = 1$
**b.** $2^3 \cdot 2^{-5} = 2^{-2} = \frac{1}{2^2} = \frac{1}{4}$
**c.** $\frac{4^9}{4^3} = 4^6 = 4096$
**d.** $\left(\frac{1}{7}\right)^3 \left(\frac{1}{7}\right)^{-4} = \left(\frac{1}{7}\right)^{-1} = 7$

**22. a.** $x^5 \cdot x^{-7} = x^{5+(-7)} = x^{-2} = \dfrac{1}{x^2}$

**b.** $\dfrac{x^8}{x^{-2}} = x^{8-(-2)} = x^{10}$

**c.** $(x^3)^3 = x^{3\cdot 3} = x^9$

**d.** $(y^4)^{-2} = y^{(4)(-2)} = y^{-8} = \dfrac{1}{y^8}$

**e.** $(-y^{-3})^{-2} = y^{(-3)(-2)} = y^6$

**There are other correct methods of working problems 23–28.**

**23.** $\dfrac{-(2xy^2)^{-2}}{(3x^{-2}y^{-3})^2} = \dfrac{(-1)(2)^{-2}x^{-2}y^{-4}}{3^2x^{-4}y^{-6}}$

$= \dfrac{(-1)x^4y^6}{2^2\cdot 3^2x^2y^4}$

$= -\dfrac{x^2y^2}{36}$

**24.** $\left(\dfrac{2}{3}x^2y^{-4}\right)^{-2} = \left(\dfrac{2}{3}\right)^{-2}\left(x^2\right)^{-2}\left(y^{-4}\right)^{-2}$

$= \left(\dfrac{3}{2}\right)^2\left(x^{-4}\right)\left(y^8\right)$

$= \left(\dfrac{9}{4}\right)\left(\dfrac{1}{x^4}\right)\left(y^8\right)$

$= \dfrac{9y^8}{4x^4}$

**25.** $\left(\dfrac{x^{-2}}{2y^{-1}}\right)^2 = \left(\dfrac{y}{2x^2}\right)^2 = \dfrac{y^2}{4x^4}$

**26.** $\dfrac{\left(-x^4y^{-2}z^2\right)^0}{-\left(x^4y^{-2}z^2\right)^{-2}} = \dfrac{1}{-\left(x^4\right)^{-2}\left(y^{-2}\right)^{-2}\left(z^2\right)^{-2}}$

$= \dfrac{1}{-x^{-8}y^4z^{-4}} = \dfrac{-x^8z^4}{y^4}$

**27.** $\left(\dfrac{x^{-3}y^4z^{-2}}{3x^{-2}y^{-3}z^{-3}}\right)^{-1} = \left(\dfrac{y^{4-(-3)}z^{-2-(-3)}}{3x^{-2-(-3)}}\right)^{-1}$

$= \left(\dfrac{y^7z}{3x}\right)^{-1} = \dfrac{3x}{y^7z}$

**28.** $\left(\dfrac{x}{2y}\right)\left(\dfrac{y}{x^2}\right)^{-2} = \left(\dfrac{x}{2y}\right)\left(\dfrac{x^2}{y}\right)^2$

$= \left(\dfrac{x}{2y}\right)\left(\dfrac{(x^2)^2}{y^2}\right) = \left(\dfrac{x}{2y}\right)\left(\dfrac{x^4}{y^2}\right) = \dfrac{x^5}{2y^3}$

**29. a.** $-\sqrt[3]{-64} = -\sqrt[3]{(-4)^3} = -(-4) = 4$

**b.** $\sqrt{\dfrac{4}{49}} = \sqrt{\dfrac{2^2}{7^2}} = \dfrac{2}{7}$

**c.** $\sqrt[7]{1.9487171} = 1.1$

**30. a.** $\sqrt{x} = x^{1/2}$

**b.** $\sqrt[3]{x^2} = x^{2/3}$

**c.** $1/\sqrt[4]{x} = \dfrac{1}{x^{1/4}} = x^{-1/4}$

**31. a.** $x^{3/7} = \sqrt[7]{x^3}$

**b.** $x^{-1/2} = \dfrac{1}{\sqrt{x}} = \dfrac{\sqrt{x}}{x}$

**c.** $-x^{3/2} = -x\sqrt{x}$

**32. a.** $\dfrac{5xy}{\sqrt{2x}}\cdot\dfrac{\sqrt{2x}}{\sqrt{2x}} = \dfrac{5xy\sqrt{2x}}{2x} = \dfrac{5y\sqrt{2x}}{2}$

**b.** $\dfrac{y}{x\sqrt[3]{xy^2}}\cdot\dfrac{\sqrt[3]{x^2y}}{\sqrt[3]{x^2y}} = \dfrac{y\sqrt[3]{x^2y}}{x\sqrt[3]{x^3y^3}}$

$= \dfrac{y\sqrt[3]{x^2y}}{x(xy)}$

$= \dfrac{y\sqrt[3]{x^2y}}{x^2y}$

$= \dfrac{\sqrt[3]{x^2y}}{x^2}$

**33.** $x^{1/2}\cdot x^{1/3} = x^{(3/6)+(2/6)} = x^{5/6}$

**34.** $\dfrac{y^{-3/4}}{y^{-7/4}} = y^{-3/4-(-7/4)} = y^{4/4} = y$

**35.** $x^4\cdot x^{1/4} = x^{(16/4)+(1/4)} = x^{17/4}$

**36.** $\dfrac{1}{x^{-4/3}\cdot x^{-7/3}} = \dfrac{1}{x^{-11/3}} = x^{11/3}$

**37.** $(x^{4/5})^{1/2} = x^{(4/5)(1/2)} = x^{2/5}$

**38.** $(x^{1/2}y^2)^4 = (x^{1/2})^4(y^2)^4 = x^2y^8$

**39.** $\sqrt{12x^3y^5} = \sqrt{4x^2y^4 \cdot 3xy} = 2xy^2\sqrt{3xy}$

**40.** $\sqrt{1250x^6y^9} = \sqrt{625x^6y^8 \cdot 2y} = 25x^3y^4\sqrt{2y}$

**41.** $\sqrt[3]{24x^4y^4} \cdot \sqrt[3]{45x^4y^{10}} = \sqrt[3]{8x^3y^3 \cdot 3xy} \cdot \sqrt[3]{9x^3y^9 \cdot 5xy}$
$= 2xy\sqrt[3]{3xy} \cdot xy^3\sqrt[3]{9 \cdot 5xy}$
$= 2x^2y^4\sqrt[3]{27 \cdot 5x^2y^2}$
$= 6x^2y^4\sqrt[3]{5x^2y^2}$

**42.** $\sqrt{16a^2b^3} \cdot \sqrt{8a^3b^5} = \sqrt{128a^5b^8}$
$= \sqrt{64a^4b^8 \cdot 2a}$
$= 8a^2b^4\sqrt{2a}$

**43.** $\dfrac{\sqrt{52x^3y^6}}{\sqrt{13xy^4}} = \sqrt{4x^2y^2} = 2xy$

**44.** $\dfrac{\sqrt{32x^4y^3}}{\sqrt{6xy^{10}}} = \sqrt{\dfrac{16x^3}{3y^7}} = \dfrac{4x\sqrt{x}}{y^3\sqrt{3y}} \cdot \dfrac{\sqrt{3y}}{\sqrt{3y}} = \dfrac{4x\sqrt{3xy}}{3y^4}$

**45.** $(3x+5)-(4x+7) = 3x+5-4x-7 = -x-2$

**46.** $x(1-x)+x[x-(2+x)] = x-x^2+x(-2) = -x^2-x$

**47.** $(3x^3-4xy-3)+(5xy+x^3+4y-1)$
$= 4x^3+xy+4y-4$

**48.** $(4xy^3)(6x^4y^2) = 24x^{1+4}y^{3+2} = 24x^5y^5$

**49.** $(3x-4)(x-1) = 3x^2-3x-4x+4 = 3x^2-7x+4$

**50.** $(3x-1)(x+2) = 3x^2+6x-x-2 = 3x^2+5x-2$

**51.** $(4x+1)(x-2) = 4x^2-8x+x-2 = 4x^2-7x-2$

**52.** $(3x-7)(2x+1) = 6x^2+3x-14x-7$
$= 6x^2-11x-7$

**53.** $(2x-3)^2 = (2x)^2-2(2x)(3)+3^2 = 4x^2-12x+9$

**54.** $(4x+3)(4x-3) = 16x^2-9$
Difference of two squares

**55.**
$$\begin{array}{r} x^2+x-3 \\ 2x^2+1 \\ \hline x^2+x-3 \\ 2x^4+2x^3-6x^2 \phantom{+x-3} \\ \hline 2x^4+2x^3-5x^2+x-3 \end{array}$$

**56.** $(2x-1)^3 = 8x^3-12x^2+6x-1$ Binomial cubed

**57.**
$$\begin{array}{r} x^2+xy+y^2 \\ x-y \\ \hline -x^2y-xy^2-y^3 \\ x^3+x^2y+xy^2 \phantom{-y^3} \\ \hline x^3 \phantom{+x^2y+xy^2} -y^3 \end{array}$$
Difference of two cubes

**58.** $\dfrac{4x^2y-3x^3y^3-6x^4y^2}{2x^2y^2} = \dfrac{2}{y} - \dfrac{3xy}{2} - 3x^2$

**59.**
$$\begin{array}{r} 3x^2+2x-3 \\ x^2+1\overline{)\,3x^4+2x^3 \phantom{-3x^2} -x+4} \\ \underline{3x^4 \phantom{+2x^3} +3x^2 \phantom{-x+4}} \\ 2x^3-3x^2-x+4 \\ \underline{2x^3 \phantom{-3x^2} +2x \phantom{+4}} \\ -3x^2-3x+4 \\ \underline{-3x^2 \phantom{-3x} -3} \\ -3x+7 \end{array}$$

Quotient is $3x^2+2x-3+\dfrac{7-3x}{x^2+1}$.

**60.**
$$\begin{array}{r} x^3-x^2+2x+7 \\ x-3\overline{)\,x^4-4x^3+5x^2+\phantom{0}x} \\ \underline{x^4-3x^3 \phantom{+5x^2+x}} \\ -x^3+5x^2 \phantom{+x} \\ \underline{-x^3+3x^2} \phantom{+x} \\ 2x^2+x \\ \underline{2x^2-6x} \\ 7x \\ \underline{7x-21} \\ 21 \end{array}$$

Quotient is $x^3-x^2+2x+7+\dfrac{21}{x-3}$.

**61.** $x^{4/3}(x^{2/3}-x^{-1/3}) = x^{6/3}-x^{3/3} = x^2-x$

**62.** $\left(\sqrt{x}+\sqrt{a-x}\right)\left(\sqrt{x}-\sqrt{a-x}\right)=\left(\sqrt{x}\right)^2-\left(\sqrt{a-x}\right)^2$
$=x-(a-x)$
$=x-a+x$
$=2x-a$

**63.** $2x^4-x^3=x^3(2x-1)$

**64.** $4(x^2+1)^2-2(x^2+1)^3=2(x^2+1)^2[2-(x^2+1)]$
$=2(x^2+1)^2(2-x^2-1)$
$=2(x^2+1)^2(1-x^2)$
$=2(x^2+1)^2(1+x)(1-x)$

**65.** $4x^2-4x+1=(2x)^2-2(2x)+1^2=(2x-1)^2$

**66.** $16-9x^2=(4+3x)(4-3x)$

**67.** $2x^4-8x^2=2x^2(x^2-4)=2x^2(x+2)(x-2)$

**68.** $x^2-4x-21=(x-7)(x+3)$

**69.** $3x^2-x-2=(3x+2)(x-1)$

**70.** $x^2-5x+6=(x-3)(x-2)$

**71.** $x^2-10x-24=(x-12)(x+2)$

**72.** $12x^2-23x-24$
Two expressions whose product is $12x^2(-24)=-288x^2$ and whose sum is $-23x$ are $-32x$ and $9x$. So,
$12x^2-23x-24=12x^2+9x-32x-24$
$=3x(4x+3)-8(4x+3)$
$=(4x+3)(3x-8).$

**73.** $16x^4-72x^2+81=(4x^2)^2-2(4x^2\cdot 9)+9^2$
$=(4x^2-9)^2$
$=[(2x+3)(2x-3)]^2$
$=(2x+3)^2(2x-3)^2$

**74.** $x^{-2/3}+x^{-4/3}=x^{-4/3}(?)$
$x^{-2/3}+x^{-4/3}=x^{-4/3}(x^{2/3}+1)$
$?=x^{2/3}+1$

**75. a.** $\dfrac{2x}{2x+4}=\dfrac{2x}{2(x+2)}=\dfrac{x}{x+2}$

**b.** $\dfrac{4x^2y^3-6x^3y^4}{2x^2y^2-3xy^3}=\dfrac{2x^2y^3(2-3xy)}{xy^2(2x-3y)}$
$=\dfrac{2xy(2-3xy)}{2x-3y}$

**76.** $\dfrac{x^2-4x}{x^2+4}\cdot\dfrac{x^4-16}{x^4-16x^2}=\dfrac{x(x-4)}{x^2+4}\cdot\dfrac{(x^2-4)(x^2+4)}{x^2(x^2-16)}$
$=\dfrac{(x-4)(x+2)(x-2)}{x(x-4)(x+4)}$
$=\dfrac{(x+2)(x-2)}{x(x+4)}$
$=\dfrac{x^2-4}{x(x+4)}$

**77.** $\dfrac{x^2+6x+9}{x^2-7x+12}\cdot\dfrac{x^2-3x-4}{x^2+4x+3}$
$=\dfrac{(x+3)(x+3)}{(x-4)(x-3)}\cdot\dfrac{(x-4)(x+1)}{(x+3)(x+1)}=\dfrac{x+3}{x-3}$

**78.** $\dfrac{x^4-2x^3}{3x^2-x-2}\div\dfrac{x(x^2-4)}{9x^2-4}$
$=\dfrac{x^3(x-2)}{(3x+2)(x-1)}\cdot\dfrac{(3x+2)(3x-2)}{x(x+2)(x-2)}$
$=\dfrac{x^2(3x-2)}{(x-1)(x+2)}$

**79.** $1+\dfrac{3}{2x}-\dfrac{1}{6x^2}=\dfrac{1}{1}\cdot\dfrac{6x^2}{6x^2}+\dfrac{3}{2x}\cdot\dfrac{3x}{3x}-\dfrac{1}{6x^2}$
$=\dfrac{6x^2+9x-1}{6x^2}$

**80.** $\dfrac{1}{x-2}-\dfrac{x-2}{4}=\dfrac{1\cdot 4-(x-2)(x-2)}{4(x-2)}=\dfrac{4x-x^2}{4(x-2)}$

**81.** $\dfrac{x+2}{x(x-1)}-\dfrac{x^2+4}{(x-1)(x-1)}+\dfrac{1}{1}$

$=\dfrac{(x+2)(x-1)-(x^2+4)x+x(x-1)(x-1)}{x(x-1)(x-1)}$

$=\dfrac{x^2+x-2-x^3-4x+x^3-2x^2+x}{x(x-1)^2}$

$=\dfrac{-(x^2+2x+2)}{x(x-1)^2}$

**82.** $\dfrac{x-1}{x^2-x-2}-\dfrac{x}{x^2-2x-3}+\dfrac{1}{x-2}=\dfrac{x-1}{(x-2)(x+1)}-\dfrac{x}{(x-3)(x+1)}+\dfrac{1}{x-2}$

$=\dfrac{(x-1)(x-3)}{(x-2)(x+1)(x-3)}-\dfrac{x(x-2)}{(x-2)(x+1)(x-3)}+\dfrac{(x+1)(x-3)}{(x-2)(x+1)(x-3)}$

$=\dfrac{x^2-4x+3-x^2+2x+x^2-2x-3}{(x-2)(x+1)(x-3)}$

$=\dfrac{x^2-4x}{(x-2)(x+1)(x-3)}$

$=\dfrac{x(x-4)}{(x-2)(x+1)(x-3)}$

**83.** $\dfrac{\frac{x-1}{1}-\frac{x-1}{x}}{\frac{1}{x-1}+1}\cdot\dfrac{x(x-1)}{x(x-1)}=\dfrac{x(x-1)^2-(x-1)^2}{x+x(x-1)}$

$=\dfrac{(x-1)^2(x-1)}{x^2}$

$=\dfrac{(x-1)^3}{x^2}$

**84.** $\dfrac{x^{-2}-x^{-1}}{x^{-2}+x^{-1}}=\dfrac{\frac{1}{x^2}-\frac{1}{x}}{\frac{1}{x^2}+\frac{1}{x}}\cdot\dfrac{x^2}{x^2}=\dfrac{1-x}{1+x}$

**85.** $\dfrac{3x-3}{\sqrt{x}-1}\cdot\dfrac{\sqrt{x}+1}{\sqrt{x}+1}=\dfrac{3(x-1)(\sqrt{x}+1)}{x-1}=3(\sqrt{x}+1)$

**86.** $\dfrac{\sqrt{x}-\sqrt{x-4}}{2}\cdot\dfrac{\sqrt{x}+\sqrt{x-4}}{\sqrt{x}+\sqrt{x-4}}=\dfrac{x-(x-4)}{2(\sqrt{x}+\sqrt{x-4})}$

$=\dfrac{x-x+4}{2(\sqrt{x}+\sqrt{x-4})}$

$=\dfrac{4}{2(\sqrt{x}+\sqrt{x-4})}$

$=\dfrac{2}{\sqrt{x}+\sqrt{x-4}}$

**87. a.** $R$: Recognized
$C$: Involved
$E$: Exercised

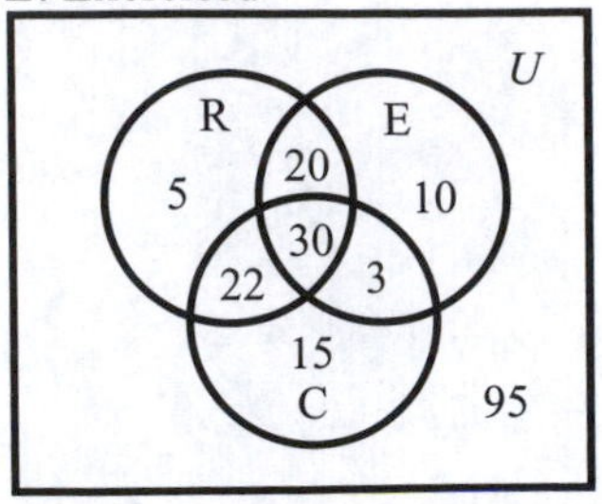


Numbered statement indicates solution for that question.

1. 30
2. $50 - 30 = 20$
3. $52 - 30 = 22$
4. $30 + 22 + 20 + \underline{5} = 77$
5. $37 - 22 = 15$
6. $77 + 15 + \underline{3} = 95$

**b.** $200-(95+5+22+30+20+3+15)=10$ So, 10 exercised only.

**c.** $63+70-(3+30)=100$ So, 100 exercised or were involved in the community.

**88.** $-0.75(15)+63.8=52.55\%$

**89.** $5^2-(5-2)^2=25-3^2=25-9=16$

**90.** $S=100\left[\dfrac{(1.0075)^n-1}{0.0075}\right]$

**a.** $S(36)=100\left[\dfrac{(1.0075)^{36}-1}{0.0075}\right]$

$\approx 100\left[\dfrac{0.30865}{0.0075}\right]\approx \$4115.27$

**b.** $S(240)=100\left[\dfrac{(1.0075)^{240}-1}{0.0075}\right]$

$\approx 100\left[\dfrac{5.00915}{0.0075}\right]\approx \$66,788.69$

**91.** $C=31.9t+310$

**a.** $t=2021-2005=16$

**b.** $C=31.9(16)+310$

$=\$820.40$

**c.** $4(820.40)=3281.60$

A family of four can expect to pay \$3281.60 for health insurance in 2021.

**92.** $h = 0.000595s^{1.922}$ or $s = 47.7h^{0.519}$

**a.** $h = 0.000595(50)^{1.922} \approx 1.1$ inch (about quarter-sized)

**b.** $s = 47.7(4.5)^{0.519} \approx 104$ mph

**93. a.** $$R = 10{,}000\left[\frac{0.0065}{1-(1.0065)^{-n}}\right] = 10{,}000\left[\frac{0.0065}{1-\frac{1}{1.0065^n}}\right]\cdot\frac{1.0065^n}{1.0065^n} = 10{,}000\left[\frac{0.0065(1.0065)^n}{1.0065^n-1}\right] = \frac{65(1.0065)^n}{1.0065^n-1}$$

**b.** $$R = 10{,}000\left[\frac{0.0065}{1-(1.0065)^{-48}}\right] \approx \$243.19$$

$$R = \frac{65(1.0065)^{48}}{1.0065^{48}-1} \approx \$243.19$$

**94.** $S = kA^{1/3}$

**a.** $S = k\sqrt[3]{A}$

**b.** Let $S_1$ be the number of species on 20,000 acres. Then $S_1 = k\sqrt[3]{20{,}000}$. Let $S_2$ be the number of species on 45,000 acres. Then

$$S_2 = k\sqrt[3]{45000} = \sqrt[3]{2.25\cdot 20{,}000} = \sqrt[3]{2.25}\cdot k\sqrt[3]{20{,}000} = \sqrt[3]{2.25}\cdot S_1$$

$$S_2 \approx 1.31S_1$$

**95.** $$\text{Profit} = 30x - 0.001x^2 - (300+4x) = -0.001x^2 + 26x - 300$$

**96.** $$\text{Value} = \$1{,}450{,}000 - 0.0025(1{,}450{,}000)x = \$1{,}450{,}000 - 3625x$$

**97.** $600 - 13x - 0.5x^2 = 0.5(1200 - 26x - x^2)$ or $0.5(50+x)(24-x)$ or $(25+0.5x)(24-x)$ or $(50+x)(12-0.5x)$

**98. a.** $$C = \frac{1{,}200{,}000}{100-p} - \frac{12{,}000}{1} = \frac{1{,}200{,}000 - 12{,}000(100-p)}{100-p} = \frac{12{,}000p}{100-p}$$

**b.** If $p = 0$, $C = \frac{12{,}000(0)}{100-0} = \frac{0}{100} = \$0$. The cost of removing no pollution is zero.

**c.** $C = \frac{12{,}000(98)}{100-98} = \$588{,}000$

**d.** The formula is not defined when $p = 100$. We are dividing by zero. The cost increases as $p$ approaches 100. It is cost prohibitive (or maybe not feasible) to remove all of the pollution.

**99.** $$\frac{1200}{1} + \frac{56x}{1} + \frac{8000}{x} = \frac{1200x}{x} + \frac{56x^2}{x} + \frac{8000}{x} = \frac{56x^2 + 1200x + 8000}{x}$$

## *Chapter 0 Test*

**1. a.** $A = \{6, 8\}$ $B' = \{3, 4, 6\}$
$A \cup B' = \{3, 4, 6, 8\}$

**b.** $\{3, 4\}$, $\{3, 6\}$, and $\{4, 6\}$ are disjoint from $B$.

**c.** $\{6\}$ and $\{8\}$ are non-empty subsets of $A$.

**2.** $$(4-2^3)^2 - 3^4\cdot 0^{15} + 12 \div 3 + 1 = (-4)^2 - 0 + 4 + 1 = 16 + 4 + 1 = 21$$

**3. a.** $x^4 \cdot x^4 = x^8$

**b.** $x^0 = 1$, if $x \neq 0$

**c.** $\sqrt{x} = x^{1/2}$

**d.** $(x^{-5})^2 = x^{-10}$ or $\frac{1}{x^{10}}$

**e.** $a^{27} \div a^{-3} = a^{27-(-3)} = a^{30}$

**f.** $x^{1/2}\cdot x^{1/3} = x^{5/6}$

**g.** $\frac{1}{\sqrt[3]{x^2}} = \frac{1}{x^{2/3}}$

**h.** $\frac{1}{x^3} = x^{-3}$

**4.** **a.** $x^{1/5} = \sqrt[5]{x}$

**b.** $x^{-3/4} = \sqrt[4]{x^{-3}}$ or $\left(\sqrt[4]{x}\right)^{-3}$ or $\dfrac{1}{\sqrt[4]{x^3}}$

**5.** **a.** $x^{-5} = \dfrac{1}{x^5}$

**b.** $\left(\dfrac{x^{-8}y^2}{x^{-1}}\right)^{-3} = \dfrac{x^{24}y^{-6}}{x^3} = \dfrac{x^{21}}{y^6}$

**6.** **a.** $\dfrac{x}{\sqrt{5x}} \cdot \dfrac{\sqrt{5x}}{\sqrt{5x}} = \dfrac{x\sqrt{5x}}{5x} = \dfrac{\sqrt{5x}}{5}$

**b.** $\sqrt{24a^2b} \cdot \sqrt{a^3b^4} = 2a\sqrt{6b} \cdot ab^2\sqrt{a}$
$= 2a^2b^2\sqrt{6ab}$

**c.** $\dfrac{1-\sqrt{x}}{1+\sqrt{x}} \cdot \dfrac{1-\sqrt{x}}{1-\sqrt{x}} = \dfrac{1-2\sqrt{x}+x}{1-x}$

**7.** $2x^3 - 7x^5 - 5x - 8$

**a.** Degree is 5.
**b.** Constant is –8.
**c.** Coefficient of $x$ is –5.

**8.** In interval notation, $(-2,\infty) \cap (-\infty, 3] = (-2, 3]$

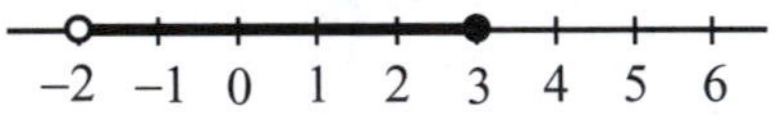


**9.** **a.** $8x^3 - 2x^2 = 2x^2(4x-1)$

**b.** $x^2 - 10x + 24 = (x-4)(x-6)$

**c.** $6x^2 - 13x + 6 = (2x-3)(3x-2)$

**d.** $2x^3 - 32x^5 = 2x^3(1-16x^2)$
$= 2x^3(1-4x)(1+4x)$

**10.** A quadratic polynomial has degree two.
(c) is the quadratic.
$4 - x - x^2 = 4 - (-3) - (-3)^2 = 4 + 3 - 9 = -2,$
when $x = -3$

**11.**
$$\begin{array}{r|l} & \quad 2x+1 \\ x^2-1 & 2x^3 + x^2 \qquad -7 \\ & \underline{2x^3 \qquad -2x} \\ & \qquad x^2+2x-7 \\ & \qquad \underline{x^2 \qquad -1} \\ & \qquad \quad 2x-6 \end{array}$$

Quotient: $2x + 1 + \dfrac{2x-6}{x^2-1}$

**12.** **a.** $4y - 5(9-3y) = 4y - 45 + 15y = 19y - 45$

**b.** $-3t^2(2t^4 - 3t^7) = -6t^6 + 9t^9$

**c.**
$$\begin{array}{r} x^2 - 5x + 2 \\ \underline{4x - 1} \\ -x^2 + 5x - 2 \\ \underline{4x^3 - 20x^2 + 8x \qquad} \\ 4x^3 - 21x^2 + 13x - 2 \end{array}$$

**d.** $(6x-1)(2-3x) = 12x - 18x^2 - 2 + 3x$
$= -18x^2 + 15x - 2$

**e.** $(2m-7)^2 = 4m^2 - 28m + 49$

**f.** $\dfrac{x^6}{x^2-9} \cdot \dfrac{x-3}{3x^2} = \dfrac{x^4}{(x+3)(x-3)} \cdot \dfrac{(x-3)}{3}$
$= \dfrac{x^4}{3(x+3)}$

**g.** $\dfrac{x^4}{9} \div \dfrac{9x^3}{x^6} = \dfrac{x^4}{9} \cdot \dfrac{x^6}{9x^3} = \dfrac{x^7}{81}$

**h.** $\dfrac{4}{x-8} - \dfrac{x-2}{x-8} = \dfrac{4-x+2}{x-8} = \dfrac{6-x}{x-8}$

**i.** $\dfrac{x-1}{x^2-2x-3} - \dfrac{3}{x^2-3x}$
$= \dfrac{x-1}{(x-3)(x+1)} - \dfrac{3}{x(x-3)}$
$= \dfrac{x(x-1) - 3(x+1)}{x(x-3)(x+1)}$
$= \dfrac{x^2 - x - 3x - 3}{x(x-3)(x+1)} = \dfrac{x^2 - 4x - 3}{x(x-3)(x+1)}$

**13.** $\dfrac{\frac{1}{x} - \frac{1}{y}}{\frac{1}{x} + y} \cdot \dfrac{xy}{xy} = \dfrac{y-x}{y+xy^2}$ or $\dfrac{y-x}{y(1+xy)}$

**14.** **a.** Construct a Venn diagram:

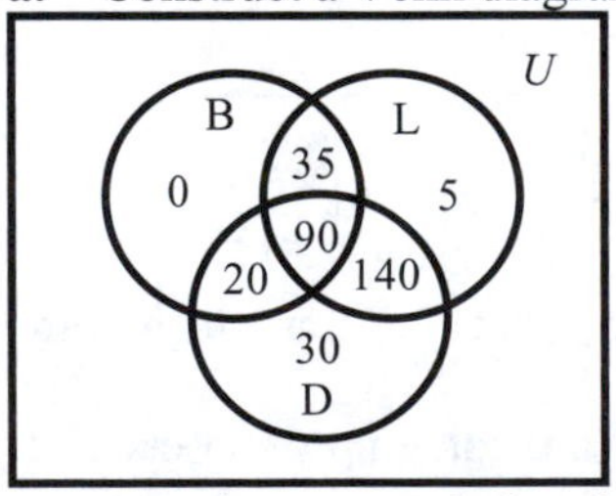


**b.** 0 students ate only breakfast.
**c.** 320 – 145 = 175. 175 students skipped breakfast.

**15.** $S = 1000\left(1+\frac{0.08}{4}\right)^{4x} = 1000\left(1+\frac{0.08}{4}\right)^{4(20)}$

$= 1000(1+0.02)^{80} = 1000(1.02)^{80} \approx 4875.44$

In 20 years, the future value will be about $4875.44.

# Chapter 1: Linear Equations and Functions

## *Exercises 1.1*

**1.** 
$$\begin{aligned} 4x-7&=8x+2\\ 4x-7+7-8x&=8x+2+7-8x\\ -4x&=9\\ x&=-\frac{9}{4}\end{aligned}$$

**3.** 
$$\begin{aligned} x+8&=8(x+1)\\ x+8&=8x+8\\ x-8x&=8-8\\ -7x&=0\\ x&=0\end{aligned}$$

**5.** 
$$\begin{aligned} -\frac{3x}{4}&=24\\ -3x&=4(24)=96\\ x&=-32\end{aligned}$$

**7.** 
$$\begin{aligned} 2(x-7)&=5(x+3)-x\\ 2x-14&=5x+15-x\\ 2x-5x+x&=15+14\\ -2x&=29\\ x&=-\frac{29}{2}\end{aligned}$$

**9.** 
$$\begin{aligned} 8-2(3x+9)-6x&=50\\ 8-6x-18-6x&=50\\ -10-12x&=50\\ -12x&=50+10\\ -12x&=60\\ x&=-5\end{aligned}$$

**11.** 
$$\begin{aligned} \frac{5x}{2}-4&=\frac{2x-7}{6}\\ 6\left(\frac{5x}{2}-4\right)&=6\left(\frac{2x-7}{6}\right)\\ 15x-24&=2x-7\\ 15x-2x&=24-7\\ 13x&=17\\ x&=\frac{17}{13}\end{aligned}$$

**13.** 
$$\begin{aligned} x+\frac{1}{3}&=2\left(x-\frac{2}{3}\right)-6x\\ x+\frac{1}{3}&=2x-\frac{4}{3}-6x\\ 3x+1&=6x-4-18x\\ 3x+18x-6x&=-4-1\\ 15x&=-5\\ x&=\frac{-5}{15}=-\frac{1}{3}\end{aligned}$$

**15.** 
$$\begin{aligned} (5x)\left(\frac{33-x}{5x}\right)&=5x(2)\\ 33-x&=10x\\ -x-10x&=-33\\ -11x&=-33\\ x&=3\end{aligned}$$

Check: $\dfrac{33-3}{5(3)}\stackrel{?}{=}2$

$\dfrac{30}{15}\stackrel{?}{=}2$

$2=2$

$x=3$ is the solution.

**17.** $\dfrac{2x}{2x+5}=\dfrac{2}{3}-\dfrac{5}{2(2x+5)}$

Multiply each term by $6(2x+5)$.

$$\begin{aligned} 12x&=(8x+20)-15\\ 12x-8x&=20-15\end{aligned}$$

$4x=5$ or $x=\dfrac{5}{4}$

Check: $\dfrac{2\left(\frac{5}{4}\right)}{2\left(\frac{5}{4}\right)+5}\stackrel{?}{=}\dfrac{2}{3}-\dfrac{5}{4\left(\frac{5}{4}\right)+10}$

$\dfrac{10}{10+20}\stackrel{?}{=}\dfrac{2}{3}-\dfrac{5}{15}$

$\dfrac{10}{30}=\dfrac{1}{3}$ and $\dfrac{2}{3}-\dfrac{5}{15}=\dfrac{1}{3}$

$x=\dfrac{5}{4}$ is the solution.

**19.** $\frac{2x}{x-1}+\frac{1}{3}=\frac{5}{6}+\frac{2}{x-1}$

$$\frac{2x-2}{x-1}+\frac{1}{3}=\frac{5}{6}$$

$$\frac{2(x-1)}{x-1}+\frac{1}{3}=\frac{5}{6}$$

$$2+\frac{1}{3}\neq\frac{5}{6}$$

There is no solution.

**21.** $3.259x-8.638=-3.8(8.625x+4.917)$

$$3.259x-8.638=-32.775x-18.6846$$

$$3.259x+32.775x=8.638-18.6846$$

$$36.034x=-10.0466$$

$$x=\frac{-10.0466}{36.034}\approx-0.279$$

**23.** $0.000316x+9.18=2.1(3.1-0.0029x)-4.68$

$$0.000316x+9.18=6.51-0.00609x-4.68$$

$$0.000316x+0.00609x=6.51-4.68-9.18$$

$$0.006406x=-7.35$$

$$x=\frac{-7.35}{0.006406}$$

$$x\approx-1147.362$$

**25.** $3x-4y=15$

$$-4y=-3x+15$$

$$y=\frac{-3x}{-4}+\frac{15}{-4}$$

$$y=\frac{3}{4}x-\frac{15}{4}$$

**27.** $2\left(9x+\frac{3}{2}y\right)=2(11)$

$$18x+3y=22$$

$$3y=-18x+22$$

$$y=-6x+\frac{22}{3}$$

**29.** $S=P+Prt$

$$Prt=S-P$$

$$t=\frac{S-P}{Pr}$$

**31.** $3(x-1)<2x-1$

$$3x-3<2x-1$$

$$x-3<-1$$

$$x<2$$

**33.** $1-2x>9$

$$-2x>8$$

$$\left(-\tfrac{1}{2}\right)(-2x)>8\left(-\tfrac{1}{2}\right)$$

$$x<-4$$

**35.** $\frac{3(x-1)}{2}\le x-2$

$$3(x-1)\le 2(x-2)$$

$$3x-3\le 2x-4$$

$$x-3\le-4$$

$$x\le-1$$

**37.** $2(x-1)-3>4x+1$

$$2x-2-3>4x+1$$

$$2x-5>4x+1$$

$$-2x-5>1$$

$$-2x>6$$

$$\left(-\tfrac{1}{2}\right)(-2x)>6\left(-\tfrac{1}{2}\right)$$

$$x<-3$$

**39.** $\frac{-3x}{2}>3-x$

$$-3x>2(3-x)$$

$$-3x>6-2x$$

$$-x>6$$

$$(-1)(-x)>6(-1)$$

$$x<-6$$

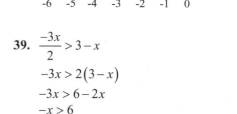

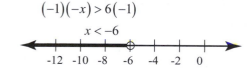

**41.** $12\left(\frac{3x}{4}-\frac{1}{6}\right)<12\left(x-\frac{2(x-1)}{3}\right)$

$$9x-2<12x-8(x-1)$$
$$9x-2<12x-8x+8$$
$$9x-2<4x+8$$
$$5x-2<8$$
$$5x<10$$
$$\left(\tfrac{1}{5}\right)(5x)<\left(\tfrac{1}{5}\right)(10)$$
$$x<2$$

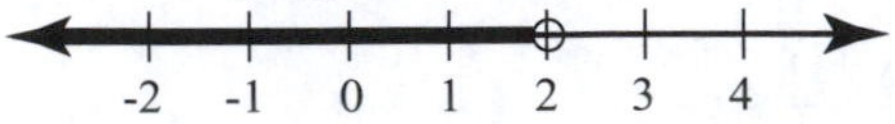

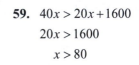

**43.**
$$y=648{,}000-1800x$$
$$387{,}000=648{,}000-1800x$$
$$1800x=648{,}000-387{,}000=261{,}000$$
$$x=\frac{261{,}000}{1800}=145 \text{ months}$$
$$x=\frac{261{,}000}{1800}=145 \text{ months}$$

**45.** $\frac{I}{175.393}+0.663=r$

$$\frac{I}{175.393}+0.663=19.8$$
$$\frac{I}{175.393}=19.8-0.663=19.137$$
$$I=19.137(175.393)$$
$$I=\$3356.50$$

**47.** $R=C$ for breakeven point

$$20x=2x+7920$$
$$18x=7920$$
$$x=440 \text{ packs or } 220{,}000 \text{ CD's}$$

**49.** $170{,}500=5.76x$

$$x=\frac{170{,}500}{5.76}=\$29{,}600$$

**51. a.**
$$25P-34t=1378$$
$$25P-34(20)=1378$$
$$25P-680=1378$$
$$25P=2058$$
$$P\approx 82.3\%$$

**b.**
$$25P-34t=1378$$
$$25(90)-34t=1378$$
$$2250-34t=1378$$
$$-34t=-872$$
$$t\approx 25.6$$

Since $1990+25.6=2015.6$, the year in which 90% of the population will have Internet access is 2016.

**53.** $\frac{93+69+89+97+FE+FE}{6}=90$

$$2FE+348=540$$
$$2FE=192$$
$$FE=96$$

A 96 is the lowest grade that can be earned on the final.

**55.** $x=$ Amount in safe fund

$120{,}000-x=$ amount in risky fund

Yield: $0.09x+0.13(120{,}000-x)=12{,}000$

$$0.09x+15{,}600-0.13x=12{,}000$$
$$-0.04x=-3600$$
$$x=90{,}000$$

$x=\$90{,}000$ in 9% fund

$120{,}000-90{,}000=\$30{,}000$ in 13% fund.

**57.** Reduced salary: $2000-0.10(2000)=\$1800$

Increased salary: $1800+0.20(1800)=\$2160$

$160=R\%$ of 2000

$$R=\frac{160}{2000}=\frac{8}{100}$$

\$160 is an 8% increase.

**59.**
$$40x>20x+1600$$
$$20x>1600$$
$$x>80$$

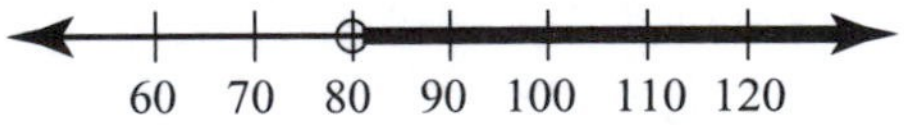


**61.**
$$695+5.75x\le 900$$
$$5.75x\le 205$$
$$x\le 35.65$$

He could buy 35 or fewer memory sticks.

**63.** **a.** $2018-1980=38$

**b.** $S=0.264t-2.57$

$10=0.264t-2.57$

$12.57=0.264t$

$t\approx 47.6$

**c.** Since $1980+47.6=2027.6$, the year in which at least 10% of adults are predicted to be obese is 2028.

**65.** $A=90.2+41.3h$

**a.** $131.5\geq 90.2+41.3h\geq 110$

$41.3\geq 41.3h\geq 19.8$

$0.479\leq h\leq 1$ ($h=1$ means 100% humidity)

**b.** $90.2\leq 90.2+41.3h<100$

$0\leq 41.3h<9.8$

$0\leq h<0.237$

## *Exercises 1.2*

**1.** **a.** For each value of $x$ there is only one $y$.

**b.** $D=\{-7,-1,0,3,4.2,9,11,14,18,22\}$

$R=\{0,1,5,9,11,22,35,60\}$

**c.** $f(0)=1,\ f(11)=35$

**3.** This is a function, since for each $x$ there is only one $y$. $D=\{1,2,3,8,9\}$, $R=\{-4,5,16\}$

**5.** The vertical-line test shows that the graph represents $y$ as a function of $x$.

**7.** The vertical-line test shows that the graph does not represent $y$ as a function of $x$.

**9.** If $y=3x^3$, then $y$ is a function of $x$.

**11.** If $y^2=3x$, then $y$ is not a function of $x$. If, for example, $x=3$, then there are two possible values for $y$.

**13.** $R(x)=8x-10$

**a.** $R(0)=8(0)-10=-10$

**b.** $R(2)=8(2)-10=6$

**c.** $R(-3)=8(-3)-10=-34$

**d.** $R(1.6)=8(1.6)-10=2.8$

**15.** $C(x)=4x^2-3$

**a.** $C(0)=4(0)^2-3=-3$

**b.** $C(-1)=4(-1)^2-3=1$

**c.** $C(-2)=4(-2)^2-3=13$

**d.** $C\left(-\frac{3}{2}\right)=4\left(-\frac{3}{2}\right)^2-3=6$

**17.** $h(x)=x-2(4-x)^3$

**a.** $h(-1)=-1-2(4-(-1))^3$

$=-1-2(4+1)^3=-1-2(5)^3$

$=-1-2(125)=-1-250=-251$

**b.** $h(0)=0-2(4-(0))^3$

$=-2(4)^3=-2(64)=128$

**c.** $h(6)=6-2(4-(6))^3$

$=6-2(4-6)^3=6-2(-2)^3$

$=6-2(-8)=6+16=22$

**d.** $h(2.5)=2.5-2(4-(2.5))^3$

$=2.5-2(1.5)^3$

$=2.5-2(3.375)=2.5-6.75$

$=-4.25$

**19.** $f(x)=x^3-\dfrac{4}{x}$

**a.** $f\left(-\frac{1}{2}\right)=\left(-\frac{1}{2}\right)^3-\frac{4}{-\frac{1}{2}}=-\frac{1}{8}+8=\frac{63}{8}$

**b.** $f(2)=2^3-\frac{4}{2}=8-2=6$

**c.** $f(-2)=(-2)^3-\frac{4}{-2}=-8+2=-6$

**21.** $f(x)=1+x+x^2$

**a.** $f(2+1)=f(3)=1+3+3^2=13$
$f(2)+f(1)=7+3=10$
$f(2)+f(1)\neq f(2+1)$

**b.** $f(x+h)=1+(x+h)+(x+h)^2$
$=1+x+h+x^2+2xh+h^2$

**c.** $f(x)+f(h)=2+x+h+x^2+h^2$
No, $f(x+h)\neq f(x)+f(h)$.

**d.** $f(x)+h=1+x+x^2+h$
No, $f(x+h)\neq f(x)+h$.

**e.** $f(x+h)=1+(x+h)+(x+h)^2$
$=1+x+h+x^2+2xh+h^2$
$f(x)=1+x+x^2$
$f(x+h)-f(x)=h+2xh+h^2$
$=h(1+2x+h)$
$\dfrac{f(x+h)-f(x)}{h}=1+2x+h$

**23.** $f(x)=x-2x^2$

**a.** $f(x+h)=(x+h)-2(x+h)^2$
$=-2x^2-4xh-2h^2+x+h$

**b.** $f(x+h)-f(x)$
$=(x+h)-2(x+h)^2-(x-2x^2)$
$=x+h-2x^2-4xh-2h^2-x+2x^2$
$=h-4xh-2h^2$

**c.** $\dfrac{f(x+h)-f(x)}{h}=\dfrac{h-4xh-2h^2}{h}$
$=1-4x-2h$

**25.** Since (9, 10) and (5, 6) are points on the graph:

**a.** $f(9)=10$

**b.** $f(5)=6$

**27. a.** The coordinates of $Q=(1,-3)$. Since the point is on the curve, the coordinates satisfy the equation.

**b.** The coordinates of $R=(3,-3)$. They satisfy the equation.

**c.** The ordered pair $(a, b)$ satisfies the equation. Thus $b=a^2-4a$.

**d.** The $x$-values are 0 and 4. These values are also solutions of $x^2-4x=0$.

**29.** $y=x^2+4$
There is no division by zero or square roots. Domain is all the reals, i.e., $\{x: x\in \text{Reals}\}$. Since $x^2\geq 0,\ x^2+4\geq 4$, the range is reals $\geq 4$ or $\{y: y\geq 4\}$

**31.** $y=\sqrt{x+4}$
There is no division by zero. To get a real number $y$, we must have $x+4\geq 0$ or $x\geq -4$.
Domain: $x\geq -4$.
The square root is always nonnegative.
Thus, the range is $\{y: y\in \text{reals},\ y\geq 0\}$.

**33.** Domain: $x\geq 1,\ x\neq 2$

**35.** Domain: $-7\leq x\leq 7$

**37.** $f(x)=3x,\quad g(x)=x^3$

**a.** $(f+g)(x)=3x+x^3$

**b.** $(f-g)(x)=3x-x^3$

**c.** $(f\cdot g)(x)=3x\cdot x^3=3x^4$

**d.** $\left(\dfrac{f}{g}\right)(x)=\dfrac{3x}{x^3}=\dfrac{3}{x^2}$

**39.** $f(x)=\sqrt{2x},\quad g(x)=x^2$

**a.** $(f+g)(x)=\sqrt{2x}+x^2$

**b.** $(f-g)(x)=\sqrt{2x}-x^2$

**c.** $(f\cdot g)(x)=\sqrt{2x}\cdot x^2=x^2\sqrt{2x}$

**d.** $\left(\dfrac{f}{g}\right)(x)=\dfrac{\sqrt{2x}}{x^2}$

**41.** $f(x)=(x-1)^3,\quad g(x)=1-2x$

**a.** $(f\circ g)(x)=f(1-2x)=(1-2x-1)^3=-8x^3$

**b.** $(g\circ f)(x)=g((x-1)^3)=1-2(x-1)^3$

**c.** $f(f(x))=f((x-1)^3)=[(x-1)^3-1]^3$

**d.** $(f\cdot f)(x)=(x-1)^3\cdot(x-1)^3=(x-1)^6$
$\left[(f\cdot f)(x)\neq f(f(x))\right]$

**43.** $f(x)=2\sqrt{x},\quad g(x)=x^4+5$

**a.** $(f\circ g)(x)=f(x^4+5)=2\sqrt{x^4+5}$

**b.** $(g\circ f)(x)=g(2\sqrt{x})=(2\sqrt{x})^4+5$
$=16x^2+5$

**c.** $f(f(x))=f(2\sqrt{x})=2\sqrt{2\sqrt{x}}$

**d.** $(f \cdot f)(x) = 2\sqrt{x} \cdot 2\sqrt{x} = 4x$
$[(f \cdot f)(x) \neq f(f(x))]$

**45. a.** $f(20) = 103,000$ means it will take 20 years to pay off a debt of \$103,000 (at \$800 per month and 7.5% compounded monthly.)

**b.**
$f(5+5) = f(10) = 69,000$;
$f(5) + f(5) = 80,000$;
No.

**c.** It will take 15 years to pay off the debt, i.e., $89,000 = f(15)$.

**47. a.** From the figure, $f(64) = \$866$ and $f(67) = \$1,080$.

**b.** $f(68) = \$1,160$. Starting benefits at age 68 provides \$1,160 per month.

**c.** $f(66) - f(62) = \$1000 - \$750 = \$250$.
Starting benefits at age 66 gives \$250 more per month than starting benefits at age 62.

**49. a.** $W(100) \approx 155$ million and $O(140) \approx 120$ million

**b.** $W(120) = 166.3$. In 2020 there are expected to be 166.3 million white, non-Hispanics in the civilian, non-institutional labor force (CN-ILF).

**c.** $O(90) = 42.6$. In 1990 there were 42.6 million non-Whites or Hispanics in the CN-ILF..

**d.** $(W-O)(120) = W(120) - O(120)$
$= 166.3 - 86.8$
$= 79.5$

In 2020 there are expected to be 79.5 million more White non-Hispanics in the CN-ILF than others.

**e.** $(W+O)(150) = W(150) + O(150)$
$= 169.4 + 143.0$
$= 312.4$
In 2050 the total size of the CN-ILF is expected to be 312.4 million.

**f.** $(W-O)(100)$ is greater than $(W-O)(140)$ because the graphs are further apart at $t = 100$ than at $t = 140$.

**51.** $C = \frac{5}{9}F - \frac{160}{9}$

**a.** $C$ is a function of $F$.

**b.** Mathematically, the domain is all reals.

**c.** Domain: $\{F : 32 \leq F \leq 212\}$
Range: $\{C : 0 \leq C \leq 100\}$

**d.** $C(40) = \frac{5}{9}(40) - \frac{160}{9}$
$= \frac{200-160}{9} = \frac{40}{9} = 4.44°C$

**53.** $C(x) = 300x + 0.1x^2 + 1200$

**a.** $C(10) = 300(10) + 0.1(10)^2 + 1200$
$= 3000 + 0.1(100) + 1200$
$= 3000 + 10 + 1200$
$= \$4210$

**b.** $C(100) = 300(100) + 0.1(100)^2 + 1200$
$= \$32,200$

**c.** The value $C(100)$ is the total cost of producing 100 items, which is \$32,200.

**55.** $C(p) = \frac{7300p}{100-p}$

**a.** Domain: $\{p : 0 \leq p < 100\}$

**b.** $C(45) = \frac{7300(45)}{100-45} = \frac{328,500}{55} = \$5972.73$

**c.** $C(90) = \frac{7300(90)}{100-90} = \frac{657,000}{10} = \$65,700$

**d.** $C(99) = \frac{7300(99)}{100-99} = \frac{722,700}{1} = \$722,700$

**e.** $C(99.6) = \frac{7300(99.6)}{100-99.6} = \frac{727,080}{0.4} = \$1,817,700$

In each case above, to remove $p$% of the particulate pollution would cost $C(p)$.

**57.** **a.** $P(q(t)) = P(1000+10t)$

$$= 180(1000+10t) - \frac{(1000+10t)^2}{100} - 200$$

$$= 169,800 + 1600t - t^2$$

**b.** $q(15) = 1000 + 10(15) = 1150$

$P(q(15)) = \$193,575$

**59.** $R = f(C) \quad C = g(A)$

**a.** $(f \circ g)(x) = f(C) = R$

**b.** $(g \circ f)(x)$ is not defined.

**c.** $A$ is the independent variable and $R$ is the dependent variable. Revenue depends on money spent for advertising.

**61.** length $= x$ width $= y$ $L = 2x + 2y$

$1600 = xy$ or $y = \frac{1600}{x}$

$$L = 2x + 2\left(\frac{1600}{x}\right) = 2x + \frac{3200}{x}$$

**63.** $R = (30+x)(100-2x)$

## *Exercises 1.3*

**1.** $3x + 4y = 12$

$x$-intercept: $y = 0$ then $x = 4$.

$y$-intercept: $x = 0$ then $y = 3$.

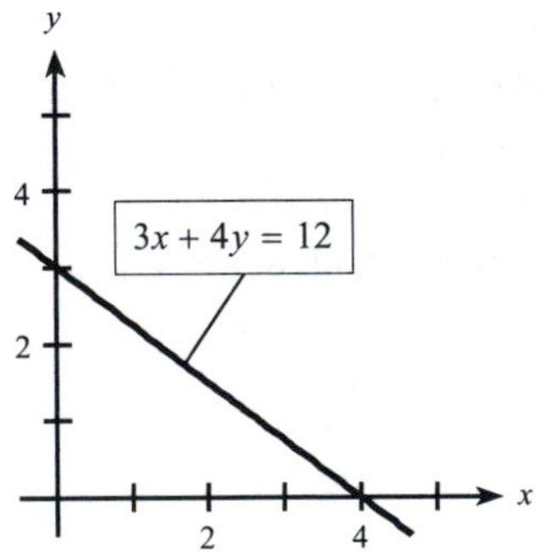


**3.** $5x - 8y = 60$

$x$-intercept: $y = 0$ then $x = 12$.

$y$-intercept: $x = 0$ then $y = -7.5$.

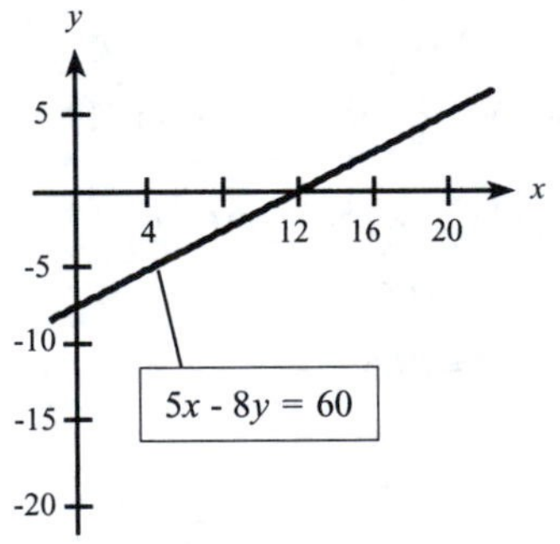


**5.** $(22,11)$ and $(15,-17)$

$$m = \frac{y_2 - y_1}{x_2 - x_1} = \frac{-17-11}{15-22} = \frac{-28}{-7} = 4$$

**7.** $(3,-1)$ and $(-1,1)$

$$m = \frac{y_2 - y_1}{x_2 - x_1} = \frac{1-(-1)}{-1-3} = \frac{2}{-4} = -\frac{1}{2}$$

**9.** (3, 2) and (–1, 2)

$$m=\frac{y_2-y_1}{x_2-x_1}=\frac{2-2}{3-(-1)}=\frac{0}{4}=0$$

**11.** A horizontal line has a slope of $0$ .

**13.** $(3,2)$ and $(-1,2)$

The rate of change is equivalent to the slope of the line.

$$m=\frac{y_2-y_1}{x_2-x_1}=\frac{2-2}{-1-3}=\frac{0}{-4}=0$$

**15. a.** The slope is negative since the line slants downward toward the right.

**b.** The slope is undefined since the line is vertical.

**17.** $y=\frac{7}{3}x-\frac{1}{4},\ m=\frac{7}{3},\ b=-\frac{1}{4}$

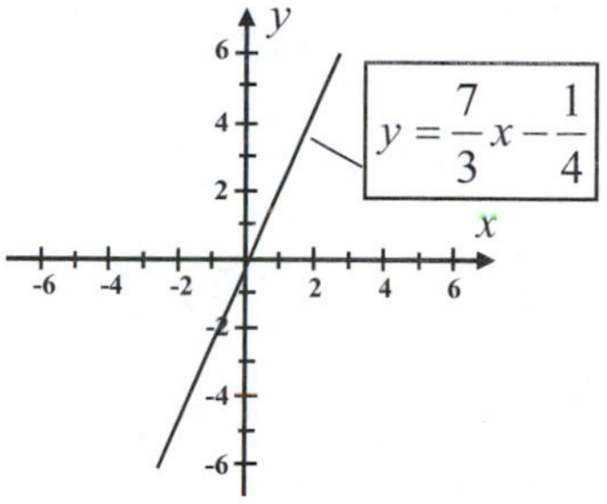


**19.** $y=3$ or $y=0x+3,\ m=0,\ b=3$

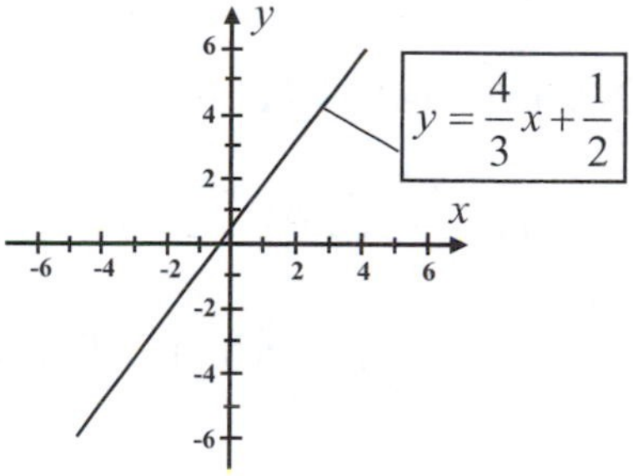


**21.** $x=-8$

Slope is undefined.

There is no $y$-intercept.

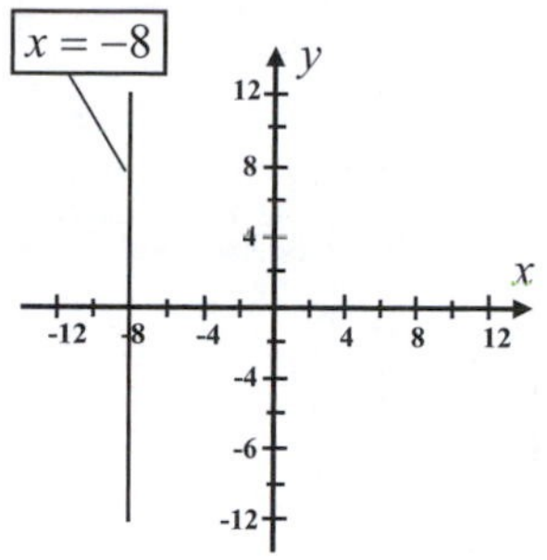


**23.** $2x+3y=6$ or $y=-\frac{2}{3}x+2,\ m=-\frac{2}{3},\ b=2$ .

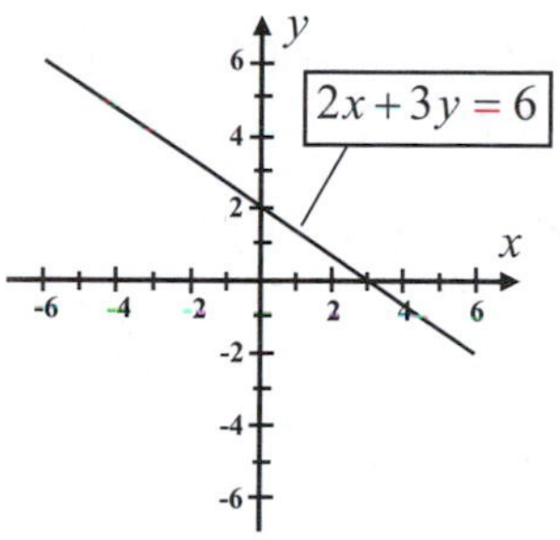


**25.** $m=\frac{1}{2},\ b=-3;\ y=\frac{1}{2}x-3$

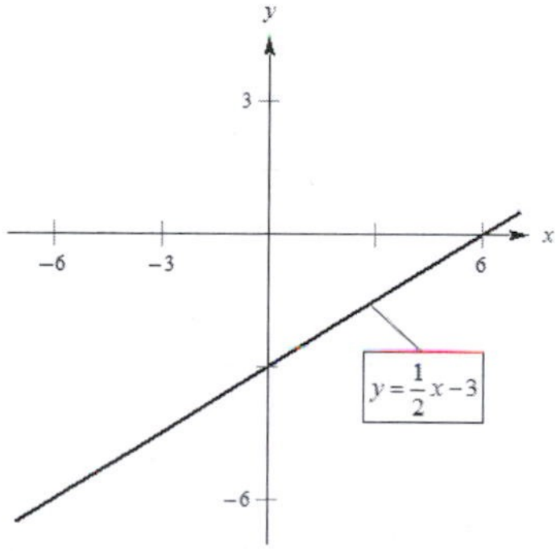


**27.**

$m=-2,\ b=\frac{1}{2},\ y=-2x+\frac{1}{2}$

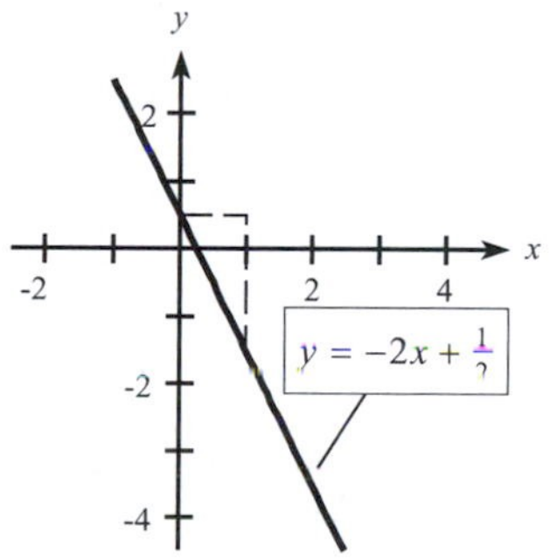

**29.** $P(2, 0),\ m=-5$

$y-0=-5(x-2)$

$y=-5x+10$

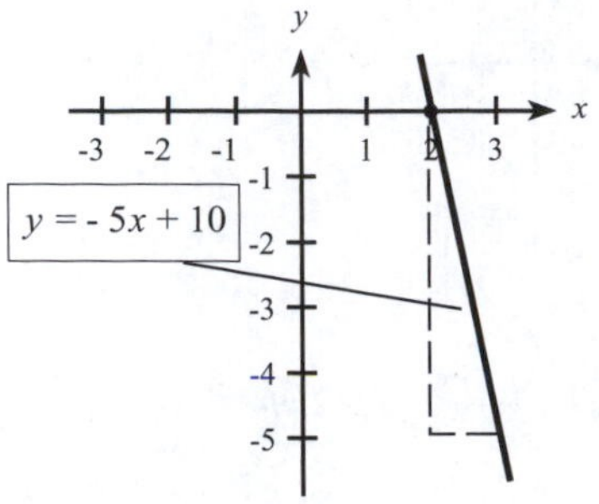


**31.** $P(-1,4),\ m=-\frac{3}{4}$

$y-4=-\frac{3}{4}(x-(-1))$

$y=-\frac{3}{4}x+\frac{13}{4}$

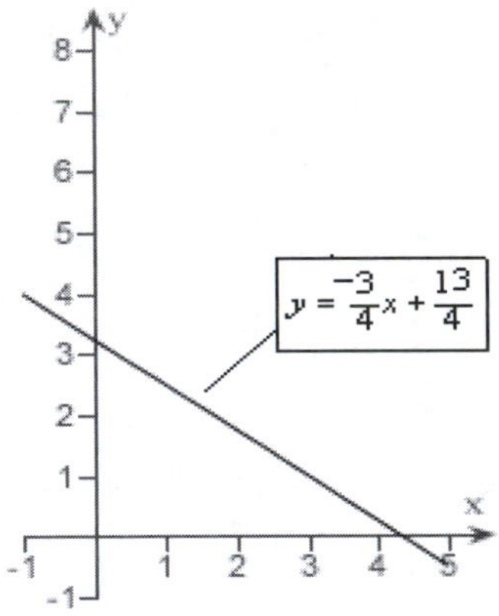


**33.** $P(-1, 1)$, $m$ is undefined

$x=-1$

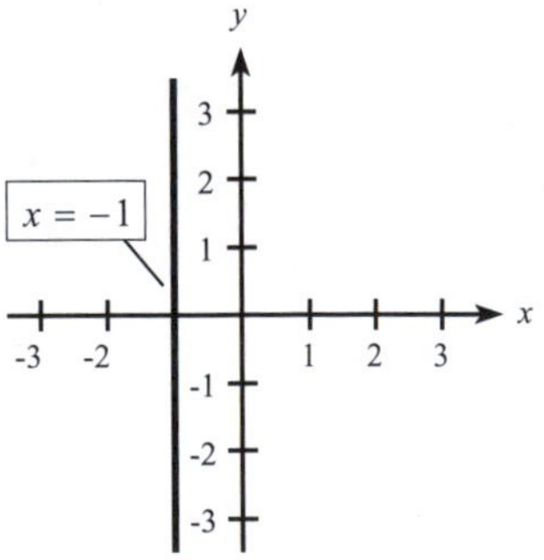


**35.** $P_1=(3,2), P_2=(-1,-6)$

$m=\frac{-6-2}{-1-3}=2$

$y-2=2(x-3)$

$y=2x-4$

**37.** $P_1=(7,3), P_2=(-6,2)$

$m=\frac{2-3}{-6-7}=\frac{-1}{-13}=\frac{1}{13}$

$y-3=\frac{1}{13}(x-7)$

$y=\frac{1}{13}x-\frac{7}{13}+3$

$y=\frac{1}{13}x+\frac{32}{13}$ or $-x+13y=32$

**39.** $P_1=(-4,0), P_2=(0,-12)$

$m=\frac{-12-0}{0-(-4)}=\frac{-12}{4}=-3$

$y=-3x-12$

**41.** $3x+2y=6$ and $2x-3y=6$

$y=-\frac{3}{2}x+3$ $\quad$ $y=\frac{2}{3}x-2$

Lines are perpendicular since $\left(-\frac{3}{2}\right)\left(\frac{2}{3}\right)=-1$.

**43.** $6x-4y=12$ and $3x-2y=6$

$y=\frac{6}{4}x-\frac{12}{4}$ $\quad$ $y=\frac{3}{2}x-3$

or $y=\frac{3}{2}x-3$

Lines are the same.

**45.** If $3x+5y=11$, then $y=-\frac{3}{5}x+\frac{11}{5}$. So, $m=-\frac{3}{5}$. A line parallel will have the same slope. Thus, $m=-\frac{3}{5}$ and $P=(-2,-7)$ gives $y-(-7)=-\frac{3}{5}(x-(-2))$ which simplifies to $y=-\frac{3}{5}x-\frac{41}{5}$.

**47.** If $5x-6y=4$, then $y=\frac{5}{6}x-\frac{4}{6}$. Slope of the perpendicular line is $-\frac{6}{5}$. Thus $m=-\frac{6}{5}$ and $P=(3, 1)$ gives $y-1=-\frac{6}{5}(x-3)$ which simplifies to $y=-\frac{6}{5}x+\frac{23}{5}$.

**49. a.**

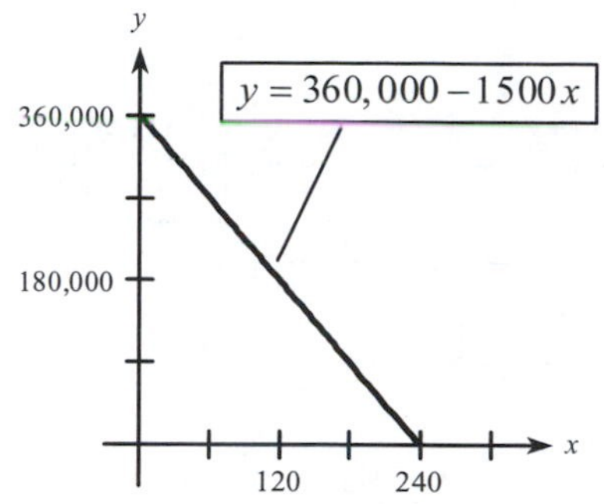


**b.** $0 = 360,000 - 1500x$

$$x = \frac{360000}{1500} = 240 \text{ months}$$

In 240 months, the building will be completely depreciated.

**c.** (60, 270,000) means that after 60 months the value of the building will be $270,000.

**51.** $y = 1.36x + 55.98$

**a.** $m = 1.36;\ b = 55.98$

**b.** The percent of the U.S. population with Internet service is changing at the rate of 1.36 percentage points per year.

**c.**

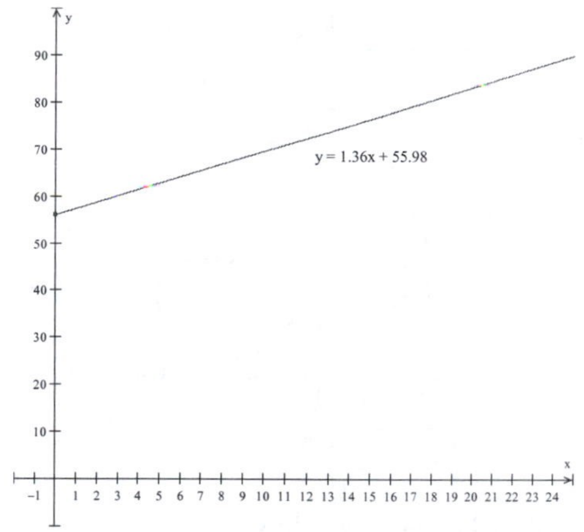


**53. a.** $m = -0.065$; $b = 31.39$

**b.** The $F$-intercept represents the percent of the world's land that was forest in 1990.

**c.** -0.065 percentage points per year. This means that after 1990, the world forest area as a percent of land area changes by -0.065 percentage points per year.

**55.** $F = 0.78M - 1.316$

**a.** $m = 0.78$

**b.** For each $1 increase in male earnings, the female's earning increases by only $0.78.

**c.** $F = 0.78(60) - 1.316$

$= 45.484$ thousand

=$48,484

**57.** $y = 16.37 + 0.0838x$

**59. a.** $(1950, 62.2)$ and $(2050, 191.8)$

$$m = \frac{191.8 - 62.2}{2050 - 1950} = \frac{129.6}{100} = 1.296$$

$$y - 62.2 = 1.296(x - 1950)$$

$$y = 1.296x - 2465$$

**b.** The slope, 1.296, indicates that the size of the U.S. civilian workforce changes at the rate of 1.296 million workers per year.

**61. a.** $(2020, 120.56)$ and $(2050, 276.05)$

$$m = \frac{276.05 - 120.56}{2050 - 2020} = \frac{155.49}{30} = 5.183$$

$$y - 120.56 = 5.183(x - 2020)$$

$$y = 5.183x - 10,349.10$$

**b.** The consumer price index increases at the rate of $5.18 per year.

**63.** $(x, p)$ is the reference. (0, 85000) is one point.

$$m = \frac{-1700}{1} = -1700$$

$p - 85,000 = -1700(x - 0)$ or

$p = -1700x + 85,000$

**65.** $(t, R)$ is the ordered pair.

$$P_1 = \left(\frac{7}{2}, 11\right),\ P_2 = (6, 19)$$

$$m = \frac{19 - 11}{6 - \frac{7}{2}} = \frac{8}{\frac{5}{2}} = \frac{16}{5} = 3.2$$

$R - 19 = 3.2(t - 6)$ or $R = 3.2t - 19.2 + 19$ or

$R = 3.2t - 0.2$

**67.** $P_1 = (200, 25)$ $P_2 = (250, 49)$

$$m = \frac{49 - 25}{250 - 200}$$

$$= \frac{24}{50}$$

$$= 0.48$$

$y - 25 = 0.48(x - 200)$ or $y = 0.48x - 71$

# Chapter 1: Linear Equations and Functions

## *Exercises 1.4*

**1.**

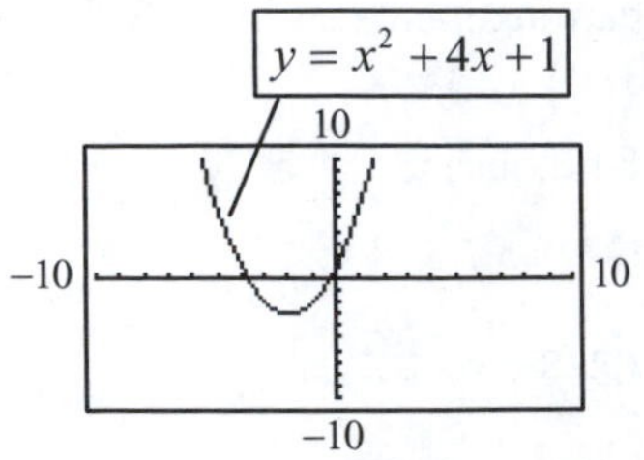


**3.**

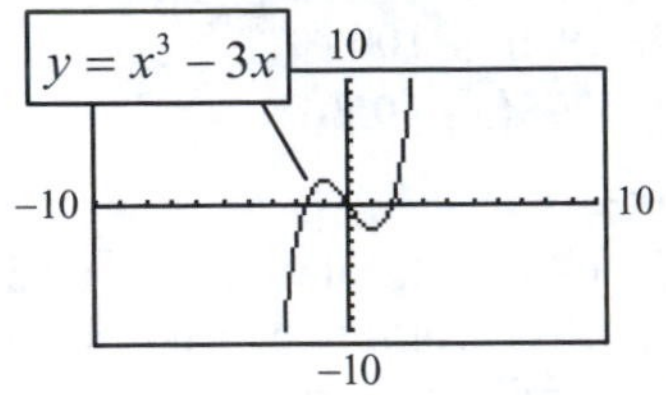


**5.**

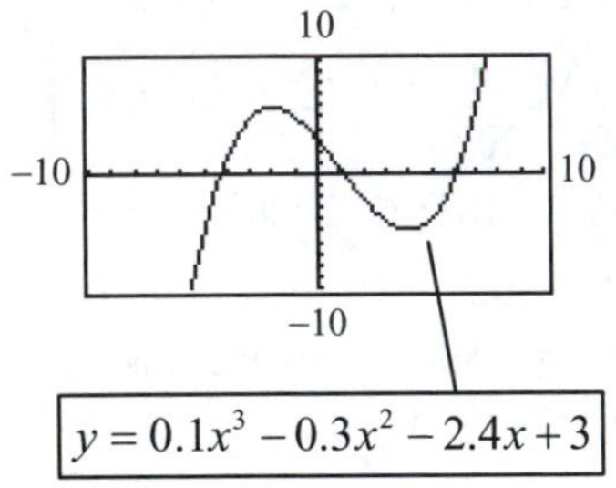


**7.**

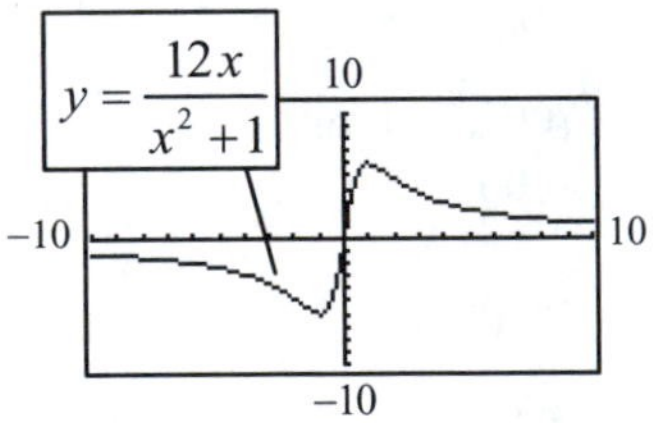


**9.**

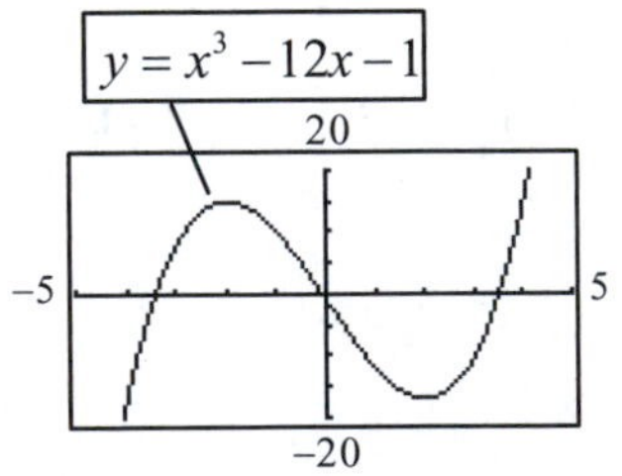


**11.**

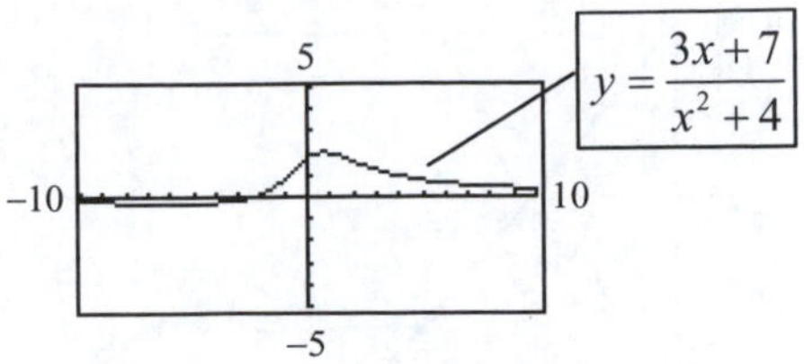


**13. a.**

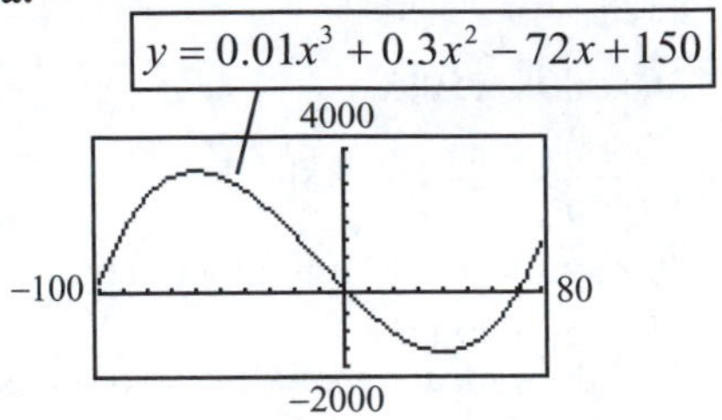


**b.** Standard viewing window

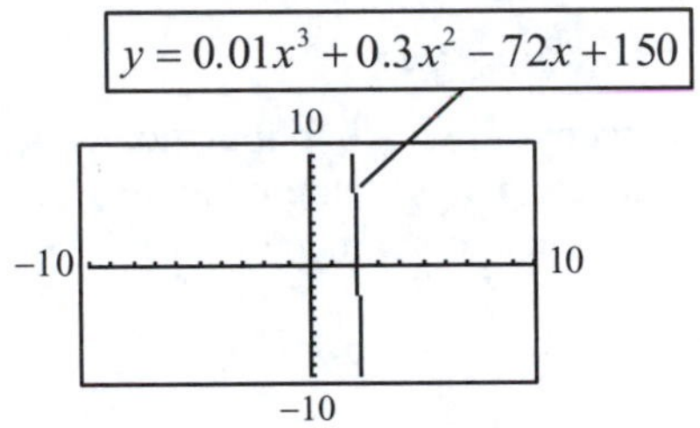


**15.**

**a.**

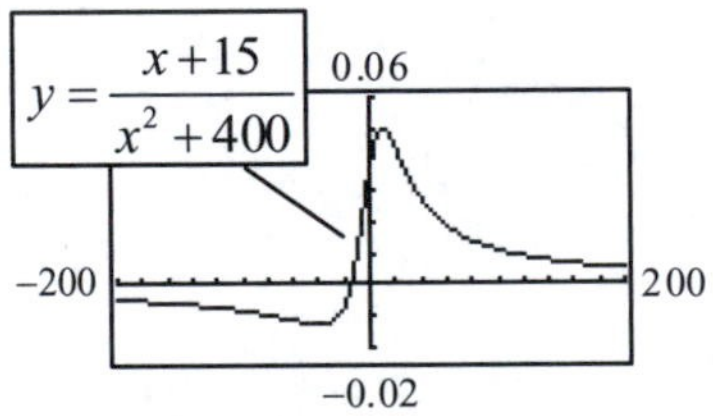


**b.** Standard Window

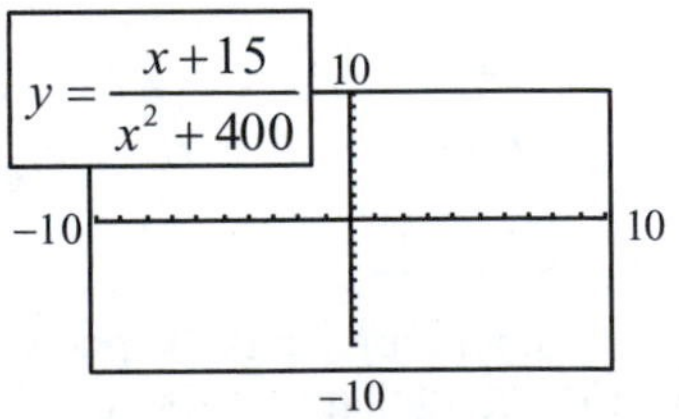


**17. a.** $y$-intercept $= -0.03$

$x$-intercept: $0.001x = 0.03$

$x = 30$

**b.**

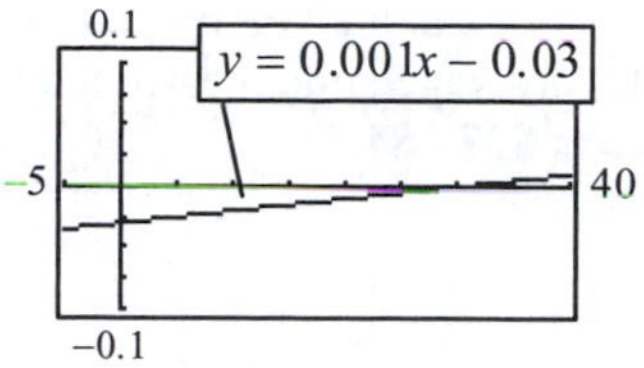


**19.** $y = -0.15(x - 10.2)^2 + 10$

There is no min.
Max value of $y = 10$.
$x$-intercepts:

$$0 = -0.15(x - 10.2)^2 + 10$$

$$(x - 10.2)^2 = \frac{10}{0.15} = 66.66$$

$$x - 10.2 = \pm\sqrt{66.66} \approx \pm 8$$

$$x = 10.2 \pm 8 \text{ or } x = 2.2 \text{ or } 18.2$$

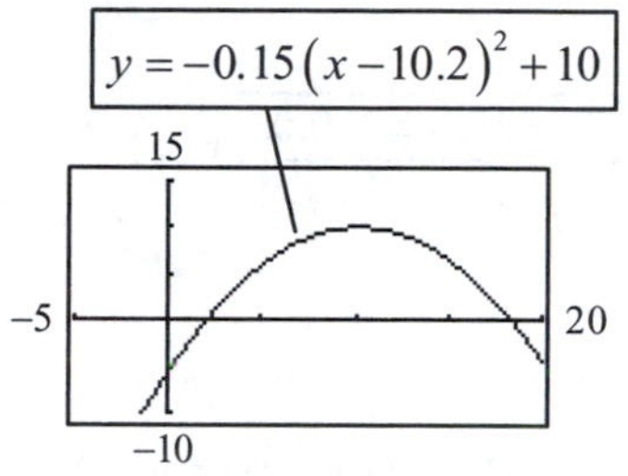


**21.** If $x = 0$, $y = -42$.
A suggested window is shown below.

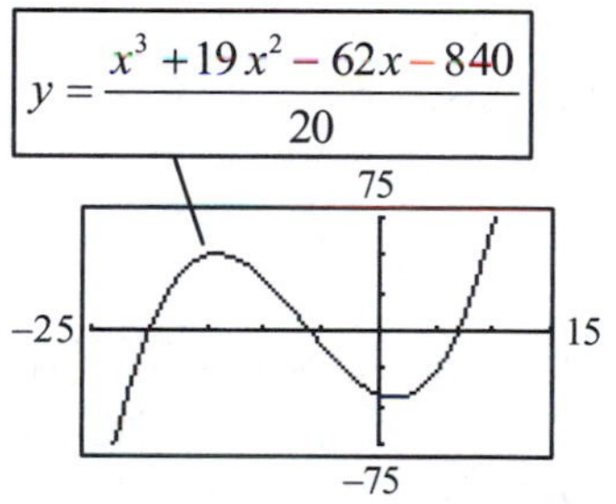


**23.** $4x - y = 8$

$y = 4x - 8$

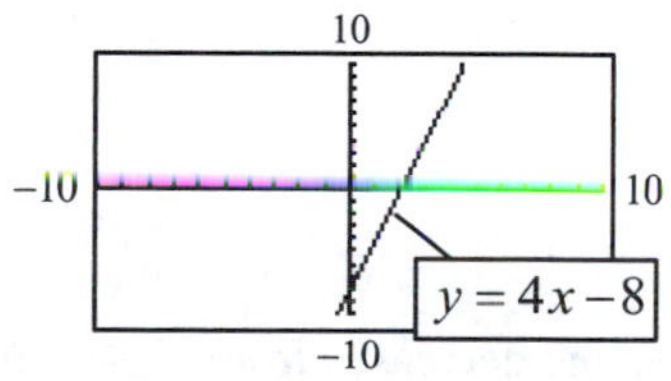


**25.** $4x^2 + 2y = 5$

$$2y = -4x^2 + 5$$

$$y = -2x^2 + \frac{5}{2}$$

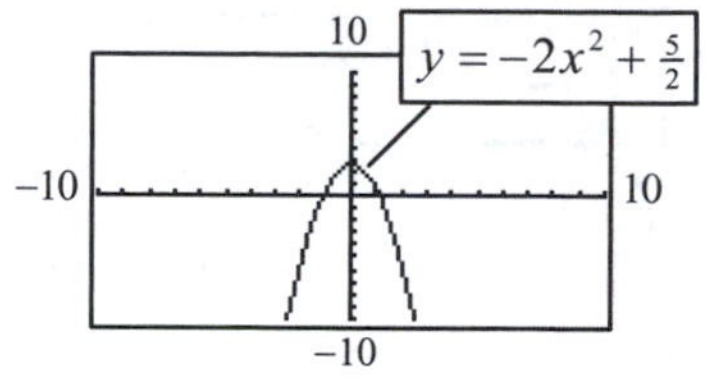


**27.** $x^2 - 6 = 2y$

$$2y = x^2 - 6$$

$$y = \frac{1}{2}x^2 - 3$$

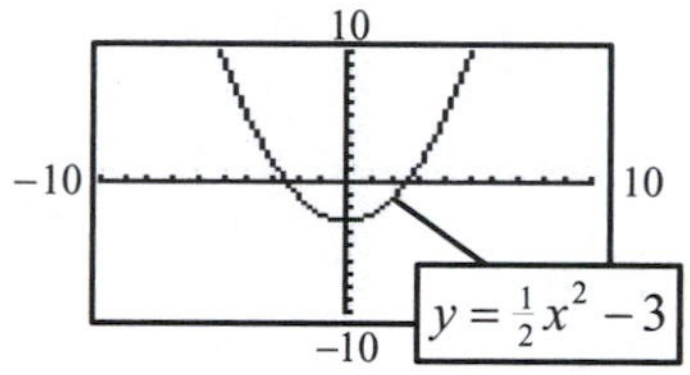


Not linear

**29.** $f(x) = x^3 - 3x^2 + 2$

$$f(-2) = (-2)^3 - 3(-2)^2 + 2 = -8 - 12 + 2 = -18$$

$$f\left(\frac{3}{4}\right) = \left(\frac{3}{4}\right)^3 - 3\left(\frac{3}{4}\right)^2 + 2 = 0.734375$$

Use your graphing calculator, and evaluate the function at these two points. If either of your answers differ, can you explain the difference?

**31.** As $x$ gets large, $y$ approaches 12.
When $x = 0$, $y = -12$, $x$ intercepts at $\pm$ 1.

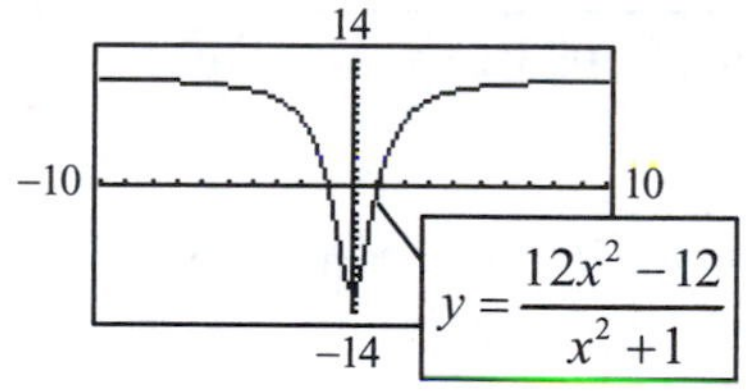

**33.** $y = \dfrac{x^2 - x - 6}{x^2 + 5x + 6} = \dfrac{(x-3)(x+2)}{(x+3)(x+2)}$

What happens to $y$ as $x$ approaches –3? –2?

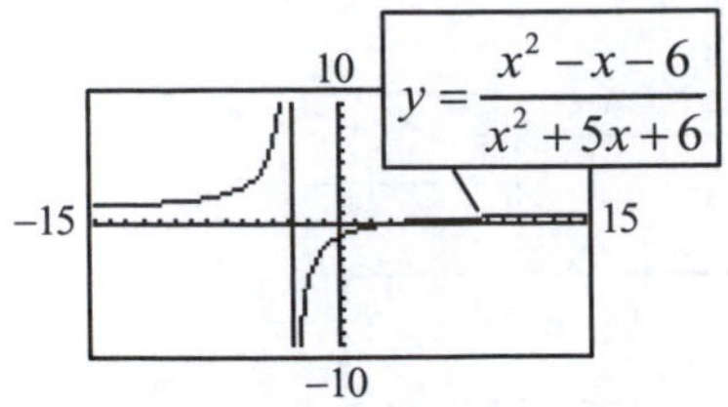


**35.** $6x - 21 = 0$

$6x = 21$

$x = \dfrac{21}{6} = \dfrac{7}{2}$

**37.** $x^2 - 3x - 10 = 0$

$(x-5)(x+2) = 0$

$x = -2$ or 5

**39. a.**

$y = x^2 - 7x - 9$

$\Rightarrow x^2 - 7x - 9 = 0$

$\Rightarrow (x + 1.1098)(x - 8.1098) = 0$

$\Rightarrow x = -1.1098$ or $x = 8.1098$

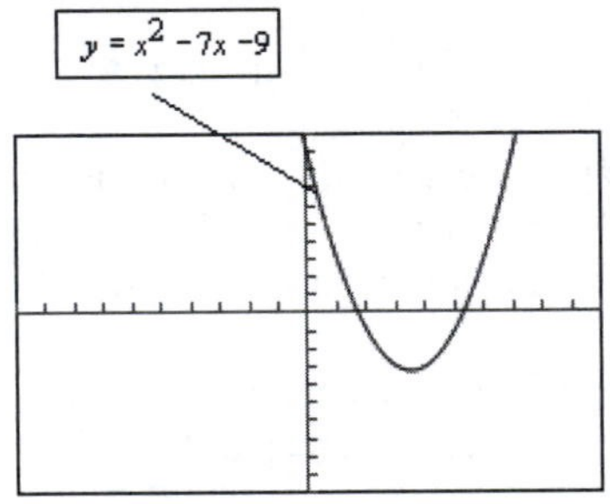


**b.** –1.1098, 8.1098

**41. a.** The graph of $F = 0.78M - 1.316$ is

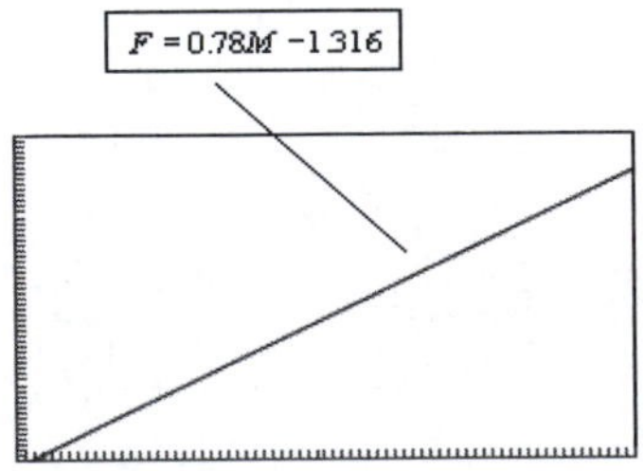


**b.** When average annual earnings for males is \$50,000, average annual earnings for females is \$37,684.

**c.** \$47,434

**43. a.**

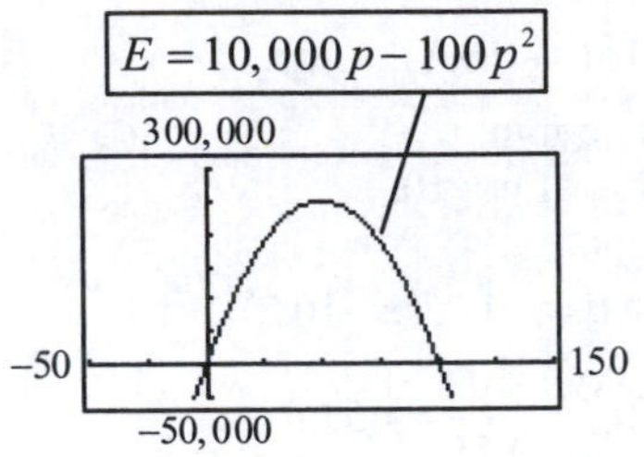


**b.** $E \geq 0$ when $0 \leq p \leq 100$.

**45. a.**

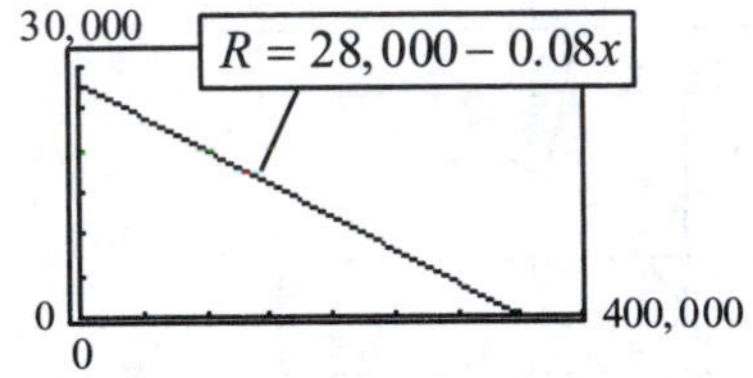


**b.** The rate is –0.08. As more people become aware of the product, there are fewer to learn about it

**47. a.** $x$-min = 0, $x$-max = 130

**b.** $P$-max = 65 is reasonable

**c.**

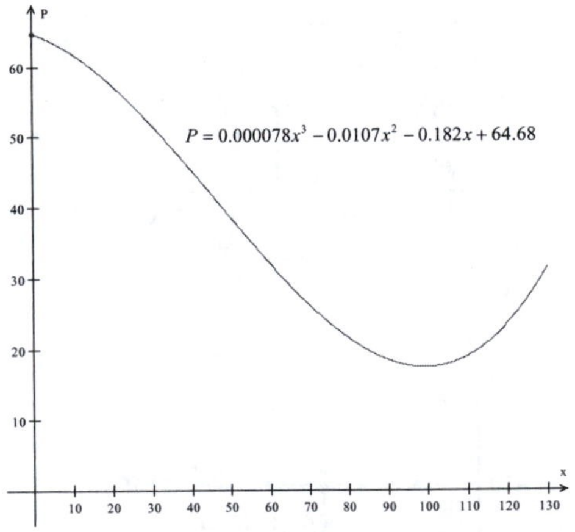


**d.** The percentage decreases from 57.4% in 1920 to 17.5% in 2000, and then it increases to 31.4% in 2030.

**49. a.**

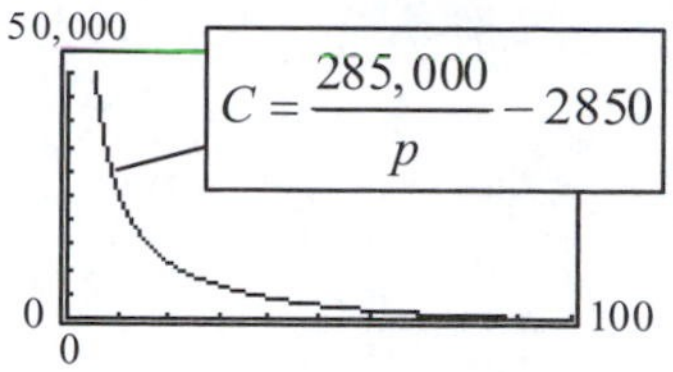


**b.** Near $p = 0$, cost grows without bound.

**c.** The coordinates of the point mean that the cost of obtaining stream water with 1% of the current pollution levels would cost \$282,150.

**d.** The $p$-intercept means that the cost of stream water with 100% of the current pollution levels would cost \$0.

**51. a.**

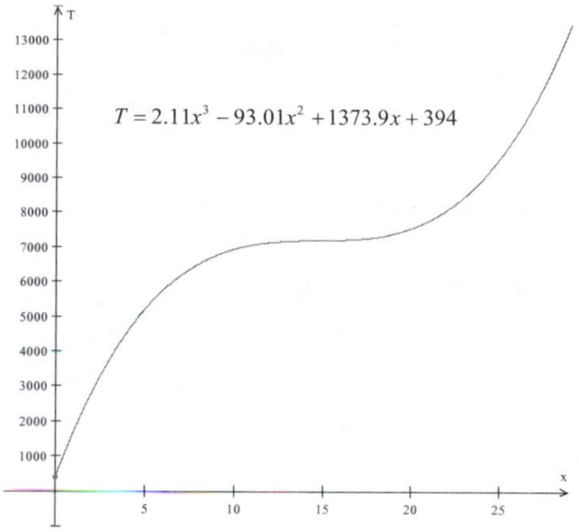


**b.** The graph is increasing. The per capita federal tax burden is increasing.

## *Exercises 1.5*

**1.** Solution: $(-1, -2)$

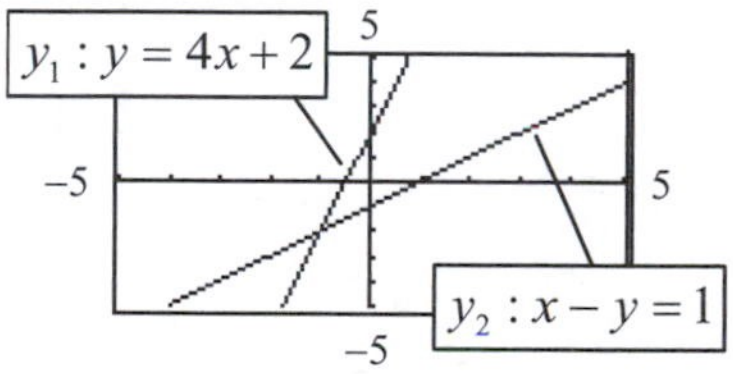


**3.** Infinitely many solutions.

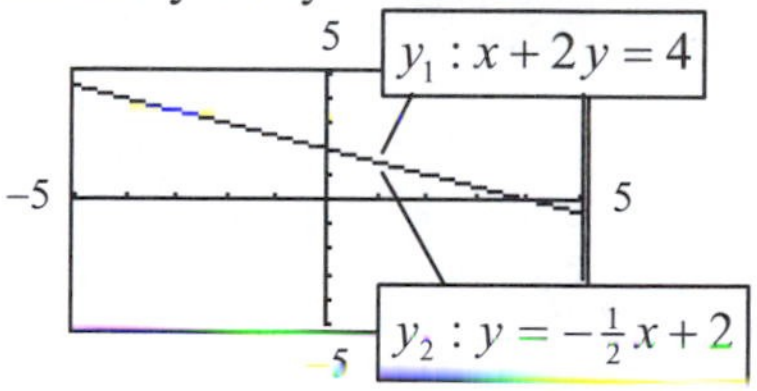


**5.** Solution: $(2, 2)$

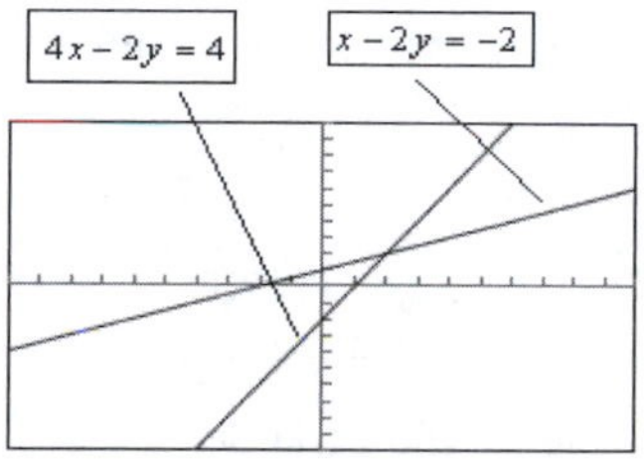


**7.** Solution: No solution, since the graphs do not intersect.

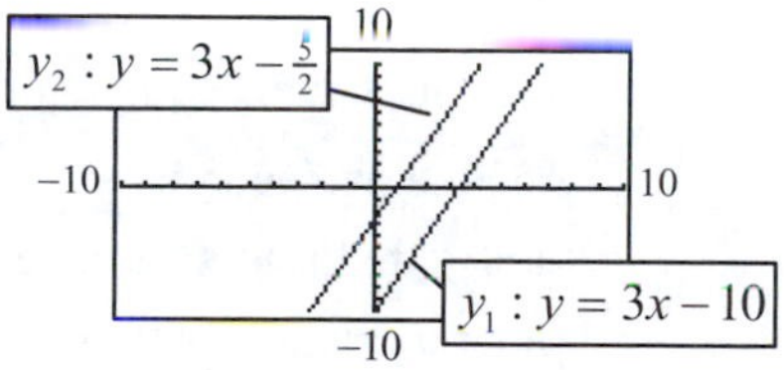

**9.** $3x-2y=6$

$4y=8$ Solve for $y$. $y=\frac{8}{4}=2$

Substitute for this variable in $3x-2(2)=6$

first equation and solve for $3x=6+4=10$

the other variable. $x=\frac{10}{3}$

The solution of the system is $x=\frac{10}{3}$ and $y=2$, or $\left(\frac{10}{3},2\right)$.

**11.** $2x-y=2$

$3x+4y=6$ Solve for $y$. $y=2x-2$

Substitute for this variable in $3x+4(2x-2)=6$

second equation and solve for $3x+8x-8=6$

the other variable. $11x=14$

$x=14/11$

Solve for $y$: $y=2\left(\frac{14}{11}\right)-2=\frac{6}{11}$

The solution of the system is $x=\frac{14}{11}$ and $y=\frac{6}{11}$, or $\left(\frac{14}{11},\frac{6}{11}\right)$.

**13.** $7x+2y=26$ Multiply 1st equation by 2. $14x+4y=52$

$3x-4y=16$ $\quad$ $3x-4y=16$

Add the two equations. $17x=68$

Solve for the variable. $x=4$

Substitute for this variable in $3(4)-4y=16$

either original equation and $4y=4$

solve for the other variable. $y=1$

The solution of the system is $x=4$ and $y=1$, or $(4,1)$.

**15.** $3x+4y=1$ Multiply 1st equation by 3. $9x+12y=3$

$2x-3y=12$ Multiply 2nd equation by 4. $8x-12y=48$

Add the two equations. $17x=51$

Solve for the variable. $x=3$

Substitute for this variable in $3(3)+4y=1$

either original equation and $4y=-8$

solve for the other variable. $y=-2$

The solution of the system is $x=3$ and $y=-2$, or $(3,-2)$.

**17.** $-4x+3y=-5$ Multiply first equation by 3. $-12x+9y=-15$

$3x-2y=4$ Multiply second equation by 4. $12x-8y=16$

Add the two equations. $y=1$

Substitute for this variable in $-4x+3(1)=-5$

either original equation and $-4x=-8$

solve for the other variable. $x=2$

The solution of the system is $x=2$ and $y=1$.

**19.** $0.2x - 0.3y = 4$ $\quad$ $0.20x - 0.3y = 4$

$2.3x - \quad y = 1.2$ $\quad$ Multiply 2nd equation by 0.3. $\quad$ $0.69x - 0.3y = 0.36$

Subtract the two equations. $\quad$ $-0.49x \quad = 3.64$

Solve for the variable. $\quad$ $x = -\frac{52}{7}$

Substitute, solve for $y$. $\quad$ $y = -\frac{128}{7}$

The solution of the system is

$x = -\dfrac{52}{7}$ and $y = -\dfrac{128}{7}$, or $\left(-\dfrac{52}{7}, -\dfrac{128}{7}\right)$.

**21.** $\frac{5}{2}x - \frac{7}{2}y = -1$ $\quad$ Multiply first equation by 6. $\quad$ $15x - 21y = -6$

$8x + 3y = 11$ $\quad$ Multiply second equation by 7. $\quad$ $56x + 21y = 77$

Add the two equations. $\quad$ $71x \quad = 71$

Substitute for this variable in $\quad$ $x = 1$

either original equation and $\quad$ $8(1) + 3y = 11$

solve for the other variable. $\quad$ $3y = 3$

$y = 1$

The solution of the system is $x = 1$ and $y = 1$, or $(1, 1)$.

**23.** $4x + 6y = 4$ $\quad$ $4x + 6y = 4$

$2x + 3y = 2$ $\quad$ Multiply second equation by $-2$. $\quad$ $-4x - 6y = -4$

Add the two equations: $\quad$ $0 = 0$

There are infinitely many solutions.
The system is dependent. Solve for one
of the variables in terms of the remaining variable:

$y = \dfrac{2}{3} - \dfrac{2}{3}x$.

Then a general solution is $\left(c, \dfrac{2}{3} - \dfrac{2}{3}c\right)$,

where any value of $c$ will give a particular solution.

**25.** $\begin{cases} y = 8 - \dfrac{3x}{2} \\ y = \dfrac{3x}{4} - 1 \end{cases}$

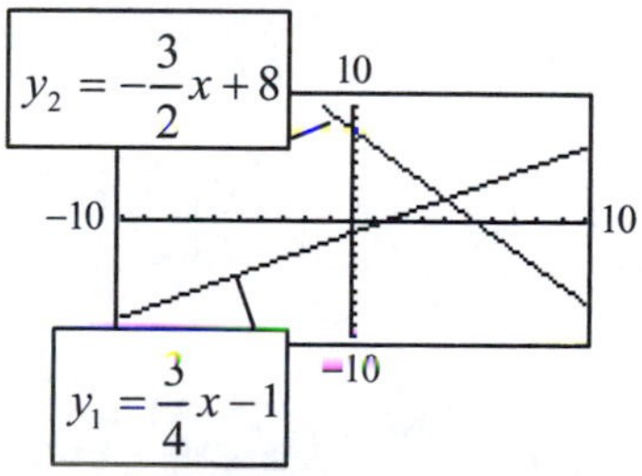


Solution: $(4, 2)$

**27.** $\begin{cases} y_1: 5x+3y=-2 \\ y_2: 3x+7y=4 \end{cases}$

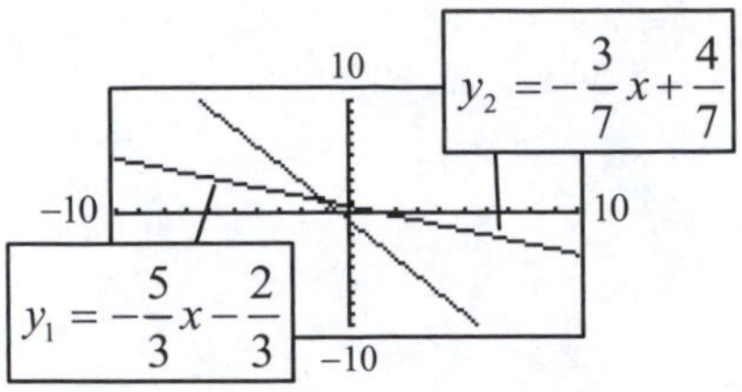


Solution: (−1, 1)

**29.** Eq. 1 $x+2y+z=2$ Steps 1, 2, and 3 of the systematic
Eq. 2 $-y+3z=8$ procedure are completed.
Eq. 3 $2z=10$ Step 4: $z=5$
From Eq. 2 $-y+3(5)=8$ or $y=7$
From Eq. 1 $x+2(7)+5=2$ or $x=-17$
The solution is $x=-17, y=7, z=5$.

**31.** Eq. 1 $x-y-8z=0$ Steps 1 and 2 of the systematic
Eq. 2 $y+4z=8$ procedure are completed.
Eq. 3 $3y+14z=22$
Step 3: $(-3)\times$ Eq 2 added to Eq. 3 gives $2z=-2$ or $z=-1$.
From Eq. 2 $y+4(-1)=8$ or $y=12$
From Eq. 1 $x-12-8(-1)=0$ or $x=4$
The solution is $x=4, y=12, z=-1$ or (4, 12, −1).

**33.** Eq. 1 $x+4y-2z=9$ Step 1 is completed.
Eq. 2 $x+5y+2z=-2$
Eq. 3 $x+4y-28z=22$
Step 2:
$x+4y-2z=9$ Eq. 1
Eq. 4 $y+4z=-11$ $(-1)\times$ Eq. 1 added to Eq. 2
Eq. 5 $-26z=13$ $(-1)\times$ Eq. 1 added to Eq. 3
Step 3 is also completed.
Step 4: $z=-\frac{1}{2}$ from Eq. 5.

From Eq. 4 $y+4\left(-\frac{1}{2}\right)=-11$ or $y=-9$

From Eq. 1 $x+4(-9)-2\left(-\frac{1}{2}\right)=9$ or $x=44$

The solution is $x=44, y=-9, z=-\frac{1}{2}$ or $\left(44, -9, -\frac{1}{2}\right)$.

**35.** **a.** If $F = 2C + 30$ and $F = 1.8C + 32$, then

$$2C + 30 = 1.8C + 32$$
$$0.2C = 2$$
$$C = 10$$

The formulas agree when the temperature is $10°$ Celsius and $50°$ Fahrenheit.

**b.** At temperatures above $10°$ Celsius, the tourist formula overestimates the actual Fahrenheit temperature. The slope of the tourist formula, 2, means that $2°$ F is added for every $1°$ C, but the actual change is $1.8°$ C.

**37.** **a.** $x + y = 1800$ Total number of tickets

**b.** $20x =$ revenue from \$20 tickets

**c.** $30y =$ revenue from \$30 tickets

**d.** $20x + 30y = 42,000$ Total Revenue

**e.** Multiply equation from part (a) by $-20$.

$$-20x - 20y = -36000$$
$$20x + 30y = 42000$$
$$10y = 6000$$
$$y = 600$$

Substitution into equation from part (a) gives $x = 1200$.
Sell 1200 of the \$20 tickets and 600 of the \$30 tickets.

**39.** $x$ = amount of safe investment.
$y$ = amount of risky investment.
$x + y = 145,600$ Total amount invested
$0.1x + 0.18y = 20,000$ Income from investments
The solution is the solution of the above system of equations.

$$x + \quad y = 145,600$$
$$x + 1.8y = 200,000 \quad (10) \times \text{ second equation}$$
$$0.8y = 54,400 \quad \text{Subtract equations}$$
$$y = 68,000 \quad \text{Solve for } y \text{ or amount of risky investment.}$$

Substituting $y = 68,000$ into one of the original equations we have $x + 68,000 = 145,600$ or $x = \$77,600$.
Solution: Put \$77,600 in a safe investment and \$68,000 in a risky investment.

**41.** $x$ = amount invested at 10%.
$y$ = amount invested at 12%.

$$x + y = 470,000 \quad \rightarrow \quad x + \quad y = 470,000 \qquad x + 200,000 = 470,000$$
$$0.10x + 0.12y = 51,000 \quad \rightarrow \quad -x - 1.2y = -510,000 \qquad x = 270,000$$
$$-0.2y = -40,000$$
$$y = \$200,000$$

**43.** $A$ = ounces of substance A.
$B$ = ounces of substance B.

Required ratio $\frac{A}{B} = \frac{3}{5}$ gives $5A - 3B = 0$.

Required nutrition is $5\%A + 12\%B = 100\%$. This gives $5A + 12B = 100$.
The % notation can be trouble. Be careful! Now we can solve the system.

$$5A - 3B = 0$$
$$5A + 12B = 100$$
$$15B = 100 \quad \text{Subtract first equation from second.}$$
$$B = \frac{100}{15} = \frac{20}{3}$$

Substituting into the original equation gives $5A - 3\left(\frac{20}{3}\right) = 0$ or $A = 4$.

The solution is 4 ounces of substance A and $6\frac{2}{3}$ ounces of substance B.

**45.** $x$ = population of species A.
$y$ = population of species B.
$2x + y = 10{,}600$ units of first nutrient
$3x + 4y = 19{,}650$ units of second nutrient

$$\begin{array}{lll} 8x+4y=42{,}400 & (4)\times \text{ first equation} \\ \underline{3x+4y=19{,}650} & \\ 5x \quad = 22{,}750 & \text{Subtract} \\ x \quad = 4550 & \text{Solve for } x \end{array}$$

Substituting $x = 4550$ into an original equation we have $2(4550) + y = 10{,}600$.
So, $y = 1500$. Solution is 4550 of species A and 1500 of species B.

**47.** $x$ = amount of 20% concentration.
$y$ = amount of 5% concentration.
$x + y = 10$ amount of solution
$0.20x + 0.05y = 0.155(10)$ concentration of medicine
Solving this system of equations:

$$\begin{array}{ll} x + \quad y = 10 & \\ \underline{x + 0.25y = 7.75} & (5)\times \text{ second equation} \\ 0.75y = 2.25 & \text{Subtract equations} \\ y = 3 & \text{Solve for } y \end{array}$$

Substituting into the first equation we have $x + 3 = 10$ or $x = 7$.
The solution is 3 cc of 5% concentration and 7 cc of 20% concentration.

**49.** $x$ = number of \$40 tickets.
$y$ = number of \$60 tickets.

$$\begin{array}{lcl} x+y=16{,}000 & \rightarrow & -40x-40y=-640{,}000 \\ 40x+60y=760{,}000 & \rightarrow & \underline{40x+60y=\ 760{,}000} \\ & & 20y = \ 120{,}000 \\ & & y = \ 6{,}000 \end{array}$$

$$x + 6{,}000 = 16{,}000$$
$$x = 10{,}000$$

**51.** $x$ = amount of 20% solution to be added.
$0.20x$ = concentration of nutrient in 20% solution.
$0.02(100) = 2$ is the concentration of nutrient in 2% solution.

$$0.20x + 2 = 0.10(x + 100)$$
$$0.20x + 2 = 0.10x + 10$$
$$0.1x = 8 \text{ or } x = 80 \text{ cc of 20\% solution is needed.}$$

**53.** $x$ = ounces of substance A,
$y$ = ounces of substance B, and
$z$ = ounces of substance C.

$$\begin{array}{rl} 5x + 15y + 12z = 100 & \text{Nutrition requirements} \\ x = z & \text{Digestive restrictions} \\ y = \frac{1}{5}z & \text{Digestive restrictions} \end{array}$$

Since both $x$ and $y$ are in terms of $z$, we can substitute in the first equation and solve for $z$.
So, $5z + 3z + 12z = 100$ or $20z = 100$. So, $z = 5$. Now, since $x = z$, we have $x = 5$.

Since $y = \frac{1}{5}z$, we have $y = 1$. The solution is 5 ounces of substance A, 1 ounce of substance B, and 5 ounces of substance C.

**55.** $A$ = number of A type clients.
$B$ = number of B type clients.
$C$ = number of C type clients.

| | |
|---|---|
| $A + B + C = 500$ | Total clients |
| $200A + 500B + 300C = 150{,}000$ | Counseling costs |
| $300A + 200B + 100C = 100{,}000$ | Food and shelter |

To find the solution we must solve the system of equations.

| | | |
|---|---|---|
| Eq. 1 | $A + B + C = 500$ | |
| Eq. 2 | $2A + 5B + 3C = 1500$ | Original equation divided by 100 |
| Eq. 3 | $3A + 2B + C = 1000$ | Original equation divided by 100 |
| | $A + B + C = 500$ | |
| Eq. 4 | $3B + C = 500$ | $(-2)\times$ Eq. 1 added to Eq. 2 |
| Eq. 5 | $-B - 2C = -500$ | $(-3)\times$ Eq. 1 added to Eq. 3 |

$A + B + C = 500$ Eq. 1

$3B + C = 500$ Eq. 4

$-\frac{5}{3}C = \frac{-1000}{3}$ $\frac{1}{3}\times$ Eq. 4 added to Eq. 5

$$C = \frac{1000}{3}\cdot\frac{3}{5} = 200$$

Substituting $C = 200$ into Eq. 4 gives $3B + 200 = 500$ or $3B = 300$. So, $B = 100$.
Substituting $C = 200$ and $B = 100$ into Eq. 1 gives $A + 100 + 200 = 500$. So, $A = 200$.
Thus, the solution is 200 type A clients, 100 type B clients, and 200 type C clients.

## *Exercises 1.6*

**1. a.** $P(x) = R(x) - C(x)$
$= 68x - (34x + 6800)$
$= 34x - 6800$

**b.** $P(3000) = 34(3000) - 6800 = \$95{,}200$

**3. a.** $P(x) = R(x) - C(x)$
$= 80x - (43x + 1850)$
$= 37x - 1850$

**b.** $P(30) = 37(30) - 1850 = -\$740$
The total costs are more than the revenue.

**c.** $P(x) = 0$ or $37x - 1850 = 0$
So, $x = \frac{1850}{37} = 50$ units is the break-even point.

**5.** $C(x) = 5x + 250$

**a.** $m = 5$, $C$-intercept: 250

**b.** $\overline{MC} = 5$ means that each additional unit produced costs \$5.

**c.** Slope = marginal cost.
$C$-intercept =fixed costs.

**d.** \$5, \$5 ($\overline{MC} = 5$ at every point)

**7.** $R = 27x$

**a.** $m = 27$

**b.** 27; each additional unit sold yields \$27 in revenue.

**c.** In each case, one more unit yields \$27.

**9.** $R(x) = 27x$, $C(x) = 5x + 250$

**a.** $P(x) = 27x - (5x + 250) = 22x - 250$

**b.** $m = 22$

**c.** Marginal profit is 22.

**d.** Each additional unit sold gives a profit of \$22. To maximize profit sell all that you can produce. Note that this is not always true.

# Chapter 1: Linear Equations and Functions

**11.** $(x, P)$ is the correct form.

$P_1 = (200, 3100)$

$P_2 = (250, 6000)$

$$m = \frac{6000-3100}{250-200} = 58$$

$P - 3100 = 58(x-200)$ or $P = 58x - 8500$

The marginal profit is 58.

**13.** **a.** $TC = 35H + 6600$

**b.** $TR = 60H$

**c.** $P = R - C$
$$= 60H - (35H + 6600)$$
$$= 25H - 6600$$

**d.** $C(200) = 35(200) + 6600$
$= \$13{,}600$ cost of 200 helmets

$R(200) = 60(200)$
$= \$12{,}000$ revenue from 200 helmets

$P(200) = R(200) - C(200)$
$= \$12{,}000 - 13{,}600$
$= -\$1600$ loss from 200 helmets

**e.** $C(300) = 35(300) + 6600$
$= \$17{,}100$ cost of 300 helmets

$R(300) = 60(300)$
$= \$18{,}000$ revenue from 300 helmets

$P(300) = R(300) - C(300)$
$= 18{,}000 - 17{,}100$
$= \$900$ profit from 300 helmets

**f.** The marginal profit is \$25. Each additional helmet sold gives a profit of \$25.

**15.** **a.** The revenue function is the graph that passes through the origin.

**b.** At a production of zero the fixed costs are \$2000.

**c.** From the graph, the break-even point is 400 units and \$3000 in revenue or costs.

**d.** Marginal cost $= \dfrac{3000-2000}{400-0} = 2.5$

Marginal revenue $= \dfrac{3000-0}{400-0} = 7.5$

**17.** $R(x) = C(x) = 85x = 35x + 1650$ or $50x = 1650$ or $x = 33$.
Thus, 33 necklaces must be sold to break even.

**19.** **a.** $R(x) = 12x$, $C(x) = 8x + 1600$

**b.** $R(x) = C(x)$ if $12x = 8x + 1600$ or $4x = 1600$ or $x = 400$.
It takes 400 units to break even.

**21.** **a.** $P(x) = R(x) - C(x)$
$$= 12x - (8x + 1600)$$
$$= 4x - 1600$$

**b.** By setting $P(x) = 0$ we get $x = 400$.
Same as 19(b).

**23.** **a.** $TC = 4.50x + 1045$

**b.** $TR = 10x$

**c.** $P = R - C$
$$= 10x - (4.50x + 1045)$$
$$= 5.50x - 1045$$

**d.** Break even also means $P = 0$.
$5.50x - 1045 = 0$
$5.50x = 1045$
$x = 190$ units to break even

**25.** **a.** $R(x) = 54.90x$

**b.** $P_1 = (2000, 50000)$

$P_2 = (800, 32120)$

$$m = \frac{32{,}120 - 50{,}000}{800-2000} = \frac{-17{,}880}{-1200} = 14.90$$

$y - 50{,}000 = 14.90(x - 2000)$ or

$y = 14.90x + 20{,}200 = C(x)$

**c.** From $54.90x = 14.90x + 20{,}200$ we have $x = 505$ units to break even.

**27.** **a.**

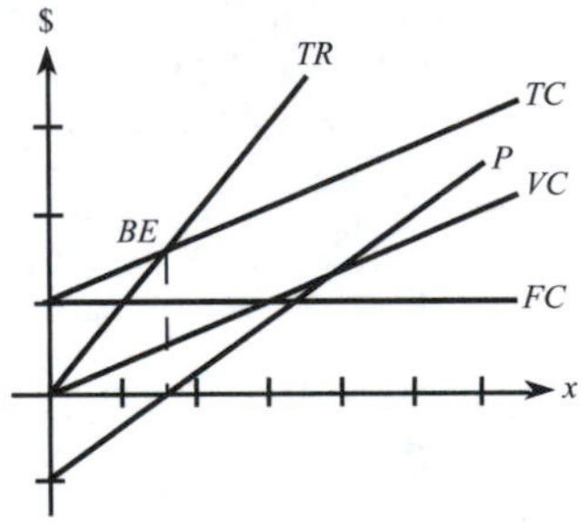


**b.** $TR$ starts at the origin and intersects $TC$ at the break-even ($BE$). $FC$ is a horizontal line from the vertical intercept of $TC$. $VC$ starts at the origin and is parallel to $TC$.

**29.** If price increases, then the demand for the product decreases.

**31.** **a.** If $p = \$100$, then $q = 600$ (approximately).

**b.** If $p = \$100$, then $q = 300$.

**c.** There is a shortage since more is demanded.

**33.** Demand: $2p+5q=200$

$$2(60)+5q=200$$
$$5q=80$$
$$q=16$$

Supply: $p-2q=10$

$$60-2q=10$$
$$2q=50$$
$$q=25$$

There will be a surplus of 9 units at a price of \$60.00.

**35.** Remember that $(q, p)$ is the correct form.

$P_1=(240, 900)$

$P_2=(315, 850)$

$$m=\frac{850-900}{315-240}=-\frac{50}{75}=-\frac{2}{3}$$

Note: $m<0$ for demand equations.

$$p-900=-\frac{2}{3}(q-240) \text{ or}$$

$$p=-\frac{2}{3}q+1060$$

**37.** $(q, p)$ is the correct form.

$P_1=(10000, 1.50)$

$P_2=(5000, 1.00)$

$$m=\frac{1-1.50}{5000-10000}=\frac{-0.50}{-5000}=0.0001$$

Note: $m>0$ for supply equations.

$p-1=0.0001(q-5000)$ or $p=0.0001q+0.5$

**39.** **a.** The decreasing function is the demand curve. The increasing function is the supply curve.

**b.** Reading the graph, we have equilibrium at $q=30$ and $p=25$.

**41.** **a.** Reading the graph, at $p=20$ we have 20 units supplied.

**b.** Reading the graph, at $p=20$ we have 40 units demanded.

**c.** At $p=20$ there is a shortage of 20 units.

**43.** By observing the graph in the figure, we see that a price below the equilibrium price results in a shortage.

**45.**

$-\frac{1}{2}q+28=\frac{1}{3}q+\frac{34}{3}$ Required condition.

$-3q+168=2q+68$ Multiply both sides by 6 to simplify.

$$-5q=-100$$
$$q=20$$

Substituting into one of the original equations gives $p=-\frac{1}{2}(20)+28=18$.

Thus, the equilibrium point is $(q, p)=(20, 18)$.

**47.** $-4q+220=15q+30$ Required condition.

$$190=19q$$

$q=10$ Solve for $q$.

Substituting $q=10$ into one of the original equations gives $p=180$.

Thus, the equilibrium point is $(q, p)=(10, 180)$.

**49.** Demand: (80, 350) and (120, 300) are two points. $m=\frac{350-300}{80-120}=-\frac{5}{4}$

$$p-p_1=m(q-q_1) \text{ or } p-300=-\frac{5}{4}(q-120) \text{ or } p=-\frac{5}{4}q+450$$

Supply: (60, 280) and (140, 370) are two points. $m=\frac{280-370}{60-140}=\frac{9}{8}$

$$p-p_1=m(q-q_1) \text{ or } p-280=\tfrac{9}{8}(q-60) \text{ or } p=\tfrac{9}{8}q+212.5$$

Now, set these two equations for $p$ equal to each other and solve for $q$.

$\frac{9}{8}q+212.5=-\frac{5}{4}q+450$ Required for equilibrium.

$9q+1700=-10q+3600$ Multiply both sides by 8 to simplify.

$$19q=1900$$
$$q=100$$

Substituting $q=100$ into one of the original equations gives $p=325$.

Thus, the equilibrium point is $(q, p)=(100, 325)$.

**51.** **a.** Reading the graph, we have that the tax is \$15.
**b.** From the graph, the original equilibrium was (100, 100).
**c.** From the graph, the new equilibrium is (50, 110).
**d.** The supplier suffers because the increased price reduces the demand.

**53.** New supply price: $p = 15q + 30 + 38 = 15q + 68$

$15q + 68 = -4q + 220$ Required condition

$$19q = 152$$

$$q = 8$$

Substituting $q = 8$ into one of the original equations gives $p = 188$.
Thus, the new equilibrium point is $(q, p) = (8, 188)$.

**55.** New supply price: $p = \frac{q}{20} + 10 + 5 = \frac{q}{20} + 15$

$\frac{q}{20} + 15 = -\frac{q}{20} + 65$ Required condition

$$q + 300 = -q + 1300$$

$$2q = 1000$$

$$q = 500$$

Thus, $p = \frac{500}{20} + 15 = 40$.

The new equilibrium point is (500, 40).

**57.** Demand: $p = \frac{-q + 2100}{60}$

Supply: $p = \frac{q + 540}{120}$

New supply:

$$p = \frac{q + 540}{120} + \frac{1}{2} = \frac{q + 540}{120} + \frac{60}{120} = \frac{q + 600}{120}$$

$\frac{q+600}{120} = \frac{-q+2100}{60}$ Required condition

$q + 600 = -2q + 4200$ Multiply both sides by 120

$$3q = 3600$$

$q = 1200$ Thus, $p = \frac{1200+600}{120} = 15$.

The new equilibrium quantity is 1200.
The new equilibrium price is \$15.

## Chapter 1 Review Exercises

For this set of exercises we will not give reasons for any steps or list any formulas.

**1.** $3x-8=23$

$3x=31$

$x=\frac{31}{3}$

**2.** $2x-8=3x+5$

$-x=13$

$x=-13$

**3.** $\frac{6x+3}{6}=\frac{5(x-2)}{9}$

$18\left(\frac{6x+3}{6}\right)=18\left(\frac{5(x-2)}{9}\right)$

$3(6x+3)=10(x-2)$

$18x+9=10x-20$

$8x=-29$

$x=-\frac{29}{8}$

**4.** $2x+\frac{1}{2}=\frac{x}{2}+\frac{1}{3}$

$12x+3=3x+2$

$9x=-1$

$x=-\frac{1}{9}$

**5.** $\frac{6}{3x-5}=\frac{6}{2x+3}$

$6(2x+3)=6(3x-5)$

$2x+3=3x-5$

$3+5=3x-2x$

$x=8$

**6.** $\frac{2x+5}{x+7}=\frac{1}{3}+\frac{x-11}{2(x+7)}$

$6(2x+5)=2(x+7)+3(x-11)$

$12x+30=2x+14+3x-33$

$12x-2x-3x=14-33-30$

$7x=-49$

$x=-7$

There is no solution since we have division by zero when $x=-7$.

**7.** $3y-6=-2x-10$

$3y=-2x-4$

$y=\frac{-2x-4}{3}$

$y=-\frac{2}{3}x-\frac{4}{3}$

**8.** $3x-9\le 4(3-x)$

$3x-9\le 12-4x$

$7x\le 21$

$x\le 3$

**9.** $\frac{2}{5}x\le x+4$

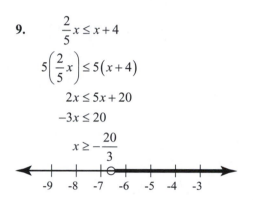

$5\left(\frac{2}{5}x\right)\le 5(x+4)$

$2x\le 5x+20$

$-3x\le 20$

$x\ge -\frac{20}{3}$

**10.** $5x+1\ge \frac{2}{3}(x-6)$

$3(5x+1)\ge 3\cdot\frac{2}{3}(x-6)$

$15x+3\ge 2(x-6)$

$15x+3\ge 2x-12$

$13x\ge -15$

$x\ge -\frac{15}{13}$

**11.** Yes.

**12.** $y^2=9x$, is not a function of $x$. If $x=1$, then $y=\pm 3$.

**13.** Yes.

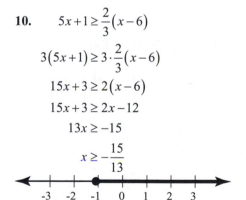

**14.** $y=\sqrt{9-x}$

Domain: $9-x\ge 0$ or $9\ge x$ or $x\le 9$.

Range: Nonnegative square root means $y\ge 0$.

**15.** $f(x)=x^2+4x+5$

**a.** $f(-3)=(-3)^2+4(-3)+5=9-12+5=2$

**b.** $f(4)=(4)^2+4(4)+5=16+16+5=37$

**c.** $f\left(\frac{1}{2}\right)=\left(\frac{1}{2}\right)^2+4\left(\frac{1}{2}\right)+5=\frac{1}{4}+2+5=\frac{29}{4}$

**16.** $g(x)=x^2+\frac{1}{x}$

**a.** $g(-1)=(-1)^2+\frac{1}{-1}=1-1=0$

**b.** $g\left(\frac{1}{2}\right)=\left(\frac{1}{2}\right)^2+\frac{1}{\frac{1}{2}}=\frac{1}{4}+2=2\frac{1}{4}$

**c.** $g(0.1)=(0.1)^2+\frac{1}{0.1}=0.01+10=10.01$

**17.** $f(x)=9x-x^2$

$$\begin{aligned}f(x+h)&=9(x+h)-(x+h)^2\\&=9x+9h-x^2-2xh-h^2\end{aligned}$$

$f(x)=9x-x^2$

$$\begin{aligned}f(x+h)-f(x)&=9h-2xh-h^2\\&=h(9-2x-h)\end{aligned}$$

$$\frac{f(x+h)-f(x)}{h}=9-2x-h$$

**18.** $y$ is a function of $x$. (Use vertical-line test.)

**19.** No, the graph fails vertical-line test.

**20.** $f(2)=4$

**21.** $x=0, x=4$

**22. a.** $D=\{-2,-1,0,1,3,4\}$, $R=\{-3,2,4,7,8\}$

**b.** $f(4)=7$

**c.** $f(x)=2$ if $x=-1, 3$

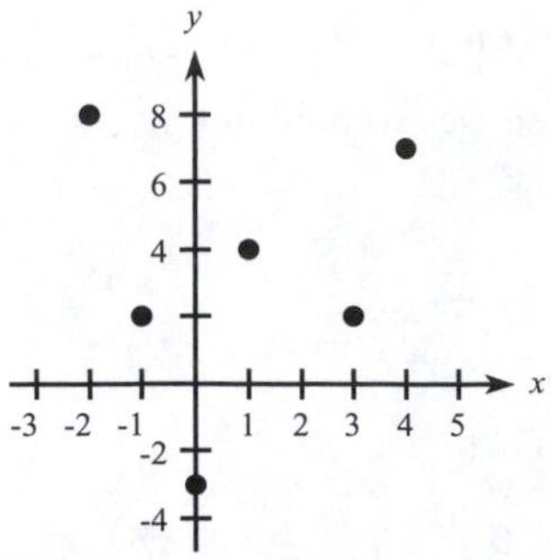


**d.** No. For $y=2$, there are two values of $x$.

**23.** $f(x)=3x+5,\ g(x)=x^2$

**a.** $(f+g)x=(3x+5)+x^2=x^2+3x+5$

**b.** $\left(\frac{f}{g}\right)x=\frac{3x+5}{x^2}$ or $\frac{3x}{x^2}+\frac{5}{x^2}=\frac{3}{x}+\frac{5}{x^2}$

**c.** $f(g(x))=f(x^2)=3x^2+5$

**d.** 
$$\begin{aligned}(f\circ f)x&=f(3x+5)\\&=3(3x+5)+5\\&=9x+20\end{aligned}$$

**24.** $5x+2y=10$

$x$-intercept: If $y=0, x=2$

$y$-intercept: If $x=0, y=5$

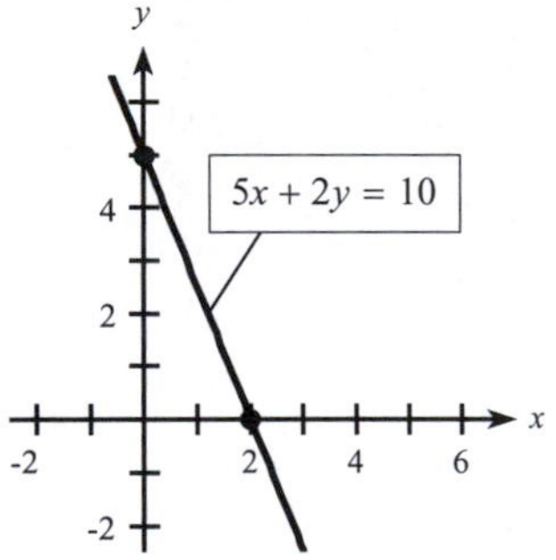


**25.** $6x+5y=9$

$x$-intercept: If $y=0$, $x=\frac{9}{6}=\frac{3}{2}$

$y$-intercept: If $x=0$ or $y=\frac{9}{5}$

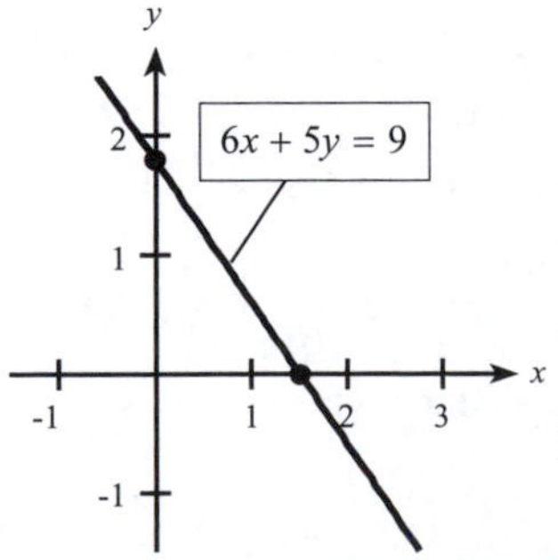

**26.** $x = -2$
$x$-intercept: $x = -2$
There is no $y$-intercept.

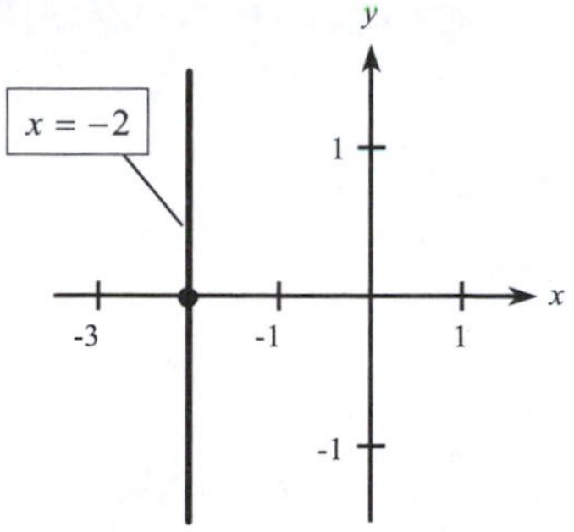


**27.** $P_1(2, -1);\ P_2(-1, -4)$

$$m = \frac{-4-(-1)}{-1-2} = \frac{-3}{-3} = 1$$

**28.** $(-3.8, -7.16)$ and $(-3.8, 1.16)$

$$m = \frac{-7.16-1.16}{-3.8-(-3.8)} = \frac{-8.32}{0}$$

Slope is undefined.

**29.** $2x + 5y = 10$

$$y = -\frac{2}{5}x + 2,\ m = -\frac{2}{5},\ b = 2$$

**30.** $x = -\frac{3}{4}y + \frac{3}{2}$ or $y = -\frac{4}{3}x + 2$

$$m = -\frac{4}{3},\ b = 2$$

**31.** $m = 4, b = 2, y = 4x + 2$

**32.** $m = -\frac{1}{2},\ b = 3,\ y = -\frac{1}{2}x + 3$

**33.** $P = (-2, 1),\ m = \frac{2}{5}$

$$y - 1 = \frac{2}{5}(x+2) \text{ or } y = \frac{2}{5}x + \frac{9}{5}$$

**34.** $(-2, 7)$ and $(6, -4)$

$$m = \frac{-4-7}{6-(-2)} = \frac{-11}{8}$$

$$y - 7 = \frac{-11}{8}(x-(-2)) \text{ or}$$

$$y = \frac{-11}{8}x + \frac{17}{4}$$

**35.** $P_1(-1, 8);\ P_2(-1, -1)$
The line is vertical since the $x$-coordinates are the same. Equation: $x = -1$ $x = -1$

**36.** Parallel to $y = 4x - 6$ means $m = 4$.
$y - 6 = 4(x-1)$ or $y = 4x + 2$

**37.** $P(-1, 2)$; $\perp$ to $3x + 4y = 12$
or

$$y = -\frac{3}{4}x + 3$$

$$m = \frac{4}{3}$$

$$y - 2 = \frac{4}{3}(x+1) \text{ or}$$

$$y = \frac{4}{3}x + \frac{10}{3}$$

**38.** $x^2 + y - 2x - 3 = 0;\ y = -x^2 + 2x + 3$

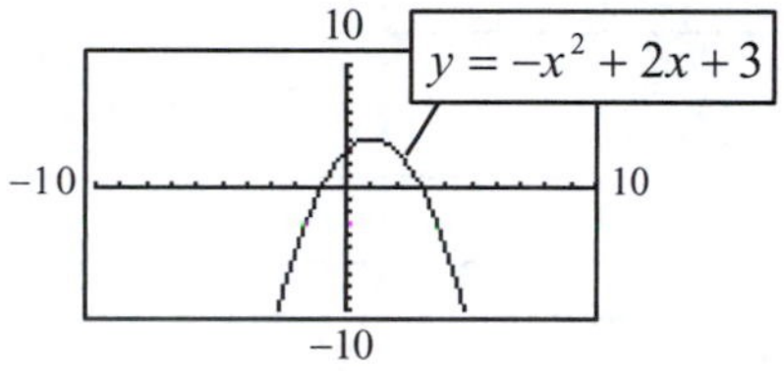


**39.**

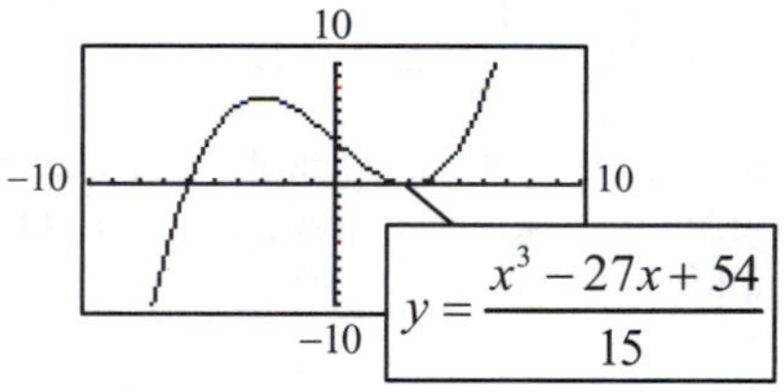


**40.** **a.**

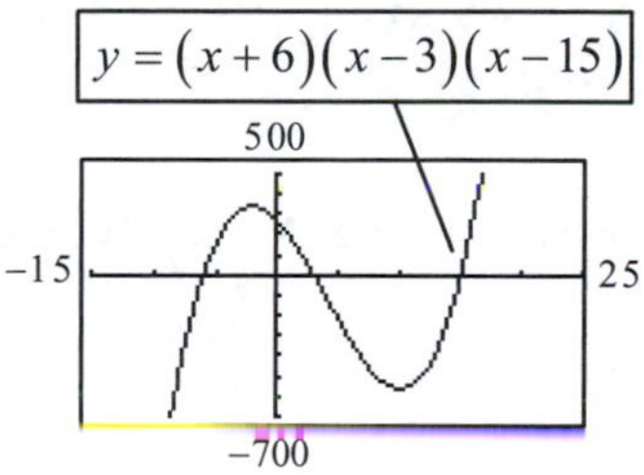

**b.**

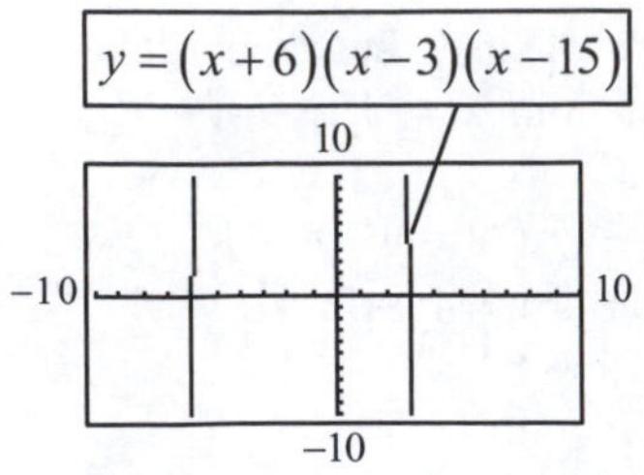


**c.** The graph in (a) shows the complete graph. The graph in (b) shows a piece that rises toward the high point and a piece between the high and low points.

**41.** $y = x^2 - x - 42$ is a parabola opening upward.

**a.**

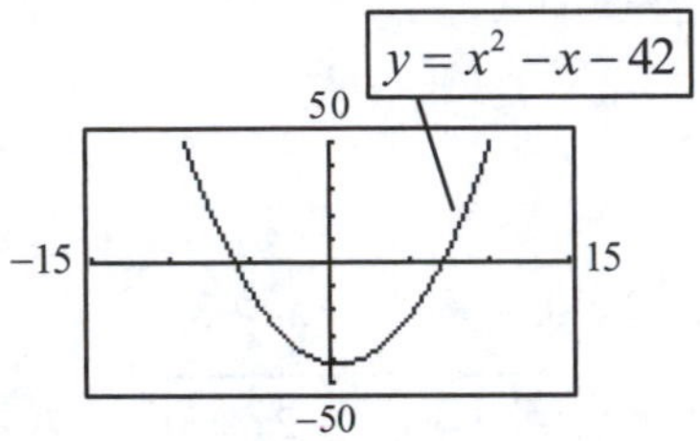


**b.**

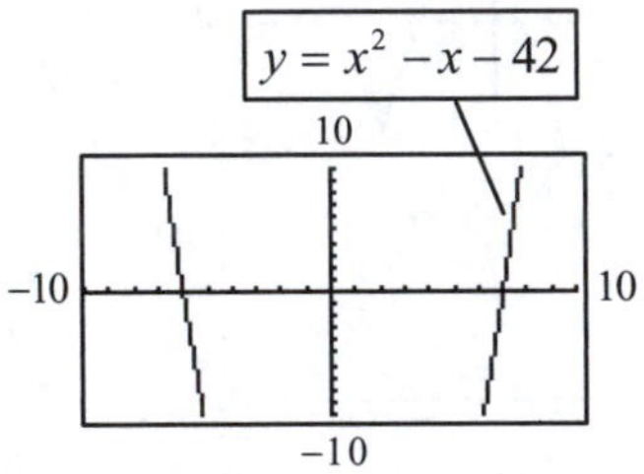


**c.** (a) shows the complete graph. The *y*-min is too large in absolute value for (b) to get a complete graph.

**42.** $y = \dfrac{\sqrt{x+3}}{x}$

$x \neq 0;\ x+3 \geq 0$ or $x \geq -3$;

Domain: $x \neq 0,\ x \geq -3$

**43.** Trace approximates $x = -7.2749,\ x = 0.2749$

**44.**
$$\begin{aligned} 4x-2y&=6\\ 3x+3y&=9\\ \text{Then,}\quad 12x-6y&=18\\ 6x+6y&=18\\ \hline 18x\qquad &=36\\ x&=2\\ 4(2)-2y&=6\\ -2y&=-2\\ y&=1 \end{aligned}$$

Solution: $(2,1)$

**45.**
$$\begin{aligned} 2x+\ y&=19\\ x-2y&=12\\ \text{Then,}\quad 4x+2y&=38\\ x-2y&=12\\ \hline 5x\qquad &=50\\ x&=10\\ 2(10)+y&=19\\ y&=-1 \end{aligned}$$

Solution: $(10,-1)$

**46.**
$$\begin{aligned} 3x+2y&=5\\ 2x-3y&=12\\ \text{Then,}\quad 9x+6y&=15\\ 4x-6y&=24\\ \hline 13x\qquad &=39\\ x&=3\\ 3(3)+2y&=5\\ 2y&=-4\\ y&=-2 \end{aligned}$$

Solution: $(3,-2)$

**47.**
$$\begin{aligned} 6x+3y&=1\\ y&=-2x+1\\ 6x+3(-2x+1)&=1\\ 6x-6x+3&=1\\ 3&=1 \end{aligned}$$

No solution.

**48.**
$$\begin{aligned} 4x-3y&=253 \qquad & 8x-6y&=506 \qquad & 4(10)-3y&=253\\ 13x+2y&=-12 \qquad & 39x+6y&=-36 \qquad & -3y&=213\\ & & \hline 47x\qquad &=470 & y&=-71\\ & & x\qquad &=10 & & \end{aligned}$$

Solution: $(10,-71)$

**49.** $x+2y+3z=5$ Steps 1 and 2: Nothing to be done.

$y+11z=21$ Step 3: $x+2y+\ 3z=5$

$5y+\ 9z=13$

$$y+11z=21$$
$$-46z=-92$$

Step 4: $z=2$ $\quad y+11(2)=21 \quad x+2(-1)+3(2)=5$

$y=-1 \quad x=1$

Solution is $x=1,\ y=-1,\ z=2$.

**50.**

$$\begin{aligned} x+\ y-\ z&=12 \\ 2y-3z&=-7 \\ 3x+3y-7z&=0 \\ x+\ y-\ z&=12 \\ 2y-3z&=-7 \\ -4z&=-36 \end{aligned}$$

Thus $z=9$

$2y-27=-7$

$2y=20$ or $y=10$

$x+10-9=12$

$x=11$

Solution: (11, 10, 9)

**51. a.** $1950+70=2020$

**b.** $y=0.077(100)+13.8=21.5$

21.5 additional years of life expectancy (to age 86.5)

**c.** $20=0.077x+13.8$

$6.2=0.077x$

$x\approx 80.5$

$1950+80.5=2030.5$

In the year 2031, the function predicts the life expectancy to be 20 years after age 65.

**52.** Student has total points of $91+82+88+50+42+42=395$.
Total of possible points is $300+150+200=650$.
To earn an A students need at least $0.9(650)=585$ points.

Student must earn $585-395=190$ points on the final. This is the same as $95\%$.

**53.** Diesel: $C=0.76x+58{,}000$

Gas: $C=0.88x+53{,}200$

$0.76x+58{,}000=0.88x+53{,}200$

$0.12x=4800$

$x=40{,}000$

Costs are equal at $40{,}000$ miles.

He probably would drive more than 40,000 miles in 7 years, so he should buy the diesel.

**54. a.** Yes

**b.** No

**c.** $f(300)=4$

**55. a.** $f(80)=565.44$

**b.** The monthly payment on a loan is \$494.75.

**56.** $P(x)=330x-0.05x^2-5000$

$x=q(t)=100+10t$

**a.** $(P\circ q)(t)=P(100+10t)$

$=330(100+10t)-0.05(100+10t)^2-5000$

**b.** $x=q(15)=100+10(15)$

$=250$ units produced

$P(250)=330(250)-0.05(250)^2-5000$

$=\$74{,}375$

**57.** $W(L)=kL^3,\ L(t)=65-0.1(t-25)^2,\ 0\le t\le 25$

$(W\circ L)(t)=W\left(65-0.1(t-25)^2\right)$

$=0.03\left(65-0.1(t-25)^2\right)^3$

**58. a.**

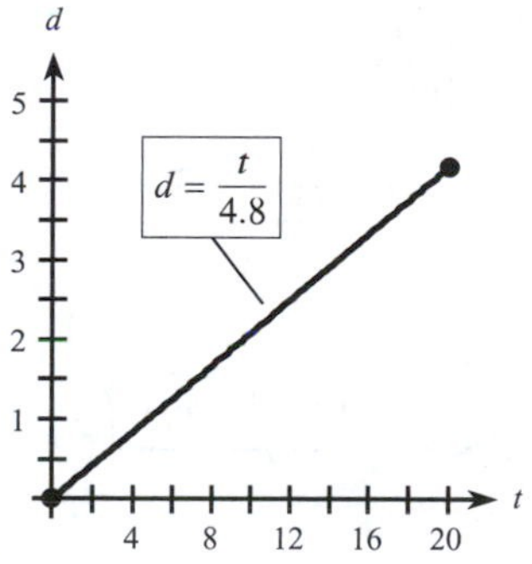


**b.** $(9.6, 2)$ means that the thunderstorm is two miles away if the flash and thunder are 9.6 seconds apart.

**59.**

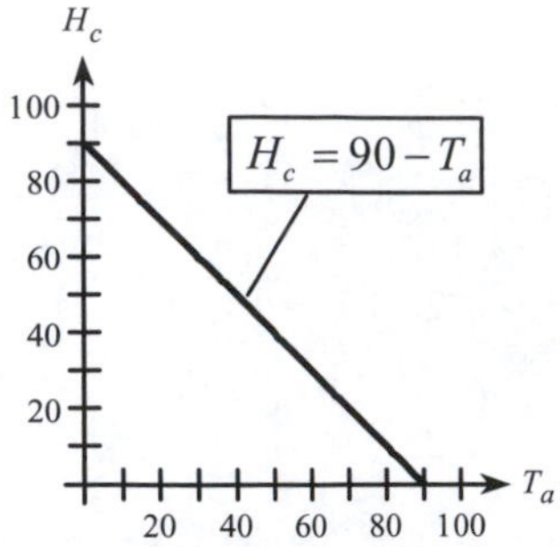


**60. a.** $(x, P)$ is the required form.

$P_1 = (200, 3100),\ P_2 = (250, 6000)$

$$m = \frac{6000-3100}{250-200} = \frac{2900}{50} = 58$$

$P - 3100 = 58(x - 200)$ or

$P(x) = 58x - 8500$

**b.** For each additional unit sold the profit increases by \$58.

**61.** $A = 427x + 4541$

**a.** Yes.

**b.** $m = 427$, $A$-intercept is $4541$

**c.** In 2000, average annual health care costs were \$4541 per customer.

**d.** The average annual cost costs are changing at the rate of \$427 per year.

**62.** $(C, F): (0, 32)$ and $(100, 212)$

$$m = \frac{212-32}{100-0} = \frac{180}{100} = \frac{9}{5}$$

Using $y = mx + b$, $F = \frac{9}{5}C + 32.$

**63. a.**

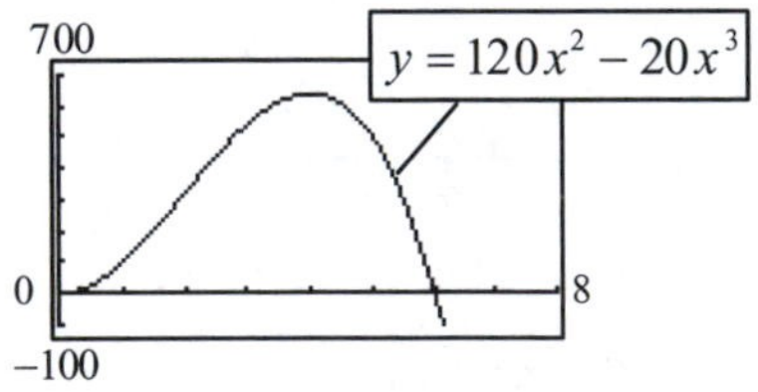


**b.** Algebraically, $y \geq 0$ if

$120x^2 - 20x^3 = 20x^2(x-6) \geq 0.$

Answer: $0 \leq x \leq 6$

**64. a.** $v^2 = 1960(h + 10)$

$$h + 10 = \frac{v^2}{1960}$$

$$h = \frac{v^2}{1960} - 10$$

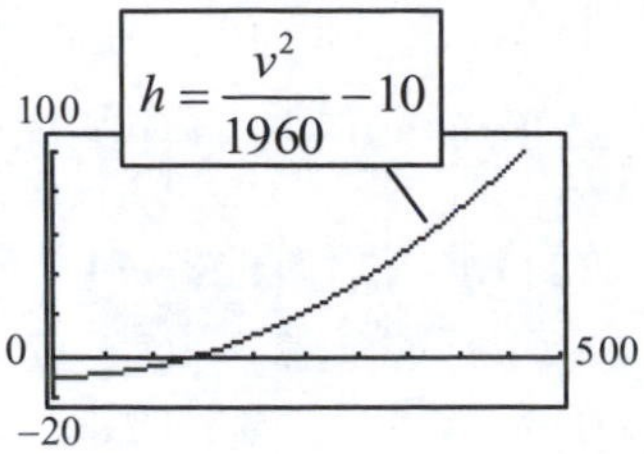


**b.** $h(210) = \frac{210^2}{1960} - 10 = 12.5$ cm

**65.** $x$ = amount of safer investment and $y$ = amount of other investment.

$x + y = 150000$

$0.095x + 0.11y = 15000$

Solving the system:

$0.11x + 0.11y = 16500$

$0.095x + 0.11y = 15000$

$0.015x = 1500$

$x = 100000$

Then $y = 50000$. Thus, invest \$100,000 at 9.5% and \$50,000 at 11%.

**66.** $x$ = liters of 20% solution

$y$ = liters of 70% solution

$x + y = 4$

$0.2x + 0.7y = 1.4$

$x + y = 4$

$x + 3.5y = 7$

$2.5y = 3 \quad y = 1.2$

$x + 1.2 = 4$

$x = 2.8$

Answer: 2.8 liters of 20%, 1.2 of 70%.

**67.** $S: p = 4q + 5,\ D: p = -2q + 81$

**a.** $S: 53 = 4q + 5 \qquad D: 53 = -2q + 81$

$4q = 48 \qquad 2q = 28$

$q = 12 \qquad q = 14$

**b.** Demand is greater. There is a shortfall.

**c.** Price is likely to increase.

**68.** **a. – c.**

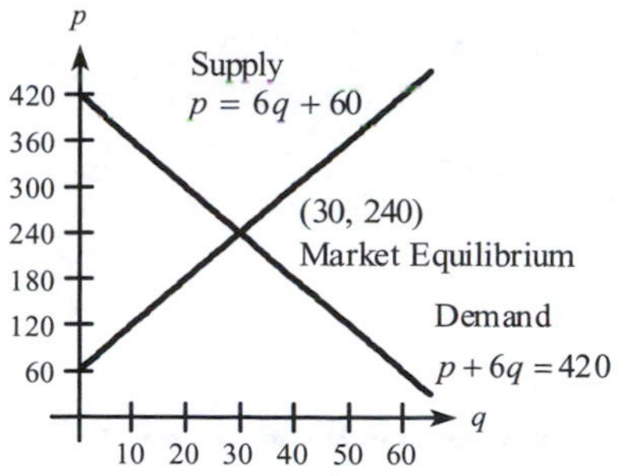


**69.** $C(x) = 38.80x + 4500, R(x) = 61.30x$

**a.** Marginal cost is \$38.80 .

**b.** Marginal revenue is \$61.30 .

**c.** Marginal profit is $\$61.30 - 38.80 = \$22.50$ .

**d.**
$$61.30x = 38.80x + 4500$$
$$22.50x = 4500$$
$$x = 200 \text{ units to break even.}$$

**70.** $FC = \$1500, VC = \$22$ per unit, $R = \$52$ per unit

**a.** $C(x) = 22x + 1500$

**b.** $R(x) = 52x$

**c.** $P = R - C = 30x - 1500$

**d.** $\overline{MC} = 22$

**e.** $\overline{MR} = 52$

**f.** $\overline{MP} = 30$

**g.** Break even means $30x - 1500 = 0$ or $x = 0$.

**71.** Supply: $m = \dfrac{200-100}{150-125} = 4$ Demand: $m = \dfrac{200-100}{330-355} = -4$

$$p - 100 = 4(q - 125) \qquad p - 100 = -4(q - 355)$$
$$p = 4q - 400 \qquad p = -4q + 1520$$

So, $4q - 400 = -4q + 1520$ or $q = 240$. With $q = 240$, $p = 4(240) - 400 = \$560$. Market equilibrium is achieved with a product quantity of 240 units at a price of \$560 per unit.

**72.** New supply equation: $p = \dfrac{q}{10} + 8 + 2 = \dfrac{q}{10} + 10$

Demand:
$$p = \frac{-q + 1500}{10} = -\frac{q}{10} + 150$$
$$\frac{q}{10} + 10 = -\frac{q}{10} + 150$$
$$\frac{2q}{10} = 140 \text{ or } q = 700$$
$$p = \frac{700}{10} + 10 = 80$$

Solution: $(700,\ 80)$

## Chapter 1 Test

**1.** $10-2(2x-9)-4(6+x)=52$

$$10-4x+18-24-4x=52$$
$$4-8x=52$$
$$-8x=48$$
$$x=-6$$

**2.** $4x-3=\dfrac{x}{2}+6$

$$8x-6=x+12$$
$$7x=18$$
$$x=\frac{18}{7}$$

**3.** $\dfrac{3}{x}+4=\dfrac{4x}{x+1}$

$$3(x+1)+4x(x+1)=4x(x)$$
$$3x+3+4x^2+4x=4x^2$$
$$7x=-3$$
$$x=-\frac{3}{7}$$

**4.** $\dfrac{3x-1}{4x-9}=\dfrac{5}{7}$

$$7(3x-1)=5(4x-9)$$
$$21x-7=20x-45$$
$$x=-38$$

**5.** $f(x)=7+5x-2x^2$

$$f(x+h)=7+5(x+h)-2(x+h)^2$$
$$=7+5x+5h-2x^2-4xh-2h^2$$
$$f(x)=7+5x-2x^2$$
$$f(x+h)-f(x)=5h-4xh-2h^2$$
$$\frac{f(x+h)-f(x)}{h}=5-4x-2h$$

**6.** $1+\dfrac{2}{3}t\le 3t+22$

$$3\left(1+\frac{2}{3}t\right)\le 3(3t+22)$$
$$3+2t\le 9t+66$$
$$-7t\le 63$$
$$t\ge -9$$

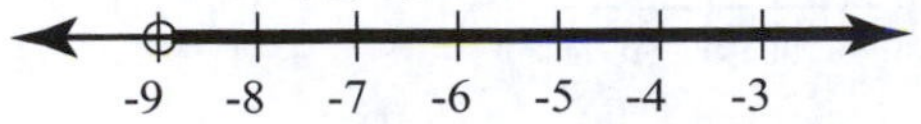


**7.** $5x-6y=30$

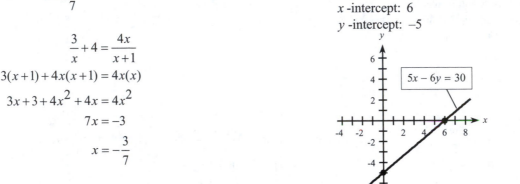

$x$-intercept: 6
$y$-intercept: $-5$

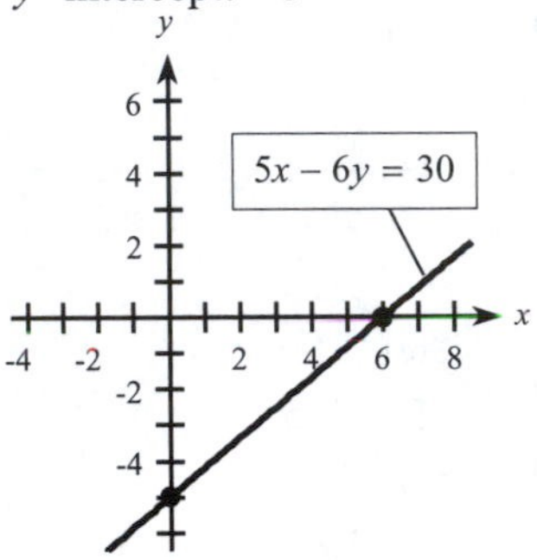


**8.** $7x+5y=21$

$x$-intercept: 3

$y$-intercept: $\dfrac{21}{5}$

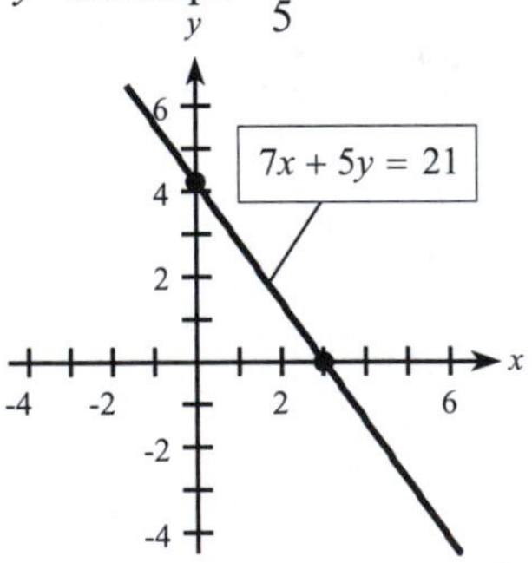


**9.** $f(x)=\sqrt{4x+16}$

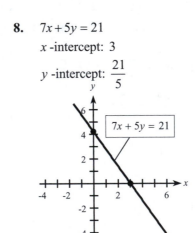

**a.** $4x+16\ge 0$
$4x\ge -16$
Domain: $x\ge -4$; Range: $y\ge 0$
For range, note square root is positive.

**b.** $f(3)=\sqrt{12+16}=2\sqrt{7}$

**c.** $f(5)=\sqrt{20+16}=6$

**10.** $(-1,2)$ and $(3,-4)$

$$m=\frac{-4-2}{3-(-1)}=\frac{-6}{4}=\frac{-3}{2}$$

$$y-2=\frac{-3}{2}(x-(-1))$$

$$y=\frac{-3}{2}x+\frac{1}{2}$$

**11.** $5x+4y=15$

$$y=-\frac{5}{4}x+\frac{15}{4}$$

$$m=-\frac{5}{4},\ b=\frac{15}{4}$$

**12.** Point $(-3,-1)$

**a.** Undefined slope means vertical line. $x=-3$

**b.** $\perp$ to $y=\frac{1}{4}x+2$ means $m=-4$.

Thus, $y+1=-4(x+3)$ or $y=-4x-13$.

**13. a.** Is not a function since for some $x$-values there are two values of $y$.

**b.** Is a function since for each $x$ there is only one $y$.

**c.** Is not a function for same reason as (a).

**14. a.**

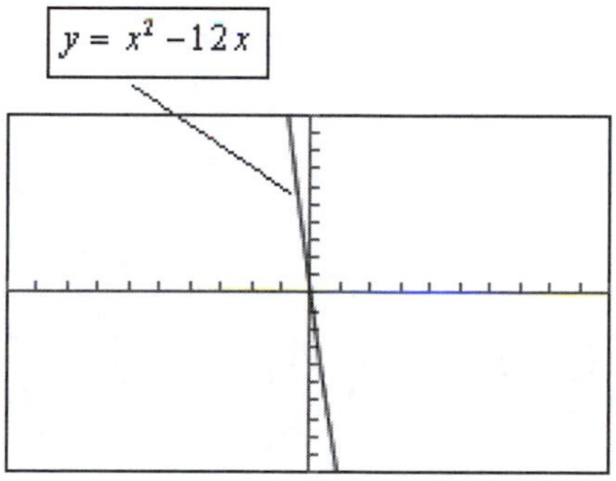


**b.**

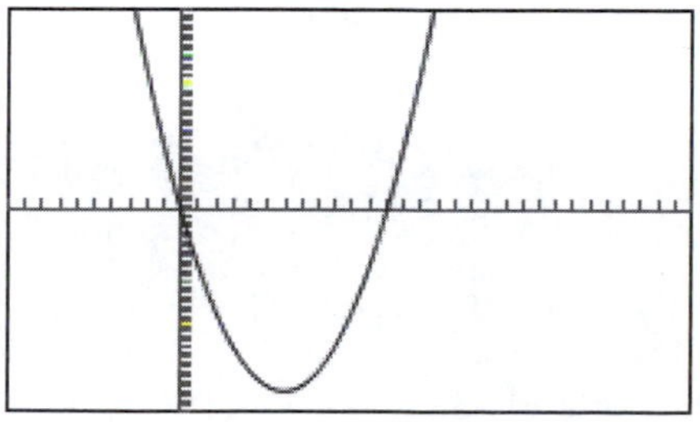

**15.** $3x+2y=-2$

$$4x+5y=2$$

$$12x+8y=-8$$

$$\underline{12x+15y=6}$$

$$-7y=-14$$

$$y=2$$

$$3x+2(2)=-2$$

$$3x=-6$$

$$x=-2$$

Solution: $(-2,2)$

**16.** $f(x)=5x^2-3x,\ g(x)=x+1$

**a.** $(fg)(x)=(5x^2-3x)(x+1)$

**b.** $g(g(x))=g(x+1)=(x+1)+1=x+2$

**c.** $(f\circ g)(x)=f(x+1)$

$$=5(x+1)^2-3(x+1)$$

$$=5x^2+10x+5-3x-3$$

$$=5x^2+7x+2$$

**17. a.**

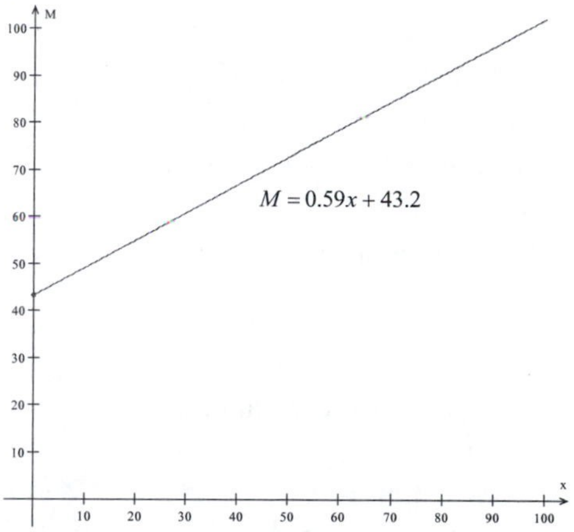


**b.** The model predicts that there will be 90.4 million men in the U.S. workforce in 2030.

**c.** $M=0.59(100)+43.2=102.2$

The model predicts that there will be 102.2 million men in the U.S. workforce in 2050.

**18. a.** $R(x) = 38x, C(x) = 30x + 1200$

, $\overline{MC} = \$30$

**b.** $P(x) = 38x - (30x + 1200)$

$= 8x - 1200$

**c.** Break-even means $P(x) = 0$.

$8x = 1200$ or $x = 150$ units

**d.** $\overline{MP} = \$8$. Each additional unit sold increases the profit by $8.

**19. a.** $R(x) = 50x$

**b.** $C(100) = 10(100) + 18000$

$= \$19,000$

It costs $19,000 to make 100 units.

**c.** $50x = 10x + 18000$

$40x = 18000$

$x = 450$ units

**20.** $S: p = 5q + 1500, D: p = -3q + 3100$

$5q + 1500 = -3q + 3100$

$8q = 1600$ or $q = 200$

$p(200) = 5(200) + 1500 = \$2500$

**21.** $y = 720,000 - 2000x$

**a.** $b = 720,000$

The original value is 720,000.

**b.** $m = -2000$.

The building is depreciating $2000 each month.

**22.** $x$ = number of reservations

$0.90x = 360$

$x = 400$

Accept 400 reservations.

**23.** $x$ = amount invested at 9%

$y$ = amount invested at 6%

$x + y = 20000$ Amount

$0.09x + 0.06y = 1560$ Interest

$0.09x + 0.09y = 1800$

$\underline{0.09x + 0.06y = 1560}$

$0.03y = 240$

$y = \$8000$

Invest $8000 at 6% and $12000 at 9%.

# Chapter 2: Quadratic and Other Special Functions

## *Exercises 2.1*

**1.** $2x^2+3=x^2-2x+4$

$x^2+2x-1=0$

**3.** $(y+1)(y+2)=4$

$y^2+3y+2=4$

$y^2+3y-2=0$

**5.** $x^2-4x=12$

$x^2-4x-12=0$

$x^2-6x+2x-12=0$

$x(x-6)+2(x-6)=0$

$(x-6)(x+2)=0$

$x-6=0$ or $x+2=0$

Solution: $x=-2, 6$

**7.** $9-4x^2=0$

$(3+2x)(3-2x)=0$

$3+2x=0$ or $3-2x=0$

Solution: $x=-\frac{3}{2}, \frac{3}{2}$

**9.** $x=x^2$

$x^2-x=0$

$x(x-1)=0$

Solution: $x=0, 1$

Never divide by a variable. A root is lost if you divide.

**11.** $4t^2-4t+1=0$

$(2t-1)(2t-1)=0$

$2t-1=0$

Solution: $t=\frac{1}{2}$

**13. a.** $x^2-4x-4=0$

$a=1, b=-4, c=-4$

$$x=\frac{-(-4)\pm\sqrt{(-4)^2-4(1)(-4)}}{2(1)}$$

$$=\frac{4\pm\sqrt{32}}{2}=\frac{4\pm4\sqrt{2}}{2}=2\pm2\sqrt{2}$$

**b.** Since $\sqrt{2}\approx1.414$, the solutions are approximately 4.83, –0.83.

**c.** $x^2-6x+7=0$

$a=1, b=-6, c=7$

$$x=\frac{6\pm\sqrt{36-28}}{2}$$

$$=\frac{6\pm\sqrt{8}}{2}=\frac{6\pm2\sqrt{2}}{2}=3\pm\sqrt{2}$$

$\sqrt{2}\approx1.414$, the solutions are approximately 4.83, –0.83.

**15.** $2w^2+w+1=0$

$a=2, b=1, c=1$

$$w=\frac{-1\pm\sqrt{1-8}}{4}=\frac{-1\pm\sqrt{-7}}{4}$$

There are no real solutions.

**17.** $y^2=7$

$y=\pm\sqrt{7}$

**19.** $5x^2=80$

$x^2=16$

$x=\pm4$

**21.** $(x+4)^2=25$

$x+4=\pm5$

$x=-4\pm5$

Solution: $x=1, -9$

**23.** $x^2+5x=21+x$

$x^2+4x-21=0$

$(x+7)(x-3)=0$

Solution: $x=-7, 3$

**25.** $\frac{w^2}{8}-\frac{w}{2}-4=0$

$w^2-4w-32=0$

$(w-8)(w+4)=0$

$w-8=0$ or $w+4=0$

Solution: $w=8, -4$

**27.** $16z^2 + 16z - 21 = 0$

$a = 16, b = 16, c = -21$

$$z = \frac{-16 \pm \sqrt{256 + 1344}}{32}$$

$$= \frac{-16 \pm 40}{32} = \frac{3}{4} \text{ or } -\frac{7}{4}$$

Solution: $z = -\frac{7}{4}, \frac{3}{4}$

**29.** $(x-1)(x+5) = 7$

$x^2 + 4x - 5 = 7$

$x^2 + 4x - 12 = 0$

$(x+6)(x-2) = 0$

Solution: $x = -6, 2$

**31.** $5x^2 = 2x + 6$ or $5x^2 - 2x - 6 = 0$

$a = 5, b = -2, c = -6$

$$x = \frac{2 \pm \sqrt{4 + 120}}{10} = \frac{1 \pm \sqrt{31}}{5}$$

Solution: $x = \frac{1 - \sqrt{31}}{5}, \frac{1 + \sqrt{31}}{5}$

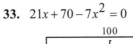

**33.** $21x + 70 - 7x^2 = 0$

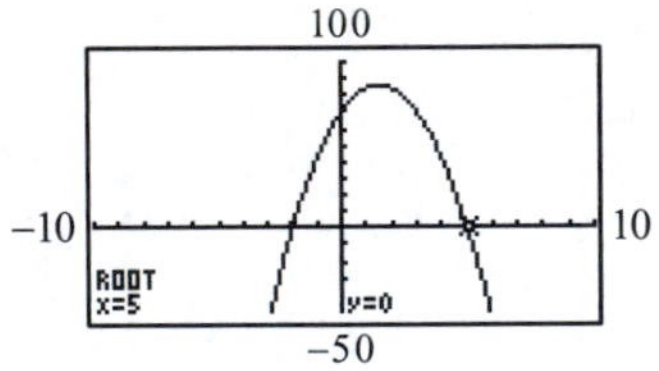


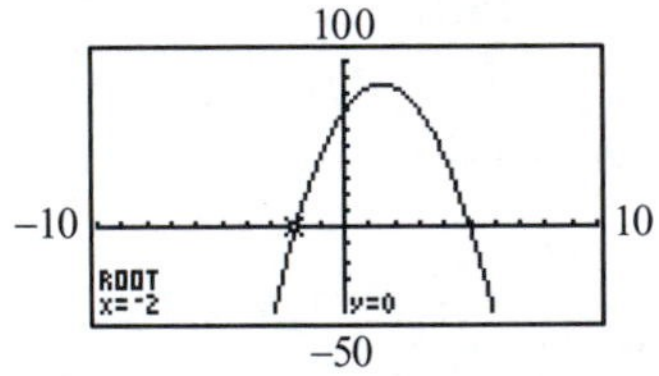


Divide by –7 and rearrange.

$x^2 - 3x - 10 = 0$

$(x-5)(x+2) = 0$

Solution: $x = -2, 5$

**35.** $300 - 2x - 0.01x^2 = 0$

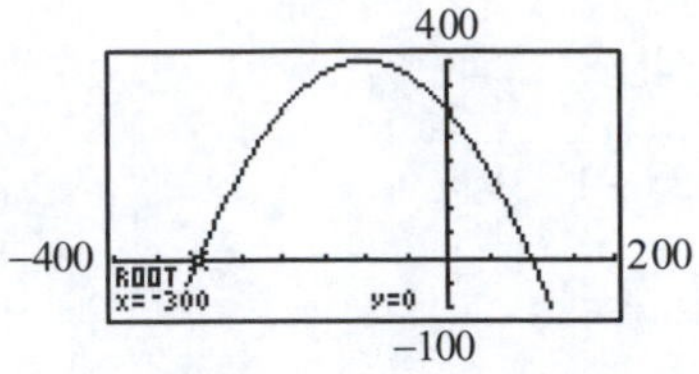


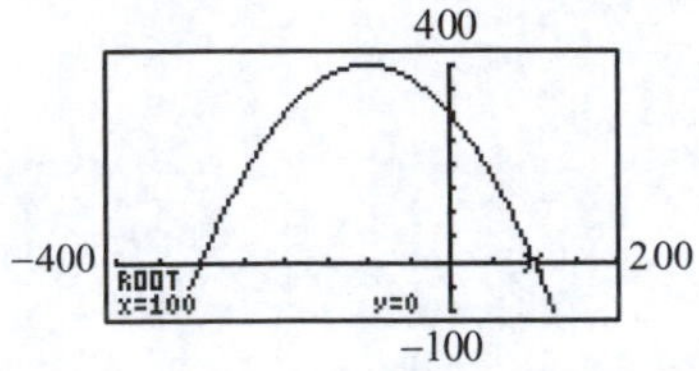


$a = -0.01, b = -2, c = 300$

$$x = \frac{2 \pm \sqrt{4 + 12}}{-0.02} = \frac{2 \pm 4}{-0.02}$$

$= -300$ or $100$

**37.** $25.6x^2 - 16.1x - 1.1 = 0$

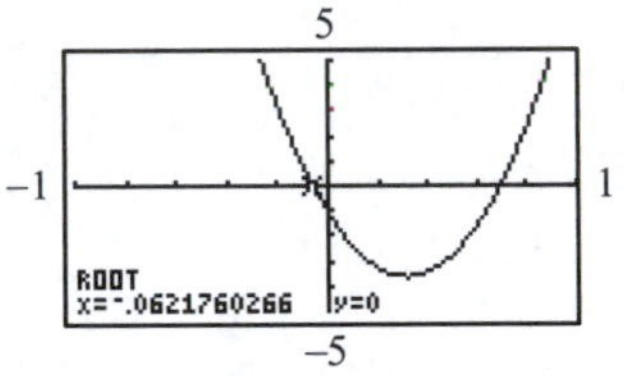


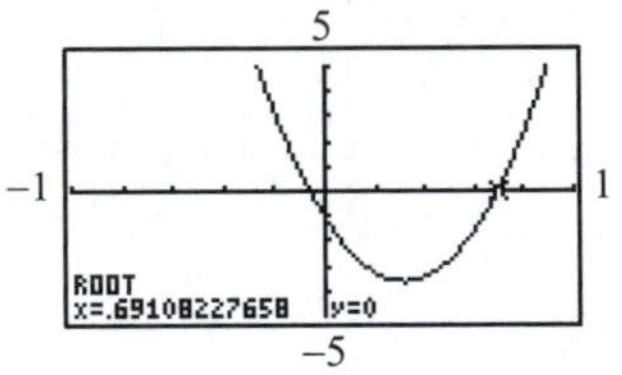


**39.** $x + \frac{8}{x} = 9$

$x^2 + 8 = 9x$

$x^2 - 9x + 8 = 0$

$(x-8)(x-1) = 0$

Solution: $x = 1, 8$

**41.** $\frac{x}{x-1} = 2x + \frac{1}{x-1}$

$x = (2x^2 - 2x) + 1$

$2x^2 - 3x + 1 = 0$

$(2x-1)(x-1) = 0$

Solution: $x = \frac{1}{2}$

1 is not a root since division by zero is not defined.

**43.** $(x+8)^2+3(x+8)+2=0$

$[(x+8)+2][(x+8)+1]=0$

$(x+8)+2=0$ or $(x+8)+1=0$

Solution: $x=-10, -9$

**45.** $P=-x^2+90x-200$

$1200=-x^2+90x-200$

$0=x^2-90x+1400$

$0=(x-20)(x-70)$

A profit of \$1200 is earned at $x = 20$ units or $x = 70$ units of production.

**47. a.** $P=-18x^2+6400x-400$

$61,800=-18x^2+6400x-400$

$18x^2-6400x+62,200=0$

Factoring appears difficult, so let us apply the quadratic formula.

$$x=\frac{6400\pm\sqrt{6400^2-4(18)(62,200)}}{36}$$

$$=\frac{6400\pm\sqrt{36,481,600}}{36}$$

$$=\frac{6400\pm 6040}{36}=10 \text{ or } 345.56$$

So, a profit of \$61,800 is earned for 10 units or for 345.56 units.

**b.** Yes. Maximum profit occurs at vertex as seen using the graphing calculator.

**49.** $S=100+96t-16t^2$

$100=100+96t-16t^2$

$0=96t-16t^2=16t(6-t)$

The ball is 100 feet high 6 seconds later.

**51.** $p=25-0.01s^2$

**a.** $0=25-0.01s^2$

$=(5+0.1s)(5-0.1s)$

$p=0$ if $5-0.1s=0$ or $s=50$.

**b.** $s\ge 0$. $p=0$ means there is no particulate pollution.

**53.** $t=0.001\left(0.732x^2+15.417x+607.738\right)$

$8.99=0.001\left(0.732x^2+15.417x+607.738\right)$

$8990=0.732x^2+15.417x+607.738$

$0=0.732x^2+15.417x-8382.262$

$$t=\frac{-15.417\pm\sqrt{(15.417)^2-4(0.732)(-8382.262)}}{2(0.732)}$$

$t\approx 96.996$ or $t\approx -118.058$

The positive answer is the one that makes sense here, 97.0 mph.

**55.** $p=0.17t^2-2.61t+52.64$

$55=0.17t^2-2.61t+52.64$

$0.17t^2-2.61t-2.36=0$

Using the quadratic formula or a graphing utility gives the positive value $t\approx 16.2$. In 2016 the percent of high school seniors who will have tried marijuana is predicted by the function to reach 55%.

**57.** $P=\left(\frac{C}{100}\right)\cdot C$

We know that the selling price is \$144 and that the selling price equals the profit plus the cost $C$ to the store.

$$144=\frac{C^2}{100}+C$$

$14400=C^2+100C$

$C^2+100C-14400=0$

$C=-180$ or $C=80$

The cost $C$ of the necklace to the store is not negative, so $C$ = \$80 is the amount the store paid for the necklace.

**59.** $E=7.94x^2+33.2x+2190$

$5000=7.94x^2+33.2x+2190$

$7.94x^2+33.2x-2810=0$

Using the quadratic formula or a graphing utility gives the positive value $x \approx 16.8$.

The model predict these expenditures will reach \$5 trillion in 2022.

**61.** $K^2 = 16v + 4$

In each case below only positive values of $K$ are reported.

**a.** $K^2 = 16(20) + 4 = 324$

$K = 18$

**b.** $K^2 = 16(60) + 4 = 964$

$K \approx 31$

**c.** Speed triples, but $K$ changes only by a factor of 1.72.

## *Exercises 2.2*

**1.** $y = \frac{1}{2}x^2 + x$

**a.** $x = \frac{-b}{2a} = \frac{-1}{2(1/2)} = -1$

$y = \frac{1}{2}(-1)^2 + (-1) = -\frac{1}{2}$

Vertex is at $\left(-1, -\frac{1}{2}\right)$.

**b.** $a > 0$, so vertex is a minimum.

**c.** $-1$

**d.** $-\frac{1}{2}$

**3.** $y = 8 + 2x - x^2$

**a.** $x = \frac{-b}{2a} = \frac{-2}{2(-1)} = 1$

$y = 8 + 2(1) - (1)^2 = 9$

Vertex is at (1, 9).

**b.** $a < 0$, so vertex is a maximum.

**c.** 1

**d.** 9

**5.** $f(x) = 6x - x^2$

**a.** $x = \frac{-b}{2a} = \frac{-6}{-2} = 3.$

$f(3) = 6(3) - (3)^2 = 9$

Vertex is at $(3, 9)$.

**b.** $a < 0$, so vertex is a maximum.

**c.** 3

**d.** 9

**7.** $y = -\frac{1}{4}x^2 + x$

Vertex is a maximum point since $a < 0$.

V: $x = \frac{-b}{2a} = \frac{-1}{2(-1/4)} = 2$

$y = -\frac{1}{4}(2)^2 + 2 = 1$

Zeros: $-\frac{1}{4}x^2 + x = 0$

$x\left(-\frac{1}{4}x + 1\right) = 0$

$x = 0, 4$

$y$-intercept = 0

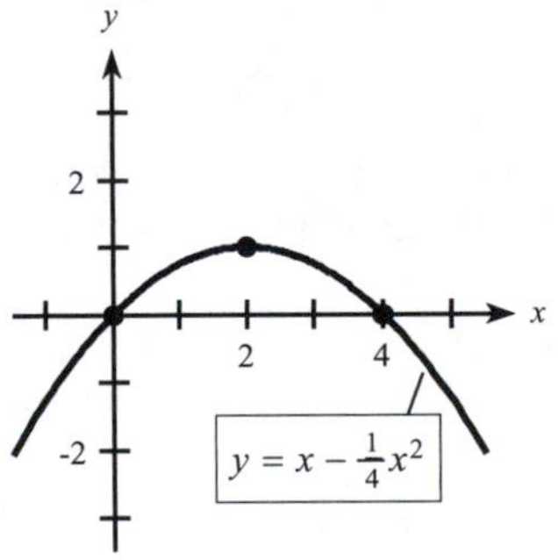

**9.** $y = x^2 + 4x + 4$

Vertex is a minimum point since $a > 0$.

V: $x = \dfrac{-b}{2a} = \dfrac{-4}{2(1)} = -2$

$y = (-2)^2 + 4(-2) + 4 = 0$

Zeros: $x^2 + 4x + 4 = (x+2)(x+2) = 0$

$x = -2$

$y$-intercept $= 4$

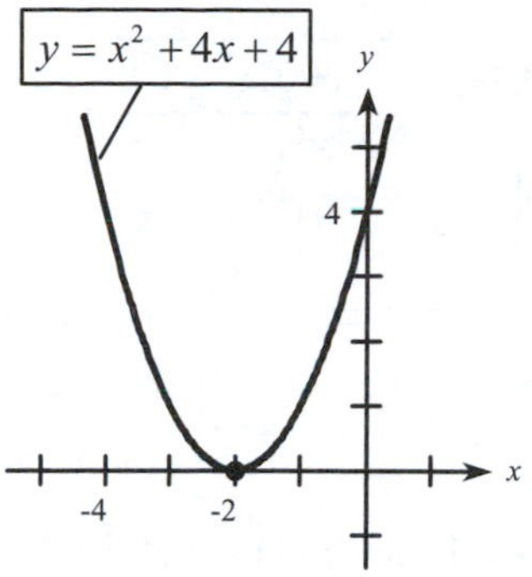


**11.** $y = \dfrac{1}{2}x^2 + x - 3$

Vertex is a minimum point since $a > 0$.

V: $x = \dfrac{-b}{2a} = \dfrac{-1}{2(1/2)} = -1$

$y = \dfrac{1}{2}(-1)^2 + (-1) - 3 = -\dfrac{7}{2}$

Zeros: $\dfrac{1}{2}x^2 + x - 3 = 0 \;\rightarrow\; x^2 + 2x - 6 = 0$

$x = \dfrac{-2 \pm \sqrt{4+24}}{2} = \dfrac{-2 \pm 2\sqrt{7}}{2} = -1 \pm \sqrt{7}$

$y$-intercept $= -3$

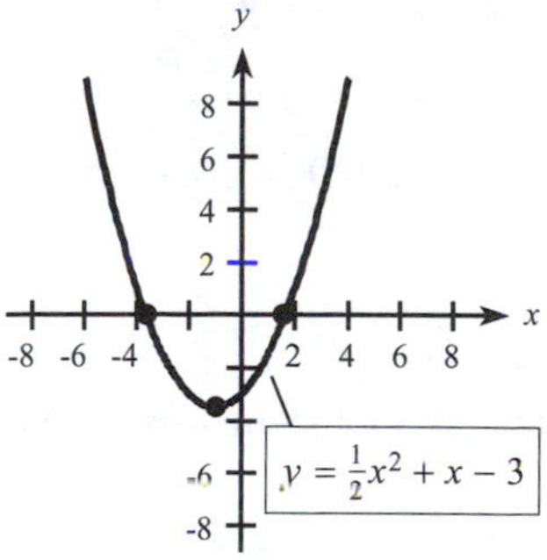


**13.** $y = (x-3)^2 + 1$

**a.** Graph is shifted 3 units to the right and 1 unit up.

**b.**

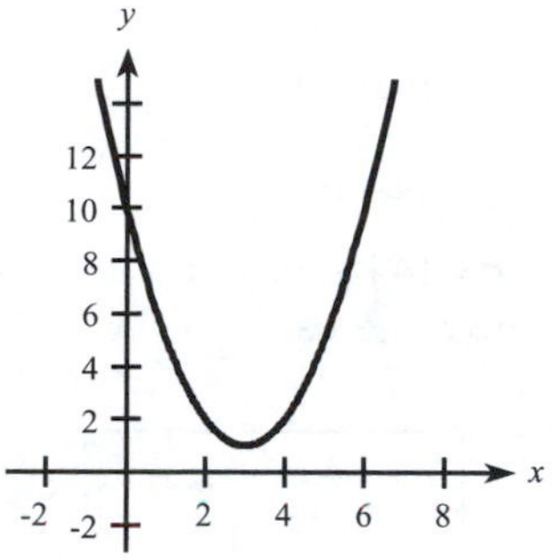


**15.** $y = (x+2)^2 - 2$

**a.** Graph is shifted 2 units to the left and 2 units down.

**b.**

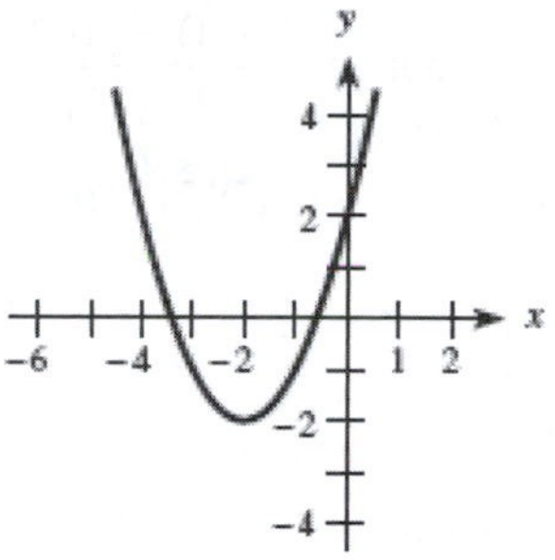


**17.** $y = \dfrac{1}{2}x^2 - x - \dfrac{15}{2}$

V: $x = \dfrac{-b}{2a} = \dfrac{-(-1)}{2(1/2)} = 1$

$y = \dfrac{1}{2}(1)^2 - 1 - \dfrac{15}{2} = -8$

Zeros: $x^2 - 2x - 15 = (x-5)(x+3) = 0$

$x = 5, -3$

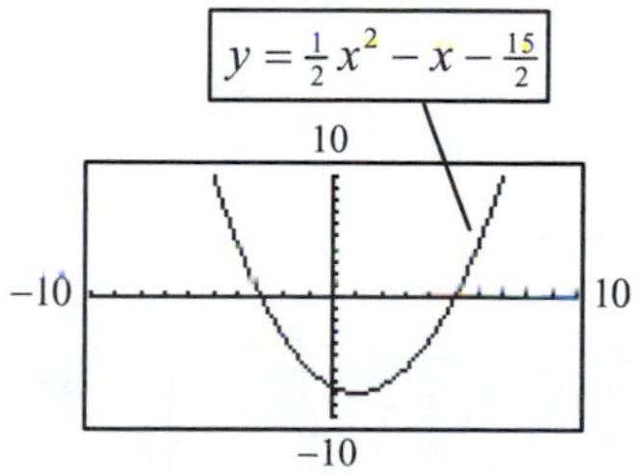

**19.** $y = \frac{1}{4}x^2 + 3x + 12$

V: $x = \frac{-b}{2a} = \frac{-3}{2\left(\frac{1}{4}\right)} = -6$

$y = \frac{1}{4}(-6)^2 + 3(-6) + 12 = 3$

Zeros: $x^2 + 12x + 48 = 0$

$b^2 - 4ac = 144 - 192 < 0$

There are no zeros.

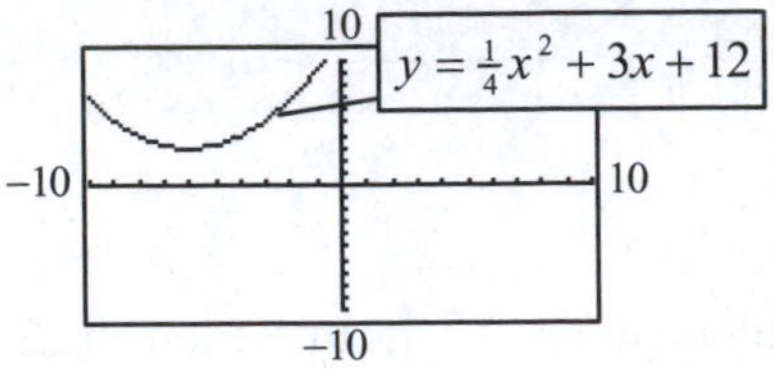


**21.** $f(x) = y = -5x - x^2$

Average Rate of Change $= \frac{f(1) - f(-1)}{1 - (-1)}$

$= \frac{-6 - 4}{2} = -\frac{10}{2} = -5$

**23.** $y = 63 + 0.2x - 0.01x^2$

V: $x = \frac{-0.2}{-0.02} = 10$

$y = 63 + 2 - 1 = 64$

Zeros: $x^2 - 20x - 6300 = (x - 90)(x + 70) = 0$

$x = 90, -70$

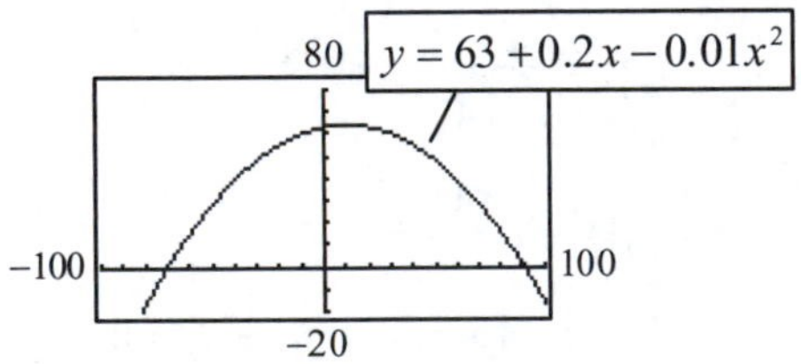


**25.** $y = 0.0001x^2 - 0.01$

V: $x = \frac{-0}{2(0.0001)} = 0$

$y = 0 - 0.01 = -0.01$

Zeros: $0.0001x^2 - 0.01 = 0.01\left(0.01x^2 - 1\right) = 0$

$0.01(0.1x + 1)(0.1x - 1) = 0$

$x = -10, 10$

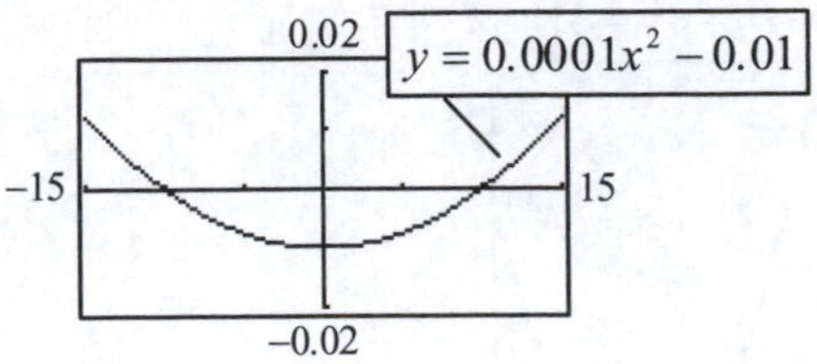


**27.** $f(x) = 8x^2 - 16x - 16$

**a.** $x = \frac{-b}{2a} = \frac{16}{16} = 1$ and $f(1) = -24$

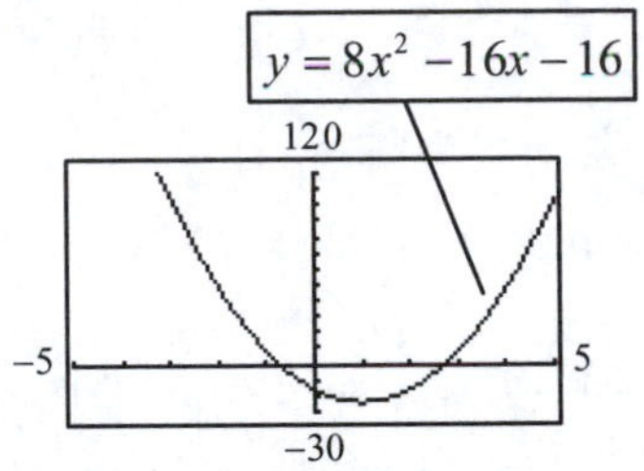


**b.** Graphical approximation gives $x = -0.73,\ 2.73$

**29.** $f(x) = 3x^2 - 8x + 4$

**a.** The TRACE gives $x = 2$ as a solution.

**b.** $(x - 2)$ is a factor.

**c.** $3x^2 - 8x + 4 = (x - 2)(3x - 2)$

**d.** $(x - 2)(3x - 2) = 0$

$x - 2 = 0$ or $3x - 2 = 0$

Solution is $x = 2,\ 2/3$.

**31.** $P = -0.1x^2 + 16x - 100$

The vertex coordinates are the answers to the questions.

**a.** $a = -0.1, b = 16$

$x = \frac{-b}{2a} = \frac{-16}{-0.2} = 80$

Profit is maximized at a production level of 80 units.

**b.** $P(80) = -0.1(80)^2 + 16(80) - 100 = \$540$ is the maximum profit.

**33.** $Y = 800x - x^2$
Opens down so maximum $Y$ is at vertex.
V: $x = \frac{-800}{-2} = 400$
Maximum yield occurs at $x = 400$ trees.

**35.** $S = 1000x - x^2$
Maximum sensitivity occurs at vertex.
V: $x = \frac{-1000}{-2} = 500$
The dosage for maximum sensitivity is 500.

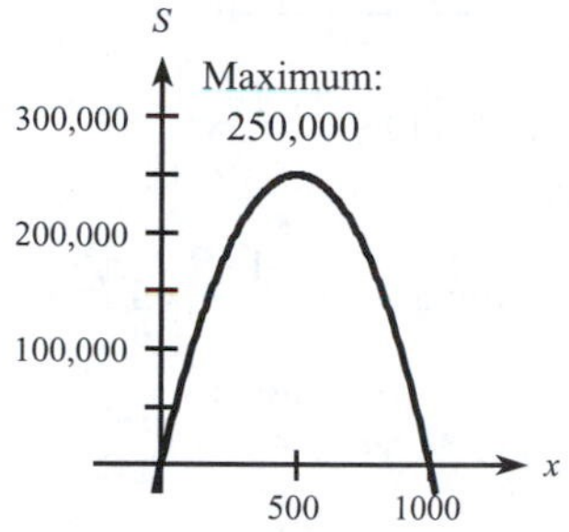


**37.** $R = 270x - 90x^2$
Maximum rate occurs at vertex.
V: $x = \frac{-270}{2(-90)} = \frac{3}{2}$ (lumens)
is the intensity for maximum rate.

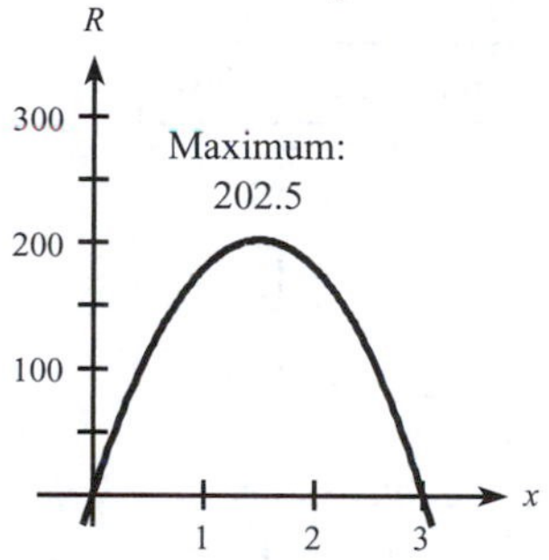


**39. a.** $y = -0.0013x^2 + x + 10$
V: $x = \frac{-1}{-0.0026} = 384.62;$
$y = -0.0013(384.62)^2 + 384.62 + 10$
$= 202.31$

**b.** $y = -\frac{1}{81}x^2 + \frac{4}{3}x + 10$
V: $x = \frac{\frac{-4}{3}}{\frac{-2}{81}} = 54$;
$y = -\frac{1}{81}(54)^2 + \frac{4}{3}(54) + 10 = 46$
Projectile **a.** goes $202.31 - 46 = 156.31$ feet higher.

**41. a.** From $b$ to $c$. The average rate of change is the same as the slope of the segment. The segment from $b$ to $c$ is steeper.
**b.** Needs to satisfy $d > b$ to make the segment from $a$ to $d$ have a greater slope.

**43. a.**

| No. of Apts | Rent | Total Revenue |
|---|---|---|
| 50 | \$600 | \$30,000 |
| 49 | \$620 | \$30,380 |
| 48 | \$640 | \$30,720 |

**b.** Revenue increases \$720
**c.** $R = (50 - x)(600 + 20x)$
**d.** $R = -20x^2 + 400x + 30{,}000$
$R$ is maximized at $x = \frac{-400}{2(-20)} = 10$.
Rent would be $\$600 + \$200 = \$800$.

**45. a.** A quadratic function or parabola.
**b.** $a < 0$ because the graph opens downward.
**c.** The vertex occurs after 2004 (or when $x > 0$), so $-\frac{b}{2a} > 0$. Hence with $a < 0$ we must have $b > 0$. The value $c = f(0)$ or the $y$-value during 2004 which is positive.

**47.** $y = 20.61x^2 - 116.4x + 7406$
For 2010, $x = 10$ gives $y = 8303$.
For 2015, $x = 15$ gives $y = 10{,}297.25$.
For 2020, $x = 20$ gives $y = 13{,}322$.
Average rate of change from 2010 to 2015:
$\frac{10{,}297.25 - 8303}{15 - 10} = 398.85$

Average rate of change from 2015 to 2020:

$$\frac{13,322-10,297.25}{20-15}=604.95$$

To the nearest dollar, the projected average rate of change of U.S. per capita health care costs from 2010 to 2015 will be \$399/year, and from 2015 to 2020 it will be \$605/year.

**49.**

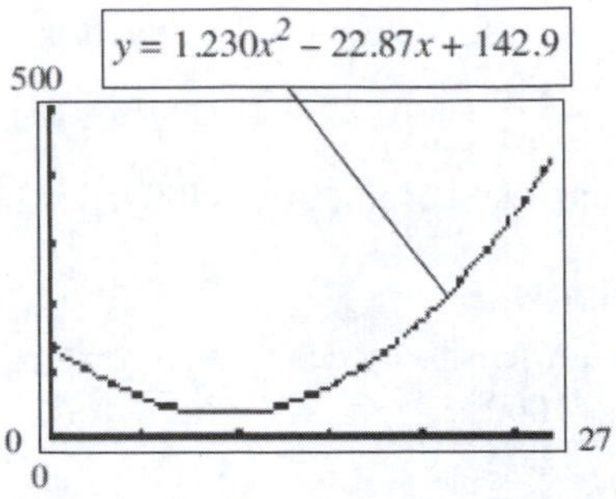


## *Exercises 2.3*

**1.** $C(x)=x^2+40x+2000$

$R(x)=130x$

$x^2+40x+2000=130x$

$x^2-90x+2000=0$

$(x-40)(x-50)=0$

$x=40$ or $x=50$

Break-even values are at $x=40$ and 50 units.

**3.** $C(x)=15,000+35x+0.1x^2$

$R(x)=385x-0.9x^2$

$15,000+35x+0.1x^2=385x-0.9x^2$

$x^2-350x+15,000=0$

$(x-300)(x-50)=0$

$x=300$ or $x=50$

**5.** $P(x)=-11.5x-0.1x^2-150$

At the break-even points, $P(x)=0$.

$0=11.5x-0.1x^2-150$

$-0.1x^2+11.5x-150=0$

$(x-15)(x-100)=0$

Since production < 75 units, $x=15$.

**7.** $R(x)=385x-0.9x^2$

$a=-0.9$, $b=385$

Maximum revenue is at the vertex.

V: $x=\frac{-385}{-1.8}=213.89$ or 214 total units

$R(214)=385(214)-0.9(214)^2=\$41,173.60$

**9.** $R(x)=x(175-0.50x)=175x-0.5x^2$

$a=-0.50$, $b=175$

Revenue is a maximum at $x=\frac{-175}{-1}=175$.

Price that will maximize revenue is $p=175-87.50=\$87.50$.

**11.** $P(x)=-x^2+110x-1000$

Maximum profit is at the vertex or when

$x=\frac{-110}{-2}=55.$

$P(55)=\$2025.$

**13. a.**

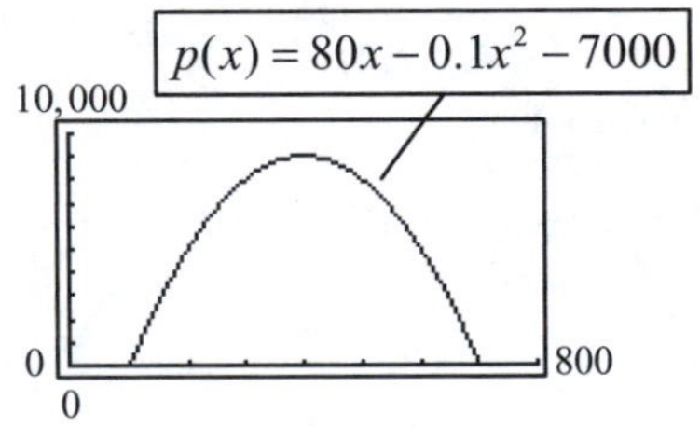


**b.** $(400,9000)$ is the maximum

**c.** positive

**d.** negative

**e.** closer to 0

**15.** $R(x) = 385x - 0.9x^2$

$C(x) = 15,000 + 35x + 0.1x^2$

**a.**

$$P(x) = 385x - 0.9x^2 - (15,000 + 35x + 0.1x^2)$$
$$= -x^2 + 350x - 15,000$$

At the vertex we have $x = \frac{-350}{-2} = 175.$

So, $P(175) = \$15,625.$

**b.** No. More units are required to maximize revenue.

**c.** The break-even values and zeros of $P(x)$ are the same.

**17. a.** $C(x) = 28,000 + \left(\frac{2}{5}x + 222\right)x$

$$= \frac{2}{5}x^2 + 222x + 28,000$$

$$R(x) = \left(1250 - \frac{3}{5}x\right)x = 1250x - \frac{3}{5}x^2$$

(The key is "per unit $x$.")

$$R(x) = C(x)$$
$$1250x - \frac{3}{5}x^2 = \frac{2}{5}x^2 + 222x + 28,000$$
$$x^2 - 1028x + 28,000 = 0$$
$$(x - 1000)(x - 28) = 0$$

Break-even values are at $x = 28$ and $x = 1000$.

**b.** Maximum revenue occurs at

$$x = \frac{-1250}{-\frac{6}{5}} = 1042 \text{ (rounded).}$$

$R(1042) = \$651,041.60$ is the maximum revenue.

**c.**

$$P(x) = 1250x - \frac{3}{5}x^2 - \left(\frac{2}{5}x^2 + 222x + 28,000\right)$$
$$= -x^2 + 1028x - 28,000$$

Maximum profit is at $x = \frac{-1028}{-2} = 514.$

$P(514) = \$236,196$ is the maximum profit.

**d.** Price that will maximize profit is

$$p = 1250 - \frac{3}{5}(514) = \$941.60.$$

**19. a.** $t \approx 5.1$, in 2012; $R \approx \$60.79$ billion

**b.** The data show a smaller revenue, R = \$60.27 billion in 2011.

**c.**

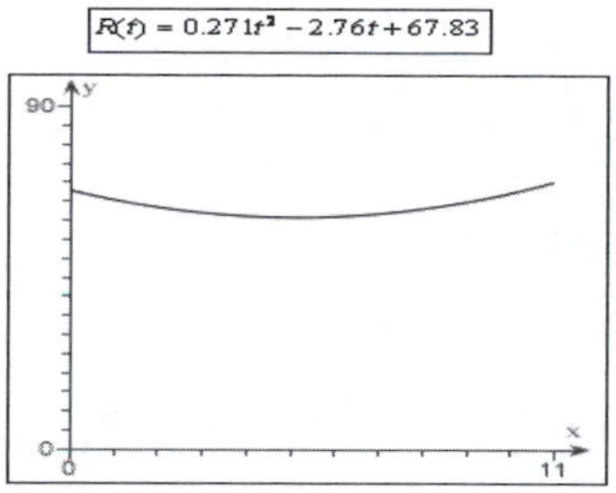


**d.** The model fits the data quite well.

**21. a.** $p(t) = -0.019t^2 + 0.284t - 0.546$

**b.** 2011

**c.**

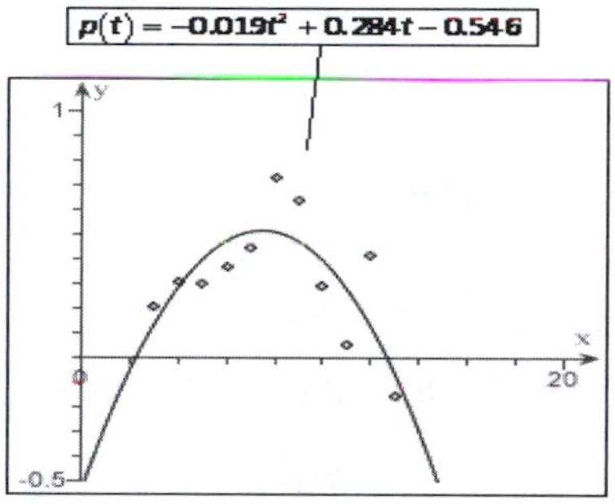


**d.** The model projects decreasing profits, and, except for 2015, the data support this.

**e.** Management would be interested in increasing revenues or reducing costs (or both) to improve profit.

**23. a.** Supply: $p = \frac{1}{4}q^2 + 10$ (see below)

Demand: $p = 86 - 6q - 3q^2$ (see below)

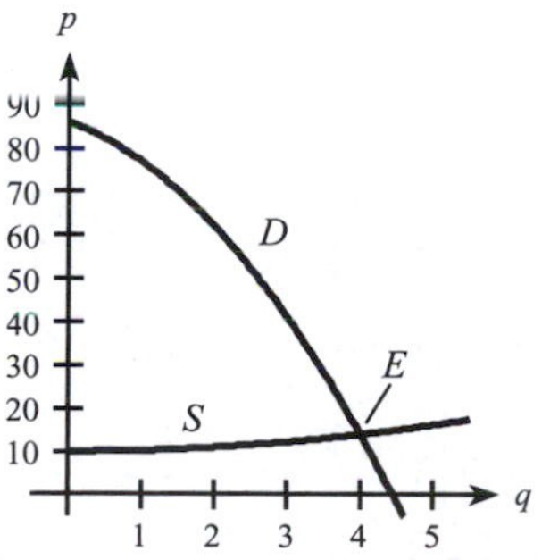

**b.** See E on graph.

**c.** $\frac{1}{4}q^2+10=86-6q-3q^2$

$$q^2+40=344-24q-12q^2$$
$$0=13q^2+24q-304$$
$$0=(q-4)(13q+76)$$

$q=4$ must be positive.

$p=\frac{1}{4}(4)^2+10=14$

E: (4, 14)

**25.** $p=q^2+8q+16$

$p=-3q^2+6q+436$

$$q^2+8q+16=-3q^2+6q+436$$
$$4q^2+2q-420=0$$
$$2q^2+q-210=0$$
$$(2q+21)(q-10)=0$$
$$q=10$$

$p=10^2+8(10)+16=196$

E: (10, 196)

**27.** $p^2+4q=1600$

$300-p^2+2q=0$

$(300+2q)+4q=1600$

$$6q=1300$$
$$q=216\frac{2}{3}$$

$p^2+4\left(\frac{1300}{6}\right)=1600$ or $p^2$

$=733.33$ or $p=27.08$

E: $\left(216\frac{2}{3}, 27.08\right)$

**29.** $p-q=10$ or $q=p-10$

$q(2p-10)=2100$

$$q=\frac{2100}{2p-10}$$
$$p-10=\frac{2100}{2p-10}$$
$$(p-10)(2p-10)=2100$$
$$2p^2-30p+100=2100$$
$$2p^2-30p-2000=0$$
$$p^2-15p-1000=0$$
$$(p-40)(p+25)=0$$

$p=40$ or $p=-25$

(only the positive answer makes sense here)

$q=40-10=30$

$E:(30,40)$

**31.** $2p-q-10=0$

$(p+10)(q+30)=7200$

So, $(p+10)(2p-10+30)=7200$

$$p^2+20p+100=3600$$
$$p^2+20p-3500=0$$
$$(p+70)(p-50)=0$$
$$p=50$$

$q=2(50)-10=90$

E: $(q, p)=(90, 50)$

**33.** $p=\frac{1}{2}q+5+22=\frac{1}{2}q+27$

So, $\left(\frac{1}{2}q+27+10\right)(q+30)=7200$

$$(q+74)(q+30)=14{,}400$$
$$q^2+104q-12{,}180=0$$
$$(q+174)(q-70)=0$$

$p=\frac{1}{2}(70)+27=62$

E: (70, 62)

## *Exercises 2.4*

**1.** b

**3.** f

**5.** j

**7.** k

**9.** a

**11.** c

**13. a.** cubic
**b.** quartic

**15.** $y=x^3-x=x(x+1)(x-1)$ : e

**17.** $y = 16x^2 - x^4 = x^2(4+x)(4-x)$ : b

**19.** $y = x^2 + 7x = x(x+7)$ : d

**21.** $y = \dfrac{x-3}{x+1}$ : g

**23.**

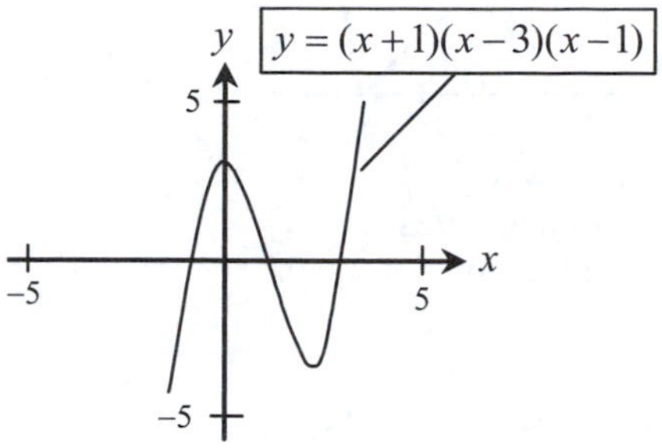


**25.**

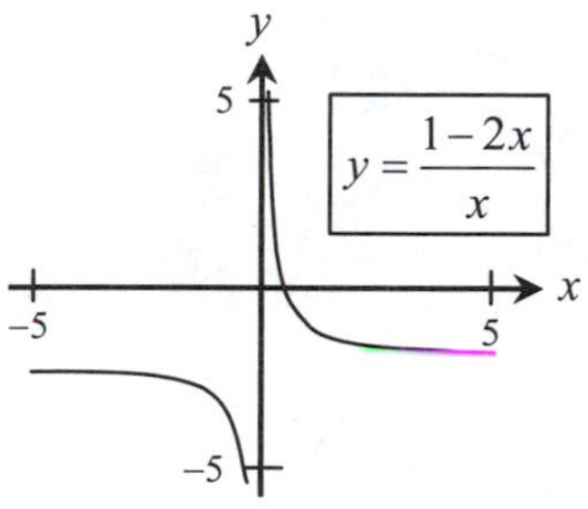


**27.**

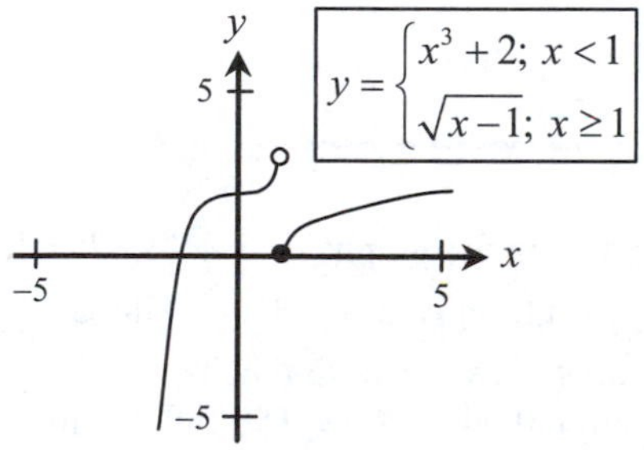


**29.** $F(x) = \dfrac{x^2 - 1}{x}$

**a.** $F\left(-\dfrac{1}{3}\right) = \dfrac{\frac{1}{9} - 1}{-\frac{1}{3}} = \dfrac{8}{3}$

**b.** $F(10) = \dfrac{100-1}{10} = \dfrac{99}{10}$

$F(x) = \dfrac{x^2 - 1}{x}$

**c.** $F\left(-\dfrac{1}{3}\right) = \dfrac{\frac{1}{9} - 1}{-\frac{1}{3}} = \dfrac{8}{3}$

**d.** $F(10) = \dfrac{100-1}{10} = \dfrac{99}{10}$

**e.** $F(0.001) = \dfrac{0.000001 - 1}{0.001} = \dfrac{-0.999999}{0.001}$
$= -999.999$

**f.** $F(0)$ is not defined–division by zero.

**31.** $f(x) = x^{3/2}$

**a.** $f(16) = (\sqrt{16})^3 = 64$

**b.** $f(1) = (\sqrt{1})^3 = 1$

**c.** $f(100) = (\sqrt{100})^3 = 1000$

**d.** $f(0.09) = (\sqrt{0.09})^3 = 0.027$

**33.** $k(x) = \begin{cases} 2 & \text{if } x < 0 \\ x+4 & \text{if } 0 \le x < 1 \\ 1-x & \text{if } x \ge 1 \end{cases}$

**a.** $k(-5) = 2$ since $x < 0$.

**b.** $k(0) = 0 + 4 = 4$

**c.** $k(1) = 1 - 1 = 0$

**d.** $k(-0.001) = 2$ since $x < 0$.

**35.** $y = 1.6x^2 - 0.1x^4$

**a.**

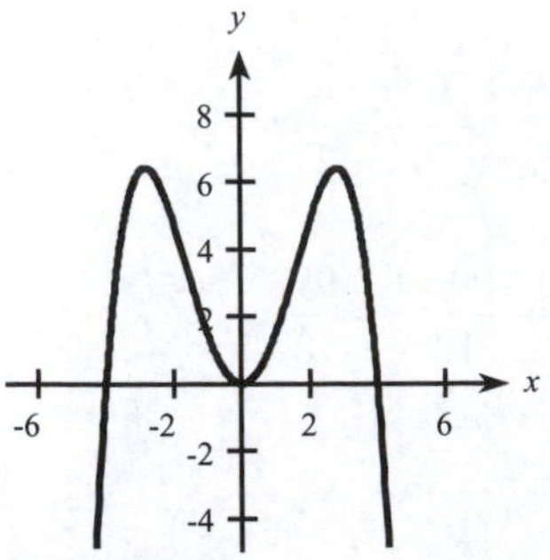


**b.** polynomial
**c.** no asymptotes
**d.** turning points at $x = 0$ and approximately $x = -2.8$ and $x = 2.8$

**37.** $y = \dfrac{2x+4}{x+1}$

**a.**

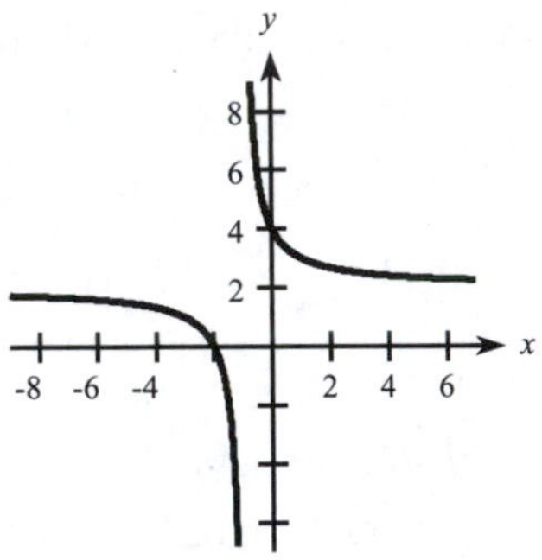


**b.** rational
**c.** vertical: $x = -1$
horizontal: $y = 2$
**d.** no turning points

**39.** $f(x) = \begin{cases} -x & \text{if } x < 0 \\ 5x & \text{if } x \geq 0 \end{cases}$

**a.**

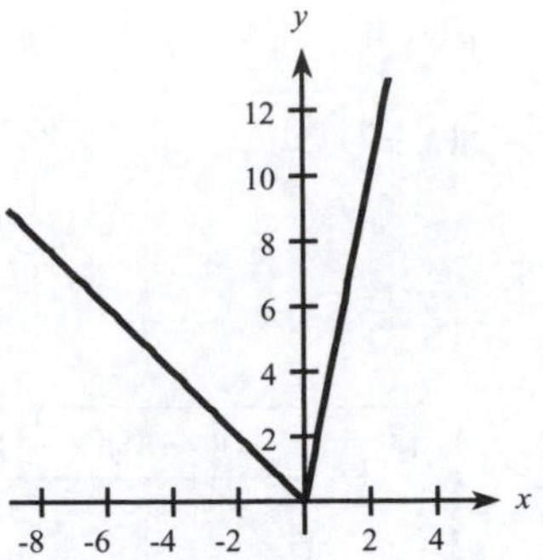


**b.** piecewise
**c.** no asymptotes
**d.** turning point at $x = 0$.

**41.** $V = V(x) = x^2(108 - 4x)$

**a.** $V(10) = 100(68) = 6800$ cubic inches
$V(20) = 400(28) = 11{,}200$ cubic inches

**b.** $108 - 4x > 0$

$$-4x > -108$$

$$0 < x < 27$$

**43.** $f(x) = 15.875x^{1.18}$

**a.** upward

**b.**

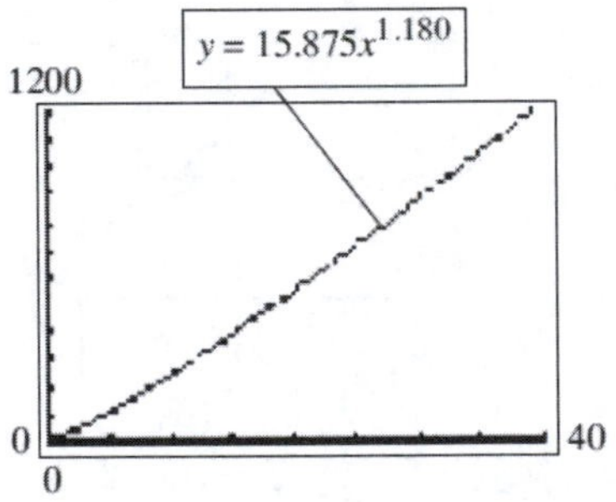


**c.** Intersecting the graphs of $y = 15.875x^{1.18}$ and $y = 1150$ gives $x \approx 37.7$. Global spending is expected to reach \$1,150,000,000,000 (\$1150 billion) in $1980 + 38 = 2018$.

**45.** $C(p) = \dfrac{7300p}{100 - p}$

**a.** $0 \le p < 100$

**b.** $C(45) = \dfrac{7300 \cdot 45}{100 - 45} = \$5972.73$

**c.** $C(90) = \dfrac{7300 \cdot 90}{100 - 90} = \$65,700$

**d.** $C(99) = \dfrac{7300 \cdot 99}{100 - 99} = \$722,700$

**e.** $C(99.6) = \dfrac{7300(99.6)}{100 - 99.6} = \$1,817,700$

**f.** To remove $p$% of the pollution would cost $C(p)$. Note how cost increases as $p$ (the percent of pollution removed) increases.

**47.** $A = A(x) = x(50 - x)$

**a.** $A(2) = 2 \cdot 48 = 96$ square feet

$A(30) = 30 \cdot 20 = 600$ square feet

**b.** $0 < x < 50$ in order to have a rectangle.

**49.** $y = \begin{cases} 5.59x^2 - 93.5x + 633 & \text{for } 0 \le x \le 55 \\ 6.56x^2 - 519x + 20,900 & \text{for } 55 < x \le 90 \end{cases}$

**a.**

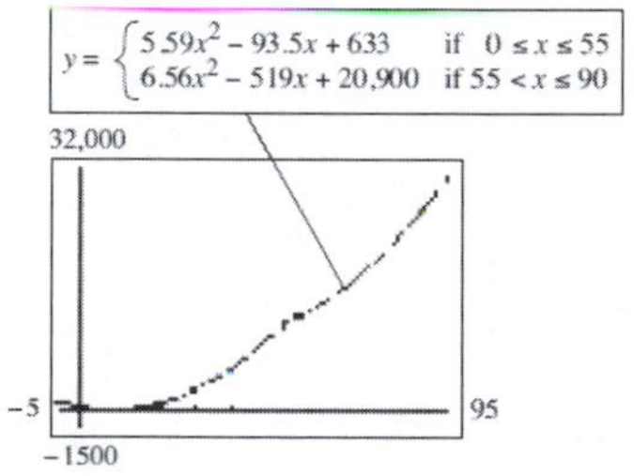


**b.** $y(50) = 5.59(50)^2 - 93.5(50) + 633 = \$9933$ billion (\$9.933 trillion)

**c.** $y(75) = 6.56(75)^2 - 519(75) + 20,900 = \$18,875$ billion (\$18.875 trillion)

**51. a.** $P(x) = \begin{cases} 49 & \text{if } 0 < x \le 1 \\ 70 & \text{if } 1 < x \le 2 \\ 91 & \text{if } 2 < x \le 3 \\ 112 & \text{if } 3 < x \le 4 \end{cases}$

**b.** $P(1.2) = 70$; it costs 70 cents to mail a 1.2-oz letter.

**c.** Domain: $0 < x \le 4$; Range: $\{49, 70, 91, 112\}$

**d.** The postage for a 2-ounce letter is 70 cents; for a 201 ounce letter, it is 91 cents

**53.** $p = \dfrac{200}{2 + 0.1x}$

**a.**

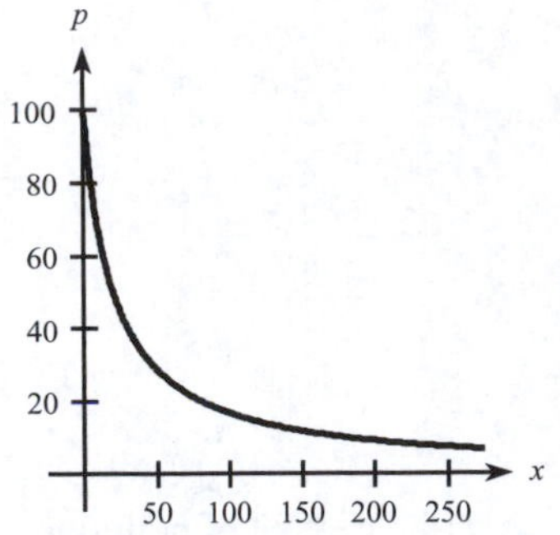


**b.** No

**a.**

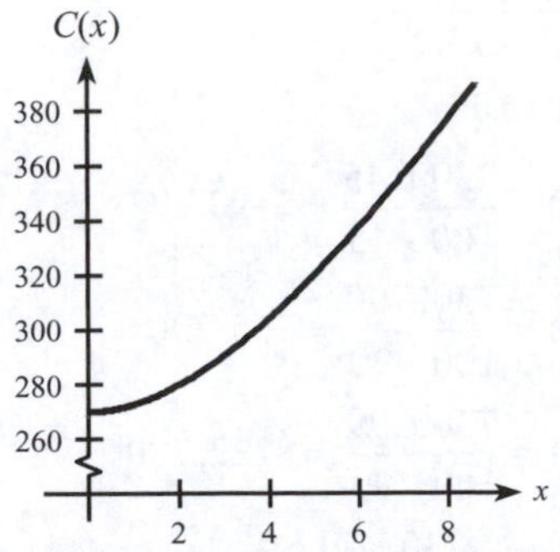


**b.** A turning point indicates a minimum or maximum cost.

**c.** This is the fixed cost of production.

**55.** $C(x) = 30(x - 1) + \dfrac{3000}{x + 10}$

## *Exercises 2.5*

**1.** Linear: The points are in a straight line.

**3.** Quadratic: The points appear to fit a parabola.

**5.** Quartic: The graph crosses the $x$-axis four times. Also there are three bends.

**7.** Quadratic: There is one bend. A parabola is the best fit.

**9.** $y = 2x - 3$ is the best fit.

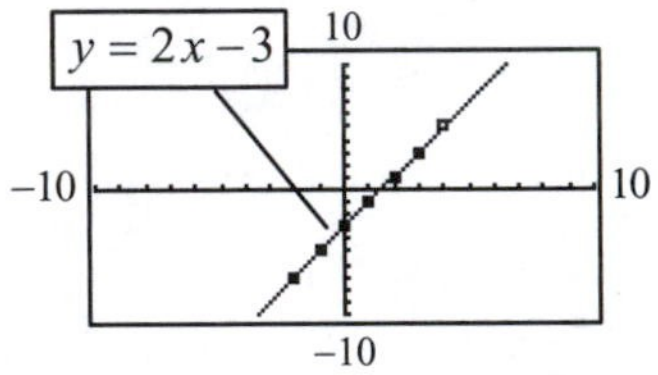


**11.** $y = 2x^2 - 1.5x - 4$ is the best fit.

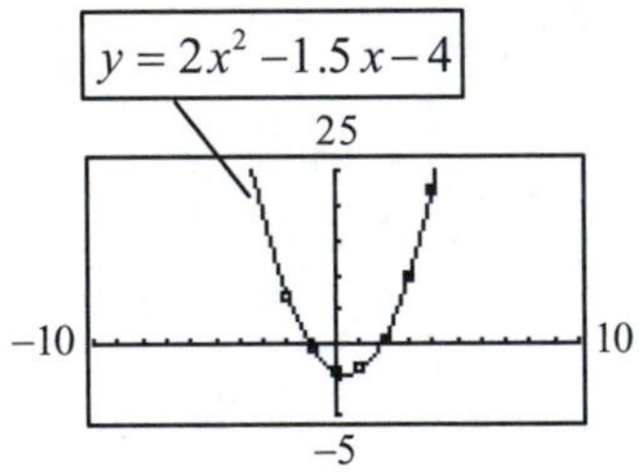


**13.** $y = x^3 - x^2 - 3x - 4$ is the best fit.

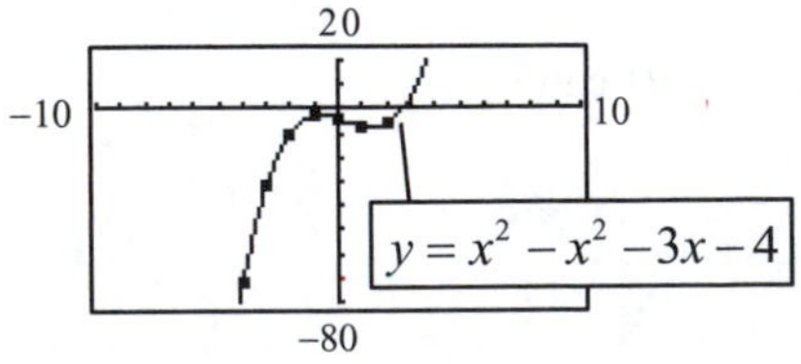


**15.** $y = 2x^{1/2}$ is the best fit.

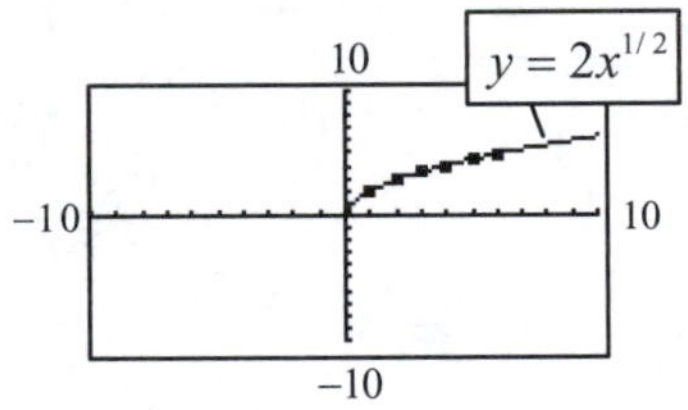


**17. a.**

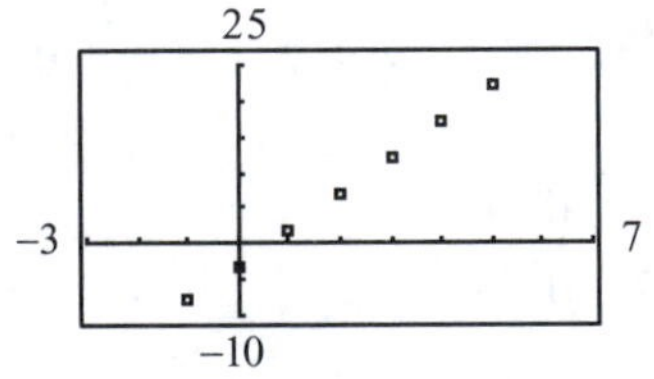


**b.** linear

**c.** $y = 5x - 3$

**19. a.**

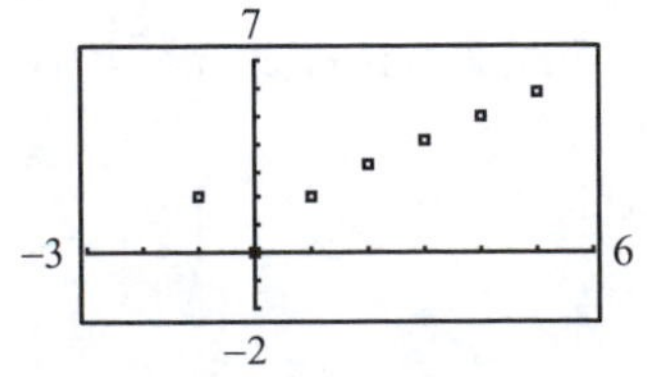


**b.** quadratic

**c.** $y = 0.0959x^2 + 0.4656x + 1.4758$

**21. a.**

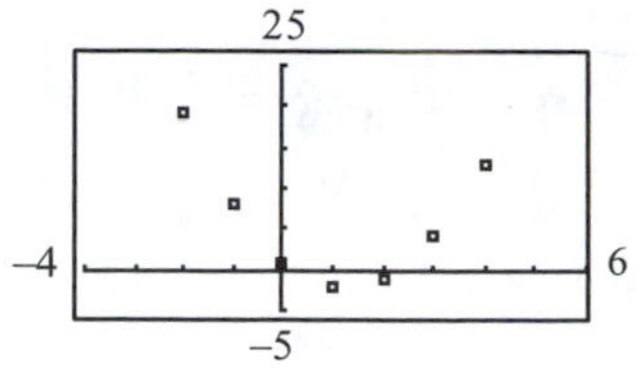


**b.** quadratic

**c.** $y = 2x^2 - 5x + 1$

**23. a.**

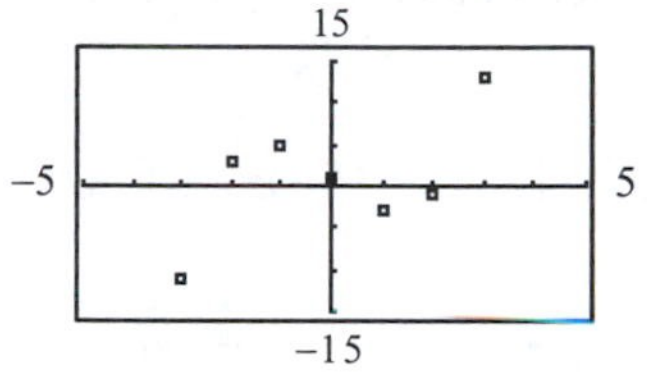


**b.** cubic

**c.** $y = x^3 - 5x + 1$

**25. a.** $y = 154.0x + 35{,}860$

**b.** $y(27) = 154.0(27) + 35{,}860 = 40{,}018$

The projected population of females under age 18 in 2037 is 40,018,000.

**c.** $45{,}000 = 154.0x + 35860 \Rightarrow x \approx 59.35$

This population will reach 45,000,000 in 2010 + 60 = 2070 according to this model.

**27. a.** A linear function is best; $y = 327.6x + 9591$

**b.** $y(17) = 327.6(17) + 9591 \approx \$15{,}160$ billion

**c.** $m = 327.6$ means the U.S. disposable income is increasing at the rate of about \$327.6 billion per year.

**29. a.** $y = 0.0052x^2 - 0.62x + 15$

**b.** $x = \dfrac{-b}{2a} = \dfrac{0.62}{2(0.0052)} \approx 59.6$

**c.** No, it is unreasonable to feel warmer for winds greater than 60 mph.

**31. a.**

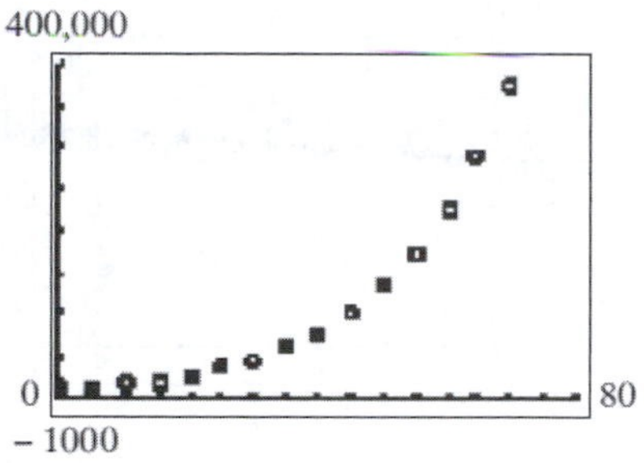


**b.** $y = 106x^2 - 2870x + 28{,}500$

**c.** $y = 1.70x^3 - 72.9x^2 + 1970x + 5270$

**d.**

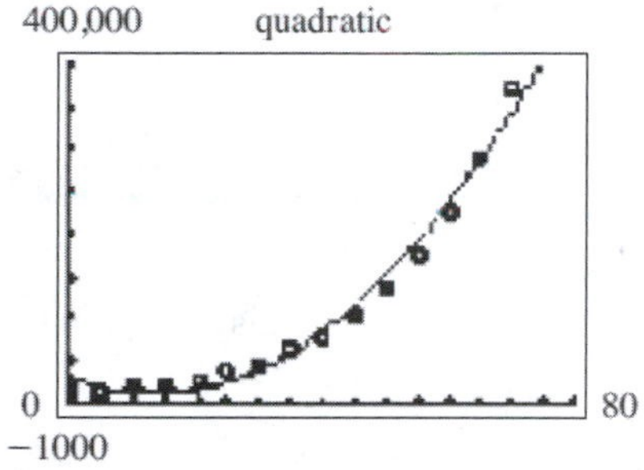


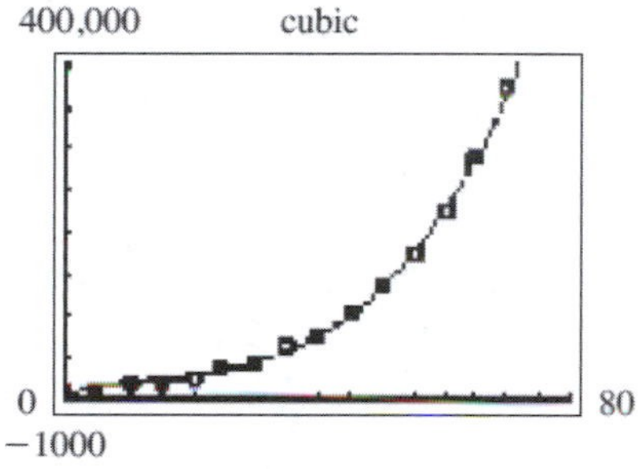


The cubic model fits better.

**33. a.** $y = 0.0157x + 2.01$

**b.** $y = -0.00105x^2 + 0.367x + 1.94$

**c.**

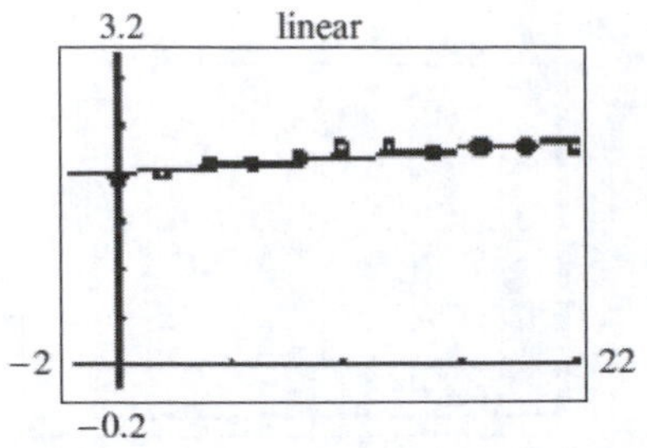


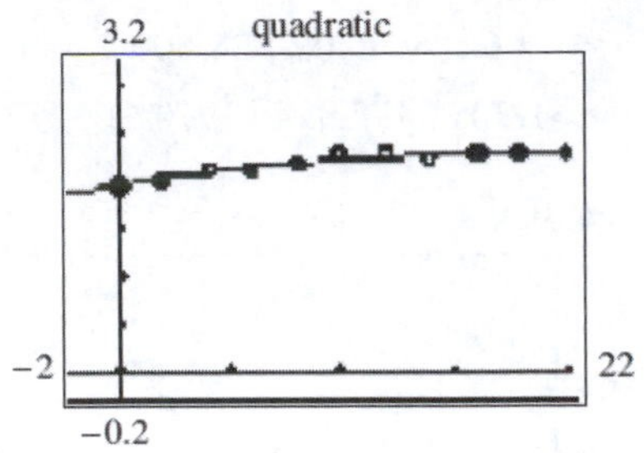


**d.** The quadratic model is a slightly better fit.

**35. a.**

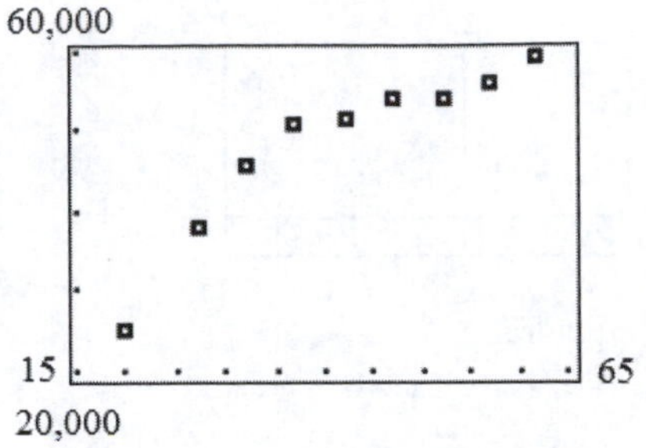


A cubic model looks best because of the two bends.

**b.** $y = 0.864x^3 - 128x^2 + 6610x - 62,600$

**c.**

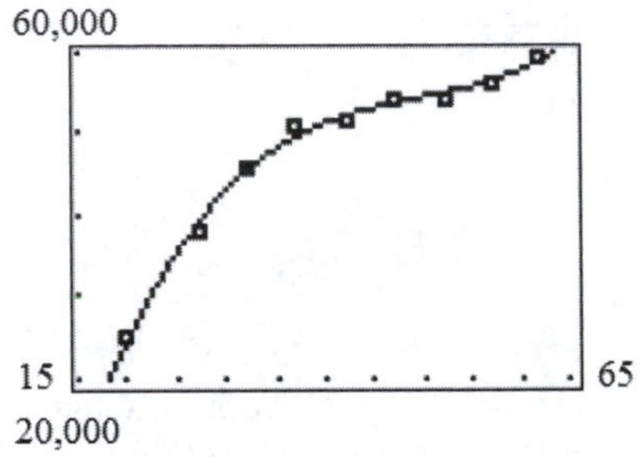


**d.** Using the coefficient values reported by the calculator, the model estimates the median income to be \$56,250 at age 57.

**37. a.** $y = 0.0514x^{2.73}$

**b.**

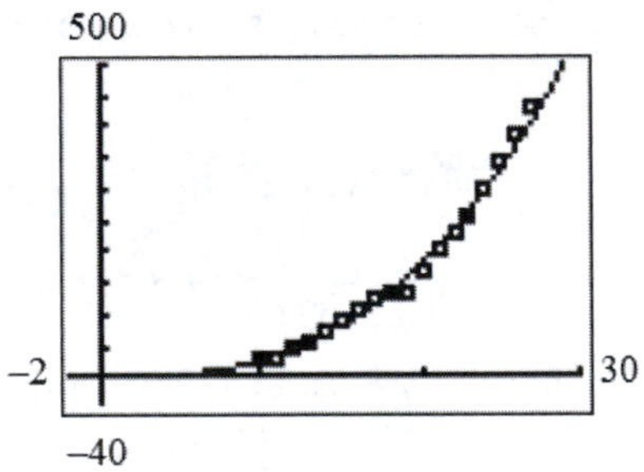


**c.** $y(30) \approx \$546$ billion

# Chapter 2: Quadratic and Other Special Functions

## *Chapter 2 Review Exercises*

**1.** $3x^2+10x=5x$

$3x^2+5x=0$

$x(3x+5)=0$

$x=0 \text{ or } x=-\dfrac{5}{3}$

**2.** $4x-3x^2=0$

$x(4-3x)=0$

$x=0 \text{ or } x=\dfrac{4}{3}$

**3.** $x^2+5x+6=0$

$(x+3)(x+2)=0$

$x=-3 \text{ or } x=-2$

**4.** $11-10x-2x^2=0$

$a=-2,\ b=-10,\ c=11$

$x=\dfrac{10\pm\sqrt{100+88}}{-4}=\dfrac{-5\pm\sqrt{47}}{2}$

**5.** $(x-1)(x+3)=-8$

$x^2+2x-3=-8$

$x^2+2x+5=0$

$b^2-4ac<0$

No real solution

**6.** $4x^2=3$

$x^2=\dfrac{3}{4}$

$x=\pm\sqrt{\dfrac{3}{4}}=\pm\dfrac{\sqrt{3}}{2}$

**7.** $20x^2+3x=20-15x^2$

$35x^2+3x-20=0$

$(7x-5)(5x+4)=0$

$x=\dfrac{5}{7} \text{ or } x=-\dfrac{4}{5}$

**8.** $8x^2+8x=1-8x^2$

$16x^2+8x-1=0$

$a=16,\ b=8,\ c=-1$

$x=\dfrac{-8\pm\sqrt{64+64}}{32}=\dfrac{-1\pm\sqrt{2}}{4}$

**9.** $7=2.07x-0.02x^2$

$0.02x^2-2.07x+7=0$

$a=0.02,\ b=-2.07,\ c=7$

$x=\dfrac{2.07\pm\sqrt{4.2849-0.56}}{0.04}=\dfrac{2.07\pm1.93}{0.04}$

$=100 \text{ or } 3.5$

**10.** $46.3x-117-0.5x^2=0$

$a=-0.5,\ b=46.3,\ c=-117$

$x=\dfrac{-46.3\pm\sqrt{2143.69+(-234)}}{-1}=\dfrac{-46.3\pm43.7}{-1}$

$=90 \text{ or } 2.6$

**11.** $4z^2+25=0$

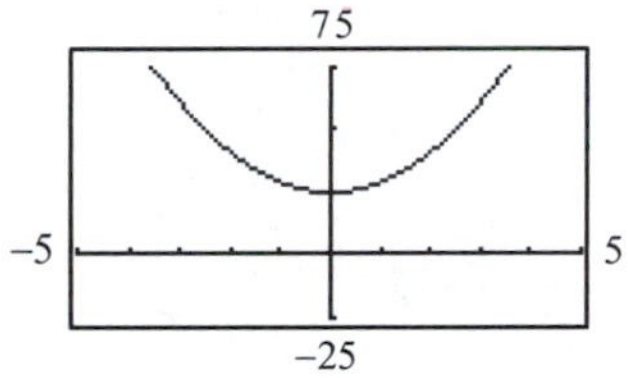


$4z^2+5^2=0$

The sum of 2 squares cannot be factored. There are no real solutions.

**12.** $f(z)=z^2+6z-27$

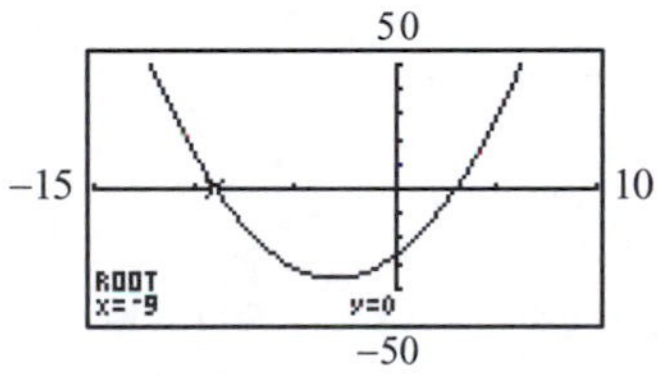


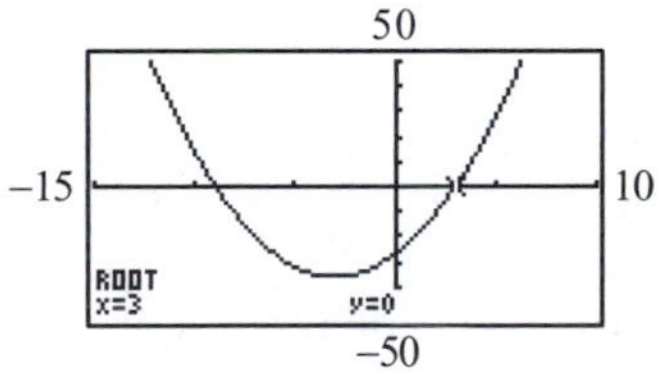


From the graph, the zeros are −9 and 3.

Algebraic solution:

$z(z+6)=27$

$z^2+6z-27=0$

$(z+9)(z-3)=0$

$z=-9 \text{ or } z=3$

**13.** $3x^2 - 18x - 48 = 0$

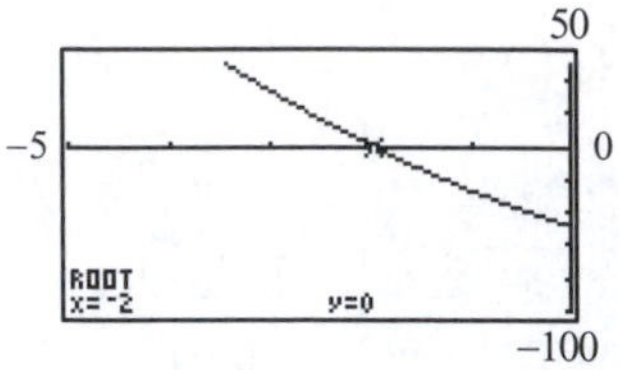


$3(x^2 - 6x - 16) = 0$

$3(x-8)(x+2) = 0$

$x = -2,\ x = 8$

**14.** $f(x) = 3x^2 - 6x - 9$

$3x^2 - 6x - 9 = 0$

$3\left(x^2 - 2x - 3\right) = 0$

$3(x-3)(x+1) = 0$

$x = 3,\ \ x = -1$

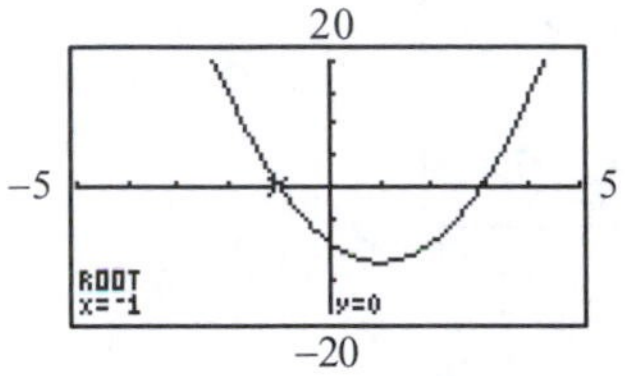


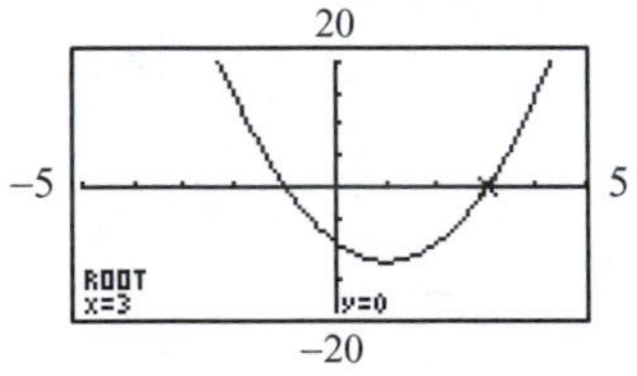


**15.** $x^2 + ax + b = 0$

To apply the quadratic formula we have "$a$" = 1, "$b$" = $a$, and "$c$" = $b$.

$$x = \frac{-a \pm \sqrt{a^2 - 4b}}{2}$$

**16.** $xr^2 - 4ar - x^2c = 0$

To solve for $r$, use the quadratic formula with "$a$" = $x$, "$b$" = $-4a$, and "$c$" = $-x^2c$.

$$r = \frac{4a \pm \sqrt{16a^2 + 4x(x^2c)}}{2x} = \frac{4a \pm \sqrt{16a^2 + 4x^3c}}{2x}$$

$$= \frac{4a \pm 2\sqrt{4a^2 + x^3c}}{2x} = \frac{2a \pm \sqrt{4a^2 + x^3c}}{x}$$

**17.** $-0.002x^2 - 14.1x + 23.1 = 0$

$$x = \frac{14.1 \pm \sqrt{198.81 + 0.1848}}{-0.004} = \frac{14.1 \pm 14.107}{-0.004}$$

$= -7051.64,\ 1.64,\ \text{or}\ 1.75$ (using 14.107)

**18.** $1.03x^2 + 2.02x - 1.015 = 0$

$a = 1.03,\ b = 2.02,\ c = -1.015$

$$x = \frac{-2.02 \pm \sqrt{4.0804 + 4.1818}}{2.06} = \frac{-2.02 \pm 2.87}{2.06}$$

$= -2.38$ or $0.41$

**19.** $y = \frac{1}{2}x^2 + 2x$

$a > 0$, thus vertex is a minimum.

V: $x = \dfrac{-2}{2\left(\frac{1}{2}\right)} = -2$

$y = \frac{1}{2}(-2)^2 + 2(-2) = -2$

Zeros: $\frac{1}{2}x^2 + 2x = 0$

$x\left(\frac{1}{2}x + 2\right) = 0$

$x = 0, -4$

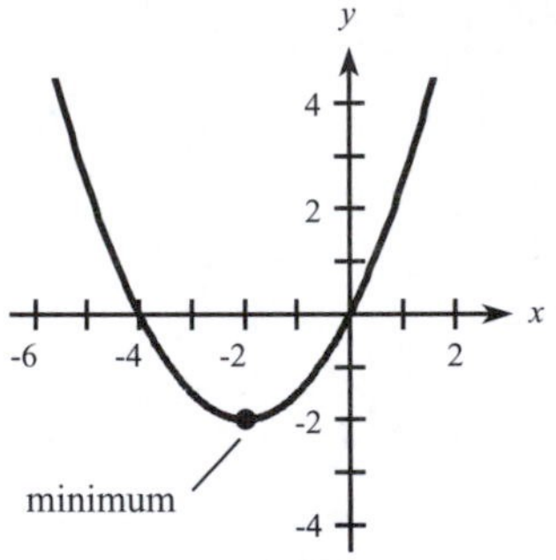


**20.** $y = 4 + \frac{1}{4}x^2$

V: $x$-coordinate = 0

$y$-coordinate = 4

(0, 4) is a maximum point

Zeros are $x = \pm 4$.

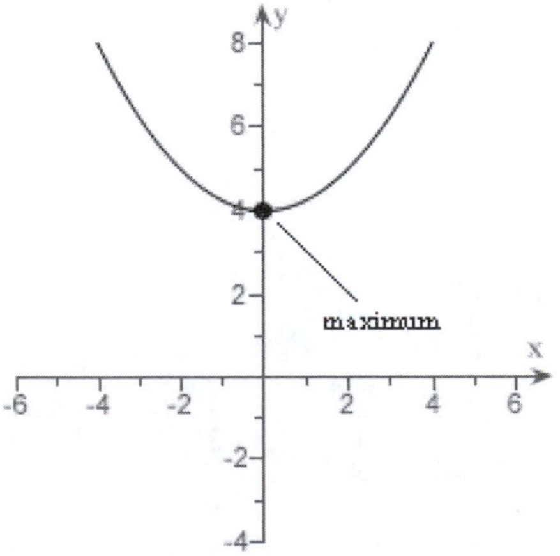

**21.** $y = 6 + x - x^2$

$a < 0$, thus vertex is a maximum.

V: $x = \frac{-1}{2(-1)} = \frac{1}{2}$

$y = 6 + \frac{1}{2} - \left(\frac{1}{2}\right)^2 = \frac{25}{4}$

Zeros: $6 + x - x^2 = 0$

$(3 - x)(2 + x) = 0$

$x = -2, 3$

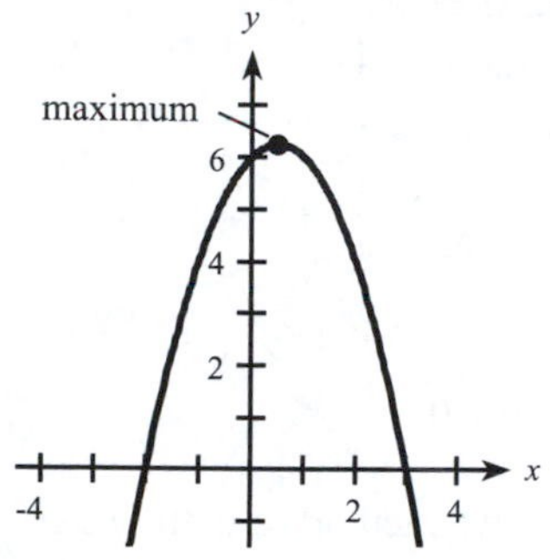


**22.** $y = x^2 - 4x + 5$

V: $x$-coordinate $= \frac{4}{2} = 2$

$y$-coordinate $= 2^2 - 4(2) + 5 = 1$

(2, 1) is a minimum point.

Zeros: Since the minimum point is above the $x$-axis, there are no zeros.

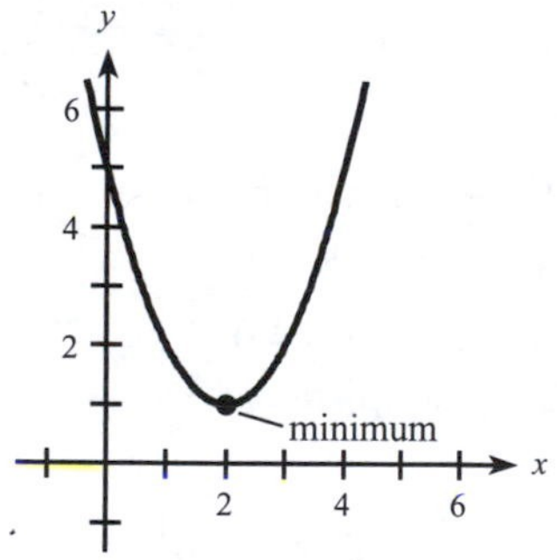


**23.** $y = x^2 + 6x + 9$

$a > 0$, thus vertex is a minimum.

V: $x = \frac{-6}{2(1)} = -3$

$y = (-3)^2 + 6(-3) + 9 = 0$

Zeros: $x^2 + 6x + 9 = 0$

$(x + 3)(x + 3) = 0$

$x = -3$

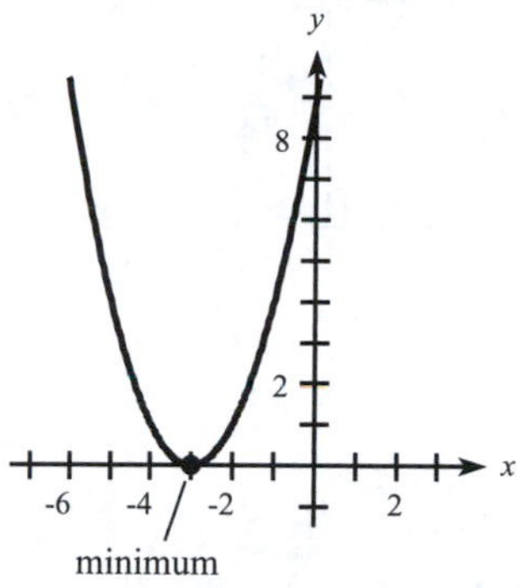


**24.** $y = 12x - 9 - 4x^2$

V: $x$-coordinate $= -\frac{12}{-8} = \frac{3}{2}$

$y$-coordinate $= 12\left(\frac{3}{2}\right) - 9 - 4\left(\frac{3}{2}\right)^2 = 0$

$\left(\frac{3}{2}, 0\right)$ is a maximum point.

Zeros: From the vertex we have that $x = \frac{3}{2}$ is the only zero.

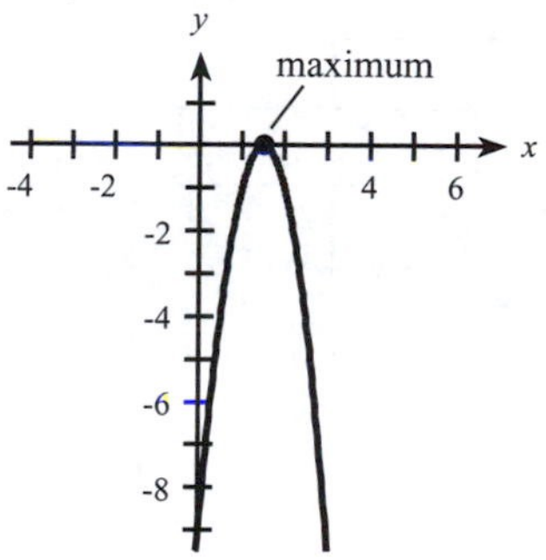

**25.** $y=\frac{1}{3}x^2-3$

V: (0, –3)

Zeros: $\frac{1}{3}x^2-3=0$

$x^2=9$

$x=\pm 3$

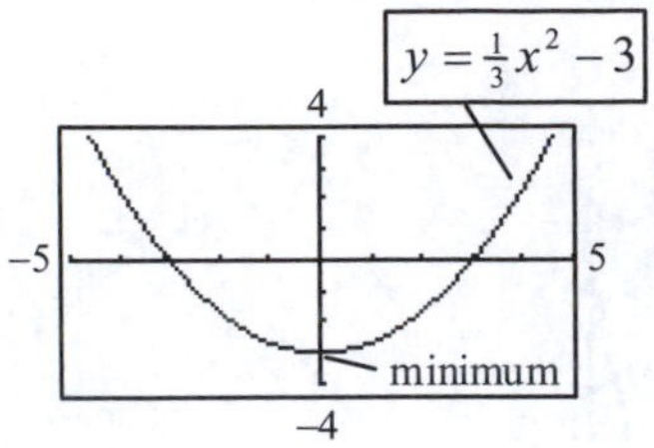


**26.** $y=\frac{1}{2}x^2+2$

Vertex: (0, 2) ← minimum

No zeros.

The graph using $x$-min = –4 $y$-min = 0
$x$-max = 4 $y$-max = 6
is shown below.

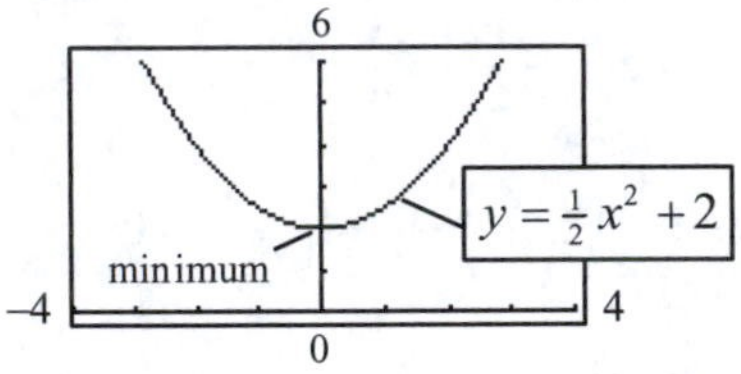


**27.** $y=x^2+2x+5$

V: (–1, 4)

There are no real zeros.

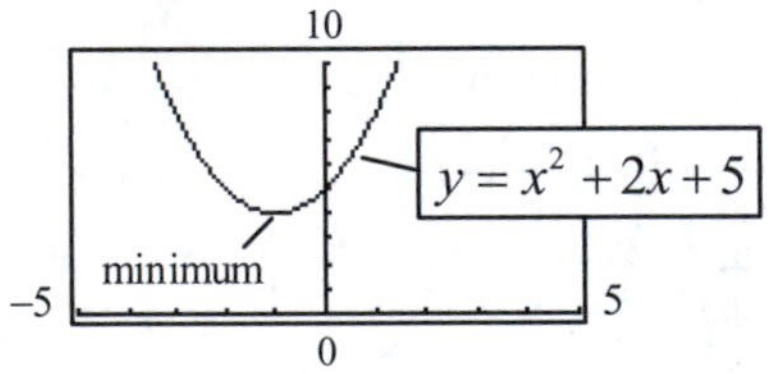


**28.** $y=-10+7x-x^2$

Vertex: $\left(\frac{7}{2},\frac{9}{4}\right)$ ← maximum

Zeros: $x^2-7x+10=0$

$(x-5)(x-2)=0$

$x=5$ or $x=2$

Graph using $x$-min = 0 $y$-min = –5
$x$-max = 8 $y$-max = 5

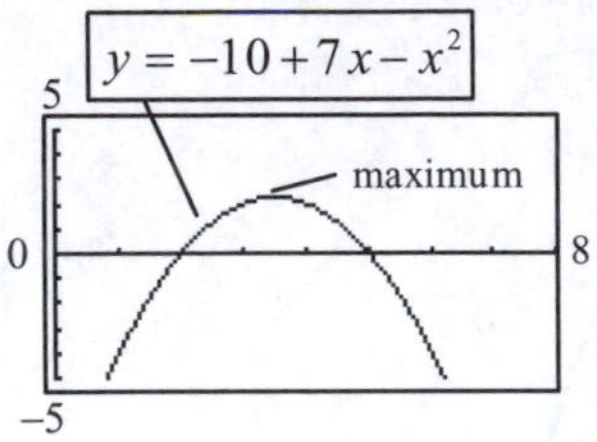


**29.** $y=20x-0.1x^2$

Zeros: $x(20-0.1x)=0$

$x=0, 200$

(This is an alternative method of getting the vertex.)

The $x$-coordinate of the vertex is halfway between the zeros.

V: (100, 1000)

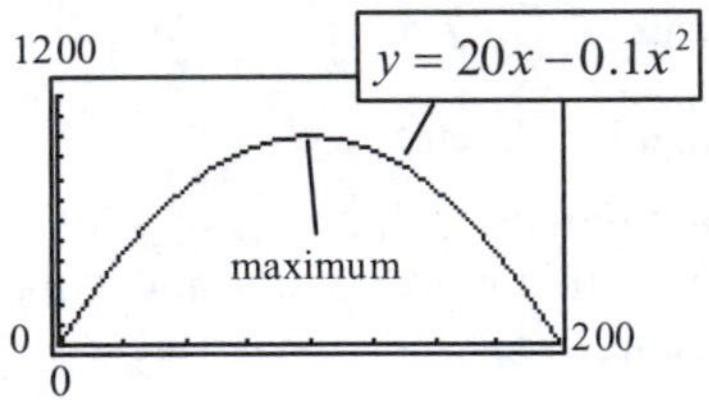


**30.** $y=50-1.5x+0.01x^2$

Vertex: (75, –6.25) ← minimum

Zeros: $0.01x^2-1.5x+50=0$

$0.01(x^2-150x+5000)=0$

$0.01(x-50)(x-100)=0$

$x=50$ or $x=100$

Graph using $x$-min = 0 $y$-min = –10
$x$-max = 125 $y$-max = 10

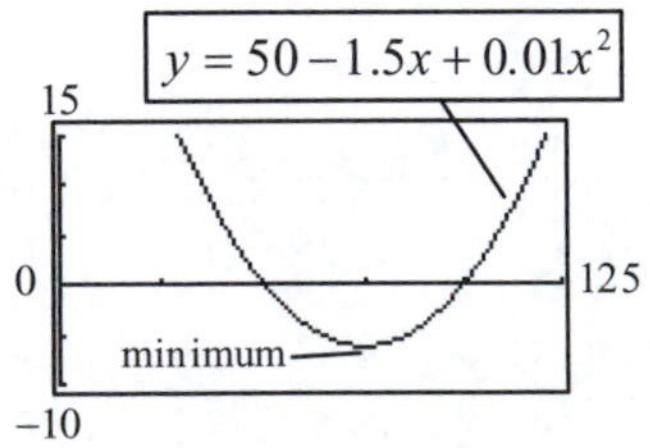


**31.** $\frac{f(50)-f(30)}{50-30}=\frac{2500-2100}{20}=\frac{400}{20}=20$

**32.** $\dfrac{f(50)-f(10)}{50-10}=\dfrac{1022+178}{40}=\dfrac{1200}{40}=30$

**33.** **a.** The vertex is halfway between the zeros. So, the vertex is $\left(1,-4\dfrac{1}{2}\right)$.

**b.** The zeros are where the graph crosses the $x$-axis. $x=-2$, 4.

**c.** The graph matches B.

**34.** From the graph,

**a.** Vertex is (0, 49)

**b.** Zeros are $x=\pm 7$.

**c.** Matches with D.

**35.** **a.** The vertex is halfway between the zeros. So, the vertex is (7, 24.5).

**b.** Zeros are $x=0$, 14.

**c.** The graph matches A.

**36.** From the graph,

**a.** Vertex is (–1, 9).

**b.** Zeros are $x=-4$ and $x=2$.

**c.** Matches with C.

**37.** **a.** $f(x)=x^2$

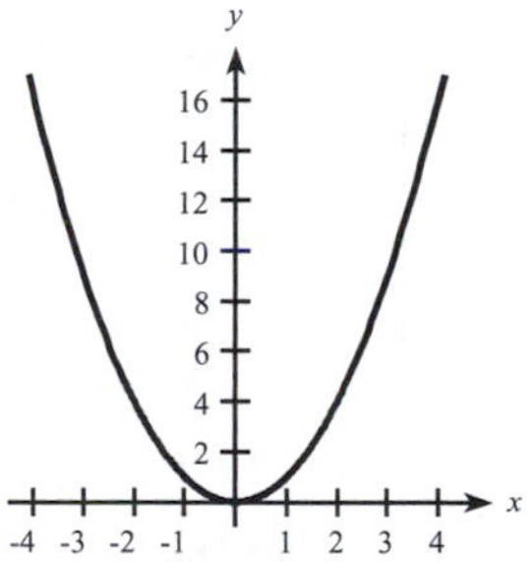


**b.** $f(x)=\dfrac{1}{x}$

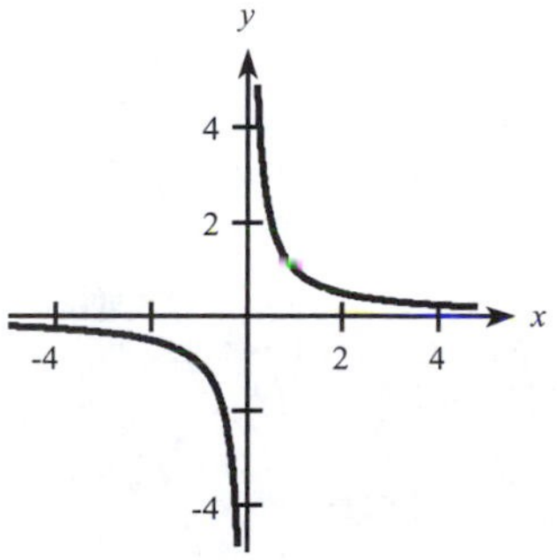


**c.** $f(x)=x^{1/4}$

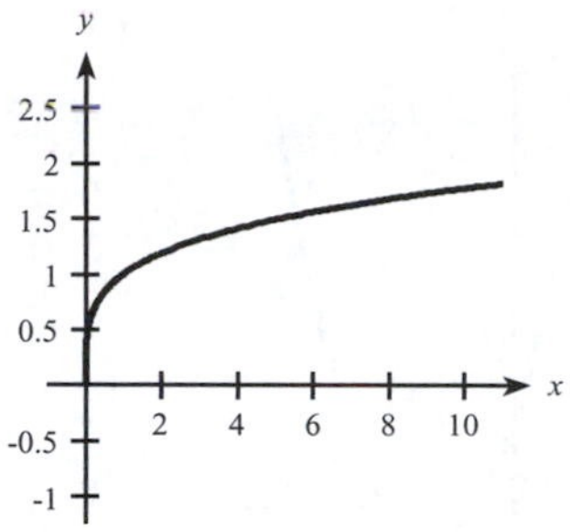


**38.** $f(x)=\begin{cases}-x^2 & \text{if } x\le 0\\ \dfrac{1}{x} & \text{if } x>0\end{cases}$

**a.** $f(0)=-(0^2)=0$

**b.** $f(0.0001)=\dfrac{1}{0.0001}=10{,}000$

**c.** $f(-5)=-(-5)^2=-25$

**d.** $f(10)=\dfrac{1}{10}=0.1$

**39.** $f(x)=\begin{cases}x & \text{if } x\le 1\\ 3x-2 & \text{if } x>1\end{cases}$

**a.** $f(-2)=-2$

**b.** $f(0)=0$

**c.** $f(1)=1$

**d.** $f(2)=3\cdot 2-2=4$

**40.** $f(x)=\begin{cases}x & \text{if } x\le 1\\ 3x-2 & \text{if } x>1\end{cases}$

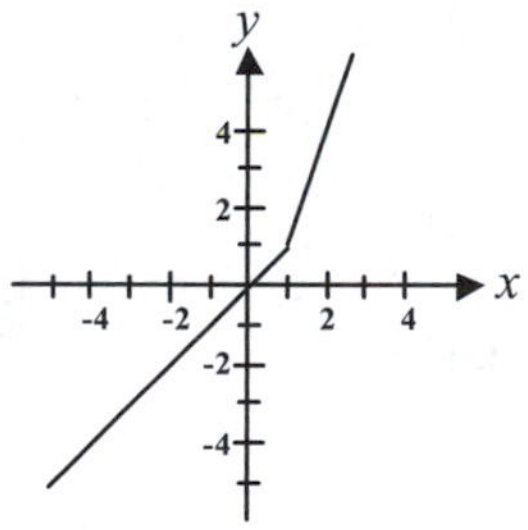

**41. a.** $f(x) = (x-2)^2$

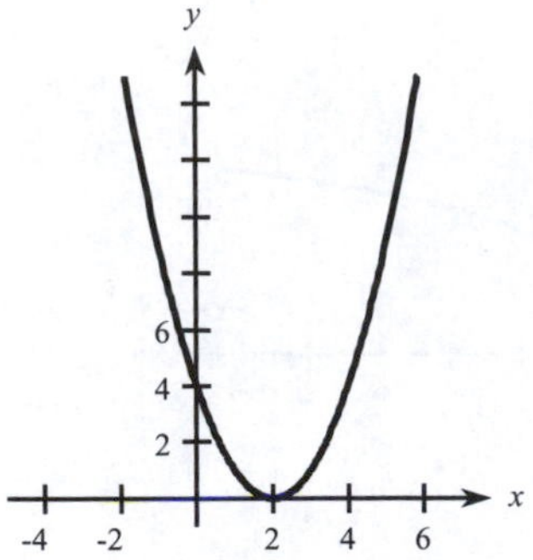


**b.** $f(x) = (x+1)^3$

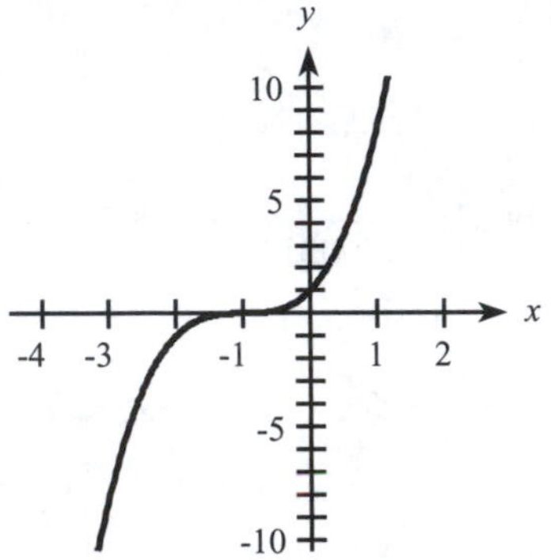


**42.** $y = x^3 + 3x^2 - 9x$

Using $x$-min = –10, $x$-max = 10, $y$-min = –10, $y$-max = 35, the turning points are at $x = -3$ and 1.

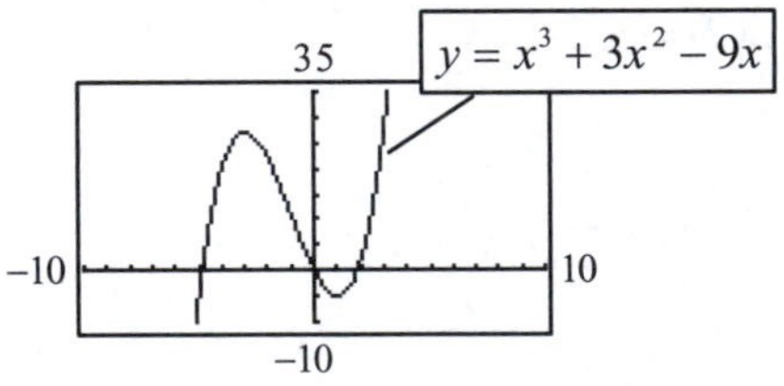


**43.** $y = x^3 - 9x$

Using $x$-min = –4.7, $x$-max = 4.7, $y$-min = –15, $y$-max = 15, the turning points are at $x = \pm 1.732$.

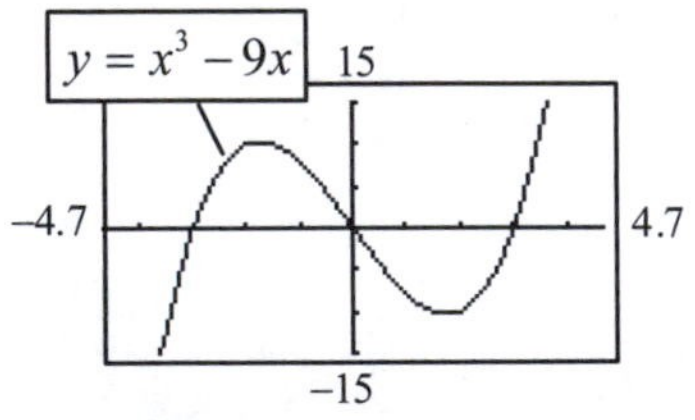


**Note: Your turning points in 42–43. may vary depending on your scale.**

**44.** $y = \dfrac{1}{x-2}$

There is a vertical asymptote $x = 2$.
There is a horizontal asymptote $y = 0$.

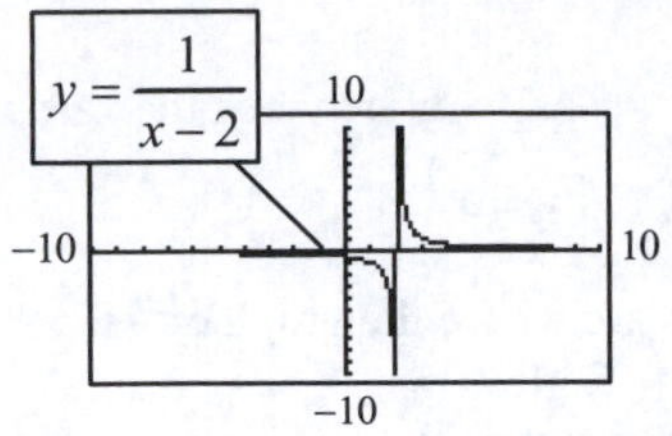


**45.** $y = \dfrac{2x-1}{x+3} = \dfrac{2 - \frac{1}{x}}{1 + \frac{3}{x}}$

Vertical asymptote is $x = -3$.
Horizontal asymptote is $y = 2$.

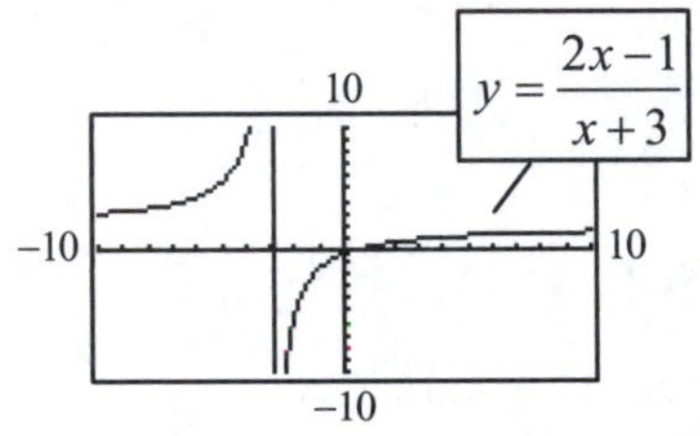


**46. a.**

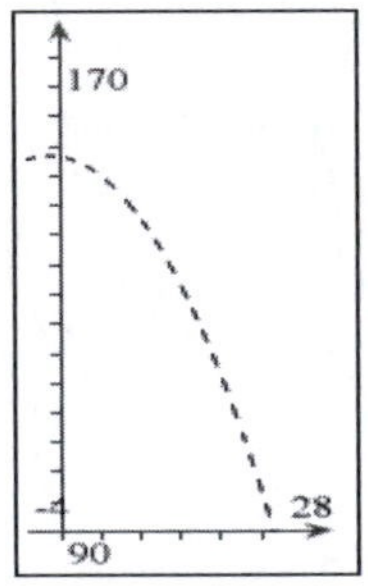


**b.** $y = -2.1786x + 159.8571$ is a good fit to the data.

**c.** $y = -0.0818x^2 - 0.2143x + 153.3095$ is a slightly better fit.

**47. a.**

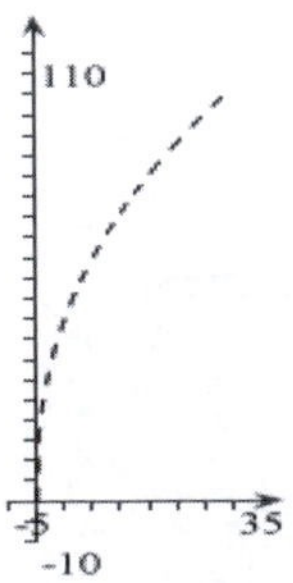


**b.** $y = 2.1413x + 34.3913$ is a good fit to the data.

**c.** $y = 22.2766x^{0.4259}$ is a slightly better fit.

**48.** $S = 96 + 32t - 16t^2$

**a.** $16(6 + 2t - t^2) = 0$

$$t = \frac{-2 \pm \sqrt{4 + 24}}{-2}$$

$t \approx -1.65$ or $t \approx 3.65$

**b.** $t \geq 0$ Use $t = 3.65$

**c.** After 3.65 seconds

**49.** $P(x) = -0.10x^2 + 82x - 1600$

$(-0.10x + 80)(x - 20) = 0$

Break-even at $x = 20, 800$

**50.** $E(t) = -0.0052t^2 + 0.080t + 12$

**a.** The employment is a maximum at

$$t = \frac{-b}{2a} = \frac{-0.080}{2(-0.0052)} \approx 7.69$$

$f(7.69) \approx 12.3$; the maximum employment in manufacturing in the U.S. is predicted to be 12.3 million in 2010 + 8 = 2018.

**b.** $11.5 = -0.0052t^2 + 0.080t + 12$

The quadratic formula gives $t \approx -4.8$ or $t \approx 20.2$. The employment in manufacturing in the U.S. will be 11.5 million in 2010 + 21 = 2031.

**51.** $A = -\frac{3}{4}x^2 + 300x$

**a.** V: $x = \frac{-300}{-\frac{3}{2}} = 200$ ft

**b.** $A = -\frac{3}{4}(200)^2 + 300(200) = 30{,}000$ sq ft

**52.**

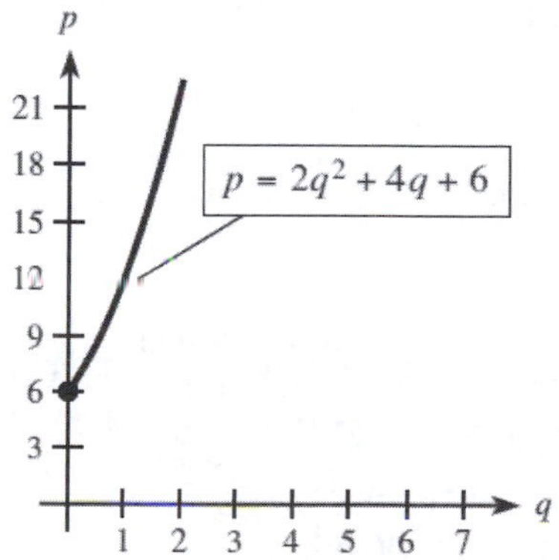


**53.**

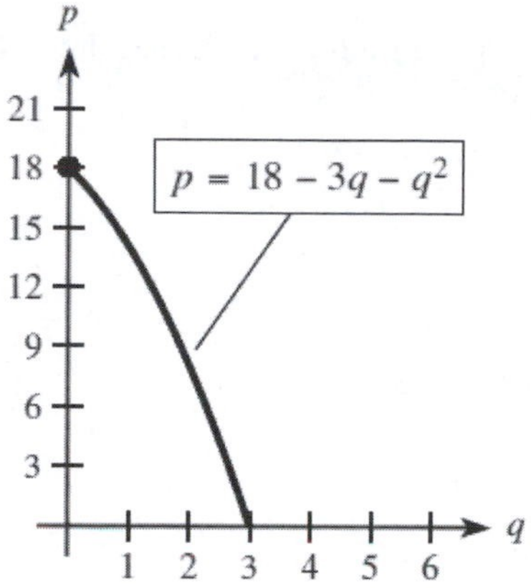


**54. a.**

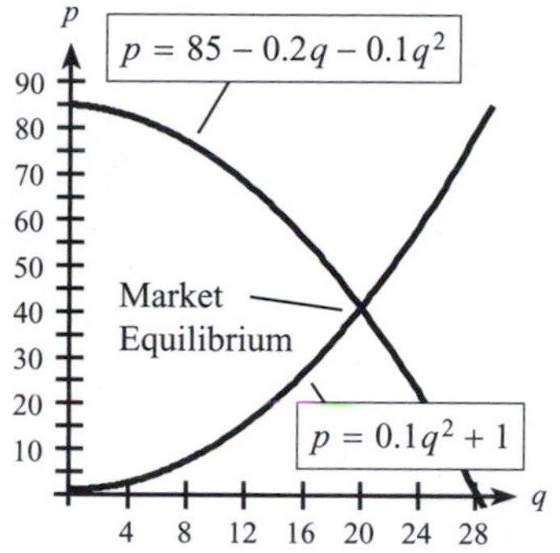


**b.** $0.1q^2 + 1 = 85 - 0.2q - 0.1q^2$

$0.2q^2 + 0.2q - 84 = 0$

$0.2(q^2 + q - 420) = 0$

$0.2(q - 20)(q + 21) = 0$

$q = 20$ (only positive value)

$p = 0.1(20)^2 + 1 = 41$

**55.** $p = q^2 + 300$

$p = -q + 410$

$q^2 + 300 = -q + 410$

$q^2 + q - 110 = 0$

$(q + 11)(q - 10) = 0$

$q = 10$

$p = -10 + 410 = 400$

So, E: (10, 400).

**56.** D: $p^2+5q=200 \rightarrow p^2=200-5q$

S: $40-p^2+3q=0$

Substitute $200-5q$ for $p^2$ in the second equation and solve for $q$.

$40-(200-5q)+3q=0$

$-160=-8q$

$q=20$

$p^2=200-5(20)$

$p^2=100$ or $p=10$

**57.** $R(x)=100x-0.4x^2$

$C(x)=1760+8x+0.6x^2$

$100x-0.4x^2=1760+8x+0.6x^2$

$x^2-92x+1760=0$

$x=\dfrac{92\pm\sqrt{1424}}{2}=46\pm2\sqrt{89}\approx 64.87, 27.13$

$(\sqrt{1424}=\sqrt{16\cdot 89})$

**58.** $C(x)=900+25x$

$R(x)=100x-x^2$

$900+25x=100x-x^2$

$x^2-75x+900=0$

$(x-60)(x-15)=0$

$x=60$ or $x=15$

$R(60)=2400$; $R(15)=1275$

(60, 2400) and (15, 1275)

**59.** $R(x)=100x-x^2$

V: $x=\dfrac{-100}{-2}=50$

$R(50)=100(50)-50^2$

$=\$2500$ max revenue

$P(x)=(100x-x^2)-(900+25x)$

$=-x^2+75x-900$

V: $x=\dfrac{-75}{-2}=37.5$

$P(37.5)=\$506.25$ max profit

**60.** $P(x)=1.3x-0.01x^2-30$

$x$-coordinate of the vertex $=\dfrac{1.3}{0.02}=65$

$P(65)=1.3(65)-0.01(65)^2-30=12.25 \leftarrow$ max

Break-even points:

$0=1.3x-0.01x^2-30$

$0=-0.01(x^2-130x+3000)$

$0=-0.01(x-30)(x-100)$

$x=30$ or $x=100$

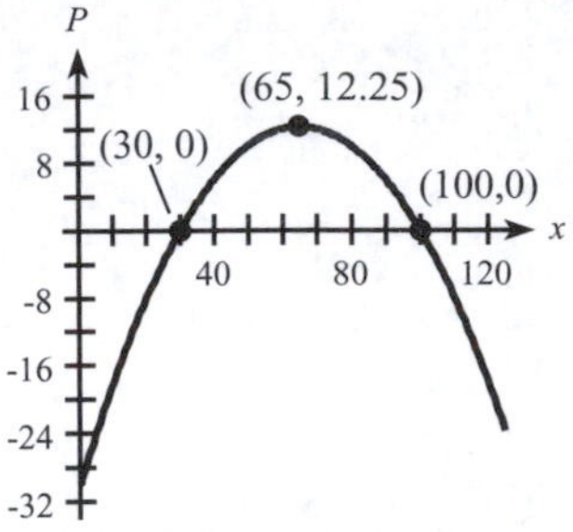


**61.** $P(x)=(50x-0.2x^2)-(360+10x+0.2x^2)$

$=-0.4x^2+40x-360$

V: $x=\dfrac{-40}{-0.8}=50$ units for maximum profit.

$P(50)=-0.4(50)^2+40(50)-360$

$=\$640$ maximum profit.

**62.** **a.** $C(x)=15,000+(140+0.04x)x$

$=15,000+140x+0.04x^2$

$R(x)=(300-0.06x)x$

$=300x-0.06x^2$

**b.** $15,000+140x+0.04x^2=300x-0.06x^2$

$0.10x^2-160x+15,000=0$

$0.1(x^2-1600x+150,000)=0$

$0.1(x-100)(x-1500)=0$

$x=100$ or $x=1500$

**c.** Maximum revenue:

$x$-coordinate: $-\dfrac{300}{-0.12}=2500$

**d.** $P(x)=R(x)-C(x)$

$=-0.10x^2+160x-15,000$

$x$-coordinate of max $=-\dfrac{160}{-0.20}=800$

**e.** $P(2500)=\$240,000$ loss

$P(800)=\$49,000$ profit

**63.** $D(t) = 4.95t^{0.495}$

**a.** power function

**b.** $D(20) \approx 21.8\%$

**c.** 24.4; in 2025 about 24.4% of U.S. adults are expected to have diabetes.

**64. a.**

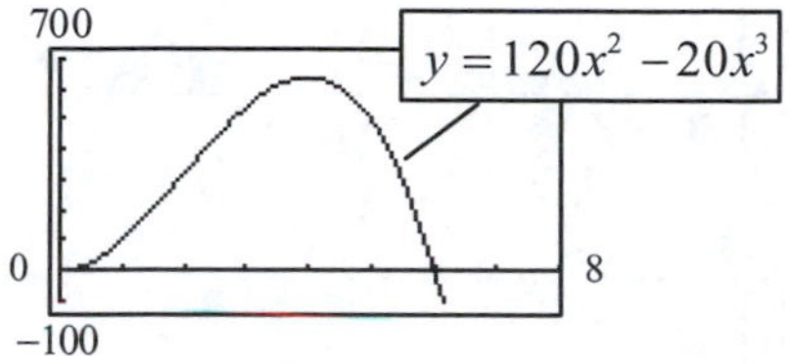


**b.** $y = 20x^2(6 - x)$

Domain: $0 \le x \le 6$

**65.** $C(p) = \dfrac{4800p}{100 - p}$

**a.** rational function

**b.** Domain: $0 \le p < 100$

**c.** $C(0) = 0$ means that there is no cost if no pollution is removed.

**d.** $C(99) = \dfrac{4800(99)}{100 - 99} = \$475,200$

**66.** $C(x) = \begin{cases} 2.557x & 0 \le x \le 100 \\ 255.70 + 2.04(x - 100) & 100 < x \le 1000 \\ 2091.7 + 1.689(x - 1000) & x > 1000 \end{cases}$

**a.** $C(12) = 2.557(12) = \$30.68$

**b.** $C(825) = 255.70 + 2.04(825 - 100) =$ $1734.70

**67. a.** Linear, quadratic, cubic, and power functions are each reasonable.

**b.** $y = 23.779x^{0.525}$

**c.**

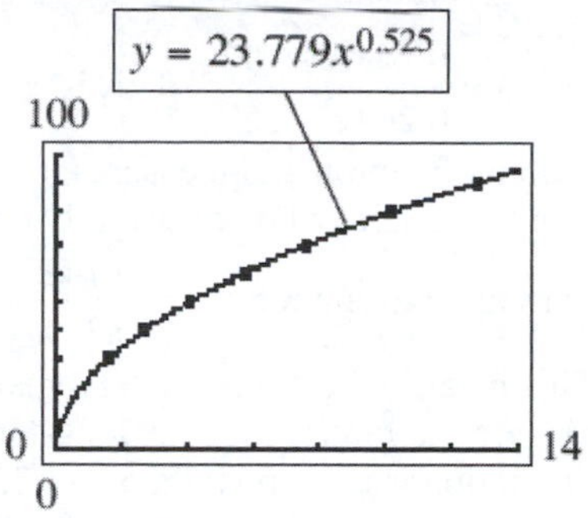


**d.** $f(5) = 23.779(5)^{0.525} \approx 55$ mph

**e.** Use the TRACE KEY. It will take 9.9 seconds.

**68. a.**

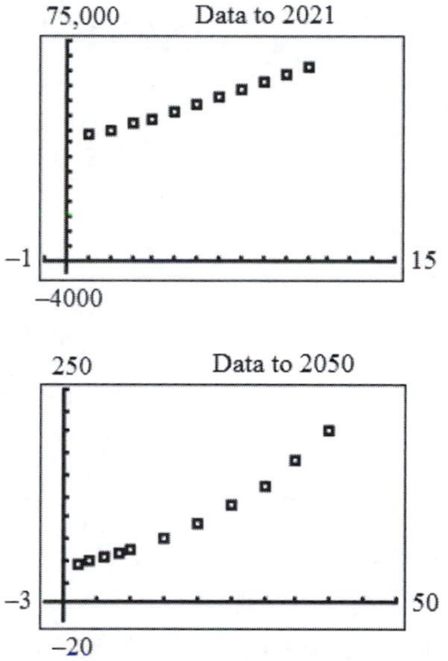


**b.** A quadratic model could be used.

$a(x) = 47.70x^2 + 1802x + 40,870;$

$A(x) = 0.07294x^2 + 0.9815x + 44.45$

**c.** 2020 data: $63,676

$a(10) \approx \$63,664$–closer;

$A(10) \approx 61.564$ ($61,564)

2050 data: 202.5 ($202,500)

$a(40) \approx \$189,292;$

$A(40) \approx 200.413$ ($200,413–closer)

**d.** $a = 150,000$ when $x \approx 32.5$, in 2043;

$A = 150,000$ when $x \approx 31.9$, in 2042

**69. a.** $O(x) = 0.743x + 6.97$

**b.** $S(x) = 0.264x - 2.57$

**c.** $F(x) = \dfrac{0.743x + 6.97}{0.264x - 2.57}$. This is called a rational function and measures the fraction of obese adults who are severely obese.

**d.** horizontal asymptote: $y \approx \dfrac{0.743}{0.264} \approx 0.355$.

This means that if this model remains valid far into the future, then the long-term projection is that about 0.355, or 35.5%, of obese adults will be severely obese.

**70. a.**

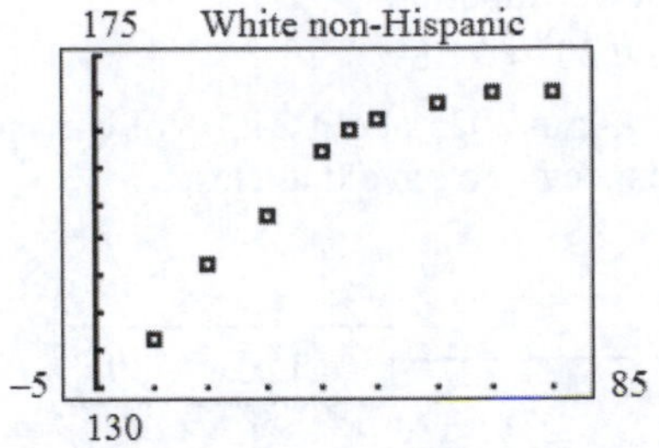


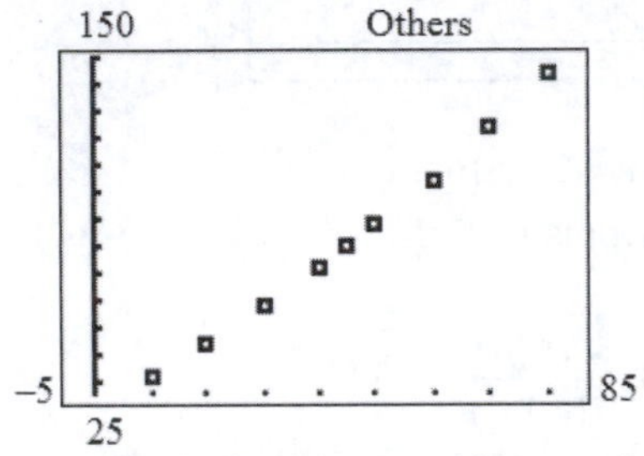


A quadratic function could be used to model each set of data.

$W(x) = -0.00903x^2 + 1.28x + 124$

$O(x) = 0.00645x^2 + 1.02x + 20.0$

**b.** At $x \approx 91.1$, $W(x) = O(x) \approx 166.4$

In 1970 + 92 = 2162, these population segments are predicted to be equal (at about 166.4 million each).

## *Chapter 2 Test*

**1. a.** $f(x) = x^4$

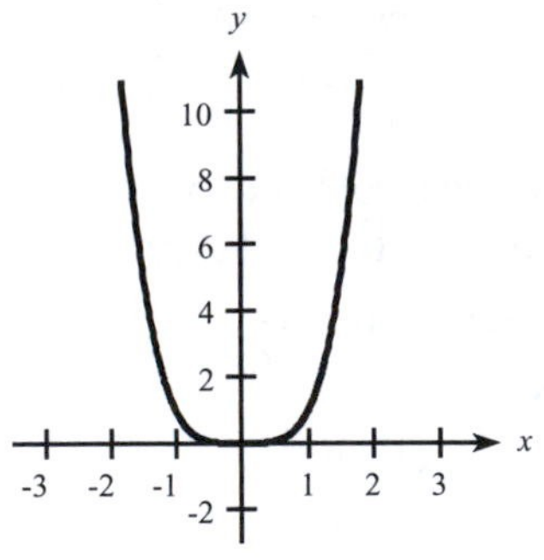


**b.** $g(x) = |x|$

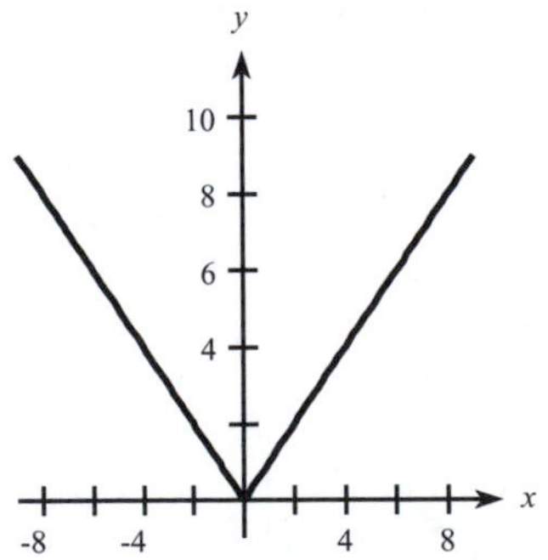

**c.** $h(x) = -1$

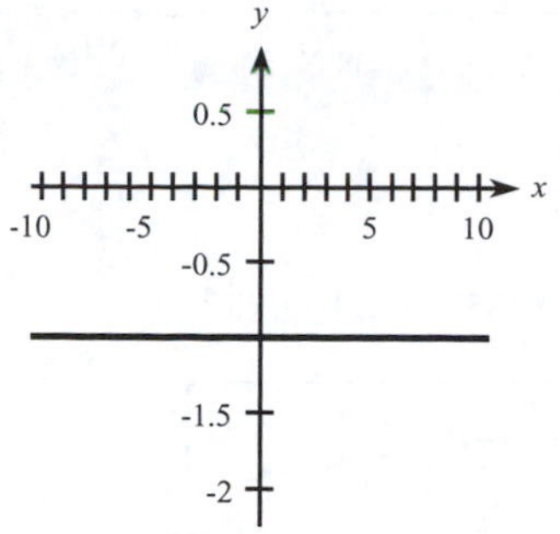


**d.** $k(x) = \sqrt{x}$

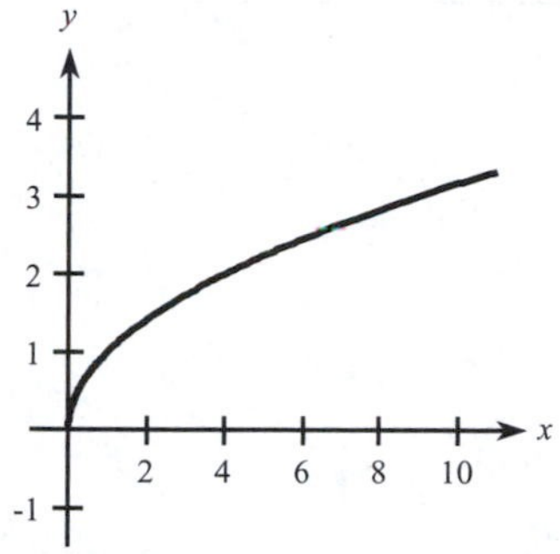


**2.** figure b is the graph for $b > 1$.
figure a is the graph for $0 < b < 1$.

**3.** $f(x) = ax^2 + bx + c$ and a < 0 is a parabola opening downward.

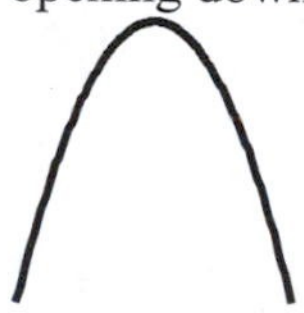

**4.** **a.** $f(x) = (x+1)^2 - 1$

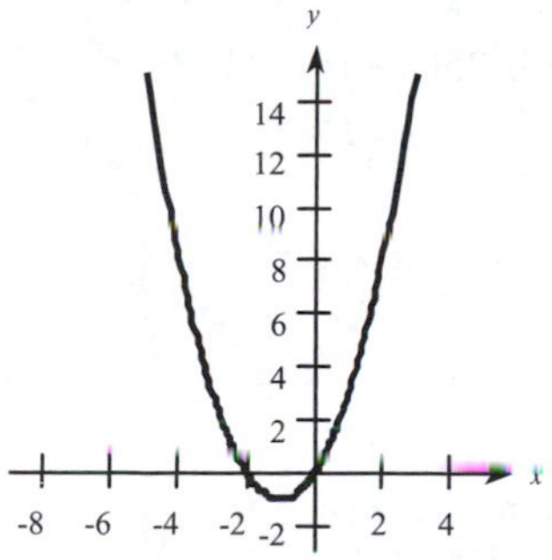


**b.** $f(x) = (x-2)^3 + 1$

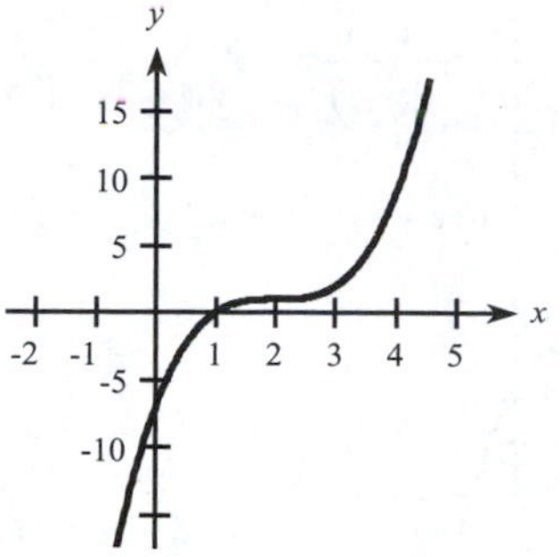


**5.** $f(x) = x^3 - 4x^2 = x^2(x-4)$.
**a.** and **b.** are the cubic choices. $f(x) < 0$ if $0 < x < 4$. Answer: b

**6.** $f(x) = \begin{cases} 8x + \dfrac{1}{x} & \text{if } x < 0 \\ 4 & \text{if } 0 \le x \le 2 \\ 6 - x & \text{if } x > 2 \end{cases}$

**a.** $f(16) = 6 - 16 = -10$

**b.** $f(-2) = 8(-2) + \dfrac{1}{-2} = -16\dfrac{1}{2}$

**c.** $f(13) = 6 - 13 = -7$

**7.** $g(x) = \begin{cases} x^2 & \text{if } x \le 1 \\ 4 - x & \text{if } x > 1 \end{cases}$

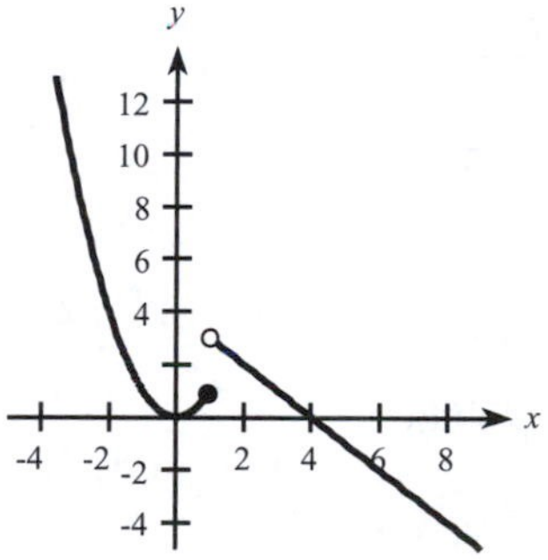

**8.** $f(x)=21-4x-x^2=(7+x)(3-x)$

Vertex: $x=\frac{-b}{2a}=\frac{-(-4)}{2(-1)}=-2$

Point: $(-2, 25)$
Zeros: $f(x)=0$ at $x=-7$ or 3.

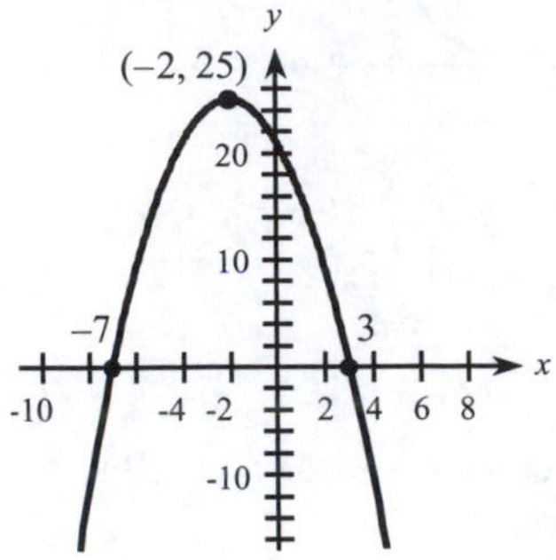


**9.** $3x^2+2=7x$

$3x^2-7x+2=0$

$(3x-1)(x-2)=0$

$3x-1=0$ or $x-2=0$

$x=\frac{1}{3}, 2$

**10.** $2x^2+6x-9=0$

$x=\frac{-6\pm\sqrt{36+72}}{4}=\frac{-6\pm 6\sqrt{3}}{4}=\frac{-3\pm 3\sqrt{3}}{2}$

**11.** $\left(\frac{1}{x}+2x=\frac{1}{3}+\frac{x+1}{x}\right)3x$

$3+6x^2=x+3x+3$

$6x^2-4x=0$

$2x(3x-2)=0$

$x=\frac{2}{3}$ is the only solution.

**12.** $g(x)=\frac{3(x-4)}{x+2}$

Vertical asymptote at $x=-2$.
$g(4)=0$
Answer: c

**13.** $f(x)=\frac{8}{2x-10}$

Horizontal: $y=0$

Vertical: $2x-10=0$

$2x=10$

$x=5$

**14.** $\frac{f(40)-f(10)}{40-10}=\frac{320-(-940)}{30}=\frac{1260}{30}=42$

**15. a.** quartic
**b.** cubic

**16. a.** $f(x)=-0.3577x+19.9227$

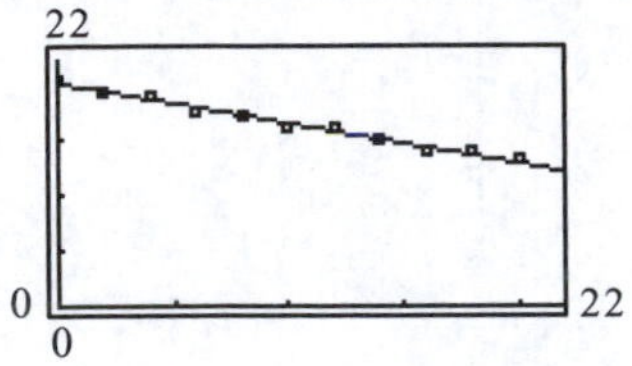


**b.** $f(40)=5.6$

**c.** $f(x)=0$ if $x=\frac{19.9227}{0.3577}\approx 55.7$

**17.** S: $p=\frac{1}{6}q+30$

D: $p=\frac{30{,}000}{q}-20$

$\left(\frac{1}{6}q+30=\frac{30{,}000}{q}-20\right)6q$

$q^2+180q=180{,}000-120q$

$q^2+300q-180{,}000=0$

$(q+600)(q-300)=0$

$E_q: q=300$

$E_p: p=50+30=80$

**18.** $R(x)=285x-0.9x^2$

$C(x)=15{,}000+35x+0.1x^2$

**a.** $P(x)=285x-0.9x^2-(15{,}000+35x+0.1x^2)$

$=-x^2+250x-15{,}000$

$=(100-x)(x-150)$

**b.** Maximum profit is at vertex.

$x=\frac{-250}{2(-1)}=125$

Maximum profit = $P(125)$ = \$625

**c.** Break-even means $P(x)=0$.
From **a.**, $x=100, 150$.

**19.** **a.** Use middle rule for $s = 15$.
$f(15) = -19.5$ means that when the air temperature is 0°F and the wind speed is 15 mph, then the air temperature feels like −19.5 °F. In winter, the TV weather report usually gives the wind chill temperature.

**b.** $f(48) = -31.4$°F

**c.** Break-even means $P(x) = 0$.
From **a.**, $x = 100, 150$.

**20.** **a.**

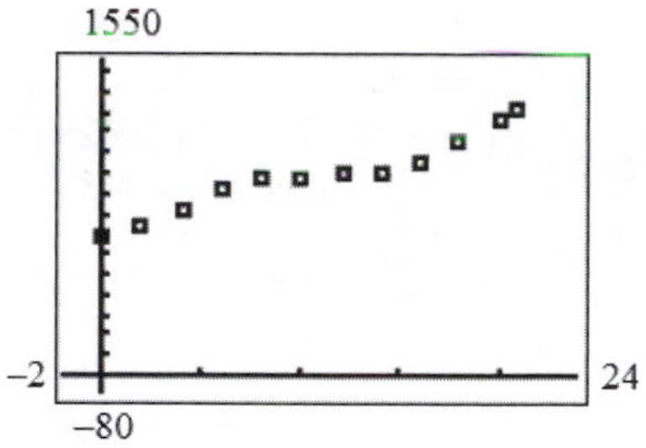


**b.** Linear: $y = 26.8x + 695$;

Cubic: $y = 0.175x^3 - 5.27x^2 + 65.7x + 654$

**c.** Linear: $y(21) \approx \$1258$;

Cubic: $y(21) \approx \$1326$

The cubic model is quite accurate, but both models are fairly close.

**d.** The linear model increases steadily, but the cubic model rises rapidly for years past 2021.

# Chapter 3: Matrices

## *Exercises 3.1*

**1.** Matrix $B$ has 3 rows.

**3.** $A$, $F$, and $Z$ have the same order as $G$.

**5.** $-F = \begin{bmatrix} -1 & -2 & -3 \\ 1 & 0 & -1 \\ -2 & 3 & 4 \end{bmatrix}$

**7.** $A$, $C$, $D$, $F$, G, and $Z$ are square.

**9.** $a_{23} = 1$

**11.** $A^T = \begin{bmatrix} 1 & 3 & 4 \\ 0 & 2 & 0 \\ -2 & 1 & 3 \end{bmatrix}$

**13.** $A+(-A) = \begin{bmatrix} 0 & 0 & 0 \\ 0 & 0 & 0 \\ 0 & 0 & 0 \end{bmatrix}$

**15.** $C + D = \begin{bmatrix} 5+4 & 3+2 \\ 1+3 & 2+5 \end{bmatrix} = \begin{bmatrix} 9 & 5 \\ 4 & 7 \end{bmatrix}$

**17.** $A - F = \begin{bmatrix} 1-1 & 0-2 & -2-3 \\ 3+1 & 2-0 & 1-1 \\ 4-2 & 0+3 & 3+4 \end{bmatrix} = \begin{bmatrix} 0 & -2 & -5 \\ 4 & 2 & 0 \\ 2 & 3 & 7 \end{bmatrix}$

**19.**

$$A + A^T = \begin{bmatrix} 1 & 0 & -2 \\ 3 & 2 & 1 \\ 4 & 0 & 3 \end{bmatrix} + \begin{bmatrix} 1 & 3 & 4 \\ 0 & 2 & 0 \\ -2 & 1 & 3 \end{bmatrix} = \begin{bmatrix} 2 & 3 & 2 \\ 3 & 4 & 1 \\ 2 & 1 & 6 \end{bmatrix}$$

**21.** Because the orders are different, $D - G$ is undefined.

**23.** $3B = 3 \cdot \begin{bmatrix} 1 & 1 & 3 & 0 \\ 4 & 2 & 1 & 1 \\ 3 & 2 & 0 & 1 \end{bmatrix} = \begin{bmatrix} 3 & 3 & 9 & 0 \\ 12 & 6 & 3 & 3 \\ 9 & 6 & 0 & 3 \end{bmatrix}$

**25.** $4C + 2D = \begin{bmatrix} 20 & 12 \\ 4 & 8 \end{bmatrix} + \begin{bmatrix} 8 & 4 \\ 6 & 10 \end{bmatrix} = \begin{bmatrix} 28 & 16 \\ 10 & 18 \end{bmatrix}$

**27.** Because the orders are different, $2A - 3B$ is undefined.

**29.** $\begin{bmatrix} x & 1 & 0 \\ 0 & y & z \\ w & 2 & 1 \end{bmatrix} = \begin{bmatrix} 3 & 1 & 0 \\ 0 & 2 & 3 \\ 4 & 2 & 1 \end{bmatrix}$

By setting corresponding elements equal we have $x = 3$, $y = 2$, $z = 3$, $w = 4$.

**31.** $\begin{bmatrix} x & 3 & 2x-1 \\ y & 4 & 4y \end{bmatrix} = \begin{bmatrix} 2x-4 & z & 7 \\ 1 & w+1 & 3y+1 \end{bmatrix}$

$2x - 4 = x$ or $x = 4$, $z = 3$, $y = 1$,
$w + 1 = 4$ or $w = 3$.

**33.** $\begin{bmatrix} 3x & 3y \\ 3y & 3z \end{bmatrix} + \begin{bmatrix} 4x & -2y \\ 6y & -8z \end{bmatrix} = \begin{bmatrix} 14 & 4-y \\ 18 & 15 \end{bmatrix}$

Setting corresponding elements equal gives the following:

$$3x + 4x = 14 \qquad 3y - 2y = 4 - y$$
$$7x = 14 \qquad 2y = 4$$
$$x = 2 \qquad y = 2$$

$$3y + 6y = 18 \qquad 3z - 8z = 15$$
$$9y = 18 \qquad -5z = 15$$
$$y = 2 \qquad z = -3$$

**35.** **a.** $A = \begin{bmatrix} 70 & 78 & 14 & 15 & 82 \\ 14 & 13 & 20 & 9 & 65 \end{bmatrix}$

$B = \begin{bmatrix} 256 & 208 & 69 & 8 & 11 \\ 16 & 15 & 17 & 1 & 1 \end{bmatrix}$

**b.** $A + B = \begin{bmatrix} 326 & 286 & 83 & 23 & 93 \\ 30 & 28 & 37 & 10 & 66 \end{bmatrix}$

**c.** $B - A = \begin{bmatrix} 186 & 130 & 55 & -7 & -71 \\ 2 & 3 & -3 & -8 & -64 \end{bmatrix}$

The absolute values of the negative entries tell us how many more endangered or threatened species are in the United States.

# Chapter 3: Matrices

**37. a.** Capital Expenditures + Gross Operating Costs = $\begin{bmatrix} 11041.7 & 8978.4 & 6461 \\ 8739.8 & 9159.6 & 6877.3 \\ 9798.1 & 9086.7 & 6448.4 \\ 9696.6 & 8926.7 & 6109.5 \end{bmatrix}$

**b.** Air pollution (R4, C1)

**39. a.** $A+B=\begin{bmatrix} 825 & 580 & 1560 \\ 810 & 650 & 350 \end{bmatrix}$ **b.** $B-A=\begin{bmatrix} -75 & 20 & -140 \\ 10 & -50 & 50 \end{bmatrix}$

**41. a.**

$$A=\begin{array}{c} \begin{matrix} \text{M} & \text{F} \end{matrix} \\ \begin{bmatrix} 74.7 & 79.9 \\ 75.7 & 80.6 \\ 76.5 & 81.3 \\ 77.1 & 81.8 \\ 77.7 & 82.4 \end{bmatrix} \end{array} \quad B=\begin{array}{c} \begin{matrix} \text{M} & \text{F} \end{matrix} \\ \begin{bmatrix} 67.8 & 74.7 \\ 69.7 & 76.5 \\ 70.2 & 77.2 \\ 71.4 & 78.2 \\ 72.6 & 79.2 \end{bmatrix} \end{array}$$

**b.**

$$A-B=\begin{array}{c} \begin{matrix} \text{M} & \text{F} \end{matrix} \\ \begin{bmatrix} 6.9 & 5.2 \\ 6 & 4.1 \\ 6.3 & 4.1 \\ 5.7 & 3.6 \\ 5.1 & 3.2 \end{bmatrix} \end{array}$$

**c.** The life expectancy difference is greatest in 2000.

**43.** All answers are in quadrillion BTUs.

**a.** $E-M=\begin{bmatrix} 0.41 & -0.18 & -0.33 \\ -1.49 & 0.09 & 1.56 \\ -13.29 & -12.05 & -12.56 \end{bmatrix}$

This represents the U.S. balance of trade for various energy types for selected years.

**b.** $\frac{1}{12}C \approx \begin{bmatrix} 3.09 & 3.13 & 3.07 \\ 2.16 & 2.23 & 2.27 \\ 2.90 & 3.06 & 3.18 \end{bmatrix}$

This represents the average monthly U.S. consumption of various energy types for selected years.

**c.** $C+E-M=\begin{bmatrix} 37.45 & 37.36 & 36.54 \\ 24.37 & 26.86 & 28.84 \\ 21.53 & 24.68 & 25.63 \end{bmatrix}$

This represents the total U.S. production of various energy types for selected years.

**45. a.** $1.05A=\begin{bmatrix} 46.20 & 84 & 210 & 10.50 \\ 42 & 84 & 42 & 0 \\ 58.80 & 147 & 94.50 & 0 \\ 31.50 & 147 & 42 & 21 \\ 42 & 0 & 210 & 10.50 \end{bmatrix}$

**b.** $1.10A=\begin{bmatrix} 48.40 & 88 & 220 & 11 \\ 44 & 88 & 44 & 0 \\ 61.60 & 154 & 99 & 0 \\ 33 & 154 & 44 & 22 \\ 44 & 0 & 220 & 11 \end{bmatrix}$

**47. a.** $\begin{bmatrix} 0 & 1 & 1 & 0 \\ 1 & 0 & 1 & 1 \\ 0 & 0 & 0 & 1 \\ 0 & 1 & 1 & 0 \end{bmatrix}$

**b.** $\begin{bmatrix} 0 & 1 & 1 & 1 \\ 1 & 0 & 1 & 1 \\ 0 & 1 & 0 & 1 \\ 1 & 1 & 1 & 0 \end{bmatrix}$

**c.** $A+B=\begin{bmatrix} 0 & 2 & 2 & 1 \\ 2 & 0 & 2 & 2 \\ 0 & 1 & 0 & 2 \\ 1 & 2 & 2 & 0 \end{bmatrix}$

Sum in Row 1 = 5, Row 2 = 6, Row 3 = 3, Row 4 = 5. Person 2 is the most active.

**49.** **a.** $A+B=\begin{bmatrix}50+30 & 30+45\\ 36+22 & 44+62\end{bmatrix}=\begin{bmatrix}80 & 75\\ 58 & 106\end{bmatrix}$

**b.**

$A+B+C=\begin{bmatrix}80+96 & 75+52\\ 58+81 & 106+37\end{bmatrix}$

$=\begin{bmatrix}176 & 127\\ 139 & 143\end{bmatrix}$

**c.** $A-D=\begin{bmatrix}50-40 & 30-26\\ 36-29 & 44-42\end{bmatrix}=\begin{bmatrix}10 & 4\\ 7 & 2\end{bmatrix}$

**d.** $B\text{–}D=\begin{bmatrix}30-40 & 45-26\\ 22-29 & 62-42\end{bmatrix}=\begin{bmatrix}-10 & 19\\ -7 & 20\end{bmatrix}$

Shortage; take from inventory.

**51.** **a**. 3,4,5,6
**b.** 1

**53.** For each row: (C1 + C2 + C3 + C4) ÷ 4 = average efficiency

| | | |
|---|---|---|
| W1: 0.9625 | W2: 0.9375 | W3: 0.9125 |
| W4: 0.8875 | W5: 0.85 | W6: 0.875 |
| W7: 0.90 | W8: 0.925 | W9: 0.95 |

W5 is least efficient. He works best at center 5.

## *Exercises 3.2*

**1.** **a.** $\begin{bmatrix}1 & 2 & 3\end{bmatrix}\begin{bmatrix}4\\5\\6\end{bmatrix}=\begin{bmatrix}1\cdot4+2\cdot5+3\cdot6\end{bmatrix}=[32]$

(not 32)

**b.** $\begin{bmatrix}1 & 2\end{bmatrix}\begin{bmatrix}3 & 5\\ 4 & 6\end{bmatrix}=\begin{bmatrix}1\cdot3+2\cdot4 & 1\cdot5+2\cdot6\end{bmatrix}$

$=\begin{bmatrix}11 & 17\end{bmatrix}$

**3.** $CD=\begin{bmatrix}5 & 3\\ 1 & 2\end{bmatrix}\cdot\begin{bmatrix}4 & 2\\ 3 & 5\end{bmatrix}$

$=\begin{bmatrix}20+9 & 10+15\\ 4+6 & 2+10\end{bmatrix}$

$=\begin{bmatrix}29 & 25\\ 10 & 12\end{bmatrix}$

**5.** $DE=\begin{bmatrix}4 & 2\\ 3 & 5\end{bmatrix}\begin{bmatrix}1 & 0 & 4\\ 5 & 1 & 0\end{bmatrix}$

$=\begin{bmatrix}4+10 & 0+2 & 16+0\\ 3+25 & 0+5 & 12+0\end{bmatrix}$

$=\begin{bmatrix}14 & 2 & 16\\ 28 & 5 & 12\end{bmatrix}$

**7.** $AB=\begin{bmatrix}1 & 0 & 2\\ 3 & 2 & 1\\ 4 & 0 & 3\end{bmatrix}\begin{bmatrix}1 & 1 & 3 & 0\\ 4 & 2 & 1 & 1\\ 3 & 2 & 0 & 1\end{bmatrix}=\begin{bmatrix}1+0+6 & 1+0+4 & 3+0+0 & 0+0+2\\ 3+8+3 & 3+4+2 & 9+2+0 & 0+2+1\\ 4+0+9 & 4+0+6 & 12+0+0 & 0+0+3\end{bmatrix}=\begin{bmatrix}7 & 5 & 3 & 2\\ 14 & 9 & 11 & 3\\ 13 & 10 & 12 & 3\end{bmatrix}$

**9.** *BA* is undefined.

**11.** $EB=\begin{bmatrix}1 & 0 & 4\\ 5 & 1 & 0\end{bmatrix}\begin{bmatrix}1 & 1 & 3 & 0\\ 4 & 2 & 1 & 1\\ 3 & 2 & 0 & 1\end{bmatrix}=\begin{bmatrix}1+0+12 & 1+0+8 & 3+0+0 & 0+0+4\\ 5+4+0 & 5+2+0 & 15+1+0 & 0+1+0\end{bmatrix}=\begin{bmatrix}13 & 9 & 3 & 4\\ 9 & 7 & 16 & 1\end{bmatrix}$

**13.** $EA^T = \begin{bmatrix} 1 & 0 & 4 \\ 5 & 1 & 0 \end{bmatrix} \begin{bmatrix} 1 & 3 & 4 \\ 0 & 2 & 0 \\ 2 & 1 & 3 \end{bmatrix} = \begin{bmatrix} 1+0+8 & 3+0+4 & 4+0+12 \\ 5+0+0 & 15+2+0 & 20+0+0 \end{bmatrix} = \begin{bmatrix} 9 & 7 & 16 \\ 5 & 17 & 20 \end{bmatrix}$

**15.** $A^2 = \begin{bmatrix} 1 & 0 & 2 \\ 3 & 2 & 1 \\ 4 & 0 & 3 \end{bmatrix} \begin{bmatrix} 1 & 0 & 2 \\ 3 & 2 & 1 \\ 4 & 0 & 3 \end{bmatrix} = \begin{bmatrix} 1+0+8 & 0+0+0 & 2+0+6 \\ 3+6+4 & 0+4+0 & 6+2+3 \\ 4+0+12 & 0+0+0 & 8+0+9 \end{bmatrix} = \begin{bmatrix} 9 & 0 & 8 \\ 13 & 4 & 11 \\ 16 & 0 & 17 \end{bmatrix}$

**17.** $C^3 = \begin{bmatrix} 5 & 3 \\ 1 & 2 \end{bmatrix} \cdot \left( \begin{bmatrix} 5 & 3 \\ 1 & 2 \end{bmatrix} \cdot \begin{bmatrix} 5 & 3 \\ 1 & 2 \end{bmatrix} \right) = \begin{bmatrix} 5 & 3 \\ 1 & 2 \end{bmatrix} \cdot \begin{bmatrix} 25+3 & 15+6 \\ 5+2 & 3+4 \end{bmatrix}$

$= \begin{bmatrix} 5 & 3 \\ 1 & 2 \end{bmatrix} \cdot \begin{bmatrix} 28 & 21 \\ 7 & 7 \end{bmatrix} = \begin{bmatrix} 140+21 & 105+21 \\ 28+14 & 21+14 \end{bmatrix} = \begin{bmatrix} 161 & 126 \\ 42 & 35 \end{bmatrix}$

**19.** $AA^T = \begin{bmatrix} 1 & 0 & 2 \\ 3 & 2 & 1 \\ 4 & 0 & 3 \end{bmatrix} \begin{bmatrix} 1 & 3 & 4 \\ 0 & 2 & 0 \\ 2 & 1 & 3 \end{bmatrix} = \begin{bmatrix} 5 & 5 & 10 \\ 5 & 14 & 15 \\ 10 & 15 & 25 \end{bmatrix}$

$(AA^T)^T = \begin{bmatrix} 5 & 5 & 10 \\ 5 & 14 & 15 \\ 10 & 15 & 25 \end{bmatrix}$ $\qquad A^T A = \begin{bmatrix} 1 & 3 & 4 \\ 0 & 2 & 0 \\ 2 & 1 & 3 \end{bmatrix} \begin{bmatrix} 1 & 0 & 2 \\ 3 & 2 & 1 \\ 4 & 0 & 3 \end{bmatrix} = \begin{bmatrix} 26 & 6 & 17 \\ 6 & 4 & 2 \\ 17 & 2 & 14 \end{bmatrix}$

No, they are not equal.

**21.** No. See problems 5 and 7

**23.** $AB = \begin{bmatrix} 2 & 5 & 4 \\ 1 & 4 & 3 \\ 1 & -3 & -2 \end{bmatrix} \begin{bmatrix} -1 & 2 & 1 \\ -5 & 8 & 2 \\ -7 & 11 & -3 \end{bmatrix} = \begin{bmatrix} -2-25-28 & 4+40+44 & 2+10-12 \\ -1-20-21 & 2+32+33 & 1+8-9 \\ -1+15+14 & 2-24-22 & 1-6+6 \end{bmatrix} = \begin{bmatrix} -55 & 88 & 0 \\ -42 & 67 & 0 \\ 28 & -44 & 1 \end{bmatrix}$

**25.** $CD = \begin{bmatrix} 3 & 0 & 4 \\ 1 & 7 & -1 \\ 3 & 0 & 4 \end{bmatrix} \begin{bmatrix} 4 & 4 & -8 \\ -1 & -1 & 2 \\ -3 & -3 & 6 \end{bmatrix} = \begin{bmatrix} 12+0-12 & 12+0-12 & -24+0+24 \\ 4-7+3 & 4-7+3 & -8+14-6 \\ 12+0-12 & 12+0-12 & -24+0+24 \end{bmatrix} = \begin{bmatrix} 0 & 0 & 0 \\ 0 & 0 & 0 \\ 0 & 0 & 0 \end{bmatrix}$

**27.** $F^2 = \begin{bmatrix} 0 & 1 & 2 \\ 0 & 0 & -4 \\ 0 & 0 & 0 \end{bmatrix} \begin{bmatrix} 0 & 1 & 2 \\ 0 & 0 & -4 \\ 0 & 0 & 0 \end{bmatrix} = \begin{bmatrix} 0+0+0 & 0+0+0 & 0-4+0 \\ 0+0+0 & 0+0+0 & 0+0+0 \\ 0+0+0 & 0+0+0 & 0+0+0 \end{bmatrix} = \begin{bmatrix} 0 & 0 & -4 \\ 0 & 0 & 0 \\ 0 & 0 & 0 \end{bmatrix}$

$F^3 = \begin{bmatrix} 0 & 0 & -4 \\ 0 & 0 & 0 \\ 0 & 0 & 0 \end{bmatrix} \begin{bmatrix} 0 & 1 & 2 \\ 0 & 0 & -4 \\ 0 & 0 & 0 \end{bmatrix} = \begin{bmatrix} 0+0+0 & 0+0+0 & 0+0+0 \\ 0+0+0 & 0+0+0 & 0+0+0 \\ 0+0+0 & 0+0+0 & 0+0+0 \end{bmatrix} = \begin{bmatrix} 0 & 0 & 0 \\ 0 & 0 & 0 \\ 0 & 0 & 0 \end{bmatrix}$

**29.** $AI = A$ See Example 6.

**31.** $Z(CI) = ZC = Z$ $\qquad (ZC)I = ZI = Z$
Product of zero matrix and A is the zero matrix.

**33.** Problems 35 and 37 show that a product can be "zero" and neither of the factors is zero.

**35. a.** We will use $\frac{1}{ad-bc}AB$ and $\frac{1}{ad-bc}BA$.

$$AB=\begin{bmatrix} a & b \\ c & d \end{bmatrix}\begin{bmatrix} d & -b \\ -c & a \end{bmatrix}=\begin{bmatrix} ad-bc & -ab+ba \\ cd-dc & -bc+ad \end{bmatrix}=\begin{bmatrix} ad-bc & 0 \\ 0 & ad-bc \end{bmatrix}$$

$$BA=\begin{bmatrix} d & -b \\ -c & a \end{bmatrix}\begin{bmatrix} a & b \\ c & d \end{bmatrix}=\begin{bmatrix} ad-bc & bd-bd \\ -ac+ac & -bc+ad \end{bmatrix}=\begin{bmatrix} ad-bc & 0 \\ 0 & ad-bc \end{bmatrix}$$

$$\frac{1}{ad-bc}AB=\frac{1}{ad-bc}BA=\begin{bmatrix} 1 & 0 \\ 0 & 1 \end{bmatrix}.$$

**b.** For $B$ to exist, we must have

**37.** $$\begin{bmatrix} 1 & 2 & 1 \\ 3 & 4 & -2 \\ 2 & 0 & -1 \end{bmatrix}\begin{bmatrix} 2 \\ -1 \\ 2 \end{bmatrix}=\begin{bmatrix} 2-2+2 \\ 6-4-4 \\ 4+0-2 \end{bmatrix}=\begin{bmatrix} 2 \\ -2 \\ 2 \end{bmatrix}$$

These values are the solution.

**39.**

$$\begin{bmatrix} 1 & 1 & 2 \\ 4 & 0 & 1 \\ 2 & 1 & 1 \end{bmatrix}\begin{bmatrix} 1 \\ 2 \\ 1 \end{bmatrix}=\begin{bmatrix} 1+2+2 \\ 4+0+1 \\ 2+2+1 \end{bmatrix}=\begin{bmatrix} 5 \\ 5 \\ 5 \end{bmatrix}$$

These values are the solution.

**41.** $$\begin{bmatrix} 0.1 & 0.0 & 0.1 & 0.1 & 0.2 \\ 0.1 & 0.2 & -0.1 & 0.1 & -0.1 \\ 0.1 & -0.1 & 0.1 & -0.2 & 0.1 \\ 0.1 & 0.2 & 0.1 & 0.1 & 0.1 \\ 0.1 & 0.0 & 0.0 & 0.0 & 0.0 \end{bmatrix}\begin{bmatrix} 0 & 0 & 0 & 0 & 10 \\ 0 & 5 & \frac{10}{3} & \frac{5}{3} & -10 \\ -10 & -15 & -\frac{20}{3} & -\frac{35}{3} & 20 \\ 0 & -5 & -\frac{20}{3} & \frac{5}{3} & 10 \\ 10 & 10 & \frac{20}{3} & -\frac{20}{3} & 20 \end{bmatrix}$$

$$=\begin{bmatrix} 1 & 0 & 0 & 0 & 0 \\ 0 & 1 & 0 & -0.002 & 0 \\ 0 & 0 & 0.999 & 0.002 & 0 \\ 0 & 0 & 0 & 0.998 & 0 \\ 0 & 0 & 0 & 0 & 1 \end{bmatrix}=\begin{bmatrix} 1 & 0 & 0 & 0 & 0 \\ 0 & 1 & 0 & 0 & 0 \\ 0 & 0 & 1 & 0 & 0 \\ 0 & 0 & 0 & 1 & 0 \\ 0 & 0 & 0 & 0 & 1 \end{bmatrix}$$

The graphing calculator should show 1's on the main diagonal and 0's everywhere else (1 may be 0.9999 or 0.998 and 0 may be 0.002 or -0.002).

**43.** $\begin{bmatrix} 0.88 & 0 \\ 0 & 0.85 \end{bmatrix}\begin{bmatrix} 36{,}000 & 42{,}000 \\ 50{,}000 & 56{,}000 \end{bmatrix} = \begin{bmatrix} 31{,}680 & 36{,}960 \\ 42{,}500 & 47{,}600 \end{bmatrix}$

The product is the price the dealer pays for the cars.

**45. a.** $A = \begin{bmatrix} 86.86 & 91.25 & 95.37 \\ 27.57 & 29.82 & 32.05 \\ 6.36 & 6.45 & 6.60 \end{bmatrix}$ and $B = \begin{bmatrix} 91.1 & 0 & 0 \\ 0 & 86.0 & 0 \\ 0 & 0 & 82.5 \end{bmatrix}$

**b.** $AB \approx \begin{bmatrix} 7913 & 7848 & 7868 \\ 2512 & 2565 & 2644 \\ 579.4 & 554.7 & 544.5 \end{bmatrix}$

**c.** The 1-1 entry is the trillions of BTUs used in single-family households in 2015. The 2-3 entry is the trillions of BTUs predicted to be used in multi-family households in 2025.

**d.** Using $AB$ as stored in a calculator, $\begin{bmatrix} 1 & 1 & 1 \end{bmatrix} AB \approx \begin{bmatrix} 11{,}004.0 & 10{,}966.7 & 11{,}056.7 \end{bmatrix}$. Left to right, each entry gives the trillions of BTUs of delivered energy consumed by all households in 2015, 2020, and 2025.

**47. a.** $\begin{bmatrix} 0.5 & 0.5 \end{bmatrix}\begin{bmatrix} 0.8 & 0.2 \\ 0.3 & 0.7 \end{bmatrix} = \begin{bmatrix} 0.55 & 0.45 \end{bmatrix}$ After 5 years, M has 55% and S has 45% of the population.

**b.** 10 years: $(PD)D = PD^2$; 15 years: $PD^3$

**c.** 60% in M and 40% in S. The population proportions are stable.

**49.**

| W | a | s | t | e | blank | n | o | t | blank | w | a | n | t | blank | n | o | t |
|---|---|---|---|---|---|---|---|---|---|---|---|---|---|---|---|---|---|
| 23 | 1 | 19 | 20 | 5 | 27 | 14 | 15 | 20 | 27 | 23 | 1 | 14 | 20 | 27 | 14 | 15 | 20 |

$\begin{bmatrix} 5 & 9 \\ 6 & 11 \end{bmatrix}\begin{bmatrix} 23 & 19 & 5 & 14 & 20 & 23 & 14 & 27 & 15 \\ 1 & 20 & 27 & 15 & 27 & 1 & 20 & 14 & 20 \end{bmatrix} = \begin{bmatrix} 124 & 275 & 268 & 205 & 343 & 124 & 250 & 261 & 255 \\ 149 & 334 & 327 & 249 & 417 & 149 & 304 & 316 & 310 \end{bmatrix}$

So 124, 149, 275, 334, 268, 327, 205, 249, 343, 417, 124, 149, 250, 304, 261, 316, 255, 310 is the message sent.

**51. a.** $B = \begin{bmatrix} \frac{3}{4} & \frac{2}{5} & \frac{1}{4} \\ \frac{1}{4} & \frac{3}{5} & \frac{3}{4} \end{bmatrix}; D = \begin{bmatrix} 22 & 30 \\ 12 & 20 \\ 8 & 11 \end{bmatrix}; BD = \begin{bmatrix} 23.3 & 33.25 \\ 18.7 & 27.75 \end{bmatrix}$ Houston needs 23,300 gallons of black crude and 18,700 gallons of gold. Gulfport needs 32,250 gallons of black and 27,750 gallons of gold.

**b.** $PBD = \begin{bmatrix} 135.945 & 197.5325 \end{bmatrix}$; Houston's cost = \$135,945; Gulfport's cost = \$197,532.50

**53. a.** $B = \begin{bmatrix} 0.7 & 8.5 & 10.2 & 1.1 & 5.6 & 3.6 \\ 0.5 & 0.2 & 6.1 & 1.3 & 0.2 & 1.0 \\ 2.2 & 0.4 & 8.8 & 1.2 & 1.2 & 4.8 \\ 251.8 & 63.4 & 81.6 & 35.2 & 54.3 & 144.2 \\ 30.0 & 1.0 & 1.0 & 1.0 & 1.0 & 1.0 \\ 788.9 & 0 & 0 & 0 & 0 & 0 \end{bmatrix}$

**b.** $A = \begin{bmatrix} 1.11 & 0 & 0 & 0 & 0 & 0 \\ 0 & 0.95 & 0 & 0 & 0 & 0 \\ 0 & 0 & 1.11 & 0 & 0 & 0 \\ 0 & 0 & 0 & 1.11 & 0 & 0 \\ 0 & 0 & 0 & 0 & 0.95 & 0 \\ 0 & 0 & 0 & 0 & 0 & 0.95 \end{bmatrix}$

## Exercises 3.3

**1.** $$\left[\begin{array}{rrr|r} 1 & -2 & -1 & -7 \\ 3 & 1 & 2 & 0 \\ 4 & 2 & 2 & 1 \end{array}\right] \begin{array}{c} \\ -3R_1 + R_2 \to R_2 \\ \to \end{array}$$
$$\left[\begin{array}{rrr|r} 1 & -2 & -1 & -7 \\ 0 & 7 & 5 & 21 \\ 4 & 2 & 2 & 1 \end{array}\right]$$

**3.** $$\left[\begin{array}{rrr|r} 1 & -3 & 4 & 2 \\ 2 & 0 & 2 & 1 \\ 1 & 2 & 1 & 1 \end{array}\right]$$

**5.** $x = 2,\ y = \dfrac{1}{2},\ z = -5$

**7.** $$\left[\begin{array}{rrr|r} 1 & 1 & 2 & -1 \\ 0 & 3 & 1 & 7 \\ 0 & -2 & 4 & 0 \end{array}\right] \begin{array}{c} \to \\ R_3 + R_2 \to R_2 \\ \\ \end{array} \left[\begin{array}{rrr|r} 1 & 1 & 2 & -1 \\ 0 & 1 & 5 & 7 \\ 0 & -2 & 4 & 0 \end{array}\right] \begin{array}{c} -R_2 + R_1 \to R_1 \\ \to \\ 2R_2 + R_3 \to R_3 \end{array}$$
$$\left[\begin{array}{rrr|r} 1 & 0 & -3 & -8 \\ 0 & 1 & 5 & 7 \\ 0 & 0 & 14 & 14 \end{array}\right] \begin{array}{c} \\ \to \\ \frac{1}{14}R_3 \to R_3 \end{array} \left[\begin{array}{rrr|r} 1 & 0 & -3 & -8 \\ 0 & 1 & 5 & 7 \\ 0 & 0 & 1 & 1 \end{array}\right] \begin{array}{c} 3R_3 + R_1 \to R_1 \\ -5R_3 + R_2 \to R_2 \\ \\ \end{array} \left[\begin{array}{rrr|r} 1 & 0 & 0 & -5 \\ 0 & 1 & 0 & 2 \\ 0 & 0 & 1 & 1 \end{array}\right]$$
Solution: $x = -5,\ y = 2,\ z = 1$

**9.** $$\left[\begin{array}{rrr|r} 1 & 2 & 5 & -4 \\ 2 & -2 & 4 & -2 \\ 0 & 1 & -3 & 7 \end{array}\right] \begin{array}{c} \to \\ -2R_1 + R_2 \to R_2 \\ \\ \end{array} \left[\begin{array}{rrr|r} 1 & 2 & 5 & -4 \\ 0 & -6 & -6 & 6 \\ 0 & 1 & -3 & 7 \end{array}\right] \begin{array}{c} R_2 \leftrightarrow R_3 \\ \to \\ \\ \end{array} \left[\begin{array}{rrr|r} 1 & 2 & 5 & -4 \\ 0 & 1 & -3 & 7 \\ 0 & -6 & -6 & 6 \end{array}\right] \begin{array}{c} -2R_2 + R_1 \to R_1 \\ \to \\ 6R_2 + R_3 \to R_3 \end{array}$$
$$\left[\begin{array}{rrr|r} 1 & 0 & 11 & -18 \\ 0 & 1 & -3 & 7 \\ 0 & 0 & -24 & 48 \end{array}\right] \begin{array}{c} \\ \to \\ -\frac{1}{24}R_3 \to R_3 \end{array} \left[\begin{array}{rrr|r} 1 & 0 & 11 & -18 \\ 0 & 1 & -3 & 7 \\ 0 & 0 & 1 & -2 \end{array}\right] \begin{array}{c} -11R_3 + R_1 \to R_1 \\ 3R_3 + R_2 \to R_2 \\ \to \end{array} \left[\begin{array}{rrr|r} 1 & 0 & 0 & 4 \\ 0 & 1 & 0 & 1 \\ 0 & 0 & 1 & -2 \end{array}\right]$$
Solution: $x = 4,\ y = 1,\ z = -2$

**11.** $$\left[\begin{array}{rrr|r} 1 & 1 & -1 & 0 \\ 1 & 2 & 3 & -5 \\ 2 & -1 & -13 & 17 \end{array}\right] \begin{array}{c} \to \\ -R_1 + R_2 \to R_2 \\ -2R_1 + R_3 \to R_3 \end{array} \left[\begin{array}{rrr|r} 1 & 1 & -1 & 0 \\ 0 & 1 & 4 & -5 \\ 0 & -3 & -11 & 17 \end{array}\right] \begin{array}{c} -R_2 + R_1 \to R_1 \\ \to \\ 3R_2 + R_3 \to R_3 \end{array}$$
$$\left[\begin{array}{rrr|r} 1 & 0 & -5 & 5 \\ 0 & 1 & 4 & -5 \\ 0 & 0 & 1 & 2 \end{array}\right] \begin{array}{c} 5R_3 + R_1 \to R_1 \\ -4R_3 + R_2 \to R_2 \\ \to \end{array} \left[\begin{array}{rrr|r} 1 & 0 & 0 & 15 \\ 0 & 1 & 0 & -13 \\ 0 & 0 & 1 & 2 \end{array}\right]$$
Solution: $x = 15,\ y = -13,\ z = 2$

**13.** $$\left[\begin{array}{ccc|c} 2 & -6 & -12 & 6 \\ 3 & -10 & -20 & 5 \\ 2 & 0 & -17 & -4 \end{array}\right] \begin{array}{c} \frac{1}{2}R_1 \to R_1 \\ \to \\ \phantom{x} \end{array} \left[\begin{array}{ccc|c} 1 & -3 & -6 & 3 \\ 3 & -10 & -20 & 5 \\ 2 & 0 & -17 & -4 \end{array}\right] \begin{array}{c} \to \\ -3R_1 + R_2 \to R_2 \\ -2R_1 + R_3 \to R_3 \end{array}$$

$$\left[\begin{array}{ccc|c} 1 & -3 & -6 & 3 \\ 0 & -1 & -2 & -4 \\ 0 & 6 & -5 & -10 \end{array}\right] \begin{array}{c} \to \\ -R_2 \to R_2 \\ \phantom{x} \end{array} \left[\begin{array}{ccc|c} 1 & -3 & -6 & 3 \\ 0 & 1 & 2 & 4 \\ 0 & 6 & -5 & -10 \end{array}\right] \begin{array}{c} 3R_2 + R_1 \to R_1 \\ \to \\ -6R_2 + R_3 \to R_3 \end{array}$$

$$\left[\begin{array}{ccc|c} 1 & 0 & 0 & 15 \\ 0 & 1 & 2 & 4 \\ 0 & 0 & -17 & -34 \end{array}\right] \begin{array}{c} \phantom{x} \\ \to \\ -\frac{1}{17}R_3 \to R_3 \end{array} \left[\begin{array}{ccc|c} 1 & 0 & 0 & 15 \\ 0 & 1 & 2 & 4 \\ 0 & 0 & 1 & 2 \end{array}\right] \begin{array}{c} \to \\ -2R_3 + R_2 \to R_2 \\ \phantom{x} \end{array} \left[\begin{array}{ccc|c} 1 & 0 & 0 & 15 \\ 0 & 1 & 0 & 0 \\ 0 & 0 & 1 & 2 \end{array}\right]$$

Solution: $x = 15,\ y = 0,\ z = 2$

**15.** $$\left[\begin{array}{cccc|c} 1 & -2 & 3 & 1 & -2 \\ 1 & -3 & 1 & -1 & -7 \\ 1 & -1 & 0 & 0 & -2 \\ 1 & 0 & 1 & 1 & 2 \end{array}\right] \begin{array}{c} \to \\ -R_1 + R_2 \to R_2 \\ -R_1 + R_3 \to R_3 \\ -R_1 + R_4 \to R_4 \end{array} \left[\begin{array}{cccc|c} 1 & -2 & 3 & 1 & -2 \\ 0 & -1 & -2 & -2 & -5 \\ 0 & 1 & -3 & -1 & 0 \\ 0 & 2 & -2 & 0 & 4 \end{array}\right] \begin{array}{c} \to \\ -1R_2 \to R_2 \\ \phantom{x} \\ \phantom{x} \end{array}$$

$$\left[\begin{array}{cccc|c} 1 & -2 & 3 & 1 & -2 \\ 0 & 1 & 2 & 2 & 5 \\ 0 & 1 & -3 & -1 & 0 \\ 0 & 2 & -2 & 0 & 4 \end{array}\right] \begin{array}{c} 2R_2 + R_1 \to R_1 \\ \to \\ -R_2 + R_3 \to R_3 \\ -2R_2 + R_4 \to R_4 \end{array} \left[\begin{array}{cccc|c} 1 & 0 & 7 & 5 & 8 \\ 0 & 1 & 2 & 2 & 5 \\ 0 & 0 & -5 & -3 & -5 \\ 0 & 0 & -6 & -4 & -6 \end{array}\right] \begin{array}{c} \phantom{x} \\ \to \\ -\frac{1}{5}R_3 \to R_3 \\ \phantom{x} \end{array}$$

$$\left[\begin{array}{cccc|c} 1 & 0 & 7 & 5 & 8 \\ 0 & 1 & 2 & 2 & 5 \\ 0 & 0 & 1 & 3/5 & 1 \\ 0 & 0 & -6 & -4 & -6 \end{array}\right] \begin{array}{c} -7R_3 + R_1 \to R_1 \\ -2R_3 + R_2 \to R_2 \\ \to \\ 6R_3 + R_4 \to R_4 \end{array} \left[\begin{array}{cccc|c} 1 & 0 & 0 & 4/5 & 1 \\ 0 & 1 & 0 & 4/5 & 3 \\ 0 & 0 & 1 & 3/5 & 1 \\ 0 & 0 & 0 & 2/5 & 0 \end{array}\right] \begin{array}{c} \text{Now take } \frac{5}{2}R_4 \\ \text{and get 0's} \\ \text{in column 4;} \\ \text{Result is} \end{array} \left[\begin{array}{cccc|c} 1 & 0 & 0 & 0 & 1 \\ 0 & 1 & 0 & 0 & 3 \\ 0 & 0 & 1 & 0 & 1 \\ 0 & 0 & 0 & 1 & 0 \end{array}\right]$$

Solution: $x = 1,\ y = 3,\ z = 1,\ w = 0$

**17.** There is no solution since the last row of the reduced matrix says $0x + 0y + 0z = 1$, which is not possible.

**19.** **a**. From first row: $x - \frac{2}{3}z = \frac{11}{3}$

From second row: $y + \frac{1}{3}z = -\frac{1}{3}$

General solution: $x = \frac{11}{3} + \frac{2}{3}z = \frac{11+2z}{3}$, $y = -\frac{1}{3} - \frac{1}{3}z = \frac{-1-z}{3}$ for any real number $z$.

**b**. Many possibilities, including $x = \frac{11}{3}, y = -\frac{1}{3}, z = 0$ and $x = \frac{13}{3}, y = -\frac{2}{3}, z = 1$

**21** If the bottom row of the reduced matrix has all zeros on the left side of the augment and a number other than zero on the right side, the system has no solutions.

**23.** $\left[\begin{array}{ccc|c} 1 & 1 & 1 & 0 \\ 2 & -1 & -1 & 0 \\ -1 & 2 & 2 & 0 \end{array}\right] \begin{array}{c} \rightarrow \\ -2R_1 + R_2 \rightarrow R_2 \\ R_1 + R_3 \rightarrow R_3 \end{array} \left[\begin{array}{ccc|c} 1 & 1 & 1 & 0 \\ 0 & -3 & -3 & 0 \\ 0 & 3 & 3 & 0 \end{array}\right] \begin{array}{c} \\ \rightarrow \\ R_2 + R_3 \rightarrow R_3 \end{array}$

$\left[\begin{array}{ccc|c} 1 & 1 & 1 & 0 \\ 0 & -3 & -3 & 0 \\ 0 & 0 & 0 & 0 \end{array}\right] \begin{array}{c} \rightarrow \\ -\frac{1}{3}R_2 \rightarrow R_2 \\ \end{array} \left[\begin{array}{ccc|c} 1 & 1 & 1 & 0 \\ 0 & 1 & 1 & 0 \\ 0 & 0 & 0 & 0 \end{array}\right] \begin{array}{c} -R_2 + R_1 \rightarrow R_1 \\ \rightarrow \\ \end{array} \left[\begin{array}{ccc|c} 1 & 0 & 0 & 0 \\ 0 & 1 & 1 & 0 \\ 0 & 0 & 0 & 0 \end{array}\right]$

General solution: $x = 0,\ \ y = -z$

**25.** $\left[\begin{array}{ccc|c} 3 & 2 & 1 & 0 \\ 1 & 1 & 2 & 2 \\ 2 & 1 & -1 & -1 \end{array}\right] \begin{array}{c} R_1 \leftrightarrow R_2 \\ \rightarrow \\ \end{array} \left[\begin{array}{ccc|c} 1 & 1 & 2 & 2 \\ 3 & 2 & 1 & 0 \\ 2 & 1 & -1 & -1 \end{array}\right] \begin{array}{c} \rightarrow \\ -3R_1 + R_2 \rightarrow R_2 \\ -2R_1 + R_3 \rightarrow R_3 \end{array}$

$\left[\begin{array}{ccc|c} 1 & 1 & 2 & 2 \\ 0 & -1 & -5 & -6 \\ 0 & -1 & -5 & -5 \end{array}\right] \begin{array}{c} \rightarrow \\ \\ -R_2 + R_3 \rightarrow R_3 \end{array} \left[\begin{array}{ccc|c} 1 & 1 & 2 & 2 \\ 0 & -1 & -5 & -6 \\ 0 & 0 & 0 & 1 \end{array}\right]$ No Solution

**27.** $\left[\begin{array}{ccc|c} 2 & 2 & 1 & 2 \\ 1 & -2 & 2 & 1 \\ -1 & 2 & -2 & -1 \end{array}\right] \begin{array}{c} R_1 \leftrightarrow R_2 \\ \rightarrow \\ \end{array} \left[\begin{array}{ccc|c} 1 & -2 & 2 & 1 \\ 2 & 2 & 1 & 2 \\ -1 & 2 & -2 & -1 \end{array}\right] \begin{array}{c} \rightarrow \\ -2R_1 + R_2 \rightarrow R_2 \\ R_1 + R_3 \rightarrow R_3 \end{array}$

$\left[\begin{array}{ccc|c} 1 & -2 & 2 & 1 \\ 0 & 6 & -3 & 0 \\ 0 & 0 & 0 & 0 \end{array}\right] \begin{array}{c} \rightarrow \\ \frac{1}{6}R_2 \rightarrow R_2 \\ \end{array} \left[\begin{array}{ccc|c} 1 & -2 & 2 & 1 \\ 0 & 1 & -1/2 & 0 \\ 0 & 0 & 0 & 0 \end{array}\right] \begin{array}{c} 2R_2 + R_1 \rightarrow R_1 \\ \rightarrow \\ \end{array} \left[\begin{array}{ccc|c} 1 & 0 & 1 & 1 \\ 0 & 1 & -1/2 & 0 \\ 0 & 0 & 0 & 0 \end{array}\right]$

General solution: $x = 1 - z,\ \ y = \dfrac{1}{2}z$

**29.** $\left[\begin{array}{ccc|c} 1 & -3 & 3 & 7 \\ 1 & 2 & -1 & -2 \\ 3 & 2 & 4 & 5 \end{array}\right] \begin{array}{c} \rightarrow \\ -R_1 + R_2 \rightarrow R_2 \\ -3R_1 + R_3 \rightarrow R_3 \end{array} \left[\begin{array}{ccc|c} 1 & -3 & 3 & 7 \\ 0 & 5 & -4 & -9 \\ 0 & 11 & -5 & -16 \end{array}\right] \begin{array}{c} \\ \rightarrow \\ -2R_2 + R_3 \rightarrow R_3 \end{array}$

$\left[\begin{array}{ccc|c} 1 & -3 & 3 & 7 \\ 0 & 5 & -4 & -9 \\ 0 & 1 & 3 & 2 \end{array}\right] \begin{array}{c} R_2 \leftrightarrow R_3 \\ \rightarrow \end{array} \left[\begin{array}{ccc|c} 1 & -3 & 3 & 7 \\ 0 & 1 & 3 & 2 \\ 0 & 5 & -4 & -9 \end{array}\right] \begin{array}{c} 3R_2 + R_1 \rightarrow R_1 \\ \rightarrow \\ -5R_2 + R_3 \rightarrow R_3 \end{array}$

$\left[\begin{array}{ccc|c} 1 & 0 & 12 & 13 \\ 0 & 1 & 3 & 2 \\ 0 & 0 & -19 & -19 \end{array}\right] \begin{array}{c} \\ \rightarrow \\ -\frac{1}{19}R_3 \rightarrow R_3 \end{array} \left[\begin{array}{ccc|c} 1 & 0 & 12 & 13 \\ 0 & 1 & 3 & 2 \\ 0 & 0 & 1 & 1 \end{array}\right] \begin{array}{c} -12R_3 + R_1 \rightarrow R_1 \\ -3R_3 + R_2 \rightarrow R_2 \\ \rightarrow \end{array} \left[\begin{array}{ccc|c} 1 & 0 & 0 & 1 \\ 0 & 1 & 0 & -1 \\ 0 & 0 & 1 & 1 \end{array}\right]$

Solution: $x = 1,\ \ y = -1,\ \ z = 1$

**31.** $\left[\begin{array}{ccc|c} 2 & -5 & 1 & -9 \\ 1 & 4 & -6 & 2 \\ 3 & -4 & -2 & -10 \end{array}\right] \begin{array}{c} R_1 \leftrightarrow R_2 \\ \rightarrow \\ \end{array} \left[\begin{array}{ccc|c} 1 & 4 & -6 & 2 \\ 2 & -5 & 1 & -9 \\ 3 & -4 & -2 & -10 \end{array}\right] \begin{array}{c} \rightarrow \\ -2R_1 + R_2 \rightarrow R_2 \\ -3R_1 + R_3 \rightarrow R_3 \end{array} \left[\begin{array}{ccc|c} 1 & 4 & -6 & 2 \\ 0 & -13 & 13 & -13 \\ 0 & -16 & 16 & -16 \end{array}\right] \begin{array}{c} \rightarrow \\ -\frac{1}{13}R_2 \rightarrow R_2 \\ \end{array}$

$\left[\begin{array}{ccc|c} 1 & 4 & -6 & 2 \\ 0 & 1 & -1 & 1 \\ 0 & -16 & 16 & -16 \end{array}\right] \begin{array}{c} -4R_2 + R_1 \rightarrow R_1 \\ \rightarrow \\ 16R_2 + R_3 \rightarrow R_3 \end{array} \left[\begin{array}{ccc|c} 1 & 0 & -2 & -2 \\ 0 & 1 & -1 & 1 \\ 0 & 0 & 0 & 0 \end{array}\right]$

General solution: $x = 2z - 2,\ \ y = z + 1$

**33.** $\left[\begin{array}{ccc|c} 1 & 1 & 1 & 3 \\ 1 & -1 & 1 & 4 \end{array}\right] \xrightarrow{-R_1+R_2 \to R_2} \left[\begin{array}{ccc|c} 1 & 1 & 1 & 3 \\ 0 & -2 & 0 & 1 \end{array}\right] \xrightarrow{-\frac{1}{2}R_2 \to R_2}$

$\left[\begin{array}{ccc|c} 1 & 1 & 1 & 3 \\ 0 & 1 & 0 & -1/2 \end{array}\right] \xrightarrow{-R_2+R_1 \to R_1} \left[\begin{array}{ccc|c} 1 & 0 & 1 & 7/2 \\ 0 & 1 & 0 & -1/2 \end{array}\right]$

General Solution: $x = \dfrac{7}{2} - z$, $y = -\dfrac{1}{2}$, $z = z$

**35.** $\left[\begin{array}{ccc|c} 3 & -2 & 5 & 14 \\ 2 & -3 & 4 & 8 \end{array}\right] \xrightarrow{-R_2+R_1 \to R_1} \left[\begin{array}{ccc|c} 1 & 1 & 1 & 6 \\ 2 & -3 & 4 & 8 \end{array}\right] \xrightarrow{-2R_1+R_2 \to R_2}$

$\left[\begin{array}{ccc|c} 1 & 1 & 1 & 6 \\ 0 & -5 & 2 & -4 \end{array}\right] \xrightarrow{-\frac{1}{5}R_2 \to R_2} \left[\begin{array}{ccc|c} 1 & 1 & 1 & 6 \\ 0 & 1 & -\frac{2}{5} & \frac{4}{5} \end{array}\right] \xrightarrow{-R_2+R_1 \to R_1} \left[\begin{array}{ccc|c} 1 & 0 & \frac{7}{5} & \frac{26}{5} \\ 0 & 1 & -\frac{2}{5} & \frac{4}{5} \end{array}\right]$

General Solution: $x = \dfrac{26}{5} - \dfrac{7}{5}z$, $y = \dfrac{4}{5} + \dfrac{2}{5}z$

**37.**

$$\left[\begin{array}{ccc|c} -0.6 & 0.1 & 0.4 & 10 \\ 0.4 & -0.7 & 0.2 & -26 \\ 0.2 & 0.6 & -0.5 & 20 \end{array}\right] \begin{array}{l} 10R_1 \to R_1 \\ 10R_2 \to R_2 \\ 10R_3 \to R_3 \end{array} \left[\begin{array}{ccc|c} -6 & 1 & 4 & 100 \\ 4 & -7 & 2 & -260 \\ 2 & 6 & -5 & 200 \end{array}\right] \begin{array}{c} \to \\ 2R_1+3R_2 \to R_2 \\ -R_2+2R_3 \to R_3 \end{array} \left[\begin{array}{ccc|c} -6 & 1 & 4 & 100 \\ 0 & -19 & 14 & -580 \\ 0 & 19 & -12 & 660 \end{array}\right]$$

$$\begin{array}{c} \\ \to \\ R_2+R_3 \to R_3 \end{array} \left[\begin{array}{ccc|c} -6 & 1 & 4 & 100 \\ 0 & -19 & 14 & -580 \\ 0 & 0 & 2 & 80 \end{array}\right] \begin{array}{c} \to \\ -7R_3+R_2 \to R_2 \\ \frac{1}{2}R_3 \to R_3 \end{array} \left[\begin{array}{ccc|c} -6 & 1 & 4 & 100 \\ 0 & -19 & 0 & -1140 \\ 0 & 0 & 1 & 40 \end{array}\right] \begin{array}{c} \frac{1}{19}R_3+R_1 \to R_1 \\ -\frac{1}{19}R_2 \to R_2 \\ \to \end{array}$$

$$\left[\begin{array}{ccc|c} -6 & 0 & 4 & 40 \\ 0 & 1 & 0 & 60 \\ 0 & 0 & 1 & 40 \end{array}\right] \begin{array}{c} -\frac{1}{6}(-4R_3+R_1) \to R_1 \\ \to \\ \to \end{array} \left[\begin{array}{ccc|c} 1 & 0 & 0 & 20 \\ 0 & 1 & 0 & 60 \\ 0 & 0 & 1 & 40 \end{array}\right]$$

General solution: $x_1 = 20$, $x_2 = 60$, $x_3 = 40$

**39.**

$$\left[\begin{array}{cccc|c} 1 & 3 & 2 & 2 & 3 \\ 1 & 1 & 3 & 0 & 4 \\ 2 & 0 & 2 & -3 & 4 \\ 1 & -3 & 0 & 0 & 1 \end{array}\right] \xrightarrow{\substack{-R_1+R_2\to R_2 \\ -2R_1+R_3\to R_3 \\ -R_1+R_4\to R_4}} \left[\begin{array}{cccc|c} 1 & 3 & 2 & 2 & 3 \\ 0 & -2 & 1 & -2 & 1 \\ 0 & -6 & -2 & -7 & -2 \\ 0 & -6 & -2 & -2 & -2 \end{array}\right] \xrightarrow{\substack{\text{*See note} \\ -3R_4+R_3\to R_3}}$$

$$\left[\begin{array}{cccc|c} 1 & 3 & 2 & 2 & 3 \\ 0 & -2 & 1 & -2 & 1 \\ 0 & 12 & 4 & -1 & 4 \\ 0 & -6 & -2 & -2 & -2 \end{array}\right] \xrightarrow{\substack{-R_3\to R_3 \\ \text{then } R_3\leftrightarrow R_4}} \left[\begin{array}{cccc|c} 1 & 3 & 2 & 2 & 3 \\ 0 & -2 & 1 & -2 & 1 \\ 0 & -6 & -2 & -2 & -2 \\ 0 & -12 & -4 & 1 & -4 \end{array}\right] \xrightarrow{\substack{-2R_4+R_1\to R_1 \\ 2R_4+R_2\to R_2 \\ 2R_4+R_3\to R_3}}$$

$$\left[\begin{array}{cccc|c} 1 & 27 & 10 & 0 & 11 \\ 0 & -26 & -7 & 0 & -7 \\ 0 & -30 & -10 & 0 & -10 \\ 0 & -12 & -4 & 1 & -4 \end{array}\right] \xrightarrow{\substack{-3R_2\to R_2 \\ -2R_3\to R_3}} \left[\begin{array}{cccc|c} 1 & 27 & 10 & 0 & 11 \\ 0 & 78 & 21 & 0 & 21 \\ 0 & 60 & 20 & 0 & 20 \\ 0 & -12 & -4 & 1 & -4 \end{array}\right] \xrightarrow{-R_3+R_2\to R_2}$$

$$\left[\begin{array}{cccc|c} 1 & 27 & 10 & 0 & 11 \\ 0 & 18 & 1 & 0 & 1 \\ 0 & 60 & 20 & 0 & 20 \\ 0 & -12 & -4 & 1 & -4 \end{array}\right] \xrightarrow{\substack{-10R_2+R_1\to R_1 \\ -20R_2+R_3\to R_3 \\ 4R_2+R_4\to R_4}} \left[\begin{array}{cccc|c} 1 & -153 & 0 & 0 & 1 \\ 0 & 18 & 1 & 0 & 1 \\ 0 & -300 & 0 & 0 & 0 \\ 0 & 60 & 0 & 1 & 0 \end{array}\right] \xrightarrow{\substack{-\frac{1}{300}R_3\to R_3 \\ \text{then } R_2\leftrightarrow R_3}}$$

$$\left[\begin{array}{cccc|c} 1 & -153 & 0 & 0 & 1 \\ 0 & 1 & 0 & 0 & 0 \\ 0 & 18 & 1 & 0 & 1 \\ 0 & 60 & 0 & 1 & 0 \end{array}\right] \xrightarrow{\substack{153R_2+R_1\to R_1 \\ -18R_2+R_3\to R_3 \\ -60R_2+R_4\to R_4}} \left[\begin{array}{cccc|c} 1 & 0 & 0 & 0 & 1 \\ 0 & 1 & 0 & 0 & 0 \\ 0 & 0 & 1 & 0 & 1 \\ 0 & 0 & 0 & 1 & 0 \end{array}\right]$$

Solution: $x_1 = 1, x_2 = 0, x_3 = 1, x_4 = 0$

*Note: The author's step by step method always works. However, it is often necessary to introduce fractions. This problem has been worked "backwards" to illustrate an alternate method and avoid fractions.

**41.**

$$\left[\begin{array}{cccc|c} 3 & 2 & 1 & -1 & 3 \\ 1 & -1 & -2 & 2 & 2 \\ 2 & 3 & -1 & 1 & 1 \\ -1 & 1 & 2 & -2 & -2 \end{array}\right] \xrightarrow{R_1\leftrightarrow R_2} \left[\begin{array}{cccc|c} 1 & -1 & -2 & 2 & 2 \\ 3 & 2 & 1 & -1 & 3 \\ 2 & 3 & -1 & 1 & 1 \\ -1 & 1 & 2 & -2 & -2 \end{array}\right] \xrightarrow{\substack{-3R_1+R_2\to R_2 \\ -2R_1+R_3\to R_3 \\ R_1+R_4\to R_4}} \left[\begin{array}{cccc|c} 1 & -1 & -2 & 2 & 2 \\ 0 & 5 & 7 & -7 & -3 \\ 0 & 5 & 3 & -3 & -3 \\ 0 & 0 & 0 & 0 & 0 \end{array}\right]$$

$$\xrightarrow{-R_2+R_3\to R_3} \left[\begin{array}{cccc|c} 1 & -1 & -2 & 2 & 2 \\ 0 & 5 & 7 & -7 & -3 \\ 0 & 0 & -4 & 4 & 0 \\ 0 & 0 & 0 & 0 & 0 \end{array}\right] \xrightarrow{\substack{\frac{1}{5}R_2\to R_2 \\ -\frac{1}{4}R_3\to R_3}} \left[\begin{array}{cccc|c} 1 & -1 & -2 & 2 & 2 \\ 0 & 1 & \frac{7}{5} & -\frac{7}{5} & -\frac{3}{5} \\ 0 & 0 & 1 & -1 & 0 \\ 0 & 0 & 0 & 0 & 0 \end{array}\right] \xrightarrow{\substack{R_2+R_1\to R_1 \\ -\frac{7}{5}R_3+R_2\to R_2}}$$

$$\left[\begin{array}{cccc|c} 1 & 0 & -\frac{3}{5} & \frac{3}{5} & \frac{7}{5} \\ 0 & 1 & 0 & 0 & -\frac{3}{5} \\ 0 & 0 & 1 & -1 & 0 \\ 0 & 0 & 0 & 0 & 0 \end{array}\right] \xrightarrow{\frac{3}{5}R_3+R_1\to R_1} \left[\begin{array}{cccc|c} 1 & 0 & 0 & 0 & \frac{7}{5} \\ 0 & 1 & 0 & 0 & -\frac{3}{5} \\ 0 & 0 & 1 & -1 & 0 \\ 0 & 0 & 0 & 0 & 0 \end{array}\right]$$

General solution: $x = \frac{7}{5},\ y = -\frac{3}{5},\ z = w$

**43.** $$\left[\begin{array}{cccc|c} 1 & 2 & -1 & 1 & 3 \\ 1 & 3 & 4 & 1 & -2 \\ 2 & 5 & 2 & 2 & 1 \\ 2 & 3 & -6 & 2 & 3 \end{array}\right] \begin{array}{c} \to \\ -R_1+R_2\to R_2 \\ -2R_1+R_3\to R_3 \\ -2R_1+R_4\to R_4 \end{array} \left[\begin{array}{cccc|c} 1 & 2 & -1 & 1 & 3 \\ 0 & 1 & 5 & 0 & -5 \\ 0 & 1 & 4 & 0 & -5 \\ 0 & -1 & -4 & 0 & -3 \end{array}\right] \begin{array}{c} -2R_2+R_1\to R_1 \\ \to \\ -R_2+R_3\to R_3 \\ R_2+R_4\to R_4 \end{array} \left[\begin{array}{cccc|c} 1 & 0 & -11 & 1 & 13 \\ 0 & 1 & 5 & 0 & -5 \\ 0 & 0 & -1 & 0 & 0 \\ 0 & 0 & 1 & 0 & -8 \end{array}\right]$$

The last two rows state that $x_3 = 0$ and $x_3 = -8$. Thus, there is no solution.

**45.** $$\left[\begin{array}{ccccc|c} 1 & 3 & 4 & -1 & 2 & 1 \\ 1 & -1 & -2 & 1 & 0 & 3 \\ 1 & 2 & 3 & 1 & 4 & 0 \\ 2 & 2 & 2 & 0 & 2 & 4 \end{array}\right] \begin{array}{c} \to \\ R_1-R_2\to R_2 \\ R_1-R_3\to R_3 \\ -2R_1+R_4\to R_4 \end{array} \left[\begin{array}{ccccc|c} 1 & 3 & 4 & -1 & 2 & 1 \\ 0 & 4 & 6 & -2 & 2 & -2 \\ 0 & 1 & 1 & -2 & -2 & 1 \\ 0 & -4 & -6 & 2 & -2 & 2 \end{array}\right] \begin{array}{c} \\ R_2 \leftrightarrow R_3 \\ \\ \\ \end{array}$$

$$\left[\begin{array}{ccccc|c} 1 & 3 & 4 & -1 & 2 & 1 \\ 0 & 1 & 1 & -2 & -2 & 1 \\ 0 & 4 & 6 & -2 & 2 & -2 \\ 0 & -4 & -6 & 2 & -2 & 2 \end{array}\right] \begin{array}{c} -3R_2+R_1\to R_1 \\ \to \\ -4R_2+R_3\to R_3 \\ 4R_2+R_4\to R_4 \end{array} \left[\begin{array}{ccccc|c} 1 & 0 & 1 & 5 & 8 & -2 \\ 0 & 1 & 1 & -2 & -2 & 1 \\ 0 & 0 & 2 & 6 & 10 & -6 \\ 0 & 0 & -2 & -6 & -10 & 6 \end{array}\right] \begin{array}{c} \to \\ \to \\ \frac{1}{2}R_3\to R_3 \\ R_4+R_3\to R_4 \end{array}$$

$$\left[\begin{array}{ccccc|c} 1 & 0 & 1 & 5 & 8 & -2 \\ 0 & 1 & 1 & -2 & -2 & 1 \\ 0 & 0 & 1 & 3 & 5 & -3 \\ 0 & 0 & 0 & 0 & 0 & 0 \end{array}\right] \begin{array}{c} -R_3+R_1\to R_1 \\ -R_3+R_2\to R_2 \\ \to \\ \\ \end{array} \left[\begin{array}{ccccc|c} 1 & 0 & 0 & 2 & 3 & 1 \\ 0 & 1 & 0 & -5 & -7 & 4 \\ 0 & 0 & 1 & 3 & 5 & -3 \\ 0 & 0 & 0 & 0 & 0 & 0 \end{array}\right]$$

There are infinitely many solutions.
General solution: $x_1 = 1 - 2x_4 - 3x_5$, $x_2 = 4 + 5x_4 + 7x_5$, $x_3 = -3 - 3x_4 - 5x_5$

**47.** $$\left[\begin{array}{cc|c} a_1 & b_1 & c_1 \\ a_2 & b_2 & c_2 \end{array}\right] \begin{array}{c} \frac{1}{a_1}R_1\to R_1 \\ \to \end{array} \left[\begin{array}{cc|c} 1 & \frac{b_1}{a_1} & \frac{c_1}{a_1} \\ a_2 & b_2 & c_2 \end{array}\right] \begin{array}{c} \to \\ -a_2R_1+R_2\to R_2 \end{array} \left[\begin{array}{cc|c} 1 & \frac{b_1}{a_1} & \frac{c_1}{a_1} \\ 0 & \frac{b_2a_1-b_1a_2}{a_1} & \frac{c_2a_1-c_1a_2}{a_1} \end{array}\right]$$

$$\begin{array}{c} -\frac{b_1}{a_1}\cdot\frac{a_1}{b_2a_1-b_1a_2}R_2+R_1\to R_1 \\ \to \end{array} \left[\begin{array}{cc|c} 1 & 0 & \frac{b_2c_1-c_2b_1}{a_1b_2-a_2b_1} \\ 0 & \frac{b_2a_1-b_1a_2}{a_1} & \frac{c_2a_1-c_1a_2}{a_1} \end{array}\right] \text{ Thus, } x = \frac{b_2c_1-c_2b_1}{a_1b_2-a_2b_1}.$$

Fractions and messy computations cannot always be avoided.

**49.** Let $x$ = cups of Beef. Let $y$ = cups of Sirloin Burger.
Fat: $2.5x + 7y = 61$, Cholesterol: $15x + 15y = 150$ or $x + y = 10$

Row reducing $\left[\begin{array}{cc|c} 2.5 & 7 & 61 \\ 1 & 1 & 10 \end{array}\right]$ gives $\left[\begin{array}{cc|c} 1 & 0 & 2 \\ 0 & 1 & 8 \end{array}\right]$. They should use 2 cups of beef and 8 cups of Sirloin Burger.

**51. a.** Let $x = 12\%$ investment amount, $y = 10\%$ investment amount, and $z = 8\%$ investment amount.

$$\begin{array}{ll} x + y + z = 235{,}000 & \text{Total investment} \\ 0.12x + 0.10y + 0.08z = 22{,}500 & \text{Investment income} \\ 2x - z = 0 & 2x = z \end{array}$$

$$\left[\begin{array}{ccc|c} 1 & 1 & 1 & 235{,}000 \\ 0.12 & 0.10 & 0.08 & 22{,}500 \\ 2 & 0 & -1 & 0 \end{array}\right] \begin{array}{c} \to \\ 100R_2 \to R_2 \\ \to \end{array} \left[\begin{array}{ccc|c} 1 & 1 & 1 & 235{,}000 \\ 12 & 10 & 8 & 2{,}250{,}000 \\ 2 & 0 & -1 & 0 \end{array}\right] \begin{array}{c} \to \\ -12R_1 + R_2 \to R_2 \\ -2R_1 + R_3 \to R_3 \end{array}$$

$$\left[\begin{array}{ccc|c} 1 & 1 & 1 & 235{,}000 \\ 0 & -2 & -4 & -570{,}000 \\ 0 & -2 & -3 & -470{,}000 \end{array}\right] \begin{array}{c} \to \\ -\frac{1}{2}R_2 \to R_2 \\ \to \end{array} \left[\begin{array}{ccc|c} 1 & 1 & 1 & 235{,}000 \\ 0 & 1 & 2 & 285{,}000 \\ 0 & -2 & -3 & -470{,}000 \end{array}\right] \begin{array}{c} -R_2 + R_1 \to R_1 \\ \to \\ 2R_2 + R_3 \to R_3 \end{array}$$

$$\left[\begin{array}{ccc|c} 1 & 0 & -1 & -50{,}000 \\ 0 & 1 & 2 & 285{,}000 \\ 0 & 0 & 1 & 100{,}000 \end{array}\right] \begin{array}{c} R_3 + R_1 \to R_1 \\ -2R_3 + R_2 \to R_2 \\ \to \end{array} \left[\begin{array}{ccc|c} 1 & 0 & 0 & 50{,}000 \\ 0 & 1 & 0 & 85{,}000 \\ 0 & 0 & 1 & 100{,}000 \end{array}\right]$$

Invest \$50,000 at 12%, \$85,000 at 10% and \$100,000 at 8%.

**b.** Income: $0.12(50{,}000) = \$6000$, $0.10(85{,}000) = \$8{,}500$, and $0.08(100{,}000) = \$8{,}000$

**53.** $AP$ = number of cars rented from Metropolis Airport
$DT$ = number of cars rented from downtown
$CA$ = number of cars rented from City Airport

$$\begin{array}{lcl} AP + DT + CA = 2200 & & AP + DT + CA = 2200 \\ AP = 0.9AP + 0.1DT + 0.1CA & \to & -0.1AP + 0.1DT + 0.1CA = 0 \\ DT = 0.05AP + 0.8DT + 0.05CA & & 0.05AP - 0.2DT + 0.05CA = 0 \end{array}$$

$$\left[\begin{array}{ccc|c} 1 & 1 & 1 & 2200 \\ -0.1 & 0.10 & 0.10 & 0 \\ 0.05 & -0.20 & 0.05 & 0 \end{array}\right] \begin{array}{c} \to \\ 10R_2 \to R_2 \\ 100R_3 \to R_3 \end{array} \left[\begin{array}{ccc|c} 1 & 1 & 1 & 2200 \\ -1 & 1 & 1 & 0 \\ 5 & -20 & 5 & 0 \end{array}\right] \begin{array}{c} \to \\ R_1 + R_2 \to R_2 \\ -5R_1 + R_3 \to R_3 \end{array} \left[\begin{array}{ccc|c} 1 & 1 & 1 & 2200 \\ 0 & 2 & 2 & 2200 \\ 0 & -25 & 0 & -11{,}000 \end{array}\right]$$

$$\begin{array}{c} \to \\ \frac{1}{2}R_2 \to R_2 \\ \to \end{array} \left[\begin{array}{ccc|c} 1 & 1 & 1 & 2200 \\ 0 & 1 & 1 & 1100 \\ 0 & -25 & 0 & -11{,}000 \end{array}\right] \begin{array}{c} -R_2 + R_1 \to R_1 \\ \to \\ 25R_2 + R_3 \to R_3 \end{array} \left[\begin{array}{ccc|c} 1 & 0 & 0 & 1100 \\ 0 & 1 & 1 & 1100 \\ 0 & 0 & 25 & 16{,}500 \end{array}\right] \begin{array}{c} \to \\ \to \\ \frac{1}{25}R_3 \to R_3 \end{array} \left[\begin{array}{ccc|c} 1 & 0 & 0 & 1100 \\ 0 & 1 & 1 & 1100 \\ 0 & 0 & 1 & 660 \end{array}\right]$$

$$\begin{array}{c} \to \\ -R_3 + R_2 \to R_2 \\ \to \end{array} \left[\begin{array}{ccc|c} 1 & 0 & 0 & 1100 \\ 0 & 1 & 0 & 440 \\ 0 & 0 & 1 & 660 \end{array}\right] \quad AP = 1100,\ DT = 440,\ CA = 660$$

**55.** Let $x$ = ounces of AF, $y$ = ounces of FP, $z$ = ounces of NMG
We start directly with the augmented matrix.

$$\begin{array}{r} \text{Calories} \\ \text{Fat} \\ \text{Carbohydrates} \end{array} \left[\begin{array}{ccc|c} 50 & 108 & 127 & 443 \\ 0 & 0.1 & 5.5 & 5.7 \\ 22 & 25.7 & 18 & 113.4 \end{array}\right] \begin{array}{c} \frac{1}{50}R_1 \to R_1 \\ 10R_2 \to R_2 \\ \to \end{array} \left[\begin{array}{ccc|c} 1 & 2.16 & 2.54 & 8.86 \\ 0 & 1 & 55 & 57 \\ 22 & 25.7 & 18 & 113.4 \end{array}\right] \begin{array}{c} \\ \to \\ -22R_1 + R_3 \to R_3 \end{array}$$

$$\left[\begin{array}{ccc|c} 1 & 2.16 & 2.54 & 8.86 \\ 0 & 1 & 55 & 57 \\ 0 & -21.82 & -37.88 & -81.52 \end{array}\right] \begin{array}{c} -2.16R_2 + R_1 \to R_1 \\ \to \\ 21.82R_2 + R_3 \to R_3 \end{array} \left[\begin{array}{ccc|c} 1 & 0 & -116.26 & -114.26 \\ 0 & 1 & 55 & 57 \\ 0 & 0 & 1162.22 & 1162.22 \end{array}\right] \begin{array}{c} \\ \to \\ \frac{1}{1162.22}R_3 \to R_3 \end{array}$$

$$\left[\begin{array}{ccc|c} 1 & 0 & -116.26 & -114.26 \\ 0 & 1 & 55 & 57 \\ 0 & 0 & 1 & 1 \end{array}\right] \begin{array}{c} 116.26R_3 + R_1 \to R_1 \\ -55R_3 + R_2 \to R_2 \\ \to \end{array} \left[\begin{array}{ccc|c} 1 & 0 & 0 & 2 \\ 0 & 1 & 0 & 2 \\ 0 & 0 & 1 & 1 \end{array}\right]$$

Use 2 ounces of AF, 2 ounces of FP, and 1 ounce of NMG.

**57.** $x = \#$ porfolio I $\quad 2x+4y+2z=12$
y= #portfolio II $\quad x+2y+2z=6$
$z = \#$portfolio III $\quad 3y+3z=6$

$$\left[\begin{array}{ccc|c} 2 & 4 & 2 & 12 \\ 1 & 2 & 2 & 6 \\ 0 & 3 & 3 & 6 \end{array}\right] \begin{array}{c} R_1 \to R_2 \\ \to \\ \phantom{x} \end{array}$$

$$\left[\begin{array}{ccc|c} 1 & 2 & 2 & 6 \\ 2 & 4 & 2 & 12 \\ 0 & 3 & 3 & 6 \end{array}\right] \begin{array}{c} \to \\ -2R_1 + R_2 \to R_2 \\ \phantom{x} \end{array} \left[\begin{array}{ccc|c} 1 & 2 & 2 & 6 \\ 0 & 0 & -2 & 0 \\ 0 & 3 & 3 & 6 \end{array}\right] \begin{array}{c} \to \\ -\tfrac{1}{2}R_2 \to R_2 \\ \tfrac{1}{3}R_3 \to R_3 \end{array} \left[\begin{array}{ccc|c} 1 & 2 & 2 & 6 \\ 0 & 0 & 1 & 0 \\ 0 & 1 & 1 & 2 \end{array}\right] \begin{array}{c} -2R_2 + R_1 \to R_1 \\ \to \\ -R_2 + R_3 \to R_3 \end{array}$$

$$\left[\begin{array}{ccc|c} 1 & 2 & 0 & 6 \\ 0 & 0 & 1 & 0 \\ 0 & 1 & 0 & 2 \end{array}\right] \begin{array}{c} R_2 \leftrightarrow R_3 \\ \to \end{array} \left[\begin{array}{ccc|c} 1 & 2 & 0 & 6 \\ 0 & 1 & 0 & 2 \\ 0 & 0 & 1 & 0 \end{array}\right] \begin{array}{c} -2R_2 + R_1 \to R_1 \\ \to \end{array} \left[\begin{array}{ccc|c} 1 & 0 & 0 & 2 \\ 0 & 1 & 0 & 2 \\ 0 & 0 & 1 & 0 \end{array}\right]$$

2 units each of portfolio I and portfolio II.

**59.** $0.1S + 3.4M + 2.2B = 12.1$ iron needed
$8.5S + 22M + 10B = 97$ protein needed
$1S + 20M + 12B = 70$ carbohydrates needed

$$\left[\begin{array}{ccc|c} 0.1 & 3.4 & 2.2 & 12.1 \\ 8.5 & 22 & 10 & 97 \\ 1 & 20 & 12 & 70 \end{array}\right] \begin{array}{c} 10R_1 \to R_1 \\ \to \\ \phantom{x} \end{array}$$

$$\left[\begin{array}{ccc|c} 1 & 34 & 22 & 121 \\ 8.5 & 22 & 10 & 97 \\ 1 & 20 & 12 & 70 \end{array}\right] \begin{array}{c} \to \\ -8.5R_1 + R_2 \to R_2 \\ -R_1 + R_3 \to R_3 \end{array} \left[\begin{array}{ccc|c} 1 & 34 & 22 & 121 \\ 0 & -267 & -177 & -931.5 \\ 0 & -14 & -10 & -51 \end{array}\right] \begin{array}{c} \to \\ -19R_3 + R_2 \to R_2 \\ \phantom{x} \end{array}$$

$$\left[\begin{array}{ccc|c} 1 & 34 & 22 & 121 \\ 0 & -1 & 13 & 37.5 \\ 0 & -14 & -10 & -51 \end{array}\right] \begin{array}{c} 34R_2 + R_1 \to R_1 \\ \to \\ -14R_2 + R_3 \to R_3 \end{array} \left[\begin{array}{ccc|c} 1 & 0 & 464 & 1396 \\ 0 & -1 & 13 & 37.5 \\ 0 & 0 & -192 & -576 \end{array}\right] \begin{array}{c} \phantom{x} \\ \to \\ -\tfrac{1}{192}R_3 \to R_3 \end{array}$$

$$\left[\begin{array}{ccc|c} 1 & 0 & 464 & 1396 \\ 0 & -1 & 13 & 37.5 \\ 0 & 0 & 1 & 3 \end{array}\right] \begin{array}{c} -464R_3 + R_1 \to R_1 \\ -13R_3 + R_2 \to R_2 \\ \to \end{array} \left[\begin{array}{ccc|c} 1 & 0 & 0 & 4 \\ 0 & -1 & 0 & -1.5 \\ 0 & 0 & 1 & 3 \end{array}\right] \begin{array}{c} -R_2 \to R_2 \\ \to \end{array} \left[\begin{array}{ccc|c} 1 & 0 & 0 & 4 \\ 0 & 1 & 0 & 1.5 \\ 0 & 0 & 1 & 3 \end{array}\right]$$

4 glasses of milk, 1.5 quarter-pound servings of meat, and 3 2-slice servings of bread. Note that this is $3/8$ pounds of meat and 6 slices of bread.

**61.** $x =$ Type I bags; $y =$ Type II bags; $z =$ Type III bags; $w =$ Type IV bags

$$\left[\begin{array}{cccc|c} 5 & 5 & 10 & 5 & 10{,}000 \\ 10 & 5 & 30 & 10 & 20{,}000 \\ 5 & 15 & 10 & 25 & 20{,}000 \end{array}\right] \begin{array}{c} \tfrac{1}{5}R_1 \to R_1 \\ \tfrac{1}{5}R_2 \to R_2 \\ \tfrac{1}{5}R_3 \to R_3 \end{array} \left[\begin{array}{cccc|c} 1 & 1 & 2 & 1 & 2000 \\ 2 & 1 & 6 & 2 & 4000 \\ 1 & 3 & 2 & 5 & 4000 \end{array}\right] \begin{array}{c} \to \\ -2R_1 + R_2 \to R_2 \\ -R_1 + R_3 \to R_3 \end{array}$$

$$\left[\begin{array}{cccc|c} 1 & 1 & 2 & 1 & 2000 \\ 0 & -1 & 2 & 0 & 0 \\ 0 & 2 & 0 & 4 & 2000 \end{array}\right] \begin{array}{c} R_2 + R_1 \to R_1 \\ \to \\ 2R_2 + R_3 \to R_3 \end{array} \left[\begin{array}{cccc|c} 1 & 0 & 4 & 1 & 2000 \\ 0 & 1 & 2 & 0 & 0 \\ 0 & 0 & 4 & 4 & 2000 \end{array}\right] \begin{array}{c} \to \\ -R_2 \to R_2 \\ \tfrac{1}{4}R_3 \to R_3 \end{array}$$

$$\left[\begin{array}{cccc|c} 1 & 0 & 4 & 1 & 2000 \\ 0 & 1 & -2 & 0 & 0 \\ 0 & 0 & 1 & 1 & 500 \end{array}\right] \begin{array}{c} -4R_3 + R_1 \to R_1 \\ 2R_3 + R_2 \to R_2 \\ \to \end{array} \left[\begin{array}{cccc|c} 1 & 0 & 0 & -3 & 0 \\ 0 & 1 & 0 & 2 & 1000 \\ 0 & 0 & 1 & 1 & 500 \end{array}\right]$$

Solution: $x = 3w$, $y = 1000 - 2w$, $z = 500 - w$, where $w$ is any non-negative amount $\le 500$.

**63.**

| Nutrient | Species $x_1$ | $x_2$ | $x_3$ | |
|---|---|---|---|---|
| A | 1 | 2 | 2 | 5100 units |
| B | 1 | 0 | 3 | 6900 units |
| C | 2 | 2 | 5 | 12,000 units |

$$\left[\begin{array}{ccc|c} 1 & 2 & 2 & 5100 \\ 1 & 0 & 3 & 6900 \\ 2 & 2 & 5 & 12{,}000 \end{array}\right] \xrightarrow[-2R_1+R_3\to R_3]{-R_1+R_2\to R_2} \left[\begin{array}{ccc|c} 1 & 2 & 2 & 5100 \\ 0 & -2 & 1 & 1800 \\ 0 & -2 & 1 & 1800 \end{array}\right] \xrightarrow[-R_2+R_3\to R_3]{R_2+R_1\to R_1} \left[\begin{array}{ccc|c} 1 & 0 & 3 & 6900 \\ 0 & -2 & 1 & 1800 \\ 0 & 0 & 0 & 0 \end{array}\right]$$

Species $x_1 = 6900 - 3x_3$ and species $x_2 = -900 + \frac{1}{2}x_3$, with any amount of $x_3$ between 1800 and 2300.

**65. a.** Let $x =$ computer shares, $y =$ utility shares, and $z =$ retail shares.

$30x + 44y + 26z = 392{,}000$ Total Cost

$6x + 6y + 2.4z = 0.15(392{,}000)$ Total Growth

$$\left[\begin{array}{ccc|c} 30 & 44 & 26 & 392{,}000 \\ 6 & 6 & 2.4 & 58{,}800 \end{array}\right] \xrightarrow{R_1 \leftrightarrow \frac{1}{6}R_2} \left[\begin{array}{ccc|c} 1 & 1 & 0.4 & 9{,}800 \\ 30 & 44 & 26 & 392{,}000 \end{array}\right] \xrightarrow{-30R_1+R_2\to R_2}$$

$$\left[\begin{array}{ccc|c} 1 & 1 & 0.4 & 9{,}800 \\ 0 & 14 & 14 & 98{,}000 \end{array}\right] \xrightarrow{\frac{1}{14}R_2\to R_2} \left[\begin{array}{ccc|c} 1 & 1 & 0.4 & 9{,}800 \\ 0 & 1 & 1 & 7{,}000 \end{array}\right] \xrightarrow{-R_2+R_1\to R_1} \left[\begin{array}{ccc|c} 1 & 0 & -0.6 & 2{,}800 \\ 0 & 1 & 1 & 7{,}000 \end{array}\right]$$

General solution: $x = 2{,}800 + 0.6z$ and $y = 7000 - z$

**b.** If $z = 1000$, then $x = 3400$ and $y = 6000$. So 3400 shares computer stock and 6000 shares utility stock.

**c.** If they buy no shares of retail stock ($z = 0$), then $x = 2{,}800$ and $y = 7000$. So, the minimum number of shares of computer stock is 2,800 and they would also buy 7,000 shares of utility stock.

**d.** The maximum number of computer shares depends on the maximum value of $z$. $y = 7000 - z$, so $z$ cannot be greater than 7000. If $z = 7000$, then $x = 2800 + 0.6(7000) = 7000$ and $y = 0$. So they would purchase 7,000 computer shares and 7,000 retail shares.

**67.** $x =$ units of Portfolio I $\quad 2x + 4y + 2z = 16$ common stock blocks
$y =$ units of Portfolio II $\quad x + 2y + z = 8$ municipal bond blocks
$z =$ units of Portfolio III $\quad 3y + 3z = 6$ preferred stock blocks

$$\begin{bmatrix} 2 & 4 & 2 & 16 \\ 1 & 2 & 1 & 8 \\ 0 & 3 & 3 & 6 \end{bmatrix} \xrightarrow{R_1 \leftrightarrow R_2}$$

$$\left[\begin{array}{ccc|c} 1 & 2 & 1 & 8 \\ 2 & 4 & 2 & 16 \\ 0 & 3 & 3 & 6 \end{array}\right] \xrightarrow[\frac{1}{3}R_3\to R_3]{-2R_1+R_2\to R_2} \left[\begin{array}{ccc|c} 1 & 2 & 1 & 8 \\ 0 & 0 & 0 & 0 \\ 0 & 1 & 1 & 2 \end{array}\right] \xrightarrow{-2R_3+R_1\to R_1} \left[\begin{array}{ccc|c} 1 & 0 & -1 & 4 \\ 0 & 0 & 0 & 0 \\ 0 & 1 & 1 & 2 \end{array}\right]$$

Thus, $x = z + 4$ and $y = -z + 2$. Note that $z = 0, 1, 2$ only.

| Possible Offerings | I | II | III |
|---|---|---|---|
| | 4 | 2 | 0 |
| | 5 | 1 | 1 |
| | 6 | 0 | 2 |

## *Exercises 3.4*

**1.** The product is the $3\times3$ identity matrix.

**3.** $$\begin{bmatrix} 1 & 2 & 1 \\ 0 & 0 & 3 \\ 1 & 0 & 1 \end{bmatrix}\begin{bmatrix} 0 & -1/3 & 1 \\ 1/2 & 0 & -1/2 \\ 0 & 1/3 & 0 \end{bmatrix} = \begin{bmatrix} 1 & 0 & 0 \\ 0 & 1 & 0 \\ 0 & 0 & 1 \end{bmatrix}$$

Yes, $B = A^{-1}$.

5. $A^{-1} = \frac{1}{(4)(2)-7(1)}\begin{bmatrix} 2 & -7 \\ -1 & 4 \end{bmatrix} = \begin{bmatrix} 2 & -7 \\ -1 & 4 \end{bmatrix}$

7. $ad - bc = 2(2) - (-4)(-1) = 0$, so no inverse exists.

9. $A^{-1} = \frac{1}{4(1)-7(2)}\begin{bmatrix} 1 & -7 \\ -2 & 4 \end{bmatrix} = \begin{bmatrix} -1/10 & 7/10 \\ 1/5 & -2/5 \end{bmatrix}$

11. $\left[\begin{array}{ccc|ccc} 3 & 0 & 0 & 1 & 0 & 0 \\ 0 & 3 & 2 & 0 & 1 & 0 \\ 0 & 0 & 1 & 0 & 0 & 1 \end{array}\right] \begin{array}{c} \frac{1}{3}R_1 \to R_1 \\ \frac{1}{3}R_2 \to R_2 \\ \to \end{array}$

$\left[\begin{array}{ccc|ccc} 1 & 0 & 0 & \frac{1}{3} & 0 & 0 \\ 0 & 1 & \frac{2}{3} & 0 & \frac{1}{3} & 0 \\ 0 & 0 & 1 & 0 & 0 & 1 \end{array}\right] \begin{array}{c} -\frac{2}{3}R_3 + R_2 \to R_2 \\ \to \end{array}$

$\left[\begin{array}{ccc|ccc} 1 & 0 & 0 & \frac{1}{3} & 0 & 0 \\ 0 & 1 & 0 & 0 & \frac{1}{3} & -\frac{2}{3} \\ 0 & 0 & 1 & 0 & 0 & 1 \end{array}\right] A^{-1}: \begin{bmatrix} \frac{1}{3} & 0 & 0 \\ 0 & \frac{1}{3} & -\frac{2}{3} \\ 0 & 0 & 1 \end{bmatrix}$

$AA^{-1} = \begin{bmatrix} 3 & 0 & 0 \\ 0 & 3 & 2 \\ 0 & 0 & 1 \end{bmatrix}\begin{bmatrix} \frac{1}{3} & 0 & 0 \\ 0 & \frac{1}{3} & -\frac{2}{3} \\ 0 & 0 & 1 \end{bmatrix}$

$= \begin{bmatrix} 1 & 0 & 0 \\ 0 & 1 & 0 \\ 0 & 0 & 1 \end{bmatrix}$

13. $\left[\begin{array}{ccc|ccc} 0 & 1 & 0 & 1 & 0 & 0 \\ 1 & 1 & 0 & 0 & 1 & 0 \\ 0 & 1 & 1 & 0 & 0 & 1 \end{array}\right] \begin{array}{c} R_1 \leftrightarrow R_2 \\ \to \end{array}$

$\left[\begin{array}{ccc|ccc} 1 & 1 & 0 & 0 & 1 & 0 \\ 0 & 1 & 0 & 1 & 0 & 0 \\ 0 & 1 & 1 & 0 & 0 & 1 \end{array}\right] \begin{array}{c} -R_2 + R_1 \to R_1 \\ \to \\ -R_2 + R_3 \to R_3 \end{array}$

$\left[\begin{array}{ccc|ccc} 1 & 0 & 0 & -1 & 1 & 0 \\ 0 & 1 & 0 & 1 & 0 & 0 \\ 0 & 0 & 1 & -1 & 0 & 1 \end{array}\right]$

Inverse $= \begin{bmatrix} -1 & 1 & 0 \\ 1 & 0 & 0 \\ -1 & 0 & 1 \end{bmatrix}$

15. $\left[\begin{array}{ccc|ccc} 3 & 1 & 2 & 1 & 0 & 0 \\ 1 & 2 & 3 & 0 & 1 & 0 \\ 1 & 1 & 1 & 0 & 0 & 1 \end{array}\right] \begin{array}{c} R_1 \leftrightarrow R_3 \\ \to \end{array}$ $\left[\begin{array}{ccc|ccc} 1 & 1 & 1 & 0 & 0 & 1 \\ 1 & 2 & 3 & 0 & 1 & 0 \\ 3 & 1 & 2 & 1 & 0 & 0 \end{array}\right] \begin{array}{c} \to \\ -R_1 + R_2 \to R_2 \\ -3R_1 + R_3 \to R_3 \end{array}$

$\left[\begin{array}{ccc|ccc} 1 & 1 & 1 & 0 & 0 & 1 \\ 0 & 1 & 2 & 0 & 1 & -1 \\ 0 & -2 & -1 & 1 & 0 & -3 \end{array}\right] \begin{array}{c} -R_2 + R_1 \to R_1 \\ \to \\ 2R_2 + R_3 \to R_3 \end{array}$ $\left[\begin{array}{ccc|ccc} 1 & 0 & -1 & 0 & -1 & 2 \\ 0 & 1 & 2 & 0 & 1 & -1 \\ 0 & 0 & 3 & 1 & 2 & -5 \end{array}\right] \begin{array}{c} \to \\ \\ \frac{1}{3}R_3 \to R_3 \end{array}$

$\left[\begin{array}{ccc|ccc} 1 & 0 & -1 & 0 & -1 & 2 \\ 0 & 1 & 2 & 0 & 1 & -1 \\ 0 & 0 & 1 & 1/3 & 2/3 & -5/3 \end{array}\right] \begin{array}{c} R_3 + R_1 \to R_1 \\ -2R_3 + R_2 \to R_2 \\ \to \end{array}$ $\left[\begin{array}{ccc|ccc} 1 & 0 & 0 & 1/3 & -1/3 & 1/3 \\ 0 & 1 & 0 & -2/3 & -1/3 & 7/3 \\ 0 & 0 & 1 & 1/3 & 2/3 & -5/3 \end{array}\right]$

Inverse $= \begin{bmatrix} 1/3 & -1/3 & 1/3 \\ -2/3 & -1/3 & 7/3 \\ 1/3 & 2/3 & -5/3 \end{bmatrix}$

17. $\left[\begin{array}{ccc|ccc} 1 & 3 & 5 & 1 & 0 & 0 \\ -1 & -1 & 2 & 0 & 1 & 0 \\ 1 & 5 & 12 & 0 & 0 & 1 \end{array}\right] \begin{array}{c} \to \\ R_1 + R_2 \to R_2 \\ -R_1 + R_3 \to R_3 \end{array}$ $\left[\begin{array}{ccc|ccc} 1 & 3 & 5 & 1 & 0 & 0 \\ 0 & 2 & 7 & 1 & 1 & 0 \\ 0 & 2 & 7 & -1 & 0 & 1 \end{array}\right] \begin{array}{c} \to \\ \\ -R_2 + R_3 \to R_3 \end{array}$ $\left[\begin{array}{ccc|ccc} 1 & 3 & 5 & 1 & 0 & 0 \\ 0 & 2 & 7 & 1 & 1 & 0 \\ 0 & 0 & 0 & -2 & -1 & 1 \end{array}\right]$

There is no inverse since there is

**19.** $\left[\begin{array}{rrr|rrr} 1 & 2 & 4 & 1 & 0 & 0 \\ 1 & -1 & -3 & 0 & 1 & 0 \\ 2 & 1 & 1 & 0 & 0 & 1 \end{array}\right] \begin{array}{l} \rightarrow \\ -R_1+R_2 \rightarrow R_2 \\ -2R_1+R_3 \rightarrow R_3 \end{array} \left[\begin{array}{rrr|rrr} 1 & 2 & 4 & 1 & 0 & 0 \\ 0 & -3 & -7 & -1 & 1 & 0 \\ 0 & -3 & -7 & -2 & 0 & 1 \end{array}\right] \begin{array}{l} \rightarrow \\ \\ -R_2+R_3 \rightarrow R_3 \end{array} \left[\begin{array}{rrr|rrr} 1 & 2 & 4 & 1 & 0 & 0 \\ 0 & -3 & -7 & -1 & 1 & 0 \\ 0 & 0 & 0 & -1 & -1 & 1 \end{array}\right]$

There is no inverse since there is a row of 0's in the original matrix.

**21.** $C^{-1} = \begin{bmatrix} 1 & 1 & 0 & 0 & -1 \\ -3 & 0 & -3 & 1 & 4 \\ 1 & -2 & 0 & 0 & 1 \\ -3 & 1 & -3 & 1 & 3 \\ -8 & 5 & -8 & 2 & 7 \end{bmatrix}$ Use a graphing calculator to find the inverse.

**23.** $AX = \begin{bmatrix} 3 \\ 2 \end{bmatrix} \quad X = A^{-1}\begin{bmatrix} 3 \\ 2 \end{bmatrix} = \begin{bmatrix} 3 & 2 \\ 1 & 1 \end{bmatrix}\begin{bmatrix} 3 \\ 2 \end{bmatrix} = \begin{bmatrix} 13 \\ 5 \end{bmatrix}$

**25.** $AX = \begin{bmatrix} 3 \\ -1 \\ 2 \end{bmatrix} \quad X = A^{-1}\begin{bmatrix} 3 \\ -1 \\ 2 \end{bmatrix} = \begin{bmatrix} 3 & 2 & 1 \\ 1 & 1 & 2 \\ 1 & 2 & 1 \end{bmatrix} \cdot \begin{bmatrix} 3 \\ -1 \\ 2 \end{bmatrix} = \begin{bmatrix} 9 \\ 6 \\ 3 \end{bmatrix}$

**27.** $A \cdot \begin{bmatrix} x \\ y \\ z \end{bmatrix} = \begin{bmatrix} 1 \\ 2 \\ 3 \end{bmatrix} \quad \begin{bmatrix} x \\ y \\ z \end{bmatrix} = A^{-1}\begin{bmatrix} 1 \\ 2 \\ 3 \end{bmatrix} = \begin{bmatrix} -1 & 1 & 0 \\ 1 & 0 & 0 \\ -1 & 0 & 1 \end{bmatrix} \cdot \begin{bmatrix} 1 \\ 2 \\ 3 \end{bmatrix} = \begin{bmatrix} 1 \\ 1 \\ 2 \end{bmatrix} \quad \begin{array}{l} x=1 \\ y=1 \\ z=2 \end{array}$

**29.** $A = \begin{bmatrix} 1 & 2 \\ 3 & 4 \end{bmatrix},\ A^{-1} = \frac{1}{-2}\begin{bmatrix} 4 & -2 \\ -3 & 1 \end{bmatrix} = \begin{bmatrix} -2 & 1 \\ 3/2 & -1/2 \end{bmatrix};\ \begin{bmatrix} x \\ y \end{bmatrix} = A^{-1}\begin{bmatrix} 4 \\ 10 \end{bmatrix} = \begin{bmatrix} -2 & 1 \\ 3/2 & -1/2 \end{bmatrix}\begin{bmatrix} 4 \\ 10 \end{bmatrix} = \begin{bmatrix} 2 \\ 1 \end{bmatrix}$

So, $x = 2, y = 1$.

**31.** $A = \begin{bmatrix} 2 & 1 \\ 3 & 1 \end{bmatrix},\ A^{-1} = \frac{1}{-1}\begin{bmatrix} 1 & -1 \\ -3 & 2 \end{bmatrix} = \begin{bmatrix} -1 & 1 \\ 3 & -2 \end{bmatrix};\ \begin{bmatrix} x \\ y \end{bmatrix} = A^{-1}\begin{bmatrix} 4 \\ 5 \end{bmatrix} = \begin{bmatrix} -1 & 1 \\ 3 & -2 \end{bmatrix}\begin{bmatrix} 4 \\ 5 \end{bmatrix} = \begin{bmatrix} 1 \\ 2 \end{bmatrix}$

So, $x = 1, y = 2$.

**33.** $\left[\begin{array}{rrr|rrr} 1 & 1 & 1 & 1 & 0 & 0 \\ 2 & 1 & 1 & 0 & 1 & 0 \\ 2 & 2 & 1 & 0 & 0 & 1 \end{array}\right] \begin{array}{l} \rightarrow \\ -2R_1+R_2 \rightarrow R_2 \\ -2R_1+R_3 \rightarrow R_3 \end{array} \left[\begin{array}{rrr|rrr} 1 & 1 & 1 & 1 & 0 & 0 \\ 0 & -1 & -1 & -2 & 1 & 0 \\ 0 & 0 & -1 & -2 & 0 & 1 \end{array}\right] \begin{array}{l} \rightarrow \\ -R_2 \rightarrow R_2 \\ -R_3 \rightarrow R_3 \end{array}$

$\left[\begin{array}{rrr|rrr} 1 & 1 & 1 & 1 & 0 & 0 \\ 0 & 1 & 1 & 2 & -1 & 0 \\ 0 & 0 & 1 & 2 & 0 & -1 \end{array}\right] \begin{array}{l} -R_2+R_1 \rightarrow R_1 \\ \rightarrow \\ \end{array} \left[\begin{array}{rrr|rrr} 1 & 0 & 0 & -1 & 1 & 0 \\ 0 & 1 & 1 & 2 & -1 & 0 \\ 0 & 0 & 1 & 2 & 0 & -1 \end{array}\right] \begin{array}{l} \rightarrow \\ -R_3+R_2 \rightarrow R_2 \\ \end{array}$

$\left[\begin{array}{rrr|rrr} 1 & 0 & 0 & -1 & 1 & 0 \\ 0 & 1 & 0 & 0 & -1 & 1 \\ 0 & 0 & 1 & 2 & 0 & -1 \end{array}\right] \qquad \begin{bmatrix} x \\ y \\ z \end{bmatrix} = \begin{bmatrix} -1 & 1 & 0 \\ 0 & -1 & 1 \\ 2 & 0 & -1 \end{bmatrix}\begin{bmatrix} 3 \\ 4 \\ 5 \end{bmatrix} = \begin{bmatrix} 1 \\ 1 \\ 1 \end{bmatrix}$

So, $x = 1,\ y = 1,\ z = 1$.

**35.** $\left[\begin{array}{ccc|ccc} 1 & 1 & 2 & 1 & 0 & 0 \\ 2 & 1 & 1 & 0 & 1 & 0 \\ 2 & 2 & 1 & 0 & 0 & 1 \end{array}\right] \begin{array}{c} \to \\ -2R_1 + R_2 \to R_2 \\ -2R_1 + R_3 \to R_3 \end{array} \left[\begin{array}{ccc|ccc} 1 & 1 & 2 & 1 & 0 & 0 \\ 0 & -1 & -3 & -2 & 1 & 0 \\ 0 & 0 & -3 & -2 & 0 & 1 \end{array}\right] \begin{array}{c} R_2 + R_1 \to R_1 \\ \to \\ -\frac{1}{3}R_3 \to R_3 \end{array}$

$\left[\begin{array}{ccc|ccc} 1 & 0 & -1 & -1 & 1 & 0 \\ 0 & -1 & -3 & -2 & 1 & 0 \\ 0 & 0 & 1 & 2/3 & 0 & -1/3 \end{array}\right] \begin{array}{c} R_3 + R_1 \to R_1 \\ \to \\ 3R_3 + R_2 \to R_2 \end{array} \left[\begin{array}{ccc|ccc} 1 & 0 & 0 & -1/3 & 1 & -1/3 \\ 0 & -1 & 0 & 0 & 1 & -1 \\ 0 & 0 & 1 & 2/3 & 0 & -1/3 \end{array}\right] \begin{array}{c} -R_2 \to R_2 \\ \to \end{array}$

$\left[\begin{array}{ccc|ccc} 1 & 0 & 0 & -1/3 & 1 & -1/3 \\ 0 & 1 & 0 & 0 & -1 & 1 \\ 0 & 0 & 1 & 2/3 & 0 & -1/3 \end{array}\right] \quad \begin{bmatrix} x \\ y \\ z \end{bmatrix} = \begin{bmatrix} -1/3 & 1 & -1/3 \\ 0 & -1 & 1 \\ 2/3 & 0 & -1/3 \end{bmatrix} \begin{bmatrix} 8 \\ 7 \\ 10 \end{bmatrix} = \begin{bmatrix} 1 \\ 3 \\ 2 \end{bmatrix}$

So, $x = 1, \ y = 3, \ z = 2$.

**37.** The graphing calculator yields

$$A^{-1} = \begin{bmatrix} 3 & 2 & 2 & 4 & 3 \\ 1 & 0 & 2 & 2 & 1 \\ 0.5 & 1 & 1 & 3 & 1.5 \\ 2 & 0 & 2 & 2 & 1 \\ 1 & 0 & 0 & 0 & 2 \end{bmatrix}; \quad X = A^{-1} \begin{bmatrix} 0.7 \\ -1.6 \\ 1.275 \\ 1.15 \\ -0.15 \end{bmatrix} = \begin{bmatrix} 5.6 \\ 5.4 \\ 3.25 \\ 6.1 \\ 0.4 \end{bmatrix}$$

**39. a.** $\begin{vmatrix} 1 & 2 \\ 3 & 4 \end{vmatrix} = 1(4) - 3(2) = -2$

**b.** Determinant does not equal zero, so matrix has an inverse.

**41. a.** $\begin{vmatrix} 3 & -1 \\ 12 & -4 \end{vmatrix} = 3(-4) - 12(-1) = 0$

**b.** Determinant equals zero, so matrix does not have an inverse.

**43. a.** Using technology, $\begin{vmatrix} 3 & 2 & 1 \\ -1 & 0 & 2 \\ 0 & 1 & 1 \end{vmatrix} = -5$

**b.** Determinant does not equal zero, so matrix has an inverse.

**45. a.** Using technology, $\begin{vmatrix} 0 & 1 & 2 \\ 3 & 1 & 1 \\ 4 & -1 & 3 \end{vmatrix} = -19$

**b.** Determinant does not equal zero, so matrix has an inverse.

**47.** $A^{-1} = \begin{bmatrix} 11 & -9 \\ -6 & 5 \end{bmatrix}, \ \begin{bmatrix} 11 & -9 \\ -6 & 5 \end{bmatrix} \cdot \begin{bmatrix} 49 & 133 & 270 & 313 \\ 59 & 161 & 327 & 381 \end{bmatrix} = \begin{bmatrix} 8 & 14 & 27 & 14 \\ 1 & 7 & 15 & 27 \end{bmatrix}$

8 1 14 7 27 15 14 27

H A N G   O N

**49.** $A^{-1} = \begin{bmatrix} 1 & 1 & -2 \\ 2 & 1 & -3 \\ -3 & -2 & 6 \end{bmatrix}, \ \begin{bmatrix} 1 & 1 & -2 \\ 2 & 1 & -3 \\ -3 & -2 & 6 \end{bmatrix} \cdot \begin{bmatrix} 47 & 28 & 63 & 56 & 17 \\ 22 & 87 & 66 & 44 & 14 \\ 34 & 46 & 55 & 43 & 15 \end{bmatrix} = \begin{bmatrix} 1 & 23 & 19 & 14 & 1 \\ 14 & 5 & 27 & 27 & 3 \\ 19 & 18 & 9 & 2 & 11 \end{bmatrix}$

1 14 19 23 5 18 19 27 9 14 27 2 1 3 11

A N S W E R S   I N   B A C K

**51.** $\begin{bmatrix}1900\\1700\end{bmatrix}=\begin{bmatrix}2/3 & 1/4\\1/3 & 3/4\end{bmatrix}\begin{bmatrix}x_0\\y_0\end{bmatrix}$ For ease in calculations, use a graphing calculator.

$$\begin{bmatrix}x_0\\y_0\end{bmatrix}=\begin{bmatrix}2/3 & 1/4\\1/3 & 3/4\end{bmatrix}^{-1}\begin{bmatrix}1900\\1700\end{bmatrix}=\begin{bmatrix}9/5 & -3/5\\-4/5 & 8/5\end{bmatrix}\begin{bmatrix}1900\\1700\end{bmatrix}=\begin{bmatrix}2400\\1200\end{bmatrix}$$

**53.** Medication A is given every 4 hours, or 6 times per day. Medication B is given 2 times per day.

The ratio of the dosage of A to the dosage of B is always 5 to 8: $\frac{A}{B}=\frac{5}{8}$ or, rearranging, $8A-5B=0$.

**a.** For Patient I, the total dosage is 50.6 mg per day: $6A+2B=50.6$, or rearranging, $3A+B=25.3$

We need to solve: $\begin{bmatrix}3 & 1\\8 & -5\end{bmatrix}\begin{bmatrix}A\\B\end{bmatrix}=\begin{bmatrix}25.3\\0\end{bmatrix}$ The inverse of $\begin{bmatrix}3 & 1\\8 & -5\end{bmatrix}$ is $-\frac{1}{23}\begin{bmatrix}-5 & -1\\-8 & 3\end{bmatrix}=\begin{bmatrix}5/23 & 1/23\\8/23 & -3/23\end{bmatrix}$.

Patient 1: $\begin{bmatrix}A\\B\end{bmatrix}=\begin{bmatrix}5/23 & 1/23\\8/23 & -3/23\end{bmatrix}\begin{bmatrix}25.3\\0.0\end{bmatrix}=\begin{bmatrix}5.5\\8.8\end{bmatrix}$ 5.5 mg of $A$, 8.8 mg of $B$

**b.** For Patient II, the total dosage is 92.0 mg per day: $6A+2B=92.0$, or rearranging, $3A+B=46.0$

We need to solve: $\begin{bmatrix}3 & 1\\8 & -5\end{bmatrix}\begin{bmatrix}A\\B\end{bmatrix}=\begin{bmatrix}46.0\\0\end{bmatrix}$

Using the inverse found in part a:

Patient II: $\begin{bmatrix}A\\B\end{bmatrix}=\begin{bmatrix}5/23 & 1/23\\8/23 & -3/23\end{bmatrix}\begin{bmatrix}46.0\\0\end{bmatrix}=\begin{bmatrix}10\\16\end{bmatrix}$ 10 mg of A, 16 mg of B

**55.** $x$ = amount invested at 10%.
$y$ = amount invested at 18%.
$x+y=145{,}600$ Total investment
$0.10x+0.18y=20{,}000$ Total income

$$\begin{bmatrix}x\\y\end{bmatrix}=\begin{bmatrix}1 & 1\\0.10 & 0.18\end{bmatrix}^{-1}\begin{bmatrix}145{,}600\\20{,}000\end{bmatrix}=\begin{bmatrix}77{,}600\\68{,}000\end{bmatrix}$$

\$77,600 at 10% and \$68,000 at 18%
Note - Use the graphing utility to find the inverse, then multiply.

**57. a.** Use the graphing utility to find the inverse.

$$\begin{bmatrix}\text{Deluxe}\\\text{Premium}\\\text{Ultimate}\end{bmatrix}=\begin{bmatrix}1.6 & 2 & 2.4\\2 & 3 & 4\\0.5 & 0.5 & 1\end{bmatrix}^{-1}\cdot\begin{bmatrix}96\\156\\37\end{bmatrix}=\begin{bmatrix}2.5 & -2 & 2\\0 & 1 & -4\\-1.25 & 0.5 & 2\end{bmatrix}\cdot\begin{bmatrix}96\\156\\37\end{bmatrix}=\begin{bmatrix}2\\8\\32\end{bmatrix}$$

Produce 2 deluxe, 8 premium, and 32 ultimate models.

**b.**
$$\begin{bmatrix}\text{Deluxe}\\\text{Premium}\\\text{Ultimate}\end{bmatrix}=\begin{bmatrix}1.6 & 2 & 2.4\\2 & 3 & 4\\0.5 & 0.5 & 1\end{bmatrix}^{-1}\cdot\begin{bmatrix}96+8\\156\\37\end{bmatrix}=\begin{bmatrix}2.5 & -2 & 2\\0 & 1 & -4\\-1.25 & 0.5 & 2\end{bmatrix}\cdot\begin{bmatrix}104\\156\\37\end{bmatrix}=\begin{bmatrix}22\\8\\22\end{bmatrix}$$

Produce 22 deluxe, 8 premium, and 22 ultimate models.

$[\text{New}]=[\text{Old}]+8[\text{Col.1 of }A^{-1}]$

**59.** $x = 8\%$ investment, $y = 10\%$ investment, $z = 6\%$ investment

$x + y + z = 1,000,000$ Total investment

$0.08x + 0.10y + 0.06z = 86,000$ Investment income

$y = x + z$ Third condition

Use the graphing utility to find the inverse.

$$\begin{bmatrix} x \\ y \\ z \end{bmatrix} = \begin{bmatrix} 1 & 1 & 1 \\ 0.08 & 0.10 & 0.06 \\ 1 & -1 & 1 \end{bmatrix}^{-1} \cdot \begin{bmatrix} 1,000,000 \\ 86,000 \\ 0 \end{bmatrix} = \begin{bmatrix} -4 & 50 & 1 \\ 0.5 & 0 & -0.5 \\ 4.5 & -50 & -0.5 \end{bmatrix} \cdot \begin{bmatrix} 1,000,000 \\ 86,000 \\ 0 \end{bmatrix} = \begin{bmatrix} 300,000 \\ 500,000 \\ 200,000 \end{bmatrix}$$

\$200,000 at 6%, \$300,000 at 8%, and \$500,000 at 10%

**61.** **a.** $$\begin{bmatrix} a & b & c \\ d & e & f \\ g & h & i \end{bmatrix} \begin{bmatrix} n_t \\ n_{t+1} \\ n_{t+2} \end{bmatrix} = \begin{bmatrix} n_{t+1} \\ n_{t+2} \\ n_t + n_{t+1} + n_{t+2} \end{bmatrix}$$

$an_t + bn_{t+1} + cn_{t+2} = n_{t+1}$, etc.

implies $a = 0,\ b = 1,\ c = 0,\ d = 0,\ e = 0,\ f = 1,\ g = h = i = 1$

$$M = \begin{bmatrix} 0 & 1 & 0 \\ 0 & 0 & 1 \\ 1 & 1 & 1 \end{bmatrix}$$

**b.** $$M^{-1} = \begin{bmatrix} -1 & -1 & 1 \\ 1 & 0 & 0 \\ 0 & 1 & 0 \end{bmatrix} \qquad \begin{bmatrix} n_t \\ n_{t+1} \\ n_{t+2} \end{bmatrix} = M^{-1} \begin{bmatrix} 191 \\ 346 \\ 645 \end{bmatrix} = \begin{bmatrix} 108 \\ 191 \\ 346 \end{bmatrix}$$

There were 108 visitors on the day before there were 191 visitors.

**63.** **a.** See exercise 61 where $M$ was found to be $\begin{bmatrix} 0 & 1 & 0 \\ 0 & 0 & 1 \\ 1 & 1 & 1 \end{bmatrix}$.

**b.** $$M^{-1} = \begin{bmatrix} -1 & -1 & 1 \\ 1 & 0 & 0 \\ 0 & 1 & 0 \end{bmatrix}$$

At the end of 3 successive 1-hour periods, the value for $N$ was $N = \begin{bmatrix} 200 \\ 370 \\ 600 \end{bmatrix}$.

$$M^{-1}N = \begin{bmatrix} -1 & -1 & 1 \\ 1 & 0 & 0 \\ 0 & 1 & 0 \end{bmatrix} \begin{bmatrix} 200 \\ 370 \\ 600 \end{bmatrix} = \begin{bmatrix} 30 \\ 200 \\ 370 \end{bmatrix}$$ So, the population was 30 one hour before it was 200.

## *Exercises 3.5*

**1.** **a.** Row 3, column 2 = 0.15 100(0.15) = 15
**b.** Row 4, column 1 = 0.10 40(0.10) = 4

**3.** 1000(0.008) = 8.

**5.** 1000(0.040) = 40.

**7.** Most dependent would be the largest entry on the main diagonal. Raw materials is the most self dependent. Likewise, Fuels is least dependent.

**9.** The largest entries in Row 2 give the industries most affected by a rise in raw material cost. These are Raw materials, Manufacturing, and Service industries.

**11.** $D=\begin{bmatrix}96\\8\end{bmatrix}$. $(I-A)X=D$ or $\begin{bmatrix}0.5 & -0.1\\-0.1 & 0.7\end{bmatrix}\begin{bmatrix}P\\M\end{bmatrix}=\begin{bmatrix}96\\8\end{bmatrix}$

$$\begin{bmatrix}P\\M\end{bmatrix}=\begin{bmatrix}0.5 & -0.1\\-0.1 & 0.7\end{bmatrix}^{-1}\cdot\begin{bmatrix}96\\8\end{bmatrix}=\frac{1}{0.34}\cdot\begin{bmatrix}0.7 & 0.1\\0.1 & 0.5\end{bmatrix}\begin{bmatrix}96\\8\end{bmatrix}=\frac{1}{0.34}\begin{bmatrix}68\\13.6\end{bmatrix}=\begin{bmatrix}200\\40\end{bmatrix}$$

$P=200$ units of farm products and $M=40$ units of machinery.

**13.** $D=\begin{bmatrix}80\\180\end{bmatrix}$. $(I-A)X=D$ or

$$\begin{bmatrix}0.7 & -0.15\\-0.3 & 0.6\end{bmatrix}\begin{bmatrix}U\\M\end{bmatrix}=\begin{bmatrix}80\\180\end{bmatrix}$$

$$\begin{bmatrix}U\\M\end{bmatrix}=\begin{bmatrix}0.7 & -0.15\\-0.3 & 0.6\end{bmatrix}^{-1}\cdot\begin{bmatrix}80\\180\end{bmatrix}=\frac{1}{0.375}\begin{bmatrix}0.6 & 0.15\\0.3 & 0.7\end{bmatrix}\begin{bmatrix}80\\180\end{bmatrix}=\begin{bmatrix}200\\400\end{bmatrix}\begin{matrix}\text{Utility}\\\text{Manufacturing}\end{matrix}$$

1.

**15. a.** $A=\begin{bmatrix}0.4 & 0.2\\0.2 & 0.1\end{bmatrix}$ $\quad D=\begin{bmatrix}0\\610\end{bmatrix}$ $\quad X=\begin{bmatrix}x_1\\x_2\end{bmatrix}$, where

$x_1$ = gross production for agriculture and $x_2$ = gross production for oil.

$$X=(I-A)^{-1}D \quad I-A=\begin{bmatrix}1 & 0\\0 & 1\end{bmatrix}-\begin{bmatrix}0.4 & 0.2\\0.2 & 0.1\end{bmatrix}=\begin{bmatrix}0.6 & -0.2\\-0.2 & 0.9\end{bmatrix}$$

$$(I-A)^{-1}=\begin{bmatrix}1.8 & 0.4\\0.4 & 1.2\end{bmatrix}$$

$$X=(I-A)^{-1}D=\begin{bmatrix}1.8 & 0.4\\0.4 & 1.2\end{bmatrix}\begin{bmatrix}0\\610\end{bmatrix}=\begin{bmatrix}244\\732\end{bmatrix}$$ Agriculture Products = 244; Oil Products = 732

**b.** The additional production needed from each industry for 1 more unit of oil surplus is

$$1\cdot\begin{bmatrix}\text{Col 2 of}\\(I-A)^{-1}\end{bmatrix}=\begin{bmatrix}0.4\\1.2\end{bmatrix}$$ Agriculture Products = 0.4; Oil Products = 1.2

**17. a.** $D=\begin{bmatrix}36\\278\end{bmatrix}$. $(I-A)X=D$ or $\begin{bmatrix}0.8 & -0.1\\-0.6 & 0.7\end{bmatrix}\begin{bmatrix}\text{MINE}\\\text{MFG}\end{bmatrix}=\begin{bmatrix}36\\278\end{bmatrix}$

$$\begin{bmatrix}\text{MINE}\\\text{MFG}\end{bmatrix}=\begin{bmatrix}0.8 & -0.1\\-0.6 & 0.7\end{bmatrix}^{-1}\cdot\begin{bmatrix}36\\278\end{bmatrix}=\frac{1}{0.5}\begin{bmatrix}0.7 & 0.1\\0.6 & 0.8\end{bmatrix}\begin{bmatrix}36\\278\end{bmatrix}=\begin{bmatrix}1.4 & 0.2\\1.2 & 1.6\end{bmatrix}\begin{bmatrix}36\\278\end{bmatrix}=\begin{bmatrix}106\\488\end{bmatrix}\begin{matrix}\text{Mining}\\\text{Manufacturing}\end{matrix}$$

**b.** The additional production needed from each industry for 1 more unit of mining surplus is

$$1\cdot\begin{bmatrix}\text{Col 1 of}\\(I-A)^{-1}\end{bmatrix}=\begin{bmatrix}1.4\\1.2\end{bmatrix}$$

**19. a.** Let $E$ = gross production for electronics and $C$ = gross production for computers.

$$\begin{matrix} & E & C\end{matrix}$$
$$A=\begin{bmatrix}0.3 & 0.6\\0.2 & 0.2\end{bmatrix}$$

**b.** $D=\begin{bmatrix}648\\16\end{bmatrix}$ $x=\begin{bmatrix}E\\C\end{bmatrix}$ $X=(I-A)^{-1}D$

$$I-A=\begin{bmatrix}1&0\\0&1\end{bmatrix}-\begin{bmatrix}0.3&0.6\\0.2&0.2\end{bmatrix}=\begin{bmatrix}0.7&-0.6\\-0.2&0.8\end{bmatrix} \qquad (I-A)^{-1}=\frac{1}{0.44}\begin{bmatrix}0.8&0.6\\0.2&0.7\end{bmatrix}=\begin{bmatrix}\frac{20}{11}&\frac{15}{11}\\\frac{5}{11}&\frac{35}{22}\end{bmatrix}$$

$$X=(I-A)^{-1}D=\begin{bmatrix}\frac{20}{11}&\frac{15}{11}\\\frac{5}{11}&\frac{35}{22}\end{bmatrix}\begin{bmatrix}648\\16\end{bmatrix}=\begin{bmatrix}1200\\320\end{bmatrix}$$

Gross production for electronics is 1200 units and for computers is 320 units.

**21. a.** $A=\begin{bmatrix}0.30&0.04\\0.35&0.10\end{bmatrix}$

**b.** $D=\begin{bmatrix}20\\1090\end{bmatrix}$. $(I-A)X=D$ or $\begin{bmatrix}0.7&-0.04\\-0.35&0.9\end{bmatrix}\begin{bmatrix}\text{Fish}\\\text{Oil}\end{bmatrix}=\begin{bmatrix}20\\1090\end{bmatrix}$

$$\begin{bmatrix}\text{Fish}\\\text{Oil}\end{bmatrix}=\begin{bmatrix}0.7&-0.04\\-0.35&0.9\end{bmatrix}^{-1}\cdot\begin{bmatrix}20\\1090\end{bmatrix}=\frac{1}{0.616}\begin{bmatrix}0.9&0.04\\0.35&0.7\end{bmatrix}\begin{bmatrix}20\\1090\end{bmatrix}=\begin{bmatrix}100\\1250\end{bmatrix}\begin{matrix}\text{Fishing}\\\text{Oil}\end{matrix}$$

**23.** $\begin{bmatrix}x\\y\end{bmatrix}=\begin{bmatrix}0&0.05\\0.1&0\end{bmatrix}\begin{bmatrix}x\\y\end{bmatrix}+\begin{bmatrix}20,400\\9900\end{bmatrix}$ or $\begin{bmatrix}1&-0.05\\-0.1&1\end{bmatrix}\begin{bmatrix}x\\y\end{bmatrix}=\begin{bmatrix}20,400\\9900\end{bmatrix}$

$$\begin{bmatrix}x\\y\end{bmatrix}=\frac{1}{1-0.005}\begin{bmatrix}1&0.05\\0.1&1\end{bmatrix}\begin{bmatrix}20,400\\9900\end{bmatrix}=\frac{1}{0.995}\begin{bmatrix}20,895\\11,940\end{bmatrix}=\begin{bmatrix}21,000\\12,000\end{bmatrix}$$

Costs for development are \$21,000. Costs for promotional are \$12,000.

**25.** $\begin{bmatrix}E\\C\end{bmatrix}=\begin{bmatrix}11,750\\10,000\end{bmatrix}+\begin{bmatrix}0&0.25\\0.2&0\end{bmatrix}\begin{bmatrix}E\\C\end{bmatrix}$

$\begin{bmatrix}1&-0.25\\-0.2&1\end{bmatrix}\begin{bmatrix}E\\C\end{bmatrix}=\begin{bmatrix}11,750\\10,000\end{bmatrix}$ or

$$\begin{bmatrix}E\\C\end{bmatrix}=\frac{1}{1-0.05}\begin{bmatrix}1&0.25\\0.2&1\end{bmatrix}\begin{bmatrix}11,750\\10,000\end{bmatrix}=\begin{bmatrix}15,000\\13,000\end{bmatrix}$$

Engineering costs are \$15,000.
Computer costs are \$13,000.

**27.** $D=\begin{bmatrix}110\\50\\50\end{bmatrix}$

$(I-A)X=D$ **or** $X=(I-A)^{-1}\cdot D$

$$X=\begin{bmatrix}0.5&-0.1&-0.1\\-0.3&0.5&-0.2\\-0.1&-0.3&0.6\end{bmatrix}^{-1}\cdot\begin{bmatrix}110\\50\\50\end{bmatrix}=\begin{bmatrix}2.79&1.05&0.81\\2.33&3.37&1.51\\1.63&1.86&2.56\end{bmatrix}\cdot\begin{bmatrix}110\\50\\50\end{bmatrix}=\begin{bmatrix}400\\500\\400\end{bmatrix}$$

400 units of fishing output, 500 units of agricultural goods and 400 units of mining goods are needed.

**29.** $D=\begin{bmatrix}100\\272\\200\end{bmatrix}$. $(I-A)X=D$ or $\left[\begin{array}{ccc|c}0.4 & -0.2 & -0.2 & 100\\ -0.1 & 0.6 & -0.5 & 272\\ -0.1 & -0.2 & 0.8 & 200\end{array}\right]$ is the required augmented matrix.

By reducing this to $\left[\begin{array}{ccc|c}1 & 0 & 0 & 1240\\ 0 & 1 & 0 & 1260\\ 0 & 0 & 1 & 720\end{array}\right]$ we have 1240 electronics, 1260 steel, and 720 autos.

**31.** $D=\begin{bmatrix}24\\62\\32\end{bmatrix}$. $(I-A)X=D$ or $X=(I-A)^{-1}\cdot D$

$$X=\begin{bmatrix}0.6 & -0.1 & -0.1\\ -0.2 & 0.5 & -0.2\\ -0.2 & -0.1 & 0.7\end{bmatrix}^{-1}\cdot\begin{bmatrix}24\\62\\32\end{bmatrix}=\begin{bmatrix}1.964 & 0.476 & 0.417\\ 1.071 & 2.381 & 0.833\\ 0.714 & 0.476 & 1.667\end{bmatrix}\cdot\begin{bmatrix}24\\62\\32\end{bmatrix}=\begin{bmatrix}90\\200\\100\end{bmatrix}$$

Service = 90; Manufacturing = 200; Agriculture = 100

**33.** 0.5P + 0.1M + 0.2H = P  
0.1P + 0.3M + 0.0H = M  
0.4P + 0.6M + 0.8H = H

or

–0.5P + 0.1M + 0.2H = 0  
0.1P – 0.7M + 0.0H = 0  
0.4P + 0.6M – 0.2H = 0

$$\left[\begin{array}{ccc|c}1 & -7 & 0 & 0\\ -5 & 1 & 2 & 0\\ 4 & 6 & -2 & 0\end{array}\right]\begin{array}{c}\to\\ 5R_1+R_2\to R_2\\ -4R_1+R_3\to R_3\end{array}\left[\begin{array}{ccc|c}1 & -7 & 0 & 0\\ 0 & -34 & 2 & 0\\ 0 & 34 & -2 & 0\end{array}\right]\begin{array}{c}\\ \to\\ R_2+R_3\to R_3\end{array}$$

$$\left[\begin{array}{ccc|c}1 & -7 & 0 & 0\\ 0 & -34 & 2 & 0\\ 0 & 0 & 0 & 0\end{array}\right]\begin{array}{c}\to\\ -\frac{1}{34}R_2\to R_2\\ \end{array}\left[\begin{array}{ccc|c}1 & -7 & 0 & 0\\ 0 & 1 & -1/17 & 0\\ 0 & 0 & 0 & 0\end{array}\right]\begin{array}{c}7R_2+R_1\to R_1\\ \to\\ \end{array}\left[\begin{array}{ccc|c}1 & 0 & -7/17 & 0\\ 0 & 1 & -1/17 & 0\\ 0 & 0 & 0 & 0\end{array}\right]$$

So, Farm Products = $\frac{7}{17}$ Households and Farm Machinery = $\frac{1}{17}$ Households.

**35.** 0.4G + 0.2I + 0.2H = G  
0.2G + 0.3I + 0.3H = I  
0.4G + 0.5I + 0.5H = H

or

–0.6G + 0.2I + 0.2H = 0  
0.2G – 0.7I + 0.3H = 0  
0.4G + 0.5I – 0.5H = 0

$$\left[\begin{array}{ccc|c}2 & -7 & 3 & 0\\ -6 & 2 & 2 & 0\\ 4 & 5 & -5 & 0\end{array}\right]\begin{array}{c}\\ 3R_1+R_2\to R_2\\ -2R_1+R_3\to R_3\end{array}\left[\begin{array}{ccc|c}2 & -7 & 3 & 0\\ 0 & -19 & 11 & 0\\ 0 & 19 & -11 & 0\end{array}\right]\begin{array}{c}\to\\ \\ R_2+R_3\to R_3\end{array}\left[\begin{array}{ccc|c}2 & -7 & 3 & 0\\ 0 & -19 & 11 & 0\\ 0 & 0 & 0 & 0\end{array}\right]\begin{array}{c}\\ -\frac{1}{19}R_2\to R_2\\ \to\end{array}$$

$$\left[\begin{array}{ccc|c}2 & -7 & 3 & 0\\ 0 & 1 & -11/19 & 0\\ 0 & 0 & 0 & 0\end{array}\right]\begin{array}{c}7R_2+R_1\to R_1\\ \to\end{array}\left[\begin{array}{ccc|c}2 & 0 & -20/19 & 0\\ 0 & 1 & -11/19 & 0\\ 0 & 0 & 0 & 0\end{array}\right]\begin{array}{c}\frac{1}{2}R_1\to R_1\\ \to\end{array}\left[\begin{array}{ccc|c}1 & 0 & -10/19 & 0\\ 0 & 1 & -11/19 & 0\\ 0 & 0 & 0 & 0\end{array}\right]$$

Government = $\frac{10}{19}$ Households. Industry = $\frac{11}{19}$ Households.

**37. a.** $A = \begin{bmatrix} 0.5 & 0.4 & 0.3 \\ 0.4 & 0.5 & 0.3 \\ 0.1 & 0.1 & 0.4 \end{bmatrix}$ $\begin{matrix} \text{Manufacturing} \\ \text{Utilities} \\ \text{Households} \end{matrix}$

**b.** $\begin{matrix} 0.5M + 0.4U + 0.3H = M \\ 0.4M + 0.5U + 0.3H = U \\ 0.1M + 0.1U + 0.4H = H \end{matrix} \quad \rightarrow \quad \begin{matrix} -0.5M + 0.4U + 0.3H = 0 \\ 0.4M - 0.5U + 0.3H = 0 \\ 0.1M + 0.1U - 0.6H = 0 \end{matrix}$

$$\left[\begin{array}{rrr|r} 1 & 1 & -6 & 0 \\ 4 & -5 & 3 & 0 \\ -5 & 4 & 3 & 0 \end{array}\right] \begin{matrix} \\ -4R_1 + R_2 \rightarrow R_2 \\ 5R_1 + R_3 \rightarrow R_3 \end{matrix} \left[\begin{array}{rrr|r} 1 & 1 & -6 & 0 \\ 0 & -9 & 27 & 0 \\ 0 & 9 & -27 & 0 \end{array}\right] \begin{matrix} \rightarrow \\ \\ R_2 + R_3 \rightarrow R_3 \end{matrix} \left[\begin{array}{rrr|r} 1 & 1 & -6 & 0 \\ 0 & -9 & 27 & 0 \\ 0 & 0 & 0 & 0 \end{array}\right]$$

$$\begin{matrix} \rightarrow \\ -\frac{1}{9}R_2 \rightarrow R_2 \\ \end{matrix} \left[\begin{array}{rrr|r} 1 & 1 & -6 & 0 \\ 0 & 1 & -3 & 0 \\ 0 & 0 & 0 & 0 \end{array}\right] \begin{matrix} -R_2 + R_1 \rightarrow R_1 \\ \rightarrow \\ \end{matrix} \left[\begin{array}{rrr|r} 1 & 0 & -3 & 0 \\ 0 & 1 & -3 & 0 \\ 0 & 0 & 0 & 0 \end{array}\right]$$

Manufacturing = 3 Households and Utilities = 3 Households.

**39.** $$\left[\begin{array}{rrrrrr|r} 1 & 0 & 0 & 0 & 0 & 0 & 24 \\ -4 & 1 & 0 & 0 & 0 & 0 & 0 \\ -1 & 0 & 1 & 0 & 0 & 0 & 0 \\ 0 & -1 & -1 & 1 & 0 & 0 & 0 \\ 0 & -4 & -4 & 0 & 1 & 0 & 12 \\ -20 & -24 & -24 & 0 & 0 & 1 & 96 \end{array}\right] \begin{matrix} \rightarrow \\ 4R_1 + R_2 \rightarrow R_2 \\ R_1 + R_3 \rightarrow R_3 \\ \\ \\ 20R_1 + R_6 \rightarrow R_6 \end{matrix} \left[\begin{array}{rrrrrr|r} 1 & 0 & 0 & 0 & 0 & 0 & 24 \\ 0 & 1 & 0 & 0 & 0 & 0 & 96 \\ 0 & 0 & 1 & 0 & 0 & 0 & 24 \\ 0 & -1 & -1 & 1 & 0 & 0 & 0 \\ 0 & -4 & -4 & 0 & 1 & 0 & 12 \\ 0 & -24 & -24 & 0 & 0 & 1 & 576 \end{array}\right] \begin{matrix} \\ \rightarrow \\ \\ R_2 + R_4 \rightarrow R_4 \\ 4R_2 + R_5 \rightarrow R_5 \\ 24R_2 + R_6 \rightarrow R_6 \end{matrix}$$

$$\left[\begin{array}{rrrrrr|r} 1 & 0 & 0 & 0 & 0 & 0 & 24 \\ 0 & 1 & 0 & 0 & 0 & 0 & 96 \\ 0 & 0 & 1 & 0 & 0 & 0 & 24 \\ 0 & 0 & -1 & 1 & 0 & 0 & 96 \\ 0 & 0 & -4 & 0 & 1 & 0 & 396 \\ 0 & 0 & -24 & 0 & 0 & 1 & 2880 \end{array}\right] \begin{matrix} \\ \rightarrow \\ \\ R_3 + R_4 \rightarrow R_4 \\ 4R_3 + R_5 \rightarrow R_5 \\ 24R_3 + R_6 \rightarrow R_6 \end{matrix} \left[\begin{array}{rrrrrr|r} 1 & 0 & 0 & 0 & 0 & 0 & 24 \\ 0 & 1 & 0 & 0 & 0 & 0 & 96 \\ 0 & 0 & 1 & 0 & 0 & 0 & 24 \\ 0 & 0 & 0 & 1 & 0 & 0 & 120 \\ 0 & 0 & 0 & 0 & 1 & 0 & 492 \\ 0 & 0 & 0 & 0 & 0 & 1 & 3456 \end{array}\right]$$

Therefore, 120 panels, 492 braces, and 3456 bolts are required to fill the order.

**41.** $D = \begin{bmatrix} 10 \\ 0 \\ 0 \\ 6 \\ 0 \\ 6 \\ 100 \end{bmatrix}$ $X = \begin{bmatrix} x_1 \\ x_2 \\ x_3 \\ x_4 \\ x_5 \\ x_6 \\ x_7 \end{bmatrix} = \begin{bmatrix} \text{total sawhorses} \\ \text{total tops} \\ \text{total leg pairs} \\ \text{total } 2 \times 4\text{s} \\ \text{total braces} \\ \text{total clamps} \\ \text{total nails} \end{bmatrix}$ Solve $(I - A)X = D$

$$\left[\begin{array}{rrrrrrr|r} 1 & 0 & 0 & 0 & 0 & 0 & 0 & 10 \\ -1 & 1 & 0 & 0 & 0 & 0 & 0 & 0 \\ -2 & 0 & 1 & 0 & 0 & 0 & 0 & 0 \\ 0 & -1 & -2 & 1 & 0 & 0 & 0 & 6 \\ 0 & 0 & -1 & 0 & 1 & 0 & 0 & 0 \\ 0 & 0 & -1 & 0 & 0 & 1 & 0 & 6 \\ -4 & 0 & -8 & 0 & 0 & 0 & 1 & 100 \end{array}\right] \begin{matrix} \\ R_1 + R_2 \rightarrow R_2 \\ 2R_1 + R_3 \rightarrow R_3 \\ \\ \rightarrow \\ \\ 4R_1 + R_7 \rightarrow R_7 \end{matrix} \left[\begin{array}{rrrrrrr|r} 1 & 0 & 0 & 0 & 0 & 0 & 0 & 10 \\ 0 & 1 & 0 & 0 & 0 & 0 & 0 & 10 \\ 0 & 0 & 1 & 0 & 0 & 0 & 0 & 20 \\ 0 & -1 & -2 & 1 & 0 & 0 & 0 & 6 \\ 0 & 0 & -1 & 0 & 1 & 0 & 0 & 0 \\ 0 & 0 & -1 & 0 & 0 & 1 & 0 & 6 \\ 0 & 0 & -8 & 0 & 0 & 0 & 1 & 140 \end{array}\right] \begin{matrix} \\ \rightarrow \\ \\ R_2 + R_4 \rightarrow R_4 \\ \\ \\ \\ \end{matrix}$$

$$\left[\begin{array}{ccccccc|c} 1 & 0 & 0 & 0 & 0 & 0 & 0 & 10 \\ 0 & 1 & 0 & 0 & 0 & 0 & 0 & 10 \\ 0 & 0 & 1 & 0 & 0 & 0 & 0 & 20 \\ 0 & 0 & -2 & 1 & 0 & 0 & 0 & 16 \\ 0 & 0 & -1 & 0 & 1 & 0 & 0 & 0 \\ 0 & 0 & -1 & 0 & 0 & 0 & 10 & 6 \\ 0 & 0 & -8 & 0 & 0 & 0 & 1 & 140 \end{array}\right] \begin{array}{c} \\ \to \\ \\ 2R_3 + R_4 \to R_4 \\ R_3 + R_5 \to R_5 \\ R_3 + R_6 \to R_6 \\ 8R_3 + R_7 \to R_7 \end{array} \left[\begin{array}{ccccccc|c} 1 & 0 & 0 & 0 & 0 & 0 & 0 & 10 \\ 0 & 1 & 0 & 0 & 0 & 0 & 0 & 10 \\ 0 & 0 & 1 & 0 & 0 & 0 & 0 & 20 \\ 0 & 0 & 0 & 1 & 0 & 0 & 0 & 56 \\ 0 & 0 & 0 & 0 & 1 & 0 & 0 & 20 \\ 0 & 0 & 0 & 0 & 0 & 1 & 0 & 26 \\ 0 & 0 & 0 & 0 & 0 & 0 & 1 & 300 \end{array}\right]$$

To fill the order, 56 $2 \times 4$s , 20 braces, 26 clamps, and 300 nails.

# Chapter 3: Matrices

## *Chapter 3 Review Exercises*

**1.** $a_{12} = 4$

**2.** $b_{23} = 0$

**3.** $A$ and $B$

**4.** None

**5.** $D, F, G, I$

**6.** The negative of $B$ is $\begin{bmatrix} -2 & 5 & 11 & -8 \\ -4 & 0 & 0 & -4 \\ 2 & 2 & -1 & -9 \end{bmatrix}$.

**7.** Zero matrix

**8.** Two matrices can be added if they have the same order.

**9.** $A + B$

$$= \begin{bmatrix} 4 & 4 & 2 & -5 \\ 6 & 3 & -1 & 0 \\ 0 & 0 & -3 & 5 \end{bmatrix} + \begin{bmatrix} 2 & -5 & -11 & 8 \\ 4 & 0 & 0 & 4 \\ -2 & -2 & 1 & 9 \end{bmatrix}$$

$$= \begin{bmatrix} 6 & -1 & -9 & 3 \\ 10 & 3 & -1 & 4 \\ -2 & -2 & -2 & 14 \end{bmatrix}$$

**10.** $C - E = \begin{bmatrix} 4 & -2 \\ 5 & 0 \\ 6 & 0 \\ 1 & 3 \end{bmatrix} - \begin{bmatrix} 1 & 1 \\ 1 & 1 \\ 4 & 6 \\ 0 & 5 \end{bmatrix} = \begin{bmatrix} 3 & -3 \\ 4 & -1 \\ 2 & -6 \\ 1 & -2 \end{bmatrix}$

**11.** $D^T - I = \begin{bmatrix} 3 & 1 \\ 5 & 2 \end{bmatrix} - \begin{bmatrix} 1 & 0 \\ 0 & 1 \end{bmatrix} = \begin{bmatrix} 2 & 1 \\ 5 & 1 \end{bmatrix}$

**12.** $3C = 3\begin{bmatrix} 4 & -2 \\ 5 & 0 \\ 6 & 0 \\ 1 & 3 \end{bmatrix} = \begin{bmatrix} 12 & -6 \\ 15 & 0 \\ 18 & 0 \\ 3 & 9 \end{bmatrix}$

**13.** $4I = \begin{bmatrix} 4 & 0 \\ 0 & 4 \end{bmatrix}$

**14.** $-2F = -2\begin{bmatrix} -1 & 6 \\ 4 & 11 \end{bmatrix} = \begin{bmatrix} 2 & -12 \\ -8 & -22 \end{bmatrix}$

**15.** $4D - 3I = \begin{bmatrix} 12 & 20 \\ 4 & 8 \end{bmatrix} - \begin{bmatrix} 3 & 0 \\ 0 & 3 \end{bmatrix} = \begin{bmatrix} 9 & 20 \\ 4 & 5 \end{bmatrix}$

**16.** $F + 2D = \begin{bmatrix} -1 & 6 \\ 4 & 11 \end{bmatrix} + \begin{bmatrix} 6 & 10 \\ 2 & 4 \end{bmatrix} = \begin{bmatrix} 5 & 16 \\ 6 & 15 \end{bmatrix}$

**17.** $3A - 5B$

$$= \begin{bmatrix} 12 & 12 & 6 & -15 \\ 18 & 9 & -3 & 0 \\ 0 & 0 & -9 & 15 \end{bmatrix} - \begin{bmatrix} 10 & -25 & -55 & 40 \\ 20 & 0 & 0 & 20 \\ -10 & -10 & 5 & 45 \end{bmatrix}$$

$$= \begin{bmatrix} 2 & 37 & 61 & -55 \\ -2 & 9 & -3 & -20 \\ 10 & 10 & -14 & -30 \end{bmatrix}$$

**18.** $AC = \begin{bmatrix} 4 & 4 & 2 & -5 \\ 6 & 3 & -1 & 0 \\ 0 & 0 & -3 & 5 \end{bmatrix}\begin{bmatrix} 4 & -2 \\ 5 & 0 \\ 6 & 0 \\ 1 & 3 \end{bmatrix} = \begin{bmatrix} 43 & -23 \\ 33 & -12 \\ -13 & 15 \end{bmatrix}$

**19.** $CD = \begin{bmatrix} 4 & -2 \\ 5 & 0 \\ 6 & 0 \\ 1 & 3 \end{bmatrix}\begin{bmatrix} 3 & 5 \\ 1 & 2 \end{bmatrix} = \begin{bmatrix} 10 & 16 \\ 15 & 25 \\ 18 & 30 \\ 6 & 11 \end{bmatrix}$

**20.** $DF = \begin{bmatrix} 3 & 5 \\ 1 & 2 \end{bmatrix}\begin{bmatrix} -1 & 6 \\ 4 & 11 \end{bmatrix} = \begin{bmatrix} 17 & 73 \\ 7 & 28 \end{bmatrix}$

**21.** $FD = \begin{bmatrix} -1 & 6 \\ 4 & 11 \end{bmatrix}\begin{bmatrix} 3 & 5 \\ 1 & 2 \end{bmatrix} = \begin{bmatrix} 3 & 7 \\ 23 & 42 \end{bmatrix}$

**22.** $FI = F$

**23.** $IF = F$

**24.** $DG^T = \begin{bmatrix} 3 & 5 \\ 1 & 2 \end{bmatrix}\begin{bmatrix} 2 & -1 \\ -5 & 3 \end{bmatrix}\begin{bmatrix} -19 & 12 \\ -8 & 5 \end{bmatrix}$

**25.**

$$DG = \begin{bmatrix} 3 & 5 \\ 1 & 2 \end{bmatrix}\begin{bmatrix} 2 & -5 \\ -1 & 3 \end{bmatrix}$$

$$= \begin{bmatrix} 1 & 0 \\ 0 & 1 \end{bmatrix} = I$$

So, $(DG)F = F$.

**26. a.** Infinitely many solutions.

**b.** The general solution is: $x = 6 + 2z$ and $y = 7 - 3z$, where $z$ is any real number. Answers will vary depending on the choice of $z$. If we let $z = 0$, then $y = 6 + 2(0) = 6$ and $y = 7 - 3(0) = 7$. If we let $z = 1$, then $y = 6 + 2(1) = 8$ and $y = 7 - 3(1) = 4$.

**27. a.** No solution. If translated into equation form, the last row would be $0x + 0y + 0z = 1$. Zero cannot equal one, so there is no solution.

**b.** No solution.

**28. a.** Unique solution, the coefficient matrix is $I_3$.

**b.** $x = 0,\ y = -10,\ z = 14$

**29.**

$$\left[\begin{array}{ccc|c} 1 & 1 & 2 & 5 \\ 4 & 0 & 1 & 5 \\ 2 & 1 & 1 & 5 \end{array}\right] \begin{array}{c} \to \\ -4R_1 + R_2 \to R_2 \\ -2R_1 + R_3 \to R_3 \end{array} \left[\begin{array}{ccc|c} 1 & 1 & 2 & 5 \\ 0 & -4 & -7 & -15 \\ 0 & -1 & -3 & -5 \end{array}\right] \begin{array}{c} R_2 \leftrightarrow R_3 \\ \to \\ \ \end{array} \left[\begin{array}{ccc|c} 1 & 1 & 2 & 5 \\ 0 & -1 & -3 & -5 \\ 0 & -4 & -7 & -15 \end{array}\right] \begin{array}{c} \to \\ -R_2 \to R_2 \\ \ \end{array}$$

$$\left[\begin{array}{ccc|c} 1 & 1 & 2 & 5 \\ 0 & 1 & 3 & 5 \\ 0 & -4 & -7 & -15 \end{array}\right] \begin{array}{c} -R_2 + R_1 \to R_1 \\ \to \\ 4R_2 + R_3 \to R_3 \end{array} \left[\begin{array}{ccc|c} 1 & 0 & -1 & 0 \\ 0 & 1 & 3 & 5 \\ 0 & 0 & 5 & 5 \end{array}\right] \begin{array}{c} \to \\ \ \\ \frac{1}{5}R_3 \to R_3 \end{array} \left[\begin{array}{ccc|c} 1 & 0 & -1 & 0 \\ 0 & 1 & 3 & 5 \\ 0 & 0 & 1 & 1 \end{array}\right] \begin{array}{c} R_3 + R_1 \to R_1 \\ -3R_3 + R_2 \to R_2 \\ \to \end{array}$$

$$\left[\begin{array}{ccc|c} 1 & 0 & 0 & 1 \\ 0 & 1 & 0 & 2 \\ 0 & 0 & 1 & 1 \end{array}\right]$$ The solution is (1, 2, 1).

**30.**

$$\left[\begin{array}{cc|c} 1 & -2 & 4 \\ -3 & 10 & 24 \end{array}\right] \begin{array}{c} \to \\ 3R_1 + R_2 \to R_2 \end{array} \left[\begin{array}{cc|c} 1 & -2 & 4 \\ 0 & 4 & 36 \end{array}\right] \begin{array}{c} \to \\ \frac{1}{4}R_2 \to R_2 \end{array} \left[\begin{array}{cc|c} 1 & -2 & 4 \\ 0 & 1 & 9 \end{array}\right] \begin{array}{c} 2R_2 + R_1 \to R_1 \\ \to \end{array} \left[\begin{array}{cc|c} 1 & 0 & 22 \\ 0 & 1 & 9 \end{array}\right]$$

Solution: $x = 22, y = 9$

**31.**

$$\left[\begin{array}{ccc|c} 1 & 1 & 1 & 4 \\ 3 & 4 & -1 & -1 \\ 2 & -1 & 3 & 3 \end{array}\right] \begin{array}{c} \to \\ -3R_1 + R_2 \to R_2 \\ -2R_1 + R_3 \to R_3 \end{array} \left[\begin{array}{ccc|c} 1 & 1 & 1 & 4 \\ 0 & 1 & -4 & -13 \\ 0 & -3 & 1 & -5 \end{array}\right] \begin{array}{c} -R_2 + R_1 \to R_1 \\ \to \\ 3R_2 + R_3 \to R_3 \end{array}$$

$$\left[\begin{array}{ccc|c} 1 & 0 & 5 & 17 \\ 0 & 1 & -4 & -13 \\ 0 & 0 & -11 & -44 \end{array}\right] \begin{array}{c} \to \\ \ \\ -\frac{1}{11}R_3 \to R_3 \end{array} \left[\begin{array}{ccc|c} 1 & 0 & 5 & 17 \\ 0 & 1 & -4 & -13 \\ 0 & 0 & 1 & 4 \end{array}\right] \begin{array}{c} -5R_3 + R_1 \to R_1 \\ 4R_3 + R_2 \to R_2 \\ \to \end{array} \left[\begin{array}{ccc|c} 1 & 0 & 0 & -3 \\ 0 & 1 & 0 & 3 \\ 0 & 0 & 1 & 4 \end{array}\right]$$

Solution: $x = -3,\ y = 3,\ z = 4$

**32.**

$$\left[\begin{array}{ccc|c} -1 & 1 & 1 & 3 \\ 3 & 0 & -1 & 1 \\ 2 & -3 & -4 & -2 \end{array}\right] \begin{array}{c} \to \\ 3R_1 + R_2 \to R_2 \\ 2R_1 + R_3 \to R_3 \end{array} \left[\begin{array}{ccc|c} -1 & 1 & 1 & 3 \\ 0 & 3 & 2 & 10 \\ 0 & -1 & -2 & 4 \end{array}\right] \begin{array}{c} R_3 + R_1 \to R_1 \\ 3R_3 + R_2 \to R_2 \\ \to \end{array} \left[\begin{array}{ccc|c} -1 & 0 & -1 & 7 \\ 0 & 0 & -4 & 22 \\ 0 & -1 & -2 & 4 \end{array}\right] \begin{array}{c} -R_1 \to R_1 \\ \to \\ R_2 \leftrightarrow R_3 \end{array}$$

$$\left[\begin{array}{ccc|c} 1 & 0 & 1 & -7 \\ 0 & -1 & -2 & 4 \\ 0 & 0 & -4 & 22 \end{array}\right] \begin{array}{c} \to \\ -R_2 \to R_2 \\ -\frac{1}{4}R_3 \to R_3 \end{array} \left[\begin{array}{ccc|c} 1 & 0 & 1 & -7 \\ 0 & 1 & 2 & -4 \\ 0 & 0 & 1 & -11/2 \end{array}\right] \begin{array}{c} -R_3 + R_1 \to R_1 \\ -2R_3 + R_2 \to R_2 \\ \to \end{array} \left[\begin{array}{ccc|c} 1 & 0 & 0 & -3/2 \\ 0 & 1 & 0 & 7 \\ 0 & 0 & 1 & -11/2 \end{array}\right]$$

Solution: $x = -\dfrac{3}{2}, y = 7, z = -\dfrac{11}{2}$

**33.** $\left[\begin{array}{rrr|r} 1 & 1 & -2 & 5 \\ 3 & 2 & 5 & 10 \\ -2 & -3 & 15 & 2 \end{array}\right] \begin{array}{c} \to \\ -3R_1 + R_2 \to R_2 \\ 2R_1 + R_3 \to R_3 \end{array} \left[\begin{array}{rrr|r} 1 & 1 & -2 & 5 \\ 0 & -1 & 11 & -5 \\ 0 & -1 & 11 & 12 \end{array}\right] \begin{array}{c} \to \\ \\ -R_2 + R_3 \to R_3 \end{array} \left[\begin{array}{rrr|r} 1 & 1 & -2 & 5 \\ 0 & -1 & 11 & -5 \\ 0 & 0 & 0 & 17 \end{array}\right]$

No solution.

**34.** $\left[\begin{array}{rrr|r} 1 & -1 & 0 & 3 \\ 1 & 1 & 4 & 1 \\ 2 & -3 & -2 & 7 \end{array}\right] \begin{array}{c} \to \\ -R_1 + R_2 \to R_2 \\ -2R_1 + R_3 \to R_3 \end{array} \left[\begin{array}{rrr|r} 1 & -1 & 0 & 3 \\ 0 & 2 & 4 & -2 \\ 0 & -1 & -2 & 1 \end{array}\right] \begin{array}{c} \to \\ \frac{1}{2}R_2 \to R_2 \\ \\ \end{array}$

$\left[\begin{array}{rrr|r} 1 & -1 & 0 & 3 \\ 0 & 1 & 2 & -1 \\ 0 & -1 & -2 & 1 \end{array}\right] \begin{array}{c} R_2 + R_1 \to R_1 \\ \to \\ R_2 + R_3 \to R_3 \end{array} \left[\begin{array}{rrr|r} 1 & 0 & 2 & 2 \\ 0 & 1 & 2 & -1 \\ 0 & 0 & 0 & 0 \end{array}\right]$

There are infinitely many solutions. The general solution is $x = 2 - 2z,\ \ y = -1 - 2z.$

**35.** $\left[\begin{array}{rrr|r} 1 & -3 & 1 & 4 \\ 2 & -5 & -1 & 6 \end{array}\right] \begin{array}{c} \to \\ -2R_1 + R_2 \to R_2 \end{array} \left[\begin{array}{rrr|r} 1 & -3 & 1 & 4 \\ 0 & 1 & -3 & -2 \end{array}\right] \begin{array}{c} 3R_2 + R_1 \to R_1 \\ \to \end{array} \left[\begin{array}{rrr|r} 1 & 0 & -8 & -2 \\ 0 & 1 & -3 & -2 \end{array}\right]$

There are infinitely many solutions. The general solution is $x = -2 + 8z$ and $y = -2 + 3z$.

**36.** $\left[\begin{array}{rrrr|r} 1 & 1 & 1 & 1 & 3 \\ 1 & -2 & 1 & -4 & -5 \\ 1 & 0 & -1 & 1 & 0 \\ 0 & 1 & 1 & 1 & 2 \end{array}\right] \begin{array}{c} \to \\ -R_1 + R_2 \to R_2 \\ -R_1 + R_3 \to R_3 \\ \\ \end{array} \left[\begin{array}{rrrr|r} 1 & 1 & 1 & 1 & 3 \\ 0 & -3 & 0 & -5 & -8 \\ 0 & -1 & -2 & 0 & -3 \\ 0 & 1 & 1 & 1 & 2 \end{array}\right] \begin{array}{c} \\ R_2 \leftrightarrow R_4 \\ \to \\ \\ \end{array}$

$\left[\begin{array}{rrrr|r} 1 & 1 & 1 & 1 & 3 \\ 0 & 1 & 1 & 1 & 2 \\ 0 & -1 & -2 & 0 & -3 \\ 0 & -3 & 0 & -5 & -8 \end{array}\right] \begin{array}{c} -R_2 + R_1 \to R_1 \\ \to \\ R_2 + R_3 \to R_3 \\ 3R_2 + R_4 \to R_4 \end{array} \left[\begin{array}{rrrr|r} 1 & 0 & 0 & 0 & 1 \\ 0 & 1 & 1 & 1 & 2 \\ 0 & 0 & -1 & 1 & -1 \\ 0 & 0 & 3 & -2 & -2 \end{array}\right] \begin{array}{c} \to \\ \\ -R_3 \to R_3 \\ \\ \end{array}$

$\left[\begin{array}{rrrr|r} 1 & 0 & 0 & 0 & 1 \\ 0 & 1 & 1 & 1 & 2 \\ 0 & 0 & 1 & -1 & 1 \\ 0 & 0 & 3 & -2 & -2 \end{array}\right] \begin{array}{c} \to \\ -R_3 + R_2 \to R_2 \\ \\ -3R_3 + R_4 \to R_4 \end{array} \left[\begin{array}{rrrr|r} 1 & 0 & 0 & 0 & 1 \\ 0 & 1 & 0 & 2 & 1 \\ 0 & 0 & 1 & -1 & 1 \\ 0 & 0 & 0 & 1 & -5 \end{array}\right] \begin{array}{c} \to \\ -2R_4 + R_2 \to R_2 \\ R_4 + R_3 \to R_3 \\ \\ \end{array} \left[\begin{array}{rrrr|r} 1 & 0 & 0 & 0 & 1 \\ 0 & 1 & 0 & 0 & 11 \\ 0 & 0 & 1 & 0 & -4 \\ 0 & 0 & 0 & 1 & -5 \end{array}\right]$

Solution: $x_1 = 1,\ \ x_2 = 11,\ \ x_3 = -4,\ \ x_4 = -5$

**37.** $DG = \begin{bmatrix} 3 & 5 \\ 1 & 2 \end{bmatrix}\begin{bmatrix} 2 & -5 \\ -1 & 3 \end{bmatrix} = \begin{bmatrix} 1 & 0 \\ 0 & 1 \end{bmatrix}$ Yes, $D$ and $G$ are inverse matrices.

**38.** $\begin{bmatrix} 7 & -1 \\ -10 & 2 \end{bmatrix}^{-1} = \dfrac{1}{14-10}\begin{bmatrix} 2 & 1 \\ 10 & 7 \end{bmatrix} = \begin{bmatrix} 1/2 & 1/4 \\ 5/2 & 7/4 \end{bmatrix}$

**39.** $\left[\begin{array}{rrr|rrr} 1 & 0 & 2 & 1 & 0 & 0 \\ 3 & 4 & -1 & 0 & 1 & 0 \\ 1 & 1 & 0 & 0 & 0 & 1 \end{array}\right] \begin{array}{c} \to \\ -3R_1 + R_2 \to R_2 \\ -R_1 + R_3 \to R_3 \end{array} \left[\begin{array}{rrr|rrr} 1 & 0 & 2 & 1 & 0 & 0 \\ 0 & 4 & -7 & -3 & 1 & 0 \\ 0 & 1 & -2 & -1 & 0 & 1 \end{array}\right] \begin{array}{c} R_2 \leftrightarrow R_3 \\ \to \\ \\ \end{array}$

$\left[\begin{array}{rrr|rrr} 1 & 0 & 2 & 1 & 0 & 0 \\ 0 & 1 & -2 & -1 & 0 & 1 \\ 0 & 4 & -7 & -3 & 1 & 0 \end{array}\right] \begin{array}{c} \to \\ \\ -4R_2 + R_3 \to R_3 \end{array} \left[\begin{array}{rrr|rrr} 1 & 0 & 2 & 1 & 0 & 0 \\ 0 & 1 & -2 & -1 & 0 & 1 \\ 0 & 0 & 1 & 1 & 1 & -4 \end{array}\right] \begin{array}{c} -2R_3 + R_1 \to R_1 \\ 2R_3 + R_2 \to R_2 \\ \to \end{array}$

$$\left[\begin{array}{ccc|ccc} 1 & 0 & 0 & -1 & -2 & 8 \\ 0 & 1 & 0 & 1 & 2 & -7 \\ 0 & 0 & 1 & 1 & 1 & -4 \end{array}\right] \qquad \text{Answer: } \begin{bmatrix} -1 & -2 & 8 \\ 1 & 2 & -7 \\ 1 & 1 & -4 \end{bmatrix}$$

**40.** $$\left[\begin{array}{ccc|ccc} 3 & 3 & 2 & 1 & 0 & 0 \\ -1 & 4 & 2 & 0 & 1 & 0 \\ 2 & 5 & 3 & 0 & 0 & 1 \end{array}\right] \begin{array}{c} 2R_2+R_1 \to R_1 \\ \to \\ 2R_2+R_3 \to R_3 \end{array} \left[\begin{array}{ccc|ccc} 1 & 11 & 6 & 1 & 2 & 0 \\ -1 & 4 & 2 & 0 & 1 & 0 \\ 0 & 13 & 7 & 0 & 2 & 1 \end{array}\right] \begin{array}{c} \to \\ R_1+R_2 \to R_2 \\ \phantom{x} \end{array}$$

$$\left[\begin{array}{ccc|ccc} 1 & 11 & 6 & 1 & 2 & 0 \\ 0 & 15 & 8 & 1 & 3 & 0 \\ 0 & 13 & 7 & 0 & 2 & 1 \end{array}\right] \begin{array}{c} \to \\ -R_3+R_2 \to R_2 \\ \phantom{x} \end{array} \left[\begin{array}{ccc|ccc} 1 & 11 & 6 & 1 & 2 & 0 \\ 0 & 2 & 1 & 1 & 1 & -1 \\ 0 & 13 & 7 & 0 & 2 & 1 \end{array}\right] \begin{array}{c} -6R_2+R_1 \to R_1 \\ \to \\ -7R_2+R_3 \to R_3 \end{array}$$

$$\left[\begin{array}{ccc|ccc} 1 & -1 & 0 & -5 & -4 & 6 \\ 0 & 2 & 1 & 1 & 1 & -1 \\ 0 & -1 & 0 & -7 & -5 & 8 \end{array}\right] \begin{array}{c} R_2 \leftrightarrow R_3 \\ \to \\ (-1)\text{ new } R_2 \end{array} \left[\begin{array}{ccc|ccc} 1 & -1 & 0 & -5 & -4 & 6 \\ 0 & 1 & 0 & 7 & 5 & -8 \\ 0 & 2 & 1 & 1 & 1 & -1 \end{array}\right] \begin{array}{c} R_2+R_1 \to R_1 \\ \to \\ -2R_2+R_3 \to R_3 \end{array}$$

$$\left[\begin{array}{ccc|ccc} 1 & 0 & 0 & 2 & 1 & -2 \\ 0 & 1 & 0 & 7 & 5 & -8 \\ 0 & 0 & 1 & -13 & -9 & 15 \end{array}\right] \qquad \text{Answer: } \begin{bmatrix} 2 & 1 & -2 \\ 7 & 5 & -8 \\ -13 & -9 & 15 \end{bmatrix}$$

**41.** Use the inverse of problem 35.

$$\begin{bmatrix} x \\ y \\ z \end{bmatrix} = \begin{bmatrix} 1 & 0 & 2 \\ 3 & 4 & -1 \\ 1 & 1 & 0 \end{bmatrix}^{-1} \begin{bmatrix} 5 \\ 2 \\ -3 \end{bmatrix} = \begin{bmatrix} -1 & -2 & 8 \\ 1 & 2 & -7 \\ 1 & 1 & -4 \end{bmatrix} \begin{bmatrix} 5 \\ 2 \\ -3 \end{bmatrix} = \begin{bmatrix} -33 \\ 30 \\ 19 \end{bmatrix}$$

**42.** Use the inverse of problem 36.

$$\begin{bmatrix} x \\ y \\ z \end{bmatrix} = \begin{bmatrix} 3 & 3 & 2 \\ -1 & 4 & 2 \\ 2 & 5 & 3 \end{bmatrix}^{-1} \begin{bmatrix} 1 \\ -10 \\ -6 \end{bmatrix} = \begin{bmatrix} 2 & 1 & -2 \\ 7 & 5 & -8 \\ -13 & -9 & 15 \end{bmatrix} \begin{bmatrix} 1 \\ -10 \\ -6 \end{bmatrix} = \begin{bmatrix} 4 \\ 5 \\ -13 \end{bmatrix}$$

**43.** $$\left[\begin{array}{ccc|ccc} 1 & 3 & 1 & 1 & 0 & 0 \\ 1 & 4 & 3 & 0 & 1 & 0 \\ 2 & -1 & -11 & 0 & 0 & 1 \end{array}\right] \begin{array}{c} \to \\ -R_1+R_2 \to R_2 \\ -2R_1+R_3 \to R_3 \end{array} \left[\begin{array}{ccc|ccc} 1 & 3 & 1 & 1 & 0 & 0 \\ 0 & 1 & 2 & -1 & 1 & 0 \\ 0 & -7 & -13 & -2 & 0 & 1 \end{array}\right] \begin{array}{c} -3R_2+R_1 \to R_1 \\ \to \\ 7R_2+R_3 \to R_3 \end{array}$$

$$\left[\begin{array}{ccc|ccc} 1 & 0 & -5 & 4 & -3 & 0 \\ 0 & 1 & 2 & -1 & 1 & 0 \\ 0 & 0 & 1 & -9 & 7 & 1 \end{array}\right] \begin{array}{c} 5R_3+R_1 \to R_1 \\ -2R_3+R_2 \to R_2 \\ \to \end{array} \left[\begin{array}{ccc|ccc} 1 & 0 & 0 & -41 & 32 & 5 \\ 0 & 1 & 0 & 17 & -13 & -2 \\ 0 & 0 & 1 & -9 & 7 & 1 \end{array}\right]$$

$$\begin{bmatrix} x \\ y \\ z \end{bmatrix} = \begin{bmatrix} -41 & 32 & 5 \\ 17 & -13 & -2 \\ -9 & 7 & 1 \end{bmatrix} \begin{bmatrix} 0 \\ 2 \\ -12 \end{bmatrix} = \begin{bmatrix} 4 \\ -2 \\ 2 \end{bmatrix}$$

**44.** The determinant of the matrix is 0. The matrix does not have an inverse.

**45. a.** $\begin{vmatrix} 4 & 4 \\ -2 & 2 \end{vmatrix} = (4)(2) - (-2)(4) = 8 - (-8) = 16$

**b.** Yes, since the determinant does not equal zero, this matrix has an inverse.

**46.** **a.** Using technology, $\begin{vmatrix} 1 & 2 & 3 \\ 4 & -1 & 8 \\ 6 & 3 & 14 \end{vmatrix} = 0$

**b.** No, since the determinant equals zero, the matrix does not have an inverse.

**47.** Total production $= N + M = \begin{bmatrix} 250 & 140 \\ 480 & 700 \end{bmatrix}$

**48.** $M + P - S = \begin{bmatrix} 1030 & 800 \\ 700 & 1200 \end{bmatrix}$

**49.** June: $\begin{bmatrix} 100 & 0 \\ 0 & 120 \end{bmatrix} \cdot M = \begin{matrix} \text{A} & \text{B} \\ \begin{bmatrix} 15{,}000 & 8000 \\ 33{,}600 & 36{,}000 \end{bmatrix} \\ \overline{48{,}600} & \overline{44{,}000} \end{matrix}$ July: $\begin{bmatrix} 100 & 0 \\ 0 & 120 \end{bmatrix} \cdot N = \begin{matrix} \text{A} & \text{B} \\ \begin{bmatrix} 10{,}000 & 6000 \\ 24{,}000 & 48{,}000 \end{bmatrix} \\ \overline{34{,}000} & \overline{54{,}000} \end{matrix}$

**a.** Production was higher at Plant A in June.
**b.** Production was higher at Plant B in July.

**50.**
$$\begin{matrix} & \text{S} & \text{M} & \text{L} \\ \text{R} & 25 & 40 & 45 \\ \text{H} & 10 & 10 & 10 \end{matrix} \quad \begin{matrix} \text{M} & \text{W} & \\ 1 & 14 & \text{S} \\ 12 & 10 & \text{M} \\ 8 & 3 & \text{L} \end{matrix} = \begin{matrix} \text{M} & \text{W} & \\ 865 & 885 & \text{R} \\ 210 & 270 & \text{H} \end{matrix}$$

*BA* gives costs of robes and hoods for men and women.

**51.**
$$\begin{bmatrix} 25 & 40 & 45 \\ 10 & 10 & 10 \end{bmatrix} \begin{bmatrix} 1 & 14 \\ 12 & 10 \\ 8 & 3 \end{bmatrix} = \begin{bmatrix} 865 & 885 \\ 210 & 270 \end{bmatrix}$$

$$\begin{bmatrix} 865 & 885 \\ 210 & 270 \end{bmatrix} \begin{bmatrix} 1 \\ 1 \end{bmatrix} = \begin{bmatrix} 1750 \\ 480 \end{bmatrix} \begin{matrix} \text{R} \\ \text{H} \end{matrix}$$

The cost of new robes is \$1750.
The cost of new hoods is \$480.

**52.** **a.**
$$\begin{bmatrix} 30 & 20 & 10 \\ 20 & 10 & 20 \end{bmatrix} \begin{bmatrix} 300 & 280 \\ 150 & 100 \\ 150 & 200 \end{bmatrix} = \begin{bmatrix} 9000+3000+1500 & 8400+2000+2000 \\ 6000+1500+3000 & 5600+1000+4000 \end{bmatrix} = \begin{bmatrix} 13{,}500 & 12{,}400 \\ 10{,}500 & 10{,}600 \end{bmatrix}$$

**b.** Column 1 is Ace's price and column 2 is Kink's price. Dept. A buys from Kink and Dept. B buys from Ace.

**53.** **a.** $W = \begin{bmatrix} 0.20 & 0.30 & 0.50 \end{bmatrix}$

**b.** $R = \begin{bmatrix} 0.013469 \\ 0.013543 \\ 0.006504 \end{bmatrix}$

**c.** $WR = \begin{bmatrix} 0.20 & 0.30 & 0.50 \end{bmatrix} \begin{bmatrix} 0.013469 \\ 0.013543 \\ 0.006504 \end{bmatrix} = 0.20(0.013469) + 0.30(0.013543) + 0.50(0.006504) = 0.0100087$

**d.** The historical return on the portfolio, 0.0100087, is the estimated expected monthly return of the portfolio. This is roughly 1% per month.

**54.** $x$ = fast food shares; $y$ = software shares; $z$ = pharmaceutical shares

$$50x+20y+80z=50{,}000$$

$$0.115(50x)+0.15(20y)+0.10(80z)=0.12(50{,}000) \quad \to \quad x=2z$$

Rearranged and simplified:

$$x-2z=0$$
$$5x+2y+8z=5000$$
$$5.75x+3y+8z=6000$$

$$\left[\begin{array}{ccc|c} 1 & 0 & -2 & 0 \\ 5 & 2 & 8 & 5000 \\ 5.75 & 3 & 8 & 6000 \end{array}\right] \begin{array}{c} \to \\ -5R_1+R_2\to R_2 \\ -5.75R_1+R_3\to R_3 \end{array} \left[\begin{array}{ccc|c} 1 & 0 & -2 & 0 \\ 0 & 2 & 18 & 5000 \\ 0 & 3 & 19.5 & 6000 \end{array}\right] \begin{array}{c} \to \\ \frac{1}{2}R_2\to R_2 \\ \phantom{x} \end{array} \left[\begin{array}{ccc|c} 1 & 0 & -2 & 0 \\ 0 & 1 & 9 & 2500 \\ 0 & 3 & 19.5 & 6000 \end{array}\right]$$

$$\begin{array}{c} \to \\ \phantom{x} \\ -3R_2+R_3\to R_3 \end{array} \left[\begin{array}{ccc|c} 1 & 0 & -2 & 0 \\ 0 & 1 & 9 & 2500 \\ 0 & 0 & -7.5 & -1500 \end{array}\right] \begin{array}{c} \to \\ \phantom{x} \\ -\frac{2}{15}R_3\to R_3 \end{array} \left[\begin{array}{ccc|c} 1 & 0 & -2 & 0 \\ 0 & 1 & 9 & 2500 \\ 0 & 0 & 1 & 200 \end{array}\right] \begin{array}{c} 2R_3+R_1\to R_1 \\ -9R_3+R_2\to R_2 \\ \to \end{array} \left[\begin{array}{ccc|c} 1 & 0 & 0 & 400 \\ 0 & 1 & 0 & 700 \\ 0 & 0 & 1 & 200 \end{array}\right]$$

Buy 400 shares of fast food company, 700 shares of software company, and 200 shares of pharmaceutical company.

**55.**

$$\begin{array}{ll} A+B+2C=2000 & \text{Units of I} \\ 3A+4B+10C=8000 & \text{Units of II} \\ A+2B+6C=4000 & \text{Units of III} \end{array}$$

**a.** $$\left[\begin{array}{ccc|c} 1 & 1 & 2 & 2000 \\ 3 & 4 & 10 & 8000 \\ 1 & 2 & 6 & 4000 \end{array}\right] \begin{array}{c} \to \\ -3R_1+R_2\to R_2 \\ -R_1+R_3\to R_3 \end{array} \left[\begin{array}{ccc|c} 1 & 1 & 2 & 2000 \\ 0 & 1 & 4 & 2000 \\ 0 & 1 & 4 & 2000 \end{array}\right] \begin{array}{c} -R_2+R_1\to R_1 \\ \to \\ -R_2+R_3\to R_3 \end{array} \left[\begin{array}{ccc|c} 1 & 0 & -2 & 0 \\ 0 & 1 & 4 & 2000 \\ 0 & 0 & 0 & 0 \end{array}\right]$$

Solution: $A=2C,\ B=2000-4C$

**b.** $500=2C$, so $C=250$

When $C=250,\ B=2000-4(250)=1000$.

Yes, it is possible to support 500 type A slugs; there would be 1000 type B slugs and 250 type C slugs.

**c.** $0=2000-4C$ so $C=500$

When $C=500$, $A=2(500)=1000$.

The maximum number of type A slugs is 1000, when there are 500 type C slugs and no type B slugs.

**56. a.** Using inverse matrices, and letting $x$ = passenger planes, $y$ = transport planes, and $z$ = jumbos:

$$\begin{bmatrix} 100 & 100 & 100 \\ 150 & 20 & 350 \\ 20 & 65 & 35 \end{bmatrix}\begin{bmatrix} x \\ y \\ z \end{bmatrix}=\begin{bmatrix} 1100 \\ 1930 \\ 460 \end{bmatrix}$$

$$\begin{bmatrix} 100 & 100 & 100 \\ 150 & 20 & 350 \\ 20 & 65 & 35 \end{bmatrix}^{-1}\begin{bmatrix} 1100 \\ 1930 \\ 460 \end{bmatrix}=\begin{bmatrix} x \\ y \\ z \end{bmatrix}$$

$$\begin{bmatrix} 0.0201 & -0.0027 & -0.0301 \\ -0.0016 & -0.0014 & 0.0183 \\ -0.0085 & 0.0041 & 0.0119 \end{bmatrix}\begin{bmatrix} 1100 \\ 1930 \\ 460 \end{bmatrix}=\begin{bmatrix} 3 \\ 4 \\ 4 \end{bmatrix}$$

Use 3 passenger, 4 transport, and 4 jumbo.

**b.** $\begin{bmatrix} 100 & 100 & 100 \\ 150 & 20 & 350 \\ 20 & 65 & 35 \end{bmatrix}^{-1} \begin{bmatrix} 1100 \\ 1930+730 \\ 460 \end{bmatrix} = \begin{bmatrix} 1 \\ 3 \\ 7 \end{bmatrix}$

Use 1 passenger, 3 transport, and 7 jumbo.

**c.** Column 2

**57. a.** $D = \begin{bmatrix} 4720 \\ 40 \end{bmatrix}$ $(I-A)X = D$ or $\begin{bmatrix} 0.9 & -0.2 \\ -0.2 & 0.6 \end{bmatrix} \begin{bmatrix} \text{S} \\ \text{A} \end{bmatrix} = \begin{bmatrix} 4720 \\ 40 \end{bmatrix}$

$\begin{bmatrix} \text{S} \\ \text{A} \end{bmatrix} = \begin{bmatrix} 0.9 & -0.2 \\ -0.2 & 0.6 \end{bmatrix}^{-1} \begin{bmatrix} 4720 \\ 40 \end{bmatrix} = \frac{1}{0.5} \begin{bmatrix} 0.6 & 0.2 \\ 0.2 & 0.9 \end{bmatrix} \begin{bmatrix} 4720 \\ 40 \end{bmatrix} = \begin{bmatrix} 1.2 & 0.4 \\ 0.4 & 1.8 \end{bmatrix} \begin{bmatrix} 4720 \\ 40 \end{bmatrix} = \begin{bmatrix} 5680 \\ 1960 \end{bmatrix}$

shipping = 5680; agriculture = 1960

**b.** $1 \begin{bmatrix} \text{Col 2 of} \\ (1-A)^{-1} \end{bmatrix} = \begin{bmatrix} 0.4 \\ 1.8 \end{bmatrix}$ shipping = 0.4; agriculture = 1.8

**58. a.**

$$A = \begin{matrix} & S & C & \\ & \begin{bmatrix} 0.1 & 0.1 \\ 0.2 & 0.05 \end{bmatrix} & & \begin{matrix} \text{Shoes} \\ \text{Cattle} \end{matrix} \end{matrix}$$

**b.** $D = \begin{bmatrix} 850 \\ 275 \end{bmatrix}$ $(I-A)X = D$ or $\begin{bmatrix} 0.9 & -0.1 \\ -0.2 & 0.95 \end{bmatrix} \begin{bmatrix} \text{S} \\ \text{C} \end{bmatrix} = \begin{bmatrix} 850 \\ 275 \end{bmatrix}$

$\begin{bmatrix} \text{S} \\ \text{C} \end{bmatrix} = \begin{bmatrix} 0.9 & -0.1 \\ -0.2 & 0.95 \end{bmatrix}^{-1} \begin{bmatrix} 850 \\ 275 \end{bmatrix} = \frac{1}{0.835} \begin{bmatrix} 0.95 & 0.1 \\ 0.2 & 0.9 \end{bmatrix} \begin{bmatrix} 850 \\ 275 \end{bmatrix} = \begin{bmatrix} 1000 \\ 500 \end{bmatrix}$

Shoes = 1000; cattle = 500

**59.** $(I-A)X = D$ or

$$\begin{bmatrix} 0.6 & -0.2 & -0.2 \\ -0.2 & 0.6 & -0.2 \\ -0.1 & -0.2 & 0.8 \end{bmatrix} \begin{bmatrix} \text{Min} \\ \text{Mfg} \\ \text{Fuel} \end{bmatrix} = \begin{bmatrix} 72 \\ 40 \\ 220 \end{bmatrix}$$

$$\begin{bmatrix} 0.6 & -0.2 & -0.2 \\ -0.2 & 0.6 & -0.2 \\ -0.1 & -0.2 & 0.8 \end{bmatrix}^{-1} \begin{bmatrix} 72 \\ 40 \\ 220 \end{bmatrix} = \begin{bmatrix} \text{Min} \\ \text{Mfg} \\ \text{Fuel} \end{bmatrix}$$

$$\begin{bmatrix} 2.1154 & 0.9615 & 0.7692 \\ 0.8654 & 2.2115 & 0.7692 \\ 0.4808 & 0.6731 & 1.5385 \end{bmatrix} \begin{bmatrix} 72 \\ 40 \\ 220 \end{bmatrix} = \begin{bmatrix} 360 \\ 320 \\ 400 \end{bmatrix}$$

Mining = 360; manufacturing = 320; fuels = 400

**60.** A closed Leontief model must be solved by the Gauss-Jordan elimination method. A graphing calculator is strongly suggested.

$\text{G} = \frac{64}{93}\text{H}, \text{A} = \frac{59}{93}\text{H}, \text{M} = \frac{40}{93}\text{H}$

# Chapter 3: Matrices

## *Chapter 3 Test*

**1.** $A^T + B = \begin{bmatrix} 1 & 3 & 4 \\ -2 & 2 & 1 \end{bmatrix} + \begin{bmatrix} 2 & -2 & 1 \\ 3 & 1 & 5 \end{bmatrix} = \begin{bmatrix} 3 & 1 & 5 \\ 1 & 3 & 6 \end{bmatrix}$

**2.** $B - C = \begin{bmatrix} 2 & -2 & 1 \\ 3 & 1 & 5 \end{bmatrix} - \begin{bmatrix} 3 & -4 & -1 \\ 2 & 2 & -1 \end{bmatrix} = \begin{bmatrix} -1 & 2 & 2 \\ 1 & -1 & 6 \end{bmatrix}$

**3.** $CD = \begin{bmatrix} 3 & -4 & -1 \\ 2 & 2 & -1 \end{bmatrix} \begin{bmatrix} 1 & 2 & 4 \\ 3 & 5 & 41 \\ 3 & 2 & 3 \end{bmatrix} = \begin{bmatrix} -12 & -16 & -155 \\ 5 & 12 & 87 \end{bmatrix}$

**4.** $DA = \begin{bmatrix} 1 & 2 & 4 \\ 3 & 5 & 41 \\ 3 & 2 & 3 \end{bmatrix} \begin{bmatrix} 1 & -2 \\ 3 & 2 \\ 4 & 1 \end{bmatrix} = \begin{bmatrix} 23 & 6 \\ 182 & 45 \\ 21 & 1 \end{bmatrix}$

**5.** $BA = \begin{bmatrix} 2 & -2 & 1 \\ 3 & 1 & 5 \end{bmatrix} \begin{bmatrix} 1 & -2 \\ 3 & 2 \\ 4 & 1 \end{bmatrix} = \begin{bmatrix} 0 & -7 \\ 26 & 1 \end{bmatrix}$

**6.** $ABD = \begin{bmatrix} 1 & -2 \\ 3 & 2 \\ 4 & 1 \end{bmatrix} \begin{bmatrix} 2 & -2 & 1 \\ 3 & 1 & 5 \end{bmatrix} \begin{bmatrix} 1 & 2 & 4 \\ 3 & 5 & 41 \\ 3 & 2 & 3 \end{bmatrix} = \begin{bmatrix} -4 & -4 & -9 \\ 12 & -4 & 13 \\ 11 & -7 & 9 \end{bmatrix} \begin{bmatrix} 1 & 2 & 4 \\ 3 & 5 & 41 \\ 3 & 2 & 3 \end{bmatrix} = \begin{bmatrix} -43 & -46 & -207 \\ 39 & 30 & -77 \\ 17 & 5 & -216 \end{bmatrix}$

**7.** $\begin{bmatrix} 1 & 3 \\ 2 & 4 \end{bmatrix}^{-1} = \frac{1}{-2} \begin{bmatrix} 4 & -3 \\ -2 & 1 \end{bmatrix} = \begin{bmatrix} -2 & 3/2 \\ 1 & -1/2 \end{bmatrix}$

**8.**
$$\left[\begin{array}{ccc|ccc} 1 & 2 & 4 & 1 & 0 & 0 \\ 1 & 2 & 2 & 0 & 1 & 0 \\ 1 & 1 & 4 & 0 & 0 & 1 \end{array}\right] \begin{array}{l} \rightarrow \\ -R_1 + R_2 \rightarrow R_2 \\ -R_1 + R_3 \rightarrow R_3 \end{array}$$

$$\left[\begin{array}{ccc|ccc} 1 & 2 & 4 & 1 & 0 & 0 \\ 0 & 0 & -2 & -1 & 1 & 0 \\ 0 & -1 & 0 & -1 & 0 & 1 \end{array}\right] \begin{array}{l} \rightarrow \\ -R_3 \rightarrow R_3 \\ R_2 \leftrightarrow \text{new } R_3 \end{array}$$

$$\left[\begin{array}{ccc|ccc} 1 & 2 & 4 & 1 & 0 & 0 \\ 0 & 1 & 0 & 1 & 0 & -1 \\ 0 & 0 & -2 & -1 & 1 & 0 \end{array}\right] \begin{array}{l} -2R_2 + R_1 \rightarrow R_1 \\ \rightarrow \\ -\tfrac{1}{2}R_3 \rightarrow R_3 \end{array}$$

$$\left[\begin{array}{ccc|ccc} 1 & 0 & 4 & -1 & 0 & 2 \\ 0 & 1 & 0 & 1 & 0 & -1 \\ 0 & 0 & 1 & 1/2 & -1/2 & 0 \end{array}\right] \begin{array}{l} -4R_3 + R_1 \rightarrow R_1 \\ \rightarrow \\ \ \end{array}$$

$$\left[\begin{array}{ccc|ccc} 1 & 0 & 0 & -3 & 2 & 2 \\ 0 & 1 & 0 & 1 & 0 & -1 \\ 0 & 0 & 1 & 1/2 & -1/2 & 0 \end{array}\right]$$

$$\text{Inverse} = \begin{bmatrix} -3 & 2 & 2 \\ 1 & 0 & -1 \\ 1/2 & -1/2 & 0 \end{bmatrix}$$

**9.**
$$AX = B$$
$$A^{-1}AX = A^{-1}B$$
$$X = A^{-1}B$$
$$X = \begin{bmatrix} 1 & 2 & 0 \\ 3 & 1 & 2 \\ 4 & 1 & 1 \end{bmatrix} \begin{bmatrix} 3 \\ 1 \\ 2 \end{bmatrix} = \begin{bmatrix} 5 \\ 14 \\ 15 \end{bmatrix}$$

**10.** Begin with the augmented matrix.

$$\left[\begin{array}{ccc|c}1&-1&2&4\\1&4&1&4\\2&2&4&10\end{array}\right]\xrightarrow{\frac{1}{2}R_3\to R_3}\left[\begin{array}{ccc|c}1&-1&2&4\\1&4&1&4\\1&1&2&5\end{array}\right]\xrightarrow[-R_1+R_3\to R_3]{-R_1+R_2\to R_2}\left[\begin{array}{ccc|c}1&-1&2&4\\0&5&-1&0\\0&2&0&1\end{array}\right]\xrightarrow[R_2\leftrightarrow \text{new } R_3]{\frac{1}{2}R_3\to R_3}$$

$$\left[\begin{array}{ccc|c}1&-1&2&4\\0&1&0&1/2\\0&5&-1&0\end{array}\right]\xrightarrow[-5R_2+R_3\to R_3]{R_2+R_1\to R_1}\left[\begin{array}{ccc|c}1&0&2&9/2\\0&1&0&1/2\\0&0&-1&-5/2\end{array}\right]\xrightarrow[-R_3\to R_3]{2R_3+R_1\to R_1}\left[\begin{array}{ccc|c}1&0&0&-1/2\\0&1&0&1/2\\0&0&1&5/2\end{array}\right]$$

Solution: $x=-\dfrac{1}{2},\ y=\dfrac{1}{2},\ z=\dfrac{5}{2}$

**11.** Begin with the augmented matrix.

$$\left[\begin{array}{ccc|c}1&-1&2&4\\1&4&1&4\\2&3&3&8\end{array}\right]\xrightarrow[-2R_1+R_3\to R_3]{-R_1+R_2\to R_2}\left[\begin{array}{ccc|c}1&-1&2&4\\0&5&-1&0\\0&5&-1&0\end{array}\right]\xrightarrow[-5R_2+R_3\to R_3]{\frac{1}{5}R_2\to R_2\text{ then}}\left[\begin{array}{ccc|c}1&-1&2&4\\0&1&-1/5&0\\0&0&0&0\end{array}\right]\xrightarrow{R_2+R_1\to R_1}$$

$$\left[\begin{array}{ccc|c}1&0&9/5&4\\0&1&-1/5&0\\0&0&0&0\end{array}\right]$$ Solution: $x=4-\dfrac{9}{5}z,\ y=\dfrac{1}{5}z,\ z=z$

**12.** Begin with augmented matrix.

$$\left[\begin{array}{ccc|c}1&-1&3&4\\1&5&2&3\\2&4&5&8\end{array}\right]\xrightarrow[-2R_1+R_3\to R_3]{-R_1+R_2\to R_2}\left[\begin{array}{ccc|c}1&-1&3&4\\0&6&-1&-1\\0&6&-1&0\end{array}\right]\xrightarrow{-R_3+R_2\to R_2}\left[\begin{array}{ccc|c}1&-1&3&4\\0&0&0&-1\\0&6&-1&0\end{array}\right]$$

There is no solution. In $R_2$ we have $0=-1$.

**13.** Your calculator will give, in decimal notation, $A^{-1}$.

$$A^{-1}=\begin{bmatrix}1&2&-1&0\\-4/3&-2/3&1&-2/3\\2/3&-2/3&0&1/3\\1&1&-1&1\end{bmatrix};\qquad \begin{bmatrix}x\\y\\z\\w\end{bmatrix}=A^{-1}\cdot\begin{bmatrix}4\\4\\10\\0\end{bmatrix}=\begin{bmatrix}2\\2\\0\\-2\end{bmatrix}$$

**14.** Your calculator will say that the coefficient matrix has no inverse. We will use the Gauss-Jordan method, give the matrices, and ask you to use the same steps with your calculator.

$$\left[\begin{array}{cccc|c}1&-1&2&-1&4\\1&4&1&1&4\\2&2&4&2&10\\0&-1&1&2&2\end{array}\right]\xrightarrow[-2R_1+R_3\to R_3]{-R_1+R_2\to R_2}\left[\begin{array}{cccc|c}1&-1&2&-1&4\\0&5&-1&2&0\\0&4&0&4&2\\0&-1&1&2&2\end{array}\right]\begin{matrix}-R_4+R_1\to R_1\\5R_4+R_2\to R_2\\4R_4+R_3\to R_3\\\to\end{matrix}\left[\begin{array}{cccc|c}1&0&1&-3&2\\0&0&4&12&10\\0&0&4&12&10\\0&-1&1&2&2\end{array}\right]$$

$$\xrightarrow[-R_4\to R_4]{-R_2+R_3\to R_3}\left[\begin{array}{cccc|c}1&0&1&-3&2\\0&0&4&12&10\\0&0&0&0&0\\0&1&-1&-2&-2\end{array}\right]\xrightarrow[\begin{subarray}{l}\text{1. Interchange } R_2 \text{ and } R_4.\\ \text{2. Interchange } R_2 \text{ and } R_3.\\ \text{3. } \frac{1}{4}R_3\to R_3\end{subarray}]{}\left[\begin{array}{cccc|c}1&0&1&-3&2\\0&1&-1&-2&-2\\0&0&1&3&5/2\\0&0&0&0&0\end{array}\right]\xrightarrow[R_3+R_2\to R_2]{-R_3+R_1\to R_1}$$

$$\left[\begin{array}{cccc|c}1&0&0&-6&-1/2\\0&1&0&1&1/2\\0&0&1&3&5/2\\0&0&0&0&0\end{array}\right]$$ Solution: $x=6w-\dfrac{1}{2},\ y=-w+\dfrac{1}{2},\ z=-3w+\dfrac{5}{2}$

**15.** **a.** $H = \$10,000;$

$B = 75,000 - 3(10,000) = \$45,000;$

$E = 20,000 + 2(10,000) = \$40,000$

**b.** $\$0 \le H \le \$25,000$ (so $B \ge 0$)

$E = 20,000 + 2(0) = \$20,000;$

**c.** $H = \$0;$

$B = 75,000 - 3(0) = \$75,000$

**16.** **a.** $AB = \begin{bmatrix} 0.08 & 0.22 & 0.12 \\ 0.10 & 0.08 & 0.19 \\ 0.05 & 0.07 & 0.09 \\ 0.10 & 0.26 & 0.15 \\ 0.12 & 0.04 & 0.24 \end{bmatrix}$

**b.** Plant type 1: 0.08, 0.22, 0.12 are consumed by carnivores 1, 2, 3, respectively.

**c.** Plant 5 by 1; Plant 4 by 2; Plant 5 by 3.

**17.** **a.** $\begin{bmatrix} 1000 & 4000 & 2000 & 1000 \end{bmatrix}$

**b.** $\begin{bmatrix} 1000 & 4000 & 2000 & 1000 \end{bmatrix} \begin{bmatrix} 10 & 5 & 0 & 0 \\ 5 & 0 & 20 & 10 \\ 5 & 20 & 0 & 10 \\ 5 & 10 & 10 & 10 \end{bmatrix}$

$= \begin{bmatrix} 45,000 & 55,000 & 90,000 & 70,000 \end{bmatrix}$

**c.** $\begin{bmatrix} 5 \\ 3 \\ 4 \\ 4 \end{bmatrix}$

**d.** $\begin{bmatrix} 45,000 & 55,000 & 90,000 & 70,000 \end{bmatrix} \begin{bmatrix} 5 \\ 3 \\ 4 \\ 4 \end{bmatrix}$

$= \begin{bmatrix} 1,030,000 \end{bmatrix}$

**e.** $\begin{bmatrix} 10 & 5 & 0 & 0 \\ 5 & 0 & 20 & 10 \\ 5 & 20 & 0 & 10 \\ 5 & 10 & 10 & 10 \end{bmatrix} \begin{bmatrix} 5 \\ 3 \\ 4 \\ 4 \end{bmatrix} = \begin{bmatrix} 65 \\ 145 \\ 125 \\ 135 \end{bmatrix}$

**18.** **a.** $\begin{matrix} 12 & 5 & 19 & 19 & 27 & 9 & 19 & 27 & 13 & 15 & 18 & 5 \\ L & E & S & S & & I & S & & M & O & R & E \end{matrix} = \begin{bmatrix} 12 & 19 & 27 & 19 & 13 & 18 \\ 5 & 19 & 9 & 27 & 15 & 5 \end{bmatrix}$

Encoding: $\begin{bmatrix} 8 & 5 \\ 3 & 2 \end{bmatrix} \begin{bmatrix} 12 & 19 & 27 & 19 & 13 & 18 \\ 5 & 19 & 9 & 27 & 15 & 5 \end{bmatrix} = \begin{bmatrix} 121 & 247 & 261 & 287 & 179 & 169 \\ 46 & 95 & 99 & 111 & 69 & 64 \end{bmatrix}$

Coded message: 121, 46, 247, 95, 261, 99, 287, 111, 179, 69, 169, 64

**b.** $A^{-1}=\frac{1}{16-15}\begin{bmatrix}2 & -5\\ -3 & 8\end{bmatrix}=\begin{bmatrix}2 & -5\\ -3 & 8\end{bmatrix}$

$$\begin{bmatrix}2 & -5\\ -3 & 8\end{bmatrix}\begin{bmatrix}138 & 140 & 255 & 141 & 201 & 287\\ 54 & 53 & 99 & 54 & 76 & 111\end{bmatrix}=\begin{bmatrix}6 & 15 & 15 & 12 & 22 & 19\\ 18 & 4 & 27 & 9 & 5 & 27\end{bmatrix}$$

| | | | | | | | | | | | |
|---|---|---|---|---|---|---|---|---|---|---|---|
| 6 | 18 | 15 | 4 | 15 | 27 | 12 | 9 | 22 | 5 | 19 | 27 |
| F | R | O | D | O | | L | I | V | E | S | |

**19.** Using the calculator and solving with the inverse is more efficient. This method shows the inefficiency of the Gauss-Jordan method.

$x$ = Growth shares $\quad 30x = 100y + 50z$
$y$ = Blue-chip shares $\quad 4.6x + 11y + 5z = 0.13(120{,}000)$
$z$ = Utility shares $\quad 30x + 100y + 50z = 120{,}000$

$$\left[\begin{array}{ccc|c}30 & -100 & -50 & 0\\ 4.6 & 11 & 5 & 15{,}600\\ 30 & 100 & 50 & 120{,}000\end{array}\right] \underset{-R_1+R_3\to R_3}{\to} \left[\begin{array}{ccc|c}30 & -100 & -50 & 0\\ 4.6 & 11 & 5 & 15{,}600\\ 0 & 200 & 100 & 120{,}000\end{array}\right] \underset{\frac{1}{30}R_1\to R_1}{\to}$$

$$\left[\begin{array}{ccc|c}1 & -10/3 & -5/3 & 0\\ 4.6 & 11 & 5 & 15{,}600\\ 0 & 200 & 100 & 120{,}000\end{array}\right] \underset{-4.6R_1+R_2\to R_2}{\to} \left[\begin{array}{ccc|c}1 & -10/3 & -5/3 & 0\\ 0 & 79/3 & 38/3 & 15{,}600\\ 0 & 200 & 100 & 120{,}000\end{array}\right] \underset{R_2\leftrightarrow R_3}{\to}$$

$$\left[\begin{array}{ccc|c}1 & -10/3 & -5/3 & 0\\ 0 & 200 & 100 & 120{,}000\\ 0 & 79/3 & 38/3 & 15{,}600\end{array}\right] \underset{\frac{1}{200}R_2\to R_2}{\to} \left[\begin{array}{ccc|c}1 & -10/3 & -5/3 & 0\\ 0 & 1 & 1/2 & 600\\ 0 & 79/3 & 38/3 & 15{,}600\end{array}\right] \underset{\frac{10}{3}R_2+R_1\to R_1,\ -\frac{79}{3}R_2+R_3\to R_3}{\to}$$

$$\left[\begin{array}{ccc|c}1 & 0 & 0 & 2000\\ 0 & 1 & 1/2 & 600\\ 0 & 0 & -1/2 & -200\end{array}\right] \underset{-2R_3\to R_3}{\to} \left[\begin{array}{ccc|c}1 & 0 & 0 & 2000\\ 0 & 1 & 1/2 & 600\\ 0 & 0 & 1 & 400\end{array}\right] \underset{-\frac{1}{2}R_3+R_2\to R_2}{\to} \left[\begin{array}{ccc|c}1 & 0 & 0 & 2000\\ 0 & 1 & 0 & 400\\ 0 & 0 & 1 & 400\end{array}\right]$$

Solution: $x = 2000,\ \ y = 400,\ \ z = 400$

**20. a.** The technological equation is $(I - A)X = D$.

$$I-A=\begin{bmatrix}0.6 & -0.2\\ -0.1 & 0.7\end{bmatrix};\qquad (I-A)^{-1}=\frac{1}{0.42-0.02}\begin{bmatrix}0.7 & 0.2\\ 0.1 & 0.6\end{bmatrix}$$

$$\begin{bmatrix}\text{Ag}\\ \text{M}\end{bmatrix}=\frac{1}{0.40}\begin{bmatrix}0.7 & 0.2\\ 0.1 & 0.6\end{bmatrix}\begin{bmatrix}100\\ 140\end{bmatrix}=\begin{bmatrix}1.75 & 0.5\\ 0.25 & 1.5\end{bmatrix}\begin{bmatrix}100\\ 140\end{bmatrix}=\begin{bmatrix}245\\ 235\end{bmatrix}$$ Agriculture = 245; minerals = 235

**b.** The additional units each industry must produce for 4 more units of agricultural surplus are given by:

$$4\begin{bmatrix}\text{Col 1 of}\\ (I-A)^{-1}\end{bmatrix}=4\begin{bmatrix}1.75\\ 0.25\end{bmatrix}=\begin{bmatrix}7\\ 1\end{bmatrix}$$ Agriculture = 7; minerals = 1

**c.** The additional units each industry must produce for 1 more unit of minerals surplus is given by:

$$1\cdot\begin{bmatrix}\text{Col 2 of}\\ (I-A)^{-1}\end{bmatrix}=1\begin{bmatrix}0.5\\ 1.5\end{bmatrix}=\begin{bmatrix}0.5\\ 1.5\end{bmatrix}$$ Agriculture = 0.5; minerals = 1.5

**21.** The technological equation is $(I - A)X = 0$.
The augmented matrix to be solved is

$$\left[\begin{array}{ccc|c}0.6 & -0.3 & -0.4 & 0\\ -0.2 & 0.6 & -0.2 & 0\\ -0.4 & -0.3 & 0.6 & 0\end{array}\right] \begin{array}{l}10R_1\to R_1\\ 5R_2\to R_2\\ 10R_3\to R_3\end{array} \left[\begin{array}{ccc|c}6 & -3 & -4 & 0\\ -1 & 3 & -1 & 0\\ -4 & -3 & 6 & 0\end{array}\right] \begin{array}{l}6R_2+R_1\to R_1\\ \to\\ -4R_2+R_3\to R_3\end{array}$$

$$\left[\begin{array}{rrr|r} 0 & 15 & -10 & 0 \\ -1 & 3 & -1 & 0 \\ 0 & -15 & 10 & 0 \end{array}\right] \begin{array}{c} \rightarrow \\ -R_2 \rightarrow R_2 \\ R_1 + R_3 \rightarrow R_3 \end{array} \left[\begin{array}{rrr|r} 0 & 15 & -10 & 0 \\ 1 & -3 & 1 & 0 \\ 0 & 0 & 0 & 0 \end{array}\right] \begin{array}{c} \frac{1}{15}R_1 \text{ and} \\ \text{then interchange} \rightarrow \\ R_1 \text{ and } R_2 \end{array}$$

$$\left[\begin{array}{rrr|r} 1 & -3 & 1 & 0 \\ 0 & 1 & -2/3 & 0 \\ 0 & 0 & 0 & 0 \end{array}\right] \begin{array}{c} 3R_2 + R_1 \rightarrow R_1 \\ \rightarrow \\ \phantom{x} \end{array} \left[\begin{array}{rrr|r} 1 & 0 & -1 & 0 \\ 0 & 1 & -2/3 & 0 \\ 0 & 0 & 0 & 0 \end{array}\right]$$

Profit = Households  Non-profit = $\frac{2}{3}$Households

**22.**

| | Ag | M | F | S |
|---|---|---|---|---|
| Ag | 0.2 | 0.1 | 0.1 | 0.1 |
| Mach | 0.3 | 0.2 | 0.2 | 0.2 |
| Fuel | 0.2 | 0.2 | 0.3 | 0.3 |
| Steel | 0.1 | 0.4 | 0.2 | 0.2 |

**23.** The technological equation is $(I-A)X = D$.

$$(I-A) = \begin{bmatrix} 0.8 & -0.1 & -0.1 & -0.1 \\ -0.3 & 0.8 & -0.2 & -0.2 \\ -0.2 & -0.2 & 0.7 & -0.3 \\ -0.1 & -0.4 & -0.2 & 0.8 \end{bmatrix} \qquad X = (I-A)^{-1}\begin{bmatrix} 1700 \\ 1900 \\ 900 \\ 300 \end{bmatrix}$$

Note: Use the graphing utility to find $(I-A)^{-1}$.

$$\text{Then } X = \begin{bmatrix} 1.74 & 0.68 & 0.62 & 0.62 \\ 1.30 & 2.30 & 1.18 & 1.18 \\ 1.39 & 1.55 & 2.50 & 1.50 \\ 1.22 & 1.62 & 1.29 & 2.29 \end{bmatrix}\begin{bmatrix} 1700 \\ 1900 \\ 900 \\ 300 \end{bmatrix} = \begin{bmatrix} 5000 \\ 8000 \\ 8000 \\ 7000 \end{bmatrix}$$

Agriculture = 5000; machinery = 8000; fuel = 8000; steel = 7000

**24.** The Gauss-Jordan method must be used since the equation is $(I-A)X = 0$. We begin with the augmented matrix and will use fractions. The student is encouraged to solve with the graphing calculator and work with decimals.

$$\left[\begin{array}{rrrr|r} 3/4 & -1/4 & -3/10 & -1/10 & 0 \\ -3/10 & 3/4 & -2/10 & -4/10 & 0 \\ -3/20 & -2/10 & 9/10 & -3/10 & 0 \\ -3/10 & -3/10 & -4/10 & 8/10 & 0 \end{array}\right] \begin{array}{c} \rightarrow \\ -4R_1 + R_2 \rightarrow R_2 \\ -3R_1 + R_3 \rightarrow R_3 \\ 8R_1 + R_4 \rightarrow R_4 \end{array} \left[\begin{array}{rrrr|r} 3/4 & -1/4 & -3/10 & -1/10 & 0 \\ -33/10 & 7/4 & 1 & 0 & 0 \\ -48/20 & 11/20 & 18/10 & 0 & 0 \\ 57/10 & -23/10 & -28/10 & 0 & 0 \end{array}\right]$$

$$\begin{array}{c} \frac{3}{10}R_2 + R_1 \rightarrow R_1 \\ \rightarrow \\ -\frac{18}{10}R_2 + R_3 \rightarrow R_3 \\ \frac{28}{10}R_2 + R_4 \rightarrow R_4 \end{array} \left[\begin{array}{rrrr|r} -24/100 & 11/40 & 0 & -1/10 & 0 \\ -33/10 & 7/4 & 1 & 0 & 0 \\ 354/100 & -104/40 & 0 & 0 & 0 \\ -354/100 & 104/40 & 0 & 0 & 0 \end{array}\right]$$

$$\begin{array}{c} \rightarrow \\ \\ \\ R_3 + R_4 \rightarrow R_4 \end{array} \left[\begin{array}{rrrr|r} -24/100 & 11/40 & 0 & -1/10 & 0 \\ -33/10 & 7/4 & 1 & 0 & 0 \\ 354/100 & -104/40 & 0 & 0 & 0 \\ 0 & 0 & 0 & 0 & 0 \end{array}\right]$$

$$\xrightarrow{-\frac{40}{104}R_3 \to R_3} \left[\begin{array}{cccc|c} -24/100 & 11/40 & 0 & -1/10 & 0 \\ -33/10 & 7/4 & 1 & 0 & 0 \\ -177/130 & 1 & 0 & 0 & 0 \\ 0 & 0 & 0 & 0 & 0 \end{array}\right] \xrightarrow{\substack{-\frac{11}{40}R_3 + R_1 \to R_1 \\ -\frac{7}{4}R_3 + R_2 \to R_2}} \left[\begin{array}{cccc|c} 699/5200 & 0 & 0 & -1/10 & 0 \\ -477/520 & 0 & 1 & 0 & 0 \\ -177/130 & 1 & 0 & 0 & 0 \\ 0 & 0 & 0 & 0 & 0 \end{array}\right]$$

Our solution yields: $\text{Hhold} = \frac{699}{520}\text{Ag}$, $\text{Fuel} = \frac{477}{520}\text{Ag}$, $\text{Steel} = \frac{177}{130}\text{Ag}$

Now we must solve in terms of Hholds to yield the desired form.

$$\text{Ag} = \frac{520}{699}\text{Hhold}$$

$$\text{Steel} = \frac{177}{130}\left(\frac{520}{699}\text{Hhold}\right) = \frac{236}{233}\text{Hhold}$$

$$\text{Fuel} = \frac{477}{520}\left(\frac{520}{699}\text{Hhold}\right) = \frac{159}{233}\text{Hhold.}$$

This was a difficult problem. The authors' step by step method always works. Sometimes it is not the shortest method.

# Chapter 4: Inequalities and Linear Programming

## *Exercises 4.1*

**1.** $y \le 2x - 1$

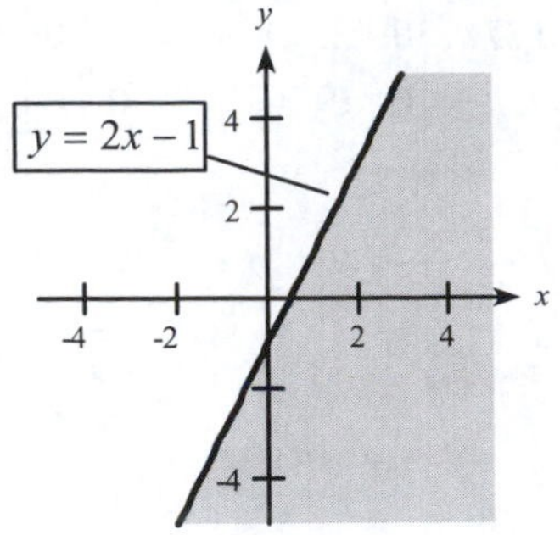


**3.** $\frac{x}{2} + \frac{y}{4} < 1$

$2x + y < 4$

$y < -2x + 4$

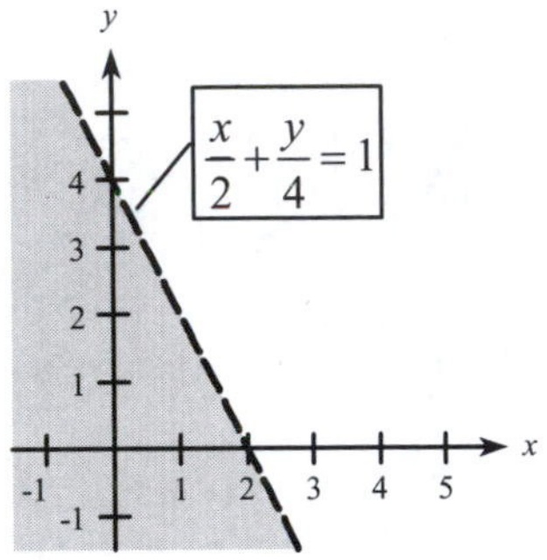


**5.** $0.4x \ge 0.8$

$x \ge 2$

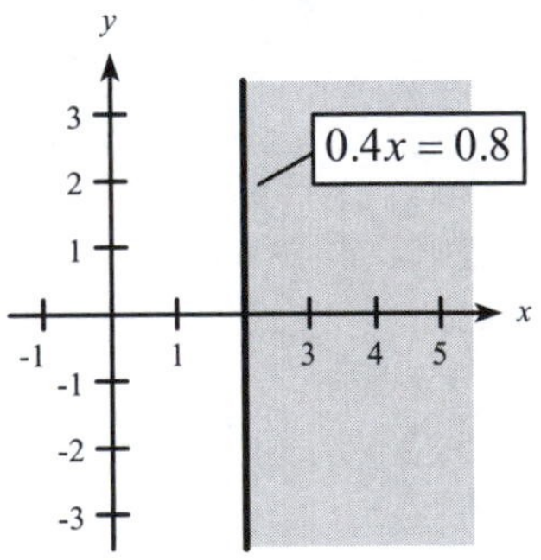


**7.** **a.**

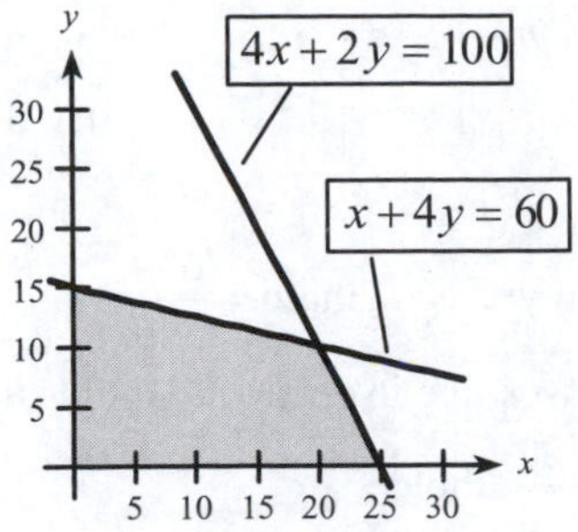


**b.** From the graph we read intercept corners (0, 15), (0, 0), (25, 0).

$4x + 2y = 100 \qquad 8x + 4y = 200$

$x + 4y = 60 \qquad \underline{x + 4y = 60}$

$7x = 140$

$x = 20$

$20 + 4y = 60$ or $y = 10$

Other corner is (20, 10).

**9.** **a.**

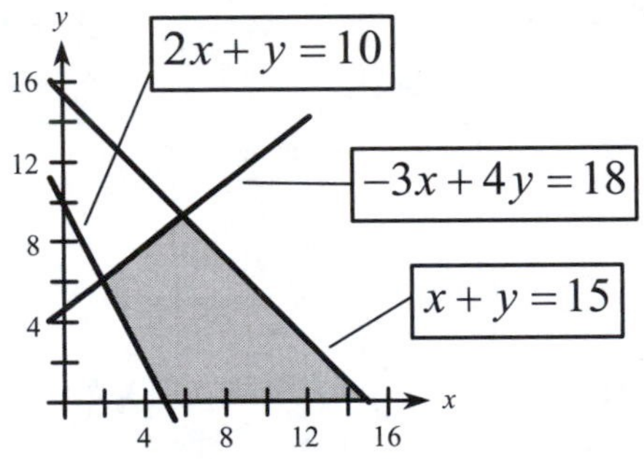


**b.** From the graph we read intercept corners (5,0) and (15,0).

$-3x + 4y = 18 \qquad -3x + 4y = 18$

$\underline{2x + y = 10} \qquad \underline{-8x - 4y = -40}$

$-11x = -22$

$x = 2$

$-3(2) + 4y = 18 \rightarrow 4y = 24 \rightarrow y = 6$

Corner is (2, 6).

$-3x + 4y = 18 \qquad -3x + 4y = 18$

$\underline{x + y = 15} \qquad \underline{3x + 3y = 45}$

$7y = 63$

$y = 9$

$x + 9 = 15 \rightarrow x = 6 \qquad$ Corner is (6, 9)

**11. a.**

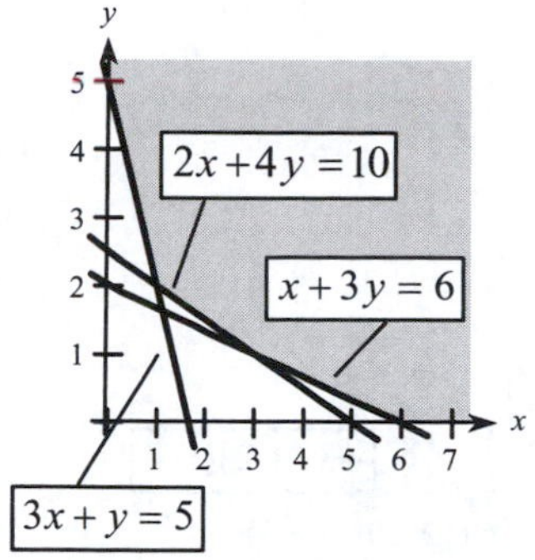


**b.** From the graph we read intercept corners (0, 5), (6, 0).

$$\begin{array}{ll} 3x+y=5 & 12x+4y=20 \\ 2x+4y=10 & \underline{2x+4y=10} \\ & 10x \quad =10 \\ & x \quad =1 \end{array}$$

$3(1)+y=5$ or $y=2$
Corner is (1, 2).

$$\begin{array}{ll} 2x+4y=10 & 2x+4y=10 \\ x+3y=6 & \underline{2x+6y=12} \\ & 2y=2 \\ & y=1 \end{array}$$

$x+3(1)=6$ or $x=3$
Corner is (3, 1)

**13.** $\begin{cases} y<2x \\ y>x-1 \end{cases}$

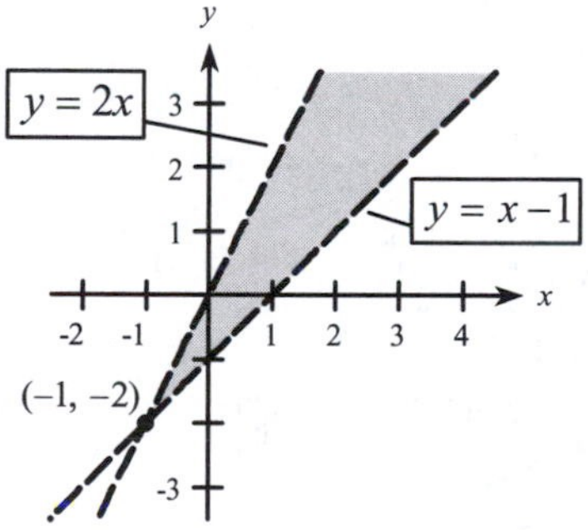


**15.** $\begin{cases} y<-2x+3 \\ y\le \dfrac{1}{2}x+\dfrac{1}{2} \end{cases}$

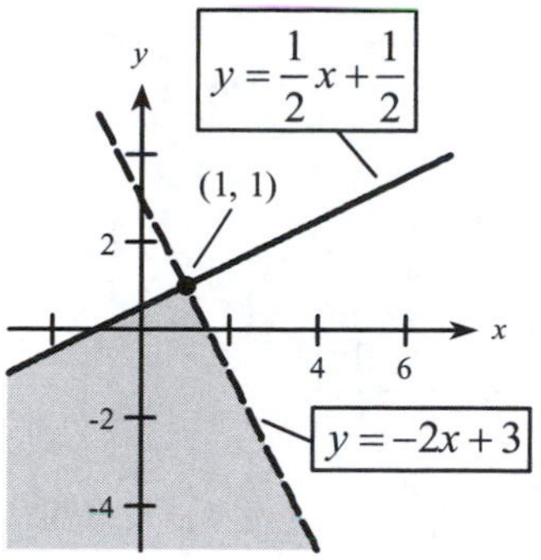


**17.** $\begin{cases} y\le -\dfrac{1}{5}x+40 \\ y\le -\dfrac{2}{3}x+\dfrac{134}{3} \\ x\ge 0,\ \ y\ge 0 \end{cases}$

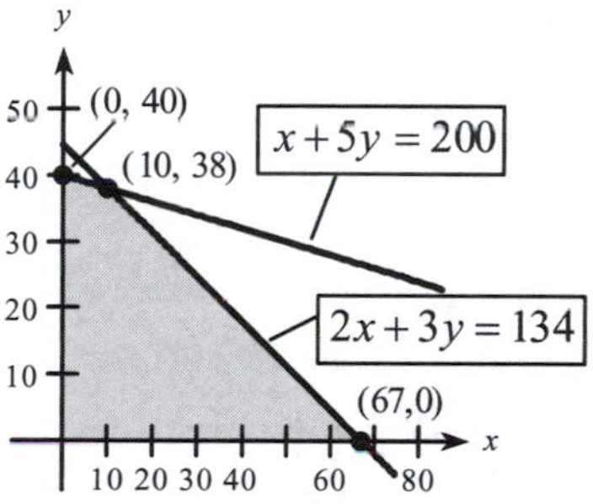


**19.** $\begin{cases} y\le -\dfrac{1}{2}x+24 \\ y\le -x+30 \\ y\le -2x+50 \\ x\ge 0,\ y\ge 0 \end{cases}$

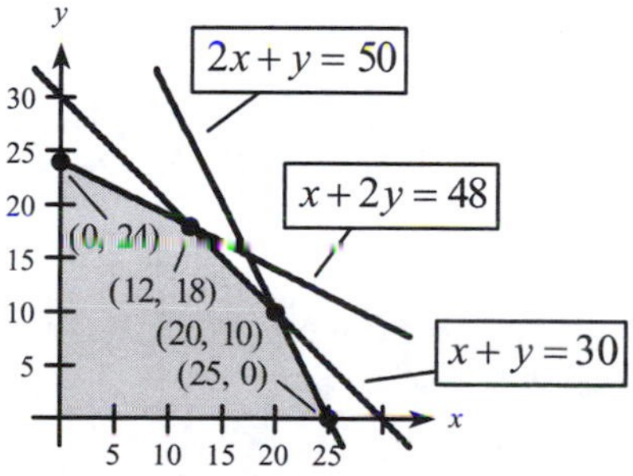

**21.** $\begin{cases} y \ge -\frac{1}{2}x + \frac{19}{2} \\ y \ge -\frac{3}{2}x + \frac{29}{2} \\ x \ge 0, y \ge 0 \end{cases}$

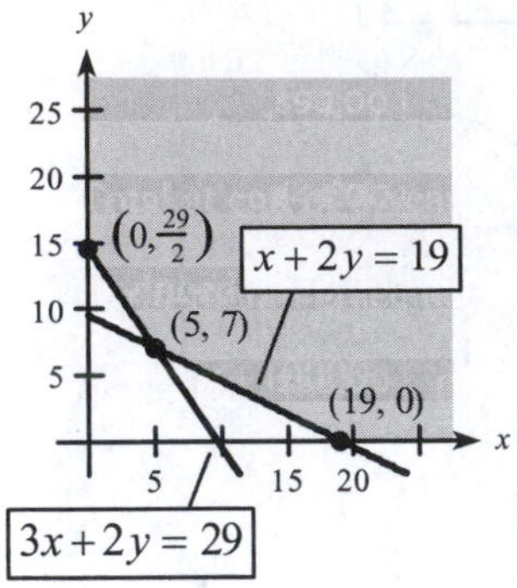


**23.** $\begin{cases} y \ge -\frac{1}{3}x + 1 \\ y \ge -\frac{2}{3}x + \frac{5}{3} \\ y \ge -2x + 3 \\ x \ge 0, y \ge 0 \end{cases}$

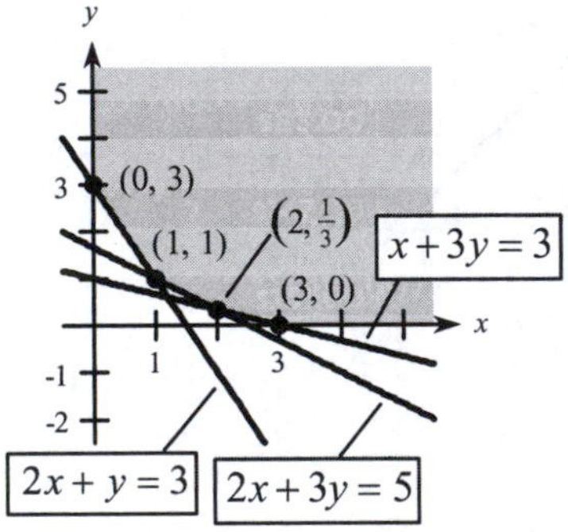


**25.** $\begin{cases} y \ge -\frac{1}{2}x + 10 \\ y \le \frac{3}{2}x + 2 \\ x \ge 12 \\ x \ge 0, \; y \ge 0 \end{cases}$

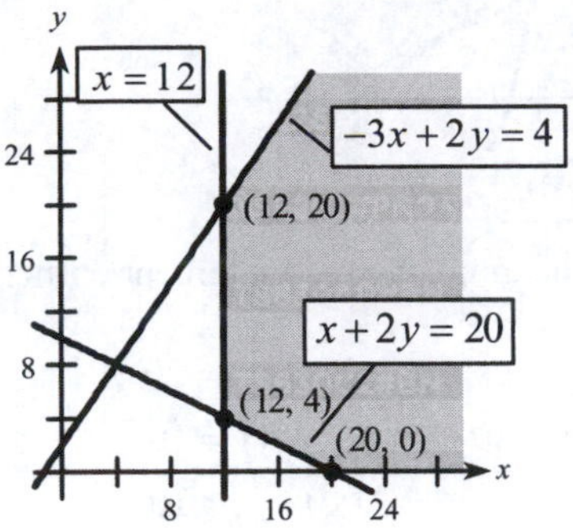


**27.** Let $x$ = number of deluxe models.
Let $y$ = number of economy models.

**a.** $x, y \ge 0 \quad 3x + 2y \le 24 \quad 0.5x + y \le 8$

**b.** Points of Intersection:

$$\begin{array}{r} 3x + 2y = 24 \\ 2(0.5x + y = 8) \\ \hline 2x \quad = 8 \\ x \quad = 4 \end{array}$$

$0.5(4) + y = 8$ or $y = 6$

Feasible corners are $(0, 0), (8, 0), (4, 6)$, and $(0, 8)$.

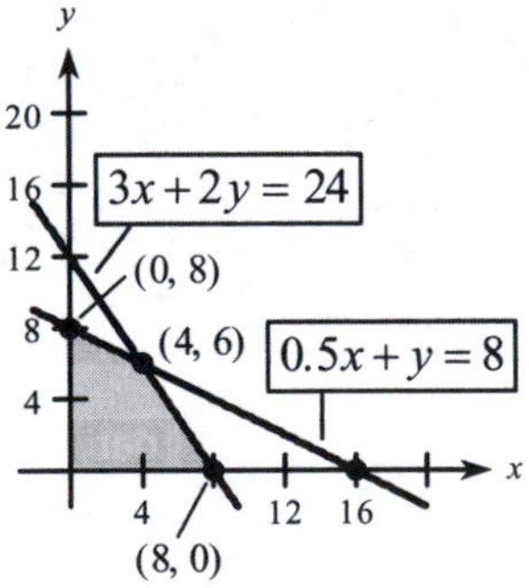


**29.** Let $x$ = number of cord type models and $y$ = number of cordless models.

**a.**

| | |
|---|---|
| $x + y \le 300$ | Packing department |
| $2x + 4y \le 800$ | Manufacturing |
| $x \ge 0$ | |
| $y \ge 0$ | |

**b.**

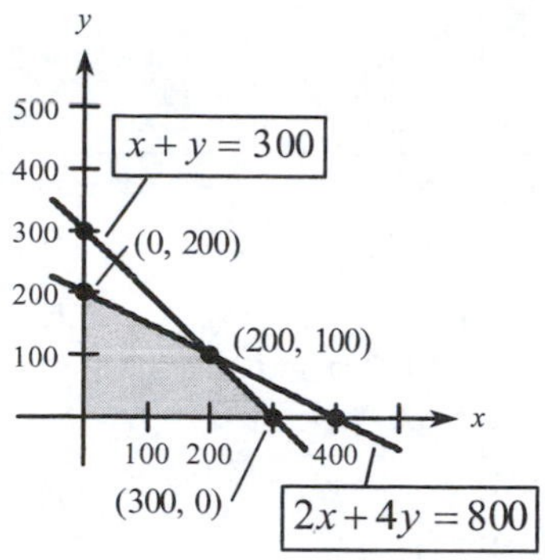


**31.** Let $x$ = minutes of business/investment program commercials
Let $y$ = minutes of sporting events commercials

**a.** $7x+2y \geq 30$ women reached

$4x+12y \geq 28$ men reached

$x \geq 0, \quad y \geq 0$

**b.**

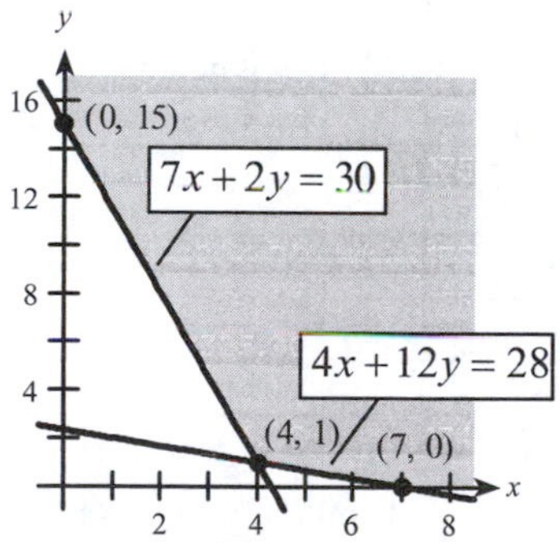


**33.** $x$ = radio minutes
$y$ = television minutes.

**a.** $x+y \geq 80$ Advertising time

$0.006x+0.09y \geq 2.16$ People reached

$x \geq 0$

$y \geq 0$

(Selecting a good scale is the most difficult part of the problem.)

**b.**

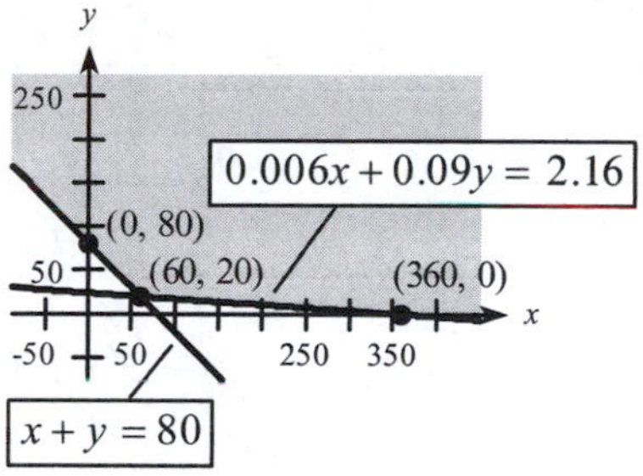


**35.** $x$ = pounds of regular meat
$y$ = pounds of all beef

**a.** $0.18x+0.75y \leq 1020$ (beef)

$0.20x+0.20y \geq 500$ (spices)

$0.30x \leq 600$ (pork)

$x, y \geq 0$

**b.**

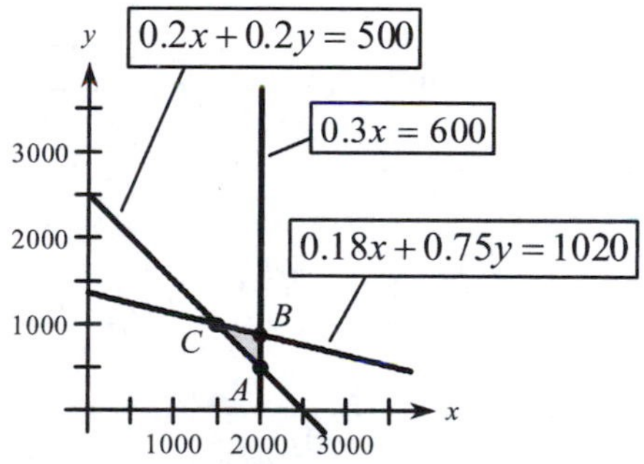


$A = (2000, 500)$
$B = (2000, 880)$
$C = (1500, 1000)$

# Chapter 4: Inequalities and Linear Programming

## *Exercises 4.2*

**1.**

| Corner | $C = 9x + 10y$ |
|---|---|
| $(0,0)$ | 0 |
| $(7,0)$ | 63 |
| $(6,2)$ | 74 |
| $(4,4)$ | 76 |
| $(0,3)$ | 30 |

Maximum $= 76$ at $(4,4)$

Minimum $= 0$ at $(0,0)$

**3.**

| Corner | $f = 5x + 2y$ |
|---|---|
| $(0,6)$ | 12 |
| $(1,3)$ | 11 |
| $(2,1)$ | 12 |
| $(4,0)$ | 20 |

Maximum: None since region continues to increase unboundedly.

Minimum $= 11$ at $(1,3)$.

**5.** $x + 5y = 100$ If $x = 0$, then $y = 20$.

$2x + y = 40$ If $y = 0$, then $x = 20$.

$8x + 5y = 170$

$$\begin{aligned} 10x + 5y &= 200 \\ 8x + 5y &= 170 \\ \hline 2x \quad &= 30 \\ x \quad &= 15 \\ y &= 10 \end{aligned} \qquad \begin{aligned} x + 5y &= 100 \\ 8x + 5y &= 170 \\ \hline 7x \quad &= 70 \\ x \quad &= 10 \\ y &= 18 \end{aligned}$$

Corners: (15, 10) and (10, 18)

| Corner | $f = 3x + 2y$ |
|---|---|
| $(0,0)$ | 0 |
| $(0,20)$ | 40 |
| $(20,0)$ | 60 |
| $(15,10)$ | 65 |
| $(10,18)$ | 66 |

Maximum $= 66$ at $(10,18)$

**7.** $3x + y = 60$

$4x + 10y = 280$

$x + y = 40$

If $x = 0$, then $y = 60$

If $y = 0$, then $x = 70$

$$\begin{aligned} 10x + 10y &= 400 \\ 4x + 10y &= 280 \\ \hline 6x \quad &= 120 \\ x \quad &= 20 \\ y &= 20 \end{aligned} \qquad \begin{aligned} x + y &= 40 \\ 3x + y &= 60 \\ \hline 2x \quad &= 20 \\ x \quad &= 10 \\ y &= 30 \end{aligned}$$

Corners: (20, 20) and (10, 30)

| Corner | $g = 2x + 3y$ |
|---|---|
| $(0,60)$ | 180 |
| $(70,0)$ | 140 |
| $(20,20)$ | 100 |
| $(10,30)$ | 100 |

Minimum $= 100$ at $(20,20)$

No Maximum

**9.** Refer to Problem 19 in Section 4.1, to obtain the corners (0, 24) and (25, 0).

$$\begin{aligned} 2x + y &= 50 \\ x + y &= 30 \\ \hline x \quad &= 20 \end{aligned}$$

(20, 10) is a point of intersection.

$$\begin{aligned} x + 2y &= 48 \\ x + y &= 30 \\ \hline y &= 18 \end{aligned}$$

(12, 18) is a feasible corner.

| Corner | $f = 30x + 50y$ |
|---|---|
| (0, 24) | 1200 |
| (25, 0) | 750 |
| (20, 10) | 1100 |
| (12, 18) | 1260 |

Maximum $= 1260$ at (12, 18).

**11.** Refer to Problem 23 in Section 4.1 to obtain the corners (3, 0) and (0, 3).

$$\begin{aligned} 2x + y &= 3 \\ 2x + 3y &= 5 \\ \hline 2y &= 2 \\ y &= 1 \end{aligned} \qquad \begin{aligned} x + 3y &= 3 \\ 2x + 3y &= 5 \\ \hline x \quad &= 2 \end{aligned}$$

Corner: (1, 1) Corner: $(2, \frac{1}{3})$

| Corner | $g = 100x + 22y$ |
|---|---|
| $(3,0)$ | 300 |
| $(0,3)$ | 66 |
| $(1,1)$ | 122 |
| $(2,\frac{1}{3})$ | $207.\overline{3}$ |

Minimum $= 66$ at (0, 3).

**13.** The graph has enough accuracy to read the feasible corners from the graph.

| Corner | $f = 3x + 4y$ |
|---|---|
| (0, 4) | 16 |
| (2, 4) | 22 |
| (4, 2) | 20 |
| (5, 0) | 15 |
| (0, 0) | 0 |

Maximum = 22 at (2, 4).

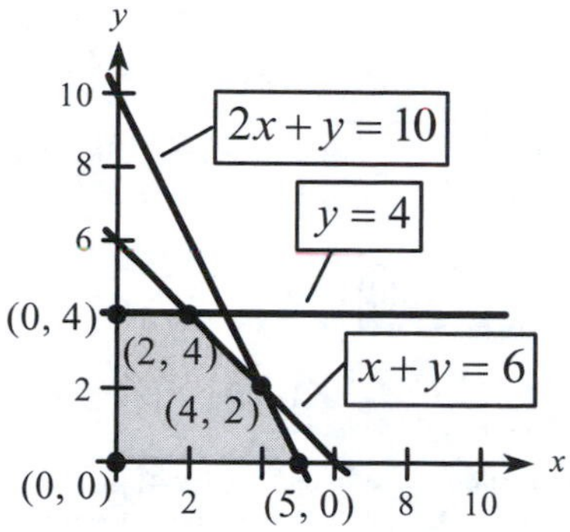


**15.** The feasible corners can be read directly from the graph.

| Corner | $f = 2x + 6y$ |
|---|---|
| (0, 5) | 30 |
| (3, 4) | 30 |
| (5, 2) | 22 |
| (6, 0) | 12 |
| (0, 0) | 0 |

Maximum = 30 at any point on line from (0, 5) to (3, 4)

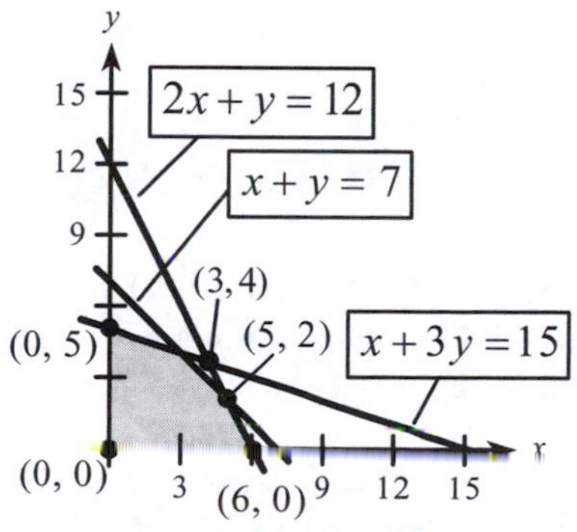


**17.** The feasible corners can be read directly from the graph.

| Corner | $g = 7x + 6y$ |
|---|---|
| (9, 0) | 63 |
| (2, 3) | 32 |
| (0, 8) | 48 |

Minimum = 32 at (3, 2).

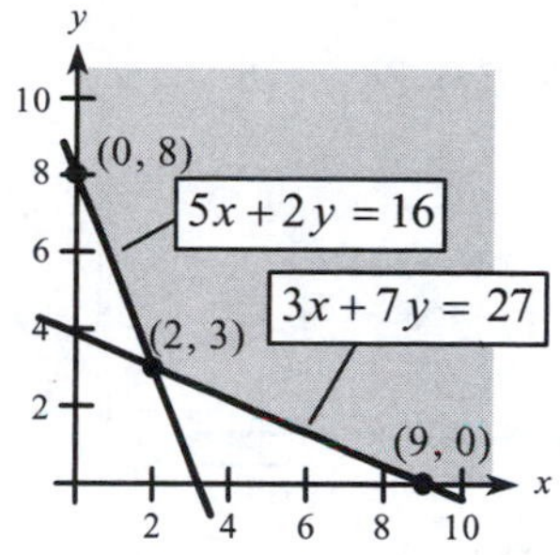


**19.**

$$\begin{aligned} 3x + 2y &= 12 \\ 4x + y &= 11 \end{aligned} \qquad \begin{aligned} 3x + 2y &= 12 \\ 8x + 2y &= 22 \\ \hline 5x &= 10 \\ x = 2,\ y &= 3 \end{aligned}$$

(2, 3) is a feasible corner.

| Corner | $g = 3x + y$ |
|---|---|
| (4, 0) | 12 |
| (0, 11) | 11 |
| (2, 3) | 9 |

Minimum = 9 at (2, 3)

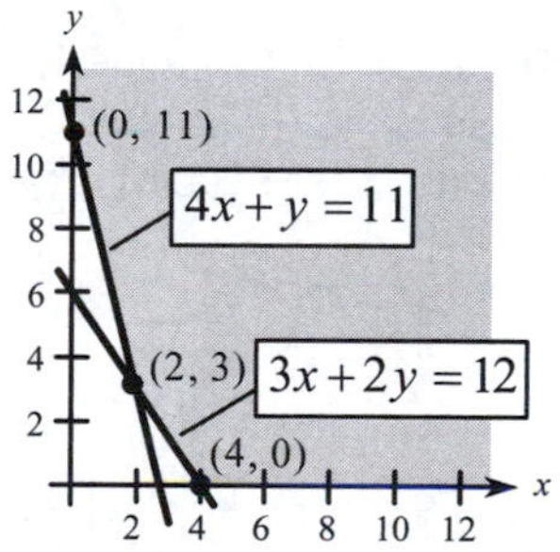

**21.** The feasible corners can be read directly from the graph.

| Corner | $f = x + 2y$ |
|---|---|
| (0, 4) | 8 |
| (2, 4) | 10 |
| (4, 0) | 4 |

Maximum = 10 at (2, 4).

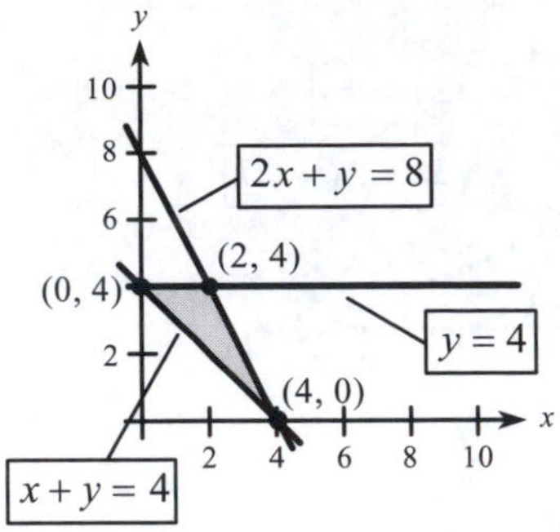


**23.**

$$\begin{aligned} x + y &= 100 \\ -x + y &= 20 \\ \hline 2y &= 120 \\ y &= 60 \\ x &= 40 \end{aligned} \qquad \begin{aligned} x + y &= 100 \\ -2x + 3y &= 30 \\ \\ 2x + 2y &= 200 \\ -2x + 3y &= 30 \\ \hline 5y &= 230 \\ y &= 46 \\ x &= 54 \end{aligned}$$

(40, 60) and (54, 46) are the feasible corners.

| Corner | $g = 40x + 25y$ | |
|---|---|---|
| (40, 60) | 3100 | ← minimum |
| (54, 46) | 3310 | |

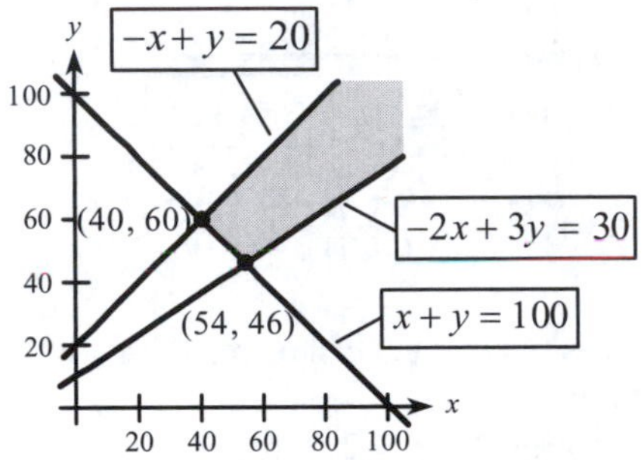


**25.** From the graph of Problem 27, Section 4.1, the corners along the axes are (0, 0), (0, 8), and (8, 0). The other corner (see graph) is (4, 6).

| Corner | $P = 90x + 72y$ | |
|---|---|---|
| (0, 8) | 576 | |
| (8, 0) | 720 | |
| (4, 6) | 792 | ← Maximum profit of \$792 at (4, 6). |
| (0, 0) | 0 | |

**27.** From the graph of Problem 29, Section 4.1, the corners are (0, 200), (0, 0), (300, 0), and (200, 100).

| Corner | $S = 22.50x + 45.00y$ |
|---|---|
| (0, 200) | 9000 |
| (300, 0) | 6750 |
| (200, 100) | 9,000 |
| (0, 0) | 0 |

Maximum profit of \$9,000 occurs at any point with integer coordinates on the segment joining (0, 200) and (200, 100). Possibilities include 0 corded and 200 cordless, 200 corded and 100 cordless, and 20 corded and 190 cordless.

**29.** From the graph of Problem 33, Section 4.1, we have corners (0, 80), (60, 20), and (360, 0).

| Corner | $C = 100x + 500y$ |
|---|---|
| (0, 80) | 40,000 |
| (360, 0) | 36,000 |
| (60, 20) | 16,000 |

← Minimum cost =$16,000 with 60 minutes of radio time and 20 minutes of TV time.

**31.** $x$ = number of inkjet printers
$y$ = number of laser printers
Maximize $P = 40x + 60y$
$x + y \leq 70$ Capacity
$x + 3y \leq 120$ Labor
$x, y \geq 0$

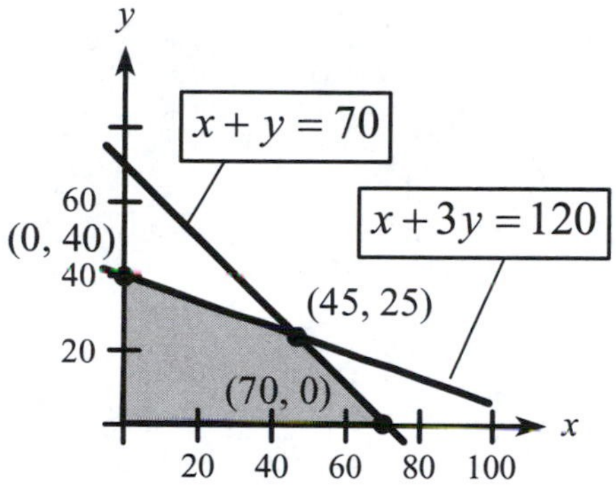


From the graph, we see that (0, 40) and (70, 0) are corners. Solve for the other corner.

$$\begin{array}{r} x + 3y = 120 \\ x + y = 70 \\ \hline \end{array} \qquad \begin{array}{r} x + 3y = 120 \\ -x - y = -70 \\ \hline 2y = 50 \\ y = 25 \end{array}$$

$x + 25 = 70 \rightarrow x = 45$ Corner: (45, 25)

| Corner | $P = 40x + 60y$ |
|---|---|
| (0,40) | \$2400 |
| (45,25) | \$3300 |
| (70,0) | \$2800 |

The maximum profit occurs with 45 inkjet printers and 25 laser printers.

**33.** $x$ = number of bass, $y$ = number of trout
Maximize $f = x + y$
$2x + 5y \leq 800$ Food A
$4x + 2y \leq 800$ Food B
$x, y \geq 0$

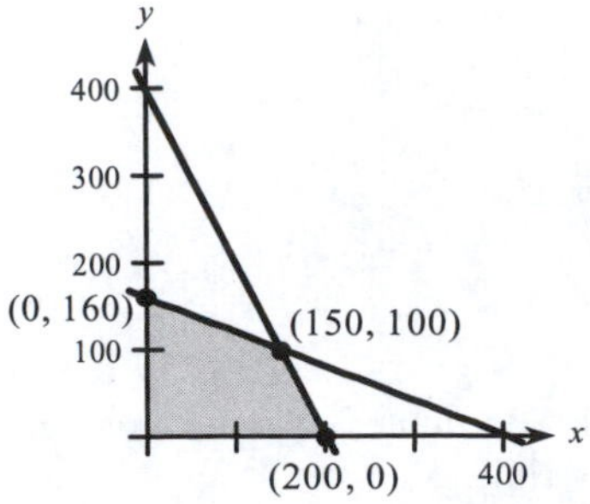


$$2x+5y=800$$
$$4x+2y=800$$

$$\begin{aligned} 4x+10y&=1600\\ 4x+2y&=800\\ \hline 8y&=800\\ y&=100\\ x&=150 \end{aligned}$$

| Corner | $f=x+y$ |
|---|---|
| (0, 160) | 160 |
| (150, 100) | 250 |
| (200, 0) | 200 |

Maximum number of fish is 250 with 150 bass and 100 trout.

**35. a.**

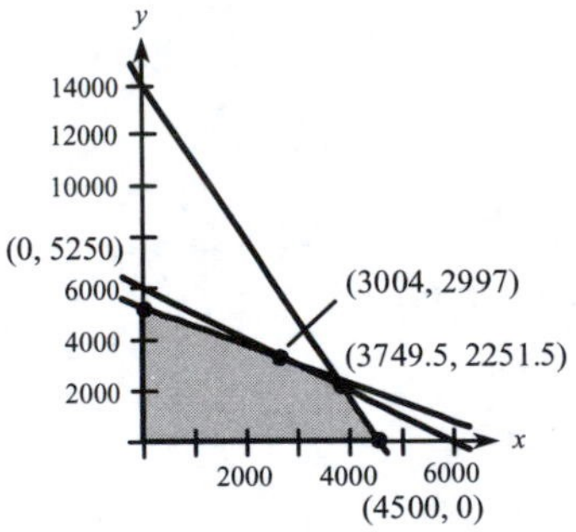


Constraints: $9x+3y \le 40,500$

$x+y \le 6001$

$\frac{3}{4}x+y \le 250$

$x \ge 0,\ y \ge 0$

| Corner | $P=60x+40y$ |
|---|---|
| (0,5250) | \$210,000 |
| (4500,0) | \$270,000 |
| (3749.5,2251.5) | \$315,030 |
| (3004,2997) | \$300,120 |

The maximum is \$315,030 when 3749.5 acres of corn are planted and 2251.5 acres of soybeans are planted.

**b.**

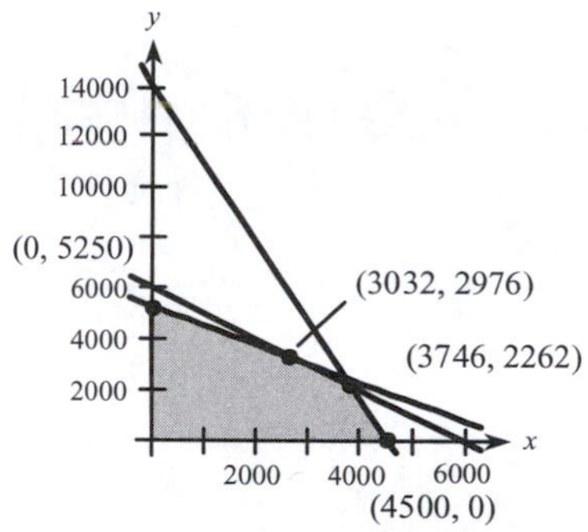


Constraints: $9x + 3y \le 40{,}500$

$x + y \le 6008$

$\frac{3}{4}x + y \le 250$

$x \ge 0,\ y \ge 0$

| Corner | $P = 60x + 40y$ |
|---|---|
| $(0, 5250)$ | $\$210{,}000$ |
| $(4500, 0)$ | $\$270{,}000$ |
| $(3746, 2262)$ | $\$315{,}240$ |
| $(3032, 2976)$ | $\$300{,}960$ |

The maximum is $315,240 when 3746 acres of corn are planted and 2260 acres of soybeans are planted.

**c.** $\$315{,}240 - \$315{,}000 = \$240$, $\$240/8 = \$30$ per acre

**37.** Let $x$ = days to keep Factory 1 open, $y$ = days to keep Factory 2 open

Minimize cost: $f = 10{,}000x + 20{,}000y$

Constraints: $80x + 20y \ge 1600$ Revised Constraints: $4x + y \ge 80$

$10x + 10y \ge 500$ $x + y \ge 50$

$20x + 70y \ge 2000$ $2x + 7y \ge 200$

$x \ge 0,\ y \ge 0$ $x \ge 0,\ y \ge 0$

| Corners | $f = 10{,}000x + 20{,}000y$ |
|---|---|
| $(0, 80)$ | $\$1{,}600{,}000$ |
| $(10, 40)$ | $\$900{,}000$ |
| $(30, 20)$ | $\$700{,}000$ ← Minimum cost is $700,000. |
| $(100, 0)$ | $\$1{,}000{,}000$ |

Solving: $\begin{cases} 4x + y = 80 \\ x + y = 50 \end{cases}$

$\begin{cases} 4x + y = 80 \\ -x - y = -50 \end{cases}$

$3x = 30$

$x = 10,\ y = 40$

Solving: $\begin{cases} x + y = 50 \\ 2x + 7y = 200 \end{cases}$

$\begin{cases} -2x + -2y = -100 \\ 2x + 7y = 200 \end{cases}$

$5y = 100$

$y = 20,\ x = 30$

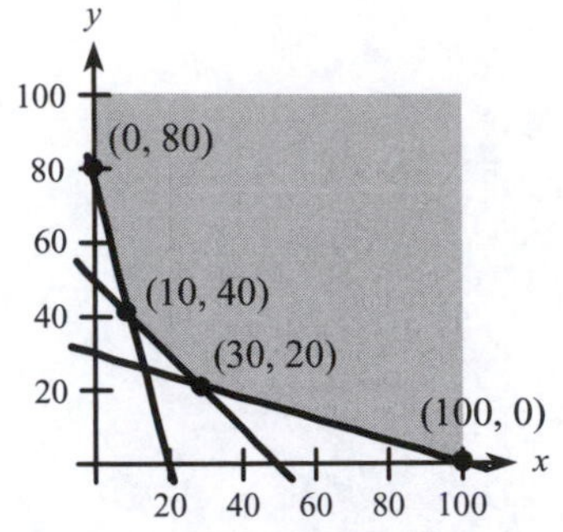


**39.** $x$ = days at location I, $y$ = days at location II
Objective function: $F = 500x + 800y$

Constraints: $x, y \geq 0$

$10x + 20y \geq 2000$ Deluxe

$20x + 50y \geq 4200$ Better

$13x + 6y \geq 1200$ Standard

$x + 2y = 200$ $13x + 6y = 1200$

$2x + 5y = 420$ $x + 2y = 200$

$A(160, 20)$ $B(60, 70)$

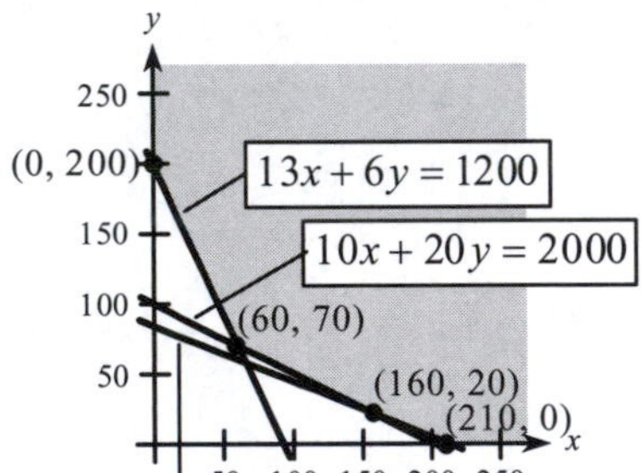


| Corner | $F = 500x + 800y$ |
|---|---|
| (160, 20) | 96,000 |
| (60, 70) | 86,000 |
| (0, 200) | 160,000 |
| (210, 0) | 105,000 |

Minimum costs of \$86,000 occurs with 60 days at location I and 70 days at location II.

**41.** See the graph for Problem 35, in Section 4.1, From the graph, feasible corners are (2000, 500), (2000, 880), and (1500, 1000).

| Corner | $P = 1.50x + y$ |
|---|---|
| (1500, 1000) | 3250 |
| (2000, 500) | 3500 |
| (2000, 880) | 3880 |

Maximum profit = \$3880 with 2000 lbs of regular and 880 lbs of all beef.

**43.** Let $x$ = number from $P$ to $B$; let $y$ = number from $P$ to $Y$.
Then $35 - x$ = number from $E$ to $B$; $40 - y$ = number from $E$ to $Y$.
Objective function: $C = 18x + 20(35 - x) + 22y + 25(40 - y)$
$= 1700 - 2x - 3y$

Constraints: $x, y \geq 0$

$x + y \leq 60$ P inventory or

$(35 - x) + (40 - y) \leq 30$ E inventory

$x + y \geq 45$

$y \leq 40$ maximum order size

$x \leq 35$ maximum order size

Drawing the graph and obtaining the feasible corners.

| Corner | $C = 1700 - 2x - 3y$ |
|---|---|
| (5, 40) | 1570 |
| (20, 40) | 1540 |
| (35, 25) | 1555 |
| (35, 10) | 1600 |

The minimum cost is $1540.

| Ship From/To | B | Y |
|---|---|---|
| P | 20 | 40 |
| E | 15 | 0 |

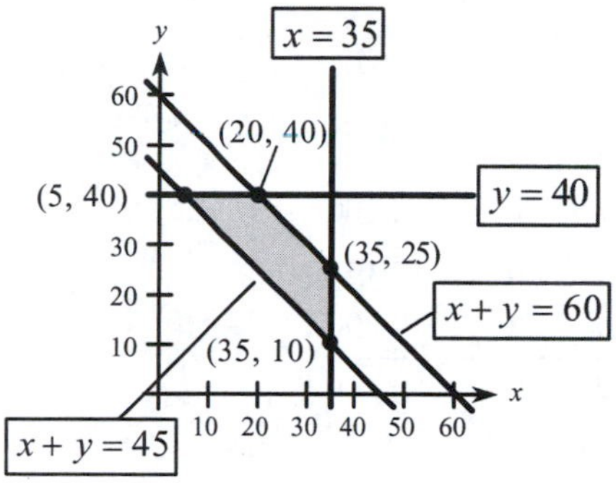


**45.** $x$ = satellite branches, $y$ = full service
Objective function: $R = 10{,}000x + 18{,}000y$

Constraints: $100{,}000x + 140{,}000y \leq 2{,}980{,}000$ Construction costs $\leftrightarrow$ $10x + 14y \leq 298$

$3x + 6y \leq 120$ Employees

$x + y \leq 25$ Number of branches

$x, y \geq 0$

**a.**
$$10x + 14y = 298 \quad 15x + 21y = 447$$
$$3x + 6y = 120 \quad 15x + 30y = 600$$
$$9y = 153$$
$$y = 17$$
$$x = 6$$

$$10x + 14y = 298 \quad 10x + 14y = 298$$
$$x + y = 25 \quad 10x + 10y = 250$$
$$4y = 48$$
$$y = 12$$
$$x = 13$$

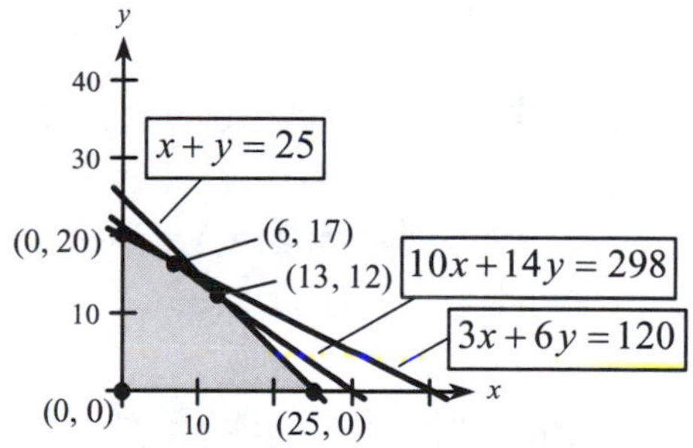


| Corner | $R = 1000(10x + 18y)$ |
|---|---|
| (13,12) | 346,000 |
| (6,17) | 366,000 |
| (25,0) | 250,000 |
| (0,20) | 360,000 |

Maximum revenue of $366,000 with 6 satellite and 17 full service branches

**b.** Branches: Used 23 of 25 possible; 2 not used (slack)
New employees: hired 120 of 120 possible; 0 not hired (slack)

Budget: used all \$2.98 million; \$0 not used (slack)

**c.** Additional new employees and additional budget. These items are completely used in the current optimal solution; more could change and improve the optimal solution.

**d.** Additional branches. The current optimal solution does not use all those allotted; more would just add to the extras.

## *Exercises 4.3*

**1.** $3x+5y+s_1=15 \quad 3x+6y+s_2=20$

**3.**
$$\begin{array}{ccccc} x & y & s_1 & s_2 & f \end{array}$$
$$\left(\begin{array}{ccccc|c} 2 & 5 & 1 & 0 & 0 & 400 \\ 1 & 2 & 0 & 1 & 0 & 175 \\ \hline -3 & -7 & 0 & 0 & 1 & 0 \end{array}\right)$$

**5.**
$$\begin{array}{ccccccc} x & y & z & s_1 & s_2 & s_3 & f \end{array}$$
$$\left(\begin{array}{ccccccc|c} 2 & 7 & 9 & 1 & 0 & 0 & 0 & 100 \\ 6 & 5 & 1 & 0 & 1 & 0 & 0 & 145 \\ 1 & 2 & 7 & 0 & 0 & 1 & 0 & 90 \\ \hline -2 & -5 & -2 & 0 & 0 & 0 & 1 & 0 \end{array}\right)$$

**7.** $f$ is maximized at 20 when $x=11$ and $y=9$.

**9.** $f$ is maximized at 525 when $y=14$, $z=11$, and $x=0$.

**11. a.** $x_1=0,\ x_2=45,\ s_1=14,\ s_2=0,\ f=75$

**b.** not complete

**c.**
$$\left[\begin{array}{ccccc|c} \textcircled{2} & 0 & 1 & -\frac{3}{4} & 0 & 14 \\ 3 & 1 & 0 & \frac{1}{3} & 0 & 45 \\ \hline -6 & 0 & 0 & 3 & 1 & 75 \end{array}\right]$$

$R_1: \dfrac{14}{2}=7 \qquad R_2=\dfrac{45}{3}=15$ Thus, pivot on 2 in $R_1C_1$.

Row Operations: $\frac{1}{2}R_1 \to R_1$, then $-3R_1+R_2 \to R_2,\ 6R_1+R_3 \to R_3$

**13. a.** $x_1=0,\ x_2=0,\ s_1=200,\ s_2=400,\ s_3=350,\ f=0$

**b.** not complete

**c.**
$$\left[\begin{array}{cccccc|c} \textcircled{10} & 27 & 1 & 0 & 0 & 0 & 200 \\ 4 & 51 & 0 & 1 & 0 & 0 & 400 \\ 15 & 27 & 0 & 0 & 1 & 0 & 350 \\ \hline -8 & -7 & 0 & 0 & 0 & 1 & 0 \end{array}\right]$$

$R_1: \dfrac{200}{10}=20 \qquad R_2: \dfrac{400}{4}=100 \qquad R_3: \dfrac{350}{15}=23.\overline{3}$ Thus, pivot on 10 in $R_1C_1$.

Row Operations: $\frac{1}{10}R_1 \to R_1$, then $-4R_1+R_2 \to R_2$, $-15R_1+R_3 \to R_3$, $8R_1+R_4 \to R_4$

**15. a.** $x_1 = 24,\ x_2 = 0,\ x_3 = 21,\ s_1 = 16,\ s_2 = 0,\ s_3 = 0,\ f = 780$

**b.** Because there are no negative indicators, the solution is complete.

**17. a.** $x_1 = 0,\ x_2 = 0,\ x_3 = 12,\ s_1 = 4,\ s_2 = 6,\ s_3 = 0,\ f = 150$

**b.** not complete

**c.** The pivot could either be in column 1 or 2.
If you choose column 1:

$$R_1 : \frac{12}{4} = 3 \qquad R_2 : \frac{4}{2} = 2 \qquad \text{Thus, pivot on the 2 in } R_2C_1.$$

If you choose column 2:

$$R_1 : \frac{12}{4} = 3 \qquad R_2 : \frac{4}{4} = 1 \qquad \text{Thus, pivot on 4 in } R_2C_2.$$

For row operations, assume we use the 4 in $R_2C_2$:

$$\frac{1}{4}R_2 \to R_2, \text{ then } -4R_2 + R_1 \to R_1,\ 11R_2 + R_3 \to R_3,\ 3R_2 + R_4 \to R_4$$

**19.** $$\left(\begin{array}{cccccc|c} 14 & \boxed{7} & 1 & 0 & 0 & & 35 \\ 5 & 5 & 0 & 1 & 0 & & 50 \\ \hline -3 & -10 & 0 & 0 & 1 & & 0 \end{array}\right) \begin{array}{c} \frac{1}{7}R_1 \to R_1 \\ \to \\ \ \end{array}$$

– 10 is the most negative. Pivot on the 7.

$$\left[\begin{array}{ccccc|c} 2 & 1 & \frac{1}{7} & 0 & 0 & 5 \\ 5 & 5 & 0 & 1 & 0 & 50 \\ \hline -3 & -10 & 0 & 0 & 1 & 0 \end{array}\right] \begin{array}{c} \to \\ -5R_1 + R_2 \to R_2 \\ 10R_1 + R_3 \to R_3 \end{array} \left[\begin{array}{ccccc|c} 2 & 1 & \frac{1}{7} & 0 & 0 & 5 \\ -5 & 0 & -\frac{5}{7} & 1 & 0 & 25 \\ \hline 17 & 0 & \frac{10}{7} & 0 & 1 & 50 \end{array}\right]$$

All indicators are $\geq 0$. The maximum is 50 when $y = 5$ and $x = 0$.

**21.** $$\left[\begin{array}{ccccc|c} 1 & \boxed{2} & 1 & 0 & 0 & 10 \\ 1 & 1 & 0 & 1 & 0 & 7 \\ \hline -2 & -3 & 0 & 0 & 1 & 0 \end{array}\right] \frac{1}{2}R_1 \to R_1 \left[\begin{array}{ccccc|c} \frac{1}{2} & 1 & \frac{1}{2} & 0 & 0 & 5 \\ 1 & 1 & 0 & 1 & 0 & 7 \\ \hline -2 & -3 & 0 & 0 & 1 & 0 \end{array}\right] \begin{array}{c} \to \\ -R_1 + R_2 \to R_2 \\ 3R_1 + R_3 \to R_3 \end{array}$$

$$\left[\begin{array}{ccccc|c} \frac{1}{2} & 1 & \frac{1}{2} & 0 & 0 & 5 \\ \boxed{\frac{1}{2}} & 0 & -\frac{1}{2} & 1 & 0 & 2 \\ \hline -\frac{1}{2} & 0 & \frac{3}{2} & 0 & 1 & 15 \end{array}\right] 2R_2 \to R_2 \left[\begin{array}{ccccc|c} \frac{1}{2} & 1 & \frac{1}{2} & 0 & 0 & 5 \\ 1 & 0 & -1 & 2 & 0 & 4 \\ \hline -\frac{1}{2} & 0 & \frac{3}{2} & 0 & 1 & 15 \end{array}\right] \begin{array}{c} -\frac{1}{2}R_2 + R_1 \to R_1 \\ \to \\ \frac{1}{2}R_2 + R_3 \to R_3 \end{array} \left[\begin{array}{ccccc|c} 0 & 1 & 1 & -1 & 0 & 3 \\ 1 & 0 & -1 & 2 & 0 & 4 \\ \hline 0 & 0 & 1 & 1 & 1 & 17 \end{array}\right]$$

Maximum is 17 at $x = 4, y = 3$.

**23.** $\left[\begin{array}{cccccc|c} -1 & 1 & 1 & 0 & 0 & 0 & 2 \\ 1 & 2 & 0 & 1 & 0 & 0 & 10 \\ \boxed{3} & 1 & 0 & 0 & 1 & 0 & 15 \\ \hline -2 & -1 & 0 & 0 & 0 & 1 & 0 \end{array}\right]$ $\begin{array}{l} \frac{2}{(-1)} < 0 \\ \frac{10}{1} = 10 \\ \frac{15}{3} = 5 \quad \frac{1}{3}R_3 \to R_3 \end{array}$ $\to$ $\left[\begin{array}{cccccc|c} -1 & 1 & 1 & 0 & 0 & 0 & 2 \\ 1 & 2 & 0 & 1 & 0 & 0 & 10 \\ 1 & \frac{1}{3} & 0 & 0 & \frac{1}{3} & 0 & 5 \\ \hline -2 & -1 & 0 & 0 & 0 & 1 & 0 \end{array}\right]$ $\begin{array}{c} R_3 + R_1 \to R_1 \\ -R_3 + R_2 \to R_2 \\ \to \\ 2R_3 + R_4 \to R_4 \end{array}$

$\left[\begin{array}{cccccc|c} 0 & \frac{4}{3} & 1 & 0 & \frac{1}{3} & 0 & 7 \\ 0 & \boxed{\frac{5}{3}} & 0 & 1 & -\frac{1}{3} & 0 & 5 \\ 1 & \frac{1}{3} & 0 & 0 & \frac{1}{3} & 0 & 5 \\ \hline 0 & -\frac{1}{3} & 0 & 0 & \frac{2}{3} & 1 & 10 \end{array}\right]$ $\begin{array}{l} 7 \div \left(\frac{4}{3}\right) = \frac{21}{4} \quad \to \\ 5 \div \left(\frac{5}{3}\right) = 3 \quad \frac{3}{5}R_2 \to R_2 \\ 5 \div \left(\frac{1}{3}\right) = 15 \end{array}$ $\left[\begin{array}{cccccc|c} 0 & \frac{4}{3} & 1 & 0 & \frac{1}{3} & 0 & 7 \\ 0 & 1 & 0 & \frac{3}{5} & -\frac{1}{5} & 0 & 3 \\ 1 & \frac{1}{3} & 0 & 0 & \frac{1}{3} & 0 & 5 \\ \hline 0 & -\frac{1}{3} & 0 & 0 & \frac{2}{3} & 1 & 10 \end{array}\right]$ $\begin{array}{c} -\frac{4}{3}R_2 + R_1 \to R_1 \\ \to \\ -\frac{1}{3}R_2 + R_3 \to R_3 \\ \frac{1}{3}R_2 + R_4 \to R_4 \end{array}$

$\left[\begin{array}{cccccc|c} 0 & 0 & 1 & -\frac{4}{5} & \frac{3}{5} & 0 & 3 \\ 0 & 1 & 0 & \frac{3}{5} & -\frac{1}{5} & 0 & 3 \\ 1 & 0 & 0 & -\frac{1}{5} & \frac{2}{5} & 0 & 4 \\ \hline 0 & 0 & 0 & \frac{1}{5} & \frac{3}{5} & 1 & 11 \end{array}\right]$

Maximum is 11 at $x = 4$, $y = 3$.

**25.** $\left[\begin{array}{ccccccc|c} 3 & 5 & 4 & 1 & 0 & 0 & 0 & 30 \\ 3 & \boxed{2} & 0 & 0 & 1 & 0 & 0 & 4 \\ 1 & 2 & 0 & 0 & 0 & 1 & 0 & 8 \\ \hline -7 & -10 & -4 & 0 & 0 & 0 & 1 & 0 \end{array}\right]$ $\begin{array}{c} \frac{1}{2}R_2 \to R_2 \\ \to \end{array}$ $\left[\begin{array}{ccccccc|c} 3 & 5 & 4 & 1 & 0 & 0 & 0 & 30 \\ \frac{3}{2} & 1 & 0 & 0 & \frac{1}{2} & 0 & 0 & 2 \\ 1 & 2 & 0 & 0 & 0 & 1 & 0 & 8 \\ \hline -7 & -10 & -4 & 0 & 0 & 0 & 1 & 0 \end{array}\right]$ $\begin{array}{c} -5R_2 + R_1 \to R_1 \\ \to \\ -2R_2 + R_3 \to R_3 \\ 10R_2 + R_4 \to R_4 \end{array}$

$\left[\begin{array}{ccccccc|c} -\frac{9}{2} & 0 & \boxed{4} & 1 & -\frac{5}{2} & 0 & 0 & 20 \\ \frac{3}{2} & 1 & 0 & 0 & \frac{1}{2} & 0 & 0 & 2 \\ -2 & 0 & 0 & 0 & -1 & 1 & 0 & 4 \\ \hline 8 & 0 & -4 & 0 & 5 & 0 & 1 & 20 \end{array}\right]$ $\begin{array}{c} \frac{1}{4}R_1 \to R_1 \\ \to \end{array}$ $\left[\begin{array}{ccccccc|c} -\frac{9}{8} & 0 & 1 & \frac{1}{4} & -\frac{5}{8} & 0 & 0 & 5 \\ \frac{3}{2} & 1 & 0 & 0 & \frac{1}{2} & 0 & 0 & 2 \\ -2 & 0 & 0 & 0 & -1 & 1 & 0 & 4 \\ \hline 8 & 0 & -4 & 0 & 5 & 0 & 1 & 20 \end{array}\right]$ $\begin{array}{c} \to \\ 4R_1 + R_4 \to R_4 \end{array}$

$\left[\begin{array}{ccccccc|c} -\frac{9}{8} & 0 & 1 & \frac{1}{4} & -\frac{5}{8} & 0 & 0 & 5 \\ \frac{3}{2} & 1 & 0 & 0 & \frac{1}{2} & 0 & 0 & 2 \\ -2 & 0 & 0 & 0 & -1 & 1 & 0 & 4 \\ \hline \frac{7}{2} & 0 & 0 & 1 & \frac{5}{2} & 0 & 1 & 40 \end{array}\right]$

Maximum is 40 at $x = 0$, $y = 2$, $z = 5$.

**27.** $\left[\begin{array}{ccccccc|c} 1 & \boxed{4} & 0 & 1 & 0 & 0 & 0 & 12 \\ 3 & 6 & 4 & 0 & 1 & 0 & 0 & 48 \\ 0 & 1 & 1 & 0 & 0 & 1 & 0 & 8 \\ \hline -1 & -3 & -1 & 0 & 0 & 0 & 1 & 0 \end{array}\right]$ $\frac{1}{4}R_1 \to R_1$ $\left[\begin{array}{ccccccc|c} \frac{1}{4} & 1 & 0 & \frac{1}{4} & 0 & 0 & 0 & 3 \\ 3 & 6 & 4 & 0 & 1 & 0 & 0 & 48 \\ 0 & 1 & 1 & 0 & 0 & 1 & 0 & 8 \\ \hline -1 & -3 & -1 & 0 & 0 & 0 & 1 & 0 \end{array}\right]$ $\to$ $-6R_1 + R_2 \to R_2$, $-R_1 + R_3 \to R_3$, $3R_1 + R_4 \to R_4$

$\left[\begin{array}{ccccccc|c} \frac{1}{4} & 1 & 0 & \frac{1}{4} & 0 & 0 & 0 & 3 \\ \frac{3}{2} & 0 & 4 & -\frac{3}{2} & 1 & 0 & 0 & 30 \\ -\frac{1}{4} & 0 & \boxed{1} & -\frac{1}{4} & 0 & 1 & 0 & 5 \\ \hline -\frac{1}{4} & 0 & -1 & \frac{3}{4} & 0 & 0 & 1 & 9 \end{array}\right]$ $\to$ $-4R_3 + R_2 \to R_2$ $\to$ $R_3 + R_4 \to R_4$ $\left[\begin{array}{ccccccc|c} \frac{1}{4} & 1 & 0 & \frac{1}{4} & 0 & 0 & 0 & 3 \\ \boxed{\frac{5}{2}} & 0 & 0 & -\frac{1}{2} & 1 & -4 & 0 & 10 \\ -\frac{1}{4} & 0 & 1 & -\frac{1}{4} & 0 & 1 & 0 & 5 \\ \hline -\frac{1}{2} & 0 & 0 & \frac{1}{2} & 0 & 1 & 1 & 14 \end{array}\right]$ $\to$ $\frac{2}{5}R_2 \to R_2$

$\left[\begin{array}{ccccccc|c} \frac{1}{4} & 1 & 0 & \frac{1}{4} & 0 & 0 & 0 & 3 \\ 1 & 0 & 0 & -\frac{1}{5} & \frac{2}{5} & -\frac{8}{5} & 0 & 4 \\ -\frac{1}{4} & 0 & 1 & -\frac{1}{4} & 0 & 1 & 0 & 5 \\ \hline -\frac{1}{2} & 0 & 0 & \frac{1}{2} & 0 & 1 & 1 & 14 \end{array}\right]$ $-\frac{1}{4}R_2 + R_1 \to R_1$ $\to$ $\frac{1}{4}R_2 + R_3 \to R_3$, $\frac{1}{2}R_2 + R_4 \to R_4$ $\left[\begin{array}{ccccccc|c} 0 & 1 & 0 & \frac{3}{10} & -\frac{1}{10} & \frac{2}{5} & 0 & 2 \\ 1 & 0 & 0 & -\frac{1}{5} & \frac{2}{5} & -\frac{8}{5} & 0 & 4 \\ 0 & 0 & 1 & -\frac{3}{10} & \frac{1}{10} & \frac{3}{5} & 0 & 6 \\ \hline 0 & 0 & 0 & \frac{2}{5} & \frac{1}{5} & \frac{1}{5} & 1 & 16 \end{array}\right]$

Solution: Maximum = 16 at $x = 4$, $y = 2$, $z = 6$

**29.** $\left[\begin{array}{cccccccc|c} 12 & 16 & 20 & 8 & 1 & 0 & 0 & 0 & 880 \\ 5 & \boxed{10} & 10 & 5 & 0 & 1 & 0 & 0 & 460 \\ 10 & -5 & 8 & 5 & 0 & 0 & 1 & 0 & 280 \\ \hline -24 & -30 & -18 & -18 & 0 & 0 & 0 & 1 & 0 \end{array}\right]$ $\frac{1}{10}R_2 \to R_2$ then $-16R_2 + R_1 \to R_1$, $5R_2 + R_3 \to R_3$, $30R_2 + R_4 \to R_4$

$\left[\begin{array}{cccccccc|c} \boxed{4} & 0 & 4 & 0 & 1 & -1.6 & 0 & 0 & 144 \\ 0.5 & 1 & 1 & 0.5 & 0 & 0.1 & 0 & 0 & 46 \\ 12.5 & 0 & 13 & 7.5 & 0 & 0.5 & 1 & 0 & 510 \\ \hline -9 & 0 & 12 & -3 & 0 & 3 & 0 & 1 & 1380 \end{array}\right]$ $\frac{1}{4}R_1 \to R_1$ then $-0.5R_1 + R_2 \to R_2$, $-12.5R_1 + R_3 \to R_3$, $9R_1 + R_4 \to R_4$

$\left[\begin{array}{cccccccc|c} 1 & 0 & 1 & 0 & 0.25 & -0.4 & 0 & 0 & 36 \\ 0 & 1 & 0.5 & 0.5 & -0.125 & 0.3 & 0 & 0 & 28 \\ 0 & 0 & 0.5 & \boxed{7.5} & -3.125 & 5.5 & 1 & 0 & 60 \\ \hline 0 & 0 & 21 & -3 & 2.25 & -0.6 & 0 & 1 & 1704 \end{array}\right]$ $\frac{2}{15}R_3 \to R_3$ then $-0.5R_3 + R_2 \to R_2$, $3R_3 + R_4 \to R_4$

$\left[\begin{array}{cccccccc|c} 1 & 0 & 1 & 0 & 0.25 & -0.4 & 0 & 0 & 36 \\ 0 & 1 & 0.4\overline{6} & 0 & 0.08\overline{3} & -0.0\overline{6} & -0.0\overline{6} & 0 & 24 \\ 0 & 0 & 0.0\overline{6} & 1 & -0.41\overline{6} & 0.7\overline{3} & 0.1\overline{3} & 0 & 8 \\ \hline 0 & 0 & 21.2 & 0 & 1 & 1.6 & 0.4 & 1 & 1728 \end{array}\right]$

Solution: Maximum = 1728 at $x_1 = 36$, $x_2 = 24$, $x_3 = 0$, $x_4 = 8$

**31.** no solution is possible because no pivot can be found

**33.** $f$ is maximized at 100 when $x = 50$ and $y = 10$. Since multiple solutions are possible, pivot on 2 in $R_2C_1$.

**35.** $\left[\begin{array}{ccccc|c} \boxed{1} & -10 & 1 & 0 & 0 & 10 \\ -1 & 1 & 0 & 1 & 0 & 40 \\ \hline -3 & -2 & 0 & 0 & 1 & 0 \end{array}\right]$ $\to$ $R_1 + R_2 \to R_2$, $3R_1 + R_3 \to R_3$ $\left[\begin{array}{ccccc|c} 1 & -10 & 1 & 0 & 0 & 10 \\ 0 & -9 & 1 & 1 & 0 & 50 \\ \hline 0 & -32 & 3 & 0 & 1 & 30 \end{array}\right]$

No nonnegative quotient exists. There is no solution.

**37.** $\left[\begin{array}{ccccc|c} 2 & 1 & 1 & 0 & 0 & 120 \\ 1 & \boxed{4} & 0 & 1 & 0 & 200 \\ \hline -3 & -12 & 0 & 0 & 1 & 0 \end{array}\right] \xrightarrow{\frac{1}{4}R_2 \to R_2} \left[\begin{array}{ccccc|c} 2 & 1 & 1 & 0 & 0 & 120 \\ \frac{1}{4} & 1 & 0 & \frac{1}{4} & 0 & 50 \\ \hline -3 & -12 & 0 & 0 & 1 & 0 \end{array}\right] \begin{array}{c} -R_2 + R_1 \to R_1 \\ \to \\ 12R_2 + R_3 \to R_3 \end{array}$

$\left[\begin{array}{ccccc|c} \boxed{\frac{7}{4}} & 0 & 1 & -\frac{1}{4} & 0 & 70 \\ \frac{1}{4} & 1 & 0 & \frac{1}{4} & 0 & 50 \\ \hline 0 & 0 & 0 & 3 & 1 & 600 \end{array}\right] \frac{4}{7}R_1 \to R_1$

Maximum is 600 at $x = 0$, $y = 50$. Continuing to find a second solution we have:

$\left[\begin{array}{ccccc|c} 1 & 0 & \frac{4}{7} & -\frac{1}{7} & 0 & 40 \\ \boxed{\frac{1}{4}} & 1 & 0 & \frac{1}{4} & 0 & 50 \\ \hline 0 & 0 & 0 & 3 & 1 & 600 \end{array}\right] \xrightarrow{-\frac{1}{4}R_1 + R_2 \to R_2} \left[\begin{array}{ccccc|c} 1 & 0 & \frac{4}{7} & -\frac{1}{7} & 0 & 40 \\ 0 & 1 & -\frac{1}{7} & \frac{2}{7} & 0 & 40 \\ \hline 0 & 0 & 0 & 3 & 1 & 600 \end{array}\right]$

Maximum is 600 at $x = 40$, $y = 40$.

**39.** Let $x$ = number of inkjet printers, $y$ = number of laser printers.
Maximize $f = 40x + 60y$
Constraints: $x + y \le 60$, $x + 3y \le 120$

**a.** $\left[\begin{array}{ccccc|c} 1 & 1 & 1 & 0 & 0 & 60 \\ 1 & \boxed{3} & 0 & 1 & 0 & 120 \\ \hline -40 & -60 & 0 & 0 & 1 & 0 \end{array}\right] \xrightarrow{\frac{1}{3}R_2 \to R_2} \left[\begin{array}{ccccc|c} 1 & 1 & 1 & 0 & 0 & 60 \\ \frac{1}{3} & 1 & 0 & \frac{1}{3} & 0 & 40 \\ \hline -40 & -60 & 0 & 0 & 1 & 0 \end{array}\right] \begin{array}{c} -R_2 + R_1 \to R_1 \\ \to \\ 60R_2 + R_3 \to R_3 \end{array}$

$\left[\begin{array}{ccccc|c} \boxed{\frac{2}{3}} & 0 & 1 & -\frac{1}{3} & 0 & 20 \\ \frac{1}{3} & 1 & 0 & \frac{1}{3} & 0 & 40 \\ \hline -20 & 0 & 0 & 20 & 1 & 2400 \end{array}\right] \frac{3}{2}R_1 \to R_1 \left[\begin{array}{ccccc|c} 1 & 0 & \frac{3}{2} & -\frac{1}{2} & 0 & 30 \\ \frac{1}{3} & 1 & 0 & \frac{1}{3} & 0 & 40 \\ \hline -20 & 0 & 0 & 20 & 1 & 2400 \end{array}\right] \begin{array}{c} \to \\ -\frac{1}{3}R_1 + R_2 \to R_2 \\ 20R_1 + R_3 \to R_3 \end{array}$

$\left[\begin{array}{ccccc|c} 1 & 0 & \frac{3}{2} & -\frac{1}{2} & 0 & 30 \\ 0 & 1 & -\frac{1}{2} & \frac{1}{2} & 0 & 30 \\ \hline 0 & 0 & 30 & 10 & 1 & 3000 \end{array}\right]$

**b.** Maximum profit is $3000 with 30 ink jet and 30 laser printers.

**41.** Let $x$ = number of pairs of style #891 jeans, $y$ = number of pairs of style #917 jeans
Maximize: $f = 18x + 22.5y$
Constraints: $10x + 10y \le 7500$, $20x + 30y \le 19500$, $x \ge 0$, $y \ge 0$

$\left[\begin{array}{ccccc|c} 10 & 10 & 1 & 0 & 0 & 7500 \\ 20 & \boxed{30} & 0 & 1 & 0 & 19500 \\ \hline -18 & -22.5 & 0 & 0 & 1 & 0 \end{array}\right] \frac{1}{30}R_2 \to R_2 \text{ then } \begin{array}{c} -10R_2 + R_1 \to R_1 \\ \to \\ 22.5R_2 + R_3 \to R_3 \end{array}$

$\left[\begin{array}{ccccc|c} \boxed{3.\overline{3}} & 0 & 1 & -0.\overline{3} & 0 & 1000 \\ 0.\overline{6} & 1 & 0 & 0.0\overline{3} & 0 & 650 \\ \hline -3 & 0 & 0 & 0.75 & 1 & 14,625 \end{array}\right] \frac{3}{10}R_1 \to R_1 \text{ then } \begin{array}{c} \to \\ -0.\overline{6}R_1 + R_2 \to R_2 \\ 3R_1 + R_3 \to R_3 \end{array}$

$\left[\begin{array}{ccccc|c} 1 & 0 & 0.3 & -0.1 & 0 & 300 \\ 0 & 1 & -0.2 & 0.1 & 0 & 450 \\ \hline 0 & 0 & 0.9 & 0.45 & 1 & 15,525 \end{array}\right]$ Solution: $x = 300$, $y = 450$, $f = 15,525$

300 pairs of style #891 jeans and 450 pairs of style #917 jeans for a maximum profit of $15,525.

**43.** Let P = hundreds of gallons of Premium ice cream, and L = hundreds of gallons of Light ice cream. Objective function: $f = 100P + 100L$

Constraints: $0.3P + 0.5L \le 140,\ 90P + 70L \le 28{,}000,\ P \ge 0,\ L \ge 0$

$$\left[\begin{array}{rrrrr|r} 0.3 & \boxed{0.5} & 1 & 0 & 0 & 140 \\ 90 & 70 & 0 & 1 & 0 & 28{,}000 \\ \hline -100 & -100 & 0 & 0 & 1 & 0 \end{array}\right] \quad 2R_1 \to R_1 \text{ then } \quad \begin{array}{c} \to \\ -70R_1 + R_2 \to R_2 \\ 100R_1 + R_3 \to R_3 \end{array}$$

$$\left[\begin{array}{rrrrr|r} 0.6 & 1 & 2 & 0 & 0 & 280 \\ \boxed{48} & 0 & -140 & 1 & 0 & 8400 \\ \hline -40 & 0 & 200 & 0 & 1 & 28{,}000 \end{array}\right] \quad \tfrac{1}{48}R_2 \to R_2 \text{ then } \quad \begin{array}{c} -0.6R_2 + R_1 \to R_1 \\ \to \\ 40R_2 + R_3 \to R_3 \end{array} \quad \left[\begin{array}{rrrrr|r} 0 & 1 & -82 & 0 & 0 & 175 \\ 1 & 0 & -\frac{35}{12} & \frac{1}{48} & 0 & 175 \\ \hline 0 & 0 & \frac{250}{3} & \frac{5}{6} & 1 & 35{,}000 \end{array}\right]$$

There is a maximum profit of \$35,000 obtained from 175 hundred gallons (17,500 gal) of each ice cream.

**45.** Objective function: $z = 20{,}000A + 25{,}000B + 30{,}000W$

Constraints:
$$3000A + 3700B + 5000W \le 47{,}500$$
$$A + B + W \le 12$$
$$205{,}000A + 279{,}600B + 350{,}000W \le 3{,}413{,}000$$

Equivalent Constraints:
$$30A + 37B + 50W \le 475$$
$$A + B + W \le 12$$
$$205A + 279.6B + 350W \le 3413$$

$$\left[\begin{array}{rrrrrrr|r} 205 & 279.6 & 350 & 1 & 0 & 0 & 0 & 3413 \\ 30 & 37 & \boxed{50} & 0 & 1 & 0 & 0 & 475 \\ 1 & 1 & 1 & 0 & 0 & 1 & 0 & 12 \\ \hline -20 & -25 & -30 & 0 & 0 & 0 & 1 & 0 \end{array}\right] \quad \tfrac{1}{50}R_2 \to R_2 \text{ then } \quad \begin{array}{c} -350R_2 + R_1 \to R_1 \\ \to \\ -R_2 + R_3 \to R_3 \\ 30R_2 + R_4 \to R_4 \end{array}$$

$$\left[\begin{array}{rrrrrrr|r} -5 & \boxed{20.6} & 0 & 1 & -70 & 0 & 0 & 88 \\ 0.6 & 0.74 & 1 & 0 & 0.2 & 0 & 0 & 9.5 \\ 0.4 & 0.26 & 0 & 0 & -0.2 & 1 & 0 & 2.5 \\ \hline -2 & -2.8 & 0 & 0 & 6 & 0 & 1 & 285 \end{array}\right] \quad \tfrac{1}{20.6}R_1 \to R_1 \text{ then } \quad \begin{array}{c} \to \\ -0.74R_1 + R_2 \to R_2 \\ -0.26R_1 + R_3 \to R_3 \\ 2.8R_1 + R_4 \to R_4 \end{array}$$

$$\left[\begin{array}{rrrrrrr|r} -0.2427 & 1 & 0 & 0.0485 & -3.3981 & 0 & 0 & 4.2718 \\ 0.7796 & 0 & 1 & -0.0359 & 2.7146 & 0 & 0 & 6.3388 \\ 0.4631 & 0 & 0 & -0.0126 & \boxed{0.6835} & 1 & 0 & 1.3893 \\ \hline -2.6796 & 0 & 0 & 0.1359 & -3.5146 & 0 & 1 & 296.9612 \end{array}\right] \quad \tfrac{1}{0.6835}R_3 \to R_3 \text{ then } \quad \begin{array}{c} 3.3981R_3 + R_1 \to R_1 \\ -2.7148R_3 + R_2 \to R_2 \\ \to \\ 3.5146R_3 + R_4 \to R_4 \end{array}$$

$$\left[\begin{array}{rrrrrrr|r} 2.0597 & 1 & 0 & -0.0142 & 0 & 4.9716 & 0 & 11.1790 \\ -1.0598 & 0 & 1 & 0.0142 & 0 & -3.9719 & 0 & 0.8206 \\ \boxed{0.6776} & 0 & 0 & -0.0185 & 1 & 1.4631 & 0 & 2.0327 \\ \hline -0.2983 & 0 & 0 & 0.0710 & 0 & 5.1421 & 1 & 304.1051 \end{array}\right] \quad \tfrac{1}{0.6776}R_3 \to R_3 \text{ then } \quad \begin{array}{c} -2.0597R_3 + R_1 \to R_1 \\ 1.0598R_3 + R_2 \to R_2 \\ \to \\ 0.2983R_3 + R_4 \to R_4 \end{array}$$

$$\left[\begin{array}{rrrrrrr|r} 0 & 1 & 0 & 0.04193 & -3.0397 & 0.5244 & 0 & 5.0004 \\ 0 & 0 & 1 & -0.0147 & 1.5638 & -1.6836 & 0 & 3.9998 \\ 1 & 0 & 0 & -0.0273 & 1.4758 & 2.1592 & 0 & 2.9998 \\ \hline 0 & 0 & 0 & 0.0629 & 0.4402 & 5.7861 & 1 & 305.0000 \end{array}\right]$$

Now round off column 8 to obtain a maximum profit of \$305,000 by building 3 A, 5 B, and 4 W homes.

**47.** Let $x_1$ = number of one-bedroom apartments, $x_2$ = number of two-bedroom apartments, and $x_3$ = number of three-bedroom apartments

Objective function: $f = 500x_1 + 800x_2 + 1150x_3$

Constraints: $x_1 + 2x_2 + 3x_3 \le 250,\ x_1 \le 26,\ x_2 \le 40,\ x_3 \le 60$

$$\left[\begin{array}{cccccccc|c} 1 & 2 & 3 & 1 & 0 & 0 & 0 & 0 & 250 \\ 1 & 0 & 0 & 0 & 1 & 0 & 0 & 0 & 26 \\ 0 & 1 & 0 & 0 & 0 & 1 & 0 & 0 & 40 \\ 0 & 0 & \boxed{1} & 0 & 0 & 0 & 1 & 0 & 60 \\ \hline -500 & -800 & -1150 & 0 & 0 & 0 & 0 & 1 & 0 \end{array}\right] \begin{array}{l} -3R_4 + R_1 \to R_1 \\ \to \\ \to \\ \to \\ 1150R_4 + R_5 \to R_5 \end{array}$$

x1 x2 x3 s1 s2 s3 s4 f

$$\left[\begin{array}{cccccccc|c} 1 & \boxed{2} & 0 & 1 & 0 & 0 & -\frac{3}{2} & 0 & 35 \\ 1 & 0 & 0 & 0 & 1 & 0 & 0 & 0 & 26 \\ 0 & 1 & 0 & 0 & 0 & 1 & 0 & 0 & 40 \\ 0 & 0 & 1 & 0 & 0 & 0 & 1 & 0 & 60 \\ \hline -500 & -800 & 0 & 0 & 0 & 0 & 1150 & 1 & 69{,}000 \end{array}\right] \begin{array}{l} \frac{1}{2}R_1 \to R_1 \text{ then } \quad \to \\ \to \\ -R_1 + R_3 \to R_3 \\ \to \\ 800R_1 + R_5 \to R_5 \end{array}$$

$$\left[\begin{array}{cccccccc|c} \frac{1}{2} & 1 & 0 & \frac{1}{2} & 0 & 0 & -\frac{3}{2} & 0 & 35 \\ \boxed{1} & 0 & 0 & 0 & 1 & 0 & 0 & 0 & 26 \\ -\frac{1}{2} & 0 & 0 & -\frac{1}{2} & 0 & 1 & \frac{3}{2} & 0 & 5 \\ 0 & 0 & 1 & 0 & 0 & 0 & 1 & 0 & 60 \\ \hline -100 & 0 & 0 & 400 & 0 & 0 & -50 & 1 & 97{,}000 \end{array}\right] \begin{array}{l} -\frac{1}{2}R_2 + R_1 \to R_1 \\ \to \\ \frac{1}{2}R_2 + R_3 \to R_3 \\ \to \\ 100R_2 + R_5 \to R_5 \end{array}$$

$$\left[\begin{array}{cccccccc|c} 0 & 1 & 0 & \frac{1}{2} & -\frac{1}{2} & 0 & -\frac{3}{2} & 0 & 22 \\ 1 & 0 & 0 & 0 & 1 & 0 & 0 & 0 & 26 \\ 0 & 0 & 0 & -\frac{1}{2} & \frac{1}{2} & 1 & \boxed{\frac{3}{2}} & 0 & 18 \\ 0 & 0 & 1 & 0 & 0 & 0 & 1 & 0 & 60 \\ \hline 0 & 0 & 0 & 400 & 100 & 0 & -50 & 1 & 99{,}600 \end{array}\right] \frac{2}{3}R_3 \to R_3 \text{ then } \begin{array}{l} \frac{3}{2}R_3 + R_1 \to R_1 \\ \to \\ \to \\ -R_3 + R_4 \to R_4 \\ 50R_3 + R_5 \to R_5 \end{array}$$

$$\left[\begin{array}{cccccccc|c} 0 & 1 & 0 & 0 & 0 & 1 & 0 & 0 & 40 \\ 1 & 0 & 0 & 0 & 1 & 0 & 0 & 0 & 26 \\ 0 & 0 & 0 & -\frac{1}{3} & \frac{1}{3} & \frac{2}{3} & 1 & 0 & 12 \\ 0 & 0 & 1 & \frac{1}{3} & -\frac{1}{3} & -\frac{2}{3} & 0 & 0 & 48 \\ \hline 0 & 0 & 0 & \frac{1150}{3} & \frac{350}{3} & \frac{100}{3} & 0 & 1 & 100{,}200 \end{array}\right]$$

**a-b.** Maximum revenue is \$100,200 by renting 26 1-BR, 40 2-BR, and 48 3-BR apartments.

**49.** Let $x$ = newspaper ads and $y$ = radio ads. Objective function: $f = 6000x + 8000y$
Constraints: $100x + 300y \le 6000,\ \ x \le 21,\ \ y \le 28$

$$\left[\begin{array}{cccccc|c} 1 & \boxed{3} & 1 & 0 & 0 & 0 & 60 \\ 1 & 0 & 0 & 1 & 0 & 0 & 21 \\ 0 & 1 & 0 & 0 & 1 & 0 & 28 \\ \hline -6000 & -8000 & 0 & 0 & 0 & 1 & 0 \end{array}\right] \begin{array}{l} \tfrac{1}{3}R_1 \to R_1 \text{ then} \\ \\ \\ \\ \end{array} \begin{array}{c} \to \\ \to \\ -R_1 + R_3 \to R_3 \\ 8000R_1 + R_4 \to R_4 \end{array}$$

$$\left[\begin{array}{cccccc|c} \frac{1}{3} & 1 & \frac{1}{3} & 0 & 0 & 0 & 20 \\ \boxed{1} & 0 & 0 & 1 & 0 & 0 & 21 \\ -\frac{1}{3} & 0 & -\frac{1}{3} & 0 & 1 & 0 & 8 \\ \hline -\frac{10{,}000}{3} & 0 & \frac{8{,}000}{3} & 0 & 0 & 1 & 160{,}000 \end{array}\right] \begin{array}{c} -\frac{1}{3}R_2 + R_1 \to R_1 \\ \to \\ \frac{1}{3}R_2 + R_3 \to R_3 \\ \frac{10{,}000}{3}R_2 + R_4 \to R_4 \end{array} \left[\begin{array}{cccccc|c} 0 & 1 & \frac{1}{3} & -\frac{1}{3} & 0 & 0 & 13 \\ 1 & 0 & 0 & 1 & 0 & 0 & 21 \\ 0 & 0 & -\frac{1}{3} & \frac{1}{3} & 1 & 0 & 15 \\ \hline 0 & 0 & -\frac{8000}{3} & \frac{10{,}000}{3} & 0 & 1 & 230{,}000 \end{array}\right]$$

Maximum exposure occurs with 21 newspaper ads and 13 radio ads. The maximum number of exposures is 230,000.

**51.** Objective function: $f = 30A + 9B + 15C$
Constraints: $9A + 3B + 0.5C \le 477$, $5A + 4B \le 350$, $3A + 2C \le 150$, $B \le 20$, $A \ge 0$, $B \ge 0$, $C \ge 0$

$$\left[\begin{array}{cccccccc|c} 9 & 3 & \frac{1}{2} & 1 & 0 & 0 & 0 & 0 & 477 \\ 5 & 4 & 0 & 0 & 1 & 0 & 0 & 0 & 350 \\ \boxed{3} & 0 & 2 & 0 & 0 & 1 & 0 & 0 & 150 \\ 0 & 1 & 0 & 0 & 0 & 0 & 1 & 0 & 20 \\ \hline -30 & -9 & -15 & 0 & 0 & 0 & 0 & 1 & 0 \end{array}\right] \begin{array}{l} \\ \\ \tfrac{1}{3}R_3 \to R_3 \text{ then} \\ \\ \\ \end{array} \begin{array}{c} -9R_3 + R_1 \to R_1 \\ -5R_3 + R_2 \to R_2 \\ \to \\ \to \\ 30R_3 + R_5 \to R_5 \end{array}$$

$$\left[\begin{array}{cccccccc|c} 0 & \boxed{3} & -\frac{11}{2} & 1 & 0 & -3 & 0 & 0 & 27 \\ 0 & 4 & -\frac{10}{3} & 0 & 1 & -\frac{5}{3} & 0 & 0 & 100 \\ 1 & 0 & \frac{2}{3} & 0 & 0 & \frac{1}{3} & 0 & 0 & 50 \\ 0 & 1 & 0 & 0 & 0 & 0 & 1 & 0 & 20 \\ \hline 0 & -9 & 5 & 0 & 0 & 10 & 0 & 1 & 1500 \end{array}\right] \begin{array}{l} \tfrac{1}{3}R_1 \to R_1 \text{ then} \\ \\ \\ \\ \\ \end{array} \begin{array}{c} \to \\ -4R_1 + R_2 \to R_2 \\ \to \\ -R_1 + R_4 \to R_4 \\ 9R_1 + R_5 \to R_5 \end{array}$$

$$\left[\begin{array}{cccccccc|c} 0 & 1 & -\frac{11}{6} & \frac{1}{3} & 0 & -1 & 0 & 0 & 9 \\ 0 & 0 & 4 & -\frac{4}{3} & 1 & \frac{7}{3} & 0 & 0 & 64 \\ 1 & 0 & \frac{2}{3} & 0 & 0 & \frac{1}{3} & 0 & 0 & 50 \\ 0 & 0 & \boxed{\frac{11}{6}} & -\frac{1}{3} & 0 & 1 & 1 & 0 & 11 \\ \hline 0 & 0 & -\frac{23}{6} & 3 & 0 & 1 & 0 & 1 & 1581 \end{array}\right] \begin{array}{l} \\ \\ \\ \tfrac{6}{11}R_4 \to R_4 \text{ then} \\ \\ \end{array} \begin{array}{c} \frac{11}{6}R_4 + R_1 \to R_1 \\ -4R_4 + R_2 \to R_2 \\ -\frac{2}{3}R_4 + R_3 \to R_3 \\ \to \\ \frac{23}{2}R_4 + R_5 \to R_5 \end{array}$$

$$\left[\begin{array}{cccccccc|c} 0 & 1 & 0 & 0 & 0 & 0 & 1 & 0 & 20 \\ 0 & 0 & 0 & -\frac{20}{33} & 1 & \frac{5}{33} & -\frac{24}{11} & 0 & 40 \\ 1 & 0 & 0 & \frac{4}{33} & 0 & -\frac{1}{33} & -\frac{4}{33} & 0 & 46 \\ 0 & 0 & 1 & -\frac{2}{11} & 0 & \frac{6}{11} & \frac{6}{11} & 0 & 6 \\ \hline 0 & 0 & 0 & \frac{10}{11} & 0 & \frac{80}{11} & \frac{69}{11} & 1 & 1650 \end{array}\right]$$

The maximum profit is \$1650 when 46 of $A$, 20 of $B$, and 6 of $C$ are produced.

**53.** Let $x_1$ = number of 20-in. LCDs, $x_2$ = number of 42-in. LCDs, $x_3$ = number of 42-in. plasma TVs, $x_4$ = number of 50-in. plasma TVs
Maximize: $f = 46x_1 + 60x_2 + 75x_3 + 100x_4$

Constraints: $x_1 + x_2 + x_3 + x_4 \le 200$

$x_1 \le 40$

$7x_1 + 10x_2 + 12x_3 + 15x_4 \le 2000$

$2x_1 + 2x_2 + 4x_3 + 5x_4 \le 500$

$x_1 \ge 0,\ x_2 \ge 0,\ x_3 \ge 0,\ x_4 \ge 0$

$$\left[\begin{array}{ccccccccc|c} 1 & 1 & 1 & 1 & 1 & 0 & 0 & 0 & 0 & 200 \\ 1 & 0 & 0 & 0 & 0 & 1 & 0 & 0 & 0 & 40 \\ 7 & 10 & 12 & 15 & 0 & 0 & 1 & 0 & 0 & 2000 \\ 2 & 2 & 4 & \boxed{5} & 0 & 0 & 0 & 1 & 0 & 500 \\ \hline -46 & -60 & -75 & -100 & 0 & 0 & 0 & 0 & 1 & 0 \end{array}\right] \begin{array}{c} \\ \\ \\ \tfrac{1}{5}R_4 \to R_4 \text{ then} \\ \\ \end{array} \begin{array}{r} -R_4 + R_1 \to R_1 \\ \to \\ -15R_4 + R_3 \to R_3 \\ \to \\ 100R_4 + R_5 \to R_5 \end{array}$$

$$\left[\begin{array}{ccccccccc|c} 0.6 & 0.6 & 0.2 & 0 & 1 & 0 & 0 & -0.2 & 0 & 100 \\ 1 & 0 & 0 & 0 & 0 & 1 & 0 & 0 & 0 & 40 \\ 1 & \boxed{4} & 0 & 0 & 0 & 0 & 1 & -3 & 0 & 500 \\ 0.4 & 0.4 & 0.8 & 1 & 0 & 0 & 0 & 0.2 & 0 & 100 \\ \hline -6 & -20 & 5 & 0 & 0 & 0 & 0 & 20 & 1 & 10,000 \end{array}\right] \begin{array}{c} \\ \\ \tfrac{1}{4}R_3 \to R_3 \text{ then} \\ \\ \\ \end{array} \begin{array}{r} -0.6R_3 + R_1 \to R_1 \\ \to \\ \to \\ -0.4R_3 + R_4 \to R_4 \\ 100R_4 + R_5 \to R_5 \end{array}$$

$$\left[\begin{array}{ccccccccc|c} 0.45 & 0 & 0.2 & 0 & 1 & 0 & -0.15 & 0.25 & 0 & 25 \\ \boxed{1} & 0 & 0 & 0 & 0 & 1 & 0 & 0 & 0 & 40 \\ 0.25 & 1 & 0 & 0 & 0 & 0 & 0.25 & -0.75 & 0 & 125 \\ 0.3 & 0 & 0.8 & 1 & 0 & 0 & -0.1 & 0.5 & 0 & 50 \\ \hline -1 & 0 & 5 & 0 & 0 & 0 & 5 & 5 & 1 & 12,500 \end{array}\right] \begin{array}{r} -0.45R_2 + R_1 \to R_1 \\ \to \\ -0.25R_2 + R_3 \to R_3 \\ -0.3R_2 + R_4 \to R_4 \\ R_2 + R_5 \to R_5 \end{array}$$

$$\left[\begin{array}{ccccccccc|c} 0 & 0 & 0.2 & 0 & 1 & -0.45 & -0.15 & 0.25 & 0 & 7 \\ 1 & 0 & 0 & 0 & 0 & 1 & 0 & 0 & 0 & 40 \\ 0 & 1 & 0 & 0 & 0 & -0.25 & 0.25 & -0.75 & 0 & 115 \\ 0 & 0 & 0.8 & 1 & 0 & -0.3 & -0.1 & 0.5 & 0 & 38 \\ \hline 0 & 0 & 5 & 0 & 0 & 1 & 5 & 5 & 1 & 12,540 \end{array}\right]$$

They should make 40 of the 20-in. LCDs, 115 of the 42-in. LCDs, and 38 of the 50-in. plasma TVs for a maximum profit of $12,540 this week.

## *Exercises 4.4*

**1.** **a.** $\left[\begin{array}{cc|c} 5 & 2 & 16 \\ 1 & 2 & 8 \\ \hline 4 & 5 & g \end{array}\right]$ transpose: $\left[\begin{array}{cc|c} 5 & 1 & 4 \\ 2 & 2 & 5 \\ \hline 16 & 8 & g \end{array}\right]$

**b.** From the transpose in (a):

Maximize $f = 16x_1 + 8x_2$

Constraints: $5x_1 + x_2 \le 4$

$2x_1 + 2x_2 \le 5$

$x_1, x_2 \ge 0$

**3.** **a.** $\left[\begin{array}{cc|c} 1 & 2 & 30 \\ 1 & 4 & 50 \\ \hline 7 & 3 & g \end{array}\right]$ transpose: $\left[\begin{array}{cc|c} 1 & 1 & 7 \\ 2 & 4 & 3 \\ \hline 30 & 50 & g \end{array}\right]$

**b.** From the transpose in (a):
Maximize $f = 30x_1 + 50x_2$
Constraints: $x_1 + x_2 \le 7$
$2x_1 + 4x_2 \le 3$
$x_1,\ x_2 \ge 0$

**5.** **a.** For the minimization problem, the last entries in the columns corresponding to the slack variables are the solution.
Minimum: $g = 452$; $y_1 = 7, y_2 = 4, y_3 = 0$

**b.** Maximum: $f = 452$; $x_1 = 15, x_2 = 0, x_3 = 29$

**7.** $\left[\begin{array}{cc|c} 2 & 1 & 11 \\ 1 & 3 & 11 \\ 1 & 4 & 16 \\ \hline 2 & 10 & g \end{array}\right]$ transpose: $\left[\begin{array}{ccc|c} 2 & 1 & 1 & 2 \\ 1 & 3 & 4 & 10 \\ \hline 11 & 11 & 16 & g \end{array}\right]$

Maximize: $f = 11x_1 + 11x_2 + 16x_3$
Constraints: $2x_1 + x_2 + x_3 \le 2$
$x_1 + 3x_2 + 4x_3 \le 10$

$$\left[\begin{array}{cccccc|c} 2 & 1 & \boxed{1} & 1 & 0 & 0 & 2 \\ 1 & 3 & 4 & 0 & 1 & 0 & 10 \\ \hline -11 & -11 & -16 & 0 & 0 & 1 & 0 \end{array}\right] \begin{array}{c} \rightarrow \\ -4R_1 + R_2 \rightarrow R_2 \\ 16R_1 + R_3 \rightarrow R_3 \end{array}$$

$$\left[\begin{array}{cccccc|c} 2 & 1 & 1 & 1 & 0 & 0 & 2 \\ -7 & -1 & 0 & -4 & 1 & 0 & 2 \\ \hline 21 & 5 & 0 & 16 & 0 & 1 & 32 \end{array}\right]$$

Primal: $y_1 = 16,\ y_2 = 0,\ g = 32$ (min)
Dual: $x_1 = 0,\ x_2 = 0,\ x_3 = 2,\ f = 32$ (max)

**9.** $\left[\begin{array}{cc|c} 4 & 1 & 11 \\ 3 & 2 & 12 \\ 3 & 1 & 6 \\ \hline 3 & 1 & g \end{array}\right]$ transpose: $\left[\begin{array}{ccc|c} 4 & 3 & 3 & 3 \\ 1 & 2 & 1 & 1 \\ \hline 11 & 12 & 6 & g \end{array}\right]$

Maximize $f = 11x_1 + 12x_2 + 6x_3$
Constraints: $4x_1 + 3x_2 + 3x_3 \le 3,\ \ x_1 + 2x_2 + x_3 \le 1,\ \ x_1, x_2, x_3 \ge 0$

$$\left[\begin{array}{cccccc|c} 4 & 3 & 3 & 1 & 0 & 0 & 3 \\ 1 & \boxed{2} & 1 & 0 & 1 & 0 & 1 \\ \hline -11 & -12 & -6 & 0 & 0 & 1 & 0 \end{array}\right] \ \tfrac{1}{2}R_2 \rightarrow R_2 \text{ then } \begin{array}{c} -3R_2 + R_1 \rightarrow R_1 \\ \rightarrow \\ 12R_2 + R_3 \rightarrow R_3 \end{array}$$

$$\left[\begin{array}{cccccc|c} \boxed{\tfrac{5}{2}} & 0 & \tfrac{3}{2} & 1 & -\tfrac{3}{2} & 0 & \tfrac{3}{2} \\ \tfrac{1}{2} & 1 & \tfrac{1}{2} & 0 & \tfrac{1}{2} & 0 & \tfrac{1}{2} \\ \hline -5 & 0 & 0 & 0 & 6 & 1 & 6 \end{array}\right] \ \tfrac{2}{5}R_1 \rightarrow R_1 \text{ then } \begin{array}{c} \rightarrow \\ -\tfrac{1}{2}R_1 + R_2 \rightarrow R_2 \\ 5R_1 + R_3 \rightarrow R_3 \end{array} \left[\begin{array}{cccccc|c} 1 & 0 & \tfrac{3}{5} & \tfrac{2}{5} & -\tfrac{3}{5} & 0 & \tfrac{3}{5} \\ 0 & 1 & \tfrac{1}{5} & -\tfrac{1}{5} & \tfrac{4}{5} & 0 & \tfrac{1}{5} \\ \hline 0 & 0 & 3 & 2 & 3 & 1 & 9 \end{array}\right]$$

Primal: $y_1 = 2,\ y_2 = 3,\ g = 9$ (min) Dual: $x_1 = \dfrac{3}{5},\ x_2 = \dfrac{1}{5},\ x_3 = 0,\ f = 9$ (max)

**11.** $\left[\begin{array}{ccc|c} 1 & 1 & 1 & 3 \\ 0 & 1 & 2 & 2 \\ 1 & 0 & 0 & 2 \\ \hline 8 & 7 & 12 & g \end{array}\right]$ transpose: $\left[\begin{array}{ccc|c} 1 & 0 & 1 & 8 \\ 1 & 1 & 0 & 7 \\ 1 & 2 & 0 & 12 \\ \hline 3 & 2 & 2 & g \end{array}\right]$

Maximize: $f = 3x_1 + 2x_2 + 2x_3$

Constraints: $x_1 + x_3 \le 8,\ x_1 + x_2 \le 7,\ x_1 + 2x_2 \le 12$

$$\left[\begin{array}{ccccccc|c} 1 & 0 & 1 & 1 & 0 & 0 & 0 & 8 \\ \boxed{1} & 1 & 0 & 0 & 1 & 0 & 0 & 7 \\ 1 & 2 & 0 & 0 & 0 & 1 & 0 & 12 \\ \hline -3 & -2 & -2 & 0 & 0 & 0 & 1 & 0 \end{array}\right] \begin{array}{c} -R_2 + R_1 \to R_1 \\ \to \\ -R_2 + R_3 \to R_3 \\ 3R_2 + R_4 \to R_4 \end{array} \left[\begin{array}{ccccccc|c} 0 & -1 & \boxed{1} & 1 & -1 & 0 & 0 & 1 \\ 1 & 1 & 0 & 0 & 1 & 0 & 0 & 7 \\ 0 & 1 & 0 & 0 & -1 & 1 & 0 & 5 \\ \hline 0 & 1 & -2 & 0 & 3 & 0 & 1 & 21 \end{array}\right] \begin{array}{c} \to \\ \\ \\ 2R_1 + R_4 \to R_4 \end{array}$$

$$\left[\begin{array}{ccccccc|c} 0 & -1 & 1 & 1 & -1 & 0 & 0 & 1 \\ 1 & 1 & 0 & 0 & 1 & 0 & 0 & 7 \\ 0 & \boxed{1} & 0 & 0 & -1 & 1 & 0 & 5 \\ \hline 0 & -1 & 0 & 2 & 1 & 0 & 1 & 23 \end{array}\right] \begin{array}{c} R_3 + R_1 \to R_1 \\ -R_3 + R_2 \to R_2 \\ \to \\ R_3 + R_4 \to R_4 \end{array} \left[\begin{array}{ccccccc|c} 0 & 0 & 1 & 1 & -2 & 1 & 0 & 6 \\ 1 & 0 & 0 & 0 & 2 & -1 & 0 & 2 \\ 0 & 1 & 0 & 0 & -1 & 1 & 0 & 5 \\ \hline 0 & 0 & 0 & 2 & 0 & 1 & 1 & 28 \end{array}\right]$$

Minimum: $g = 28$ when $x = 2,\ y = 0,\ z = 1$

**13.** $\left[\begin{array}{ccc|c} 1 & 3 & 0 & 1 \\ 4 & 6 & 1 & 3 \\ 0 & 4 & 1 & 1 \\ \hline 12 & 48 & 8 & g \end{array}\right]$ transpose: $\left[\begin{array}{ccc|c} 1 & 4 & 0 & 12 \\ 3 & 6 & 4 & 48 \\ 0 & 1 & 1 & 8 \\ \hline 1 & 3 & 1 & g \end{array}\right]$ Maximize: $f = x_1 + 3x_2 + x_3$

Constraints: $x_1 + 4x_2 \le 12,\ 3x_1 + 6x_2 + 4x_3 \le 48,\ x_2 + x_3 \le 8,\ \ x_1, x_2, x_3 \ge 0$

$$\left[\begin{array}{ccccccc|c} 1 & \boxed{4} & 0 & 1 & 0 & 0 & 0 & 12 \\ 3 & 6 & 4 & 0 & 1 & 0 & 0 & 48 \\ 0 & 1 & 1 & 0 & 0 & 1 & 0 & 8 \\ \hline -1 & -3 & -1 & 0 & 0 & 0 & 1 & 0 \end{array}\right] \begin{array}{l} \tfrac{1}{4}R_1 \to R_1 \text{ then } \quad \to \\ \qquad -6R_1 + R_2 \to R_2 \\ \qquad -R_1 + R_3 \to R_3 \\ \qquad 3R_1 + R_4 \to R_4 \end{array}$$

$$\left[\begin{array}{ccccccc|c} \frac{1}{4} & 1 & 0 & \frac{1}{4} & 0 & 0 & 0 & 3 \\ \frac{3}{2} & 0 & 4 & -\frac{3}{2} & 1 & 0 & 0 & 30 \\ -\frac{1}{4} & 0 & \boxed{1} & -\frac{1}{4} & 0 & 1 & 0 & 5 \\ \hline -\frac{1}{4} & 0 & -1 & \frac{3}{4} & 0 & 0 & 1 & 9 \end{array}\right] \begin{array}{c} -4R_3 + R_2 \to R_2 \\ \to \\ R_3 + R_4 \to R_4 \end{array}$$

$$\left[\begin{array}{ccccccc|c} \frac{1}{4} & 1 & 0 & \frac{1}{4} & 0 & 0 & 0 & 3 \\ \boxed{\frac{5}{2}} & 0 & 0 & -\frac{1}{2} & 1 & -4 & 0 & 10 \\ -\frac{1}{4} & 0 & 1 & -\frac{1}{4} & 0 & 1 & 0 & 5 \\ \hline -\frac{1}{2} & 0 & 0 & \frac{1}{2} & 0 & 1 & 1 & 14 \end{array}\right] \tfrac{2}{5}R_2 \to R_2 \text{ then } \begin{array}{c} -\frac{1}{4}R_2 + R_1 \to R_1 \\ \to \\ \frac{1}{4}R_2 + R_3 \to R_3 \\ \frac{1}{2}R_2 + R_4 \to R_4 \end{array} \left[\begin{array}{ccccccc|c} 0 & 1 & 0 & \frac{3}{10} & -\frac{1}{10} & \frac{2}{5} & 0 & 2 \\ 1 & 0 & 0 & -\frac{1}{5} & \frac{2}{5} & -\frac{8}{5} & 0 & 4 \\ 0 & 0 & 1 & -\frac{3}{10} & \frac{1}{10} & \frac{3}{5} & 0 & 6 \\ \hline 0 & 0 & 0 & \frac{2}{5} & \frac{1}{5} & \frac{1}{5} & 1 & 16 \end{array}\right]$$

Minimum: $g = 16$ when $y_1 = \dfrac{2}{5}, y_2 = \dfrac{1}{5}$, and $y_3 = \dfrac{1}{5}$.

**15. a.** Minimize: $g = 120y_1 + 50y_2$

Constraints: $3y_1 + y_2 \geq 40,\quad 2y_1 + y_2 \geq 20$

**b.** $$\left[\begin{array}{cccccc|c} \boxed{3} & 2 & 1 & 0 & 0 & 120 \\ 1 & 1 & 0 & 1 & 0 & 50 \\ \hline -40 & -20 & 0 & 0 & 1 & 0 \end{array}\right] \begin{array}{c} \frac{1}{3}R_1 \to R_1 \\ \to \\ \phantom{0} \end{array}$$

$$\left[\begin{array}{ccccc|c} \boxed{1} & \frac{2}{3} & \frac{1}{3} & 0 & 0 & 40 \\ 1 & 1 & 0 & 1 & 0 & 50 \\ \hline -40 & -20 & 0 & 0 & 1 & 0 \end{array}\right] \begin{array}{c} \to \\ -R_1 + R_2 \to R_2 \\ 40R_1 + R_3 \to R_3 \end{array} \left[\begin{array}{ccccc|c} 1 & \frac{2}{3} & \frac{1}{3} & 0 & 0 & 40 \\ 0 & \frac{1}{3} & -\frac{1}{3} & 1 & 0 & 10 \\ \hline 0 & \frac{20}{3} & \frac{40}{3} & 0 & 1 & 1600 \end{array}\right]$$

Primal: $f = 1600$ (max) at $x_1 = 40, x_2 = 0$ Dual: $g = 1600$ (min) at $y_1 = \dfrac{40}{3}, y_2 = 0$

**17.** $$\left[\begin{array}{ccc|c} 1 & 2 & 1 & 16 \\ 1 & 5 & 2 & 18 \\ 2 & 5 & 3 & 38 \\ \hline 40 & 90 & 30 & g \end{array}\right] \quad \text{transpose: } \left[\begin{array}{ccc|c} 1 & 1 & 2 & 40 \\ 2 & 5 & 5 & 90 \\ 1 & 2 & 3 & 30 \\ \hline 16 & 18 & 38 & g \end{array}\right]$$

$$\left[\begin{array}{ccccccc|c} 1 & 1 & 2 & 1 & 0 & 0 & 0 & 40 \\ 2 & 5 & 5 & 0 & 1 & 0 & 0 & 90 \\ 1 & 2 & \boxed{3} & 0 & 0 & 1 & 0 & 30 \\ \hline -16 & -18 & -38 & 0 & 0 & 0 & 1 & 0 \end{array}\right] \begin{array}{c} \\ \\ \frac{1}{3}R_3 \to R_3 \text{ then} \\ \\ \end{array} \begin{array}{c} -2R_3 + R_1 \to R_1 \\ -5R_3 + R_2 \to R_2 \\ \to \\ 38R_3 + R_4 \to R_4 \end{array}$$

$$\left[\begin{array}{ccccccc|c} \frac{1}{3} & -\frac{1}{3} & 0 & 1 & 0 & -\frac{2}{3} & 0 & 20 \\ \frac{1}{3} & \frac{5}{3} & 0 & 0 & 1 & -\frac{5}{3} & 0 & 40 \\ \boxed{\frac{1}{3}} & \frac{2}{3} & 1 & 0 & 0 & \frac{1}{3} & 0 & 10 \\ \hline -\frac{10}{3} & \frac{22}{3} & 0 & 0 & 0 & \frac{38}{3} & 1 & 380 \end{array}\right] \begin{array}{c} \\ \\ 3R_3 \to R_3 \text{ then} \\ \\ \end{array} \begin{array}{c} -\frac{1}{3}R_3 + R_1 \to R_1 \\ -\frac{1}{3}R_3 + R_2 \to R_2 \\ \to \\ \frac{10}{3}R_3 + R_4 \to R_4 \end{array} \left[\begin{array}{ccccccc|c} 0 & -1 & -1 & 1 & 0 & -1 & 0 & 10 \\ 0 & 1 & -1 & 0 & 1 & -2 & 0 & 30 \\ 1 & 2 & 3 & 0 & 0 & 1 & 0 & 30 \\ \hline 0 & 14 & 10 & 0 & 0 & 16 & 1 & 480 \end{array}\right]$$

Minimum is 480 at $y_1 = 0, y_2 = 0, y_3 = 16$.

**19.** $$\left[\begin{array}{cccc|c} 40 & 20 & 15 & 50 & 50 \\ 3 & 2 & 0 & 5 & 6 \\ 2 & 2 & 4 & 4 & 10 \\ 2 & 4 & 1 & 5 & 8 \\ \hline 50 & 20 & 30 & 80 & g \end{array}\right] \quad \text{transpose: } \left[\begin{array}{cccc|c} 40 & 3 & 2 & 2 & 50 \\ 20 & 2 & 2 & 4 & 20 \\ 15 & 0 & 4 & 1 & 30 \\ 50 & 5 & 4 & 5 & 80 \\ \hline 50 & 6 & 10 & 8 & g \end{array}\right]$$

$$\left[\begin{array}{ccccccccc|c} 40 & 3 & 2 & 2 & 1 & 0 & 0 & 0 & 0 & 50 \\ \boxed{20} & 2 & 2 & 4 & 0 & 1 & 0 & 0 & 0 & 20 \\ 15 & 0 & 4 & 1 & 0 & 0 & 1 & 0 & 0 & 30 \\ 50 & 5 & 4 & 5 & 0 & 0 & 0 & 1 & 0 & 80 \\ \hline -50 & -6 & -10 & -8 & 0 & 0 & 0 & 0 & 1 & 0 \end{array}\right] \begin{array}{c} \\ \frac{1}{20}R_2 \to R_2 \text{ then} \\ \\ \\ \\ \end{array} \begin{array}{c} -40R_2 + R_1 \to R_1 \\ \to \\ -15R_2 + R_3 \to R_3 \\ -50R_2 + R_4 \to R_4 \\ 50R_2 + R_5 \to R_5 \end{array}$$

$$\left[\begin{array}{ccccccccc|c} 0 & -1 & -2 & -6 & 1 & -2 & 0 & 0 & 0 & 10 \\ 1 & 0.1 & 0.1 & 0.2 & 0 & 0.05 & 0 & 0 & 0 & 1 \\ 0 & -1.5 & \boxed{2.5} & -2 & 0 & -0.75 & 1 & 0 & 0 & 15 \\ 0 & 0 & -1 & -5 & 0 & -2.5 & 0 & 1 & 0 & 30 \\ \hline 0 & -1 & -5 & 2 & 0 & 2.5 & 0 & 0 & 1 & 50 \end{array}\right] \quad \tfrac{2}{5}R_3 \to R_3 \text{ then } \quad \begin{array}{c} 2R_3 + R_1 \to R_1 \\ -0.1R_3 + R_2 \to R_2 \\ \to \\ R_3 + R_4 \to R_4 \\ 5R_3 + R_5 \to R_5 \end{array}$$

$$\left[\begin{array}{ccccccccc|c} 0 & -2.2 & 0 & -7.6 & 1 & -2.6 & 0.8 & 0 & 0 & 22 \\ 1 & \boxed{0.16} & 0 & 0.28 & 0 & 0.08 & -0.04 & 0 & 0 & 0.4 \\ 0 & -0.6 & 1 & -0.8 & 0 & -0.3 & 0.4 & 0 & 0 & 6 \\ 0 & -0.6 & 0 & -5.8 & 0 & -2.8 & 0.4 & 1 & 0 & 36 \\ \hline 0 & -4 & 0 & -2 & 0 & 1 & 2 & 0 & 1 & 80 \end{array}\right] \quad \tfrac{25}{4}R_2 \to R_2 \text{ then } \quad \begin{array}{c} 2.2R_2 + R_1 \to R_1 \\ \to \\ 0.6R_2 + R_3 \to R_3 \\ 0.6R_2 + R_4 \to R_4 \\ 4R_2 + R_5 \to R_5 \end{array}$$

$$\left[\begin{array}{ccccccccc|c} 13.75 & 0 & 0 & -3.75 & 1 & -1.5 & 0.25 & 0 & 0 & 27.5 \\ 6.25 & 1 & 0 & 1.75 & 0 & 0.5 & -0.25 & 0 & 0 & 2.5 \\ 3.75 & 0 & 1 & 0.25 & 0 & 0 & 0.25 & 0 & 0 & 7.5 \\ 3.75 & 0 & 0 & -4.75 & 0 & -2.5 & 0.25 & 1 & 0 & 37.5 \\ \hline 25 & 0 & 0 & 5 & 0 & 3 & 1 & 0 & 1 & 90 \end{array}\right]$$

Minimum is 90 at $y_1 = 0,\ y_2 = 3,\ y_3 = 1,\ y_4 = 0$.

**21.** Let $y_1$ = hours per week to operate the Atlanta plant, $y_2$ = hours per week to operate the Ft. Worth plant

Minimize: $g = 700y_1 + 2100y_2$

Constraints: $160y_1 + 800y_2 \geq 64000,\ 200y_1 + 200y_2 \geq 40000,\ y_1 \geq 0,\ y_2 \geq 0$

$$\left[\begin{array}{cc|c} 160 & 800 & 64000 \\ 200 & 200 & 40000 \\ \hline 700 & 2100 & g \end{array}\right] \quad \text{transpose:} \quad \left[\begin{array}{cc|c} 160 & 200 & 700 \\ 800 & 200 & 2100 \\ \hline 64000 & 40000 & g \end{array}\right]$$

$$\left[\begin{array}{ccccc|c} 160 & 200 & 1 & 0 & 0 & 700 \\ \boxed{800} & 200 & 0 & 1 & 0 & 2100 \\ \hline -64000 & -40000 & 0 & 0 & 1 & 0 \end{array}\right] \quad \tfrac{1}{800}R_2 \to R_2 \text{ then } \quad \begin{array}{c} -160R_2 + R_1 \to R_1 \\ \to \\ 64000R_2 + R_3 \to R_3 \end{array}$$

$$\left[\begin{array}{ccccc|c} 0 & \boxed{160} & 1 & -0.2 & 0 & 280 \\ 1 & 0.25 & 0 & 0.00125 & 0 & 2.625 \\ \hline 0 & -24000 & 0 & 80 & 1 & 168000 \end{array}\right] \quad \tfrac{1}{160}R_1 \to R_1 \text{ then } \quad \begin{array}{c} \to \\ -0.25R_1 + R_2 \to R_2 \\ 24000R_1 + R_3 \to R_3 \end{array}$$

$$\left[\begin{array}{ccccc|c} 0 & 1 & 0.00625 & -0.00125 & 0 & 1.75 \\ 1 & 0 & -0.0015625 & 0.0015625 & 0 & 2.1875 \\ \hline 0 & 0 & 150 & 50 & 1 & 210{,}000 \end{array}\right]$$

Minimum cost is \$210,000 per week, when the Atlanta plant is operated for 150 hours and the Ft. Worth plant is operated for 50 hours.

**23.** $x$ = hours for Line 1, $y$ = hours for Line 2

Minimize: $C = 200x + 400y$

Constraints: $30x + 150y \geq 270,\ \ 40x + 40y \geq 200$

Equivalent constraints: $x + 5y \geq 9,\ \ x + y \geq 5$

$$\left[\begin{array}{cc|c} 1 & 5 & 9 \\ 1 & 1 & 5 \\ \hline 200 & 400 & g \end{array}\right] \quad \text{transpose: } \left[\begin{array}{cc|c} 1 & 1 & 200 \\ 5 & 1 & 400 \\ \hline 9 & 5 & g \end{array}\right]$$

$$\left[\begin{array}{ccccc|c} 1 & 1 & 1 & 0 & 0 & 200 \\ \boxed{5} & 1 & 0 & 1 & 0 & 400 \\ \hline -9 & -5 & 0 & 0 & 1 & 0 \end{array}\right] \quad \tfrac{1}{5}R_2 \to R_2 \text{ then } \quad \begin{array}{c} -R_2 + R_1 \to R_1 \\ \to \\ 9R_2 + R_3 \to R_3 \end{array}$$

$$\left[\begin{array}{ccccc|c} 0 & \boxed{\tfrac{4}{5}} & 1 & -\tfrac{1}{5} & 0 & 120 \\ 1 & \tfrac{1}{5} & 0 & \tfrac{1}{5} & 0 & 80 \\ \hline 0 & -\tfrac{16}{5} & 0 & \tfrac{9}{5} & 1 & 720 \end{array}\right] \quad -\tfrac{5}{4}R_1 \to R_1 \text{ then } \quad \begin{array}{c} \to \\ -\tfrac{1}{5}R_1 + R_2 \to R_2 \\ \tfrac{16}{5}R_1 + R_3 \to R_3 \end{array} \quad \left[\begin{array}{ccccc|c} 0 & 1 & \tfrac{5}{4} & -\tfrac{1}{4} & 0 & 150 \\ 1 & 0 & -\tfrac{1}{4} & \tfrac{1}{4} & 0 & 50 \\ \hline 0 & 0 & 4 & 1 & 1 & 1200 \end{array}\right]$$

The minimum cost is \$1200 for 4 hours on Line 1, and 1 hour on Line 2.

**25.** Minimize: $\text{Cost} = 1000A + 3000B + 4000C$

Constraints: $200A + 200B + 400C \geq 2000 \ \to\ A + B + 2C \geq 10$

$100A + 200B + 100C \geq 1200 \ \to\ A + 2B + C \geq 12$

$$\left[\begin{array}{ccc|c} 1 & 1 & 2 & 10 \\ 1 & 2 & 1 & 12 \\ \hline 1000 & 3000 & 4000 & g \end{array}\right] \quad \text{transpose: } \left[\begin{array}{cc|c} 1 & 1 & 1000 \\ 1 & 2 & 3000 \\ 2 & 1 & 4000 \\ \hline 10 & 12 & g \end{array}\right]$$

$$\left[\begin{array}{cccccc|c} 1 & \boxed{1} & 1 & 0 & 0 & 0 & 1000 \\ 1 & 2 & 0 & 1 & 0 & 0 & 3000 \\ 2 & 1 & 0 & 0 & 1 & 0 & 4000 \\ \hline -10 & -12 & 0 & 0 & 0 & 1 & 0 \end{array}\right] \quad \begin{array}{c} \to \\ -2R_1 + R_2 \to R_2 \\ -R_1 + R_3 \to R_3 \\ 12R_1 + R_4 \to R_4 \end{array} \quad \left[\begin{array}{cccccc|c} 1 & 1 & 1 & 0 & 0 & 0 & 1000 \\ -1 & 0 & -2 & 1 & 0 & 0 & 1000 \\ 1 & 0 & -1 & 0 & 1 & 0 & 3000 \\ \hline 2 & 0 & 12 & 0 & 0 & 1 & 12000 \end{array}\right]$$

The minimum cost of \$12,000 is obtained by using facility $A$ for 12 weeks, and 0 weeks for $B$ and $C$.

**27.** $x$ = days for Factory 1; $y$ = days for Factory 2

Minimize: $C = 10{,}000x + 20{,}000y$

Constraints: $80x + 20y \geq 1600 \ \to\ 4x + y \geq 80$

$10x + 10y \geq 500 \ \to\ x + y \geq 50$

$50x + 20y \geq 1900 \ \to\ 5x + 2y \geq 190$

$$\left[\begin{array}{cc|c} 4 & 1 & 80 \\ 1 & 1 & 50 \\ 5 & 2 & 190 \\ \hline 10000 & 20000 & g \end{array}\right] \quad \text{transpose: } \left[\begin{array}{ccc|c} 4 & 1 & 5 & 10000 \\ 1 & 1 & 2 & 20000 \\ \hline 80 & 50 & 190 & g \end{array}\right]$$

$$\left[\begin{array}{cccccc|c} 4 & 1 & \boxed{5} & 1 & 0 & 0 & 10{,}000 \\ 1 & 1 & 2 & 0 & 1 & 0 & 20{,}000 \\ \hline -80 & -50 & -190 & 0 & 0 & 1 & 0 \end{array}\right] \quad \tfrac{1}{5}R_1 \to R_1 \text{ then } \quad \begin{array}{c} \to \\ -2R_1 + R_2 \to R_2 \\ 190R_1 + R_3 \to R_3 \end{array}$$

$$\left[\begin{array}{cccccc|c} \frac{4}{5} & \boxed{\frac{1}{5}} & 1 & \frac{1}{5} & 0 & 0 & 2000 \\ -\frac{3}{5} & \frac{3}{5} & 0 & -\frac{2}{5} & 1 & 0 & 16,000 \\ \hline 72 & -12 & 0 & 38 & 0 & 1 & 380,000 \end{array}\right] \quad 5R_1 \to R_1 \text{ then } \begin{array}{c} -\frac{3}{5}R_1 + R_2 \to R_2 \\ \to \\ 12R_1 + R_3 \to R_3 \end{array}$$

$$\left[\begin{array}{cccccc|c} 4 & 1 & 5 & 1 & 0 & 0 & 10,000 \\ -3 & 0 & -3 & -1 & 1 & 0 & 10,000 \\ \hline 120 & 0 & 60 & 50 & 0 & 1 & 500,000 \end{array}\right]$$

Minimum cost is \$500,000 using 50 days at Factory 1 and 0 days at Factory 2.

**29.** Minimize: $C = 500T + 100R$
Constraints: $0.9T + 0.6R \geq 63$
$T + R \geq 90$

$$\left[\begin{array}{cc|c} 0.9 & 0.6 & 63 \\ 1 & 1 & 90 \\ \hline 500 & 100 & g \end{array}\right] \quad \text{transpose:} \quad \left[\begin{array}{cc|c} 0.9 & 1 & 500 \\ 0.6 & 1 & 100 \\ \hline 63 & 90 & g \end{array}\right]$$

$$\left[\begin{array}{ccccc|c} 0.9 & 1 & 1 & 0 & 0 & 500 \\ 0.6 & \boxed{1} & 0 & 1 & 0 & 100 \\ \hline -63 & -90 & 0 & 0 & 1 & 0 \end{array}\right] \quad \begin{array}{c} -R_2 + R_1 \to R_1 \\ \to \\ 90R_2 + R_3 \to R_3 \end{array}$$

$$\left[\begin{array}{ccccc|c} 0.3 & 0 & 1 & -1 & 0 & 400 \\ \boxed{0.6} & 1 & 0 & 1 & 0 & 100 \\ \hline -9 & 0 & 0 & 90 & 1 & 9000 \end{array}\right] \quad \tfrac{5}{3}R_2 \to R_2 \text{ then } \begin{array}{c} -\frac{3}{10}R_2 + R_1 \to R_1 \\ \to \\ 9R_2 + R_3 \to R_3 \end{array} \quad \left[\begin{array}{ccccc|c} 0 & -\frac{1}{2} & 1 & -\frac{3}{2} & 0 & 350 \\ 1 & \frac{5}{3} & 0 & \frac{5}{3} & 0 & \frac{500}{3} \\ \hline 0 & 15 & 0 & 105 & 1 & 10,500 \end{array}\right]$$

Minimum cost is \$10,500 using 0 minutes of TV time and 105 minutes of radio time.

**31.** Minimize: $\text{Cost} = 120x_1 + 140x_2 + 126x_3$
Constraints: $10x_1 + 5x_2 + 6x_3 \geq 230$
$5x_1 + 8x_2 + 6x_3 \geq 240$
$6x_1 + 6x_2 + 6x_3 \geq 210 \quad \to \quad x_1 + x_2 + x_3 \geq 35$

$$\left[\begin{array}{ccccccc|c} 10 & 5 & 1 & 1 & 0 & 0 & 0 & 120 \\ 5 & \mathbf{8} & 1 & 0 & 1 & 0 & 0 & 140 \\ 6 & 6 & 1 & 0 & 0 & 1 & 0 & 126 \\ \hline -230 & -240 & -35 & 0 & 0 & 0 & 1 & 0 \end{array}\right] \quad \begin{array}{c} \to \\ \frac{1}{8}R_2 \to R_2 \end{array}$$

$$\left[\begin{array}{ccccccc|c} 10 & 5 & 1 & 1 & 0 & 0 & 0 & 120 \\ 0.625 & \mathbf{1} & 0.125 & 0 & 0.125 & 0 & 0 & 17.5 \\ 6 & 6 & 1 & 0 & 0 & 1 & 0 & 126 \\ \hline -230 & -240 & -35 & 0 & 0 & 0 & 1 & 0 \end{array}\right] \quad \begin{array}{c} -5R_2 + R_1 \to R_1 \\ \to \\ -6R_2 + R_3 \to R_3 \\ 240R_2 + R_4 \to R_4 \end{array}$$

$$\left[\begin{array}{ccccccc|c} \mathbf{6.875} & 0 & 0.375 & 1 & -0.625 & 0 & 0 & 32.5 \\ 0.625 & 1 & 0.125 & 0 & 0.125 & 0 & 0 & 17.5 \\ 2.250 & 0 & 0.250 & 0 & -0.750 & 1 & 0 & 21 \\ \hline -80 & 0 & -5 & 0 & 30 & 0 & 1 & 4200 \end{array}\right] \quad \begin{array}{c} \frac{1}{6.875}R_1 \to R_1 \\ \\ \to \end{array}$$

$$\left[\begin{array}{ccccccc|c} \mathbf{1} & 0 & 0.0545 & 0.1455 & -0.0909 & 0 & 0 & 4.7273 \\ 0.625 & 1 & 0.125 & 0 & 0.125 & 0 & 0 & 17.5 \\ 2.250 & 0 & 0.250 & 0 & -0.750 & 1 & 0 & 21 \\ \hline -80 & 0 & -5 & 0 & 30 & 0 & 1 & 4200 \end{array}\right] \begin{array}{c} \rightarrow \\ -0.625R_1 + R_2 \rightarrow R_2 \\ -2.25R_1 + R_3 \rightarrow R_3 \\ 80R_1 + R_4 \rightarrow R_4 \end{array}$$

$$\left[\begin{array}{ccccccc|c} 1 & 0 & 0.0545 & 0.1455 & -0.0909 & 0 & 0 & 4.7273 \\ 0 & 1 & 0.0909 & -0.0909 & 0.1818 & 0 & 0 & 14.5455 \\ 0 & 0 & \mathbf{0.1273} & -0.3273 & -0.5455 & 1 & 0 & 10.3636 \\ \hline 0 & 0 & -0.6363 & 11.6364 & 22.7273 & 0 & 1 & 4578.1818 \end{array}\right] \begin{array}{c} \rightarrow \\ \\ \frac{1}{0.1273}R_3 \rightarrow R_3 \\ \\ \end{array}$$

$$\left[\begin{array}{ccccccc|c} 1 & 0 & 0.0545 & 0.1455 & -0.0909 & 0 & 0 & 4.7273 \\ 0 & 1 & 0.0909 & -0.0909 & 0.1818 & 0 & 0 & 14.5455 \\ 0 & 0 & \mathbf{1} & -2.5709 & -4.2848 & 7.8558 & 0 & 81.4111 \\ \hline 0 & 0 & -0.6363 & 11.6364 & 22.7273 & 0 & 1 & 4578.1818 \end{array}\right] \begin{array}{c} -0.0545R_3 + R_1 \rightarrow R_1 \\ -0.0909R_3 + R_2 \rightarrow R_2 \\ \rightarrow \\ 0.6363R_3 + R_4 \rightarrow R_4 \end{array}$$

$$\left[\begin{array}{ccccccc|c} 1 & 0 & 0 & 0.2856 & 0.1426 & -0.4281 & 0 & 0.2904 \\ 0 & 1 & 0 & 0.1428 & 0.5713 & -0.7141 & 0 & 7.1452 \\ 0 & 0 & 1 & -2.5709 & -4.2848 & 7.8555 & 0 & 81.4111 \\ \hline 0 & 0 & 0 & 10.0005 & 20.0009 & 4.9984 & 1 & 4629.98 \end{array}\right]$$

**a.** DeTurris should buy 10 Georgia packages, 20 Union packages, and 5 Pacific packages.
**b.** The minimum possible cost is \$4630.

**33.** $x$ = ounces of Food I, $y$ = ounces of Food II, $z$ = ounces of Food III
Minimize: $g = x + 5y + 2z$
Constraints: $2x + 2y + 2z \ge 24, \quad x + 5y + z \ge 16$
Equivalent constraints: $x + y + z \ge 12, \quad x + 5y + z \ge 16$

$$\left[\begin{array}{ccc|c} 1 & 1 & 1 & 12 \\ 1 & 5 & 1 & 16 \\ \hline 1 & 5 & 2 & g \end{array}\right] \quad \text{transpose:} \quad \left[\begin{array}{cc|c} 1 & 1 & 1 \\ 1 & 5 & 5 \\ 1 & 1 & 2 \\ \hline 12 & 16 & g \end{array}\right]$$

$$\left[\begin{array}{cccccc|c} 1 & \boxed{1} & 1 & 0 & 0 & 0 & 1 \\ 1 & 5 & 0 & 1 & 0 & 0 & 5 \\ 1 & 1 & 0 & 0 & 1 & 0 & 2 \\ \hline -12 & -16 & 0 & 0 & 0 & 1 & 0 \end{array}\right] \begin{array}{c} \rightarrow \\ -5R_1 + R_2 \rightarrow R_2 \\ -R_1 + R_3 \rightarrow R_3 \\ 16R_1 + R_4 \rightarrow R_4 \end{array} \left[\begin{array}{cccccc|c} 1 & 1 & 1 & 0 & 0 & 0 & 1 \\ -4 & 0 & -5 & 1 & 0 & 0 & 0 \\ 0 & 0 & -1 & 0 & 1 & 0 & 1 \\ \hline 4 & 0 & 16 & 0 & 0 & 1 & 16 \end{array}\right]$$

**a.** Minimum possible cost is \$16.
**b.** 16 ounces of Food I and 0 ounces of Food II and Food III should be served to give the minimum cost (NOTE: This is one of several possible solutions, another is 11 oz of Food I, 1 oz of Food II, and 0 oz of Food III.)

**35.** Let $y_1$ = number of workers who begin on Monday, $y_2$ = number of workers who begin on Tuesday, etc.
Minimize: $g = y_1 + y_2 + y_3 + y_4 + y_5 + y_6 + y_7$

Constraints: $y_1 + y_2 + y_3 + y_4 + y_5 \geq 22$

$y_2 + y_3 + y_4 + y_5 + y_6 \geq 12$

$y_3 + y_4 + y_5 + y_6 + y_7 \geq 17$

$y_1 + y_4 + y_5 + y_6 + y_7 \geq 20$

$y_1 + y_2 + y_5 + y_6 + y_7 \geq 16$

$y_1 + y_2 + y_3 + y_6 + y_7 \geq 16$

$y_1 + y_2 + y_3 + y_4 + y_7 \geq 20$

$$\left[\begin{array}{ccccccccccccccc|c} 1 & 0 & 0 & 1 & 1 & 1 & 1 & 1 & 0 & 0 & 0 & 0 & 0 & 0 & 0 & 1 \\ 1 & 1 & 0 & 0 & 1 & 1 & 1 & 0 & 1 & 0 & 0 & 0 & 0 & 0 & 0 & 1 \\ 1 & 1 & 1 & 0 & 0 & 1 & 1 & 0 & 0 & 1 & 0 & 0 & 0 & 0 & 0 & 1 \\ 1 & 1 & 1 & 1 & 0 & 0 & 1 & 0 & 0 & 0 & 1 & 0 & 0 & 0 & 0 & 1 \\ 1 & 1 & 1 & 1 & 1 & 0 & 0 & 0 & 0 & 0 & 0 & 1 & 0 & 0 & 0 & 1 \\ 0 & 1 & 1 & 1 & 1 & 1 & 0 & 0 & 0 & 0 & 0 & 0 & 1 & 0 & 0 & 1 \\ 0 & 0 & 1 & 1 & 1 & 1 & 1 & 0 & 0 & 0 & 0 & 0 & 0 & 1 & 0 & 1 \\ \hline -22 & -12 & -17 & -20 & -16 & -16 & -20 & 0 & 0 & 0 & 0 & 0 & 0 & 0 & 1 & 0 \end{array}\right]$$

Solve using technology: The minimum number of workers to staff the depot is 25 when 8 employees begin their shift on Monday, none on Tuesday, 5 on Wednesday, 4 on Thursday, 5 on Friday, none on Saturday, and 3 on Sunday.

## *Exercises 4.5*

**1.** $3x - y \geq 5$

$-3x + y \leq -5$

**3.** $y \geq 40 - 6x$

$-6x - y \leq -40$

**5.** Maximize: $f = 2x + 3y$

Constraints: $7x + 4y \leq 28$

$-3x + y \geq 2$

$x,\ y \geq 0$

**a.** $f = 2x + 3y$

$7x + 4y \leq 28$

$3x - y \leq -2$

$x,\ y \geq 0$

**b.**

$$\begin{array}{c} \begin{matrix} x & y & s_1 & s_2 & f \end{matrix} \\ \left[\begin{array}{ccccc|c} 7 & 4 & 1 & 0 & 0 & 28 \\ 3 & \boxed{-1} & 0 & 1 & 0 & -2 \\ \hline -2 & -3 & 0 & 0 & 1 & 0 \end{array}\right] \end{array}$$

**7.** Minimize: $g = 3x + 8y$

Constraints: $4x - 5y \leq 50$

$x + y \leq 80$

$-x + 2y \geq 4$

$x,\ y \geq 0$

**a.** $-g = -3x - 8y$ (to maximize)

$4x - 5y \leq 50$

$x + y \leq 80$

$x - 2y \leq -4$

$x,\ y \geq 0$

**b.**

$$\begin{array}{c} \begin{matrix} x & y & s_1 & s_2 & s_3 & -g \end{matrix} \\ \left[\begin{array}{cccccc|c} 4 & -5 & 1 & 0 & 0 & 0 & 50 \\ 1 & 1 & 0 & 1 & 0 & 0 & 80 \\ 1 & \boxed{-2} & 0 & 0 & 1 & 0 & -4 \\ \hline 3 & 8 & 0 & 0 & 0 & 1 & 0 \end{array}\right] \end{array}$$

**9.** The minimum is $f = 120$ at $x = 6$, $y = 8$, $z = 12$.

**11.** Maximize: $f = 4x + y$

Constraints: $5x + 2y \le 84,\ 3x - 2y \le -4,\ x, y \ge 0$

$$\left[\begin{array}{ccccc|c} 5 & 2 & 1 & 0 & 0 & 84 \\ 3 & \boxed{-2} & 0 & 1 & 0 & -4 \\ \hline -4 & -1 & 0 & 0 & 1 & 0 \end{array}\right] \quad -\tfrac{1}{2}R_2 \to R_2 \text{ then} \quad \begin{array}{c} -2R_2 + R_1 \to R_1 \\ \to \\ R_2 + R_3 \to R_3 \end{array}$$

$$\left[\begin{array}{ccccc|c} \boxed{8} & 0 & 1 & 1 & 0 & 80 \\ -\frac{3}{2} & 1 & 0 & -\frac{1}{2} & 0 & 2 \\ \hline -\frac{11}{2} & 0 & 0 & -\frac{1}{2} & 1 & 2 \end{array}\right] \quad \tfrac{1}{8}R_1 \to R_1 \text{ then} \quad \begin{array}{c} \to \\ \frac{3}{2}R_1 + R_2 \to R_2 \\ \frac{11}{2}R_1 + R_3 \to R_3 \end{array} \quad \left[\begin{array}{ccccc|c} 1 & 0 & \frac{1}{8} & \frac{1}{8} & 0 & 10 \\ 0 & 1 & \frac{3}{16} & -\frac{5}{16} & 0 & 17 \\ \hline 0 & 0 & \frac{11}{16} & \frac{3}{16} & 1 & 57 \end{array}\right]$$

A maximum of 57 occurs when $x = 10$ and $y = 17$.

**13.** Maximize: $-f = -2x - 3y$

Constraints: $-x \le -5,\ y \le 13,\ x - y \le -2,\ x \ge 0,\ y \ge 0$

$$\left[\begin{array}{cccccc|c} \boxed{-1} & 0 & 1 & 0 & 0 & 0 & -5 \\ 0 & 1 & 0 & 1 & 0 & 0 & 13 \\ 1 & -1 & 0 & 0 & 1 & 0 & -2 \\ \hline 2 & 3 & 0 & 0 & 0 & 1 & 0 \end{array}\right] \quad -R_1 \to R_1 \text{ then} \quad \begin{array}{c} \to \\ \to \\ -R_1 + R_3 \to R_3 \\ -2R_1 + R_4 \to R_4 \end{array}$$

$$\left[\begin{array}{cccccc|c} 1 & 0 & -1 & 0 & 0 & 0 & 5 \\ 0 & 1 & 0 & 1 & 0 & 0 & 13 \\ 0 & \boxed{-1} & 1 & 0 & 1 & 0 & -7 \\ \hline 0 & 3 & 2 & 0 & 0 & 1 & -10 \end{array}\right] \quad -R_3 \to R_3 \quad \begin{array}{c} \to \\ -R_3 + R_2 \to R_2 \\ \to \\ -3R_3 + R_4 \to R_4 \end{array} \quad \left[\begin{array}{cccccc|c} 1 & 0 & -1 & 0 & 0 & 0 & 5 \\ 0 & 0 & 1 & 1 & 1 & 0 & 6 \\ 0 & 1 & -1 & 0 & -1 & 0 & 7 \\ \hline 0 & 0 & 5 & 0 & 3 & 1 & -31 \end{array}\right]$$

Minimum is 31 at $x = 5$ and $y = 7$.

**15.** Minimize: $f = 3x + 2y$

Subject to:
$$\begin{aligned} -x + y &\le 10 \\ x + y &\ge 20 \\ x + y &\le 35 \\ x, y &\ge 0 \end{aligned}$$

Maximize: $-f = -3x - 2y$

Subject to:
$$\begin{aligned} -x + y &\le 10 \\ -x - y &\le -20 \\ x + y &\le 35 \\ x, y &\ge 0 \end{aligned}$$

$$\left[\begin{array}{cccccc|c} -1 & 1 & 1 & 0 & 0 & 0 & 10 \\ \boxed{-1} & -1 & 0 & 1 & 0 & 0 & -20 \\ 1 & 1 & 0 & 0 & 1 & 0 & 35 \\ \hline -3 & -2 & 0 & 0 & 0 & 1 & 0 \end{array}\right] \quad -R_2 \to R_2 \text{ then} \quad \begin{array}{c} R_2 + R_1 \to R_1 \\ \to \\ -R_2 + R_3 \to R_3 \\ -3R_2 + R_4 \to R_4 \end{array}$$

$$\left[\begin{array}{cccccc|c} 0 & \boxed{2} & 1 & -1 & 0 & 0 & 30 \\ 1 & 1 & 0 & -1 & 0 & 0 & 20 \\ 0 & 0 & 0 & 1 & 1 & 0 & 15 \\ \hline 0 & -1 & 0 & 3 & 0 & 1 & -60 \end{array}\right] \quad \tfrac{1}{2}R_1 \to R_1 \text{ then} \quad \begin{array}{c} \to \\ -R_1 + R_2 \to R_2 \\ \to \\ R_1 + R_4 \to R_4 \end{array} \quad \left[\begin{array}{cccccc|c} 0 & 1 & \frac{1}{2} & -\frac{1}{2} & 0 & 0 & 15 \\ 1 & 0 & -\frac{1}{2} & -\frac{1}{2} & 0 & 0 & 5 \\ 0 & 0 & 0 & 1 & 1 & 0 & 15 \\ \hline 0 & 0 & \frac{1}{2} & \frac{5}{2} & 0 & 1 & -45 \end{array}\right]$$

Minimum of 45 occurs when $x = 5$ and $y = 15$.

**17.** Maximize: $f = 2x + 5y$

Constraints: $-x + y \le 10,\ x + y \le 30,\ 2x - y \le 24,\ x, y \ge 0$

$$\left[\begin{array}{cccccc|c} -1 & \boxed{1} & 1 & 0 & 0 & 0 & 10 \\ 1 & 1 & 0 & 1 & 0 & 0 & 30 \\ 2 & -1 & 0 & 0 & 1 & 0 & 24 \\ \hline -2 & -5 & 0 & 0 & 0 & 1 & 0 \end{array}\right] \begin{array}{l} \to \\ -R_1 + R_2 \to R_2 \\ R_1 + R_3 \to R_3 \\ 5R_1 + R_4 \to R_4 \end{array}$$

$$\left[\begin{array}{cccccc|c} -1 & 1 & 1 & 0 & 0 & 0 & 10 \\ \boxed{2} & 0 & -1 & 1 & 0 & 0 & 20 \\ 1 & 0 & 1 & 0 & 1 & 0 & 34 \\ \hline -7 & 0 & 5 & 0 & 0 & 1 & 50 \end{array}\right] \tfrac{1}{2}R_2 \to R_2 \text{ then } \begin{array}{l} R_2 + R_1 \to R_1 \\ \to \\ -R_2 + R_3 \to R_3 \\ 7R_2 + R_4 \to R_4 \end{array} \left[\begin{array}{cccccc|c} 0 & 1 & \frac{1}{2} & \frac{1}{2} & 0 & 0 & 20 \\ 1 & 0 & -\frac{1}{2} & \frac{1}{2} & 0 & 0 & 10 \\ 0 & 0 & \frac{3}{2} & -\frac{1}{2} & 1 & 0 & 24 \\ \hline 0 & 0 & \frac{3}{2} & \frac{7}{2} & 0 & 1 & 120 \end{array}\right]$$

Maximum is 120 at $x = 10, y = 20$.

**19.** Maximize: $-f = -x - 2y - 3z$

Constraints: $x + z \le 20,\ -x - y \le -30,\ y + z \le 20,\ x \ge 0,\ y \ge 0,\ z \ge 0$

$$\left[\begin{array}{ccccccc|c} \boxed{1} & 0 & 1 & 1 & 0 & 0 & 0 & 20 \\ -1 & -1 & 0 & 0 & 1 & 0 & 0 & -30 \\ 0 & 1 & 1 & 0 & 0 & 1 & 0 & 20 \\ \hline 1 & 2 & 3 & 0 & 0 & 0 & 1 & 0 \end{array}\right] \begin{array}{l} \to \\ R_1 + R_2 \to R_2 \\ \to \\ -R_1 + R_4 \to R_4 \end{array}$$

$$\left[\begin{array}{ccccccc|c} 1 & 0 & 1 & 1 & 0 & 0 & 0 & 20 \\ 0 & \boxed{-1} & 1 & 1 & 1 & 0 & 0 & -10 \\ 0 & 1 & 1 & 0 & 0 & 1 & 0 & 20 \\ \hline 0 & 2 & 2 & -1 & 0 & 0 & 1 & -20 \end{array}\right] -R_2 \to R_2 \text{ then } \begin{array}{l} \to \\ \\ -R_2 + R_3 \to R_3 \\ -2R_2 + R_4 \to R_4 \end{array} \left[\begin{array}{ccccccc|c} 1 & 0 & 1 & 1 & 0 & 0 & 0 & 20 \\ 0 & 1 & -1 & -1 & -1 & 0 & 0 & 10 \\ 0 & 0 & 2 & 1 & 1 & 1 & 0 & 10 \\ \hline 0 & 0 & 4 & 1 & 2 & 0 & 1 & -40 \end{array}\right]$$

Minimum is 40 at $x = 20, y = 10, z = 0$.

Note: recall basic variables have columns with a single entry of 1 and other entries, 0.

**21.** Maximize: $f = 2x - y + 4z$

Constraints: $x + y + z \le 8,\ -x + y - z \le -4,\ -x - y + z \le -2,\ x, y, z \ge 0$

$$\left[\begin{array}{ccccccc|c} 1 & 1 & 1 & 1 & 0 & 0 & 0 & 8 \\ -1 & 1 & -1 & 0 & 1 & 0 & 0 & -4 \\ \boxed{-1} & -1 & 1 & 0 & 0 & 1 & 0 & -2 \\ \hline -2 & 1 & -4 & 0 & 0 & 0 & 1 & 0 \end{array}\right] -R_3 \to R_3 \text{ then } \begin{array}{l} -R_3 + R_1 \to R_1 \\ R_3 + R_2 \to R_2 \\ \to \\ 2R_3 + R_4 \to R_4 \end{array}$$

$$\left[\begin{array}{ccccccc|c} 0 & 0 & 2 & 1 & 0 & 1 & 0 & 6 \\ 0 & 2 & \boxed{-2} & 0 & 1 & -1 & 0 & -2 \\ 1 & 1 & -1 & 0 & 0 & -1 & 0 & 2 \\ \hline 0 & 3 & -6 & 0 & 0 & -2 & 1 & 4 \end{array}\right] -\tfrac{1}{2}R_2 \to R_2 \text{ then } \begin{array}{l} -2R_2 + R_1 \to R_1 \\ \to \\ R_2 + R_3 \to R_3 \\ 6R_2 + R_4 \to R_4 \end{array}$$

$$\left[\begin{array}{ccccccc|c} 0 & \boxed{2} & 0 & 1 & 1 & 0 & 0 & 4 \\ 0 & -1 & 1 & 0 & -\frac{1}{2} & \frac{1}{2} & 0 & 1 \\ 1 & 0 & 0 & 0 & -\frac{1}{2} & -\frac{1}{2} & 0 & 3 \\ \hline 0 & -3 & 0 & 0 & -3 & 1 & 1 & 10 \end{array}\right] \tfrac{1}{2}R_1 \to R_1 \text{ then } \begin{array}{l} \to \\ R_1 + R_2 \to R_2 \\ \to \\ 3R_1 + R_4 \to R_4 \end{array}$$

$$\left[\begin{array}{ccccccc|c} 0 & 1 & 0 & \frac{1}{2} & \boxed{\frac{1}{2}} & 0 & 0 & 2 \\ 0 & 0 & 1 & \frac{1}{2} & 0 & \frac{1}{2} & 0 & 3 \\ 1 & 0 & 0 & 0 & -\frac{1}{2} & -\frac{1}{2} & 0 & 3 \\ \hline 0 & 0 & 0 & \frac{3}{2} & -\frac{3}{2} & 1 & 1 & 16 \end{array}\right] \begin{array}{l} 2R_1 \to R_1 \text{ then} \\ \to \\ \frac{1}{2}R_1 + R_3 \to R_3 \\ \frac{3}{2}R_1 + R_4 \to R_4 \end{array} \left[\begin{array}{ccccccc|c} 0 & 2 & 0 & 1 & 1 & 0 & 0 & 4 \\ 0 & 0 & 1 & \frac{1}{2} & 0 & \frac{1}{2} & 0 & 3 \\ 1 & 1 & 0 & \frac{1}{2} & 0 & -\frac{1}{2} & 0 & 5 \\ \hline 0 & 3 & 0 & 3 & 0 & 1 & 1 & 22 \end{array}\right]$$

Maximum is 22 at $x = 5, y = 0, z = 3$.

**23.** Maximize: $-f = -10x - 30y - 35z$

Constraints: $x + y + z \le 250,\ -x - y - 2z \le -150,\ 2x + y + z \le 180,\ x \ge 0,\ y \ge 0,\ z \ge 0$

$$\left[\begin{array}{ccccccc|c} 1 & 1 & 1 & 1 & 0 & 0 & 0 & 250 \\ -1 & -1 & \boxed{-2} & 0 & 1 & 0 & 0 & -150 \\ 2 & 1 & 1 & 0 & 0 & 1 & 0 & 180 \\ \hline 10 & 30 & 35 & 0 & 0 & 0 & 1 & 0 \end{array}\right] -\tfrac{1}{2}R_2 \to R_2 \text{ then} \begin{array}{c} -R_2 + R_1 \to R_1 \\ \to \\ -R_2 + R_3 \to R_3 \\ -35R_2 + R_4 \to R_4 \end{array}$$

$$\left[\begin{array}{ccccccc|c} \frac{1}{2} & \frac{1}{2} & 0 & 1 & \frac{1}{2} & 0 & 0 & 175 \\ \frac{1}{2} & \frac{1}{2} & 1 & 0 & -\frac{1}{2} & 0 & 0 & 75 \\ \boxed{\frac{3}{2}} & \frac{1}{2} & 0 & 0 & \frac{1}{2} & 1 & 0 & 105 \\ \hline -\frac{15}{2} & \frac{25}{2} & 0 & 0 & \frac{35}{2} & 0 & 1 & -2625 \end{array}\right] \tfrac{2}{3}R_3 \to R_3 \text{ then} \begin{array}{c} -\frac{1}{2}R_3 + R_1 \to R_1 \\ -\frac{1}{2}R_3 + R_2 \to R_2 \\ \to \\ \frac{15}{2}R_3 + R_4 \to R_4 \end{array} \left[\begin{array}{ccccccc|c} 0 & \frac{1}{3} & 0 & 1 & \frac{1}{3} & -\frac{1}{3} & 0 & 140 \\ 0 & \frac{1}{3} & 1 & 0 & -\frac{2}{3} & -\frac{1}{3} & 0 & 40 \\ 1 & \frac{1}{3} & 0 & 0 & \frac{1}{3} & \frac{2}{3} & 0 & 70 \\ \hline 0 & 15 & 0 & 0 & 20 & 5 & 1 & -2100 \end{array}\right]$$

Minimum is 2100 at $x = 70$, $y = 0$, $z = 40$.

**25.** Maximize: $f = 25x_1 + 25x_2 + 25x_3 + 20x_4 + 20x_5 + 20x_6$

Constraints: $x_1 + x_4 \le 100,\quad x_2 + x_5 \le 20,\ x_3 + x_6 \le 30$

$-0.8x_2 + 0.2x_1 + 0.2x_3 \le 0$

$-0.9x_5 + 0.1x_4 + 0.1x_6 \le 0$

$-0.9x_6 + 0.1x_4 + 0.1x_5 \le 0$

$$\left[\begin{array}{ccccccccccccc|c} 1 & 0 & 0 & 1 & 0 & 0 & 1 & 0 & 0 & 0 & 0 & 0 & 0 & 100 \\ 0 & 1 & 0 & 0 & 1 & 0 & 0 & 1 & 0 & 0 & 0 & 0 & 0 & 20 \\ 0 & 0 & 1 & 0 & 0 & 1 & 0 & 0 & 1 & 0 & 0 & 0 & 0 & 30 \\ 0.2 & -0.8 & 0.2 & 0 & 0 & 0 & 0 & 0 & 0 & 1 & 0 & 0 & 0 & 0 \\ 0 & 0 & 0 & 0.1 & -0.9 & 0.1 & 0 & 0 & 0 & 0 & 1 & 0 & 0 & 0 \\ 0 & 0 & 0 & 0.1 & 0.1 & -0.9 & 0 & 0 & 0 & 0 & 0 & 1 & 0 & 0 \\ \hline -25 & -25 & -25 & -20 & -20 & -20 & 0 & 0 & 0 & 0 & 0 & 0 & 1 & 0 \end{array}\right]$$

**Steps:** Pivot on 0.2 in $R_4C_1$. Row Ops: $5R_4 \to R_4$, then $-R + R_1 \to R_1$, $25R_4 + R_7 \to R_7$

Pivot on 1 in $R_2C_2$. Row Ops: $-4R_2 + R_1 \to R_1$, $4R_2 + R_4 \to R_4$, $125R_2 + R_7 \to R_7$

Pivot on 1 in $R_5C_6$. Row Ops: $10R_6 \to R_6$, then $-R_6 + R_1 \to R_1$, $-0.1R_6 + R_5 \to R_5$, $20R_6 + R_7 \to R_7$

Pivot on 4 in $R_1C_5$. Row Ops: $\frac{1}{4}R_1 \to R_1$, then $-R_1 + R_2 \to R_2$, $-R_1 + R_3 \to R_3$, $-4R_1 + R_4 \to R_4$

$R_1 + R_5 \to R_5$, $8R_1 + R_6 \to R_6$, $75R_1 + R_7 \to R_7$

Pivot on 1.25 in $R_3C_3$. Row Ops: $\frac{4}{3}R_3 \to R_3$, then $0.25R_3 + R_1 \to R_1$, $-0.25R_3 + R_2 \to R_2$, $-2R_3 + R_4 \to R_4$, $0.25R_3 + R_5 \to R_5$, $2R_3 + R_6 \to R_6$, $18.75R_3 + R_7 \to R_7$

$$\left[\begin{array}{ccccccccccccc|c} 0 & 0 & 0 & 0 & 1 & 0 & 0.2 & -0.8 & 0.2 & -1 & -2 & 0 & 0 & 10 \\ 0 & 1 & 0 & 0 & 0 & 0 & -0.2 & 1.8 & -0.2 & 1 & 2 & 0 & 0 & 10 \\ 0 & 0 & 1 & 0 & 0 & 0 & -0.2 & 0.8 & 0.8 & 1 & 1 & 1 & 0 & 20 \\ 1 & 0 & 0 & 0 & 0 & 0 & -0.6 & 6.4 & -1.6 & 8 & 7 & -1 & 0 & 20 \\ 0 & 0 & 0 & 0 & 0 & 1 & 0.2 & -0.8 & 0.2 & -1 & -1 & -1 & 0 & 10 \\ 0 & 0 & 0 & 1 & 0 & 0 & 1.6 & -6 & 1.6 & -8 & -7 & 1 & 0 & 80 \\ \hline 0 & 0 & 0 & 0 & 0 & 0 & 15 & 65 & 15 & 50 & 50 & 0 & 1 & 3250 \end{array}\right]$$

Maximum is 3250 when $x_1 = 20$, $x_2 = 10$, $x_3 = 20$, $x_4 = 80$, $x_5 = 10$, $x_6 = 10$.

**27.**

| | All Beef | Regular | Available |
|---|---|---|---|
| Beef | $0.75x$ | $0.18y$ | $\le 1020$ |
| Spices | $0.2x$ | $0.2y$ | $\ge 500$ |
| Pork | -- | $0.3y$ | $\le 600$ |

Objective function: $f = 0.6x + 0.4y$

$$\left[\begin{array}{cccccc|c} \boxed{0.75} & 0.18 & 1 & 0 & 0 & 0 & 1020 \\ -0.20 & -0.20 & 0 & 1 & 0 & 0 & -500 \\ 0 & 0.3 & 0 & 0 & 1 & 0 & 600 \\ \hline -1.50 & -1.00 & 0 & 0 & 0 & 1 & 0 \end{array}\right] \quad \frac{4}{3}R_1 \to R_1 \text{ then } \quad \begin{array}{c} \to \\ 0.2R_1 + R_2 \to R_2 \\ \to \\ 1.5R_1 + R_4 \to R_4 \end{array}$$

$$\left[\begin{array}{cccccc|c} 1 & 0.24 & 1.\bar{3} & 0 & 0 & 0 & 1360 \\ 0 & \boxed{-0.152} & 0.2\bar{6} & 1 & 0 & 0 & -228 \\ 0 & 0.3 & 0 & 0 & 1 & 0 & 600 \\ \hline 0 & -0.64 & 2 & 0 & 0 & 1 & 2040 \end{array}\right]$$

$$-\frac{1}{0.152}R_2 \to R_2 \text{ then } \begin{array}{c} -0.24R_2 + R_1 \to R_1 \\ \to \\ -0.3R_2 + R_3 \to R_3 \\ 0.64R_2 + R_4 \to R_4 \end{array} \approx \left[\begin{array}{cccccc|c} 1 & 0 & 1.7544 & 1.5789 & 0 & 0 & 1000 \\ 0 & 1 & -1.7544 & -6.5789 & 0 & 0 & 1500 \\ 0 & 0 & 0.5263 & \boxed{1.9737} & 1 & 0 & 150 \\ \hline 0 & 0 & 0.8772 & -4.2105 & 0 & 1 & 3000 \end{array}\right]$$

$$\frac{1}{1.9737}R_3 \to R_3 \text{ then } \begin{array}{c} -1.5789R_3 + R_1 \to R_1 \\ 6.5789R_3 + R_2 \to R_2 \\ \to \\ 4.2105R_3 + R_4 \to R_4 \end{array} \left[\begin{array}{cccccc|c} 1 & 0 & 1.\bar{3} & 0 & -0.8 & 0 & 880 \\ 0 & 1 & 0 & 0 & 3.\bar{3} & 0 & 2000 \\ 0 & 0 & 0.2\bar{6} & 1 & 0.50\bar{6} & 0 & 76 \\ \hline 0 & 0 & 2 & 0 & 2.1\bar{3} & 1 & 3320 \end{array}\right]$$

Maximum profit is \$3320 with 880 lbs of all beef and 2000 lbs of regular.

**29.**

| | Total Needed | Monaca | Hamburg |
|---|---|---|---|
| Heating Components | 500 | -- | -- |
| Produced | -- | $x$ | $500-x$ |
| Profit | -- | $400x$ | $390(500-x)$ |
| Domestic Furnaces | 750 | -- | -- |
| Produced | -- | $y$ | $750-y$ |
| Profit | -- | $200y$ | $215(750-y)$ |

Maximize: $f = 400x + 390(500-x) + 200y + 215(750-y) = 10x - 15y + 356,250$

Constraints: $x, y \ge 0$

$x + y \le 1000$ — Capacity at Monaca

$(500-x)+(750-y) \le 850$ or $x+y \ge 400$ — Capacity at Hamburg

$x \le 100 + \frac{1}{2}y$ or $2x - y \le 200$ — Monaca capacity restriction

$$\left[\begin{array}{cccccc|c} 1 & 1 & 1 & 0 & 0 & 0 & 1000 \\ -1 & \boxed{-1} & 0 & 1 & 0 & 0 & -400 \\ 2 & -1 & 0 & 0 & 1 & 0 & 200 \\ \hline -10 & 15 & 0 & 0 & 0 & 1 & 356,250 \end{array}\right] \begin{array}{l} \\ -R_2 \to R_2 \text{ then} \\ \\ \end{array} \begin{array}{c} -R_2 + R_1 \to R_1 \\ \to \\ R_2 + R_3 \to R_3 \\ -15R_2 + R_4 \to R_4 \end{array}$$

$$\left[\begin{array}{cccccc|c} 0 & 0 & 1 & 1 & 0 & 0 & 600 \\ 1 & 1 & 0 & -1 & 0 & 0 & 400 \\ \boxed{3} & 0 & 0 & -1 & 1 & 0 & 600 \\ \hline -25 & 0 & 0 & 15 & 0 & 1 & 350,250 \end{array}\right] \begin{array}{l} \\ \\ \frac{1}{3}R_3 \to R_3 \text{ then} \\ \\ \end{array} \begin{array}{c} \to \\ -R_3 + R_2 \to R_2 \\ \to \\ 25R_3 + R_4 \to R_4 \end{array} \left[\begin{array}{cccccc|c} 0 & 0 & 1 & 1 & 0 & 0 & 600 \\ 0 & 1 & 0 & -0.\overline{6} & -0.\overline{3} & 0 & 200 \\ 1 & 0 & 0 & -0.\overline{3} & 0.\overline{3} & 0 & 200 \\ \hline 0 & 0 & 0 & 6.\overline{6} & 8.\overline{3} & 1 & 355,250 \end{array}\right]$$

Maximum profit is $355,250.
$x = 200$ heating components at Monaca and $500 - 200 = 300$ at Hamburg.
$y = 200$ domestic furnaces at Monaca and $750 - 200 = 550$ at Hamburg.

**31.** Let $x_1$ = number of filters and $x_2$ = number of housings.

Minimize: $f = 6.60x_1 + 8.35x_2$

Constraints: $x_1 \ge 400$

$0 \le x_2 \le 400$

$20x_1 + 40x_2 \le 20,000$

Maximize: $-f = -6.60x_1 - 8.35x_2$

Constraints: $-x_1 \le -400$

$x_2 \le 400$

$20x_1 + 40x_2 \le 20,000$

$-6.6x_1 - 8.35x_2 + f = 0$

$$\left[\begin{array}{cccccc|c} \boxed{-1} & 0 & 1 & 0 & 0 & 0 & -400 \\ 0 & 1 & 0 & 1 & 0 & 0 & 400 \\ 20 & 40 & 0 & 0 & 1 & 0 & 20,000 \\ \hline -6.60 & -8.35 & 0 & 0 & 0 & 1 & 0 \end{array}\right] \begin{array}{l} -R_1 \to R_1 \\ \\ \\ \\ \end{array} \begin{array}{c} \to \\ \to \\ -20R_1 + R_3 \to R_3 \\ 6.60R_1 + R_4 \to R_4 \end{array}$$

$$\left[\begin{array}{cccccc|c} 1 & 0 & -1 & 0 & 0 & 0 & 400 \\ 0 & 1 & 0 & 1 & 0 & 0 & 400 \\ 0 & \boxed{40} & 20 & 0 & 1 & 0 & 12,000 \\ \hline 0 & -8.35 & -6.60 & 0 & 0 & 1 & 2640 \end{array}\right] \begin{array}{c} \to \\ \to \\ \frac{1}{40}R_3 \to R_3 \text{ then} \\ \to \end{array} \begin{array}{c} \to \\ -R_3 + R_2 \to R_2 \\ \to \\ 8.35R_3 + R_4 \to R_4 \end{array}$$

$$\left[\begin{array}{cccccc|c} 1 & 0 & -1 & 0 & 0 & 0 & 400 \\ 0 & 0 & -\frac{1}{2} & 1 & -\frac{1}{40} & 0 & 100 \\ 0 & 1 & \frac{1}{2} & 0 & \frac{1}{40} & 0 & 300 \\ \hline 0 & 0 & -2.425 & 0 & 0.20875 & 1 & 5145 \end{array}\right]$$

The minimum cost is \$5145 when 400 filters and 300 housing units are produced.

**33.**

| | Total Needed | Monaca | Hamburg |
|---|---|---|---|
| Heating Components | 500 | -- | -- |
| Produced | -- | $x$ | $500-x$ |
| Cost | -- | $380x$ | $400(500-x)$ |
| Domestic Furnaces | 750 | -- | -- |
| Produced | -- | $y$ | $750-y$ |
| Cost | -- | $200y$ | $185(750-y)$ |

Cost = objective function = $C = 380x + 400(500-x) + 200y + 185(750-y)$

$= -20x + 15y + 338,750$

Constraints: $x, y \geq 0$

$x + y \leq 1000$ Capacity at Monaca

$(500-x) + (750y) \leq 850$ or $x + y \geq 400$ Capacity at Hamburg

$x \leq 100 + \frac{1}{2}y$ or $2x - y \leq 200$ Monaca capacity restriction

Need to maximize $-C = 20x - 15y - 338,750$

$$\left[\begin{array}{cccccc|c} 1 & 1 & 1 & 0 & 0 & 0 & 1000 \\ -1 & \boxed{-1} & 0 & 1 & 0 & 0 & -400 \\ 2 & -1 & 0 & 0 & 1 & 0 & 200 \\ \hline -20 & 15 & 0 & 0 & 0 & 1 & -338,750 \end{array}\right] -R_2 \to R_2 \text{ then} \begin{array}{c} -R_2 + R_1 \to R_1 \\ \to \\ R_2 + R_3 \to R_3 \\ -15R_2 + R_4 \to R_4 \end{array}$$

$$\left[\begin{array}{cccccc|c} 0 & 0 & 1 & 1 & 0 & 0 & 600 \\ 1 & 1 & 0 & -1 & 0 & 0 & 400 \\ \boxed{3} & 0 & 0 & -1 & 1 & 0 & 600 \\ \hline -35 & 0 & 0 & 15 & 0 & 1 & -344,750 \end{array}\right] \frac{1}{3}R_3 \to R_3 \text{ then} \begin{array}{c} \to \\ -R_3 + R_2 \to R_2 \\ \to \\ 35R_3 + R_4 \to R_4 \end{array} \left[\begin{array}{cccccc|c} 0 & 0 & 1 & 1 & 0 & 0 & 600 \\ 0 & 1 & 0 & -\frac{2}{3} & -\frac{1}{3} & 0 & 200 \\ 1 & 0 & 0 & -\frac{1}{3} & \frac{1}{3} & 0 & 200 \\ \hline 0 & 0 & 0 & \frac{10}{3} & \frac{35}{3} & 1 & -337,750 \end{array}\right]$$

Minimum cost is \$337,750.

$x = 200$ heating components at Monaca and 500 – 200 = 300 at Hamburg.

$y = 200$ domestic furnaces at Monaca and 750 – 200 = 550 at Hamburg.

**35.**

| | Plant 1 | Plant 2 | Plant 3 | Totals | |
|---|---|---|---|---|---|
| Million gallons | $x$ | $y$ | $z$ | $\le 10$ | |
| Impurities | $0.20x$ | $0.15y$ | $0.10z$ | $\le 0.15(x+y+z)$ | $\Rightarrow\ 0.05x - 0.05z \le 0$ |
| Cost (Thousands) | $20x$ | $30y$ | $40z$ | $= C$ | |
| Other constraints | $x$ | | $z$ | $\ge 6$ | |

$$\left[\begin{array}{ccccccc|c} 1 & 1 & 1 & 1 & 0 & 0 & 0 & 10 \\ 0.05 & 0 & -0.05 & 0 & 1 & 0 & 0 & 0 \\ \boxed{-1} & 0 & -1 & 0 & 0 & 1 & 0 & -6 \\ \hline 20 & 30 & 40 & 0 & 0 & 0 & 1 & 0 \end{array}\right] \begin{array}{c} \rightarrow \\ 100R_2 \rightarrow R_2 \\ -R_3 \rightarrow R_3 \\ \rightarrow \end{array} \left[\begin{array}{ccccccc|c} 1 & 1 & 1 & 1 & 0 & 0 & 0 & 10 \\ 5 & 0 & -5 & 0 & 100 & 0 & 0 & 0 \\ \boxed{1} & 0 & 1 & 0 & 0 & -1 & 0 & 6 \\ \hline 20 & 30 & 40 & 0 & 0 & 0 & 1 & 0 \end{array}\right] \begin{array}{c} -R_3 + R_1 \rightarrow R_1 \\ -5R_3 + R_2 \rightarrow R_2 \\ \rightarrow \\ -20R_3 + R_4 \rightarrow R_4 \end{array}$$

$$\left[\begin{array}{ccccccc|c} 0 & 1 & 0 & 1 & 0 & 1 & 0 & 4 \\ 0 & 0 & \boxed{-10} & 0 & 100 & 5 & 0 & -30 \\ 1 & 0 & 1 & 0 & 0 & -1 & 0 & 6 \\ \hline 0 & 30 & 20 & 0 & 0 & 20 & 1 & -120 \end{array}\right] \begin{array}{c} \\ -\tfrac{1}{10}R_2 \rightarrow R_2 \text{ then} \\ \\ \\ \end{array} \begin{array}{c} \rightarrow \\ \rightarrow \\ -R_2 + R_3 \rightarrow R_3 \\ -20R_2 + R_4 \rightarrow R_4 \end{array}$$

$$\left[\begin{array}{ccccccc|c} 0 & 1 & 0 & 1 & 0 & 1 & 0 & 4 \\ 0 & 0 & 1 & 0 & -10 & -\tfrac{1}{2} & 0 & 3 \\ 1 & 0 & 0 & 0 & 10 & -\tfrac{1}{2} & 0 & 3 \\ \hline 0 & 30 & 0 & 0 & 200 & 30 & 1 & -180 \end{array}\right]$$

Minimum cost is \$180,000 with plants 1 and 3 each handling 3,000,000 gallons. Plant 2 is zero, not 4, because the $y$-column still represents a nonbasic variable.

**37.** $x =$ number of footballs, $y =$ number of soccer balls, $z =$ number of volleyballs

Maximize: $f = 30x + 25y + 20z$

Subject to: $10x + 8y + 8z \ge 10{,}000 \quad \rightarrow \quad -10x - 8y - 8z \le -10{,}000$

$10x + 12y + 8z \le 20{,}000$

$$\left[\begin{array}{cccccc|c} \boxed{-10} & -8 & -8 & 1 & 0 & 0 & -10{,}000 \\ 10 & 12 & 8 & 0 & 1 & 0 & 20{,}000 \\ \hline -30 & -25 & -20 & 0 & 0 & 1 & 0 \end{array}\right] \begin{array}{c} -\tfrac{1}{10}R_1 \rightarrow R_1 \text{ then} \\ \\ \\ \end{array} \begin{array}{c} \rightarrow \\ -10R_1 + R_2 \rightarrow R_2 \\ 30R_1 + R_3 \rightarrow R_3 \end{array}$$

$$\left[\begin{array}{cccccc|c} 1 & 0.8 & 0.8 & -0.1 & 0 & 0 & 1000 \\ 0 & 4 & 0 & \boxed{1} & 1 & 0 & 10{,}000 \\ \hline 0 & -1 & 4 & -3 & 0 & 1 & 30{,}000 \end{array}\right] \begin{array}{c} 0.1R_2 + R_1 \rightarrow R_1 \\ \rightarrow \\ 3R_2 + R_3 \rightarrow R_3 \end{array} \left[\begin{array}{cccccc|c} 1 & 1.2 & 0.8 & 0 & 0.1 & 0 & 2000 \\ 0 & 4 & 0 & 1 & 1 & 0 & 10{,}000 \\ \hline 0 & 11 & 4 & 0 & 3 & 1 & 60{,}000 \end{array}\right]$$

The maximum revenue is \$60,000 when 2000 footballs and no soccer balls or volleyballs are produced.

## Chapter 4 Review Exercises

**1.**

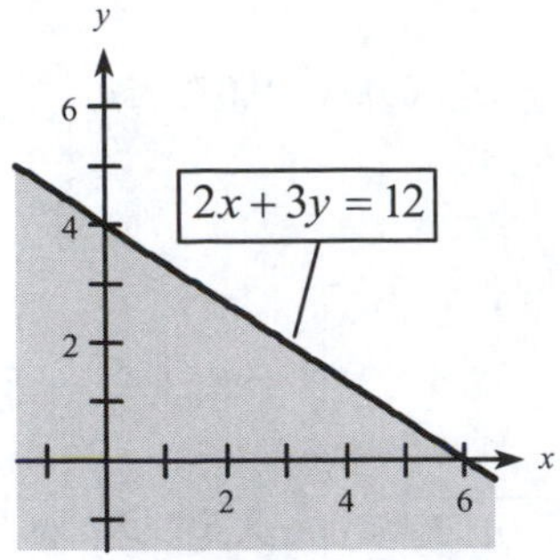


**2.**

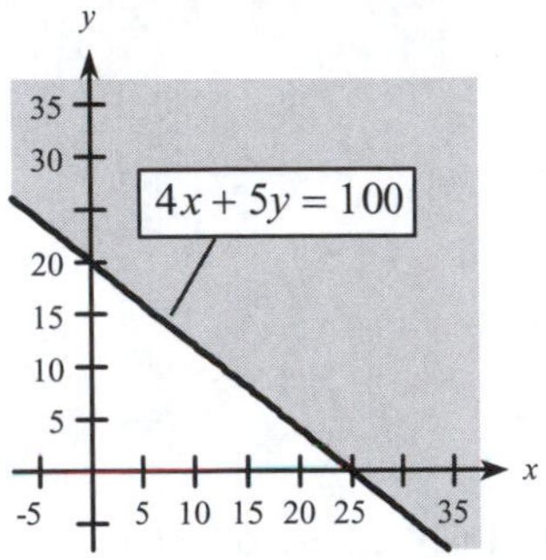


**3.**

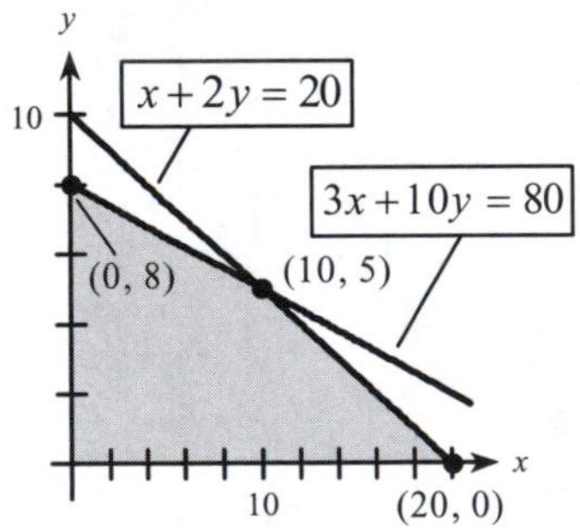


**4.**

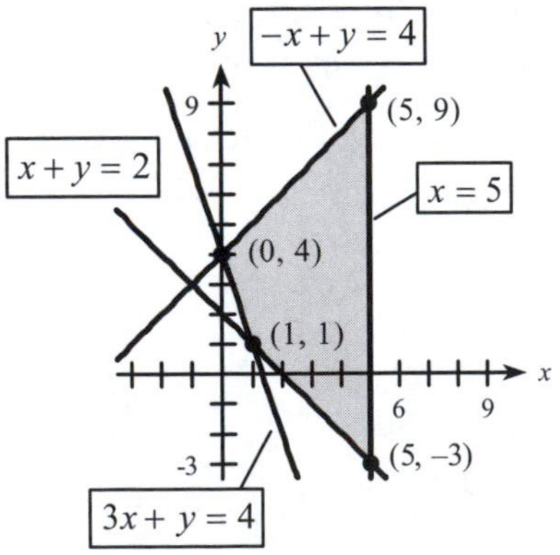


**5.**

| Corners | $f=-x+3y$ |
|---|---|
| $(0,5)$ | 15 |
| $(5,10)$ | 25 |
| $(10,6)$ | 8 |
| $(12,0)$ | −12 |
| $(0,0)$ | 0 |

Maximum is 25 at (5,10).
Minimum is −12 at (12,0).

**6.**

| Corners | $f=6x+4y$ |
|---|---|
| $(0,30)$ | 120 |
| $(0,40)$ | 160 |
| $(17,23)$ | 194 |
| $(8,14)$ | 104 |

Maximum is 194 at (17,23).
Minimum is 104 at (8,14).

**7.** Solving: $\begin{cases}-3x+2y=4\\ x+2y=20\end{cases}$

$$\begin{cases}-3x+2y=4\\ -x-2y=-20\end{cases}$$

$$-4x=-16$$

$$x=4$$

Corner: $(4,8)$

Solving: $\begin{cases}-3x+2y=4\\ x=20\end{cases}$

$$-3(20)+2y=4$$

$$2y=64$$

$$y=32$$

Corner: $(20,32)$

| Corners | $f=7x-6y$ |
|---|---|
| $(20,0)$ | 140 |
| $(4,8)$ | −20 |
| $(20,32)$ | −52 |

Maximum is 140 at $(20,0)$.

Minimum is −52 at $(20,32)$.

**8.** Solving: $\begin{cases} x+2y=19 \\ 3x+2y=29 \end{cases}$

$\begin{cases} -x-2y=-19 \\ 3x+2y=29 \end{cases}$

$2x=10$

$x=10$

$10+2y=19$

$2y=9$

$y=4.5$

Corner: $(10,4.5)$

| Corners | $f=9x+10y$ |
|---|---|
| $(10,4.5)$ | 135 |
| $(19,0)$ | 171 |
| $(0,14.5)$ | 145 |

Minimum is 135 at (10,4.5)
There is no maximum because the region is unbounded.

**9.** Maximize $f=5x+6y$

Subject to: $x+3y\le 24$

$4x+3y\le 42$

$2x+y\le 20$

Solving $\begin{cases} 4x+3y=42 \\ x+3y=24 \end{cases}$ we get $A(6, 6)$.

Solving $\begin{cases} 2x+y=20 \\ 4x+3y=42 \end{cases}$ we get $B(9, 2)$.

| Feasible Corners | $f=5x+6y$ |
|---|---|
| $(0,8)$ | 48 |
| $(6,6)$ | 66 |
| $(9,2)$ | 57 |
| $(10,0)$ | 50 |

Maximum is 66 at (6,6).

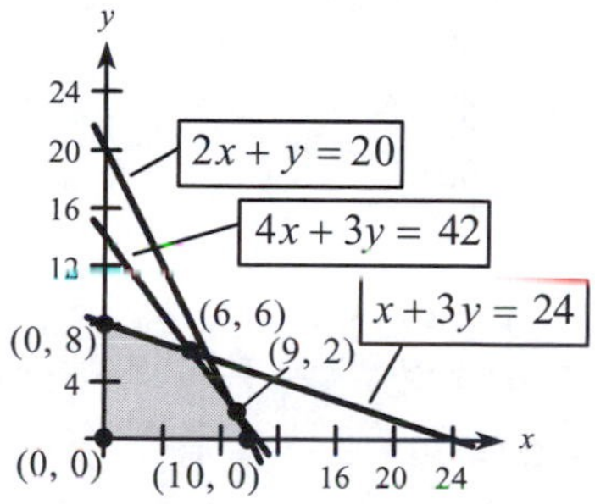


**10.** Maximize $f=x+4y$

Subject to: $7x+3y\le 105$

$2x+5y\le 59$

$x+7y\le 70$

Solving $\begin{cases} 7x+3y=105 \\ 2x+5y=59 \end{cases}$ we get $B$ (12, 7).

Solving $\begin{cases} x+7y=70 \\ 2x+5y=59 \end{cases}$ we get $A$ (7, 9).

| Feasible Corners | $f=x+4y$ | |
|---|---|---|
| $(0,10)$ | 40 | |
| $(7,9)$ | 43 | ← Minimum |
| $(12,7)$ | 40 | |
| $(15,0)$ | 15 | |

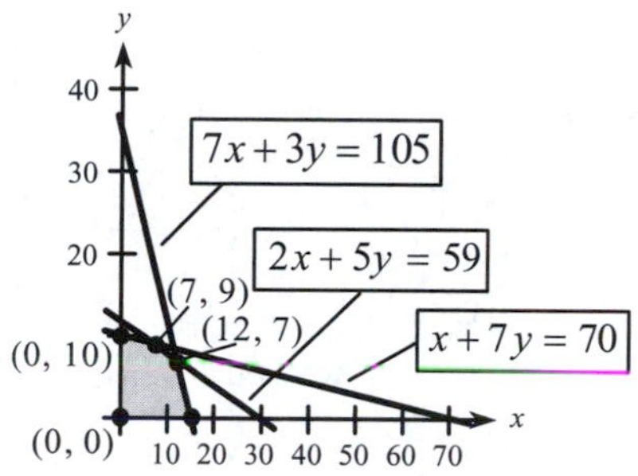


**11.** Maximize $f=2x+5y$

Subject to: $4x+5y\le 400$

$x\le 70$

$6x+15y\le 900$

Solving $\begin{cases} 4x+5y=400 \\ 6x+15y=900 \end{cases}$ we get $(50,40)$.

Solving $\begin{cases} 4x+5y=400 \\ x=70 \end{cases}$ we get $(70,24)$.

| Feasible Corners | $f=2x+5y$ |
|---|---|
| $(0,0)$ | 0 |
| $(0,60)$ | 300 |
| $(50,40)$ | 300 |
| $(70,24)$ | 260 |
| $(70,0)$ | 140 |

The maximum is $f = 300$ at any point on the segment joining $(0,60)$ and $(50,40)$.

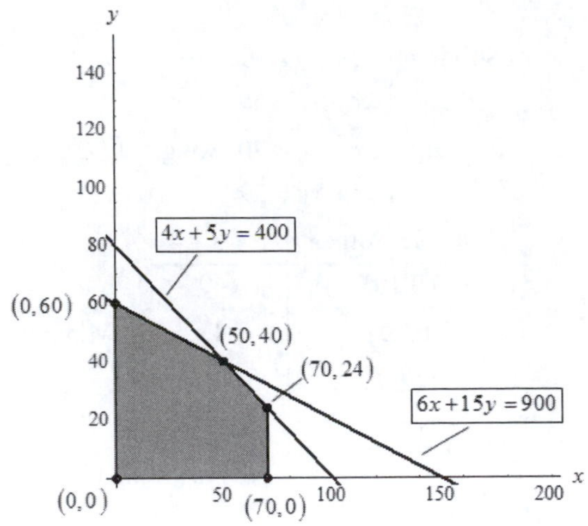


**12.** Minimize $g = 5x + 3y$

Subject to: $3x + y \geq 12$

$x + y \geq 6$

$x + 6y \geq 11$

Solving $\begin{cases} 3x + y = 12 \\ x + y = 6 \end{cases}$ we get $A(3, 3)$.

Solving $\begin{cases} x + 6y = 11 \\ x + y = 6 \end{cases}$ we get $B(5, 1)$.

| Feasible Corners | $g = 5x + 3y$ | |
|---|---|---|
| $(0,12)$ | 36 | |
| $(3,3)$ | 24 | ←Minimum |
| $(5,1)$ | 28 | |
| $(11,0)$ | 55 | |

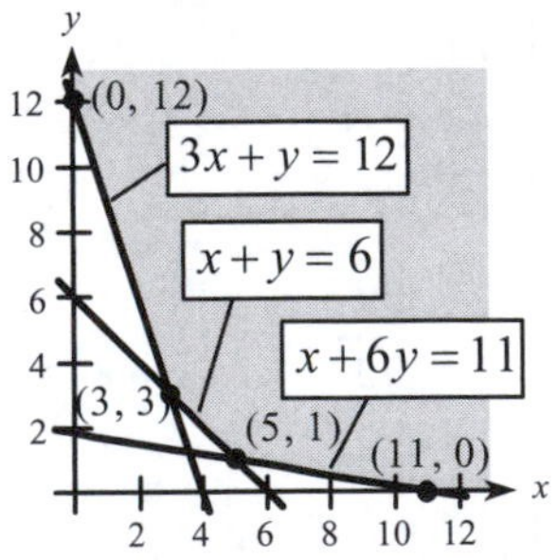


**13.** Minimize $g = x + 5y$

Subject to: $8x + y \geq 85$, $x + y \geq 50$

$x + 4y \geq 80$, $x + 10y \geq 104$

Solving $\begin{cases} 8x + y = 85 \\ x + y = 50 \end{cases}$ we get $A$ (5, 45).

Solving $\begin{cases} x + y = 50 \\ x + 4y = 80 \end{cases}$ we get $B$ (40, 10).

Solving $\begin{cases} x + 4y = 80 \\ x + 10y = 104 \end{cases}$ we get $C$ $(64, 4)$.

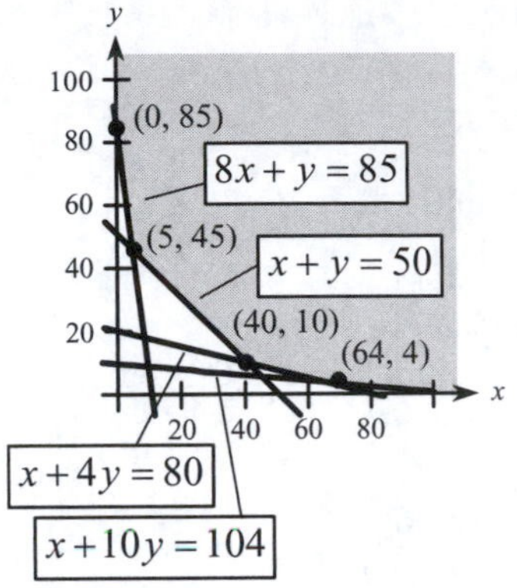


| Feasible Corners | $g = x + 5y$ | |
|---|---|---|
| $(0,85)$ | 425 | |
| $(5,45)$ | 230 | |
| $(40,10)$ | 90 | |
| $(64,4)$ | 84 | ← Minimum |
| $(94,0)$ | 94 | |

**14.** Maximize $f = 5x + 2y$

Subject to: $y \leq 20$, $2x + y \leq 32$

$-x + 2y \geq 4$, $x \geq 0$, $y \geq 0$

Solving: $\begin{cases} 2x + y = 32 \\ -x + 2y = 4 \end{cases}$

$\begin{cases} -4x - 2y = -64 \\ -x + 2y = 4 \end{cases}$

$-5x = -60$

$x = 12$, $y = 8$

Solving: $\begin{cases} y = 20 \\ 2x + y = 32 \end{cases}$

$2x + 20 = 32$

$2x = 12$

$x = 6$

Corners: $(6,20)$ and $(12,8)$

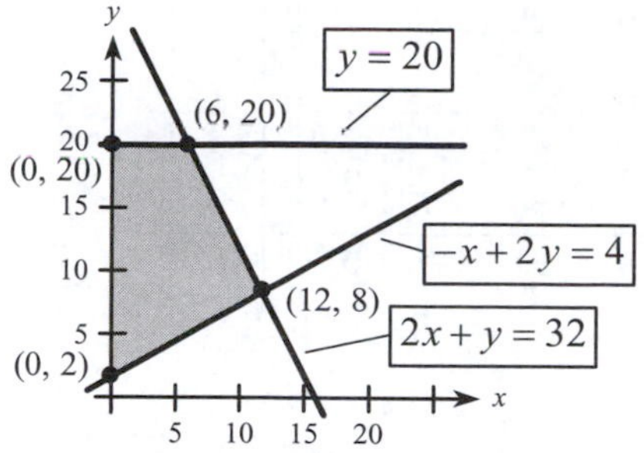


| Feasible Corners | $g = 5x + 2y$ | |
|---|---|---|
| $(0, 2)$ | 4 | |
| $(12, 8)$ | 76 | ← Maximum |
| $(6, 20)$ | 70 | |
| $(0, 20)$ | 40 | |

**15.** Maximize $f = x + 4y$

Subject to:

$$\begin{aligned} y &\le 30 \\ 3x + 2y &\le 75 \\ -3x + 5y &\ge 30 \\ x,\ y &\ge 0 \end{aligned}$$

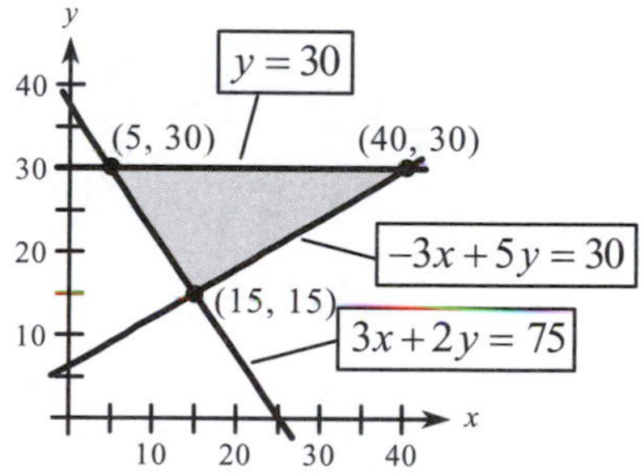


$$\begin{array}{r} y = 30 \\ \underline{3x + 2y = 75} \\ 3x = 15 \\ x = 5 \end{array} \qquad \begin{array}{r} y = 30 \\ \underline{-3x + 5y = 30} \\ -3x = -120 \\ x = 40 \end{array}$$

(5, 30) and (40, 30) are feasible corners.

$$\begin{array}{r} 3x + 2y = 75 \\ \underline{-3x + 5y = 30} \\ 7y = 105 \\ y = 15 \end{array}$$

(15, 15) is a feasible corner.

| Corner | $f = x + 4y$ |
|---|---|
| (5, 30) | 125 |
| (40, 30) | 160 |
| (15, 15) | 75 |

Minimum = 75 at (15, 15).

**16.**

$$\left[\begin{array}{cccccc|c} 7 & 3 & 1 & 0 & 0 & 0 & 105 \\ 2 & 5 & 0 & 1 & 0 & 0 & 59 \\ 1 & \boxed{7} & 0 & 0 & 1 & 0 & 70 \\ \hline -7 & -12 & 0 & 0 & 0 & 1 & 0 \end{array}\right] \quad \tfrac{1}{7}R_3 \to R_3 \text{ then } \quad \begin{array}{l} -3R_3 + R_1 \to R_1 \\ -5R_3 + R_2 \to R_2 \\ \to \\ 12R_3 + R_4 \to R_4 \end{array}$$

$$\left[\begin{array}{cccccc|c} \frac{46}{7} & 0 & 1 & 0 & -\frac{3}{7} & 0 & 75 \\ \boxed{\frac{9}{7}} & 0 & 0 & 1 & -\frac{5}{7} & 0 & 9 \\ \frac{1}{7} & 1 & 0 & 0 & \frac{1}{7} & 0 & 10 \\ \hline -\frac{37}{7} & 0 & 0 & 0 & \frac{12}{7} & 1 & 120 \end{array}\right] \quad \tfrac{7}{9}R_2 \to R_2 \text{ then } \quad \begin{array}{l} -\frac{46}{7}R_2 + R_1 \to R_1 \\ -\frac{1}{7}R_2 + R_3 \to R_3 \\ \frac{37}{7}R_2 + R_4 \to R_4 \end{array}$$

$$\left[\begin{array}{cccccc|c} 0 & 0 & 1 & -\frac{46}{9} & \boxed{\frac{29}{9}} & 0 & 29 \\ 1 & 0 & 0 & \frac{7}{9} & -\frac{5}{9} & 0 & 7 \\ 0 & 1 & 0 & -\frac{1}{9} & \frac{2}{9} & 0 & 9 \\ \hline 0 & 0 & 0 & \frac{37}{9} & -\frac{11}{9} & 1 & 157 \end{array}\right] \begin{array}{l} \frac{9}{29}R_1 \to R_1 \text{ then} \\ \frac{5}{9}R_1 + R_2 \to R_2 \\ -\frac{2}{9}R_1 + R_3 \to R_3 \\ \frac{11}{9}R_1 + R_4 \to R_4 \end{array} \to \left[\begin{array}{cccccc|c} 0 & 0 & \frac{9}{29} & -\frac{46}{29} & 1 & 0 & 9 \\ 1 & 0 & \frac{5}{29} & -\frac{3}{29} & 0 & 0 & 12 \\ 0 & 1 & -\frac{2}{29} & \frac{7}{29} & 0 & 0 & 7 \\ \hline 0 & 0 & \frac{11}{29} & \frac{63}{29} & 0 & 1 & 168 \end{array}\right]$$

Maximum is 168 at $x = 12, y = 7$.

**17.**
$$\left[\begin{array}{cccccc|c} 1 & \boxed{4} & 1 & 0 & 0 & 0 & 160 \\ 1 & 2 & 0 & 1 & 0 & 0 & 100 \\ 4 & 3 & 0 & 0 & 1 & 0 & 300 \\ \hline -3 & -4 & 0 & 0 & 0 & 1 & 0 \end{array}\right] \begin{array}{l} \frac{1}{4}R_1 \to R_1 \text{ then} \\ -2R_1 + R_2 \to R_2 \\ -3R_1 + R_3 \to R_3 \\ 4R_1 + R_4 \to R_4 \end{array} \to$$

$$\left[\begin{array}{cccccc|c} \frac{1}{4} & 1 & \frac{1}{4} & 0 & 0 & 0 & 40 \\ \boxed{\frac{1}{2}} & 0 & -\frac{1}{2} & 1 & 0 & 0 & 20 \\ \frac{13}{4} & 0 & -\frac{3}{4} & 0 & 1 & 0 & 180 \\ \hline -2 & 0 & 1 & 0 & 0 & 1 & 160 \end{array}\right] \; 2R_2 \to R_2 \text{ then} \begin{array}{l} -\frac{1}{4}R_2 + R_1 \to R_1 \\ -\frac{13}{4}R_2 + R_3 \to R_3 \\ 2R_2 + R_4 \to R_4 \end{array} \to$$

$$\left[\begin{array}{cccccc|c} 0 & 1 & \frac{1}{2} & -\frac{1}{2} & 0 & 0 & 30 \\ 1 & 0 & -1 & 2 & 0 & 0 & 40 \\ 0 & 0 & \boxed{\frac{5}{2}} & -\frac{13}{2} & 1 & 0 & 50 \\ \hline 0 & 0 & -1 & 4 & 0 & 1 & 240 \end{array}\right] \; \frac{2}{5}R_3 \to R_3 \text{ then} \begin{array}{l} -\frac{1}{2}R_3 + R_1 \to R_1 \\ R_3 + R_2 \to R_2 \\ R_3 + R_4 \to R_4 \end{array} \to \left[\begin{array}{cccccc|c} 0 & 1 & 0 & \frac{4}{5} & -\frac{1}{5} & 0 & 20 \\ 1 & 0 & 0 & -\frac{3}{5} & \frac{2}{5} & 0 & 60 \\ 0 & 0 & 1 & -\frac{13}{5} & \frac{2}{5} & 0 & 20 \\ \hline 0 & 0 & 0 & \frac{7}{5} & \frac{2}{5} & 1 & 260 \end{array}\right]$$

Solution: $x = 60, y = 20;\; f = 260$

**18.**
$$\left[\begin{array}{cccccc|c} 1 & \boxed{4} & 1 & 0 & 0 & 0 & 160 \\ 1 & 2 & 0 & 1 & 0 & 0 & 100 \\ 4 & 3 & 0 & 0 & 1 & 0 & 300 \\ \hline -3 & -8 & 0 & 0 & 0 & 1 & 0 \end{array}\right] \begin{array}{l} \frac{1}{4}R_1 \to R_1 \text{ then} \\ -2R_1 + R_2 \to R_2 \\ -3R_1 + R_3 \to R_3 \\ 8R_1 + R_4 \to R_4 \end{array} \to$$

$$\left[\begin{array}{cccccc|c} \frac{1}{4} & 1 & \frac{1}{4} & 0 & 0 & 0 & 40 \\ \boxed{\frac{1}{2}} & 0 & -\frac{1}{2} & 1 & 0 & 0 & 20 \\ \frac{13}{4} & 0 & -\frac{3}{4} & 0 & 1 & 0 & 180 \\ \hline -1 & 0 & 2 & 0 & 0 & 1 & 320 \end{array}\right] \; 2R_2 \to R_2 \text{ then} \begin{array}{l} -\frac{1}{4}R_2 + R_1 \to R_1 \\ -\frac{13}{4}R_2 + R_3 \to R_3 \\ R_2 + R_4 \to R_4 \end{array} \to \left[\begin{array}{cccccc|c} 0 & 1 & \frac{1}{2} & -\frac{1}{2} & 0 & 0 & 30 \\ 1 & 0 & -1 & 2 & 0 & 0 & 40 \\ 0 & 0 & \frac{5}{2} & -\frac{13}{2} & 1 & 0 & 50 \\ \hline 0 & 0 & 1 & 2 & 0 & 1 & 360 \end{array}\right]$$

Maximum is 360 at $x = 40, y = 30$.

**19.**
$$\left[\begin{array}{ccccccc|c} 1 & 2 & 1 & 0 & 0 & 0 & 0 & 48 \\ 1 & 1 & 0 & 1 & 0 & 0 & 0 & 30 \\ \boxed{2} & 1 & 0 & 0 & 1 & 0 & 0 & 50 \\ 1 & 10 & 0 & 0 & 0 & 1 & 0 & 200 \\ \hline -3 & -2 & 0 & 0 & 0 & 0 & 1 & 0 \end{array}\right] \; \frac{1}{2}R_3 \to R_3 \text{ then} \begin{array}{l} -R_3 + R_1 \to R_1 \\ -R_3 + R_2 \to R_2 \\ -R_3 + R_4 \to R_4 \\ 3R_3 + R_5 \to R_5 \end{array} \to$$

$$\left[\begin{array}{ccccccc|c} 0 & 1.5 & 1 & 0 & -0.5 & 0 & 0 & 23 \\ 0 & \boxed{0.5} & 0 & 1 & -0.5 & 0 & 0 & 5 \\ 1 & 0.5 & 0 & 0 & 0.5 & 0 & 0 & 25 \\ 0 & 9.5 & 0 & 0 & -0.5 & 1 & 0 & 175 \\ \hline 0 & -0.5 & 0 & 0 & 1.5 & 0 & 1 & 75 \end{array}\right] \begin{array}{c} \\ 2R_2 \to R_2 \text{ then} \\ \\ \\ \\ \end{array} \begin{array}{c} -1.5R_2 + R_1 \to R_1 \\ \to \\ -0.5R_2 + R_3 \to R_3 \\ -9.5R_2 + R_4 \to R_4 \\ 0.5R_2 + R_5 \to R_5 \end{array} \left[\begin{array}{ccccccc|c} 0 & 0 & 1 & -3 & 1 & 0 & 0 & 8 \\ 0 & 1 & 0 & 2 & -1 & 0 & 0 & 10 \\ 1 & 0 & 0 & -1 & 1 & 0 & 0 & 20 \\ 0 & 0 & 0 & -19 & 9 & 1 & 0 & 80 \\ \hline 0 & 0 & 0 & 1 & 1 & 0 & 1 & 80 \end{array}\right]$$

Maximum is 80 at $x = 20$ and $y = 10$.

**20.**

$$\left[\begin{array}{cccccc|c} 1 & \boxed{5} & 1 & 0 & 0 & 0 & 500 \\ 1 & 2 & 0 & 1 & 0 & 0 & 230 \\ 1 & 1 & 0 & 0 & 1 & 0 & 160 \\ \hline -4 & -4 & 0 & 0 & 0 & 1 & 0 \end{array}\right] \begin{array}{c} \frac{1}{5}R_1 \to R_1 \text{ then} \\ \\ \\ \\ \end{array} \begin{array}{c} \to \\ -2R_1 + R_2 \to R_2 \\ -R_1 + R_3 \to R_3 \\ 4R_1 + R_4 \to R_4 \end{array}$$

$$\left[\begin{array}{cccccc|c} \frac{1}{5} & 1 & \frac{1}{5} & 0 & 0 & 0 & 100 \\ \boxed{\frac{3}{5}} & 0 & -\frac{2}{5} & 1 & 0 & 0 & 30 \\ \frac{4}{5} & 0 & -\frac{1}{5} & 0 & 1 & 0 & 60 \\ \hline -\frac{16}{5} & 0 & \frac{4}{5} & 0 & 0 & 1 & 400 \end{array}\right] \begin{array}{c} \\ \frac{5}{3}R_2 \to R_2 \text{ then} \\ \\ \\ \end{array} \begin{array}{c} -\frac{1}{5}R_2 + R_1 \to R_1 \\ \to \\ -\frac{4}{5}R_2 + R_3 \to R_3 \\ \frac{16}{5}R_2 + R_4 \to R_4 \end{array}$$

$$\left[\begin{array}{cccccc|c} 0 & 1 & \frac{1}{3} & -\frac{1}{3} & 0 & 0 & 90 \\ 1 & 0 & -\frac{2}{3} & \frac{5}{3} & 0 & 0 & 50 \\ 0 & 0 & \boxed{\frac{1}{3}} & -\frac{4}{3} & 1 & 0 & 20 \\ \hline 0 & 0 & -\frac{4}{3} & \frac{16}{3} & 0 & 1 & 560 \end{array}\right] \begin{array}{c} \\ 3R_3 \to R_3 \text{ then} \\ \\ \\ \end{array} \begin{array}{c} -\frac{1}{3}R_3 + R_1 \to R_1 \\ \frac{2}{3}R_3 + R_2 \to R_2 \\ \to \\ \frac{4}{3}R_3 + R_4 \to R_4 \end{array} \left[\begin{array}{cccccc|c} 0 & 1 & 0 & 1 & -1 & 0 & 70 \\ 1 & 0 & 0 & -1 & 2 & 0 & 90 \\ 0 & 0 & 1 & -4 & 3 & 0 & 60 \\ \hline 0 & 0 & 0 & 0 & 4 & 1 & 640 \end{array}\right]$$

There are multiple solutions. Maximum is 640 at $x = 90$, $y = 70$.

$$\left[\begin{array}{cccccc|c} 0 & 1 & 0 & \boxed{1} & -1 & 0 & 70 \\ 1 & 0 & 0 & -1 & 2 & 0 & 90 \\ 0 & 0 & 1 & -4 & 3 & 0 & 60 \\ \hline 0 & 0 & 0 & 0 & 4 & 1 & 640 \end{array}\right] \begin{array}{c} \to \\ R_1 + R_2 \to R_2 \\ 4R_1 + R_3 \to R_3 \\ \to \end{array} \left[\begin{array}{cccccc|c} 0 & 1 & 0 & 1 & -1 & 0 & 70 \\ 1 & 1 & 0 & 0 & 1 & 0 & 160 \\ 0 & 4 & 1 & 0 & -1 & 0 & 340 \\ \hline 0 & 0 & 0 & 0 & 4 & 1 & 640 \end{array}\right]$$

Second maximum occurs at $x = 160$, $y = 0$.

$f = 640$ on the line between $(160, 0)$ and $(90, 70)$.

**21.**

$$\left[\begin{array}{ccccc|c} -4 & \boxed{1} & 1 & 0 & 0 & 40 \\ 1 & -7 & 0 & 1 & 0 & 70 \\ \hline -2 & -5 & 0 & 0 & 1 & 0 \end{array}\right] \begin{array}{c} \to \\ 7R_1 + R_2 \to R_2 \\ 5R_1 + R_3 \to R_3 \end{array} \left[\begin{array}{ccccc|c} -4 & 1 & 1 & 0 & 0 & 40 \\ -27 & 0 & 7 & 1 & 0 & 350 \\ \hline -22 & 0 & 5 & 0 & 1 & 200 \end{array}\right]$$

No solution

**22.** $\left[\begin{array}{cc|c} 5 & 2 & 16 \\ 3 & 7 & 27 \\ \hline 7 & 6 & g \end{array}\right]$ Dual: $\left[\begin{array}{cc|c} 5 & 3 & 7 \\ 2 & 7 & 6 \\ \hline 16 & 27 & g \end{array}\right]$

$$\left[\begin{array}{ccccc|c} 5 & 3 & 1 & 0 & 0 & 7 \\ 2 & \boxed{7} & 0 & 1 & 0 & 6 \\ \hline -16 & -27 & 0 & 0 & 1 & 0 \end{array}\right] \tfrac{1}{7}R_2 \to R_2 \text{ then } \begin{array}{c} -3R_2 + R_1 \to R_1 \\ \to \\ 27R_2 + R_3 \to R_3 \end{array}$$

$$\left[\begin{array}{ccccc|c} \boxed{\frac{29}{7}} & 0 & 1 & -\frac{3}{7} & 0 & \frac{31}{7} \\ \frac{2}{7} & 1 & 0 & \frac{1}{7} & 0 & \frac{6}{7} \\ \hline -\frac{58}{7} & 0 & 0 & \frac{27}{7} & 1 & \frac{162}{7} \end{array}\right] \tfrac{7}{29}R_1 \to R_1 \text{ then } \begin{array}{c} \to \\ -\frac{2}{7}R_1 + R_2 \to R_2 \\ \frac{58}{7}R_1 + R_3 \to R_3 \end{array} \left[\begin{array}{ccccc|c} 1 & 0 & \frac{7}{29} & -\frac{3}{29} & 0 & \frac{31}{29} \\ 0 & 1 & -\frac{2}{29} & \frac{5}{29} & 0 & \frac{16}{29} \\ \hline 0 & 0 & 2 & 3 & 1 & 32 \end{array}\right]$$

Minimum is 32 at $y_1 = 2, y_2 = 3$.

**23.** $\left[\begin{array}{cc|c} 3 & 1 & 8 \\ 1 & 1 & 6 \\ 2 & 5 & 18 \\ \hline 3 & 4 & g \end{array}\right] \to \left[\begin{array}{ccc|c} 3 & 1 & 2 & 3 \\ 1 & 1 & 5 & 4 \\ \hline 8 & 6 & 18 & g \end{array}\right]$

Maximize $g = 8x_1 + 6x_2 + 18x_3$

Constraints: $3x_1 + x_2 + 2x_3 \le 3$

$x_1 + x_2 + 5x_3 \le 4$

$x_1, x_2, x_3 \ge 0$

$$\left[\begin{array}{cccccc|c} 3 & 1 & 2 & 1 & 0 & 0 & 3 \\ 1 & 1 & \boxed{5} & 0 & 1 & 0 & 4 \\ \hline -8 & -6 & -18 & 0 & 0 & 1 & 0 \end{array}\right] \tfrac{1}{5}R_2 \to R_2 \text{ then } \begin{array}{c} -2R_2 + R_1 \to R_1 \\ \to \\ 18R_2 + R_3 \to R_3 \end{array}$$

$$\left[\begin{array}{cccccc|c} \boxed{\frac{13}{5}} & \frac{3}{5} & 0 & 1 & -\frac{2}{5} & 0 & \frac{7}{5} \\ \frac{1}{5} & \frac{1}{5} & 1 & 0 & \frac{1}{5} & 0 & \frac{4}{5} \\ \hline -\frac{22}{5} & -\frac{12}{5} & 0 & 0 & \frac{18}{5} & 1 & \frac{72}{5} \end{array}\right] \tfrac{5}{13}R_1 \to R_1 \text{ then}$$

$$\begin{array}{c} \to \\ -\frac{1}{5}R_1 + R_2 \to R_2 \\ \frac{22}{5}R_1 + R_3 \to R_3 \end{array} \left[\begin{array}{cccccc|c} 1 & \boxed{\frac{3}{13}} & 0 & \frac{5}{13} & -\frac{2}{13} & 0 & \frac{7}{13} \\ 0 & \frac{2}{13} & 1 & -\frac{1}{13} & \frac{3}{13} & 0 & \frac{9}{13} \\ \hline 0 & -\frac{18}{13} & 0 & \frac{22}{13} & \frac{38}{13} & 1 & \frac{218}{13} \end{array}\right] \tfrac{13}{3}R_1 \to R_1 \text{ then}$$

$$\begin{array}{c} \to \\ -\frac{2}{13}R_1 + R_2 \to R_2 \\ \frac{18}{13}R_1 + R_3 \to R_3 \end{array} \left[\begin{array}{cccccc|c} \frac{13}{3} & 1 & 0 & \frac{5}{3} & -\frac{2}{3} & 0 & \frac{7}{3} \\ -\frac{2}{3} & 0 & 1 & -\frac{1}{3} & \frac{1}{3} & 0 & \frac{1}{3} \\ \hline 6 & 0 & 0 & 4 & 2 & 1 & 20 \end{array}\right]$$

Solution: $y_1 = 4, y_2 = 2; g = 20$

**24.** $\left[\begin{array}{cc|c} 3 & 1 & 8 \\ 1 & 1 & 6 \\ 2 & 5 & 18 \\ \hline 2 & 1 & g \end{array}\right]$ Dual: $\left[\begin{array}{ccc|c} 3 & 1 & 1 & 2 \\ 1 & 1 & 5 & 1 \\ \hline 8 & 6 & 18 & g \end{array}\right]$

$$\left[\begin{array}{cccccc|c} 3 & 1 & 2 & 1 & 0 & 0 & 2 \\ 1 & 1 & \boxed{5} & 0 & 1 & 0 & 1 \\ \hline -8 & -6 & -18 & 0 & 0 & 1 & 0 \end{array}\right] \quad \tfrac{1}{5}R_2 \to R_2 \text{ then } \begin{array}{c} -2R_2 + R_1 \to R_1 \\ \to \\ 18R_2 + R_3 \to R_3 \end{array}$$

$$\left[\begin{array}{cccccc|c} \boxed{\frac{13}{5}} & \frac{3}{5} & 0 & 1 & -\frac{2}{5} & 0 & \frac{8}{5} \\ \frac{1}{5} & \frac{1}{5} & 1 & 0 & \frac{1}{5} & 0 & \frac{1}{5} \\ \hline -\frac{22}{5} & -\frac{12}{5} & 0 & 0 & \frac{18}{5} & 1 & \frac{18}{5} \end{array}\right] \quad \tfrac{5}{13}R_1 \to R_1 \text{ then } \begin{array}{c} \to \\ -\frac{1}{5}R_1 + R_2 \to R_2 \\ \frac{22}{5}R_1 + R_3 \to R_3 \end{array}$$

$$\left[\begin{array}{cccccc|c} 1 & \frac{3}{13} & 0 & \frac{5}{13} & -\frac{2}{13} & 0 & \frac{8}{13} \\ 0 & \boxed{\frac{2}{13}} & 1 & -\frac{1}{13} & \frac{3}{13} & 0 & \frac{1}{13} \\ \hline 0 & -\frac{18}{13} & 0 & \frac{22}{13} & \frac{38}{13} & 1 & \frac{82}{13} \end{array}\right] \quad \tfrac{13}{2}R_2 \to R_2 \text{ then } \begin{array}{c} -\frac{3}{13}R_2 + R_1 \to R_1 \\ \to \\ \frac{18}{13}R_2 + R_3 \to R_3 \end{array} \left[\begin{array}{cccccc|c} 1 & 0 & -\frac{3}{2} & \frac{1}{2} & -\frac{1}{2} & 0 & \frac{1}{2} \\ 0 & 1 & \frac{13}{2} & -\frac{1}{2} & \frac{3}{2} & 0 & \frac{1}{2} \\ \hline 0 & 0 & 9 & 1 & 5 & 1 & 7 \end{array}\right]$$

Minimum is 7 at $y_1 = 1, y_2 = 5$.

**25.** $\left[\begin{array}{cc|c} 1 & 1 & 100 \\ 2 & 1 & 140 \\ 6 & 5 & 580 \\ \hline 12 & 11 & g \end{array}\right]$ transpose: $\left[\begin{array}{ccc|c} 1 & 2 & 6 & 12 \\ 1 & 1 & 5 & 11 \\ \hline 100 & 140 & 580 & g \end{array}\right]$

$$\left[\begin{array}{cccccc|c} 1 & 2 & \boxed{6} & 1 & 0 & 0 & 12 \\ 1 & 1 & 5 & 0 & 1 & 0 & 11 \\ \hline -100 & -140 & -580 & 0 & 0 & 1 & 0 \end{array}\right] \quad \tfrac{1}{6}R_1 \to R_1 \text{ then } \begin{array}{c} \to \\ -5R_1 + R_2 \to R_2 \\ 580R_1 + R_3 \to R_3 \end{array}$$

$$\left[\begin{array}{cccccc|c} \frac{1}{6} & \frac{1}{3} & 1 & \frac{1}{6} & 0 & 0 & 2 \\ \boxed{\frac{1}{6}} & -\frac{2}{3} & 0 & -\frac{5}{6} & 1 & 0 & 1 \\ \hline -\frac{10}{3} & 53\frac{1}{3} & 0 & 96\frac{2}{3} & 0 & 1 & 1160 \end{array}\right] \quad 6R_2 \to R_2 \text{ then } \begin{array}{c} -\frac{1}{6}R_2 + R_1 \to R_1 \\ \to \\ \frac{10}{3}R_2 + R_3 \to R_3 \end{array} \left[\begin{array}{cccccc|c} 0 & 1 & 1 & 1 & -1 & 0 & 1 \\ 1 & -4 & 0 & -5 & 6 & 0 & 6 \\ \hline 0 & 40 & 0 & 80 & 20 & 1 & 1180 \end{array}\right]$$

The minimum is 1180 when $y_1 = 80$ and $y_2 = 20$.

**26.** Maximize $f = 3x + 5y$

Constraints: $-x - y \le -19,\ x - y \le -1,\ -x + 10y \le 190,\ x, y \ge 0$

$$\left[\begin{array}{cccccc|c} -1 & -1 & 1 & 0 & 0 & 0 & -19 \\ 1 & \boxed{-1} & 0 & 1 & 0 & 0 & -1 \\ -1 & 10 & 0 & 0 & 1 & 0 & 190 \\ \hline -3 & -5 & 0 & 0 & 0 & 1 & 0 \end{array}\right] \quad -R_2 \to R_2 \text{ then } \begin{array}{c} R_2 + R_1 \to R_1 \\ \to \\ -10R_2 + R_3 \to R_3 \\ 5R_2 + R_4 \to R_4 \end{array}$$

$$\left[\begin{array}{cccccc|c} \boxed{-2} & -0 & 1 & -1 & 0 & 0 & -18 \\ -1 & 1 & 0 & -1 & 0 & 0 & 1 \\ 9 & 0 & 0 & 10 & 1 & 0 & 180 \\ \hline -8 & 0 & 0 & -5 & 0 & 1 & 5 \end{array}\right] \begin{array}{l} -\frac{1}{2}R_1 \to R_1 \text{ then } \to \\ R_1 + R_2 \to R_2 \\ -9R_1 + R_3 \to R_3 \\ 8R_1 + R_4 \to R_4 \end{array}$$

$$\left[\begin{array}{cccccc|c} 1 & 0 & -\frac{1}{2} & \frac{1}{2} & 0 & 0 & 9 \\ 0 & 1 & -\frac{1}{2} & -\frac{1}{2} & 0 & 0 & 10 \\ 0 & 0 & \boxed{\frac{9}{2}} & \frac{11}{2} & 1 & 0 & 99 \\ \hline 0 & 0 & -4 & -1 & 0 & 1 & 77 \end{array}\right] \frac{2}{9}R_3 \to R_3 \text{ then } \begin{array}{l} \frac{1}{2}R_3 + R_1 \to R_1 \\ \frac{1}{2}R_3 + R_2 \to R_2 \\ \to \\ 4R_3 + R_4 \to R_4 \end{array} \left[\begin{array}{cccccc|c} 1 & 0 & 0 & \frac{10}{9} & \frac{1}{9} & 0 & 20 \\ 0 & 1 & 0 & \frac{1}{9} & \frac{1}{9} & 0 & 21 \\ 0 & 0 & 1 & \frac{11}{9} & \frac{2}{9} & 0 & 22 \\ \hline 0 & 0 & 0 & \frac{35}{9} & \frac{8}{9} & 1 & 165 \end{array}\right]$$

Maximum is 165 at $x = 20$, $y = 21$.

**27.**

$$\left[\begin{array}{cccccc|c} 2 & 5 & 1 & 0 & 0 & 0 & 37 \\ 5 & -1 & 0 & 1 & 0 & 0 & 34 \\ 1 & \boxed{-2} & 0 & 0 & 1 & 0 & -4 \\ \hline -4 & -6 & 0 & 0 & 0 & 1 & 0 \end{array}\right] -\frac{1}{2}R_3 \to R_3 \text{ then } \begin{array}{l} -5R_3 + R_1 \to R_1 \\ R_3 + R_2 \to R_2 \\ \to \\ 6R_3 + R_4 \to R_4 \end{array}$$

$$\left[\begin{array}{cccccc|c} \boxed{\frac{9}{2}} & 0 & 1 & 0 & \frac{5}{2} & 0 & 27 \\ \frac{9}{2} & 0 & 0 & 1 & -\frac{1}{2} & 0 & 36 \\ -\frac{1}{2} & 1 & 0 & 0 & -\frac{1}{2} & 0 & 2 \\ \hline -7 & 0 & 0 & 0 & -3 & 1 & 12 \end{array}\right] s \quad \frac{2}{9}R_1 \to R_1 \text{ then } \begin{array}{l} \to \\ -\frac{9}{2}R_1 + R_2 \to R_2 \\ \frac{1}{2}R_1 + R_3 \to R_3 \\ 7R_1 + R_4 \to R_4 \end{array} \left[\begin{array}{cccccc|c} 1 & 0 & \frac{2}{9} & 0 & \frac{5}{9} & 0 & 6 \\ 0 & 0 & -1 & 1 & -3 & 0 & 9 \\ 0 & 1 & \frac{1}{9} & 0 & -\frac{2}{9} & 0 & 5 \\ \hline 0 & 0 & \frac{14}{9} & 0 & \frac{8}{9} & 1 & 54 \end{array}\right]$$

Solution: $x = 6, y = 5; f = 54$

**28.**

$$\left[\begin{array}{ccccccc|c} 1 & 0 & 1 & 1 & 0 & 0 & 0 & 7 \\ 3 & 5 & 0 & 0 & 1 & 0 & 0 & 30 \\ \boxed{3} & 1 & 0 & 0 & 0 & 1 & 0 & 18 \\ \hline -39 & -5 & -30 & 0 & 0 & 0 & 1 & 0 \end{array}\right] \frac{1}{3}R_3 \to R_3 \text{ then } \begin{array}{l} -R_3 + R_1 \to R_1 \\ -3R_3 + R_2 \to R_2 \\ \to \\ 39R_3 + R_4 \to R_4 \end{array}$$

$$\left[\begin{array}{ccccccc|c} 0 & -\frac{1}{3} & \boxed{1} & 1 & 0 & -\frac{1}{3} & 0 & 1 \\ 0 & 4 & 0 & 0 & 1 & -1 & 0 & 12 \\ 1 & \frac{1}{3} & 0 & 0 & 0 & \frac{1}{3} & 0 & 6 \\ \hline 0 & 8 & -30 & 0 & 0 & 13 & 1 & 234 \end{array}\right] \begin{array}{l} \to \\ \to \\ \to \\ 30R_1 + R_4 \to R_4 \end{array} \left[\begin{array}{ccccccc|c} 0 & -\frac{1}{3} & 1 & 1 & 0 & -\frac{1}{3} & 0 & 1 \\ 0 & \boxed{4} & 0 & 0 & 1 & -1 & 0 & 12 \\ 1 & \frac{1}{3} & 0 & 0 & 0 & \frac{1}{3} & 0 & 6 \\ \hline 0 & -2 & 0 & 30 & 0 & 3 & 1 & 264 \end{array}\right]$$

$$\frac{1}{4}R_2 \to R_2 \text{ then } \begin{array}{l} \frac{1}{3}R_2 + R_1 \to R_1 \\ \to \\ -\frac{1}{3}R_2 + R_3 \to R_3 \\ 2R_2 + R_4 \to R_4 \end{array} \left[\begin{array}{ccccccc|c} 0 & 0 & 1 & 1 & \frac{1}{12} & -\frac{5}{12} & 0 & 2 \\ 0 & 1 & 0 & 0 & \frac{1}{4} & -\frac{1}{4} & 0 & 3 \\ 1 & 0 & 0 & 0 & -\frac{1}{12} & \frac{5}{12} & 0 & 5 \\ \hline 0 & 0 & 0 & 30 & \frac{1}{2} & \frac{5}{2} & 1 & 270 \end{array}\right]$$

Solution: $x = 5, y = 3, z = 2; f = 270$

**29.** $\left[\begin{array}{ccc|c} 1 & 2 & 1 & 60 \\ 12 & 4 & 3 & 120 \\ 2 & 3 & 1 & 80 \\ \hline 12 & 5 & 2 & g \end{array}\right]$ transpose: $\left[\begin{array}{ccc|c} 1 & 12 & 2 & 12 \\ 2 & 4 & 3 & 5 \\ 1 & 3 & 1 & 2 \\ \hline 60 & 120 & 80 & g \end{array}\right]$

$$\left[\begin{array}{ccccccc|c} 1 & 12 & 2 & 1 & 0 & 0 & 0 & 12 \\ 2 & 4 & 3 & 0 & 1 & 0 & 0 & 5 \\ 1 & \boxed{3} & 1 & 0 & 0 & 1 & 0 & 2 \\ \hline -60 & -120 & -80 & 0 & 0 & 0 & 1 & 0 \end{array}\right] \quad \tfrac{1}{3}R_3 \to R_3 \text{ then } \begin{array}{r} -12R_3 + R_1 \to R_1 \\ -4R_3 + R_2 \to R_2 \\ \to \\ 120R_3 + R_4 \to R_4 \end{array}$$

$$\left[\begin{array}{ccccccc|c} -3 & 0 & -2 & 1 & 0 & -4 & 0 & 4 \\ \frac{2}{3} & 0 & \boxed{\frac{5}{3}} & 0 & 1 & -\frac{4}{3} & 0 & \frac{7}{3} \\ \frac{1}{3} & 1 & \frac{1}{3} & 0 & 0 & \frac{1}{3} & 0 & \frac{2}{3} \\ \hline -20 & 0 & -40 & 0 & 0 & 40 & 1 & 80 \end{array}\right] \quad \tfrac{3}{5}R_2 \to R_2 \text{ then } \begin{array}{r} 2R_2 + R_1 \to R_1 \\ \to \\ -\frac{1}{3}R_2 + R_3 \to R_3 \\ 40R_2 + R_4 \to R_4 \end{array}$$

$$\left[\begin{array}{ccccccc|c} -2.2 & 0 & 0 & 1 & 1.2 & -5.6 & 0 & 6.8 \\ 0.4 & 0 & 1 & 0 & 0.6 & -0.8 & 0 & 1.4 \\ \boxed{0.2} & 1 & 0 & 0 & -0.2 & 0.6 & 0 & 0.2 \\ \hline -4 & 0 & 0 & 0 & 24 & 8 & 1 & 136 \end{array}\right] \quad 5R_3 \to R_3 \text{ then } \begin{array}{r} 2.2R_3 + R_1 \to R_1 \\ -0.4R_3 + R_2 \to R_2 \\ \to \\ 4R_3 + R_4 \to R_4 \end{array}$$

$$\left[\begin{array}{ccccccc|c} 0 & 11 & 0 & 1 & -1 & 1 & 0 & 9 \\ 0 & -2 & 1 & 0 & 1 & -2 & 0 & 1 \\ 1 & 5 & 0 & 0 & -1 & 3 & 0 & 1 \\ \hline 0 & 20 & 0 & 0 & 20 & 20 & 1 & 140 \end{array}\right]$$

The minimum is 140 at $y_1 = 0,\ y_2 = 20,\ y_3 = 30.$

**30.** $\left[\begin{array}{ccc|c} 7.5 & 4.5 & 2 & 650 \\ 6.5 & 3 & 1.5 & 400 \\ 1 & 1.5 & 0.5 & 200 \\ \hline 25 & 10 & 4 & g \end{array}\right]$ transpose: $\left[\begin{array}{ccc|c} 7.5 & 6.5 & 1 & 25 \\ 4.5 & 3 & 1.5 & 10 \\ 2 & 1.5 & 0.5 & 4 \\ \hline 650 & 400 & 200 & g \end{array}\right]$

$$\left[\begin{array}{ccccccc|c} 7.5 & 6.5 & 1 & 1 & 0 & 0 & 0 & 25 \\ 4.5 & 3 & 1.5 & 0 & 1 & 0 & 0 & 10 \\ \boxed{2} & 1.5 & 0.5 & 0 & 0 & 1 & 0 & 4 \\ \hline -650 & -400 & -200 & 0 & 0 & 0 & 1 & 0 \end{array}\right] \quad 0.5R_3 \to R_3 \text{ then } \begin{array}{r} -7.5R_3 + R_1 \to R_1 \\ -4.5R_3 + R_2 \to R_2 \\ \to \\ 650R_3 + R_4 \to R_4 \end{array}$$

$$\left[\begin{array}{ccccccc|c} 0 & 0.875 & -0.875 & 1 & 0 & -3.75 & 0 & 10 \\ 0 & -0.375 & \boxed{0.375} & 0 & 1 & -2.25 & 0 & 1 \\ 1 & 0.75 & 0.25 & 0 & 0 & 0.5 & 0 & 2 \\ \hline 0 & 87.5 & -37.5 & 0 & 0 & 325 & 1 & 1300 \end{array}\right] \quad \tfrac{8}{3}R_2 \to R_2 \text{ then } \begin{array}{r} 0.875R_2 + R_1 \to R_1 \\ \to \\ -0.25R_2 + R_3 \to R_3 \\ 37.5R_2 + R_4 \to R_4 \end{array}$$

$$\left[\begin{array}{ccccccc|c} 0 & 0 & 0 & 1 & \frac{7}{3} & -9 & 0 & 12\frac{1}{3} \\ 0 & -1 & 1 & 0 & \frac{8}{3} & -6 & 0 & 2\frac{2}{3} \\ 1 & 1 & 0 & 0 & -\frac{2}{3} & 2 & 0 & 1\frac{1}{3} \\ \hline 0 & 50 & 0 & 00 & 100 & 100 & 1 & 1400 \end{array}\right]$$

Minimum is 1400 when $y_1 = 0,\ y_2 = 100,\ y_3 = 100.$

**31.** Maximize : $-f = -10x - 3y$

Constraints: $x - 10y \le -5,\ -4x - y \le -62,\ x + y \le 50,\ x, y \ge 0$

$$\left[\begin{array}{cccccc|c} 1 & \boxed{-10} & 1 & 0 & 0 & 0 & -5 \\ -4 & -1 & 0 & 1 & 0 & 0 & -62 \\ 1 & 1 & 0 & 0 & 1 & 0 & 50 \\ \hline 10 & 3 & 0 & 0 & 0 & 1 & 0 \end{array}\right] \quad \begin{array}{l} -\tfrac{1}{10}R_1 \to R_1 \text{ then } \to \\ R_1 + R_2 \to R_2 \\ -R_1 + R_3 \to R_3 \\ -3R_1 + R_4 \to R_4 \end{array}$$

$$\left[\begin{array}{cccccc|c} -\frac{1}{10} & 1 & -\frac{1}{10} & 0 & 0 & 0 & \frac{1}{2} \\ \boxed{-\frac{41}{10}} & 0 & -\frac{1}{10} & 1 & 0 & 0 & -\frac{123}{2} \\ \frac{11}{10} & 0 & \frac{1}{10} & 0 & 1 & 0 & \frac{99}{2} \\ \hline \frac{103}{10} & 0 & \frac{3}{10} & 0 & 0 & 1 & -\frac{3}{2} \end{array}\right] \quad -\tfrac{10}{41}R_2 \to R_2 \text{ then } \begin{array}{l} \tfrac{1}{10}R_2 + R_1 \to R_1 \\ \to \\ -\tfrac{11}{10}R_2 + R_3 \to R_3 \\ -\tfrac{103}{10}R_2 + R_4 \to R_4 \end{array} \left[\begin{array}{cccccc|c} 0 & 1 & -\frac{4}{41} & -\frac{1}{41} & 0 & 0 & 2 \\ 1 & 0 & \frac{1}{41} & -\frac{10}{41} & 0 & 0 & 15 \\ 0 & 0 & \frac{3}{41} & \frac{11}{41} & 1 & 0 & 33 \\ \hline 0 & 0 & \frac{2}{41} & \frac{103}{41} & 0 & 1 & -156 \end{array}\right]$$

Minimum is 156 at $x = 15, y = 2$.

**32.** Maximize: $-f = -4x - 3y$

Subject to: $x - y \le -1,\ x + y \le 45,\ -10x - y \le -45$

$$\left[\begin{array}{cccccc|c} 1 & \boxed{-1} & 1 & 0 & 0 & 0 & -1 \\ 1 & 1 & 0 & 1 & 0 & 0 & 45 \\ -10 & -1 & 0 & 0 & 1 & 0 & -45 \\ \hline 4 & 3 & 0 & 0 & 0 & 1 & 0 \end{array}\right] \quad \begin{array}{l} -R_1 \to R_1 \text{ then } \to \\ -R_1 + R_2 \to R_2 \\ R_1 + R_3 \to R_3 \\ -3R_1 + R_4 \to R_4 \end{array}$$

$$\left[\begin{array}{cccccc|c} -1 & 1 & -1 & 0 & 0 & 0 & 1 \\ 2 & 0 & 1 & 1 & 0 & 0 & 44 \\ \boxed{-11} & 0 & -1 & 0 & 1 & 0 & -44 \\ \hline 7 & 0 & 3 & 0 & 0 & 1 & -3 \end{array}\right] \quad -\tfrac{1}{11}R_3 \to R_3 \text{ then } \begin{array}{l} R_3 + R_1 \to R_1 \\ -2R_3 + R_2 \to R_2 \\ \to \\ -7R_3 + R_4 \to R_4 \end{array} \left[\begin{array}{cccccc|c} 0 & 1 & -\frac{10}{11} & 0 & -\frac{1}{11} & 0 & 5 \\ 0 & 0 & \frac{9}{11} & 1 & \frac{2}{11} & 0 & 36 \\ 1 & 0 & \frac{1}{11} & 0 & -\frac{1}{11} & 0 & 4 \\ \hline 0 & 0 & \frac{26}{11} & 0 & \frac{7}{11} & 1 & -31 \end{array}\right]$$

Minimum is 31 at $x = 4, y = 5$.

**33.**

$$\left[\begin{array}{ccccccccc|c} 3 & 2 & 2 & 5 & 1 & 0 & 0 & 0 & 0 & 200 \\ 2 & 2 & \boxed{4} & 5 & 0 & 1 & 0 & 0 & 0 & 100 \\ 1 & 1 & 1 & 1 & 0 & 0 & 1 & 0 & 0 & 200 \\ 1 & 0 & 0 & 0 & 0 & 0 & 0 & 1 & 0 & 40 \\ \hline -88 & -86 & -100 & -100 & 0 & 0 & 0 & 0 & 1 & 0 \end{array}\right] \quad \tfrac{1}{4}R_2 \to R_2 \text{ then } \begin{array}{l} -2R_2 + R_1 \to R_1 \\ \to \\ -R_2 + R_3 \to R_3 \\ \to \\ 100R_2 + R_5 \to R_5 \end{array}$$

$$\left[\begin{array}{ccccccccc|c} 2 & 1 & 0 & 2.1 & 1 & -0.5 & 0 & 0 & 0 & 150 \\ 0.5 & 0.5 & 1 & 1.25 & 0 & 0.25 & 0 & 0 & 0 & 25 \\ 0.5 & 0.5 & 0 & -0.25 & 0 & -0.25 & 1 & 0 & 0 & 175 \\ \boxed{1} & 0 & 0 & 0 & 0 & 0 & 0 & 1 & 0 & 40 \\ \hline -38 & -36 & 0 & 25 & 0 & 25 & 0 & 0 & 1 & 2500 \end{array}\right] \quad \begin{array}{l} -2R_4 + R_1 \to R_1 \\ -0.5R_4 + R_2 \to R_2 \\ -0.5R_4 + R_3 \to R_3 \\ \to \\ 38R_4 + R_5 \to R_5 \end{array}$$

$$\left[\begin{array}{ccccccccc|c} 0 & 1 & 0 & 2.5 & 1 & -0.5 & 0 & -2 & 0 & 70 \\ 0 & \boxed{0.5} & 1 & 1.25 & 0 & 0.25 & 0 & -0.5 & 0 & 5 \\ 0 & 0.5 & 0 & -0.25 & 0 & -0.25 & 1 & -0.5 & 0 & 155 \\ 1 & 0 & 0 & 0 & 0 & 0 & 0 & 1 & 0 & 40 \\ \hline 0 & -36 & 0 & 25 & 0 & 25 & 0 & 38 & 1 & 4020 \end{array}\right] \quad 2R_2 \to R_2 \text{ then} \quad \begin{array}{c} -R_2 + R_1 \to R_1 \\ \to \\ -0.5R_2 + R_3 \to R_3 \\ \to \\ 36R_2 + R_5 \to R_5 \end{array}$$

$$\left[\begin{array}{ccccccccc|c} 0 & 0 & -2 & 0 & 1 & -1 & 0 & -1 & 0 & 60 \\ 0 & 1 & 2 & 2.5 & 0 & 0.5 & 0 & -1 & 0 & 10 \\ 0 & 0 & -1 & -1.5 & 0 & -0.5 & 1 & 0 & 0 & 150 \\ 1 & 0 & 0 & 0 & 0 & 0 & 0 & 1 & 0 & 40 \\ \hline 0 & 0 & 72 & 115 & 0 & 43 & 0 & 2 & 1 & 4380 \end{array}\right]$$

Maximum is 4380 at $x_1 = 40,\ x_2 = 10,\ x_3 = 0,$ and $x_4 = 0.$

**34.** Maximize: $f = 8x_1 + 10x_2 + 12x_3 + 14x_4$

Constraints: $-6x_1 - 3x_2 - 2x_3 - x_4 \le -350$

$3x_1 + 2x_2 + 5x_3 + 6x_4 \le 300$

$8x_1 + 3x_2 + 2x_3 + x_4 \le 400$

$x_1 + x_2 + x_3 + x_4 \le 100$

$$\left[\begin{array}{ccccccccc|c} -6 & -3 & -2 & -1 & 1 & 0 & 0 & 0 & 0 & -350 \\ 3 & 2 & 5 & 6 & 0 & 1 & 0 & 0 & 0 & 300 \\ 8 & 3 & 2 & 1 & 0 & 0 & 1 & 0 & 0 & 400 \\ 1 & 1 & 1 & 1 & 0 & 0 & 0 & 1 & 0 & 100 \\ \hline -8 & -10 & -12 & -14 & 0 & 0 & 0 & 0 & 1 & 0 \end{array}\right]$$

Using technology, the maximum is 1000 at $x_1 = 25,\ x_2 = 62.5,\ x_3 = 0,\ x_4 = 12.5.$

**35.** Maximize: $-g = -10x_1 - 9x_2 - 12x_3 - 8x_4$

Constraints: $-45x_1 - 58.5x_3 \le -4680$

$-36x_2 - 31.5x_4 \le -4230$

$x_1 + x_2 \le 100$

$x_3 + x_4 \le 100$

$$\left[\begin{array}{ccccccccc|c} -45 & 0 & -58.5 & 0 & 1 & 0 & 0 & 0 & 0 & -4680 \\ 0 & -36 & 0 & -31.5 & 0 & 1 & 0 & 0 & 0 & -4230 \\ 1 & 1 & 0 & 0 & 0 & 0 & 1 & 0 & 0 & 100 \\ 0 & 0 & 1 & 1 & 0 & 0 & 0 & 1 & 0 & 100 \\ \hline 10 & 9 & 12 & 8 & 0 & 0 & 0 & 0 & 1 & 0 \end{array}\right]$$

Using technology, the minimum is 2020 at $x_1 = 0,\ x_2 = 100,\ x_3 = 80,\ x_4 = 20.$

**36.** Let $x$ = number of large sets and $y$ = number of small sets.

Maximize $f = 200x + 100y$

Constraints: $5x + 2y \le 700$, $x + y \le 185$

Solving the two equations we get $A(110, 75)$.

| Feasible Corners | $f = 200x + 100y$ |
|---|---|
| $(0, 185)$ | 18,500 |
| $(110, 75)$ | 29,500 |
| $(140, 0)$ | 28,000 |

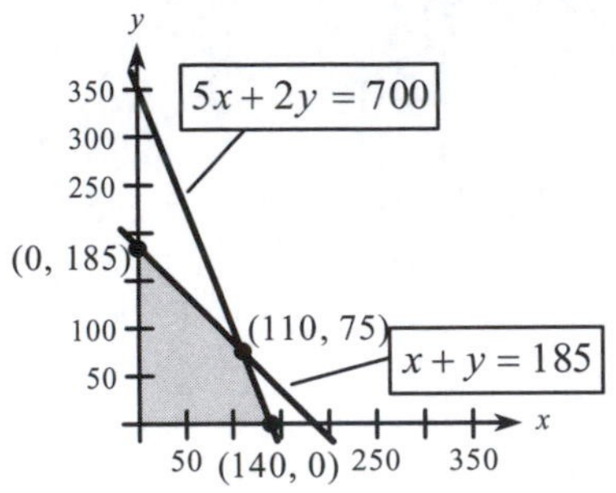


Maximum profit is $29,500 with 110 large sets and 75 small sets.

**37.** $x_1$ = days for factory 1, $x_2$ = days for factory 2

Minimize: $g = 5000x_1 + 6000x_2$

Subject to: $x_1 + 2x_2 \ge 80$

$3x_1 + 2x_2 \ge 140$

$x_1, x_2 \ge 0$

$x_1 + 2x_2 = 80$ and $3x_1 + 2x_2 = 140$ intersect at $x_1 = 30, x_2 = 25$.

The feasible corners are (80, 0), (30, 25), and (0, 70).

At (80, 0), $g = 400{,}000$

At (30, 25), $g = 300{,}000$ ← Minimum

At (0, 70), $g = 420{,}000$

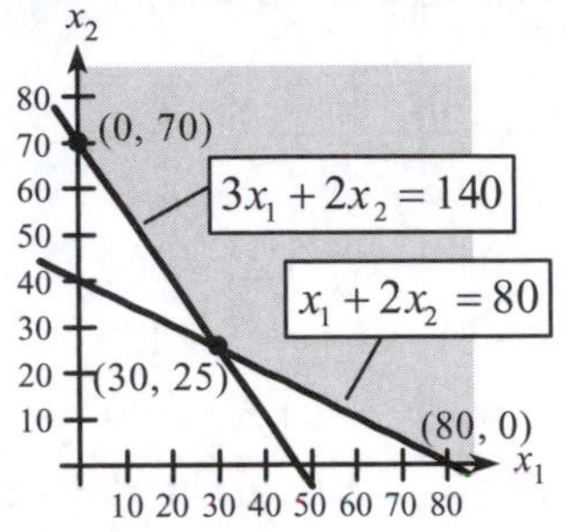


The minimum cost is $300,000 when factory 1 operates 30 days and factory 2 operates 25 days.

**38.** Let $x$ = number of standard chairs and $y$ = number of plush chairs.

Maximize $f = 89x + 133.5y$

Constraints: $2x + 3y \le 240$, $x + 3y \le 150$

Solving the system given by $2x + 3y = 240$ and $x + 3y = 150$, we get $(90, 20)$.

| Feasible Corners | $f = 89x + 133.5y$ |
|---|---|
| $(0, 0)$ | 0 |
| $(0, 50)$ | 6675 |
| $(90, 20)$ | 10,680 |
| $(120, 0)$ | 10,680 |

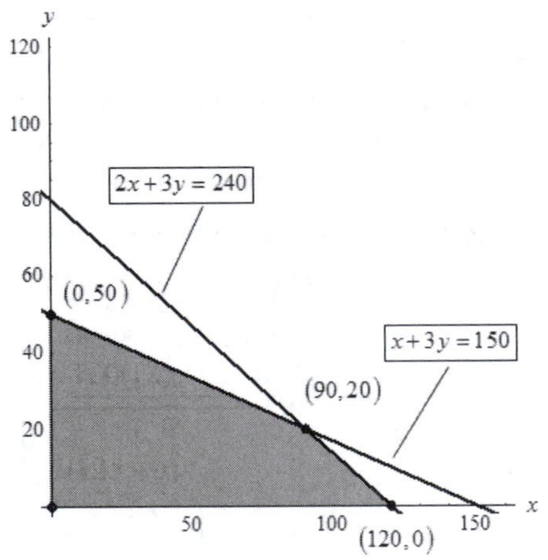


The profit will be maximized at $10,680 at any pair of integer values along the segment joining $(120, 0)$ and $(90, 20)$.

**39.** Let $x$ = number of Is and $y$ = number of IIs.
Maximize $P = 18x + 12y$
Constraints: $2x + y \le 100,\ \ x + y \le 60$

$$\left[\begin{array}{ccccc|c} \boxed{2} & 1 & 1 & 0 & 0 & 100 \\ 1 & 1 & 0 & 1 & 0 & 60 \\ \hline -18 & -12 & 0 & 0 & 1 & 0 \end{array}\right] \quad \begin{array}{l} \frac{1}{2}R_1 \to R_1 \text{ then} \quad \to \\ -R_1 + R_2 \to R_2 \\ 18R_1 + R_3 \to R_3 \end{array}$$

$$\left[\begin{array}{ccccc|c} 1 & \frac{1}{2} & \frac{1}{2} & 0 & 0 & 50 \\ 0 & \boxed{\frac{1}{2}} & -\frac{1}{2} & 1 & 0 & 10 \\ \hline 0 & -3 & 9 & 0 & 1 & 900 \end{array}\right] \quad 2R_2 \to R_2 \text{ then} \quad \begin{array}{l} -\frac{1}{2}R_2 + R_1 \to R_1 \\ \to \\ 3R_2 + R_3 \to R_3 \end{array} \quad \left[\begin{array}{ccccc|c} 1 & 0 & 1 & -1 & 0 & 40 \\ 0 & 1 & -1 & 2 & 0 & 20 \\ \hline 0 & 0 & 6 & 6 & 1 & 960 \end{array}\right]$$

Maximum profit is \$960 with 40 of the Is and 20 of the IIs.

**40.** Let $x$ = number of Jacob's Ladders, let $y$ = number of locomotive engines.
Maximize $P = 9x + 15y$
Constraints:

| | |
|---|---|
| $x + y \le 120$ | finishing hours |
| $\frac{1}{2}x + y \le 75$ | carpentry hours |
| $x \le 100$ | |
| $x, y \ge 0$ | |

$$\left[\begin{array}{cccccc|c} 1 & 1 & 1 & 0 & 0 & 0 & 120 \\ \frac{1}{2} & \boxed{1} & 0 & 1 & 0 & 0 & 75 \\ 1 & 0 & 0 & 0 & 1 & 0 & 100 \\ \hline -9 & -15 & 0 & 0 & 0 & 1 & 0 \end{array}\right] \quad \begin{array}{l} -R_2 + R_1 \to R_1 \\ \to \\ \to \\ 15R_2 + R_4 \to R_4 \end{array}$$

$$\left[\begin{array}{cccccc|c} \boxed{\frac{1}{2}} & 0 & 1 & -1 & 0 & 0 & 45 \\ \frac{1}{2} & 1 & 0 & 1 & 0 & 0 & 75 \\ 1 & 0 & 0 & 0 & 1 & 0 & 100 \\ \hline -\frac{3}{2} & 0 & 0 & 15 & 0 & 1 & 1125 \end{array}\right] \quad 2R_1 \to R_1 \text{ then} \quad \begin{array}{l} -\frac{1}{2}R_1 + R_2 \to R_2 \\ -R_1 + R_3 \to R_3 \\ \frac{3}{2}R_1 + R_4 \to R_4 \end{array} \quad \left[\begin{array}{cccccc|c} 1 & 0 & 2 & -2 & 0 & 0 & 90 \\ 0 & 1 & -1 & 2 & 0 & 0 & 30 \\ 0 & 0 & -2 & 2 & 1 & 0 & 10 \\ \hline 0 & 0 & 3 & 12 & 0 & 1 & 1260 \end{array}\right]$$

Maximum profit is \$1260, obtained by producing 90 Jacob's Ladders and 30 locomotive engines.

**41.** **a.** $x_1$ = number of 27-in. LCD models, $x_2$ = number of 32-in. LCD models, $x_3$ = number of 42-in. LCD models, $x_4$ = number of 42-in. plasma models

**b.** Maximize: $f = 80x_1 + 120x_2 + 160x_3 + 200x_4$

Constraints: 
$$\begin{aligned} 8x_1 + 10x_2 + 12x_3 + 15x_4 &\le 1870 \\ 2x_1 + 4x_2 + 4x_3 + 4x_4 &\le 530 \\ x_1 + x_2 + x_3 + x_4 &\le 200 \\ x_3 + x_4 &\le 100 \\ x_2 &\le 120 \end{aligned}$$

**c.** 
$$\left[\begin{array}{cccccccccc|c} 8 & 10 & 12 & 15 & 1 & 0 & 0 & 0 & 0 & 0 & 1870 \\ 2 & 4 & 4 & 4 & 0 & 1 & 0 & 0 & 0 & 0 & 530 \\ 1 & 1 & 1 & 1 & 0 & 0 & 1 & 0 & 0 & 0 & 200 \\ 0 & 0 & 1 & 1 & 0 & 0 & 0 & 1 & 0 & 0 & 100 \\ 0 & 1 & 0 & 0 & 0 & 0 & 0 & 0 & 1 & 0 & 120 \\ \hline -80 & -120 & -160 & -200 & 0 & 0 & 0 & 0 & 0 & 1 & 0 \end{array}\right]$$

Using technology to solve the problem, Nolmaur Electronics can manufacture 15 27-in. LCD models, 25 32-in. LCD models, and 100 42-in. plasma models for a maximum profit of $24,200.

**42.** We display the information in a table.

| | $x$ food I | $y$ food II | Requirement |
|---|---|---|---|
| A | 2 | 10 | 5 |
| B | 1 | 10 | 30 |
| cost | 30 cents/ ounce | 20 cents/ ounce | |

Minimize: $g = 30x + 20y$

Subject to: 
$$\begin{aligned} 2x + 10y &\ge 5 \\ x + 10y &\ge 30 \\ x, y &\ge 0 \end{aligned}$$

$$\left[\begin{array}{cc|c} 2 & 10 & 5 \\ 1 & 10 & 30 \\ \hline 30 & 20 & g \end{array}\right] \quad \text{transpose: } \left[\begin{array}{cc|c} 2 & 1 & 30 \\ 10 & 10 & 20 \\ \hline 5 & 30 & g \end{array}\right]$$

Maximization dual problem:

$$\left[\begin{array}{ccccc|c} 2 & 1 & 1 & 0 & 0 & 30 \\ 10 & \boxed{10} & 0 & 1 & 0 & 20 \\ \hline -5 & -30 & 0 & 0 & 1 & 0 \end{array}\right] \quad \tfrac{1}{10}R_2 \to R_2 \text{ then } \begin{array}{c} -R_2 + R_1 \to R_1 \\ \to \\ 30R_2 + R_3 \to R_3 \end{array} \left[\begin{array}{ccccc|c} 1 & 0 & 1 & -\frac{1}{10} & 0 & 28 \\ 1 & 1 & 0 & \frac{1}{10} & 0 & 2 \\ \hline 25 & 0 & 0 & 3 & 1 & 60 \end{array}\right]$$

The nutritionist should use none of food I and 3 ounces of food II for a minimum cost of $0.60.

**43.** Minimize $C = 14A + 16B$

Constraints: $A + 4B \geq 40$

$2A + B \geq 80$

$$\left[\begin{array}{cc|c} 1 & 4 & 40 \\ 2 & 1 & 80 \\ \hline 14 & 16 & C \end{array}\right] \quad \text{Dual: } \left[\begin{array}{cc|c} 1 & 2 & 14 \\ 4 & 1 & 16 \\ \hline 40 & 80 & C \end{array}\right]$$

$$\left[\begin{array}{ccccc|c} 1 & \boxed{2} & 1 & 0 & 0 & 14 \\ 4 & 1 & 0 & 1 & 0 & 16 \\ \hline -40 & -80 & 0 & 0 & 1 & 0 \end{array}\right] \begin{array}{l} \tfrac{1}{2}R_1 \to R_1 \text{ then } \\ \phantom{x} \end{array} \begin{array}{r} -R_1 + R_2 \to R_2 \\ 80R_1 + R_3 \to R_3 \end{array} \left[\begin{array}{ccccc|c} \frac{1}{2} & 1 & \frac{1}{2} & 0 & 0 & 7 \\ \frac{7}{2} & 0 & -\frac{1}{2} & 1 & 0 & 9 \\ \hline 0 & 0 & 40 & 0 & 1 & 560 \end{array}\right]$$

The laboratory should purchase 40 pounds of Feed A and none of Feed B for a minimum cost of $5.60.

**44.** $x =$ days at factory $A$, $y =$ days of factory $B$, $z =$ days at factory $C$

Minimize $g = 2000x + 3000y + 5000z$

Subject to: $10x + 20z \geq 200 \quad \to \quad x + 2z \geq 20$

$10x + 20y + 20z \geq 500 \quad\quad x + 2y + 2z \geq 50$

$10x + 20y + 10z \geq 300 \quad\quad x + 2y + z \geq 30$

$x, y, z \geq 0$

$$\left[\begin{array}{ccc|c} 1 & 0 & 2 & 20 \\ 1 & 2 & 2 & 50 \\ 1 & 2 & 1 & 30 \\ \hline 2000 & 3000 & 5000 & g \end{array}\right] \quad \text{Dual: } \left[\begin{array}{ccc|c} 1 & 1 & 1 & 2000 \\ 0 & 2 & 2 & 3000 \\ 2 & 2 & 1 & 5000 \\ \hline 20 & 50 & 30 & g \end{array}\right]$$

$$\left[\begin{array}{ccccccc|c} 1 & 1 & 1 & 1 & 0 & 0 & 0 & 2000 \\ 0 & \boxed{2} & 2 & 0 & 1 & 0 & 0 & 3000 \\ 2 & 2 & 1 & 0 & 0 & 1 & 0 & 5000 \\ \hline -20 & -50 & -30 & 0 & 0 & 0 & 1 & 0 \end{array}\right] \tfrac{1}{2}R_2 \to R_2 \text{ then } \begin{array}{r} -R_2 + R_1 \to R_1 \\ \to \\ -2R_2 + R_3 \to R_3 \\ 50R_2 + R_4 \to R_4 \end{array}$$

$$\left[\begin{array}{ccccccc|c} 1 & 0 & 0 & 1 & -\frac{1}{2} & 0 & 0 & 500 \\ 0 & 1 & 1 & 0 & \frac{1}{2} & 0 & 0 & 1500 \\ 2 & 0 & -1 & 0 & -1 & 1 & 0 & 2000 \\ \hline -20 & 0 & 20 & 0 & 25 & 0 & 1 & 75{,}000 \end{array}\right] \begin{array}{r} \to \\ \to \\ -2R_1 + R_3 \to R_3 \\ 20R_1 + R_4 \to R_4 \end{array} \left[\begin{array}{ccccccc|c} 1 & 0 & 0 & 1 & -\frac{1}{2} & 0 & 0 & 500 \\ 0 & 1 & 1 & 0 & \frac{1}{2} & 0 & 0 & 1500 \\ 0 & 0 & -1 & -2 & 0 & 1 & 0 & 1000 \\ \hline 0 & 0 & 20 & 20 & 15 & 0 & 1 & 85{,}000 \end{array}\right]$$

The company should operate factory A for 20 days and Factory B for 15 days for a minimum cost of $85,000.

**45.**

| | Flour | Shortening | Sugar | Objective Function |
|---|---|---|---|---|
| Pancake: $x$ | $0.6x$ | $0.1x$ | $--$ | $P = 0.35x + 0.25y$ |
| Cake: $y$ | $0.4y$ | $0.1y$ | $0.4y$ | |

Constraints:

$$\begin{array}{ll} 0.6x + 0.4y \le 6000 & \text{Flour} \\ 0.1x + 0.1y \ge 500 & \text{Shortening} \\ 0.4y \le 1200 & \text{Sugar} \\ x,\ y \ge 0 & \end{array}$$

$$\left[\begin{array}{cccccc|c} 0.6 & 0.4 & 1 & 0 & 0 & 0 & 6000 \\ \boxed{-0.1} & -0.1 & 0 & 1 & 0 & 0 & -500 \\ 0 & 0.4 & 0 & 0 & 1 & 0 & 1200 \\ \hline -0.35 & -0.25 & 0 & 0 & 0 & 1 & 0 \end{array}\right] \begin{array}{l} \\ -10R_2 \to R_2 \text{ then} \\ \\ \\ \end{array} \begin{array}{r} -0.6R_2 + R_1 \to R_1 \\ \to \\ \\ 0.35R_2 + R_4 \to R_4 \end{array}$$

$$\left[\begin{array}{cccccc|c} 0 & -0.2 & 1 & \boxed{6} & 0 & 0 & 3000 \\ 1 & 1 & 0 & -10 & 0 & 0 & 5000 \\ 0 & 0.4 & 0 & 0 & 1 & 0 & 1200 \\ \hline 0 & 0.1 & 0 & -3.5 & 0 & 1 & 1750 \end{array}\right] \begin{array}{l} \frac{1}{6}R_1 \to R_1 \text{ then} \\ \\ \\ \\ \end{array} \begin{array}{r} \to \\ 10R_1 + R_2 \to R_2 \\ \to \\ 3.5R_1 + R_4 \to R_4 \end{array}$$

$$\left[\begin{array}{cccccc|c} 0 & -0.0\overline{3} & 0.1\overline{6} & 1 & 0 & 0 & 500 \\ 1 & 0.\overline{6} & 1.\overline{6} & 0 & 0 & 0 & 10{,}000 \\ 0 & \boxed{0.4} & 0 & 0 & 1 & 0 & 1200 \\ \hline 0 & -0.01\overline{6} & 0.58\overline{3} & 0 & 0 & 1 & 3500 \end{array}\right] \begin{array}{l} \\ \\ 2.5R_3 \to R_3 \text{ then} \\ \\ \end{array} \begin{array}{r} 0.0\overline{3}R_3 + R_1 \to R_1 \\ -0.\overline{6}R_3 + R_2 \to R_2 \\ \to \\ 0.01\overline{6}R_3 + R_4 \to R_4 \end{array}$$

$$\left[\begin{array}{cccccc|c} 0 & 0 & 0.1\overline{6} & 1 & 0.08\overline{3} & 0 & 600 \\ 1 & 0 & 1.\overline{6} & 0 & -1.\overline{6} & 0 & 8000 \\ 0 & 1 & 0 & 0 & 2.5 & 0 & 3000 \\ \hline 0 & 0 & 0.58\overline{3} & 0 & 0.041\overline{6} & 1 & 3550 \end{array}\right]$$

The company should make 8000 lbs of pancake mix and 3000 lbs of cake mix for a maximum profit of $3550.

**46.** $x_1$ = number of desks at Texas, $x_2$ = number of tables at Texas,
$x_3$ = number of desks at Louisiana, $x_4$ = number of tables at Louisiana

Minimize $g = 36x_1 + 60x_2 + 42x_3 + 57x_4$

Subject to: $x_1 + x_2 \le 120$, $x_3 + x_4 \le 150$, $x_1 + x_3 \ge 130$, $x_2 + x_4 \ge 130$, $-x_1 + x_2 \ge 10$, $x_1, x_2, x_3, x_4 \ge 0$

$$\left[\begin{array}{cccccccccc|c} 1 & 1 & 0 & 0 & 1 & 0 & 0 & 0 & 0 & 0 & 120 \\ 0 & 0 & 1 & 1 & 0 & 1 & 0 & 0 & 0 & 0 & 150 \\ \boxed{-1} & 0 & -1 & 0 & 0 & 0 & 1 & 0 & 0 & 0 & -130 \\ 0 & -1 & 0 & -1 & 0 & 0 & 0 & 1 & 0 & 0 & -130 \\ 1 & -1 & 0 & 0 & 0 & 0 & 0 & 0 & 1 & 0 & -10 \\ \hline 36 & 60 & 42 & 57 & 0 & 0 & 0 & 0 & 0 & 1 & 0 \end{array}\right] \begin{array}{l} \\ \\ -R_3 \to R_3 \text{ then} \\ \\ \\ \\ \end{array} \begin{array}{r} -R_3 + R_1 \to R_1 \\ \to \\ \to \\ \to \\ -R_3 + R_5 \to R_5 \\ -36R_3 + R_6 \to R_6 \end{array}$$

$$\left[\begin{array}{cccccccccc|c} 0 & 1 & -1 & 0 & 1 & 0 & 1 & 0 & 0 & 0 & -10 \\ 0 & 0 & 1 & 1 & 0 & 1 & 0 & 0 & 0 & 0 & 150 \\ 1 & 0 & 1 & 0 & 0 & 0 & -1 & 0 & 0 & 0 & 130 \\ 0 & -1 & 0 & -1 & 0 & 0 & 0 & 1 & 0 & 0 & -130 \\ 0 & \boxed{-1} & -1 & 0 & 0 & 0 & 1 & 0 & 1 & 0 & -140 \\ \hline 0 & 60 & 6 & 57 & 0 & 0 & 36 & 0 & 0 & 1 & -4680 \end{array}\right] \begin{array}{c} R_5 + R_1 \to R_1 \\ \to \\ \to \\ -R_5 + R_4 \to R_4 \\ \to \\ 60R_5 + R_6 \to R_6 \end{array} \text{ then } -R_5 \to R_5$$

$$\left[\begin{array}{cccccccccc|c} 0 & 0 & -2 & 0 & 1 & 0 & 2 & 0 & 1 & 0 & -150 \\ 0 & 0 & 1 & 1 & 0 & 1 & 0 & 0 & 0 & 0 & 150 \\ 1 & 0 & 1 & 0 & 0 & 0 & -1 & 0 & 0 & 0 & 130 \\ 0 & 0 & \boxed{1} & -1 & 0 & 0 & -1 & 1 & -1 & 0 & 10 \\ 0 & 1 & 1 & 0 & 0 & 0 & -1 & 0 & -1 & 0 & 140 \\ \hline 0 & 0 & -54 & 57 & 0 & 0 & 96 & 0 & 60 & 1 & -13,080 \end{array}\right] \begin{array}{c} 2R_4 + R_1 \to R_1 \\ -R_4 + R_2 \to R_2 \\ -R_4 + R_3 \to R_3 \\ \to \\ -R_4 + R_5 \to R_5 \\ 54R_4 + R_6 \to R_6 \end{array}$$

$$\left[\begin{array}{cccccccccc|c} 0 & 0 & 0 & \boxed{-2} & 1 & 0 & 0 & 2 & -1 & 0 & -130 \\ 0 & 0 & 0 & 2 & 0 & 1 & 1 & -1 & 1 & 0 & 140 \\ 1 & 0 & 0 & 1 & 0 & 0 & 0 & -1 & 1 & 0 & 120 \\ 0 & 0 & 1 & -1 & 0 & 0 & -1 & 1 & -1 & 0 & 10 \\ 0 & 1 & 0 & 1 & 0 & 0 & 0 & -1 & 0 & 0 & 130 \\ \hline 0 & 0 & 0 & 3 & 0 & 0 & 42 & 54 & 6 & 1 & -12540 \end{array}\right] \begin{array}{c} \\ R_1 + R_2 \to R_2 \\ \frac{1}{2}R_1 + R_3 \to R_3 \\ -\frac{1}{2}R_1 + R_4 \to R_4 \\ \frac{1}{2}R_1 + R_5 \to R_5 \\ \frac{3}{2}R_1 + R_6 \to R_6 \end{array} \text{ then } -\tfrac{1}{2}R_1 \to R_1$$

$$\left[\begin{array}{cccccccccc|c} 0 & 0 & 0 & 1 & -\frac{1}{2} & 0 & 0 & -1 & \frac{1}{2} & 0 & 65 \\ 0 & 0 & 0 & 0 & 1 & 1 & 1 & 1 & 0 & 0 & 10 \\ 1 & 0 & 0 & 0 & \frac{1}{2} & 0 & 0 & 0 & \frac{1}{2} & 0 & 55 \\ 0 & 0 & 1 & 0 & -\frac{1}{2} & 0 & -1 & 0 & -\frac{1}{2} & 0 & 75 \\ 0 & 1 & 0 & 0 & \frac{1}{2} & 0 & 0 & 0 & -\frac{1}{2} & 0 & 65 \\ \hline 0 & 0 & 0 & 0 & \frac{1}{2} & 0 & 42 & 57 & \frac{9}{2} & 1 & -12,735 \end{array}\right]$$

The company should schedule the plant in Texas to make 55 desks and 65 tables, and the plant in Louisiana to make 75 desks and 65 tables for a minimum cost of $12,735.

**47.** $x_1$ = number of tons of Grade 1 steel made each week in Midland

$x_2$ = number of tons of Grade 2 steel made each week in Midland

$y_1$ = number of tons of Grade 1 steel made each week in Donora

$y_2$ = number of tons of Grade 2 steel made each week in Donora

Minimize: $f = 300x_1 + 360x_2 + 330y_1 + 270y_2$

Maximize: $-f = -300x_1 - 360x_2 - 330y_1 - 270y_2$

Constraints: $40x_1 + 42x_2 \le 19{,}460$
$44y_1 + 45y_2 \le 21{,}380$
$x_1 + y_1 \ge 500$
$x_2 + y_2 \ge 450$

Revised Constraints: $40x_1 + 42x_2 \le 19{,}460$
$44y_1 + 45y_2 \le 21{,}380$
$-x_1 - y_1 \le -500$
$-x_2 - y_2 \le -450$

$$\left[\begin{array}{cccccccc|c} 40 & 42 & 0 & 0 & 1 & 0 & 0 & 0 & 0 & 19{,}460 \\ 0 & 0 & 44 & 45 & 0 & 1 & 0 & 0 & 0 & 21{,}380 \\ -1 & 0 & -1 & 0 & 0 & 0 & 1 & 0 & 0 & -500 \\ 0 & -1 & 0 & -1 & 0 & 0 & 0 & 1 & 0 & -450 \\ \hline 300 & 360 & 330 & 270 & 0 & 0 & 0 & 0 & 1 & 0 \end{array}\right]$$

Using technology to solve the problem, Armstrong Industries should instruct the Midland plant to make 486.5 tons of Grade 1 steel, and instruct the Donora plant to make 13.5 tons of Grade 1 steel and 450 tons of Grade 2 steel for a minimum cost of $271,905.

## Chapter 4 Test

**1.**

| Corners | $f = 3x + 5y$ |
|---|---|
| $(0,24)$ | 120 |
| $(8,18)$ | 114 |
| $(20,0)$ | 60 |
| $(0,0)$ | 0 |

Maximum is 120 at (0,24).

**2.** **a.** $C$, since there is a zero indicator in a nonbasic variable column. Pivot on the 1 in $R_2C_2$.

Row Ops: $-2R_2 + R_1 \to R_1; -3R_2 + R_3 \to R_3$

**b.** $A$; Pivot column is column 3, but no pivot element is defined.

**c.** $B$ $x_1 = 40,\ x_2 = 12,\ x_3 = 0,$

$s_1 = 0,\ s_2 = 20,\ s_3 = 0,\ f = 170$

**3.** **a.**

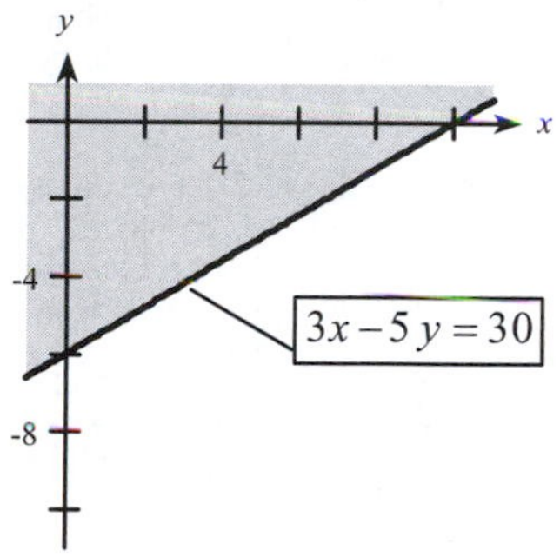


**b.**

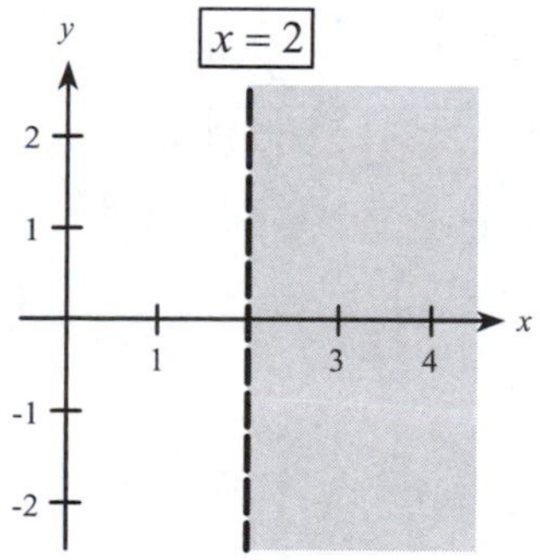


**c.**

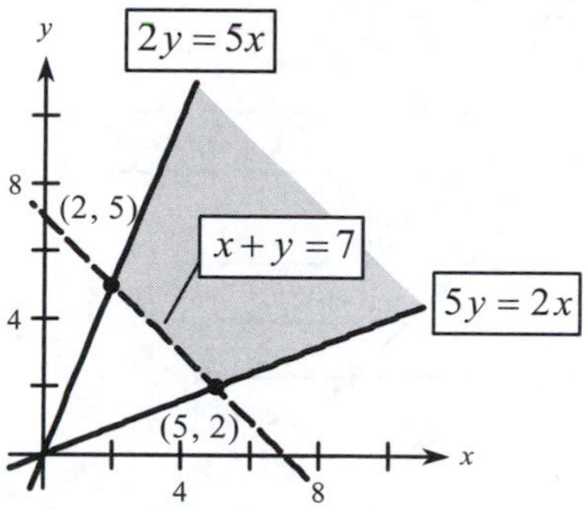


**4.** $\left[\begin{array}{ccc|c} 3 & 5 & 1 & 100 \\ 4 & 6 & 3 & 120 \\ \hline 2 & 3 & 5 & g \end{array}\right]$

Take the transpose to obtain:

Maximize: $f = 100x_1 + 120x_2$

Subject to: $3x_1 + 4x_2 \le 2$

$5x_1 + 6x_2 \le 3$

$x_1 + 3x_2 \le 5$

$x_1, x_2 \ge 0$

**5.** Solving: $\begin{cases} 4x + y = 12 \\ x + y = 9 \end{cases}$

$\begin{cases} 4x + y = 12 \\ -x - y = -9 \end{cases}$

$3x = 3$

$x = 1,\ y = 9$

Solving: $\begin{cases} x + y = 9 \\ x + 3y = 15 \end{cases}$

$\begin{cases} -x - y = -9 \\ x + 3y = 15 \end{cases}$

$2y = 6$

$y = 3,\ x = 6$

| Corners | $f = 5x + 2y$ |
|---|---|
| $(0,12)$ | 24 |
| $(1,9)$ | 23 |
| $(6,3)$ | 36 |
| $(15,0)$ | 75 |

*Minimum is 23 at* $(1,9)$.

*There is no maximum since the region is unbounded.*

**6.**

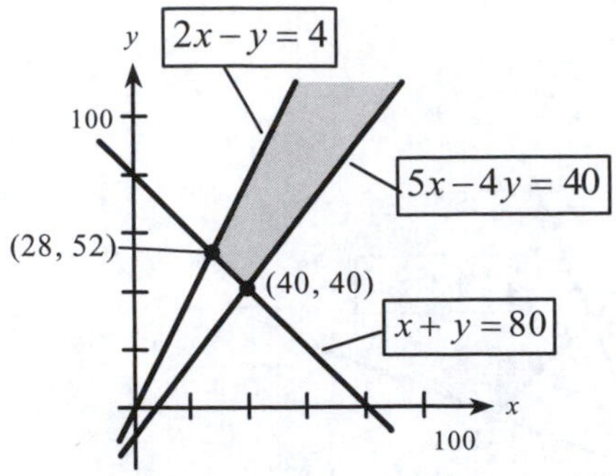


| Corners | $g = 3x + y$ |
|---|---|
| $(28,52)$ | 136 |
| $(1,9)$ | 23 |
| $(6,3)$ | 36 |

Minimum is 136 at (28, 52). There are only two feasible corners.

**7.** Maximize (change signs) $-g = -7x - 3y$

Subject to: $x - 4y \le -4$

$x - y \le 5$

$2x + 3y \le 30$

**8.** Maximum is $f = 658$ at $x_1 = 17, x_2 = 15, x_3 = 0.$

Minimum is $g = 658$ at $y_1 = 4, y_2 = 18, y_3 = 0.$

**9.** Maximize: $f = 70x + 5y$

Subject to: $x + 1.5y \le 150$

$x + 0.5y \le 90$

$x, y \ge 0$

Clear the decimals by multiplying each inequality by 2.

$$\left[\begin{array}{ccccc|c} 2 & 3 & 1 & 0 & 0 & 300 \\ \boxed{2} & 1 & 0 & 1 & 0 & 180 \\ \hline -70 & -5 & 0 & 0 & 1 & 0 \end{array}\right] \quad \tfrac{1}{2}R_2 \to R_2 \text{ then } \quad \begin{array}{c} -2R_2 + R_1 \to R_1 \\ \to \\ 70R_2 + R_3 \to R_3 \end{array} \left[\begin{array}{ccccc|c} 0 & 2 & 1 & -1 & 0 & 120 \\ 1 & \frac{1}{2} & 0 & \frac{1}{2} & 0 & 90 \\ \hline 0 & 30 & 0 & 35 & 1 & 6300 \end{array}\right]$$

Maximum is 6300 at $x = 90, y = 0$.

**10.** Maximize $f = 12x + 20y$

Subject to: $3x + 5y \le 1500$

$9x + 5y \le 3600$

$x + 5y \le 1000$

Solving $\begin{cases} 3x + 5y = 1500 \\ x + 5y = 1000 \end{cases}$ we get $(250, 150)$.

Solving $\begin{cases} 3x + 5y = 1500 \\ 9x + 5y = 3600 \end{cases}$ we get $(350, 90)$.

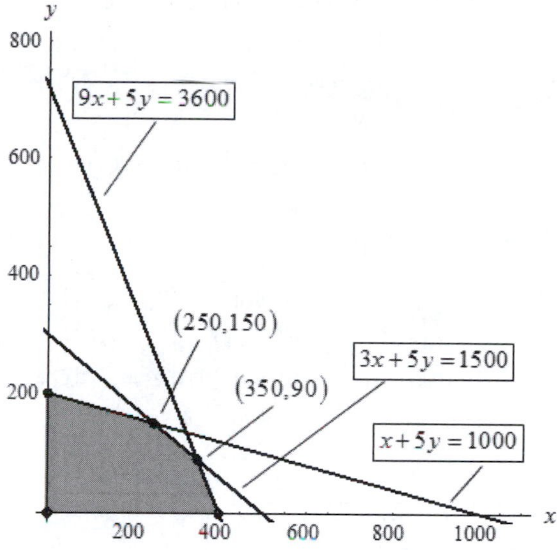


| Feasible Corners | $f = 12x + 20y$ |
|---|---|
| $(0, 0)$ | 0 |
| $(0, 200)$ | 4000 |
| $(250, 150)$ | 6000 |
| $(350, 90)$ | 6000 |
| $(400, 0)$ | 4800 |

The maximum of $f = 6000$ occurs at any point on the segment joining $(250, 150)$ and $(350, 90)$.

**11.** $$\left[\begin{array}{ccccccc|c} 8 & 6 & 1 & 1 & 0 & 0 & 0 & 0 & 108 \\ \boxed{4} & 2 & 1.5 & 0 & 1 & 0 & 0 & 0 & 50 \\ 2 & 1.5 & 0.5 & 0 & 0 & 1 & 0 & 0 & 40 \\ 0 & 0 & 1 & 0 & 0 & 0 & 1 & 0 & 16 \\ \hline -60 & -48 & -36 & 0 & 0 & 0 & 0 & 1 & 0 \end{array}\right] \quad \tfrac{1}{4}R_2 \to R_2 \text{ then } \begin{array}{c} -8R_2 + R_1 \to R_1 \\ \to \\ -2R_2 + R_3 \to R_3 \\ \to \\ 60R_2 + R_5 \to R_5 \end{array}$$

$$\left[\begin{array}{cccccccc|c} 0 & \boxed{2} & -2 & 1 & -2 & 0 & 0 & 0 & 8 \\ 1 & 0.5 & 0.375 & 0 & 0.25 & 0 & 0 & 0 & 12.5 \\ 0 & 0.5 & -0.25 & 0 & -0.5 & 1 & 0 & 0 & 15 \\ 0 & 0 & 1 & 0 & 0 & 0 & 1 & 0 & 16 \\ \hline 0 & -18 & -13.5 & 0 & 15 & 0 & 0 & 1 & 750 \end{array}\right] \quad \tfrac{1}{2}R_1 \to R_1 \text{ then } \begin{array}{c} \to \\ -0.5R_1 + R_2 \to R_2 \\ -0.5R_1 + R_3 \to R_3 \\ \to \\ 18R_1 + R_5 \to R_5 \end{array}$$

$$\left[\begin{array}{cccccccc|c} 0 & 1 & -1 & 0.5 & -1 & 0 & 0 & 0 & 4 \\ 1 & 0 & \boxed{0.875} & -0.25 & 0.75 & 0 & 0 & 0 & 10.5 \\ 0 & 0 & 0.25 & -0.25 & 0 & 1 & 0 & 0 & 13 \\ 0 & 0 & 1 & 0 & 0 & 0 & 1 & 0 & 16 \\ \hline 0 & 0 & -31.5 & 9 & -3 & 0 & 0 & 1 & 822 \end{array}\right] \quad \tfrac{8}{7}R_2 \to R_2 \text{ then } \begin{array}{c} R_2 + R_1 \to R_1 \\ \to \\ -0.25R_2 + R_3 \to R_3 \\ -R_2 + R_4 \to R_4 \\ 31.5R_2 + R_5 \to R_5 \end{array}$$

$$\left[\begin{array}{cccccccc|c} \frac{8}{7} & 1 & 0 & \frac{3}{14} & -\frac{2}{7} & 0 & 0 & 0 & 16 \\ \frac{8}{7} & 0 & 1 & -\frac{2}{7} & \frac{7}{8} & 0 & 0 & 0 & 12 \\ -\frac{2}{7} & 0 & 0 & -\frac{5}{28} & -\frac{3}{14} & 1 & 0 & 0 & 10 \\ -\frac{8}{7} & 0 & 0 & \frac{2}{7} & -\frac{7}{8} & 0 & 1 & 0 & 4 \\ \hline 36 & 0 & 0 & 0 & 24 & 0 & 0 & 1 & 1200 \end{array}\right]$$

The maximum is 1200 at $x = 0$, $y = 16$, $z = 12$.

**12.** Let $x$ = barrels of lager , $y$ = barrels of ale.

Maximize: $P = 35x + 30y$

Subject to: $3x + 2y \le 1200$

$2x + 2y \le 1000$

$$\left[\begin{array}{ccccc|c} 3 & 2 & 1 & 0 & 0 & 1200 \\ 2 & 2 & 0 & 1 & 0 & 1000 \\ \hline -35 & -30 & 0 & 0 & 1 & 0 \end{array}\right] \tfrac{1}{3}R_1 \to R_1 \left[\begin{array}{ccccc|c} 1 & \tfrac{2}{3} & \tfrac{1}{3} & 0 & 0 & 400 \\ 2 & 2 & 0 & 1 & 0 & 1000 \\ \hline -35 & -30 & 0 & 0 & 1 & 0 \end{array}\right] \begin{array}{c} \to \\ -2R_1 + R_2 \to R_2 \\ 35R_1 + R_3 \to R_3 \end{array}$$

$$\left[\begin{array}{ccccc|c} 1 & \tfrac{2}{3} & \tfrac{1}{3} & 0 & 0 & 400 \\ 0 & \tfrac{2}{3} & -\tfrac{2}{3} & 1 & 0 & 200 \\ \hline 0 & -\tfrac{20}{3} & \tfrac{35}{3} & 0 & 1 & 14{,}000 \end{array}\right] \begin{array}{c} -R_2 + R_1 \to R_1 \\ \to \\ 10R_2 + R_3 \to R_3 \end{array} \left[\begin{array}{ccccc|c} 1 & 0 & 1 & -1 & 0 & 200 \\ 0 & \tfrac{2}{3} & -\tfrac{2}{3} & 1 & 0 & 200 \\ \hline 0 & 0 & 5 & 10 & 1 & 16{,}000 \end{array}\right] \tfrac{3}{2}R_2 \to R_2$$

$$\left[\begin{array}{ccccc|c} 1 & 0 & 1 & -1 & 0 & 200 \\ 0 & 1 & -1 & \tfrac{3}{2} & 0 & 300 \\ \hline 0 & 0 & 5 & 10 & 1 & 16{,}000 \end{array}\right]$$

River Brewery should produce 200 barrels of lager and 300 barrels of ale for a maximum profit of \$16,000.

**13.** Let $x$ = number of day calls, let $y$ = number of evening calls.

Minimize $C = 3x + 4y$

Subject to: $0.3x + 0.3y \ge 150$

$0.1x + 0.3y \ge 120$

$x, y \ge 0$

| Corners | $C = 3x + 4y$ |
|---|---|
| $(0, 500)$ | 2000 |
| $(150, 350)$ | 1850 |
| $(1200, 0)$ | 3600 |

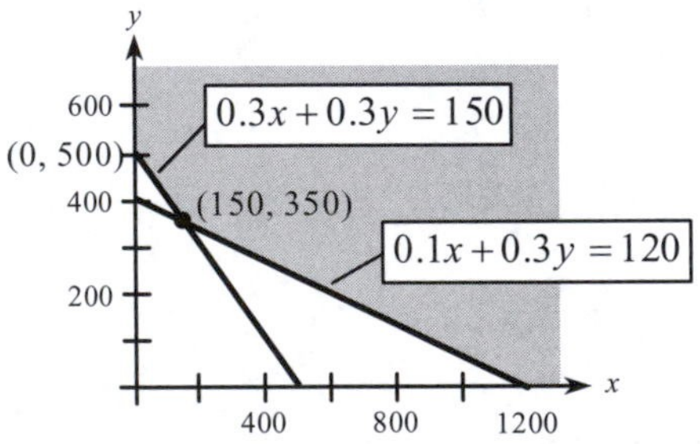


The marketing research group should make 150 day calls and 350 evening calls for a minimum cost of \$1850.

**14.** $x_1$ = tons of Product 1, $x_2$ = tons of Product 2, $x_3$ = tons of Product 3, and $x_4$ = tons of Product 4

Maximize: $f = 120x_1 + 150x_2 + 180x_3 + 210x_4$

| Constraints: | Revised: |
|---|---|
| $0.60x_1 + 0.30x_2 + 0.20x_3 + 0.10x_4 \ge 35$ | $-6x_1 - 3x_2 - 2x_3 - x_4 \ge -350$ |
| $0.20x_1 + 0.20x_2 + 0.20x_3 + 0.20x_4 \le 20$ | $x_1 + x_2 + x_3 + x_4 \le 100$ |
| $0.09x_1 + 0.06x_2 + 0.15x_3 + 0.18x_4 \le 9$ | $9x_1 + 6x_2 + 15x_3 + 18x_4 \le 900$ |
| $0.08x_1 + 0.03x_2 + 0.02x_3 + 0.01x_4 \le 4$ | $8x_1 + 3x_2 + 2x_3 + x_4 \le 400$ |

$$\left[\begin{array}{cccccccccc|c} -6 & -3 & -2 & -1 & 1 & 0 & 0 & 0 & 0 & -350 \\ 1 & 1 & 1 & 1 & 0 & 1 & 0 & 0 & 0 & 100 \\ 9 & 6 & 15 & 18 & 0 & 0 & 1 & 0 & 0 & 900 \\ 8 & 3 & 2 & 1 & 0 & 0 & 0 & 1 & 0 & 400 \\ \hline -120 & -150 & -180 & -210 & 0 & 0 & 0 & 0 & 1 & 0 \end{array}\right]$$

Using technology to solve the problem, Lawn Rich should use 25 tons of product 1, 62.5 tons of product 2, and 12.5 tons of product 4 for a maximum profit of \$15000.

# Chapter 5: Exponential and Logarithmic Functions

## *Exercises 5.1*

**1.** **a.** $10^{0.5}$

$10\boxed{y^x}0.5\boxed{=}3.1623$

**b.** $5^{-2.7}$

$5\boxed{y^x}2.7\boxed{\pm}\boxed{=}0.012965$

**3.** **a.** $3^{1/3}$

$3\boxed{y^x}0.3333\boxed{=}1.44225$

**b.** $e^2$

$e\boxed{y^x}2\boxed{=}7.3891$

**5.**

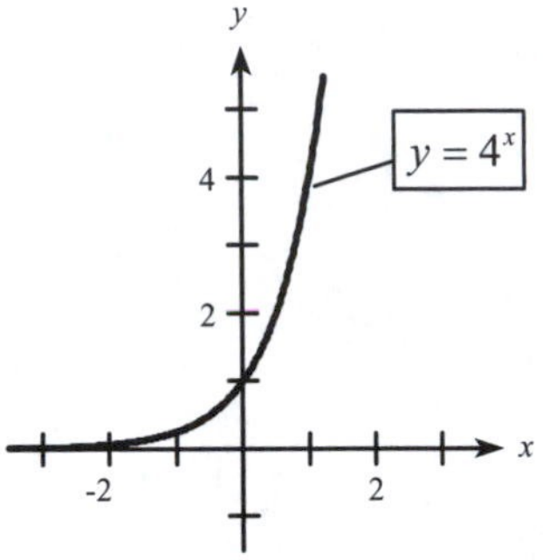


**7.**

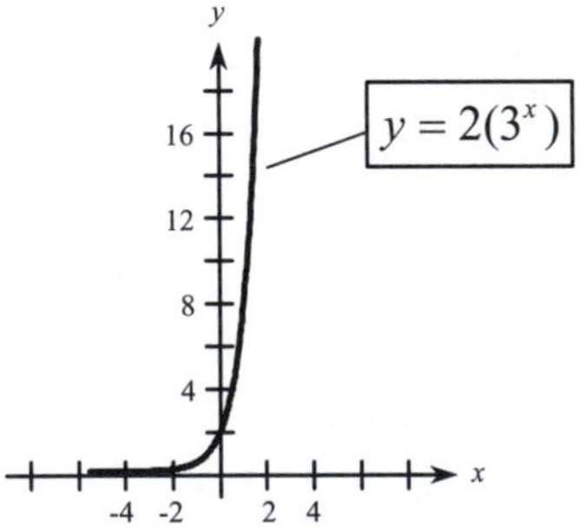


**9.**

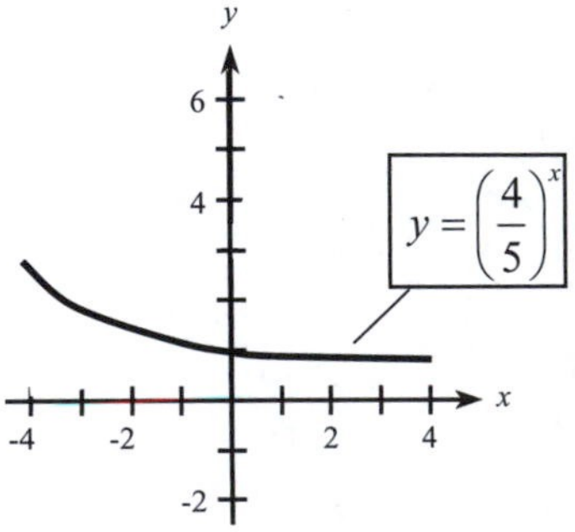


**11.**

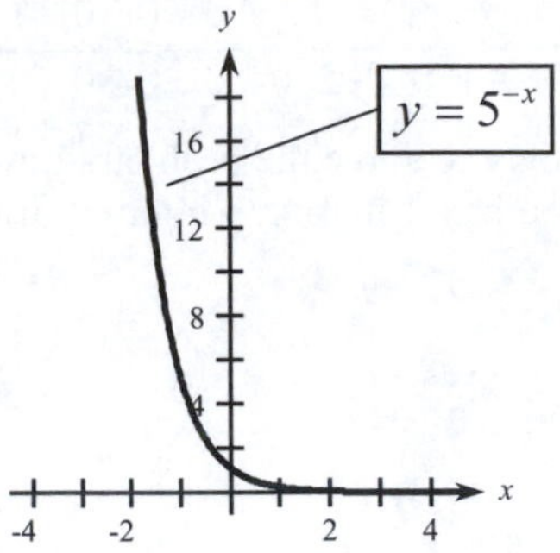


**13.**

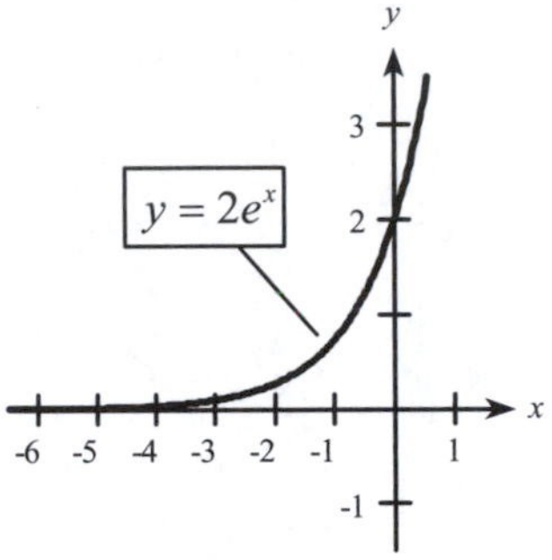


**15.**

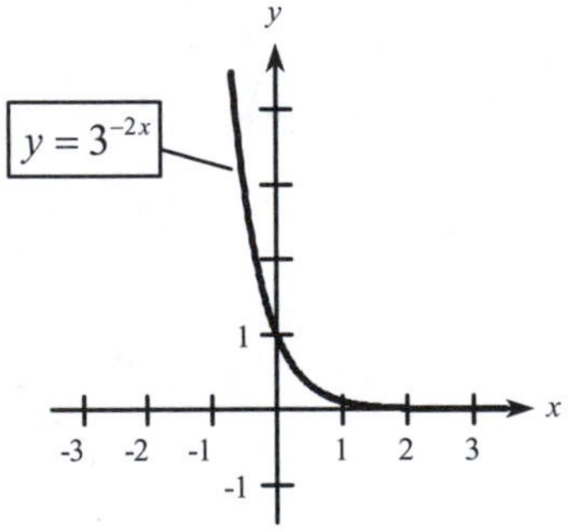


**17.** **a.** $y = 3\left(\frac{2}{5}\right)^x = 3\left(\frac{5}{2}\right)^{-x}$

**b.** These functions are decay exponentials. They are algebraically equivalent and an exponential decay function is one of the form $f(x) = cb^{-x}$ where $b > 1$. In this form $b = 5/2$ which is greater than 1.

**c.** Graphs are identical and falling.

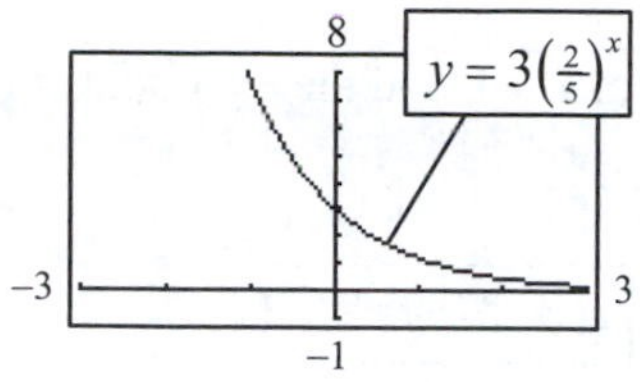


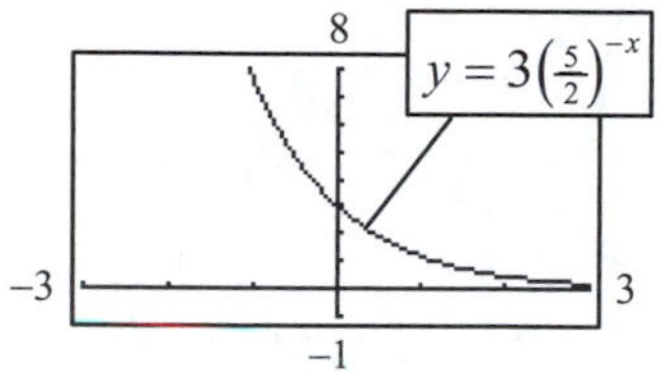


**19. a.**

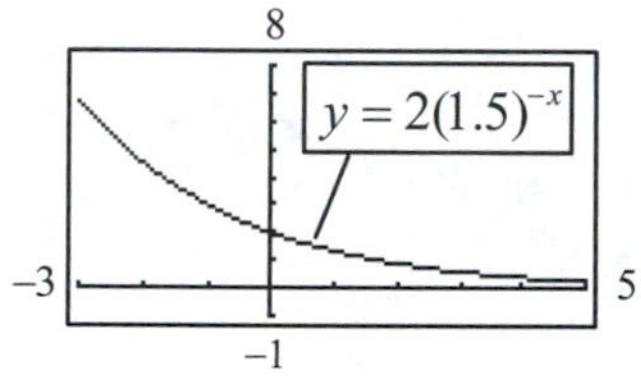


**b.**

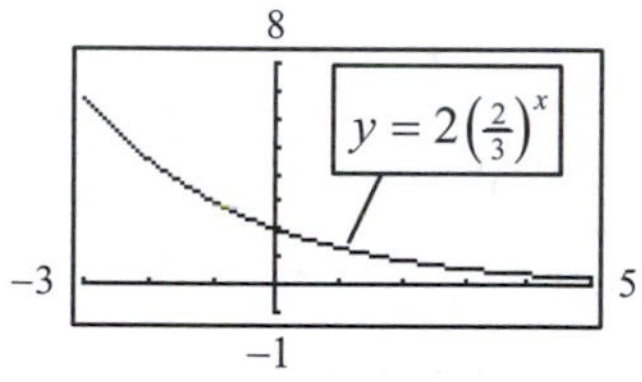


**c.** $y = 2(1.5)^{-x} = 2\left(\dfrac{3}{2}\right)^{-x} = 2\left(\dfrac{2}{3}\right)^x$

**21.** $y = \left(\dfrac{4}{5}\right)^x = \dfrac{4^x}{5^x} = \dfrac{5^{-x}}{4^{-x}} = \left(\dfrac{5}{4}\right)^{-x}$

**23.** $f(x) = e^{-x}$, $f(kx) = e^{-kx}$

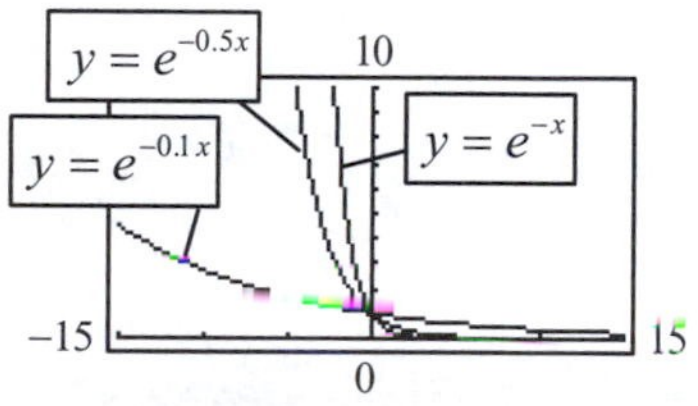


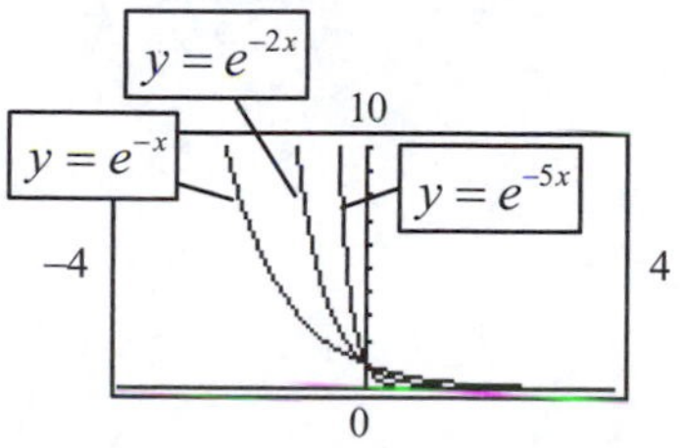


For $k > 1$ the graphs rise and fall more sharply than $y = e^{-x}$. For $k < 1$ the graphs rise and fall more slowly than $y = e^{-x}$.

**25.** $f(x) = 4^x \qquad f(x) + C = 4^x + C$

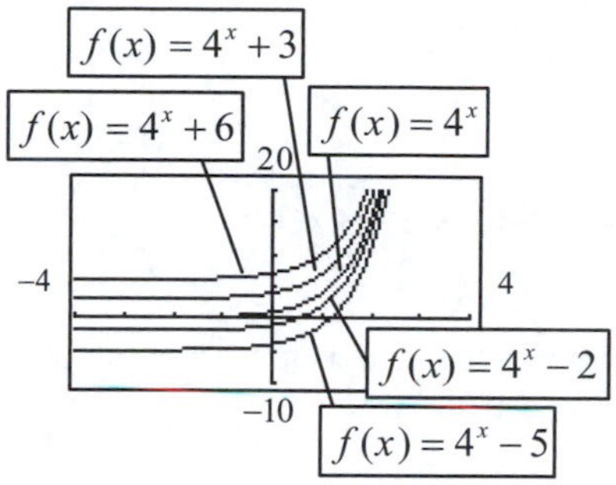


The graphs are the same but $f(x) + C$ is shifted $C$ units on the $y$-axis.

**27. a.**

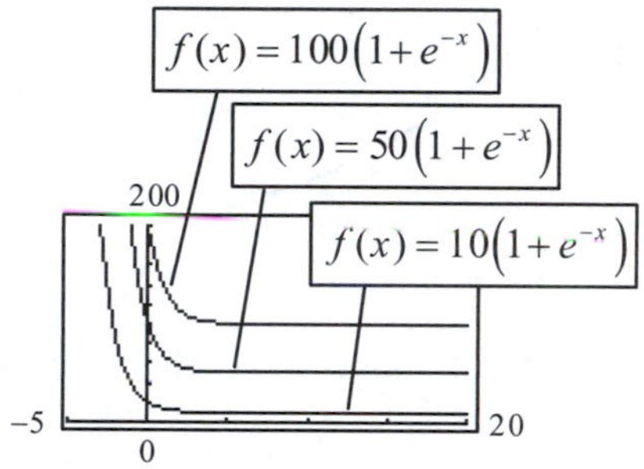


**b.** As $c$ changes, the $y$-intercept and the horizontal asymptote change.

**29.** $S(x) = 1000(1.02)^{4x}$

$S(8) = 1.02$ $\boxed{y^x}$ $32$ $\boxed{\times}$ $1000$ $\boxed{=}$ \$1884.54

**31.**

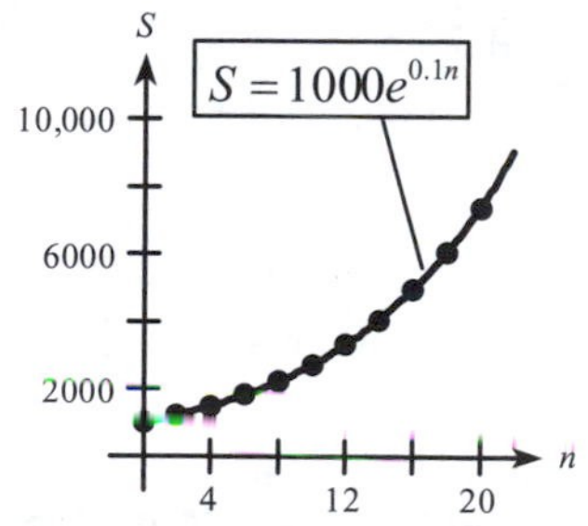

**33.**

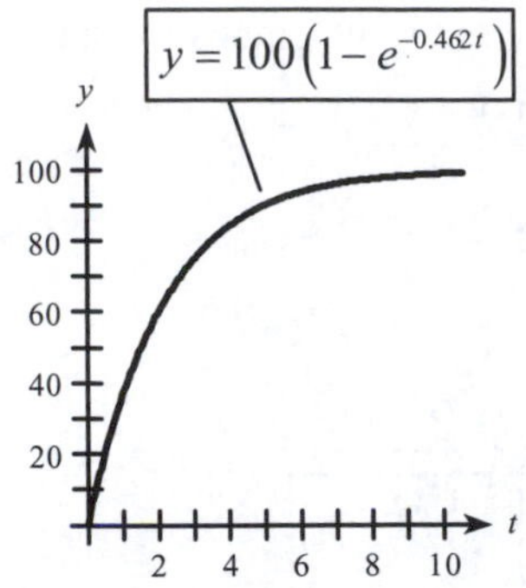


After 10 hours the drug is almost completely in the bloodstream.

**35.**

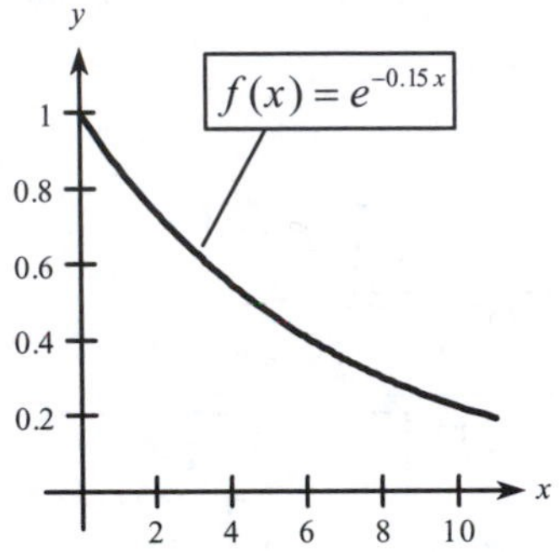


After 10 years about 20% of the television sets are still in service.

**37.**

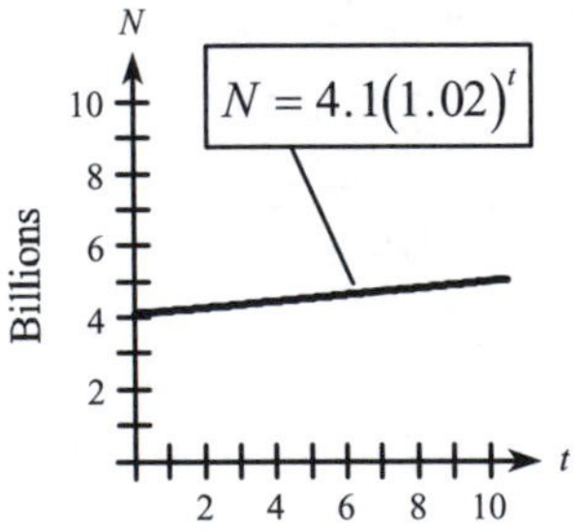


**39.**

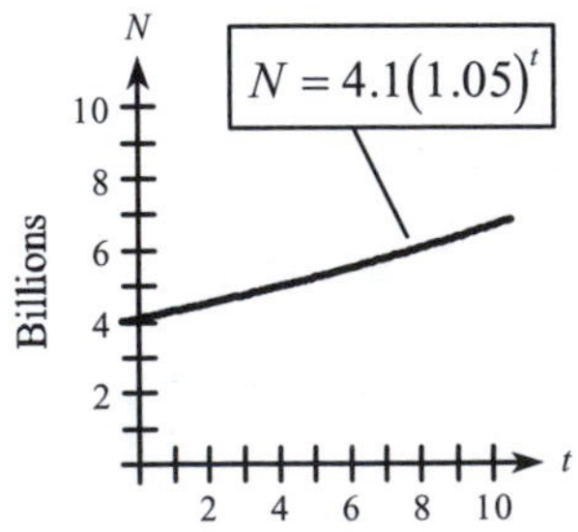


**41. a.** Growth; $e > 1$ and the exponent is positive for $t > 0$.

**b.**

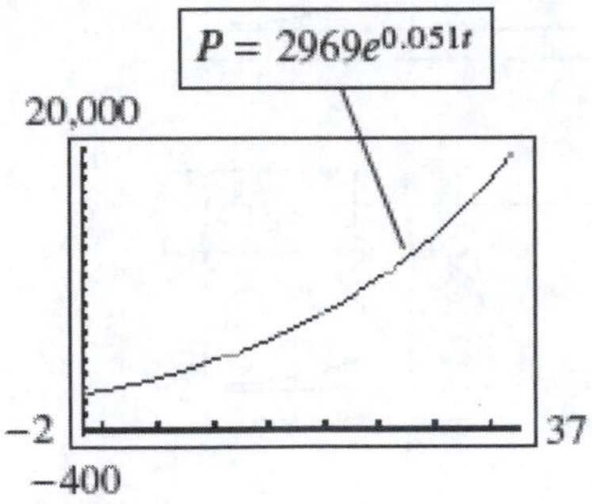


**43. a.**

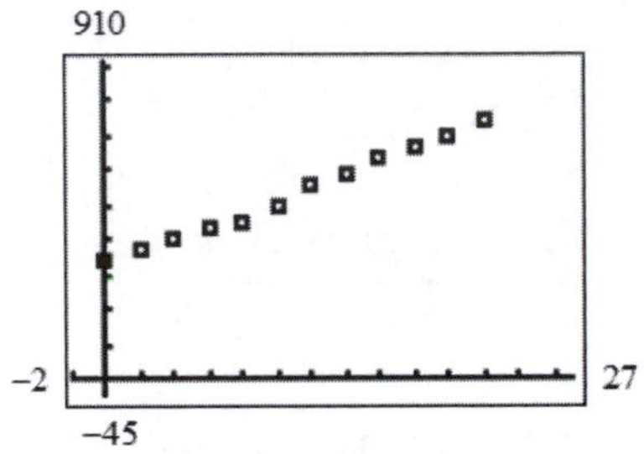


**b.** $y = 342.8(1.037)^x$

**c.**

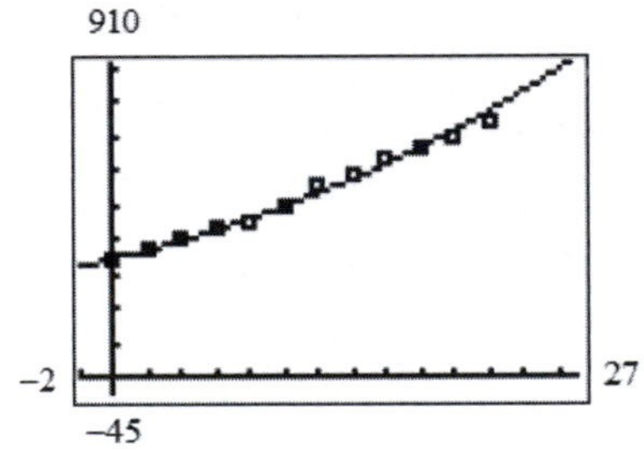

**45. a.** Letting $x = 0$ correspond to the year 1960, enter the data values $(0, 411.5)$, $(10, 838.8)$, etc. and obtain the exponential regression equation $y = 492.37(1.070)^x$

**b.** $x = 58$ in 2018, so $y = 492.37(1.070)^{58} \approx 24920$ If model is accurate, the total U.S. personal income in 2018 will be $24,920 billion which is an overestimate.

**c.** Use a graph of $y = 492.37(1.0697)^x$ and $y = 30000$. The intersection point is the point where the income will be $30 trillion (or $30,000 billion). Thus the model predicts that the total personal income will be $30 trillion when $x \approx 61$, or in 2021.

**47. a.** $y = 92.750(1.0275)^x$

**b.**

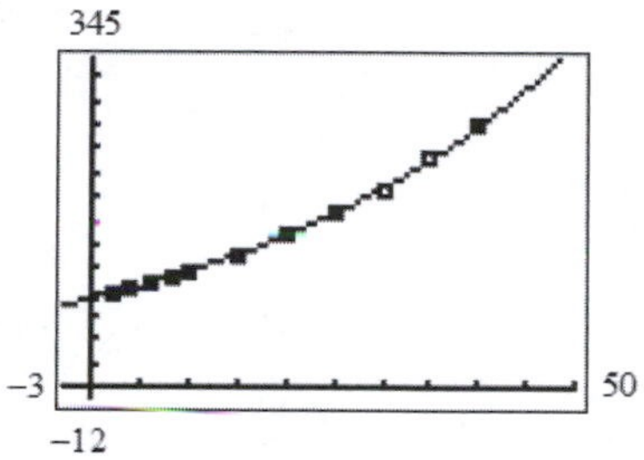


**c.** For the year 2038, $x = 28$. Substituting 28 into the unrounded function from a calculator gives $y \approx 198.23$. The model predicts a CPI of 198.23 in 2038.

**d.** Use a graph of the unrounded version of $y = 92.750(1.0275)^x$ and $y = 250$. The intersection point is the point where the CPI will be 250. Thus the model predicts that the CPI will be 250 when $x \approx 36.6$, or during 2047.

**49. a.** $y = 4.10(1.02)^x$

**b.** This model is an exponential growth function. $C > 0$ and $a > 1$.

**c.**

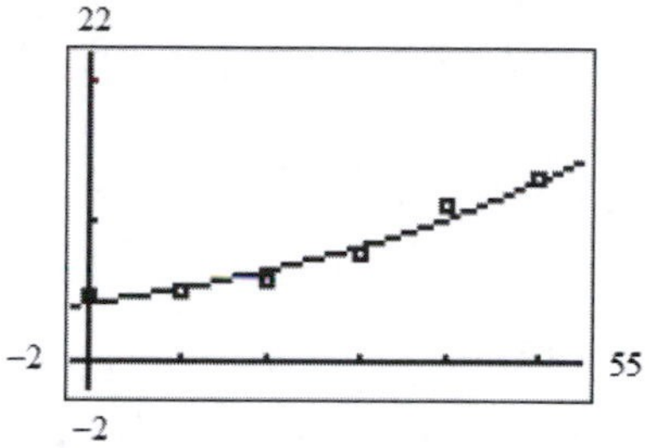


**d.** For the year 2060, $x = 60$. Substituting 60 into the unrounded function from a calculator gives $y \approx 16.13$. The model projects that 16.1 million Americans will have Alzheimer's in 2060.

## *Exercises 5.2*

NOTE: $a^y = x$ means $y = \log_a x$.

**1.** $4 = \log_2 16$

$y = 4, a = 2, x = 16$

$2^4 = 16$

**3.** $\frac{1}{2} = \log_4 2$

$y = \frac{1}{2}, a = 4, x = 2$

$4^{1/2} = 2$

**5.** $\log_3 x = 4$

$x = 3^4 = 81$

**7.** $\log_{16} x = -\frac{1}{2}$

$x = 16^{-\frac{1}{2}}$

$x = (2^4)^{-\frac{1}{2}}$

$x = 2^{-2}$

$x = \frac{1}{4}$

**9.** $\log_7(3x + 1) = 2$

$3x + 1 = 7^2$

$3x = 49 - 1$

$x = \frac{48}{3}$

$x = 16$

**11.** $\log(4x-7)=2$

$$4x-7=10^2$$
$$4x=100-7$$
$$x=\frac{93}{4}$$
$$x=23.25$$

**13.** $\log(2x+5)=2.2$

$$2x+5=e^{2.2}$$
$$2x=2.71828^{2.2}-5$$
$$2x=9.025-5$$
$$x=\frac{4.025}{2}$$
$$x=2.0125$$
$$x\approx 2.013$$

**15.** $2^5=32$

$y=5, a=2, x=32$

$5=\log_2 32$

**17.** $4^{-1}=\frac{1}{4}$

$y=-1, a=4, x=\frac{1}{4}$

$-1=\log_4\left(\frac{1}{4}\right)$

**19.** $3x+5=\ln(0.55)$

$$3x=\ln(0.55)-5$$
$$x=\frac{\ln(0.55)-5}{3}\approx -1.866$$

**21.**

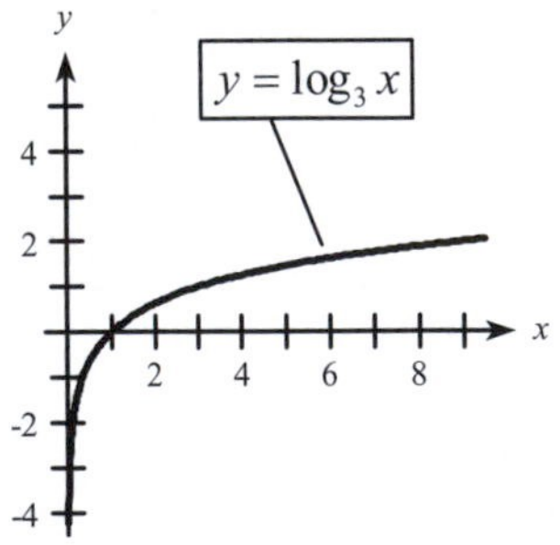


**23.**

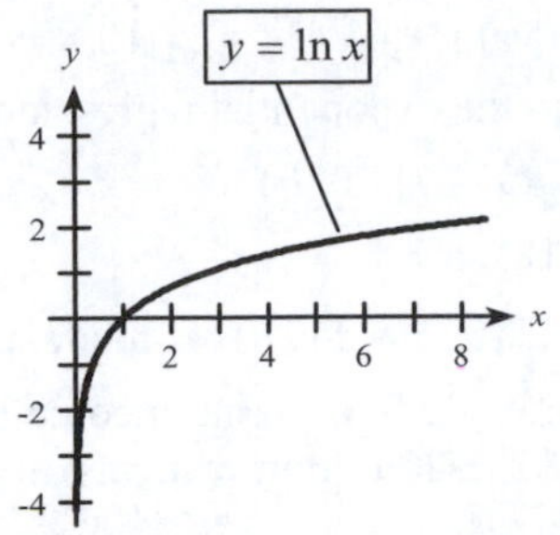


**25.**

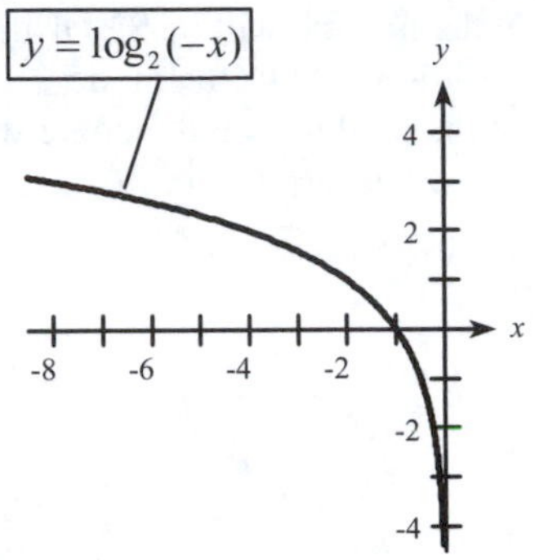


**27.** **a.** $\log_3 27=\log_3 3^3=3\log_3 3=3\cdot 1=3$

**b.** $\log_5\left(\frac{1}{5}\right)=\log_5 5^{-1}=-1\log_5 5=-1\cdot 1=-1$

**29.** $f(x)=\ln(x)$

$f\left(e^x\right)=\ln\left(e^x\right)=x\ln e=x\cdot 1=x$

**31.** If $f(x)=\log(x)$ then

$f\left(10^{-7}\right)=\log\left(10^{-7}\right)=-7$

**33.** $f(x)=e^x$

$f(\ln 3)=e^{\ln 3}=3 \quad (e^{\ln 3}=e^{\log_e 3})$

**35.** **a.** $\log_a(xy)=\log_a x+\log_a y=3.1+1.8=4.9$

**b.** $\log_a\left(\frac{x}{z}\right)=\log_a x-\log_a z=3.1-2.7=0.4$

**c.** $\log_a x^4=4\log_a x=4(3.1)=12.4$

**d.** $\log_a\sqrt{y}=\frac{1}{2}\log_a y=\frac{1}{2}(1.8)=0.9$

**37.** $\log\left(\frac{x}{x+1}\right)=\log x-\log(x+1)$

**39.** $\log_7 x\sqrt[3]{x+4} = \log_7 x + \log_7 (x+4)^{1/3}$
$= \log_7 x + \frac{1}{3}\log_7 (x+4)$

**41.** $\ln x - \ln y = \ln\left(\frac{x}{y}\right)$

**43.** $\log_5 (x+1) + \frac{1}{2}\log_5 x = \log_5 (x+1) + \log_5 x^{1/2}$
$= \log_5 \left[x^{1/2}(x+1)\right]$

**45.** **a.** $\ln(4\cdot 6)^{1/2} = 0.7630$ i.e. equivalent

**b.** $\frac{1}{2}(\ln 4 + \ln 6)$
The expressions are equivalent. Properties V and III are illustrated.

**47.** **a.** $\log\left(\frac{8}{5}\right)^{1/3} = \log(8 \div 5)^{1/3} \approx 0.06804$
i.e. not equivalent

**b.** $\frac{1}{3}\log 8 - \log 5 = \frac{1}{3}(0.9031) - 0.6990$
$\neq 0.06804$
The expressions are not equivalent.
Change (a) to $\log\frac{\sqrt[3]{8}}{5}$ to get (a) = (b).

**49.** **a.**

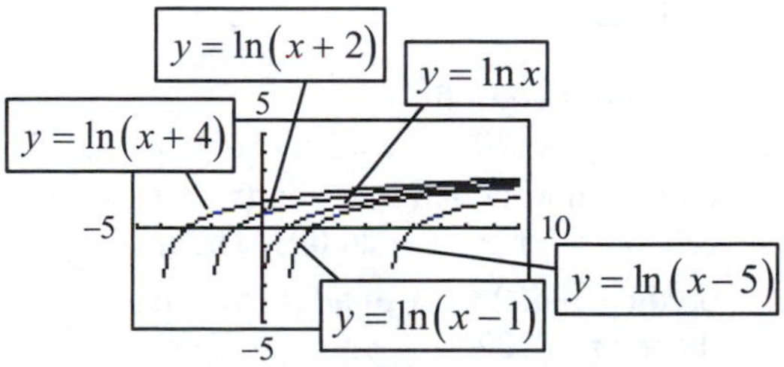


**b.** For each $c$, the domain is $x > c$ and the vertical asymptote is at $x = c$.

**c.** Each $x$-intercept is at $x = c + 1$.
$\ln(x-c) = \ln(c+1-c) = \ln 1 = 0$.

**d.** The graph of $f(x-c)$ is the graph of $f(x)$ shifted $c$ units on the $x$-axis.

**51.** **a.** $\log_2 17 = \frac{\ln 17}{\ln 2} = \frac{2.8332}{0.6931} \approx 4.0875$

**b.** $\log_5 (0.78) = \frac{\ln(0.78)}{\ln 5} = \frac{-0.2485}{1.6094} \approx -0.1544$

**53.** $y = \log_5 x$
$y = \frac{\ln x}{\ln 5} = \frac{1}{1.6094}\ln x$

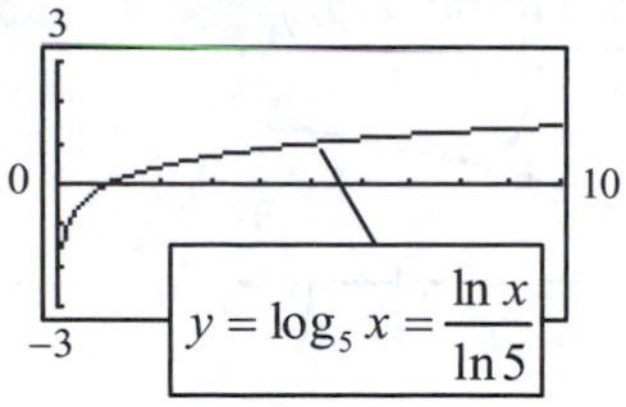


**55.** $y = \log_{13} x$ $\quad y = \frac{\ln x}{\ln 13} = \frac{1}{2.5649}\ln x$

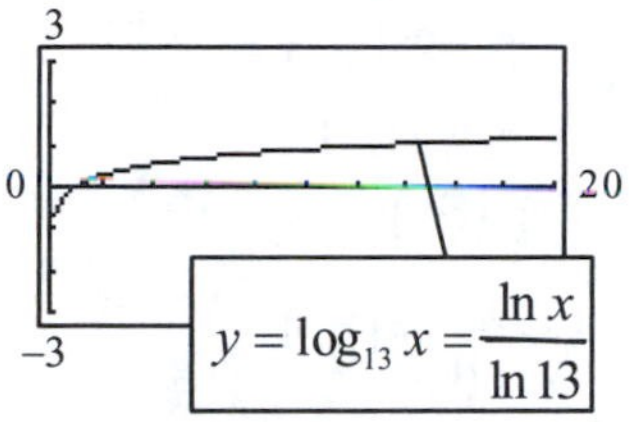


**57.** Let $u = \log_a M$ and $v = \log_a N$. Then, $M = a^u$ and $N = a^v$. So,
$\log_a\left(\frac{M}{N}\right) = \log_a\left(\frac{a^u}{a^v}\right) = \log_a(a^{u-v})$
$= (u-v)\log_a a$
$= u - v = \log_a M - \log_a N.$

**59.** $8.8 = \log\left(\frac{I}{I_0}\right)$ and $7.0 = \log\left(\frac{I}{I_0}\right)$
$10^{8.8} = \frac{I}{I_0}$ $\quad 10^{7.0} = \frac{I}{I_0}$
$\frac{10^{8.8}}{10^{7.0}} = 10^{1.8} \approx 63.1$ So the earthquake was approximately 63.1 times as severe.

**61.** Measurement is $7.8 - 7.3 = 0.5$ units larger. Intensity is $10^{0.5} \approx 3.2$ times greater.

**63.** $L = 10\log(10{,}000) = 10\cdot 4 = 40$

**65.** $L = 10\log\left(\frac{I}{I_0}\right)$

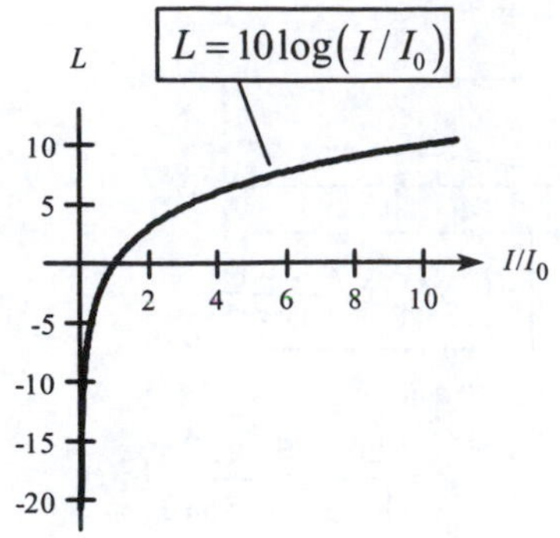


**67.** When pH = 1, $1 = -\log[H^+]$

$$-1 = \log[H^+]$$
$$10^{-1} = [H^+]$$
$$0.1 = \frac{1}{10} = [H^+]$$

When pH = 14, $[H^+] = 10^{-14}$

**69.** $\text{pH} = \log\left(\frac{1}{[H^+]}\right)$

Since $\log\left(\frac{1}{[H^+]}\right) = \log 1 - \log[H^+]$ and

$\log 1 = 0$,

we have $\text{pH} = -\log[H^+]$.

**71.**
$$2 = \left(1 + \frac{8}{100(4)}\right)^{4t}$$
$$2 = (1.02)^{4t}$$
$$\log_{1.02} 2 = 4t$$
$$\frac{\ln 2}{\ln 1.02} = 4t$$
$$t = \frac{\ln 2}{4\ln 1.02} \approx 8.75 \text{ years}$$

**73. a.**

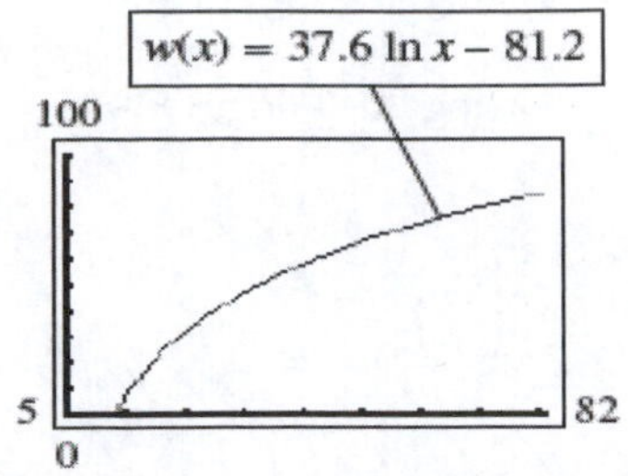


**b.** In 2030, $x = 80$, so

$w(80) = 37.6\ln(80) - 81.2 \approx 83.6$ million

In 2030, 83.6 million women will be in the workforce.

**c.** Graph $y_1 = 37.6\ln(80) - 81.2$ and $y_2 = 80$ and

find where they intersect. The point of intersection is (72.24, 80).

$72.24 \approx 73$, which corresponds to the year 2023.

The number of women in the workforce reached 80 million in 2023.

**75. a.** $y = -13.0 + 11.9\ln x$

**b.**

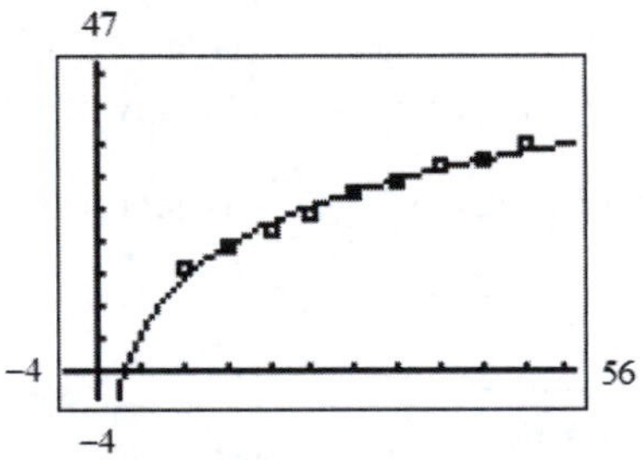


There is a good fit.

**c.** For the year 2027, $x = 27$. Substituting 27 into the unrounded function from a calculator gives $y \approx 26.09$. The model predicts that 26.1% of adults will have diabetes in 2027.

**77. a.** $y = 20.0 + 20.8\ln x$

**b.**

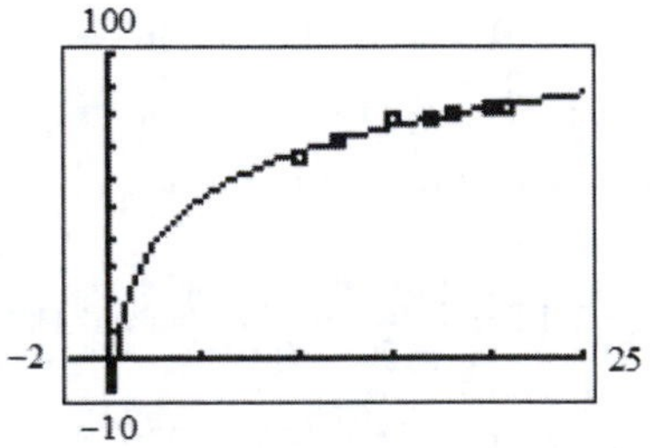


There is a good fit.

KISHWAUKEE COLLEGE BOOKSTORE
21193 MALTA RD. MALTA, IL 60150
815-825-9445
BOOKSTORE.KISH.EDU

Sale

Receipt: KCB-696080 — 001 002 Reg_4
Cashier: ENRIQUE — 08/21/18 14:49

---

| | | | Sale |
|---|---|---|---|
| 1 | GASKIN / MY ITLAB ACCESS CODE<br>10258217<br>978-0-13-449791-4 | Y | $119.25 |
| 1 | HARSHBARGER / SOLUTIONS MANUA<br>10242414<br>1-305-10806-X | Y | $63.95 |

Subtotal: $183.20

Tax:
STATE $11.45

Total: $194.65

Tender:

FINANCIAL AID $194.65
0197056

Change Due: $0.00

FULL REFUNDS ON FALL TEXTBOOKS WILL BE GIVEN THROUGH 08/31/18 WITH A COPY OF YOUR ORIGINAL RECEIPT AND CLASS SCHEDULE. NO RETURNS ON ELECTRONICS. REFUNDS WILL BE PROCESSED AS ORIGINAL TENDER GENERAL MERCHANDISE RETURNS & EXCHANGES MUST BE RETURNED WITHIN 10 BUSINESS DAYS, WITH THE RECEIPT. MERCHANDISE MUST STILL BE IN NEW CONDITION.

*KCB-696080*

c. For the year 2020, $x = 30$. Substituting 30 into the unrounded function from a calculator gives $y \approx 90.65$. The model predicts that 91% of the U.S. population will use the Internet in 2020.

## Exercises 5.3

**NOTE:** Remember $\ln e = 1$.

**1.**
$$8^{3x} = 32,768$$
$$\ln 8^{3x} = \ln 32768$$
$$3x \ln 8 = \ln 32768$$
$$x = \frac{\ln 32768}{3\ln 8} \approx 1.667$$

**3.**
$$0.13P = P(2^{-x})$$
$$\frac{0.13P}{P} = \frac{P(2^{-x})}{P}$$
$$0.13 = 2^{-x}$$
$$\ln 0.13 = \ln 2^{-x}$$
$$\ln 0.13 = -x\ln 2$$
$$x = \frac{\ln 0.13}{-\ln 2} \approx 2.943$$

**5.**
$$25000 = 10000(1.05)^{2x}$$
$$\frac{25000}{10000} = (1.05)^{2x}$$
$$2.5 = (1.05)^{2x}$$
$$\ln 2.5 = \ln(1.05)^{2x}$$
$$\ln 2.5 = 2x\ln 1.05$$
$$x = \frac{\ln 2.5}{\ln 1.05} \approx 9.3901$$

**7.**
$$10000 = 1500e^{0.10x}$$
$$\tfrac{20}{3} = e^{0.10x}$$
$$\ln\tfrac{20}{3} = \ln e^{0.10x}$$
$$\ln\tfrac{20}{3} = 0.10x$$
$$x = \frac{\ln\frac{20}{3}}{0.10} \approx 18.971$$

**9.**
$$78 = 100 - 100e^{-0.01x}$$
$$-22 = -100e^{-0.01x}$$
$$0.22 = e^{-0.01x}$$
$$\ln 0.22 = \ln e^{-0.01x}$$
$$\ln 0.22 = -0.01x$$
$$x = \frac{\ln 0.22}{-0.01} \approx 151.413$$

**11.**
$$55 = \frac{60}{1+5e^{-0.6x}}$$
$$55\left(1+5e^{-0.6x}\right) = 60$$
$$1+5e^{-0.6x} = \tfrac{12}{11}$$
$$5e^{-0.6x} = \tfrac{1}{11}$$
$$e^{-0.6x} = \tfrac{1}{55}$$
$$\ln e^{-0.6x} = \ln\tfrac{1}{55}$$
$$-0.6x = \ln\tfrac{1}{55}$$
$$x = \frac{\ln\frac{1}{55}}{-0.6} \approx 6.679$$

**13.**
$$\log x = 5$$
$$10^{\log x} = 10^5$$
$$x = 10^5 = 100,000$$

**15.**
$$\log_4(9x+1) = 3$$
$$4^{\log_4(9x+1)} = 4^3$$
$$9x+1 = 64$$
$$9x = 63$$
$$x = 7$$

**17.** $\ln x - \ln 5 = 10$

$\ln\left(\frac{x}{5}\right) = 10$

$e^{\ln\left(\frac{x}{5}\right)} = e^{10}$

$\frac{x}{5} = e^{10}$

$x = 5e^{10}$

**19.** $7 + \log(8x) = 25 - 2\log x$

$\log(8x) + 2\log x = 25 - 7$

$\log(8x) + \log(x^2) = 18$

$\log(8x^3) = 18$

$10^{\log 8x^3} = 10^{18}$

$8x^3 = 10^{18}$

$x^3 = \frac{10^{18}}{8}$

$x = \sqrt[3]{\frac{10^{18}}{8}} = \frac{10^6}{2} = 5\cdot 10^5$

**21.** $\ln(x+2) + \ln x = \ln(x+12)$

$\ln(x(x+2)) = \ln(x+12)$

$x(x+2) = x+12$

$x^2 + 2x = x + 12$

$x^2 + x - 12 = 0$

$(x+4)(x-3) = 0$

$x = -4$ or $x = 3$

But $\ln(-4)$ is undefined, so the only solution is $x = 3$.

**23. a.**

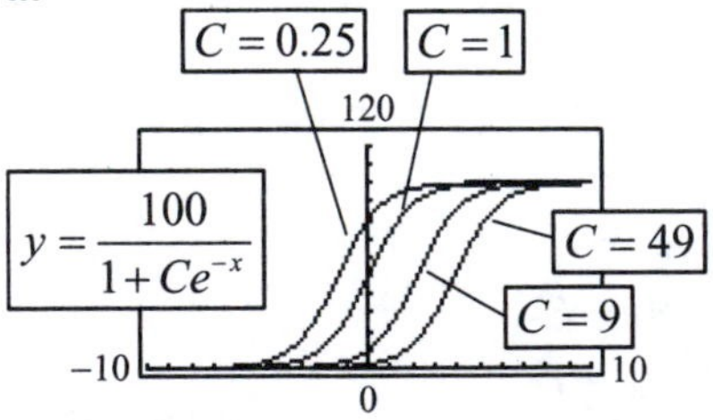


**b.** Different $c$ values change the $y$-intercept and how the graph approaches the asymptote.

**25.** $S = 50,000e^{-0.8x}$

**a.** $x = 4$

$S = 50,000e^{-3.2}$

$= 50,000(0.0408) = \$2038.11$

**b.** $S = 1000$

$1000 = 50,000e^{-0.8x}$

$e^{-0.8x} = 0.02$

$-0.8x = \ln 0.02$

$x = \frac{\ln 0.02}{-0.8} = \frac{-3.912}{-0.8} = 4.89$

In the fifth month sales will drop below \$1000.

**27. a.** $30,000 = 60,000e^{-0.05t}$

$e^{-0.05t} = \frac{30,000}{60,000}$

$e^{-0.05t} = 0.5$ take ln of both sides

$-0.05t \approx -0.6931$

$t = \frac{-0.6931}{-0.05} \approx 13.86$ years

**b.** $50,000 = A\cdot e^{-0.05(15)}$

$50,000 = Ae^{-0.75}$

$A = \frac{50,000}{e^{-0.75}} = \frac{50,000}{0.47237} \approx \$105,849.22$

**29.** $Q(t) = 100e^{-0.02828t}$

At $t = 0$ the initial amount is 100 gm. Thus,

$50 = 100e^{-0.02828t}$

$0.5 = e^{-0.02828t}$

$\ln 0.5 = -0.02828t$ take ln of both sides

$t = \frac{\ln 0.5}{-0.02828} = \frac{-0.6931}{-0.02828} = 24.5$ years

**31.** $y = P_0 e^{ht}$

$110,517 = 100,000e^{10h}$

$e^{10h} = 1.10517$

$10h = \ln 1.10517$ (Take ln of both sides.)

$10h = 0.099999$

$h = 0.009999 = 0.01$

2015 gives $t = 15$ and $P_0 = 110,517$.

So, $y = 110,517e^{0.01(15)} = 110,517e^{0.15}$

$= 110,517(1.161834) = 128,402$

**33.** $H = 2009e^{0.05194t}$

$4000 = 2009e^{0.05194t}$

$1.9910 = e^{0.05194t}$

$\ln 1.9910 = \ln e^{0.05194t}$

$\ln 1.9910 = 0.05194t$

$t = \dfrac{\ln 1.9910}{0.05194} \approx 13.3$, in the year 2019

**35.** $p = 100e^{-q/2}$

**a.** $q = 6$

$p = 100e^{-3} = 100(0.04978) = \$4.98$

**b.** $p = \$1.83$

$1.83 = 100e^{-q/2}$

$0.0183 = e^{-q/2}$

$-q/2 = \ln(0.0183)$

$q = -2(\ln 0.0183)$

$= -2(-4.000) = 8$ units

**37.** $p = \dfrac{100e^q}{q+1}$

$q = \dfrac{300}{100} = 3$

$p = \dfrac{100e^3}{4} = 25e^3 = \$502.14$

**39.** $C(x) = 2500\ln(2x+1) + 1500$

**a.** $C(80) = 2500\ln(2\cdot 80+1) + 1500$

$\approx \$14,203.51$

**b.** $16,000 = 2500\ln(2x+1) + 1500$

$14,500 = 2500\ln(2x+1)$

$5.8 = \ln(2x+1)$

$e^{5.8} = e^{\ln(2x+1)}$

$e^{5.8} = 2x+1$

$e^{5.8} - 1 = 2x$

$x = \dfrac{e^{5.8}-1}{2} \approx 164.6$

**41.** $166.5 = 96.12 + 17.43\ln x$

$70.38 = 17.43\ln x$

$\ln x = \dfrac{70.38}{17.43}$

$e^{\ln x} = e^{\frac{70.38}{17.43}}$

$x \approx 56.7$

$1970 + 56.7 = 2026.7$

The year in which the number of White non-Hispanics is expected to reach 166.5 million is 2027.

**43.** $S(t) = 8500e^{0.115t}$

**a.** $S(1.5) = 8500e^{0.1725} = 8500(1.1883)$

$= \$10,100.31$

**b.** $17,000 = 8500e^{0.115t}$

$2 = e^{0.115t}$

$0.115t = \ln 2$

$t = \dfrac{\ln 2}{0.115} = \dfrac{0.6931}{0.115} = 6.03$

The account will double in 6.03 years.

**45.** $S = 5000(1.0075)^t$

**a.** $t = 12$

$S = 5000(1.0075)^{12} = 5000(1.0938)$

$= \$5469.03$

**b.** $S = 10,000$

$10,000 = 5000(1.0075)^t$

$1.0075^t = 2$

$t \ln 1.0075 = \ln 2$

$$t = \frac{\ln 2}{\ln 1.0075} = \frac{0.6931}{0.00747}$$

$= 92.8$ months

Thus, $t = 7$ years, 9 months (approximately)

**47. a.** Profit

$= R(t) - C(t) = 21.4e^{0.131t} - 18.6e^{0.131t}$

$= 2.8e^{0.131t}$

$P(30) = 2.8e^{0.131(30)} \approx \$142.5$ million

**b.** $285 = 2.8e^{0.131t}$

$52.89 = e^{0.131t}$

$\ln 101.79 = 0.131t$

$$t = \frac{\ln 101.79}{0.131} \approx 35.3; \text{ the year 2026.}$$

It will only take 36 years after 1990 to double.

**49. a.** $P(18) = 1.078(1.028)^{-18} \approx 0.66$

This means that in 2028, \$1.00 is expected to purchase 66% of what it did in 2012.

**b.**

$0.48 = 1.078(1.028)^{-x}$

$$\frac{0.48}{1.078} = 1.028^{-x}$$

$$\ln\left(\frac{0.48}{1.078}\right) = \ln\left(1.028^{-x}\right)$$

$$\ln\left(\frac{0.48}{1.078}\right) = -x \ln 1.028$$

$$x = -\frac{\ln\left(\frac{0.48}{1.078}\right)}{\ln 1.028} \approx 29.3$$

$2010 + 29.3 = 2039.3$

The year when the purchasing power of a 2012 dollar is expected to reach \$0.48 is 2040.

**51.** $50 = 10 + 5\ln(3x+1)$

$40 = 5\ln(3x+1)$

$8 = \ln(3x+1)$

$e^8 = e^{\ln(3x+1)}$

$e^8 = 3x+1$

$e^8 - 1 = 3x$

$$x = \frac{e^8 - 1}{3} \approx 993.3$$

**53.** It will take 52.5 years before it will cost \$1 to purchase goods that cost \$0.25 in 1983. The year is 2013.

$N = 3000(0.2)^{0.6^t}$

**a.** $t = 0$ $\quad N = 3000(0.2)^1$

$= 600$

**b.** $t = 3$ $\quad N = 3000(0.2)^{0.6^t}$

$= 3000(0.2)^{0.216}$

$= 3000(0.706354)$

$= 2119$

**c.** The largest value of $(0.2)^{0.6^t}$ occurs when $0.6^t = 0$. Then, $(0.2)^0 = 1$ and the maximum sales are 3000.

**d.**

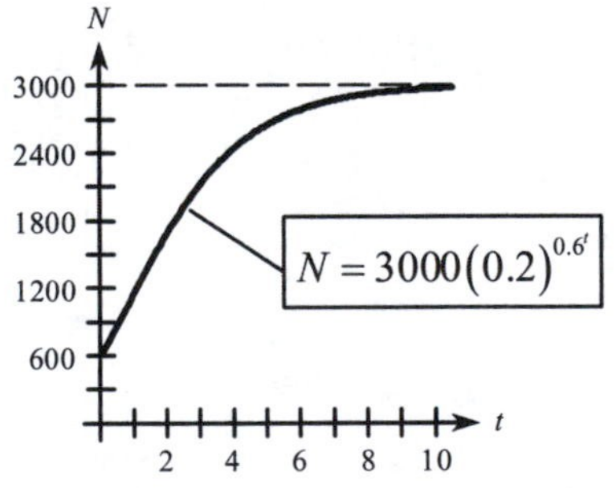

**55.** $N = 500(0.02)^{0.7^t}$

**a.** If $t = 0$, $N = 500(0.02) = 10$ employees when company opens.

**b.** $N = 100$

$$100 = 500(0.02)^{0.7^t}$$
$$(0.02)^{0.7^t} = 0.2$$
$$0.7^t \log 0.02 = \log 0.2$$
$$0.7^t = \frac{\log 0.2}{\log 0.02} = \frac{-0.69897}{-1.69897} = 0.41141$$
$$t \log 0.7 = \log 0.41141$$
$$t = \frac{\log 0.41141}{\log 0.7} = \frac{-0.38573}{-0.15490} = 2.49$$

At least 100 employees are working $2\frac{1}{2}$ years later.

**57.** $y = 100\left(1 - e^{-0.462t}\right)$

**a.** $t = 1$

$$y = 100\left(1 - e^{-0.462}\right) = 100(1 - 0.6300) = 37$$

**b.** $y = 50$

$$50 = 100\left(1 - e^{-0.462t}\right)$$
$$1 - e^{-0.462t} = 0.5$$
$$e^{-0.462t} = 0.5$$
$$-0.462t = \ln 0.5 = -0.6931$$
$$t = \frac{-0.6931}{-0.462} = 1.5 \text{ hours}$$

**59.** $y = \dfrac{10{,}000}{1 + 9999e^{-0.99t}}$

**a.** $t = 4$

$$y = \frac{10{,}000}{1 + 9999e^{-3.96}} = \frac{10{,}000}{1 + 190.61} \approx 52$$

**b.** $y = 5000$

$$5000 = \frac{10{,}000}{1 + 9999e^{-0.99t}}$$
$$1 + 9999e^{-0.99t} = 2$$
$$e^{-0.99t} = \frac{1}{9999}$$
$$-0.99t = \ln 1 - \ln 9999$$
$$t = \frac{-\ln 9999}{-0.99} = \frac{9.2102}{0.99} = 9.30$$

School will close during the tenth day.

**61.** $y = 40 - 40e^{-0.05t}$

**a.** $t = 1$

$$y = 40 - 40e^{-0.05} = 40 - 40(0.95123)$$
$$\approx 1.95\% \approx 2\%$$

**b.** $y = 25$

$$25 = 40 - 40e^{-0.05t}$$
$$40e^{-0.05t} = 15$$
$$e^{-0.05t} = \frac{15}{40} = 0.375$$
$$t = \frac{\ln 0.375}{-0.05} = \frac{-0.9808}{-0.05}$$
$$t = 19.6 \text{ or } 20 \text{ months}$$

**63.** $x = 0.05 + 0.18e^{-0.38t}$

**a.** $t = 0 \quad x = 0.05 + 0.18(1) = 0.23\text{km}^3$

**b.** 30% of 0.23 = 0.069.
So, for $x = 0.069$ we have

$$0.069 = 0.05 + 0.18e^{-0.38t}$$
$$e^{-0.38t} = \frac{0.069 - 0.05}{0.18} = 0.10556$$
$$t = \frac{\ln 0.10556}{-0.38} = \frac{-2.2485}{-0.38} = 5.9 \text{ years}$$

**65.** $L = \dfrac{197}{1+3.6e^{-0.037t}}$,

$t$ = number of years after 1940

**a.** $L = \dfrac{197}{1+3.6e^{-0.037\times 75}} \approx 160.89$ millon

**b.** $170 = \dfrac{197}{1+3.6e^{-0.037t}}$

$$170\left(1+3.6e^{-0.037t}\right) = 197$$
$$1+3.6e^{-0.037t} = 1.1588$$
$$3.6e^{-0.037t} = 0.1588$$
$$e^{-0.037t} = 0.04411$$
$$\ln e^{-0.037t} = \ln 0.04411$$
$$-0.037t = \ln 0.04411$$
$$t = \frac{\ln 0.04411}{-0.037} \approx 84.35$$
$$= \text{the year 2025}$$

**67. a.** $y(x) = \dfrac{120.}{1+5.25e^{-0.0637x}}$

**b.** $y(28) = \dfrac{120.}{1+5.25e^{-0.0637(28)}} \approx 63.9$ million

**c.** 120 million

**d.**

$$80 = \frac{120.}{1+5.25e^{-0.0637x}}$$
$$120 = 80\left(1+5.25e^{-0.0637x}\right)$$
$$1.5 = 1+5.25e^{-0.0637x}$$
$$0.5 = 5.25e^{-0.0637x}$$
$$e^{-0.0637x} = \frac{0.5}{5.25}$$
$$-0.0637x = \ln\left(\frac{0.5}{5.25}\right)$$
$$x = \frac{\ln\left(\frac{0.5}{5.25}\right)}{-0.637} \approx 36.9;\ \text{in 2037}$$

# Chapter 5: Exponential and Logarithmic Functions

## *Chapter 5 Review Exercises*

**1.** **a.** $2^x = y \to x = \log_2 y$

**b.** $3^y = 2x \to y = \log_3(2x)$

**2.** **a.** $\log_7\left(\frac{1}{49}\right) = -2 \to 7^{-2} = \frac{1}{49}$

**b.** $\log_4 x = -1 \to 4^{-1} = x$

**3.**

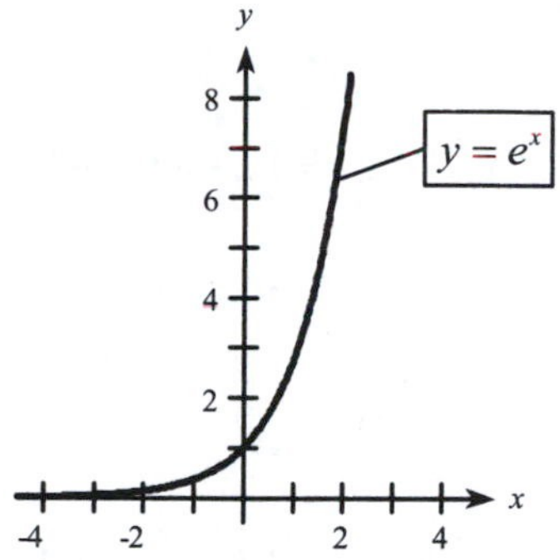


**4.**

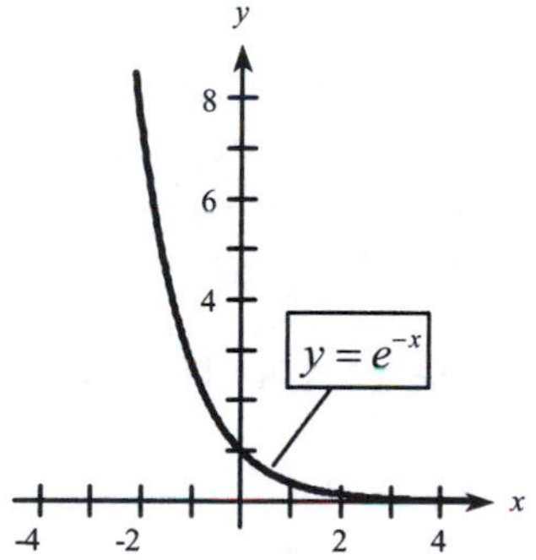


**5.**

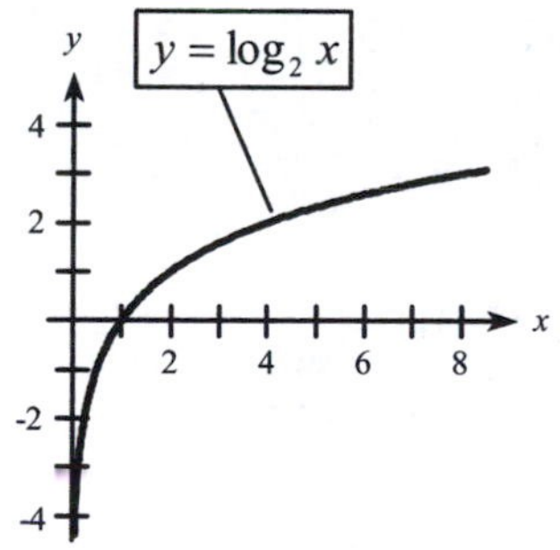


**6.**

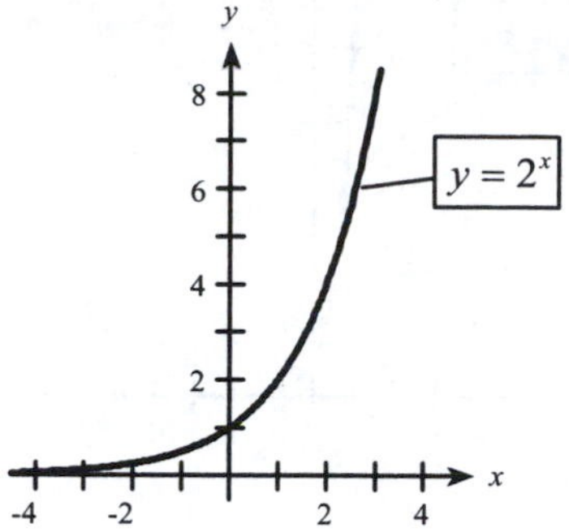


**7.**

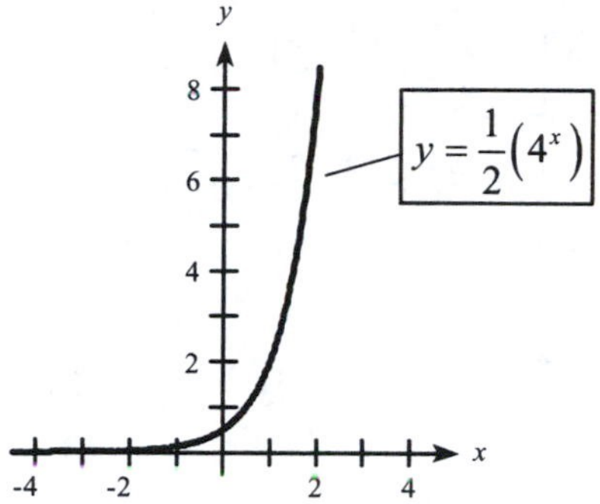


**8.**

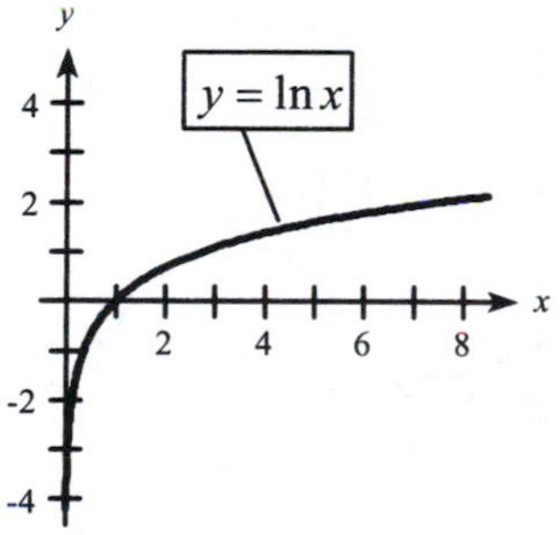


**9.**

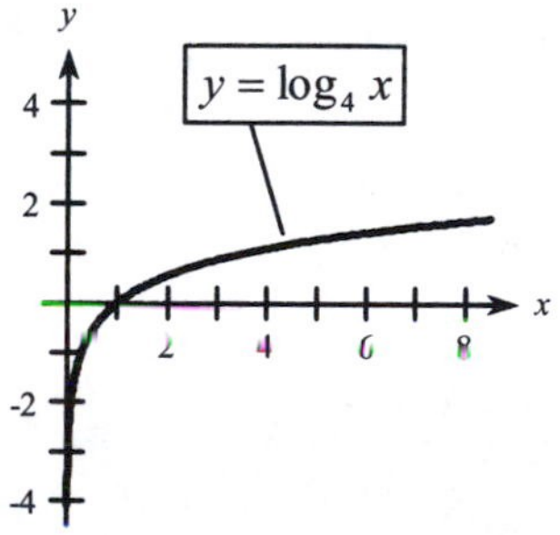

**10.**

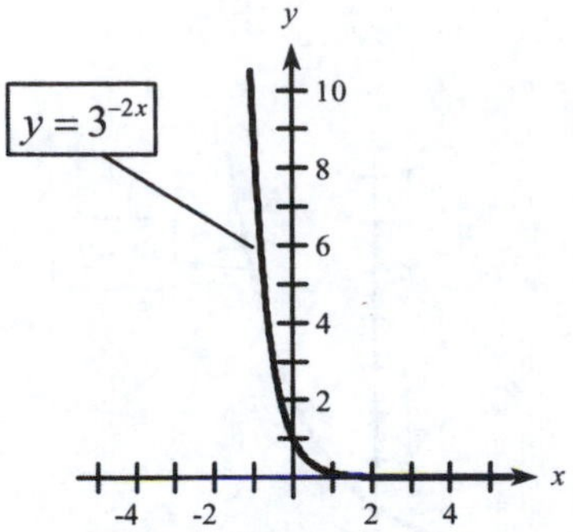


**11.**

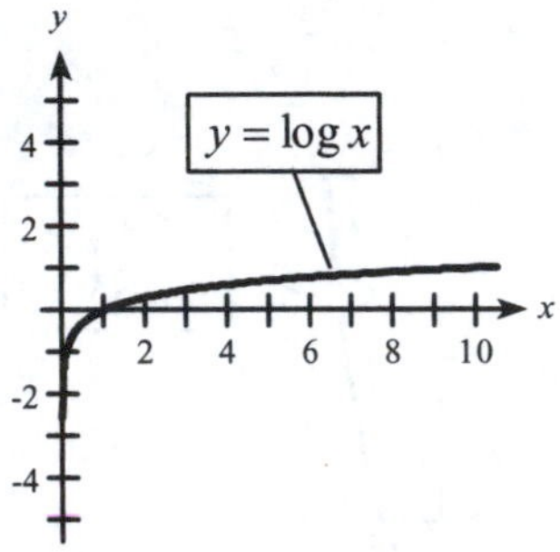


**12.**

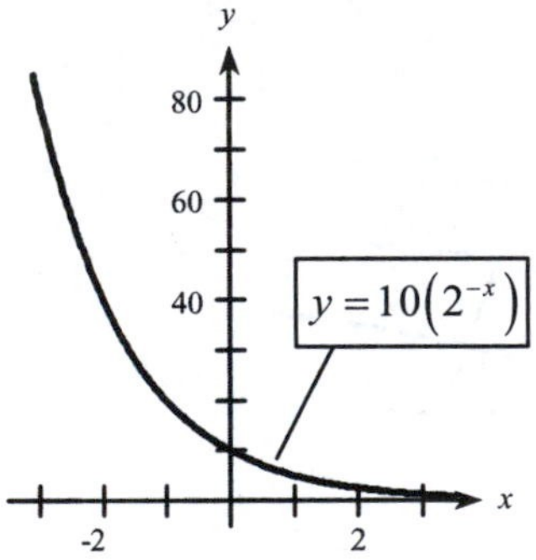


**13.** $\log_5 1 = 0$

$$\log_5 1 = \frac{\log 1}{\log 5} = \frac{0}{\log 5} = 0$$

**14.** $\log_2 8 = \log_2 2^3 = 3$

$$\log_2 8 = \frac{\log 8}{\log 2} = \frac{3\log 2}{\log 2} = 3$$

**15.** $\log_{25} 5 = \log_{25}(25)^{1/2} = \frac{1}{2}$

**16.** $\log_3\left(\frac{1}{3}\right) = \log_3 3^{-1} = -1$

**17.** $\log_3 3^8 = 8\log_3 3 = 8\cdot 1 = 8$

**18.** $\ln e = \ln e^1 = 1$

**19.** $e^{\ln 5} = 5$

**20.** $10^{\log 3.15} = 3.15$

**21.** $\log_a\left(\frac{x}{y}\right) = \log_a x - \log_a y = 1.2 - 3.9 = -2.7$

**22.** $\log_a x = 1.2$

$$\log_a \sqrt{x} = \frac{1}{2}\log_a x = \frac{1}{2}(1.2) = 0.6$$

**23.** $\log_a(xy) = \log_a x + \log_a y = 1.2 + 3.9 = 5.1$

**24.** $\log_a y = 3.9$

$$\log_a(y^4) = 4\log_a y = 4(3.9) = 15.6$$

**25.** $\log(yz) = \log y + \log z$

**26.** $\ln\sqrt{\frac{x+1}{x}} = \frac{1}{2}\ln\frac{x+1}{x} = \frac{1}{2}\left[\ln(x+1) - \ln x\right]$

$$= \frac{1}{2}\ln(x+1) - \frac{1}{2}\ln x$$

**27.** No, Let $x = y = 1$. Then $\ln 1 + \ln 1 = 0 + 0 = 0$ and $\ln(1+1) = \ln 2 \approx 0.6931$.

**28.** $f(e^{-2}) = \ln\left(e^{-2}\right) = -2$

**29.** $f(x) = 2^x + \log(7x-4)$

$$f(2) = 2^2 + \log(14-4) = 4 + 1 = 5$$

**30.** $f(0) = e^0 + \ln 1 = 1 + 0 = 1$

**31.** $f(x) = \ln\left(3e^x - 5\right)$

$$f(\ln 2) = \ln\left(3e^{\ln 2} - 5\right)$$
$$= \ln(3\cdot 2 - 5)$$
$$= \ln 1 = 0$$

**32.** $\log_9 2158 = \dfrac{\log 2158}{\log 9} \approx 3.494$

**33.** $\log_{12} 0.0195 = \dfrac{\ln 0.0195}{\ln 12} = \dfrac{-3.9373}{2.4849} = -1.5845$

**34.**

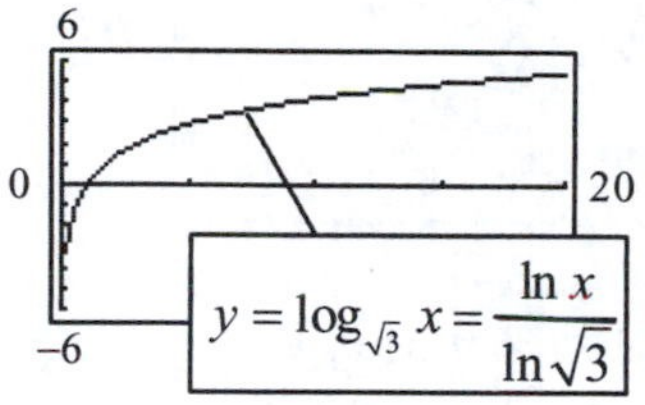


**35.** $f(x) = \log_{11}(2x-5)$

$$= \frac{\ln(2x-5)}{\ln 11} = \frac{\ln(2x-5)}{2.3979}$$

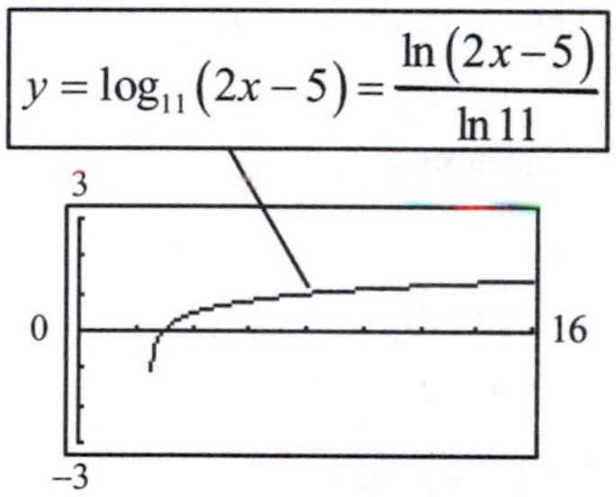


**36.**
$$6^{4x} = 46{,}656$$
$$\ln 6^{4x} = \ln 46656$$
$$4x \ln 6 = \ln 46656$$
$$x = \frac{\ln 46656}{4 \ln 6} = 1.5$$

**37.**
$$8000 = 250(1.07)^x$$
$$32 = 1.07^x$$
$$\ln 32 = \ln 1.07^x$$
$$\ln 32 = x \ln 1.07$$
$$x = \frac{\ln 32}{\ln 1.07} \approx 51.224$$

**38.**
$$11{,}000 = 45{,}000e^{-0.05x}$$
$$0.244 = e^{-0.05x}$$
$$\ln 0.244 = \ln e^{-0.05x}$$
$$\ln 0.244 = -0.05x$$
$$x = \frac{\ln 0.244}{-0.05} \approx 28.212$$

**39.**
$$312 = 300 + 300e^{-0.08x}$$
$$12 = 300e^{-0.08x}$$
$$0.04 = e^{-0.08x}$$
$$\ln 0.04 = \ln e^{-0.08x}$$
$$\ln 0.04 = -0.08x$$
$$x = \frac{\ln 0.04}{-0.08} \approx 40.236$$

**40.**
$$\log_3(4x-5) = 3$$
$$3^{\log_3(4x-5)} = 3^3$$
$$4x - 5 = 27$$
$$4x = 32$$
$$x = 8$$

**41.**
$$\ln(2x-1) - \ln 3 = \ln 9$$
$$\ln\left(\frac{2x-1}{3}\right) = \ln 9$$
$$\frac{2x-1}{3} = 9$$
$$2x - 1 = 27$$
$$2x = 28$$
$$x = 14$$

**42.**
$$3 + 2\log_2 x = \log_2(x+3) + 5$$
$$\log_2 x^2 - \log_2(x+3) = 2$$
$$\log_2\left(\frac{x^2}{x+3}\right) = 2$$
$$2^{\log_2\left(\frac{x^2}{x+3}\right)} = 2^2$$
$$\frac{x^2}{x+3} = 4$$
$$x^2 = 4x + 12$$
$$x^2 - 4x - 12 = 0$$
$$x = 6 \text{ or } x = -2$$

But a negative answer is invalid, so the solution is $x = 6$.

**43.** Growth exponential, because the general outline has the same shape as growth exponential.

**44. a.** $C(15) = 92.7e^{0.0271(15)} \approx 139.19$

This means that goods and services that cost \$100.00 in 2012 are expected to cost \$139.19 in 2025.

**b.**

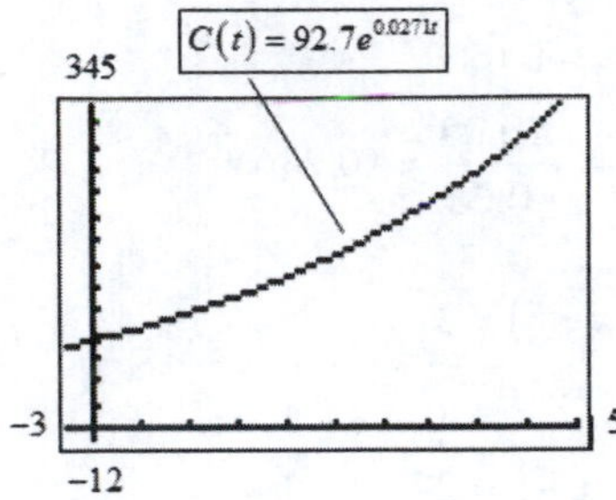


**45. a.** $P(20) = 60{,}000(0.97)^{20} \approx \$32{,}627.66$

**b.**

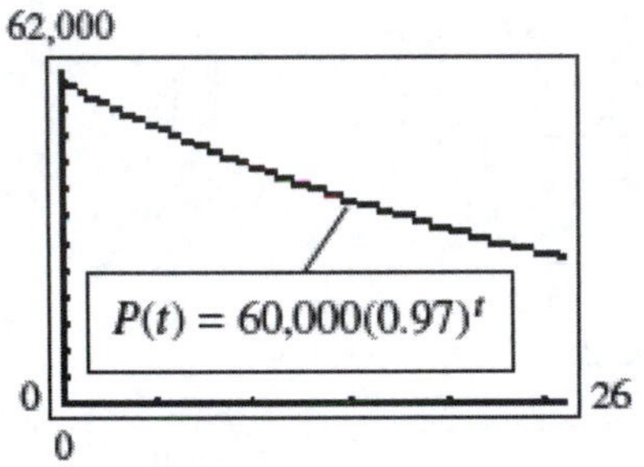


**46. a.** $y = 42.1(1.04^x)$

**b.**

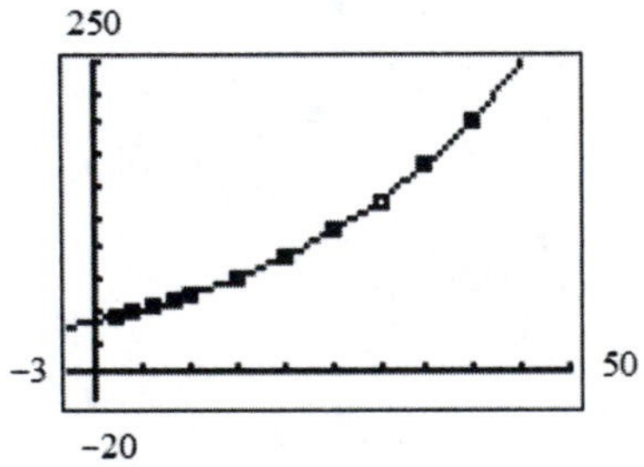


**c.** $y(44) \approx 239.2$; \$239.2 thousand

**47. a.** $y(38) = -4199.9 + 4436.3\ln(38)$

$\approx \$11{,}938$

**b.**

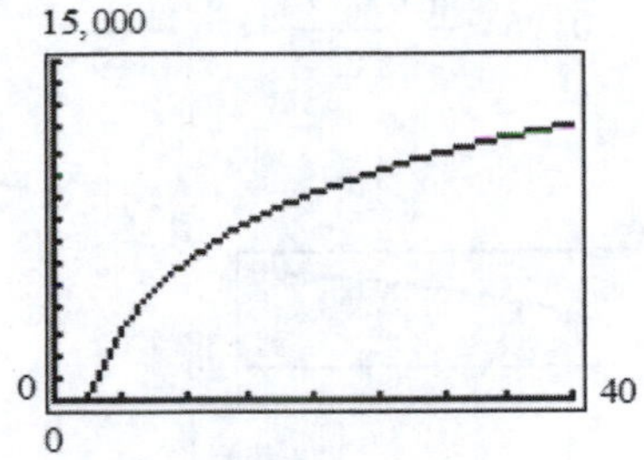


**48. a.** $y = -363.3 + 166.9\ln(x)$

**b.** $y(68) = -363.3 + 166.9\ln(68)$

$\approx 340.9$ million

**c.** $400 = -363.3 + 166.9\ln x$

$763.3 = 166.9\ln x$

$\dfrac{763.3}{166.9} = \ln x$

$x = e^{\frac{763.3}{166.9}} \approx 96.9$; in 2047

**49. a.** $M = -\dfrac{5}{2}\log\left(\dfrac{36.3B_0}{B_0}\right) = -\dfrac{5}{2}\log 36.3$

$= -\dfrac{5}{2}(1.5599) = -3.8998$

**b.** $2.1 = -\dfrac{5}{2}\log\left(\dfrac{B}{B_0}\right)$

$\log\dfrac{B}{B_0} = \dfrac{-4.2}{5} = -0.84$

$\dfrac{B}{B_0} = 10^{-0.84}$

$B = 10^{-0.84}B_0 = 0.14B_0$

**c.** $6 = -\dfrac{5}{2}\log\left(\dfrac{B}{B_0}\right)$

$\log\dfrac{B}{B_0} = \dfrac{-12}{5} = -2.4$

$\dfrac{B}{B_0} = 10^{-2.4}$

$B = 10^{-2.4}B_0 = 0.004B_0$

**d.** Yes

**50. a.** $S = 50{,}000e^{-0.1(6)} = 50{,}000(0.5488)$
$= \$27{,}440.58$

**b.** $15{,}000 = 50{,}000e^{-0.1x}$

$e^{-0.1x} = \dfrac{15{,}000}{50{,}000}$ Take ln of each side.

$-0.1x = \ln 15{,}000 - \ln 50{,}000$

$x = \dfrac{9.6158 - 10.8198}{-0.1} = 12$ weeks

**51.** $S = 50{,}000e^{-0.6x}$

$x = 6$

$S = 50{,}000e^{-3.6} = 50{,}000(0.02732) = \$1366.19$

**52.** $2000 = 1000(1.01)^{12t}$

$(1.01)^{12t} = 2$

$12t = \dfrac{\ln 2}{\ln 1.01} = 69.66$

$t = \dfrac{69.66}{12} = 5.8$ years

**53.** $S = 5000e^{0.135t}$

**a.** $t = 0.75$ 9 months

$S = 5000e^{0.10125} = 5000(1.10655)$
$= \$5532.77$

**b.** $S = 10{,}000$

$10{,}000 = 5000e^{0.135t}$

$2 = e^{0.135t}$

$t = \dfrac{\ln 2}{0.135} = \dfrac{0.6931}{0.135} = 5.13$ years

**54.** Logistic – begins like an exponential function, then grows at a slower rate.

**55.** $N = 10{,}000(0.3)^{0.5^t}$

**a.** $t = 0$

$N = 10{,}000(0.3)^1 = \$3000$

**b.** $t = 3$

$N = 10{,}000(0.3)^{0.125} = 10{,}000(0.86028)$
$\approx \$8603$

**c.** Sales are a maximum when $(0.3)^{0.5^t}$ is a maximum. Since 0.3 is between 0 and 1 we know that 0.3 to a positive power cannot be greater than 1. Thus, maximum sales will be \$10,000.

**56. a.**

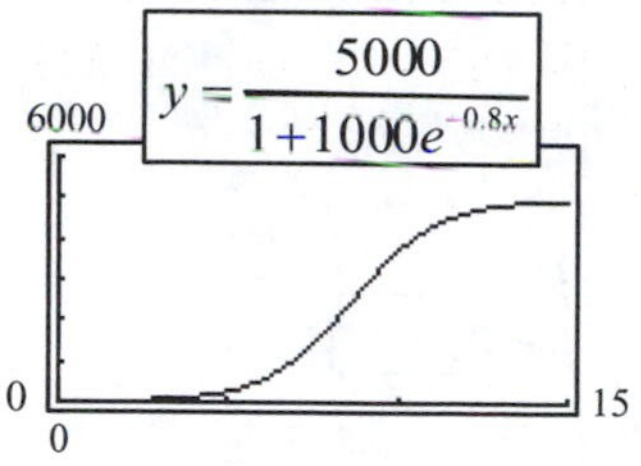


**b.** $x = 0$ when first discovered.

$y = \dfrac{5000}{1+1000e^{-0.8(0)}} = \dfrac{5000}{1001} = 4.995 \approx 5$

5 students had the virus when it was first discovered.

**c.** When $x = 15$,

$y = \dfrac{5000}{1+1000e^{-0.8(15)}} = 4969.4665$.

The total number infected by the virus during the first 15 days is 4970.

**d.** $3744 = \dfrac{5000}{1+1000e^{-0.8x}}$

$3744\left(1+1000e^{-0.8x}\right) = 5000$

$3744 + 3744000e^{-0.8x} = 5000$

$3744000e^{-0.8x} = 1256$

$e^{-0.8x} = \dfrac{1256}{3744000}$

$e^{-0.8x} \approx 0.0003547$

take ln of both sides

$-0.8x = \ln 0.0003547$

$-0.8x = -7.9442$

$x = \dfrac{-7.9442}{-0.8} \approx 9.9$

$= 10$ to the nearest whole number

In 10 days the number infected will reach 3744.

# Chapter 5: Exponential and Logarithmic Functions

***Chapter 5 Test***

1.

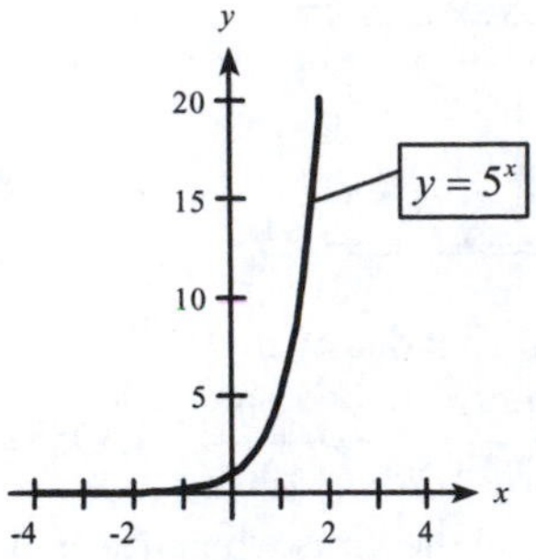


2.

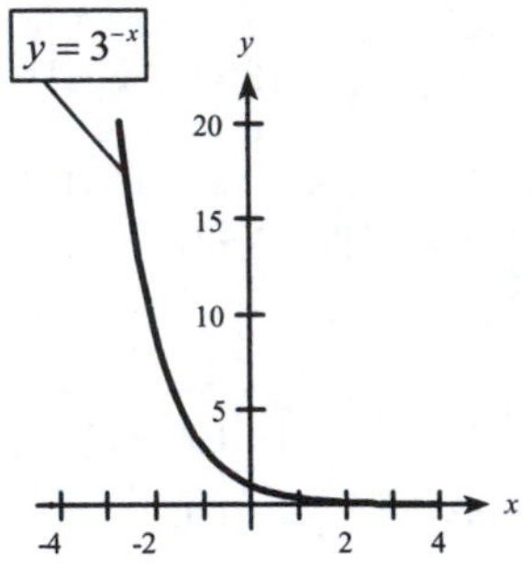


3.

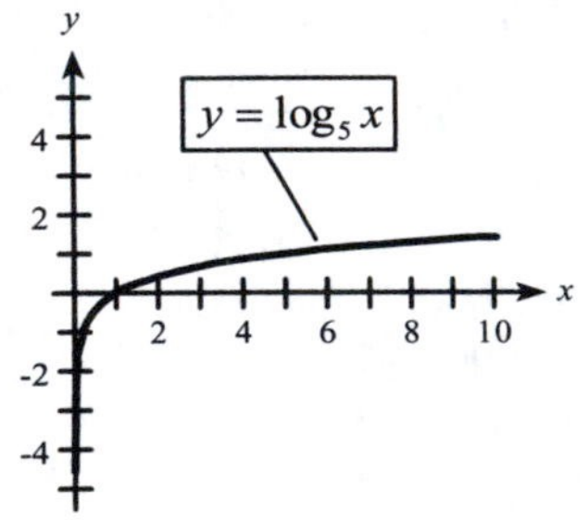


4.

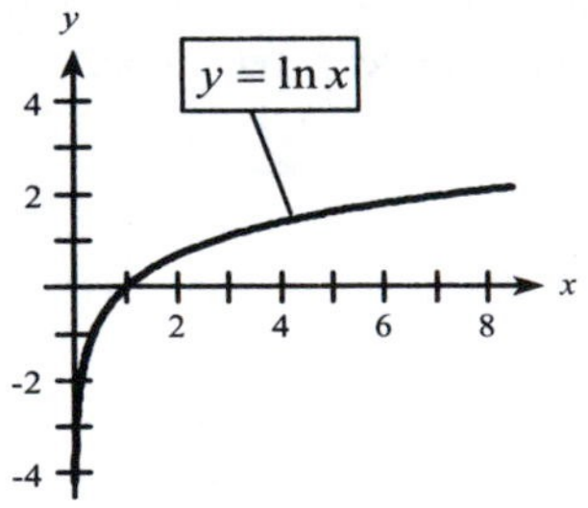


5.

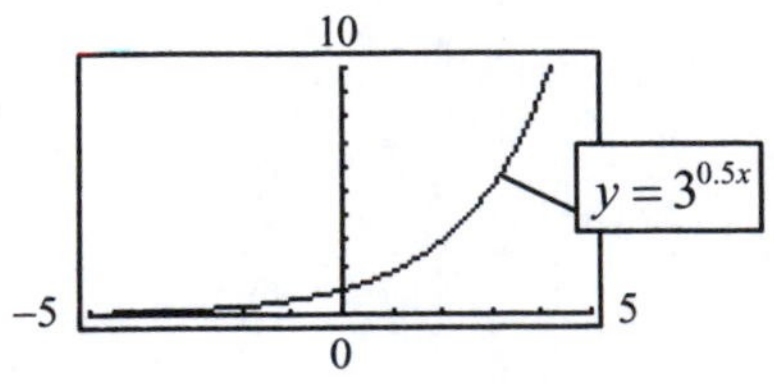


6.

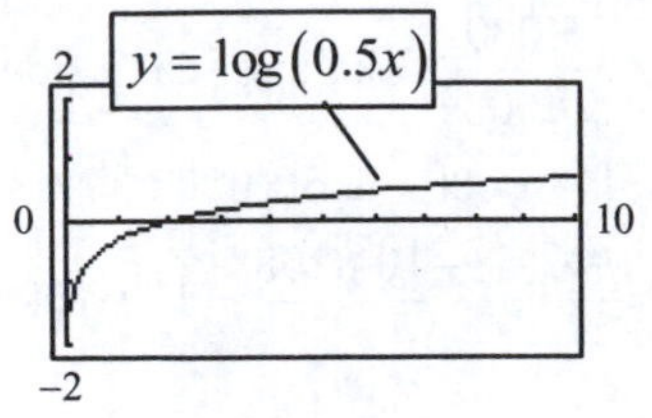


7.

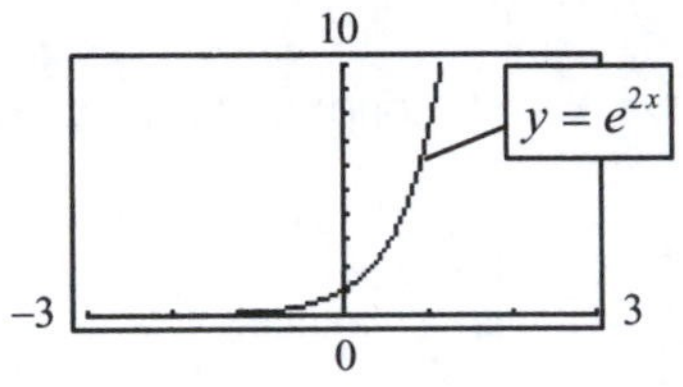


8.

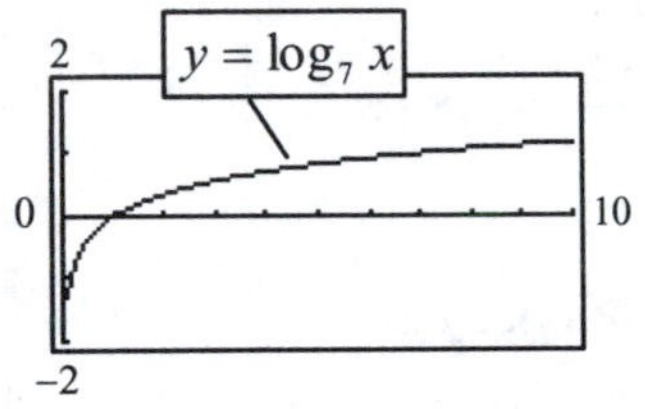


9. [2nd] [ln] 4 [=] 54.5982

10. 3 [^] [(−)] 2.1 [=] 0.100

11. [ln] 4 [=] 1.3863

12. [log] 21 [=] 1.3222

13. $\log_7 x = 3.1$

$x = 7^{3.1}$

$x \approx 416.681$

14. $3^{2x} = 27$

$2x = \log_3 27$

$x = \frac{1}{2}\log_3(3^3) = \frac{1}{2}\cdot 3 = \frac{3}{2} = 1.5$

**15.** $3+6e^{-2x}=7$

$6e^{-2x}=4$

$e^{-2x}=\frac{2}{3}$

$\ln e^{-2x}=\ln\left(\frac{2}{3}\right)$

$-2x=\ln\left(\frac{2}{3}\right)$

$x=-\frac{1}{2}\ln\left(\frac{2}{3}\right)\approx 0.203$

**16.** $8-\log_3(5x-13)=5$

$\log_3(5x-13)=3$

$3^{\log_3(5x-13)}=3^3$

$5x-13=27$

$5x=40$

$x=8$

**17.** $\log_2 8=\log_2 2^3=3\log_2 2=3\cdot 1=3$

**18.** $e^{\ln x^4}=x^4$ (Property II)

**19.** $\log_7 7^3=3$ (Property I)

**20.** $\ln e^{x^2}=x^2$ (Property I)

**21.** $\ln(M\cdot N)=\ln M+\ln N$

**22.** $\ln\left(\frac{x^3-1}{x+2}\right)=\ln\left(x^3-1\right)-\ln(x+2)$

**23.** $\log_4\left(x^3+1\right)=\frac{\ln(x^3+1)}{\ln 4}$ or $0.721\ln\left(x^3+1\right)$

**24.** $47,500=1500(1.015)^{6x}$

$31.667=(1.015)^{6x}$

$\ln 31.667=\ln 1.015^{6x}$

$\ln 31.667=6x\ln 1.015$

$x=\frac{\ln 31.667}{6\ln 1.015}\approx 38.679$

**25.** A decay exponential is more appropriate.

**26. a.** Decay exponential

**b.** In 2030, the expected tons of $CO_2$ emissions per person is 14.8 tons.

**27. a.** $H(13)=6791e^{0.04343(13)}$

$\approx \$11,943.51$

**b.** $23,887.02=6791e^{0.04343t}$

$3.51745=e^{0.04343t}$

$\ln 3.51745=\ln e^{0.04343t}$

$\ln 3.51745=0.04343t$

$t=\frac{\ln 3.51745}{0.04343}\approx 29.0$

$29.0-13=$ about 16 years (in 2034)

**28.** $2500=22,000e^{-0.35t}$

$0.1136=e^{-0.35t}$

$\ln 0.1136=\ln e^{-0.35t}$

$\ln 0.1136=-0.35t$

$t=\frac{\ln 0.1136}{-0.35}\approx 6.21$ months

**29. a.** $I(33)=\frac{70,290}{1+18.26e^{-0.058(30)}}$

$\approx \$16715.80$

$\approx \$16,716$ billion

**b.** $25,000=\frac{70290}{1+18.26e^{-0.0058t}}$

$25,000\left(1+18.26e^{-0.0058t}\right)=70290$

$1+18.26e^{-0.0058t}=2.8116$

$18.26e^{-0.0058t}=1.8116$

$e^{-0.0058t}=0.0992113$

$\ln e^{-0.0058t}=\ln 0.0992113$

$-0.058t=\ln 0.0992113$

$t=\frac{\ln 0.0992113}{-0.058}\approx 39.83$

In the year 2025.

**30. a.** $y = 63.31(1.048^x)$

**b.** $y = 63.31(1.048^{42}) \approx 461$; about 461,000

**c.**

$$1030 = 63.31(1.048^x)$$

$$\frac{1030}{63.31} = 1.048^x$$

$$\ln\left(\frac{1030}{63.31}\right) = \ln(1.048^x)$$

$$\ln\left(\frac{1030}{63.31}\right) = x\ln(1.048)$$

$$x = \frac{\ln\left(\frac{1030}{63.31}\right)}{\ln(1.048)} \approx 59.5; \text{ in } 2070$$

# Chapter 6: Mathematics of Finance

## Exercises 6.1

**1.** $250 = 1000(r)(4)$

$250 = 4000r$

$0.0625 = r$

$P = 1000,\ t = 4,\ I = 250$

**3.** $9600 = P + P(0.05)(4)$

$9600 = P + 0.2P$

$9600 = 1.2P$

$8000 = P$

$A = 9600,\ I = 1600,\ r = 0.05,\ t = 4$

**5.** $P = 10{,}000,\ t = 6,\ r = 0.16$

**a.** $I = Prt = 10{,}000(0.16)(6) = \$9600$

**b.** $S = P + I = 10{,}000 + 9600 = \$19{,}600$

**7.** $P = 1000,\ t = \frac{1}{4},\ r = 0.12$

**a.** $I = Prt = 1000(0.12)\left(\frac{1}{4}\right) = \$30$

**b.** $S = P + I = 1000 + 30 = \$1030$

**9.** $P = 800,\ t = \frac{1}{2},\ r = 0.16$

$I = Prt = 800(0.16)\left(\frac{1}{2}\right) = \$64$

$S = P + I = \$864$

**11.** $P = 3500,\ t = \frac{15}{12} = \frac{5}{4},\ r = 0.08$

$I = 3500(0.08)\left(\frac{5}{4}\right) = \$350$

$S = P + I = 3500 + 350 = \$3850$

**13.** $P = 30,\ t = 1,\ I = 0.90$

From $I = Prt$ we have

$r = \frac{I}{Pt} = \frac{0.90}{30(1)} = 0.03$. Thus, the couple earned 3% from the dividend

$P = 30,\ t = 1,\ I = 3.90$

So, $r = \frac{3.90}{30(1)} = 0.13$. The total rate earned on the investment is 13%.

**15.** $S = 10{,}000,\ P = 9750,\ t = \frac{1}{2}$

**a.** $I = S - P = 250$

$r = \frac{I}{Pt} = \frac{250}{(9750)\left(\frac{1}{2}\right)} = 0.051$

The interest rate is 5.1%.

**b.** $P = 9750 + 40 = \$9790$

$I = S - P = 210$

$r = \frac{210}{(9750)\left(\frac{1}{2}\right)} = 0.0429$

The interest rate is 4.29%.

**17.** $S = 12(140) = 1680,\ t = \frac{1}{4},\ r = 0.12$, find $P$

$S = P + I = P + Prt = P(1 + rt)$

Thus,

$P = \frac{S}{1+rt} = \frac{1680}{1+(0.12)\left(\frac{1}{4}\right)} = \frac{1680}{1.03} = \$1631.07$

The firm must deposit \$1631.07 now in order to pay the bill in 90 days.

**19.** $S = 13{,}500,\ P = ?\quad t = \frac{10}{12} = \frac{5}{6},\ r = 0.15$

$S = P + Prt = P + P(0.15)\left(\frac{5}{6}\right) = 1.125P$

$P = 13500 \div 1.125 = \$12{,}000$

**21.** $S = 9000\quad P = \$5000\quad t = ??\ r = 0.08$

$I = S - P = \$4000$

$4000 = 5000\ (0.08)(t)$

$t = \frac{4000}{5000(0.08)} = 10$ years

**23.** 45 day note: $I = 500{,}000(0.06)\left(\frac{45}{360}\right) = \$37500$

60 day note:

$I = 500{,}000(0.07)\left(\frac{60}{360}\right) = \$58333.33$

Penalty is \$5000. By paying on time the retailer will earn \$37500 versus \$58333.33 if he waits.

**25. a.** $P = 2000,\ t = \frac{9}{12},\ r = 0.08$

$S = P(1 + rt) =$

$2000\left[1+(0.08)\left(\frac{9}{12}\right)\right] = \$2120$

**b.** $S = 2120,\ t = \frac{3}{12},\ r = 0.10$

$$P = \frac{S}{1+rt} = \frac{2120}{1+\left(\frac{1}{10}\right)\left(\frac{3}{12}\right)} = \$2068.29$$

**27.**

| $n$ | 1 | 2 | 3 | 4 | 5 | 6 | 7 | 8 | 9 | 10 |
|---|---|---|---|---|---|---|---|---|---|---|
| $a_n$ | 3 | 6 | 9 | 12 | 15 | 18 | 21 | 24 | 27 | 30 |

**29.** $a_n = \frac{(-1)^n}{2n+1}$

| $n$ | 1 | 2 | 3 | 4 | 5 | 6 |
|---|---|---|---|---|---|---|
| $a_n$ | $\frac{-1}{3}$ | $\frac{1}{5}$ | $\frac{-1}{7}$ | $\frac{1}{9}$ | $-\frac{1}{11}$ | $\frac{1}{13}$ |

**31.** $a_n = \frac{n-4}{n(n+2)}$

$$a_1 = \frac{1-4}{1(1+2)} = -1 \quad a_2 = \frac{2-4}{2(2+2)} = -\frac{1}{4}$$

$$a_3 = \frac{3-4}{3(5)} = -\frac{1}{15} \quad a_4 = \frac{0}{4(6)} = 0$$

$$a_{10} = \frac{10-4}{10(12)} = \frac{1}{20}$$

**33.** **a.** $d = 3,\ a_1 = 2$

**b.** 11, 14, 17

**35.** **a.** $d = \frac{9}{2} - \frac{6}{2} = \frac{3}{2},\ a_1 = 3$

**b.** $\frac{15}{2}, 9, \frac{21}{2}$

**37.** $a_1 = 6,\ d = -\frac{1}{2},\ n = 83$

$$a_{83} = 6 + (83-1)\left(-\frac{1}{2}\right) = -35$$

**39.** $a_1 = 5,\ a_8 = 19,\ n = 100$

$$19 = 5 + (8-1)d \text{ or } 14 = 7d \text{ or } d = 2$$

$$a_{100} = 5 + (100-1)2 = 203$$

**41.** $a_1 = 2,\ d = 3,\ n = 38,\ a_{38} = 113$

$$s_{38} = \frac{38}{2}(2+113) = 2185$$

$a_n = a_1 + (n-1)d$
$113 = 2 + (38-1)d$
$3 = d$

**43.** $a_1 = 10, d = \frac{1}{2},\ n = 70$

$$a_{70} = 10 + (70-1)\left(\frac{1}{2}\right) = \frac{89}{2}$$

$$s_{70} = \frac{70}{2}\left(10 + \frac{89}{2}\right) = \frac{3815}{2} = 1907.5$$

**45.** $a_1 = 6, d = -\frac{3}{2}, n = 150$

$$a_{150} = 6 + (150-1)\left(-\frac{3}{2}\right) = -\frac{435}{2}$$

$$s_{150} = \frac{150}{2}\left(6 - \frac{435}{2}\right) = -\frac{31,725}{2} = -15862.5$$

**47.** Beginning with the third term and following terms: The value of the term is the sum of the values of the preceding two terms. 21, 34, 55

**49.** $a_1 = -4000, d = 800, n = 12,$ Find $a_{12}$

$$a_{12} = -4000 + (12-1)(800) = \$4800$$

**51.** $a_1 = 40,000, d = 2000, n = 10,$ Find $a_{10}$

$$a_{10} = 40,000 + (10-1)2000 = \$58,000$$

$$a_1 = 36,000, d = 2400, n = 10, \text{ Find } a_{10}$$

$$a_{10} = 36,000 + (10-1)2400 = \$57,600$$

Through 10 years of work, the \$40,000 beginning salary is better.

**53.** **a.** The total base salary for 3 years is \$120,000. The total salary received for the 3 years is \$126,000. Thus, the sum of the raises is \$6000.

**b.** The total salary received for 3 years is \$129,000. Thus, the sum of the raises is \$9000.

**c.** Plan II is better by a total of \$3000.

**d.** Total salary for Plan I for 5 years is

$$126,000 + 23,000 + 23,000$$
$$+24,000 + 24,000 = \$220,000$$

Total base salary is 5(40,000) = \$200,000. The sum of raises is \$20,000.

**e.** Total salary for Plan II for 5 years is:

$$129,000 + 23,600 + 24,200 + 24,800$$
$$+25,400 = \$227,000$$

Sum of the raises is \$27,000.

**f.** Plan II is better by \$7000.

**g.** Choose Plan II. As the years increase, its pay advantage also increases.

# Chapter 6: Mathematics of Finance

## *Exercises 6.2*

**1.** $S = 2000(1+0.02)^{24}$

$S = 3216.87$ (future value)

Periodic rate $= 0.02 = 2\%$

Number of periods $= 24$

Principal $= 2{,}000$

**3.** $25{,}000 = P(1+0.03)^{48}$

$\dfrac{25{,}000}{(1.03)^{48}} = P$

$P = 6049.97 =$ Principal

Periodic rate $= 0.03 = 3\%$

Number of periods $= 48$

Future value $= 25{,}000$

**5.** **a.** 8%

**b.** 7

**c.** $\dfrac{8\%}{4} = 2\% = 0.02$

**d.** $4 \cdot 7 = 28$

**7.** **a.** 9%

**b.** 5

**c.** $\dfrac{9\%}{12} = \dfrac{3}{4}\% = 0.0075$

**d.** $12 \cdot 5 = 60$

**9.** $P = 8{,}000,\ \dfrac{r}{m} = \dfrac{0.12}{1} = 0.12,\ mt = 1(10) = 10$

$S = 8000(1+0.12)^{10}$

$= 8{,}000(3.105848) = \$24{,}846.79$

**11.** $P = 3{,}200, \dfrac{r}{m} = \dfrac{0.08}{4} = 0.02, mt = 4(5) = 20$

$S = 3{,}200(1+0.02)^{20} = 3200(1.485947)$

$= \$4{,}755.03$

**13.** $S = 80{,}000, P = ??\ \dfrac{r}{m} = \dfrac{0.10}{12} = 0.008333$

$mt = 12(18) = 216$

$P = \dfrac{80{,}000}{(1+0.008333)^{216}} = \$13{,}323.86$

**15.** $S = 10{,}000,\ \dfrac{r}{m} = \dfrac{0.06}{1} = 0.06$

$mt = 1(10) = 10$

$P = \dfrac{10{,}000}{(1+0.06)^{10}} = \$5{,}583.95$

**17.** $P = 5100,\ r = 0.09,\ t = 4,\ rt = 0.36$

$S = 5100e^{0.36}$

$= 5100(1.433329) = \$7309.98$

**19.** $P = 410, r = 0.08, t = 10, rt = 0.8$

$S = 410e^{0.8} = 410(2.225541) = \$912.47$

$I = S - P = \$502.47$

**21.** $S = 100{,}000,\ P = ??,\ t = 20$

**a.** $r = 0.105,\ rt = 2.1$

$P = \dfrac{S}{e^{rt}} = \dfrac{100{,}000}{e^{2.1}} = \dfrac{100{,}000}{8.16617} = \$12{,}245.64$

**b.** $r = 0.11,\ rt = 2.2$

$P = \dfrac{S}{e^{rt}} = \dfrac{100{,}000}{e^{2.2}} = \dfrac{100{,}000}{9.02501} = \$11{,}080.32$

**c.** A $\dfrac{1}{2}\%$ increase in the interest rate reduces the investment by \$1165.32

**23.** $P = 1000,\ \dfrac{r}{m} = 0.08,\ mt = 5$

$S = 1000(1+0.08)^5$

$= 1000(1.469328) = \$1469.33$

$P = 1000,\ r = 0.07,\ t = 5,\ rt = 0.35$

$S = 1000e^{0.35} = 1000(1.419068) = \$1419.07$

The 8% compounded annually yields \$50.26 more interest.

**25.** **a.** $APY = \left(1 + \dfrac{0.073}{12}\right)^{12} - 1$

$= 1.0755 - 1 = 7.55\%$

**b.** $S = Pe^{0.06} = P(1.0618)$

$APY = 1.0618 - 1 = 0.0618 = 6.18\%$

**27.** The highest yield occurs with the most compounding periods. Rank: 8% compounded monthly, 8% compounded quarterly, 8% compounded annually.

**29.** $S_1 = Pe^r = Pe^{0.082} = P(1.08546)$

$S_2 = P\left(1+\frac{0.084}{4}\right)^4 = P(1.08668)$

$APY_1 = 1.08546 - 1 = 0.08546 = 8.546\%$

$APY_2 = 1.08668 - 1 = 0.08668 = 8.668\%$

The 8.4% investment is better.

**31.** The higher graph represents the continuous compounding since it has the higher (its effective annual rate) yield.

**33.** $S = \$600,000,\ P = \$10,000,\ n = 13,\ i = ?$

$S = P(1+i)^n$

$600,000 = 10,000(1+i)^{13}$

$60 = (1+i)^{13}$

$60^{1/13} = 1+i$

$i \approx 1.3702 - 1 \approx 0.3702 \approx 37.02\%$

**35.** $P = 700,\ S = 700+300 = 1000,\ r = 0.119,\ t = ?$

$1000 = 700e^{0.119t}$

$e^{0.119t} = \dfrac{1000}{700}$ Take ln of both sides.

$0.119t = \ln 1000 - \ln 700 = 6.9078 - 6.5511$

$t = 2.997$ or approximately 3 years.

**37.** $P = 20,000,\ S = 26,425.82,\ \dfrac{r}{m} = \dfrac{r}{4},\ t = 7$

$$26,425.82 = 20,000\left(1+\frac{r}{4}\right)^{28}$$

$$\left(1+\frac{r}{4}\right) = (1.32129)^{1/28}$$

$$= 1.00999 = 1.01$$

$$\frac{r}{4} = 0.01$$

$$r = 0.04 = 4\%$$

**39.** $P = 1000,\ \dfrac{r}{m} = \dfrac{0.08}{1} = 0.08,\ mt = 1(18) = 18$

$S = 1000(1+0.08)^{18} = \$3996.02$

**41.** **a.** $P = 432,860,\ i = 7.5\%,\ n = 22,\ S = ?$

$S = P(1+i)^n$

$S = 432,860(1+0.075)^{22} = \$2,124,876.38$

**b.** $i = 8.5\%$

$S = 432,860(1+0.085)^{22} = \$2,604,963.82$

At the higher rate of interest there will be an additional \$480,087.44 available for retirement.

**43.** $P = 30,000,\ S = 45,000,$

$\dfrac{r}{m} = \dfrac{0.08}{4} = 0.02,\ mt = 4t$

$$45,000 = 30,000(1+0.02)^{4t}$$

$$(1+0.02)^{4t} = 1.5000$$

$$4t \cdot \ln(1+0.02) = \ln(1.500)$$

$$t = \frac{\ln 1.5}{4\ln 1.02} \approx 5.12 \text{ years}$$

**45.** $P_1 = 2500,\ t = 3,\ r = 0.085$

$S_1 = 2500\left[1+3(0.085)\right] = \$3137.50$

$S_1 = P_2 = 3137.50,\ n = 9,\ r = 0.18$

$S_2 = 3137.50(1+0.18)^9 = \$13,916.24$

**47.** This problem is more meaningful if a spreadsheet is used.

**a.** From a table, the investments will be worth more than \$7500 during the sixth year.

**b.** $P = 5000$ $\quad i = \frac{0.063}{4}$ $\quad i = \frac{0.063}{12}$

3 years $\quad 5000\left(1+\frac{.063}{4}\right)^{12} = \$6031.31 \quad 5000\left(1+\frac{.063}{12}\right)^{36} = \$6037.22$

7 years $\quad 5000\left(1+\frac{.063}{4}\right)^{28} = \$7744.634 \quad 5000\left(1+\frac{.063}{12}\right)^{84} = 7762.34$

10 years $\quad 5000\left(1+\frac{.063}{4}\right)^{40} = 9342.07 \quad 5000\left(1+\frac{.063}{12}\right)^{120} = \$9372.59$

**49. a.** 3, 6, 12 Ratio is 2.
Next three terms are 24, 48, 96.

**b.** 81, 54, 36

$81 \cdot r = 54$ or $r = \frac{54}{81} = \frac{2}{3}$

Next three terms are 24, 16, $\frac{32}{3}$.

**51.** $a_1 = 10, r = 2, n = 13$

$a_{13} = 10 \cdot 2^{13-1} = 10 \cdot 4096 = 40,960$

**53.** $a_1 = 4, r = \frac{3}{2}, n = 16$

$a_{16} = 4\left(\frac{3}{2}\right)^{16-1} = 4\left(\frac{3}{2}\right)^{15}$

**55.** $a_1 = 6, r = 3, n = 17$

$s_{17} = \frac{6\left(1-3^{17}\right)}{1-3} = -3\left(1-3^{17}\right)$

**57.** $a_1 = 1, r = 3, n = 35$

$s_{35} = \frac{1-3^{35}}{1-3} = \frac{3^{35}-1}{2}$

**59.** $a_1 = 6, r = \frac{2}{3}, n = 18$

$s_{18} = \frac{6\left[1-\left(\frac{2}{3}\right)^{18}\right]}{1-\frac{2}{3}} = 18\left[1-\left(\frac{2}{3}\right)^{18}\right]$

**61.** $a_1 = 160,000, r = 1.04,$

$n = 21$ (at the end of 20 years)

$a_{21} = 160,000(1.04)^{20} = \$350,580$

**63.** $a_1 = 20(\text{million}), r = 1.02, n = 11$

$a_{11} = 20(1.02)^{10}$

$= 20(1.218994) = 24.38$ million

$n$ is 11 since 20 million is the value at the beginning of the year.

**65.** $a_1 = P, a_n = 2P, r = 1.02$

$2P = P(1.02)^n$

$(1.02)^n = 2$

$n \ln 1.02 = \ln 2$

$n = \frac{\ln 2}{\ln 1.02} = \frac{0.6931}{0.0198} = 35$ years

**67.** Down: 128 96 72 54

Up: 96 72 54 $\frac{81}{2}$

Note that there are two separate sequences.

Ball will bounce up $\frac{81}{2}$ or 40.5 ft.

**69.** Value now: 10,000

| Value end of year | 1 | 2 | 3 | 4 |
|---|---|---|---|---|
| | 8000 | 6400 | 5120 | 4096 |

**71.** Number now: 5000

| End of Hour | 1 | 2 | 3 | 4 | 5 |
|---|---|---|---|---|---|
| Number | 10,000 | 20,000 | 40,000 | 80,000 | 160,000 |

At the end of the 6th hour, 320,000 bacteria are present. $N = 10,000 \cdot 2^{6-1}$

**73.** Profit now: $8,000,000

| End of Year | 2006 | 2007 | 2008 | 2009 | 2010 |
|---|---|---|---|---|---|
| Profit | $7,840,000 | $7,683,200 | $7,529,536 | $7,378,945 | $7,231,366 |

**75.** Five people receive your letter with your name in position 6. They cross out the top name and each sends out 5 letters. Your name is now in place 5 on $25 = 5^2$ letters. Each of these sends out 5 letters and now your name is in position 4 on $25 \cdot 5 = 5^3$ letters. Continuing this pattern we have: position 3 $5^4$ letters
position 2 $5^5$ letters
position 1 $5^6$ letters

You will receive $5^6 = 15,625$ dimes. There are six sets of mailings.

**77.** Total number of letters mailed: $5 + 5^2 + 5^3 + \cdots + 5^{12}$ $\quad s_{12} = \dfrac{5(1-5^{12})}{1-5} = 305,175,780$ letters

## *Exercises 6.3*

**1.** $S = 2500\left[\dfrac{(1+0.02)^{60}-1}{0.02}\right] \approx 2500\left[\dfrac{2.281}{0.02}\right]$

$\approx 285,128.85$ (future value)

$R = 2500,\ i = 0.02,$

$n = 60$

**3.** $80,000 = R\left[\dfrac{(1+0.04)^{30}-1}{0.04}\right]$

$80,000 \approx R\left[\dfrac{2.24}{0.04}\right]$

$R \approx 1,426.43$ (payment)

$S = 80,000,\ i = 0.04,\ n = 30$

**5.** **a.** The higher graph is the $1120 per year annuity.

**b.** $S = 1000\left[\dfrac{(1+0.08)^{25}-1}{0.08}\right] = \$73,105.94$

$S = 1120\left[\dfrac{(1+0.08)^{25}-1}{0.08}\right] = \$81,878.65$

The difference is
$\$81,878.65 - 73,105.94 = \$8772.71$

**7.** $R = 1300,\quad i = 0.06,\quad n = 5$

$S = 1300\left[\dfrac{(1+0.06)^{5}-1}{0.06}\right] \approx \$7328.22$

**9.** $R = 80,\ n = 4 \cdot 3 = 12,\ i = \dfrac{0.08}{4} = 0.02$

$S = 80\left[\dfrac{(1+0.02)^{12}-1}{0.02}\right] \approx \$1072.97$

**11.** $S = 40,000,\ n = 2(12) = 24,\ i = \dfrac{0.12}{24} = 0.005$

$R = 40,000\left[\dfrac{0.005}{(1+0.005)^{24}-1}\right] \approx 1572.82$

**13.** $S = 80,000,\ R = 2500,\ i = \dfrac{0.05}{4} = 0.0125$

$80000 = 2500\left[\dfrac{(1+0.0125)^{n}-1}{0.0125}\right]$

$32 = \dfrac{(1.0125)^{n}-1}{0.0125}$

$0.4 = (1.0125)^{n} - 1$

$1.4 = (1.0125)^{n}$

$\ln 1.4 = n \ln 1.0125$

$n = \dfrac{\ln 1.4}{\ln 1.0125} \approx 27.1$ quarters

# Chapter 6: Mathematics of Finance

**15.** Twin 1: $R = 2000, n = 10, i = 0.08$

$$S_1 = 2000\left[\frac{(1+0.08)^{10}-1}{0.08}\right] \approx \$28,973.12$$

$P_2 = 28,973.12, i = 0.08, n = 32$ (age 65)

$S_2 = 28,973.12(1+0.08)^{32} = 340,060$ (rounded)

Twin 2: $S = 340,060, i = 0.08, n = 25$ (age 65)

$$R = 340,060\left[\frac{0.08}{(1+0.08)^{25}-1}\right] \approx \$4651.61$$

**17.** $R = 100, n = \frac{5}{2}\cdot 4 = 10, i = \frac{0.12}{4} = 0.03$

$$S_{due} = 100\left[\frac{(1+0.3)^{10}-1}{0.03}\right](1+0.03) \approx \$1180.78$$

**19.** $R = 200, n = 8\cdot 2 = 16, i = \frac{0.06}{2} = 0.03$

$$S_{due} = 200\left[\frac{(1+.03)^{16}-1}{0.03}\right](1+0.03) \approx \$4152.32$$

**21.** $S_{due} = 24000, n = 5, i = 0.08$

$$R = 24000\left[\frac{0.08}{(1+.08)^5-1}\right]\left(\frac{1}{1+0.08}\right) \approx \$3787.92$$

**23.** $R = 40,000,\ S_{due} = 800,000, i = \frac{0.052}{4} = 0.013$

$$80,000 = 40,000\left[\frac{(1+0.013)^n-1}{0.013}\right]\cdot(1+0.013)$$

$$\frac{20(0.013)}{1.013} = (1+0.013)^n - 1$$

$$\frac{20(0.013)}{1.013} + 1 \approx 1.256663 \approx (1.013)^n$$

$$\ln(1.256663) \approx \ln(1.013)^n = n[\ln(1.013)]$$

$$n \approx \frac{\ln(1.256663)}{\ln(1.013)} \approx 17.7 \text{ quarters}$$

**25. a.** Ordinary annuity

**b.** $R = 500,\ n = (4)(2) = 8,\ i = \frac{0.10}{2} = 0.05$

$$S = 500\left[\frac{(1+0.05)^8-1}{0.05}\right] \approx \$4774.55$$

**27. a.** Annuity due

**b.** $S_{due} = 50,000,\ n = 8,\ i = 0.10$

$$50,000 = R\left[\frac{(1+0.10)^8-1}{0.10}\right](1+0.10)$$

$$50,000 \approx R\left[\frac{1.1436}{0.10}\right](1.10)$$

$R \approx \$3974.73$

**29. a.** Ordinary annuity

**b.** $S = 200,000,\ n = 80,\ i = \frac{0.076}{4} = 0.019$

$$200,000 = R\left[\frac{(1+0.019)^{80}-1}{0.019}\right]$$

$$200,000 \approx R\left[\frac{3.5075}{0.019}\right]$$

$R \approx \$1083.40$

**31. a.** Ordinary annuity

**b.** $S = 1,500,000,\ \ n = 25\cdot 12 = 300,\ \ i = \frac{0.10}{12}$

$$R = 1,500,000\left(\frac{0.10/12}{(1+0.10/12)^{300}-1}\right)$$

$\approx \$1130.51$

**33. a.** Annuity due

**b.** $R = 12,000,\ n = 5\cdot 4 = 20,\ i = \frac{0.072}{4} = 0.018$

$$S_{due} = 12,000\left[\frac{(1+0.018)^{20}-1}{0.018}\right](1+0.018)$$

$\approx \$290,976.81$

**35. a.** Ordinary annuity

**b.** $S = \$50,000,\ R = \$300,\ i = 0.09/12,\ n = ?$

$$50000 = 300\left[\frac{(1+0.0075)^n - 1}{0.0075}\right]$$

$$166.67 = \frac{(1.0075)^n - 1}{0.0075}$$

$$1.25 = (1.0075)^n - 1$$

$$2.25 = (1.0075)^n$$

$$\ln 2.25 = n \ln 1.0075$$

$$n = \frac{\ln 2.25}{\ln 1.0075} \approx 108.53 \text{ months}$$

**37. a.** Annuity due

**b.** $S_{due} = 180{,}000$, $n = 18 \cdot 12 = 216$,

$$i = \frac{0.12}{12} = 0.01$$

$$R = 180{,}000\left[\frac{0.01}{(1+0.1)^{216} - 1}\right]\left(\frac{1}{1+0.01}\right)$$

$$\approx \$235.16$$

**39. a.** Annuity due

**b.** $R = 500$, $n = 9 \cdot 4 = 36$, $i = \frac{0.08}{4} = 0.02$

$$S_{due} = 500\left[\frac{(1+0.02)^{36} - 1}{0.02}\right](1+0.02)$$

$$\approx \$26{,}517.13$$

**41. a.** Ordinary annuity

**b.** $S = 60{,}000$, $n = 40$, $i = \frac{0.12}{4} = 0.03$

$$60{,}000 = R\left[\frac{(1+0.03)^{40} - 1}{0.03}\right]$$

$$60{,}000 \approx R\left[\frac{2.262}{0.03}\right]$$

$$R \approx \$795.75$$

**43.** $R = 100$, $n = 8 \cdot 12 = 96$, $i = \frac{0.09}{12} = 0.0075$

$$S_1 = 100\left[\frac{(1+0.0075)^{96} - 1}{0.0075}\right] \approx \$13{,}985.62$$

$P = 13{,}985.62$, $n = 15 \cdot 12 = 180$, $i = 0.0075$

$$S_2 = 13{,}985.62(1+0.0075)^{180} \approx 53{,}677.41$$

**45. a.** $S = 150{,}000$, $R = 3000$, $i = \frac{0.08}{4} = 0.02$, $n = ?$, use the formula $S = R\left[\frac{(1+i)^n - 1}{i}\right]$.

$$150{,}000 = 3000\left[\frac{(1+0.02)^n - 1}{0.02}\right]$$

$$50 = \frac{(1.02)^n - 1}{0.02}$$

$$1 = 1.02^n - 1$$

$$2 = 1.02^n$$

$$\ln 2 = n \ln 1.02$$

$$n = \frac{\ln 2}{\ln 1.02} \approx 35 \text{ quarters}$$

**b.** Treat this as two problems. The \$150,000 will grow at 8% compounded quarterly for 15 years. The second part is the deposits of \$5000 each quarter at 8% compounded quarterly.

Part I: $S = P(1+i)^n$, where $P = 150{,}000$, $i = 0.02$, $n = 15 \cdot 4 = 60$

$$S = 150{,}000(1+0.02)^{60} = \$492{,}154.62$$

Part II: $S = R\left[\frac{(1+i)^n - 1}{i}\right]$, where $R = 5000$, $i = 0.02$, $n = 60$

$$S = 5000\left[\frac{(1+0.02)^{60} - 1}{0.02}\right] \approx \$570,257.70$$

So, his total will be $492,154.62 + $570,257.70 = $1,062,412.32

## *Exercises 6.4*

Answers for all problems in this section will depend on whether the numbers were rounded during the calculations.

**1.** $A_n = 1300\left[\frac{1-(1+0.04)^{-30}}{0.04}\right]$

$\approx \$22,480$ (present value)

Payment $= R = 1300$

Periodic rate $= i = 0.04$

Number of periods $= n = 30$

**3.** $135,000 = R\left[\frac{1-(1+0.005)^{-360}}{0.005}\right]$

$135,000 \approx R[166.79]$

$R \approx \$809$ (payment)

Present value $= A_n = 135,000$

Periodic rate $= i = 0.005$

Number of periods $= 360$

**5.** $R = 6000, n = 8 \cdot 2 = 16, i = \frac{0.08}{2} = 0.04$

$$A_n = 6000\left[\frac{1-(1+.04)^{-16}}{0.04}\right] = \$69,913.77$$

**7.** $R = 250,000\ n = 20, i = 0.10$

$$A_n = 250,000\left[\frac{1-(1+.10)^{-20}}{0.10}\right] = \$2,128,390.93$$

**9.** $A = 135,000, n = 10 \cdot 4 = 40, i = \frac{0.064}{4} = 0.016$

$$R = 135,000\left[\frac{0.016}{1-(1+.016)^{-40}}\right] = \$4595.46$$

**11.** $A_n = \$200,000,\ i = 0.015,\ R = 4500,\ n = ?$

$$200,000 = 4500\left[\frac{1-(1+0.015)^{-n}}{0.015}\right]$$

$$44.44 = \frac{1-(1.015)^{-n}}{0.015}$$

$$0.666 = 1-(1.015)^{-n}$$

$$-0.333 = -(1.015)^{-n}$$

$$0.333 = 1.015^{-n}$$

$$\ln 0.333 = -n \ln 1.015$$

$$n = -\frac{\ln 0.333}{\ln 1.015} \approx 73.8 \approx 74 \text{ quarters}$$

**13.** **a.** $S = 1000,\ n = 50,\ i = \frac{0.05}{2} = 0.025$

$$R = (1000)\left(\frac{0.06}{2}\right) = \$30$$

$$S = P(1+i)^n$$

$$1000 = P(1+0.025)^{50}$$

$$1000 \approx P(3.4371)$$

$$P \approx \$290.94$$

$$A_n = 30\left[\frac{1-(1+0.025)^{-50}}{0.025}\right]$$

$$\approx 30[28.3623] \approx \$850.87$$

Market Price $= A_n + P = \$1141.81$

**b.** Selling at premium

**15.** **a.** The higher graph corresponds to 8%.
**b.** $1500 (approximately)
**c.** A 10% interest rate has a present value of $9000. An 8% interest rate has a present value of $10,500.

**17.** The payments are at the end of each period for an ordinary annuity. The payments are at the beginning of each period for an annuity due.

**19.** $R = 3000, n = 7 \cdot 4 = 28, i = \frac{0.058}{4} = 0.0145$

$$A_{(n,due)} = 3000\left[\frac{1-(1+.0145)^{-28}}{0.0145}\right](1+.0145)$$
$$= \$69,632.02$$

**21.** $R = 50,000, n = 12, i = 0.0592$

$$A_{(n,due)} = 50,000\left[\frac{1-(1+.0592)^{-12}}{0.0592}\right](1+.0592)$$
$$= \$445,962.23$$

**23.** $A_{(n,due)} = 25,000, n = 12, i = \frac{0.0648}{12} = 0.0054$

$$R = 25,000\left[\frac{0.0054}{1-(1+0.0054)^{-12}}\right]\left(\frac{1}{1+0.0054}\right)$$
$$= \$2,145.59$$

**25. a.** Ordinary annuity

**b.** $A_n = 1,500,000$, $n = 40 \cdot 12 = 480$, $i = 0.007$

$$R = 1,500,000\left[\frac{0.007}{1-(1+0.007)^{-480}}\right] = \$10,882.46$$

**27. a.** Annuity due

**b.** $R = 40,000, n = \frac{9}{2} \cdot 2 = 9,$

$$i = \frac{0.0668}{2} = 0.0334$$

$$A_{(n,due)} = 40,000\left[\frac{1-(1+.0334)^{-9}}{0.0334}\right](1+.0334)$$
$$= \$316,803.61$$

**29. a.** Ordinary annuity

**b.** $R = 140,000, n = 10, i = 0.065$

$$A_n = 140,000\left[\frac{1-(1+0.065)^{-10}}{0.065}\right]$$
$$\approx \$1,006,436.24$$
$$1,006,436.24 + 500,000 = \$1,506,436.24$$

Taking \$500,000 now and \$140,000 payments for the next 10 years has a slightly higher present value: \$1,506,436.24.

**31. a.** Annuity due

**b.** $A_n = 800,000, n = 6, i = \frac{0.077}{2} = 0.0335$

$$800,000 = R\left[\frac{1-(1+0.0385)^{-6}}{0.0385}\right](1+0.0385)$$

$$R = \frac{800,000(0.0385)}{1.0385\left[1-(1.0385)^{-6}\right]}$$

$$R \approx \$146,235.06$$

**33. a.** Ordinary annuity

**b.** $A_n = 2,200,000, R = 10,000,$

$$i = \frac{0.054}{12} = 0.0045$$

$$2,200,000 = 10,000\left[\frac{1-(1+0.0045)^{-n}}{0.0045}\right]$$

$$220(0.0045) = 1-(1.0045)^{-n}$$
$$(1.0045)^{-n} = 0.01$$
$$-n\ln(1.0045) = \ln(0.01)$$
$$n = \frac{\ln(0.01)}{-\ln(1.0045)} \approx 1025.7$$

1026 months; about 85.5 years

**35. a.** Annuity due

**b.** $R = 800, n = 2.5(12) = 30, i = \frac{0.048}{12} = 0.004$

$$A_n = 800\left[\frac{1-(1+0.004)^{-30}}{0.004}\right](1+0.004)$$
$$\approx 800(28.2168)(1.004) \approx \$22,663.74$$

**37. a.** Ordinary annuity

**b.** $A_n = 1,500,000, n = 240, i = \frac{0.072}{12} = 0.006$

$$1,500,000 = R\left[\frac{1-(1+0.006)^{-240}}{0.006}\right]$$

$$R = \frac{1,500,000(0.006)}{\left[1-(1.006)^{-240}\right]}$$

$$R \approx \$11,810.24$$

**39.** **a.** Ordinary annuity

**b.** $R = 2000$, $n = 16$, $i = \dfrac{0.072}{4} = 0.018$

$$A_n = 2000\left[\frac{1-(1+0.018)^{-16}}{0.018}\right] = \$27{,}590.62$$

**41.** **a.** $S = 10{,}000,\ n = 20,\ i = 0.05$

$$R = 10{,}000\left(\tfrac{0.078}{2}\right) = 390$$

$$10{,}000 = P(1+0.05)^{20}$$

$$P \approx \$3768.89$$

$$A_n = 390\left[\frac{1-(1+0.05)^{-20}}{0.05}\right]$$

$$A_n \approx 390[12.4622] \approx \$4860.26$$

Market price $= A_n + P = \$8{,}629.15$

**b.** $R = 10{,}000,\ n = 16,\ i = 0.04$

$$10{,}000 = P(1+0.04)^{16}$$

$$P \approx \$5339.08$$

$$A_n = 390\left[\frac{1-(1+0.04)^{-16}}{0.04}\right]$$

$$A_n \approx 390[11.6523] \approx \$4544.40$$

Selling price $= P + A_n = \$9{,}883.48$

**43.** $P = 2500$, $n = 12\cdot 12 = 144$, $i = 0.0065$

$$S_1 = 2500(1+0.0065)^{144} = \$6355.13$$

$$S_2 = 100\left[\frac{(1+0.0065)^{144}-1}{0.0065}\right] = \$23{,}723.86$$

**a.** After the last deposit, $S_1 + S_2 = \$30{,}078.99$ is in the account.

**b.** $\$2500 + \$14{,}400 = \$16{,}900$ was deposited.

**c.** $A_n = 30{,}078.99$, $n = 5\cdot 12 = 60$, $i = 0.0065$

$$R = 30{,}078.99\left[\frac{0.0065}{1-(1+0.0065)^{-60}}\right]$$

$$= \$607.02$$

**d.** The total amount withdrawn is $60(607.02) = \$36{,}421.20$.

**45.** The couple wants an annuity with $R = 30{,}000$ and $n = 8\cdot 2 = 16$. We assume from part **a** that $i = 0.04$.

$$A_n = 30{,}000\left[\frac{1-(1+0.04)^{-16}}{0.04}\right] = \$349{,}568.87$$

They need a fund with \$349,568.87.

**a.** $S = 349{,}568.87,\ i = 0.04,\ n = 18\cdot 2 = 36$

$$S = R\left[\frac{(1+i)^n - 1}{i}\right]$$

$$349{,}568.87 = R\left[\frac{(1+0.04)^{36}-1}{0.04}\right]$$

$$349{,}568.87 = R(77.598)$$

$$R = \$4{,}504.87$$

**b.** When the extra \$38,000 is invested, there are 10 years left to compound the interest.

$$S = 38{,}000(1.04)^{20} = \$83{,}262.68$$

Now the account will have \$349,568.87 + \$83,262.68 = \$432,831.55 in it when the withdrawals begin. We need to calculate a new value of $n$.

$i = 0.04,\ R = 30{,}000,\ A_n = 432{,}831.55$

$$432{,}831.55 = 30{,}000\left[\frac{1-(1+0.04)^{-n}}{0.04}\right]$$

$$14.4277 = \frac{1-(1.04)^{-n}}{0.04}$$

$$0.5771 = 1-(1.04)^{-n}$$

$$-0.4229 = -1.04^{-n}$$

$$0.4229 = 1.04^{-n}$$

$$\ln 0.4229 = -n\ln 1.04$$

$$n = -\frac{\ln 0.4229}{\ln 1.04}$$

$$\approx 21.9 \approx 22 \text{ withdrawals}$$

**47.** $R = \$2000, n = (4)(5) = 20, k = (4)(3) = 12,$

$$i = \frac{0.08}{4} = 0.02$$

$$A_{(n,k)} = 2000\left[\frac{1-(1+0.02)^{-20}}{0.02}\right](1+0.02)^{-12}$$

$$= \$25{,}785.99$$

**49.** $R = 16{,}000, n = 7\left(\text{difficult part}\right),$

$k = 3, i = 0.06$

$$A_{(n,k)} = 16{,}000\left[\frac{1-(1+.06)^{-7}}{0.06}\right](1+.06)^{-3}$$

$$= \$74{,}993.20$$

**51.** $R = 30{,}000, n = 8, k = 36,$

$i = \frac{0.07}{2} = 0.035$

$$A_{(8,36)} = 30{,}000\left[\frac{1-(1+.035)^{-8}}{0.035}\right](1+.035)^{-36}$$

$$\approx \$59{,}768.92$$

**53.** $A_{(4,18)} = 1600,\ \ n = 4,\ \ k = 18,\ \ i = 0.06$

$$1600 = R\left[\frac{1-(1+.06)^{-4}}{0.06}\right](1+.06)^{-18}$$

$$= R(1.21398)$$

$$R = \frac{1600}{1.21398} = \$1{,}317.98\left(\text{Alternate method}\right)$$

**55.**

$A_{(n,k)} = 150{,}000{,}000,\ n = 12\cdot 50 = 600,$

$k = 12\cdot 16 = 192, i = \frac{0.045}{12} = 0.00375$

$$150{,}000{,}000 = R\left[\frac{1-(1+0.00375)^{-600}}{0.00375}\right](1+0.00375)^{-192}$$

$$R = \frac{150{,}000{,}000}{(1.0375)^{-192}}\left[\frac{0.00375}{1-(1.00375)^{-600}}\right]$$

$$R \approx \$1{,}290{,}673.16$$

**57. a.**

| | A | B | C | D |
|---|---|---|---|---|
| 1 | End of Month | Acct. Value | Payment | New Balance |
| 2 | 0 | \$100,000.00 | \$0.00 | \$100,000.00 |
| 3 | 1 | \$100,650.00 | \$1,000.00 | \$99,650.00 |
| 4 | 2 | \$100,297.73 | \$1,000.00 | \$99,297.73 |
| 5 | 3 | \$99,943.16 | \$1,000.00 | \$98,943.16 |
| 6 | 4 | \$99,586.29 | \$1,000.00 | \$98,586.29 |
| 7 | 5 | \$99,227.10 | \$1,000.00 | \$98,227.10 |
| 8 | 6 | \$98,865.58 | \$1,000.00 | \$97,865.58 |
| 9 | 7 | \$98,501.70 | \$1,000.00 | \$97,501.70 |
| 10 | 8 | \$98,135.47 | \$1,000.00 | \$97,135.47 |
| 11 | 9 | \$97,766.85 | \$1,000.00 | \$96,766.85 |
| 12 | 10 | \$97,395.83 | \$1,000.00 | \$96,395.83 |
| ⋮ | ⋮ | ⋮ | ⋮ | ⋮ |
| 159 | 157 | \$5,938.16 | \$1,000.00 | \$4,938.16 |
| 160 | 158 | \$4,970.26 | \$1,000.00 | \$3,970.26 |
| 161 | 159 | \$3,996.06 | \$1,000.00 | \$2,996.06 |
| 162 | 160 | \$3,015.54 | \$1,000.00 | \$2,015.54 |
| 163 | 161 | \$2,028.64 | \$1,000.00 | \$1,028.64 |
| 164 | 162 | \$1,035.32 | \$1,000.00 | \$35.32 |
| 164 | 163 | \$35.55 | \$35.55 | \$0.00 |

**b.**

| | A | B | C | D |
|---|---|---|---|---|
| 1 | End of Month | Acct. Value | Payment | New Balance |
| 2 | 0 | $100,000.00 | $0.00 | $100,000.00 |
| 3 | 1 | $100,650.00 | $2,500.00 | $98,150.00 |
| 4 | 2 | $98,787.98 | $2,500.00 | $96,287.98 |
| 5 | 3 | $96,913.85 | $2,500.00 | $94,413.85 |
| 6 | 4 | $95,027.54 | $2,500.00 | $92,527.54 |
| 7 | 5 | $93,128.97 | $2,500.00 | $90,628.97 |
| 8 | 6 | $91,218.05 | $2,500.00 | $88,718.05 |
| 9 | 7 | $89,294.72 | $2,500.00 | $86,794.72 |
| 10 | 8 | $87,358.89 | $2,500.00 | $84,858.89 |
| 11 | 9 | $85,410.47 | $2,500.00 | $82,910.47 |
| 12 | 10 | $83,449.39 | $2,500.00 | $80,949.39 |
| ⋮ | ⋮ | ⋮ | ⋮ | ⋮ |
| 43 | 41 | $15,902.26 | $2,500.00 | $13,402.26 |
| 44 | 42 | $13,489.37 | $2,500.00 | $10,989.37 |
| 45 | 43 | $11,060.81 | $2,500.00 | $8,560.81 |
| 46 | 44 | $8,616.45 | $2,500.00 | $6,116.45 |
| 47 | 45 | $6,156.21 | $2,500.00 | $3,656.21 |
| 48 | 46 | $3,679.98 | $2,500.00 | $1,179.98 |
| 49 | 47 | $1,187.65 | $1,187.65 | $0.00 |

## *Exercises 6.5*

**1.** **a.** The 10 year loan requires more payment to principal since the loan must be paid more quickly.

**b.** The 25 year loan requires lower payment period since the loan is paid more slowly.

**3.** $A_n = 8000,\ n = 8,\ i = \dfrac{0.12}{2} = 0.06,$

$$R = 8000\left[\frac{0.06}{1-(1+0.06)^{-8}}\right] \approx \$1288.29$$

**5.** $A_n = 18{,}000,\ n = 40,\ i = \dfrac{0.042}{4} = 0.0105$

$$R = 18{,}000\left[\frac{0.0105}{1-(1+0.0105)^{-40}}\right] \approx \$553.34$$

**7.** $R = 600,\ n = 5\cdot 12 = 60,\ i = \dfrac{0.06}{12} = 0.005$

$$A_n = 600\left[\frac{1-(1+0.005)^{-60}}{0.005}\right] \approx \$31{,}035.34$$

**9.** $A_n = 100{,}000,\ n = 3,\ i = 0.09,$

$$R = 100{,}000\left[\frac{0.09}{1-(1+0.09)^{-3}}\right] \approx \$39{,}505.48$$

Payment – Interest = Balance Reduction

| Period | Payment | Interest | Balance Reduction | Unpaid Balance |
|---|---|---|---|---|
| | | | | 100,000.00 |
| 1 | 39,505.48 | 9000.00 | 30,505.48 | 69,494.52 |
| 2 | 39,505.48 | 6254.51 | 33,250.98 | 36,243.55 |
| 3 | 39,505.47 | 3261.92 | 36,243.55 | 0.00 |

Last payment usually leaves a few cents on balance. Also, method used can affect answer by 2 or 3 cents.

**11.** $A_n = 20{,}000,\ n = 4,\ i = \frac{0.12}{4} = 0.03$

$$R = 20{,}000\left[\frac{0.03}{1-(1+0.03)^{-4}}\right] \approx \$5380.54$$

| Period | Payment | Interest | Balance Reduction | Unpaid Balance |
|---|---|---|---|---|
| | | | | 20,000.00 |
| 1 | 5380.54 | 600.00 | 4780.54 | 15,219.46 |
| 2 | 5380.54 | 456.58 | 4923.96 | 10,295.50 |
| 3 | 5380.54 | 308.87 | 5071.67 | 5223.83 |
| 4 | 5380.54 | 156.71 | 5223.83 | 0.00 |

**13.** $R = 334.27,\quad n = 40,\quad k = 6,\quad i = \frac{0.06}{4} = 0.015$

Unpaid balance: $334.27\left[\frac{1-(1+.015)^{-(40-6)}}{0.015}\right] \approx \$8{,}852.05$

**15.**

$R = 684.88,\ n = 48,\ k = 30,\ i = \frac{0.081}{12} = 0.00675$

Unpaid balance:

$$684.88\left[\frac{1-(1+.00675)^{-(48-30)}}{0.00675}\right] \approx \$11{,}571.67$$

**17.** $A_n = 350{,}000 - 150{,}000 = \$200{,}000$

$n = 10(2) = 20,\ i = \frac{0.12}{2} = 0.06$

**a.** $R = 200{,}000\left[\frac{0.06}{1-(1+.06)^{-20}}\right] \approx \$17{,}436.91$

**b.** Total paid $= R \cdot n + \text{downpayment}$
$= 348{,}738.20 + 150{,}000$
$= \$498{,}738.20$

**c.** Interest paid
$= \text{Total paid} - A_n - \text{downpayment} = \$148{,}738.20$

**19.** $A_n = 1{,}000{,}000, \quad n = 120, \quad i = \dfrac{0.072}{12} = 0.006$

**a.** $R = 1{,}000{,}000\left[\dfrac{0.006}{1-(1+0.006)^{-120}}\right]$

$\approx \$11{,}714.19$

**b.** Total paid:

$120(\$11{,}714.19) + \$250{,}000$

**c.** $= \$1{,}405{,}702.80 + \$250{,}000$

$= \$1{,}655{,}702.80$

**d.** Interest paid:

$\$1{,}655{,}702.80 - \$1{,}250{,}000$

$= \$405{,}702.80$

**21.** $R = 900, n = 36, i = 0.01$

$A_n = 900\left[\dfrac{1-(1+.01)^{-36}}{0.01}\right] \approx \$27{,}096.75$

Down payment is

$\$36{,}000 - \$27096.75 = \$8903.25$

**23.** $R = 1800, n = 25 \cdot 12 = 300$

**a.** $i = \dfrac{0.069}{12} = 0.00575$

$A_n = 1800\left[\dfrac{1-(1+0.00575)^{-300}}{0.00575}\right]$

$\approx \$256{,}991.33$

$\$256{,}991.33 + 20{,}000 = \$276{,}991.33$

**b.** $i = \dfrac{0.075}{12} = 0.00625$

$A_n = 1800\left[\dfrac{1-(1+0.00625)^{-300}}{0.00625}\right]$

$\approx \$243{,}575.30$

$\$243{,}575.30 + 20{,}000 = \$263{,}575.30$

**25.** $R = 610.91, n = 360, k = 12, i = 0.006$

**a.** Unpaid balance:

$610.91\left[\dfrac{1-(1+.006)^{-(360-12)}}{0.006}\right] \approx \$89{,}120.53$

**b.** Interest paid: 12(610.91) – 879.47

= \$6451.45

**27. a.** $R = 1199.10, i = \dfrac{0.06}{12} = 0.005,$

$n = 12 \cdot 30 = 360, k = 120$

$A_{n-k} = 1199.10\left[\dfrac{1-(1+.005)^{-(360-120)}}{0.005}\right]$

$\approx \$167{,}371.30$

$\$167{,}371.30 + \$750 = \$168{,}121.30$

**b.** $A_n = \$168{,}121.30, n = 12 \cdot 15 = 180,$

$i = \dfrac{0.045}{12} = 0.00375$

$R = 168{,}121.30\left[\dfrac{0.00375}{1-(1+0.00375)^{-180}}\right]$

$\approx \$1286.12$

**c.** Original remaining:

$240(1199.10) = \$287{,}784.00$

Refinance:

$180(1286.12) = \$231{,}501.60$

$\$287{,}784.00 - \$231{,}501.60$

$= \$56{,}282.40$ saved

**29. a.** $A_n = 18{,}000, \quad n = 60, \quad i = \dfrac{0.084}{12} = 0.007$

$R = 18{,}000\left[\dfrac{0.007}{1-(1+0.007)^{-60}}\right] \approx \$368.43$

\$368.43 is required by the loan, she decides to pay \$383.43 a month.

**b.** $A_n = 18{,}000,\ R = 383.43,\ i = 0.007$

$$18{,}000 = 383.43\left[\frac{1-(1+0.007)^{-n}}{0.007}\right]$$

$$46.94 = \frac{1-(1.007)^{-n}}{0.007}$$

$$0.3286 = 1 - 1.007^{-n}$$

$$-0.6714 = -1.007^{-n}$$

$$0.6714 = 1.007^{-n}$$

$$\ln 0.6714 = -n\ln 1.007$$

$$n = -\frac{\ln 0.6714}{\ln 1.007} = 57.1$$

It will take a little over 57 months.

**c.** Without the extra, she would pay $\$368.43 \cdot 60 = \$22{,}105.80$.

With the extra, she pays $\$383.43 \cdot 57.1 = \$21{,}893.85$ for a savings of \$211.95.

**31. a.** $R = 261{,}094.80,\ i = \frac{0.072}{4} = 0.018,\ n = 4\cdot 30 = 120,\ k = 42$

$$A_{n-k} = 261{,}094.80\left[\frac{1-(1+.018)^{-(120-42)}}{0.018}\right] \approx \$10{,}897{,}827.36$$

$$\$10{,}897{,}827.36 + \$1{,}100{,}000 + \$10{,}000 = \$12{,}007{,}827.36$$

**b.** $A_n = 12{,}007{,}827.36,\ n = 4\cdot 25 = 100,\ i = \frac{0.066}{4} = 0.0165$

$$R = 12{,}007{,}827.36\left[\frac{0.0165}{1-(1+0.0165)^{-100}}\right] \approx \$246{,}017.20$$

**c.**

$$12{,}007{,}827.36 = 261{,}094.80\left[\frac{1-(1+0.0165)^{-n}}{0.0165}\right]$$

$$\frac{12{,}007{,}827.36(0.0165)}{261{,}094.80} = 1-(1.0165)^{-n}$$

$$(1.0165)^{-n} = 1 - \frac{12{,}007{,}827.36(0.0165)}{261{,}094.80}$$

$$-n\ln(1.0165) \approx \ln(0.2411601018)$$

$n \approx 86.9$; 87 payments (rather than 100)

**33. a.** The line is the total amount paid ($644.30 per month × the number of months).

The other curve is the amount paid toward the principal.

b. The length of the vertical line segment from the lower curve to the line at $x = 250$ represents the total interest paid after 250 months.

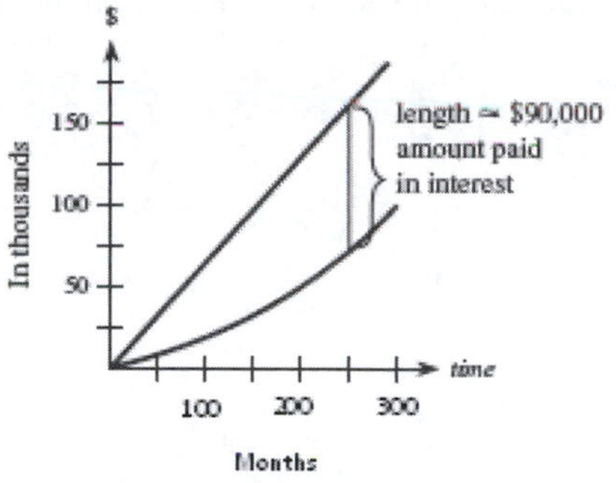


**35. a.** $A_n = 15{,}000, \quad n = 12 \cdot 4 = 48$

$$i = \frac{0.08}{12} = 0.00\overline{6}$$

$$R = 15{,}000\left[\frac{0.00\overline{6}}{1-\left(1+0.00\overline{6}\right)^{-48}}\right] \approx \$366.19$$

$$i = \frac{0.085}{12} = 0.007083\overline{3}$$

$$R = 15{,}000\left[\frac{0.007083\overline{3}}{1-\left(1+0.007083\overline{3}\right)^{-48}}\right]$$

$\approx \$369.72$

Total interest at:

8% $: 48(366.19) - 15{,}000 = \$2577.12$

8.5% $: 48(369.72) - 15{,}000 = \$2746.56$

**b.** $A_n = 80{,}000, \quad n = 12 \cdot 30 = 360$

$$i = \frac{0.0675}{12} = 0.005625 \quad \rightarrow$$

$$R = 80{,}000\left[\frac{0.005625}{1-\left(1+0.005625\right)^{-360}}\right]$$

$\approx \$518.88$

$$i = \frac{0.0725}{12} \approx 0.0060417$$

$$R = 80000\left[\frac{0.0060417}{1-\left(1+0.0060417\right)^{-360}}\right]$$

$\approx \$545.74$

Total interest at:

6.75%:

$360(518.88) - 80{,}000 = \$106{,}796.80$

7.25%:

$360(545.74) - 80{,}000 = \$116.466.40$

**c.** The duration of the loan has the greatest effect on the borrower.
It affects $R$ and the total interest paid.

**37. i.** $A_n = 100{,}000, \quad n = 25 \cdot 12 = 300$ **i.**

**ii.** $$R = 100{,}000\left[\frac{\left(\frac{.0725}{12}\right)}{1-\left(1+\frac{.0725}{12}\right)^{-300}}\right] \approx \$722.81$$

**iii.** $$R = 100{,}000\left[\frac{\left(\frac{.07}{12}\right)}{1-\left(1+\frac{.07}{12}\right)^{-300}}\right] \approx \$706.78$$

**a.**

| Payment | Points | Total Paid $\left[(R \cdot 300) + \text{Points}\right]$ |
|---|---|---|
| \$738.99 | 0 | \$221,697 |
| 722.81 | \$1000 | 217,843 |
| 706.78 | \$2000 | 214,034 |

**b.** The lower total cost is the 7%, 2 points loan.

**39.** **a.** To find the loan balance after the grace period, we need to calculate the accrued interest on the four loans.

$i = \frac{0.01}{4} = 0.0025$ during the in-school period of four years.

$n_1 = 4 \cdot 4 - 1 = 15,\ S_1 = 4000(1+0.0025)^{15} = \$4152.65$

$n_2 = 3 \cdot 4 - 1 = 11,\ S_2 = 3500(1+0.0025)^{11} = \$3597.46$

$n_3 = 2 \cdot 4 - 1 = 7,\ S_3 = 4400(1+0.0025)^{7} = \$4477.58$

$n_4 = 1 \cdot 4 - 1 = 3,\ S_4 = 5000(1+0.0025)^{3} = \$5037.59$

Total at end of 4 years is \$17,265.28.

For 6-month grace period, $i = \frac{0.03}{4} = 0.0075.\quad S = 17{,}265.28(1+0.0075)^2 = \$17{,}525.23$

**b.** $A_n = 17{,}525.23,\ i = \frac{0.03}{4} = 0.0075,\ n = 10 \cdot 4 = 40,\ R = ?$

$$R = 17525.23\left[\frac{0.0075}{1-(1+0.0075)^{-40}}\right] \approx \$508.76$$

**c.** $R = \$598.76,\ i = 0.0075,\ A_n = 17{,}525.23,\ n = ?$

$$17525.23 = 598.76\left[\frac{1-(1+0.0075)^{-n}}{0.0075}\right]$$

$$29.2692 = \frac{1-(1.0075)^{-n}}{0.0075}$$

$$0.2195 = 1 - 1.0075^{-n}$$

$$-0.7805 = -1.0075^{-n}$$

$$0.7805 = 1.0075^{-n}$$

$$\ln 0.7805 = -n \ln 1.0075$$

$$n = -\frac{\ln 0.7805}{\ln 1.0075} \approx 33.2 \text{ quarters}$$

**d.** Without the extra payment, Nolan would've paid $\$508.76 \cdot 40 = \$20{,}350.40$. With the extra payment, he will pay $\$598.76 \cdot 33.2 = \$19{,}878.83$, so the savings will be \$471.57.

**41.** $A_n = 16{,}700,\ i = \frac{0.082}{12},\ n = (4)(12) = 48$

$$R = 16{,}700\left[\frac{\frac{0.082}{12}}{1-\left(1+\frac{0.082}{12}\right)^{-48}}\right] \approx \$409.27$$

| | A | B | C | D | E |
|---|---|---|---|---|---|
| 1 | **Period** | **Payment** | **Interest** | **Bal. Reduction** | **Unpaid Bal.** |
| 2 | 0 | $0.00 | $0.00 | $0.00 | $16,700.00 |
| 3 | 1 | $409.27 | $114.12 | $295.15 | $16,404.85 |
| 4 | 2 | $409.27 | $112.10 | $297.17 | $16,107.68 |
| 5 | 3 | $409.27 | $110.07 | $299.20 | $15,808.48 |
| 6 | 4 | $409.27 | $108.02 | $301.25 | $15,507.23 |
| 7 | 5 | $409.27 | $105.97 | $303.30 | $15,203.93 |
| 8 | 6 | $409.27 | $103.89 | $305.38 | $14,898.55 |
| 9 | 7 | $409.27 | $101.81 | $307.46 | $14,591.09 |
| 10 | 8 | $409.27 | $99.71 | $309.56 | $14,281.52 |
| ⋮ | ⋮ | ⋮ | ⋮ | ⋮ | ⋮ |
| 45 | 43 | $409.27 | $16.38 | $392.89 | $2,004.81 |
| 46 | 44 | $409.27 | $13.70 | $395.57 | $1,609.24 |
| 47 | 45 | $409.27 | $11.00 | $398.27 | $1,210.97 |
| 48 | 46 | $409.27 | $8.27 | $401.00 | $809.98 |
| 49 | 47 | $409.27 | $5.53 | $403.74 | $406.24 |
| 50 | 48 | $409.02 | $2.78 | $406.69 | $0.00 |

# Chapter 6: Mathematics of Finance

## *Chapter 6 Review Exercises*

**1.** $a_n = \frac{1}{n^2}$ $a_1 = \frac{1}{1^2} = 1$ $a_2 = \frac{1}{2^2} = \frac{1}{4}$

$a_3 = \frac{1}{3^2} = \frac{1}{9}$ $a_4 = \frac{1}{4^2} = \frac{1}{16}$

**2.** 12, 7, 2, –3, … arithmetic with $d = -5$.

$\frac{1}{6}, \frac{2}{6}, \frac{3}{6}, \frac{4}{6}$, … arithmetic with $d = \frac{1}{6}$.

**3.** $a_1 = -2, d = 3, n = 80, a_{80} = -2 + (80-1)3 = 235$

**4.** $a_3 = 10, a_8 = 25, n = 36$

$$\begin{Bmatrix} 25 = a_1 + 7d \\ 10 = a_1 + 2d \end{Bmatrix} \quad \begin{matrix} d = 3 \\ a_1 = 4 \end{matrix}$$

$a_{36} = 4 + (36-1)3 = 109$

**5.** $a_1 = \frac{1}{3}, d = \frac{1}{6}, n = 60$

$a_{60} = \frac{1}{3} + (60-1)\frac{1}{6} = \frac{61}{6}$

$s_{60} = \frac{60}{2}\left(\frac{1}{3} + \frac{61}{6}\right) = 315$

**6.** $\frac{1}{4}, 2, 16, 128,$ … geometric with $r = 8$.

$16, -12, 9, -\frac{27}{4}$, … geometric with $r = -\frac{3}{4}$.

$\left(16r = -12, r = -\frac{3}{4}\right)$

**7.** $a_1 = 64, a_8 = \frac{1}{2}, n = 4$

$\frac{1}{2} = 64r^7, r^7 = \frac{1}{128}$

or $r = \frac{1}{2}, a_4 = 64\left(\frac{1}{2}\right)^3 = 8$

**8.** $a_1 = \frac{1}{9}, r = 3, n = 16,$

$s_{16} = \left(\frac{1}{9}\right)\frac{\left(1-3^{16}\right)}{1-3} = 2{,}391{,}484\frac{4}{9}$

**9.** $P = 8000, r = 0.12, t = 3$

$S = P + Prt$

$= 8000 + 8000(0.12)(3) = \$10{,}880$

**10.** $P = 2000, S = 2100, t = 0.75, r = ?$

$I = S - P = 100;\ 100 = 2000 \cdot r(0.75)$

$r = \frac{100}{2000(0.75)} = 6\frac{2}{3}\%$

**11.** $S = 3000, r = 0.06, t = \frac{4}{12} = \frac{1}{3}$

$3000 = P\left(1 + (.06)\frac{1}{3}\right),$

$P = \frac{3000}{1.02} = \$2941.18$

**12.** $a_1 = 10, a_2 = 20, a_3 = 30, \ldots, a_{30} = 300$

$s_{30} = \frac{30}{2}(10 + 300) = \$4650$

**13.** First job:

$a_1 = 40{,}000,\ d = 2000,\ n = 10$

$a_{10} = 40{,}000 + (10-1)(2000) = 58{,}000$

$S_n = \frac{n}{2}(a_1 + a_n)$

$= \frac{10}{2}(40{,}000 + 58{,}000) = 490{,}000$

Second job:

$a_1 = 36{,}000,\ d = 2500,\ n = 10$

$a_{10} = 36{,}000 + (10-1)(2500) = 58{,}500$

$S_n = \frac{10}{2}(36{,}00 + 58{,}500) = 472{,}500$

The \$40,000 job pays more money over 10 years.

**14. a.** $n = 4 \cdot 10 = 40$

**b.** $i = \frac{0.08}{4} = 0.02 = 2\%$

**15. a.** $S = P\left(1 + \frac{r}{m}\right)^{mt} = P(1+i)^n$

**b.** $S = Pe^{rt}$

**16.** Monthly compounding will earn more.

**17.** Interest = $1000(1.02)^{16} - 1000 = \$372.79$

**18.** $S = 18000, n = 4 \cdot 12 = 48, i = \frac{0.054}{12} = 0.0045$

$P = \frac{18000}{(1+0.0045)^{48}} = \$14{,}510.26$

**19.** $S = 1000e^{(0.08)(6)} = \$1616.07$

**20.** $S = 100{,}000, t = 15, r = 0.1031$

$$P = \frac{100{,}000}{e^{(.1031)(15)}} = \$21{,}299.21$$

**Note: In 21–22 we can use either log or ln.**

**21.** $25{,}000 = 15{,}000\left(1+\frac{0.06}{4}\right)^{4t}$

$$\frac{5}{3} = (1.015)^{4t}$$

$$\ln\left(\frac{5}{3}\right) = 4t\ln(1.015)$$

$$t = \frac{\ln(5/3)}{4\ln(1.015)} \approx 8.577 \text{ years}$$

$$= (8.577 \times 3 \text{ quarters})$$

$$\approx 34.3 \text{ quarters}$$

**22. a.**

$$257{,}000 = 35{,}000e^{15r}$$

$$\ln\left(\frac{257{,}000}{35{,}000}\right) = 15r$$

$$\frac{\ln 257{,}000 - \ln 35{,}000}{15} = r$$

$$r \approx 0.1329 = 13.29\%$$

**b.** $APY = e^{0.1329} - 1 = 0.1421 = 14.21\%$

**23. a.** $APY = \left(1+\frac{r}{m}\right)^m - 1, \quad m = 4, \quad r = 0.072$

$$APY = \left(1+\frac{0.072}{4}\right)^4 - 1$$

$$= 1.0740 - 1 = 0.0740 = 7.40\%$$

**b.** $APY = e^r - 1 = e^{.072} - 1$

$$= 0.07466 = 7.47\%$$

**24.** $a_1 = 1, a_2 = 2, \ldots, r = 2;$

$$a_{64} = 1 \cdot 2^{64-1} = 2^{63}$$

**25.** $a_1 = 1, r = 2, a_{32} = 1 \cdot 2^{32-1} = 2^{31}$

$$s_{32} = \frac{1(1-2^{32})}{1-2} = 2^{32} - 1 \approx 4.295 \times 10^9$$

**26.** $R = 800, n = 20, \frac{0.12}{2} = 0.06$

$$S = 800 \cdot s_{\overline{20}|0.06} = 800(36.785592)$$

$$= \$29{,}428.47$$

**27.** $S = 80{,}000, n = 10, i = 0.06;$

$$R = S \cdot \frac{1}{s_{\overline{10}|0.06}} = 80{,}000(0.075868) = \$6069.44$$

**28.** $R = 800, n = 10 \cdot 2 = 20, i = \frac{0.12}{2} = 0.06$

$$S_{due} = 800\left[\frac{(1+.06)^{20}-1}{0.06}\right](1+.06) = \$31{,}194.18$$

**29.** $S = 250{,}000, n = \frac{9}{2} \cdot 4 = 18, i = \frac{0.102}{4} = 0.0255$

$$R = 250{,}000\left[\frac{0.0255}{(1+.0255)^{18}-1}\right]\left(\frac{1}{1+.0255}\right)$$

$$= \$10{,}841.24$$

**30.** $S = 60{,}000,\ n = ?,\ i = \frac{0.072}{4} = 0.018,\ R = 1200$

$$60{,}000 = 1200\left[\frac{(1+0.018)^n - 1}{0.018}\right]$$

$$50 = \frac{(1+0.018)^n - 1}{0.018}$$

$$0.9 = (1.018)^n - 1$$

$$1.9 = 1.018^n$$

$$\ln 1.9 = n \ln 1.018$$

$$n = -\frac{\ln 1.9}{\ln 1.018} \approx 36 \text{ quarters}$$

**31.** $R = 10{,}000,\ n = 10 \cdot 2 = 20, i = \frac{0.09}{2} = 0.045$

$$A_n = 10{,}000\left[\frac{1-(1+.045)^{-20}}{0.045}\right] = \$130{,}079.36$$

**32.** $R = 3000,\ n = 6,\ k = 5 \cdot 12 - 1 = 59,$

$$i = \frac{0.078}{12} = 0.0065$$

$$A_{(n,k)} = 3000\left[\frac{1-(1+.0065)^{-6}}{0.0065}\right](1+.0065)^{-59}$$

$$= \$12{,}007.09$$

First payment is before they leave.

**33. a.** $R = \frac{295.7}{25}$ million $= \$11{,}828{,}000$

**b.** $R = 11{,}828{,}000, n = 24, i = 0.0591$

$$A_n = 11{,}828{,}000\left[\frac{1-(1+.0591)^{-24}}{0.0591}\right]$$
$$= \$149{,}688{,}218$$

Now add first payment at beginning:
$A = \$161{,}516{,}218$. This is also an annuity due with $n = 25$.

**34.** $A_n = 20{,}000,\ n = 12,\ i = \frac{0.066}{12} = 0.0055$

$$R = 20{,}000\left[\frac{0.0055}{1-(1+.0055)^{-12}}\right] = \$1{,}726.85$$

**35.** $A_{due} = 250{,}000,\ n = 20 \cdot 4 = 80,$

$i = \frac{0.062}{4} = 0.0155$

$$A_{due} : R = 250{,}000\left[\frac{0.0155}{1-(1+.0155)^{-80}}\right]\left(\frac{1}{1+.0155}\right)$$
$$= \$5390.77$$

**36.** Each semi annual coupon payment is

$$(5000)\left(\frac{0.08}{2}\right) = \$200$$

Since the desired rate of return is 10% semiannually and the bond is for 30 years, we set $i = 0.10/2 = 0.05$ and $n = (30)(2) = 60$.

Then the market price of this bond is the sum of the present values found in (1) and (2) below.

(1) the principal (or present value) of a compound interest investment at $i = 0.05$ for $n = 60$ periods with future value $S = \$5{,}000$.

$$S = P(1+i)^n$$
$$\$5{,}000 = P(1+0.05)^{60} = P(1.05)^{60}$$
$$P = \frac{\$5{,}000}{(1.05)^{60}} \approx \$267.68$$

(2) the present value of the ordinary annuity formed by the coupon payment of $R = \$200$ at $i = 0.05$ for $n = 60$ periods

$$A_n = R\left[\frac{1-(1+i)^{-n}}{i}\right]$$
$$= \$200\left[\frac{1-(1.05)^{-60}}{0.05}\right] \approx \$3{,}785.86$$

Thus, to earn the desired 10 yield, the market price an investor should pay for this bond is

Price = $\$3{,}785.86 + \$267.68 = \$4053.54$

Hence the bond is selling at a discount.

**37.** $R = 40{,}000,\ i = \frac{0.068}{2} = 0.034,$

$A_n = \$488{,}000,\ n = ?$

$$488{,}000 = 40{,}000\left[\frac{1-(1+0.034)^{-n}}{0.034}\right]$$
$$12.2 = \frac{1-(1.034)^{-n}}{0.034}$$
$$0.4148 = 1-(1.034)^{-n}$$
$$-0.5852 = -1.034^{-n}$$
$$0.5852 = 1.034^{-n}$$
$$\ln 0.5852 = -n \ln 1.034$$
$$n = -\frac{\ln 0.5852}{\ln 1.034} \approx 16 \text{ half-years, or 8 years}$$

**38.** $A_n = 1000,\ n = 12,\ i = 0.01$

$$R = 1000\left[\frac{0.01}{1-(1+.01)^{-12}}\right] = \$88.85$$

**39.** $R = 1288.29, n = 8, k = 5, i = 0.06$

$$A_{n-k} = 1288.29\left[\frac{1-(1+.06)^{-(8-5)}}{0.06}\right] = \$3443.61$$

**40.** $R = 4500, n = 18, i = \frac{0.04}{4} = 0.01$

$$A_n = 4500\left[\frac{1-(1+0.04)^{-18}}{0.04}\right]$$
$$\approx 56{,}966.84 \text{ (borrowed amount)}$$

Adding the \$90,000 down payment, the house would have cost \$146,966.84 if she paid cash.

**41.** $i = 0.00625$
Interest = (Unpaid Balance)(.00625)(1)
Balance Reduction = Payment – Interest
57: Interest = (95042.20)(.00625) = 594.01

| Payment | Amount | Interest | Balance Reduction | Unpaid Balance |
|---|---|---|---|---|
| 57 | $699.22 | $594.01 | $105.21 | $94,936.99 |
| 58 | $699.22 | $593.36 | $105.86 | $94,831.13 |

**42. a.** $R = 36,795.60,\ i = \dfrac{0.054}{2} = 0.027,$
$n = 2 \cdot 15 = 30,\ k = 17$

$$A_{n-k} = 36,795.60\left[\frac{1-(1+.027)^{-(30-17)}}{0.027}\right]$$
$$\approx \$398,934.67$$
$$\$398,934.67 + \$2,200 = \$401,134.67$$

**b.** $A_n = 401,134.67,\ n = 2 \cdot 5 = 10,$
$i = \dfrac{0.048}{2} = 0.024$

$$R = 401,134.67\left[\frac{0.024}{1-(1+0.024)^{-10}}\right]$$
$$\approx \$45,596.63$$

**c.** Original remaining:
$13(36,795.60) = \$478,342.80$
Refinance:
$10(45,596.63) = \$455,966.30$
$\$478,342.80 - \$455,966.30$
$= \$22,376.50$ saved

**43.** $S = R\left[\dfrac{(1+i)^n - 1}{i}\right]$

**44.** $I = Prt$

**45.** $A_n = R\left[\dfrac{1-(1+i)^{-n}}{i}\right]$

**46.** $S = P(1+i)^n$

**47.** $A_n = R\left[\dfrac{1-(1+i)^{-n}}{i}\right]$, solved for $R$

**48.** $S = Pe^{rt}$

**49.** $P = 2500, S = 38,000, n = 18$
$$38,000 = 2500(1+i)^{18}$$
$$(1+i)^{18} = \frac{38,000}{2500}$$
$$18\ln(1+i) = \ln 38,000 - \ln 2500$$
$$18\ln(1+i) = 10.5453 - 7.8240 = 2.7213$$
$$\ln(1+i) = \frac{2.7213}{18} = 0.1512$$
$$1 + i = e^{0.1512} = 1.1632$$
$$i = 0.1632 = 16.32\%$$

**50.** $S = 40,000, n = 10 \cdot 12 = 120, i = \dfrac{0.084}{12} = 0.007$
$$R = 40,000\left[\frac{0.007}{(1+.007)^{120} - 1}\right] = \$213.81$$

**51. a.** $P = 1000,\ r = 8\%,\ t = 6$
$S = P + Prt = 1000 + 1000(0.08)(6) = \$1,480$

**b.** $P = 1000,\ i = \dfrac{0.08}{2} = 0.04,\ n = 6 \cdot 2 = 12$
$S = P(1+i)^n = 1000(1+0.04)^{12} = \$1601.03$

**52.** $R = 500,\ i = \dfrac{0.08}{4} = 0.02,\ n = 4 \cdot 4 = 16$
$$S = R\left[\frac{(1+i)^n - 1}{i}\right] = 500\left[\frac{(1+0.02)^{16} - 1}{0.02}\right]$$
$$= 500(18.6393) = \$9319.64$$

**53.** $P = 8000, S = 22,000, r = 0.07$
$$22,000 = 8000e^{0.07t}$$
$$e^{0.07t} = \frac{22,000}{8000}$$
$$0.07t = \ln 22,000 - \ln 8000$$
$$0.07t = 9.9988 - 8.9872 = 1.0116$$
$$t = 14.45$$
$$\approx 14.5 \text{ years}$$

**54.** $P = 87.89$, $S = 105.34$, $t = 1/4$
$I = S - P = 105.34 - 87.89 = 17.45$
$17.45 = 87.89 \cdot r \cdot \frac{1}{4}$
$$r = \frac{4(17.45)}{87.89} = 0.794 = 79.4\%$$

**55.** $R = 2000,\ n = 30 \cdot 12 = 360,\ i = \dfrac{0.048}{12} = 0.004$
$$A_n = 2000\left[\frac{1-(1+0.004)^{-360}}{0.004}\right] \approx \$381{,}195$$

**56. a.** Each semi annual coupon payment is
$$(5000)\left(\frac{0.072}{2}\right) = \$180$$
Since the desired rate of return is 8% semiannually and the bond is for 10 years, we set $i = 0.08/2 = 0.04$ and $n = (10)(2) = 20$.
Then the market price of this bond is the sum of the present values found in (1) ands (2) below.

(1) the principal (or present value) of a compound interest investment at $i = 0.04$ for $n = 20$ periods with future value $S = \$5,000$
$$S = P(1+i)^n$$
$$\$5,000 = P(1+0.04)^{20} = P(1.04)^{20}$$
$$P = \frac{\$5,000}{(1.04)^{20}} \approx \$2281.93 \text{ (nearest cent)}$$

(2) the present value of the ordinary annuity formed by the coupon payment of $R = \$180$ at $i = 0.04$ for $n = 20$ periods
$$A_n = R\left[\frac{1-(1+i)^{-n}}{i}\right]$$
$$= \$180\left[\frac{1-(1.04)^{-20}}{0.04}\right] \approx \$2,446.26 \text{ (nearest cent)}$$
Thus, to earn the desired 10 yield, the market price an investor should pay for this bond is
Price = $\$2,281.93 + \$2,446.26 = \$4,728.19$
Hence the bond is selling at a discount.

**b.** $5398.07

**c.** $1749.88

**d.** 10.78%

**57.** $R = 400$, $n = 4 \cdot 12 = 48$, $i = \dfrac{0.054}{12} = 0.0045$
$$S_{due} = R\left[\frac{(1+i)^n - 1}{i}\right](1+i)$$
$$= 400\left[\frac{(1+0.0045)^{48} - 1}{0.0045}\right](1+0.0045)$$
$$= \$21,474.08$$

**58.** $A_{(n,k)} = 72{,}000,\ \ n = 9 \cdot 2 = 18,$

$k = 11 \cdot 2 = 22,\ \ i = 0.0365$

$$A_{(n,k)} : R = 72{,}000\left[\frac{0.0365}{1-(1+.0365)^{-18}}\right](1+.0365)^{22}$$

$= \$12{,}162.06$

**59.** $APY = \left(1+\frac{r}{m}\right)^m - 1 = \left(1+\frac{0.0652}{4}\right)^4 - 1 \approx 0.066811\ \ APY = e^r - 1 = e^{0.0648} - 1 = 0.066946$

The compounded continuous offer is the higher rate.

**60. a.** $R = 1653.21,\ i = \frac{0.051}{12} = 0.00425,\ n = 12 \cdot 25 = 300,\ k = 70$

$$A_{n-k} = 1653.21\left[\frac{1-(1+.00425)^{-(300-70)}}{0.00425}\right] \approx \$242{,}329.20$$

$\$242{,}329.20 + \$1100 = \$243{,}429.20$

**b.** $A_n = 243{,}429.20,\ n = 12 \cdot 15 = 180,\ i = \frac{0.042}{12} = 0.0035$

$$R = 243{,}429.20\left[\frac{0.0035}{1-(1+0.0035)^{-180}}\right] \approx \$1825.11$$

**c.** $$243{,}429.20 = 2000\left[\frac{1-(1+0.0035)^{-n}}{0.0035}\right]$$

$$\frac{243{,}429.20(0.0035)}{2000} = 1-(1.0035)^{-n}$$

$$(1.0035)^{-n} = 1 - \frac{243{,}429.20(0.0035)}{2000}$$

$$-n\ln(1.0035) \approx \ln(0.5739989)$$

$n \approx 158.9$; 159 payments (rather than 180)

**d.** Original remaining:

$230(1653.21) = \$380{,}238.30$

Refinance:

$158.9(2000) = \$317{,}800$

$\$380{,}238.30 - \$317{,}800$

$= \$62{,}438.30$ saved

**61.** $R = 1000,\ n = 40,\ i = \frac{0.12}{12} = 0.01$

$$A_n = 1000\left[\frac{1-(1+0.01)^{-40}}{0.01}\right] \approx \$32{,}834.69$$

**62.** $P = 8000,\ i = 0.12,\ n = 3$

$S = Pe^{rt} = 8000e^{0.36} = 8000(1.433329415) = \$11{,}466.64$

$S - P = \$3{,}466.64$

**63.** **a.** Here $A_n = 184,000,\ n = 25 \cdot 12 = 300, i = \dfrac{0.06}{12} = 0.005$

The monthly payment size of the loan will be

$$184,000 = R\left[\frac{1-(1+0.005)^{-300}}{0.005}\right]$$

$$184,000 = R(155.207)$$

$$R = \$1185.51$$

Hence the monthly payment is $\$1185.51$.

**b.** In 25 years the couple made 300 payments, so the total amount paid after 25 years is $(25)(1185.51) = \$355,653$.

**c.** Of the \$355,653 paid, \$184,000 was paid toward the principal (for the condominium itself). The remaining \$171,653 is the total amount of interest paid.

**d.** $R = 1185.51,\ i = 0.005,\ n = 300 = 120,\ k = 12 \cdot 7 = 84$

$$A_{n-k} = 1185.51\left[\frac{1-(1+0.005)^{-(300-84)}}{0.005}\right] \approx \$156,366.25$$

**64.** Bonus: $P = 12,500$, $n = 150$, $i = \dfrac{10.8\%}{12} = 0.009$

$$S_1 = 12,500(1+0.009)^{150} = \$47,927.52$$

Annuity: $S_2 = 150\left[\dfrac{(1+0.009)^{150}-1}{0.009}\right] = \$47,236.69$

**a.** $S = S_1 + S_2 = \$95,164.21$

**b.** $R = \$95,164.21\left[\dfrac{0.009}{1-(1+0.009)^{-120}}\right] = \$1300.14$

**65.** The withdrawals begin one month after the last deposit. Look at the problem this way. Amount needed in account at first withdrawal is $A = 10,000 + 10,000\left[\dfrac{1-(1+0.0055)^{-3}}{0.0055}\right] = \$39,673.00$. Now we have

$$39,673.00 = R\left[\frac{(1+0.0055)^{36}-1}{0.0055}\right](1+0.0055).$$

Solving we have $R = \$994.08$.

**66.** We'll assume that Aruam invests the $4000 IRA at 8.4% compounded monthly from age 22 to age 67, a total of 45 years. This would give her $S = 4000(1+0.007)^{12\times 45} = 4000(1.007)^{540} = \$172{,}971.32$. At age 30, she has 37 years till retirement. She wants to receive $20,000 a month for 20 years ($n = 12\cdot 20 = 240$). First we need to figure out what the present value of the annuity will be when she turns 67.

$$R = 10{,}000,\ i = 0.007,\ n = 240,\ \text{so}\ A_n = 20{,}000\left[\frac{1-(1+0.007)^{-240}}{0.007}\right] \approx \$2{,}321{,}520.10$$

She will need to have $2,321,520.10 when she turns 67. She will need an additional $\$2{,}321{,}520.10 - \$172{,}971.32 = \$2{,}148{,}548.68$ from the monthly payments. So now we calculate the necessary monthly payment. $A_n = 1{,}074{,}274.39,\ R = ?,\ i = 0.007,\ n = 37\times 12 = 444$.

$$2{,}148{,}548.78 = R\left[\frac{(1+0.007)^{444}-1}{0.007}\right]$$

$$2{,}148{,}548.78 = R(3019.325)$$

$$R = \$711.60$$

She should make a monthly payment of $711.60 from age 30 to age 67.

**67.** First we calculate what the regular payment would be on $2,600,000 borrowed for 15 years at 5.6% compounded quarterly. $A_n = 2{,}600{,}000,\ i = 0.056/4 = 0.014,\ R = ?,\ n = 15\cdot 4 = 60$

$$R = 2{,}600{,}000\left[\frac{0.014}{1-(1+0.014)^{-60}}\right] = \$64{,}337.43$$

Next we find the unpaid balance after 2 years, at this time there will be $60 - 8 = 52$ payments remaining.

$$A_{n-k} = \$64{,}337.43\left[\frac{1-(1+0.014)^{-52}}{0.014}\right] = \$2{,}365{,}237.24$$

Increasing to $R = 70{,}000$, we solve for $n$.

$$2{,}365{,}237.24 = 70{,}000\left[\frac{1-(1+0.014)^{-n}}{0.014}\right]$$

$$33.7891 = \frac{1-(1.014)^{-n}}{0.014}$$

$$0.473 = 1 - 1.014^{-n}$$

$$-0.527 = -1.014^{-n}$$

$$0.527 = 1.014^{-n}$$

$$\ln 0.527 = -n\ln 1.014$$

$$n = -\frac{\ln 0.527}{\ln 1.014} = 46.1 \text{ quarterly payments}$$

Now we calculate the savings. Without altering the monthly payment, the company would have made 60 payments of $64,337.43 for a total of $3,860,245.80. Really, they made 8 payments of $64,337.43 and 46.1 payments of $70,000 for a total of $3,741,699.44. Their savings was $118,546.36.

# Chapter 6: Mathematics of Finance

## Chapter 6 Test

**1.** $S = 47{,}000, P = 8000, r = 0.07,$

$$S = Pe^{rt}$$
$$47{,}000 = 8000e^{0.07t}$$
$$e^{0.07t} = \frac{47{,}000}{8000} \approx 5.875$$
$$0.07t \ln e = \ln 5.875$$
$$t = \frac{\ln 5.875}{0.07} \approx 25.3 \text{ years}$$

**2.** $R = 100, n = \frac{11}{2} \cdot 12 = 66, i = \frac{0.069}{12} = 0.00575$

$$S_n = 100\left[\frac{(1+.00575)^{66} - 1}{0.00575}\right] \approx \$7999.41$$

**3.** $S = 10{,}000,\ P = 9510,\ t = 0.75$

$$I = S - P = 490$$
$$r = \frac{I}{Pt} = \frac{490}{9510(0.75)} \approx 0.0687 = 6.87\%$$

**4.** $R = 14{,}357.78, n = 40, k = 25, i = .041$

Unpaid balance;

$$= 14{,}357.78\left[\frac{1-(1+.041)^{-(40-25)}}{0.041}\right]$$
$$\approx \$158{,}524.90$$

**5.** $P = 1000, S = 13{,}500, n = 9$

$$13{,}500 = 1000(1+i)^9$$
$$1+i = \left(\frac{13{,}500}{1000}\right)^{1/9} \approx 1.3353$$
$$i \approx 0.3353 \approx 33.53\%$$

This problem can also be solved by taking ln of both sides.

**6.** $A_n = 97{,}000,\ n = 25 \cdot 12 = 300,\ i = 0.006$

**a.** $R = 97{,}000\left[\frac{0.006}{1-(1+.006)^{-300}}\right] \approx \$698$

**b.** Interest paid = 300(698) – 97,000 = \$112,400

**7.** $P = 2500, r = 0.04, t = \frac{15}{12} = 1.25$

$$S = 2500 + 2500(.04)(1.25) = \$2625$$

**8.** $S = 12{,}000, n = 6 \cdot 2 = 12, i = 0.031$

$$12{,}000 = R\left[\frac{(1+0.031)^{12} - 1}{0.031}\right](1+0.031)$$
$$R = \frac{12{,}000}{1.031}\left[\frac{0.031}{(1+.031)^{12} - 1}\right] \approx \$815.47$$

**9.** $i = \frac{0.084}{12} = 0.007, APY = (1+.007)^{12} - 1$

$$= 0.0873 = 8.73\%$$

**10.** $R = 3000, n = 15 \cdot 4 = 60, i = 0.015$

$$A_{due} = 3000\left[\frac{1-(1+.015)^{-60}}{0.015}\right](1+.015)$$
$$\approx \$119{,}912.92$$

**11.** $P = 10{,}000, r = 0.07, t = 20$

$$S = 10{,}000e^{(.07)(20)} = \$40{,}552$$

**12.** $S = 780{,}000,\ n = 4 \cdot 15 = 60,\ i = \frac{0.08}{4} = 0.02$

$$780{,}000 = R\left[\frac{1-(1+0.02)^{-60}}{0.02}\right]$$
$$R = 780{,}000\left[\frac{0.02}{1-(1+0.02)^{-60}}\right]$$
$$R \approx \$22{,}439.01$$

**13.** $S = 9{,}500, n = 5 \cdot 4 = 20, i = 0.017$

$$P = \frac{9500}{(1+.017)^{20}} \approx \$6781.17$$

**14.** $S = 500{,}000,\ R = 1500,\ i = 0.041,\ n = ?$

$$500{,}000 = 1500\left[\frac{(1+0.041)^n - 1}{0.041}\right]$$
$$333.33 = \frac{(1.041)^n - 1}{0.041}$$
$$13.666 = 1.041^n - 1$$
$$14.666 = 1.041^n$$
$$\ln 14.666 = n \ln 1.041$$
$$n = \frac{\ln 14.666}{\ln 1.041} \approx 66.8;\ 67 \text{ half years}$$

**15.** $R = 400, n = 15 \cdot 12 = 180, i = \dfrac{0.06}{12} = 0.005$

$$S_{due} = 400\left[\frac{(1+.005)^{180}-1}{0.005}\right](1+.005)$$

$$\approx \$116{,}909.12$$

**16.** Deferred Annuity: $R = 4000$, $n = 16$, $k = 40$, $i = 0.016$

$$A = 4000\left[\frac{1-(1+.016)^{-16}}{0.016}\right](1+.016)^{-40}$$

$$\approx \$29{,}716.47$$

**17. a.** $R = 1010.76,\ i = \dfrac{0.054}{12} = 0.0045,$

$n = 12 \cdot 30 = 360,\ k = 80$

$$A_{n-k} = 1010.76\left[\frac{1-(1+.0045)^{-(360-80)}}{0.0045}\right]$$

$$\approx \$160{,}720.52$$

$\$160{,}720.52 + \$550 = \$161{,}270.52$

**b.** $A_n = \$161{,}270.52,\ n = 12 \cdot 25 = 300,$

$i = \dfrac{0.048}{12} = 0.004$

$$R = 161{,}270.52\left[\frac{0.004}{1-(1+0.004)^{-300}}\right]$$

$$\approx \$924.08$$

**c.** Original remaining:

$280(1010.76) = \$283{,}012.80$

Refinance:

$300(924.08) = \$277{,}224.00$

$\$283{,}012.80 - \$277{,}224.00$

$= \$5{,}788.80$ saved

**18.** Each semiannual coupon payment is

$$(\$10{,}000)(0.10/2) = (\$10{,}000)(0.05) = \$500$$

Since the desired rate of return is 6% semiannually and the bond is for 10 years, we set

$i = \dfrac{0.06}{2} = 0.03$ and $n = 10 \times 2 = 20$

Then the market price of this bond is the sum of the present values found in (1) and (2) below.

(1) the principal (or present value) of a interest investment at $i = 0.03$ for $n = 20$ periods with future value $S = \$10000$

$$S = P(1+i)^n$$

$$\$10000 = P(1+0.03)^{20}$$

$$P = \frac{\$10000}{(1.03)^{20}} \approx 5536.76$$

(2) The present value of the ordinary annuity formed by the coupon payments of $R = \$500$ at $i = 0.03$ for $n = 20$ periods

$$A_n = R\left[\frac{1-(1+i)^{-n}}{i}\right]$$

$$A_n = 500\left[\frac{1-(1.03)^{-20}}{0.03}\right] \approx 7438.74$$

Thus, to earn the desired 7.2% yield, the market price an investor should pay for this bond is

Price $= \$5536.76 + \$7438.74$

$= \$12975.49$; premium

**19. a.** The difference between successive terms is $-5.5$.

**b.** $a_{51} = 298.8 + (51-1)(-5.5) = 23.8$

**c.** $s_{51} = \dfrac{51}{2}(298.8 + 23.8) = 8226.3$

**20.** $a_1 = 400, r = 0.6, n = 31$ (Look at sequence.)

$$s_{31} = 400\frac{(1-0.6^{31})}{1-0.6} = 1000 \text{ mg}$$

**21. a.** After 24 payments there are 336 payments remaining.
$i = 0.084/12 = 0.007$, $R = \$1142.76$, and $n - k = 336$.

$$A_{n-k} = 1142.76\left[\frac{1-(1+0.007)^{-336}}{0.007}\right] \approx \$147,585.55\text{(without \$2000)}$$

The extra \$2000 payment would essentially reduce the unpaid balance to \$145,585.55.

**b.** $R = 1142.76,\ A_n = 145,585.55,\ i = 0.007,\ n = ?$

$$145,585.55 = 1142.76\left[\frac{1-(1+0.007)^{-n}}{0.007}\right]$$

$$127.3982 = \frac{1-(1.007)^{-n}}{0.007}$$

$$0.8918 = 1-(1.007)^{-n}$$

$$-0.1082 = -1.007^{-n}$$

$$0.1082 = 1.007^{-n}$$

$$\ln 0.1082 = -n \ln 1.007$$

$$n = -\frac{\ln 0.1082}{\ln 1.007} \approx 318.8 \text{ monthly payments}$$

Total payment of loan is $\$1142.76(318.8+24)+2000 = 393,738.13$, so \$243,738.13 is interest.

**c.** $R = 1160,\ A_n = 147,585.55,\ i = 0.007,\ n = ?$

$$147,585.55 = 1160\left[\frac{1-(1+0.007)^{-n}}{0.007}\right]$$

$$127.2289 = \frac{1-(1.007)^{-n}}{0.007}$$

$$0.8906 = 1-(1.007)^{-n}$$

$$-0.1094 = -1.007^{-n}$$

$$0.1094 = 1.007^{-n}$$

$$\ln 0.1094 = -n \ln 1.007$$

$$n = -\frac{\ln 0.1094}{\ln 1.007} \approx 317.2 \text{ monthly payments}$$

Total payment of loan is $\$1142.75(24)+\$1160(317.2) = \$395,378$, so \$245,378 is interest.

**d.** Paying the \$2000 is slightly better; it saves approximately \$1640 in interest.

**22.** $P = 10,000,\ n = \frac{9}{2}\cdot 4 = 18,\ i = 0.01925$

$$S_1 = 10,000(1+.01925)^{18} = 14,094.61$$

$$S_2 = 50,000 - S_1 = 35,905.39$$

$$R = 35,905.39\left[\frac{0.01925}{(1+.01925)^{18}-1}\right] \approx \$1688.02$$

**23.** **a.** $R = 3000, n = 8 \cdot 2 = 16, i = 0.0375$

$$S_1 = 3000\left[\frac{(1+.0375)^{16}-1}{0.0375}\right] = 64{,}178.22$$

$$S = 64{,}178.22(1+0.0375)^{40} \approx \$279{,}841.35$$

**b.** $A = 279{,}841.35, n = 20 \cdot 2 = 40, i = 0.0375$

$$A_{due} : R = 279{,}841.35\left[\frac{0.0375}{1-(1+.0375)^{-40}}\right](1+.0375)^{-1}$$

$$\approx \$13{,}124.75$$

# Chapter 7: Introduction to Probability

## *Exercises 7.1*

**1. a.** $\Pr(\text{R}) = \frac{4}{10} = \frac{2}{5}$

**b.** $\Pr(\text{G}) = \frac{0}{10} = 0$

**c.** $\Pr(\text{R or W}) = \frac{10}{10} = 1$

**3.** $\Pr(4,8,12) = \frac{3}{12} = \frac{1}{4}$

**5.** $\Pr(\text{greater than } 0) = \frac{6}{6} = 1$

**7. a.** $\Pr(\text{Red}) = \frac{3}{10}$

**b.** $\Pr(\text{Odd}) = \frac{5}{10} = \frac{1}{2}$

**c.** $\Pr(\text{Red and Odd}) = \Pr(1 \text{ or } 3) = \frac{2}{10} = \frac{1}{5}$

**d.** $\Pr(\text{Red or Odd})$

$= \Pr(1,2,3,5,7,\text{or } 9) = \frac{6}{10} = \frac{3}{5}$

**e.** $\Pr(\text{Not Black})$

$= 1 - \Pr(\text{Black}) = 1 - \frac{3}{10} = \frac{7}{10}$

**9. a.** $\Pr(\text{Queen}) = \frac{4}{52} = \frac{1}{13}$

**b.** $\Pr(\text{Heart or Diamond}) = \frac{1}{2}$

**c.** $\Pr(\text{Spade}) = \frac{1}{4}$

**11.** Sample space = {HH, HT, TH, TT}

**a.** $\Pr(0\text{H}) = \frac{1}{4}$

**b.** $\Pr(1\text{H}) = \frac{2}{4} = \frac{1}{2}$

**c.** $\Pr(2\text{H}) = \frac{1}{4}$

**13. a.** $\Pr(\text{Sum} = 4) = \frac{3}{36} = \frac{1}{12}$

**b.** $\Pr(\text{Sum} = 10) = \frac{3}{36} = \frac{1}{12}$

**c.** $\Pr(\text{Sum} = 12) = \frac{1}{36}$

**15. a.** $\Pr(4 \le S \le 7) = \frac{3+4+5+6}{36} = \frac{1}{2}$

(3 ways to roll a 4, etc.)

**b.** $\Pr(8 \le S \le 12) = \frac{5+4+3+2+1}{36} = \frac{15}{36} = \frac{5}{12}$

(5 ways to roll 8, etc.)

**17. a.** $\Pr(6) = \frac{431}{1200}$

**b.** The die is biased since Pr(6) should be about $1/6$.

**19. a.** $2:3$ or $\frac{2}{3}$

**b.** $3:2$ or $\frac{3}{2}$

**21. a.** $\Pr(\text{Win}) = \frac{1}{20+1} = \frac{1}{21}$

**b.** $\Pr(\text{Lose}) = \frac{20}{20+1} = \frac{20}{21}$

**23.** $\Pr(\text{inner city}) = 0.46$

**25. a.** $\Pr(\text{defective turn signal}) = \frac{63}{425} = 0.1482$

**b.** $\Pr(\text{defective tires}) = \frac{32}{425} = 0.753$

**27. a.** $\Pr(\text{Republican will vote})$

$= \frac{2835}{4500} = \frac{63}{100} = 0.63$

$\Pr(\text{Democrat will vote})$

$= \frac{2501}{6100} = \frac{41}{100} = 0.41$

$\Pr(\text{Independent will vote})$

$= \frac{1122}{2200} = \frac{51}{100} = 0.51$

**b.** Probability is highest that a Republican will vote in the next election.

**29. a.** $\Pr(100\% \text{ discount}) = \frac{1}{3601}$

**b.** $\Pr(50\% \text{ discount}) = \frac{100}{3601}$

**c.** $\Pr(< 50\% \text{ discount}) = \dfrac{3000 + 500}{3601} = \dfrac{3500}{3601}$

**d.** $\Pr(30\% \text{ discount}) = \dfrac{500}{3601}$;

$\Pr(> 30\% \text{ discount}) = \dfrac{101}{3601}$.

So, more likely to get a 30% discount than a higher discount.

**31. a.** $\Pr(\text{black \& recommended activity})$

$= \dfrac{40.2}{100} = 0.402$

**b.** $\Pr(\text{white \& not recommended activity})$

$= \dfrac{100 - 50.9}{100} = \dfrac{49.1}{100} = 0.491$

**c.** $\text{Sum} = 41.1 + 36.8 + 22.1 = 100$

This means that all categories of physical activity for Hispanics are included in the table.

**33.** $S = \{A^+, A^-, B^+, B^-, AB^+, AB^-, O^+, O^-\}$

**35. a.** $\Pr(AB) = 0.04$

**b.** $\Pr(\text{not } AB) = 0.96$

**37. a.** $\Pr(\text{Asian descent}) = 0.13$

**b.** $\Pr(\text{not Asian descent}) = 1 - 0.13 = 0.87$

**39.** $\Pr(\text{Empty}) = \dfrac{60,000}{2,000,000} = \dfrac{3}{100}$

**41.** $\Pr(\text{has lactose intolerance}) = 0.75$

**43.** $\Pr(\text{Woman}) = \dfrac{3}{9} = \dfrac{1}{3}$

**45.** $\Pr(\text{grade}) = \dfrac{2}{6} = \dfrac{1}{3}$

**47.** $\Pr(\text{No V and DS}) = 0.22$; yes

**49.** Sample space = {GGG, GGB, GBG, BGG, BBG, BGB, GBB, BBB}; $\Pr(2G) = \dfrac{3}{8}$

**51. a.** No

**b.** Sample space = {GG, GB, BG, BB}

**c.** $\Pr(1B \text{ and } 1G) = \dfrac{2}{4} = \dfrac{1}{2}$

**53.** $\Pr(\text{Q1 is correct}) = \dfrac{1}{5}$

**55.** $\Pr(\text{Defective}) = \dfrac{6}{250} = \dfrac{3}{125}$

**57.** $\Pr(A) = \dfrac{182}{25,500(365)} \approx 0.0000196$

$\Pr(B) = \dfrac{51}{3890(365)} \approx 0.0000359$

$\Pr(C) = \dfrac{118}{8580(365)} \approx 0.0000377$

Intersection C is the most dangerous.

**59. a.** $\Pr(A) = \dfrac{557}{1200}$

**b.** $\Pr(\text{No Vote}) = \dfrac{110}{1200} = \dfrac{11}{120}$

**61. a.** $\Pr(B) = \dfrac{2}{10} = \dfrac{1}{5}$; $\Pr(G) = \dfrac{8}{10} = \dfrac{4}{5}$

**b.** $\Pr(B) = \dfrac{1194}{1220 + 1194} \approx 0.4946$

$\Pr(G) = \dfrac{1220}{2414} \approx 0.5054$

**c.** Reality is most likely that $\Pr(B) = \Pr(G) = \dfrac{1}{2}$, as reflected in part b.

**63.** Odds in favor: 3 to 1

$\Pr(\text{Marriage will last a lifetime}) = \dfrac{3}{3+1} = \dfrac{3}{4}$

**65.** Odds against dying in accident: 3992 to 3

$\Pr(\text{dying in accident}) = \dfrac{3}{3 + 3992} \approx 0.00075$

**67.** Pr(newspaper) = 0.25; Pr (not newspaper) = 0.75
Odds against newspaper: 75 to 25 = 3 to 1

# Chapter 7: Introduction to Probability Functions

## *Exercises 7.2*

**1.** $\Pr(6, 12) = \frac{2}{12} = \frac{1}{6}$

**3.** $\Pr(1, 3, 5) = \frac{1}{2}$

**5.** $\Pr(E') = 1 - \Pr(E) = 1 - \frac{3}{5} = \frac{2}{5}$

**7. a.** $\Pr(R \cap \text{Even}) = \Pr(2, 4) = \frac{2}{14} = \frac{1}{7}$

**b.** $\Pr(R \cup \text{Even})$

$= \Pr(R) + \Pr(\text{Even}) - \Pr(R \cap \text{Even})$

$= \frac{5}{14} + \frac{7}{14} - \frac{2}{14} = \frac{10}{14} = \frac{5}{7}$

**9.** $\Pr(\text{odd or } \{4, 8, 12\})$

$= \frac{6}{12} + \frac{3}{12} - 0 = \frac{9}{12} = \frac{3}{4}$

**11.** $\Pr(\text{odd or } \{3, 6, 9, 12\})$

$= \frac{6}{12} + \frac{4}{12} - \frac{2}{12} = \frac{8}{12} = \frac{2}{3}$

**13.** $\Pr(W \cup G) = \frac{4}{17} + \frac{6}{17} - 0 = \frac{10}{17}$

**15.** $\Pr(R \cup W) = \frac{2}{6} + \frac{2}{6} - 0 = \frac{4}{6} = \frac{2}{3}$

**17. a.** $\Pr(\text{not } E) = \frac{9}{18} = \frac{1}{2}$

**b.** $\Pr(R \cap E) = \frac{6}{18} = \frac{1}{3}$

**c.** $\Pr(R \cup E) = \frac{13}{18} + \frac{9}{18} - \frac{6}{18} = \frac{16}{18} = \frac{8}{9}$

**d.** $\Pr(R' \cap E') = \Pr(15, 17) = \frac{2}{18} = \frac{1}{9}$

**19.** $\Pr(IC') = 1 - \Pr(IC) = 1 - 0.46 = 0.54$

**21. a.** $\Pr(DT') = 1 - \Pr(DT) = 1 - \frac{63}{425} = \frac{362}{425}$

**b.** $\Pr(\text{No Defects}) = 1 - \frac{63+32}{425} = \frac{330}{425} = \frac{66}{85}$

**23.** $\Pr(F \cup G) = \frac{24}{100} + \frac{18}{100} - \frac{8}{100} = \frac{34}{100} = \frac{17}{50}$

**25. a.** $\Pr(S \cup R) = \frac{1}{3} + \frac{1}{2} - 0 = \frac{5}{6}$

**b.** $\Pr(S' \cap R') = 1 - \frac{5}{6} = \frac{1}{6}$

**27. a.** $\Pr(\text{female} \cap \text{less than } \$30{,}000) = 0.35$

**b.** $\Pr(\text{at least } \$50{,}000) = 0.08$

**c.** $\Pr(\text{male } \cup \text{ less than } \$30{,}000)$

$= \Pr(\text{male}) + \Pr(\text{less than } \$30{,}000) - \Pr(\text{male} \cap \text{less than } \$30{,}000) = 0.48 + 0.60 - 0.25 = 0.83$

**29. a.** $\Pr(\text{Latino } \cup \text{ female}) = \Pr(\text{Latino}) + \Pr(\text{female}) - \Pr(\text{Latino } \cap \text{ female})$

$$= \frac{15715}{119151} + \frac{36026 + 6664 + 2078 + 5671}{119151} - \frac{5671}{119151} = \frac{60483}{119151} \approx 0.508$$

**b.** $\Pr(\text{male } \cup \text{ black}) = \Pr(\text{male}) + \Pr(\text{black}) - \Pr(\text{male } \cap \text{ black})$

$$= \frac{49804 + 6084 + 2780 + 10044}{119151} + \frac{12748}{119151} - \frac{6084}{119151} = \frac{75376}{119151} \approx 0.633$$

**c.** $\Pr(\text{Asian } \cup \text{ white}) = \Pr(\text{Asian}) + \Pr(\text{white}) - \Pr(\text{Asian } \cap \text{ white})$

$$= \frac{4858}{119151} + \frac{85830}{119151} - \frac{0}{119151} = \frac{90688}{119151} \approx 0.761$$

**31. a.** $\Pr(\text{female}) = \frac{\text{\# of females}}{\text{Total}} = \frac{8978}{42,676} \approx 0.210$

**b.** $\Pr(\text{female} \cup \text{injected drug user}) = \Pr(\text{female}) + \Pr(\text{injected drug user}) - \Pr(\text{female} \cap \text{injected drug user})$

$$= \frac{8978}{42,676} + \frac{1132+8978}{42,676} - \frac{8978}{42,676} = \frac{10,110}{42,676} \approx 0.237$$

**c.** $\Pr(\text{male} \cup \text{hetrosexual}) = \Pr(\text{male}) + \Pr(\text{hetrosexual}) - \Pr(\text{male} \cap \text{hetrosexual})$

$$= \frac{33,698}{42,676} + \frac{2712+3835}{42,676} - \frac{2712}{42,676} = \frac{37,533}{42,676} \approx 0.879$$

**33. a.** $\Pr(A \cup B) = \frac{7}{12} + \frac{7}{12} - \frac{3}{12} = \frac{11}{12}$

**b.** $\Pr(B \cup C) = \frac{7}{12} + \frac{8}{12} - \frac{5}{12} = \frac{10}{12} = \frac{5}{6}$

**b.** $\Pr(E \cup S) = \frac{16}{32} + \frac{12}{32} - 0 = \frac{28}{32} = \frac{7}{8}$

**c.** $\Pr(E \cup F) = \frac{16}{32} + \frac{13}{32} - \frac{5}{32} = \frac{24}{32} = \frac{3}{4}$

**35. a.** $\Pr(S \cup LA) = \frac{12}{32} + \frac{4}{32} - 0 = \frac{16}{32} = \frac{1}{2}$

**37.** $\Pr(R \cup \text{Favors}) = \Pr(R) + \Pr(\text{Favors}) - \Pr(\text{R and Favors}) = \frac{400+50}{1000} + \frac{435}{1000} - \frac{300+25}{1000} = \frac{560}{1000} = 0.56$

**39.**

$\Pr(W \cup \text{Opposes}) = \Pr(W) + \Pr(\text{Opposes}) - \Pr(\text{W and Opposes}) = \frac{400+350}{1000} + \frac{565}{1000} - \frac{100+250}{1000} = \frac{965}{1000} = 0.965$

**41. a.** $\Pr(F \cup RA) = \frac{390+160}{1000} + \frac{560}{1000} - \frac{390}{1000} = \frac{720}{1000} = 0.72$

**b.** $\Pr(M \cup RA) = \frac{450}{1000} + \frac{560}{1000} - \frac{170}{1000} = \frac{840}{1000} = 0.84$

**c.** $\Pr(M \cup RA') = \frac{450}{1000} + \frac{440}{1000} - \frac{280}{1000} = \frac{610}{1000} = 0.61$

**43.** $\Pr(M \cup B') = \Pr(M) + \Pr(B') - \Pr(M \cap B') = \frac{50}{80} + \frac{42}{80} - \frac{30}{80} = \frac{62}{80} = \frac{31}{40}$

**45.** $\Pr(PC') = 1 - 0.87 = 0.13$

## *Exercises 7.3*

**1. a.** $\Pr(H \mid R) = \frac{\text{13 Hearts}}{\text{26 Red}} = \frac{1}{2}$

**b.** $\Pr(K \mid R) = \frac{\text{2 Kings}}{\text{26 Red}} = \frac{1}{13}$

**3. a.** $\Pr(6 \mid \text{even})$

$$= \frac{\Pr(6 \text{ and even})}{\Pr(\text{even})} = \frac{\frac{2}{9}}{\frac{6}{9}} = \frac{2}{6} = \frac{1}{3}$$

b. $\Pr(3 \mid 3 \text{ or } 6)$

$$= \frac{\Pr(3 \text{ and } (3 \text{ or } 6))}{\Pr(3 \text{ or } 6)} = \frac{\frac{1}{9}}{\frac{3}{9}} = \frac{1}{3}$$

5. $\Pr(R \mid E) = \frac{\Pr(R \text{ and } E)}{\Pr(E)} = \frac{\frac{4}{15}}{\frac{7}{15}} = \frac{4}{7}$

7. a. $\Pr(W \mid R) = \frac{6(W)}{9(\text{Total})} = \frac{2}{3}$

b. $\Pr(R \mid W) = \frac{4(R)}{9(\text{Total})} = \frac{4}{9}$

c. First draw has no effect on second draw.

$\Pr(W) = \frac{6}{10} = \frac{3}{5}$

9. a. $\Pr(H2 \text{ and } H3 \mid H1)$

$$= \frac{\Pr(H1H2H3)}{\Pr(H1)} = \frac{\frac{1}{8}}{\frac{1}{2}} = \frac{1}{4}$$

b. $\Pr(H3 \mid H1 \text{ and } H2)$

$$= \frac{\Pr(H1H2H3)}{\Pr(H1 \text{ and } H2)} = \frac{\frac{1}{8}}{\frac{1}{4}} = \frac{1}{2}$$

11. $\Pr(3 \text{ first} \cap 6 \text{ second}) = \frac{1}{6} \cdot \frac{1}{6} = \frac{1}{36}$

13. a. $\Pr(HHH) = \frac{1}{2} \cdot \frac{1}{2} \cdot \frac{1}{2} = \frac{1}{8}$

b. $\Pr(\text{at least 1 tail})$

$$= 1 - \Pr(\text{no tails}) = 1 - \frac{1}{8} = \frac{7}{8}$$

15. a. $\Pr(R1 \text{ and } W2) = \Pr(R) \cdot \Pr(W) = \frac{3}{10} \cdot \frac{2}{10} = \frac{3}{50}$

b. $\Pr(R1 \text{ and } W2) = \Pr(R) \cdot \Pr(W \mid R) = \frac{3}{10} \cdot \frac{2}{9} = \frac{1}{15}$

c. The events in (a) are independent. Since the ball was replaced, it did not affect the probability of getting a white ball the second time. The events in (b) were dependent. Not replacing the first ball changed the sample space which affected the probability of getting a white ball the second time.

17. a. $\Pr(RR) = \Pr(R) \cdot \Pr(R) = \frac{2}{5} \cdot \frac{2}{5} = \frac{4}{25}$

b. $\Pr(WW) = \Pr(W) \cdot \Pr(W) = \frac{3}{5} \cdot \frac{3}{5} = \frac{9}{25}$

c. $\Pr(R1 \text{ and } W2) = \Pr(R) \cdot \Pr(W) = \frac{2}{5} \cdot \frac{3}{5} = \frac{6}{25}$

d. $\Pr(\text{Black}) = 0$

19. $\Pr(N1Q2N3) = \Pr(N) \cdot \Pr(Q \mid N) \cdot \Pr(N \mid N \cap Q) = \frac{9}{18} \cdot \frac{5}{17} \cdot \frac{8}{16} = \frac{5}{68}$

21. a. $\Pr(R1 \text{ and } W2) = \Pr(R) \cdot \Pr(W \mid R) = \frac{1}{5} \cdot \frac{4}{4} = \frac{1}{5}$

b. $\Pr(W1 \text{ and } W2) = \Pr(W) \cdot \Pr(W \mid W) = \frac{4}{5} \cdot \frac{3}{4} = \frac{3}{5}$

c. $\Pr(R1 \text{ and } R2) = 0$ (Only one red ball in box)

23. a. $\Pr(S1 \text{ and } S2) = \Pr(S) \cdot \Pr(S \mid S) = \frac{13}{52} \cdot \frac{12}{51} = \frac{1}{17}$

b. $\Pr(H1 \text{ and } C2) = \Pr(H) \cdot \Pr(C \mid H) = \frac{13}{52} \cdot \frac{13}{51} = \frac{13}{204}$

**25.** **a.** $\Pr(R|W)=\frac{13}{17}$

**b.** $\Pr(E1 \text{ and } E2)=\Pr(E)\cdot\Pr(E \mid E)=\frac{9}{18}\cdot\frac{8}{17}=\frac{4}{17}$

**c.** $\Pr(ER1 \text{ and } E2)=\Pr(E \text{ and } R)\cdot\Pr(E \mid E\cap R)=\frac{6}{18}\cdot\frac{8}{17}=\frac{8}{51}$

**27.** $\Pr(\text{Favors} \mid \text{Democrat})=\frac{\Pr(\text{Favors and Democrat})}{\Pr(\text{Democrat})}=\frac{\frac{310}{1000}}{\frac{520}{1000}}=\frac{310}{520}=\frac{31}{52}$

**29.** $\Pr(\text{Favors} \mid \text{Republican})=\frac{\Pr(\text{Favors and Republican})}{\Pr(\text{Republican})}=\frac{\frac{125}{1000}}{\frac{480}{1000}}=\frac{125}{480}=\frac{25}{96}$

**31.** $\Pr(\text{Opposes} \mid \text{Nonwhite})=\frac{\Pr(\text{Opposes and NonWhite})}{\Pr(\text{Democrat})}=\frac{\frac{25+190}{1000}}{\frac{50+200}{1000}}=\frac{215}{250}=\frac{43}{50}$

**33.** $\Pr(\text{Favors} \mid \text{Republican})=\frac{\Pr(\text{Favors and Republican})}{\Pr(\text{Republican})}=\frac{300+25}{435}=\frac{325}{435}=\frac{65}{87}$

**35.** $\Pr(\text{Nonwhite} \mid \text{Favors})=\frac{\Pr(\text{Nonwhite and Favors})}{\Pr(\text{Favors})}=\frac{\frac{25+10}{1000}}{\frac{435}{1000}}=\frac{35}{435}=\frac{7}{87}$

**37.** $\Pr(\text{WR and O})=\frac{100}{1000}=\frac{1}{10}$ (No "given that" in this problem)

**39.** Events are independent. $\Pr(\text{H1 and H1})=\frac{1}{12{,}000}\cdot\frac{1}{12{,}000}=\frac{1}{144{,}000{,}000}$

**41.** $\Pr(\text{A and SD and PGM})=(0.337)(0.796)(0.016)=0.004292032$

**43.** $\Pr(\text{not identified on test 1} \mid \text{defective})=0.3$

$\Pr(\text{not identified on test 2} \mid \text{defective})=0.2$

$\Pr(\text{not identified on either test} \mid \text{defective})=(0.3)(0.2)=0.06$

**45.** $\Pr(\text{H and LI})=\Pr(H)\cdot\Pr(LI \mid H)=(0.09)(0.5)=0.045$

**47.** $\Pr(\text{Good})=0.95$; $\Pr(\text{GGGGG})=(0.95)^5$ (Independent events)

**49.** $\Pr(\text{See and Buy})=\Pr(\text{See})\cdot\Pr(\text{Buy} \mid \text{See})=(0.3)(0.2)=0.06$

**51.** Events (each woman) are independent.

**a.** $\Pr(0 \text{ pregnancies}) = (0.99)^{100} \approx 0.366$

**b.** $\Pr(\text{at least 1 pregnancy}) \approx 1 - 0.366 = 0.634$

**53. a.** $\Pr(0 \text{ defects}) = (0.77)^3 \approx 0.456$

**b.** $\Pr(\text{at least 1 defect}) \approx 1 - 0.456 = 0.544$

**55.** Events are independent.

**a.** $\Pr(\text{all correct}) = \frac{1}{3}\cdot\frac{1}{3}\cdot\frac{1}{3}\cdot\frac{1}{5}\cdot\frac{1}{5}\cdot\frac{1}{5}\cdot\frac{1}{5} = \frac{1}{16{,}875}$

**b.** $\Pr(\text{no correct}) = \frac{2}{3}\cdot\frac{2}{3}\cdot\frac{2}{3}\cdot\frac{4}{5}\cdot\frac{4}{5}\cdot\frac{4}{5}\cdot\frac{4}{5} = \frac{2048}{16{,}875}$

**c.** $\Pr(\text{at least 1 correct}) = 1 - \frac{2048}{16{,}875} = \frac{14{,}827}{16{,}875}$

**57.** $\Pr(\text{Portia}) = \frac{3}{11}$ $\quad\Pr(\text{Trinka}) = \frac{1}{11}$

$\Pr(\text{P or T}) = \Pr(\text{P}) + \Pr(\text{T}) = \frac{3}{11} + \frac{1}{11} = \frac{4}{11}$; Odds for Portia or Trinka $= \frac{4}{7}$ or $= 4:7$

**59.** Choose one person. Now choose the other person.

**a.** $\Pr(\text{different from birthday \#1}) = \frac{364}{365}$

**b.** $\Pr(\text{same birthday as \#1}) = \frac{1}{365}$

**61. a.** $$\Pr(\text{Different days}) = \underset{\substack{\updownarrow \\ 1}}{1}\ \underset{\substack{\updownarrow \\ 2}}{\left(\frac{364}{365}\right)}\ \underset{\substack{\updownarrow \\ 3}}{\left(\frac{363}{365}\right)} \cdots \underset{\substack{\updownarrow \\ 20}}{\left(\frac{346}{365}\right)} \approx 0.588 \approx 0.59$$

**b.** $\Pr(\text{at least 2 same}) = 1 - \Pr(\text{different days}) \approx 1 - 0.59 = 0.41$

## *Exercises 7.4*

**1.**

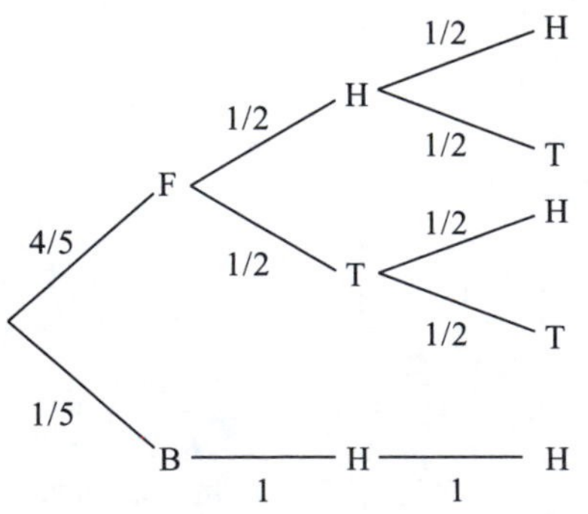


$\Pr(2\text{H}) = \frac{4}{5}\cdot\frac{1}{2}\cdot\frac{1}{2} + \frac{1}{5}\cdot 1\cdot 1 = \frac{2}{5}$

**3.**

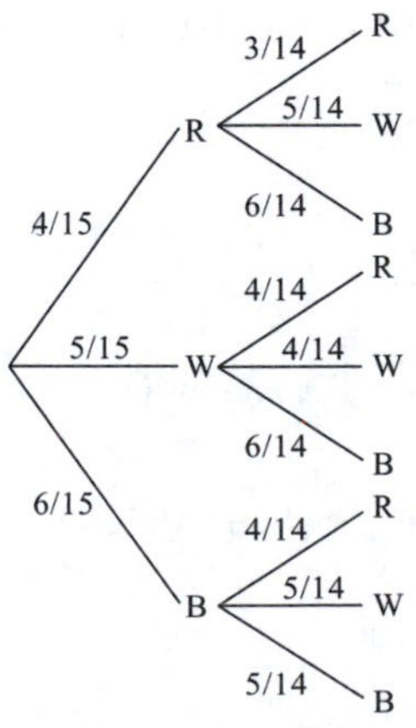

# Chapter 7: Introduction to Probability Functions

**a.** $\Pr(\text{WW}) = \frac{5}{15}\cdot\frac{4}{14} = \frac{2}{21}$

**b.** $\Pr(\text{RW or WR}) = \frac{4}{15}\cdot\frac{5}{14} + \frac{5}{15}\cdot\frac{4}{14} = \frac{4}{21}$

**c.** $\Pr(\text{at least one black})$

$$= \frac{6}{15} + \frac{4}{15}\cdot\frac{6}{14} + \frac{5}{15}\cdot\frac{6}{14}$$
$$= \frac{84}{210} + \frac{24}{210} + \frac{30}{210}$$
$$= \frac{138}{210} = \frac{69}{105} = \frac{23}{35}$$

**5.**

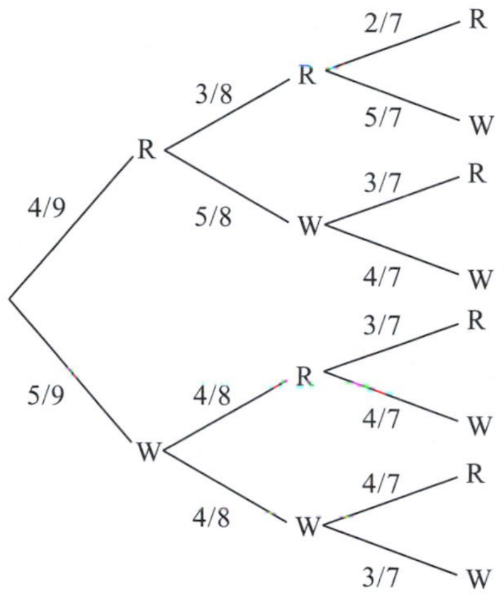


**a.** $\Pr(\text{WWW}) = \frac{5}{9}\cdot\frac{4}{8}\cdot\frac{3}{7} = \frac{5}{42}$

**b.** $\Pr(\text{2W and 1R})$

$$= \Pr(\text{WWR}) + \Pr(\text{WRW}) + \Pr(\text{RWW})$$
$$= \frac{5}{9}\cdot\frac{4}{8}\cdot\frac{4}{7} + \frac{5}{9}\cdot\frac{4}{8}\cdot\frac{4}{7} + \frac{4}{9}\cdot\frac{5}{8}\cdot\frac{4}{7}$$
$$= \frac{10}{63} + \frac{10}{63} + \frac{10}{63} = \frac{30}{63} = \frac{10}{21}$$

**c.**

$\Pr(\text{B3 is R})$

$$= \Pr(\text{RRR}) + \Pr(\text{RWR}) + \Pr(\text{WRR}) + \Pr(\text{WWR})$$
$$= \frac{4}{9}\cdot\frac{3}{8}\cdot\frac{2}{7} + \frac{4}{9}\cdot\frac{5}{8}\cdot\frac{3}{7} + \frac{5}{9}\cdot\frac{4}{8}\cdot\frac{3}{7} + \frac{5}{9}\cdot\frac{4}{8}\cdot\frac{4}{7} = \frac{4}{9}$$

**7.**

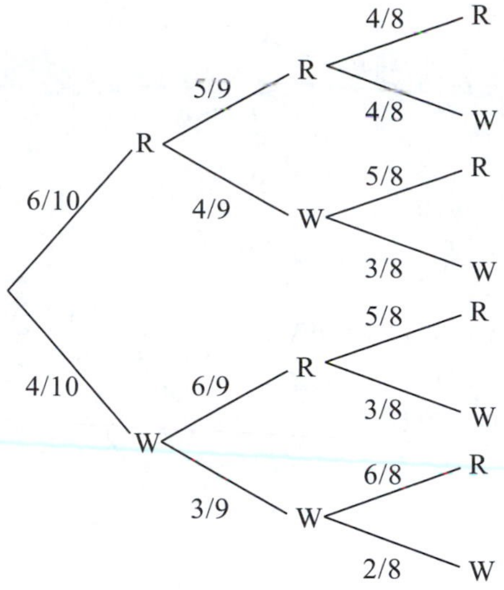


**a.** $\Pr(\text{WWW}) = \frac{4}{10}\cdot\frac{3}{9}\cdot\frac{2}{8} = \frac{1}{30}$

**b.** $\Pr(\text{WRR or RWR or RRW})$

$$= \frac{4}{10}\cdot\frac{6}{9}\cdot\frac{5}{8} + \frac{6}{10}\cdot\frac{4}{9}\cdot\frac{5}{8} + \frac{6}{10}\cdot\frac{5}{9}\cdot\frac{4}{8}$$
$$= \frac{3(4\cdot5\cdot6)}{8\cdot9\cdot10} = \frac{1}{2}$$

**c.** $\Pr(\text{at least 1W}) = 1 - \Pr(\text{No W})$

$$= 1 - \frac{6}{10}\cdot\frac{5}{9}\cdot\frac{4}{8} = \frac{5}{6}$$

**9.**

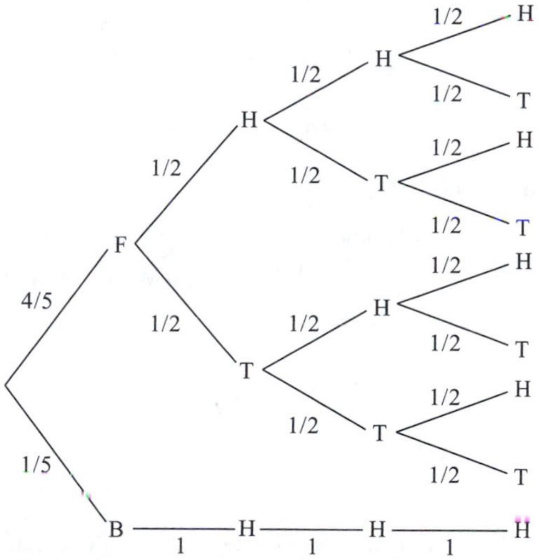


$\Pr(\text{HHT or HTH or THH or HHH})$

$$= 3\left(\frac{4}{5}\cdot\frac{1}{2}\cdot\frac{1}{2}\cdot\frac{1}{2}\right) + \left(\frac{4}{5}\cdot\frac{1}{2}\cdot\frac{1}{2}\cdot\frac{1}{2} + \frac{1}{5}\cdot1\cdot1\cdot1\right) = \frac{3}{5}$$

**11.**

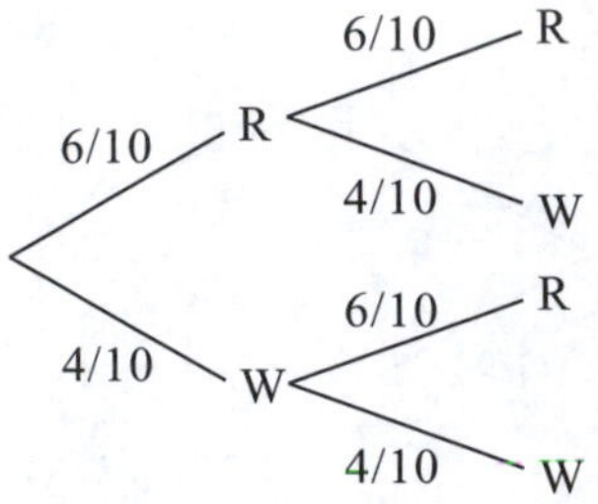


**a.** $\Pr(\text{R1 and W2}) = \dfrac{6}{10}\cdot\dfrac{4}{10} = \dfrac{24}{100} = \dfrac{6}{25}$

**b.** $\Pr(\text{RR}) = \dfrac{6}{10}\cdot\dfrac{6}{10} = \dfrac{36}{100} = \dfrac{9}{25}$

**c.** $\Pr(\text{R1W2 or W1R2}) = \dfrac{6}{10}\cdot\dfrac{4}{10} + \dfrac{4}{10}\cdot\dfrac{6}{10}$

$= \dfrac{48}{100} = \dfrac{12}{25}$

**d.** $\Pr(\text{RR or RW or WW})$

$= \dfrac{6}{10}\cdot\dfrac{6}{10} + \dfrac{6}{10}\cdot\dfrac{4}{10} + \dfrac{4}{10}\cdot\dfrac{4}{10} = \dfrac{76}{100} = \dfrac{19}{25}$

**13.**

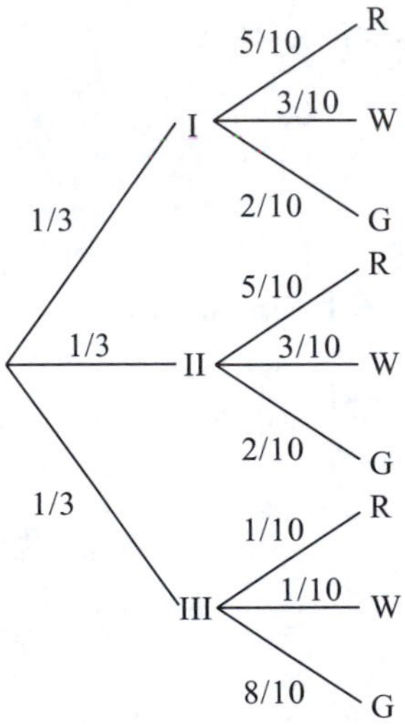


**a.** $\Pr(\text{III} \mid \text{G})$

$= \dfrac{\text{product of branch probabilities on III-G path}}{\text{sum of all branch products leading to G}}$

$= \dfrac{\frac{1}{3}\cdot\frac{8}{10}}{\frac{2}{30}+\frac{2}{30}+\frac{8}{30}} = \dfrac{\frac{8}{30}}{\frac{12}{30}} = \dfrac{2}{3}$

**b.** $\Pr(\text{III})|\text{G})$

$= \dfrac{\Pr(\text{III})\cdot\Pr(\text{G}|\text{III})}{\Pr(\text{III})\cdot\Pr(\text{G}|\text{III}) + \Pr(\text{II})\cdot\Pr(\text{G}|\text{II}) + \Pr(\text{I})\cdot\Pr(\text{G}|\text{I})}$

$= \dfrac{\frac{1}{3}\cdot\frac{8}{10}}{\frac{1}{3}\cdot\frac{8}{10}+\frac{1}{3}\cdot\frac{2}{10}+\frac{1}{3}\cdot\frac{2}{10}} = \dfrac{\frac{8}{30}}{\frac{12}{30}} = \dfrac{2}{3}$

**15.**

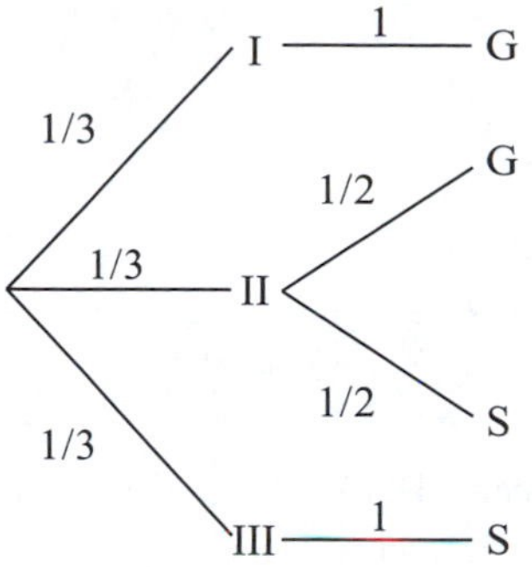


**a.** $\Pr(I \mid G)$

$$= \frac{\text{product of branch probabilities on I-G path}}{\text{sum of all branch probabilities leading to G}}$$

$$= \frac{\frac{1}{3}\cdot 1}{\frac{1}{3}+\frac{1}{6}} = \frac{\frac{1}{3}}{\frac{3}{6}} = \frac{2}{3}$$

**b.** $\Pr(I|G)$

$$= \frac{\Pr(I)\cdot \Pr(G|I)}{\Pr(I)\cdot \Pr(G|I)+\Pr(II)\cdot \Pr(G|II)+\Pr(III)\cdot \Pr(G|III)}$$

$$= \frac{\frac{1}{3}\cdot 1}{\frac{1}{3}\cdot 1+\frac{1}{3}\cdot\frac{1}{2}+\frac{1}{3}\cdot 0} = \frac{\frac{1}{3}}{\frac{3}{6}} = \frac{2}{3}$$

**17.** $\Pr(W) = 0.76$ $\Pr(H) = 0.09$ $\Pr(AANA) = 0.15$

$\Pr(I \mid W) = 0.20$ $\Pr(I \mid H) = 0.50$ $\Pr(I \mid AANA) = 0.75$

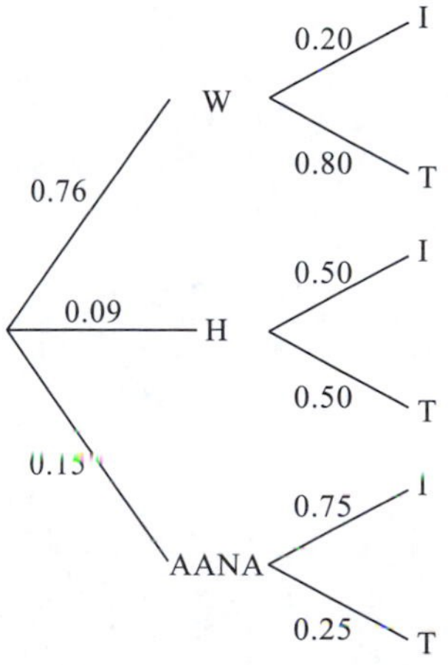


$\Pr(I) = 0.76(0.20) + 0.09(0.50) + 0.15(0.75) = 0.3095$

**19. a.** $\Pr(\text{HHHH}) = \frac{15}{50} \cdot \frac{15}{50} \cdot \frac{15}{50} \cdot \frac{15}{50} = \frac{81}{10,000}$

**b.** There are 6 successful events, each with the same probability.
Success: HHMM, HMHM, HMMH, MHMH, MMHH, MHHM

$\Pr(\text{Success}) = 6\left(\frac{15}{50} \cdot \frac{15}{50} \cdot \frac{35}{50} \cdot \frac{35}{50}\right) = \frac{1323}{5000}$

**21.**

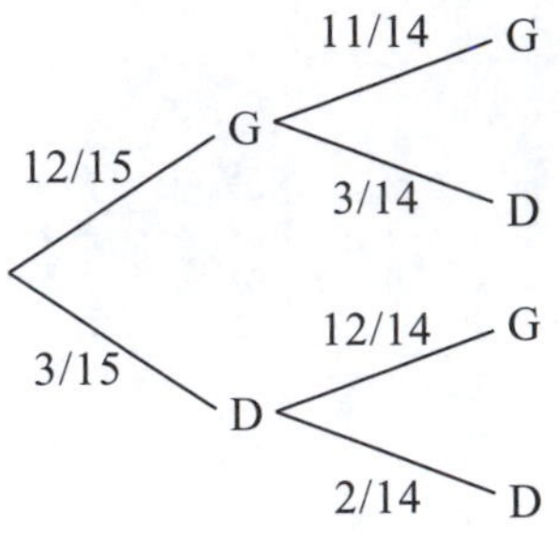


**a.** $\Pr(\text{G1D2}) = \frac{12}{15} \cdot \frac{3}{14} = \frac{6}{35}$

**b.** $\Pr(\text{D1G2}) = \frac{3}{15} \cdot \frac{12}{14} = \frac{6}{35}$

**c.** $\Pr(\text{G1D2 or D1G2}) = \frac{6}{35} + \frac{6}{35} = \frac{12}{35}$

**23.**

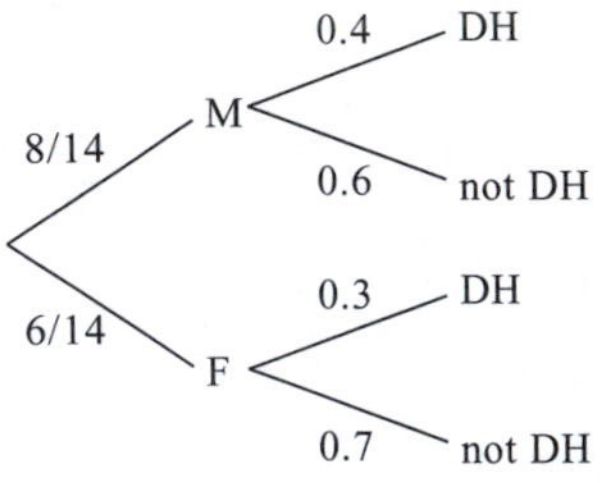


**a.** $\Pr(\text{Male}) = \frac{8}{14} = \frac{4}{7}$

**b.** 3200 males and 1800 females drink heavily.

$\Pr(\text{DH}) = \frac{3200 + 1800}{14,000} = \frac{5}{14}$

**c.** $\Pr(\text{M or DH}) = \frac{8}{14} + \frac{5}{14} - \frac{8}{14} \cdot (0.4) = \frac{7}{10}$

**d.** Using Bayes' formula and the probability tree, we have

$$\Pr(\text{M} \mid \text{DH}) = \frac{(0.4)\left(\frac{8}{14}\right)}{(0.4)\left(\frac{8}{14}\right) + (0.3)\left(\frac{6}{14}\right)} = \frac{16}{25}$$

**25.** $\Pr(\text{F} \mid \text{M}) = \frac{170}{170 + 280} = \frac{170}{450} = \frac{17}{45}$

**27.** T = Test indicates pregnant (P)
not T = Test shows not pregnant

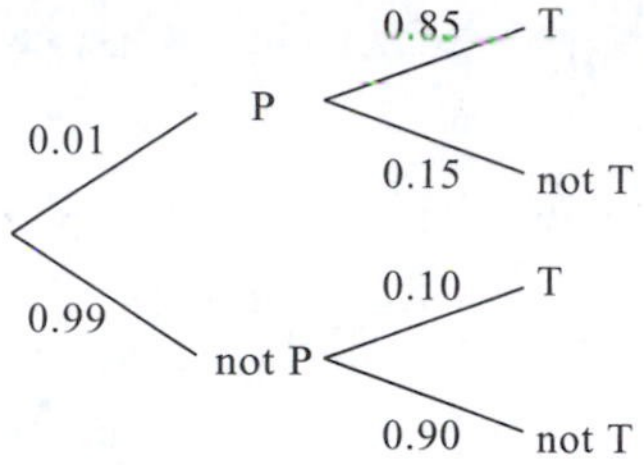


Using Bayes' formula and the probability tree, we have $\Pr(\mathrm{P} \mid \mathrm{T}) = \dfrac{(0.01)(0.85)}{(0.01)(0.85)+(0.99)(0.1)} \approx 0.079$.

Alternate Method:

| | T | not T | |
|---|---|---|---|
| P | 0.0085 | | 0.01 |
| not P | 0.099 | | 0.99 |
| | | | 1 |

| | T | not T | |
|---|---|---|---|
| P | 0.0085 | 0.0015 | 0.01 |
| not P | 0.099 | 0.891 | 0.99 |
| | 0.1075 | 0.8925 | 1 |

$$\Pr(\mathrm{P} \mid \mathrm{T}) = \frac{\Pr(\mathrm{P\ and\ T})}{\Pr(\mathrm{T})} = \frac{0.0085}{0.1075} \approx 0.079$$

**29. a.** $\Pr(\text{Female}) = \dfrac{10+12+6+13+8}{30+25+20+15+10} = \dfrac{49}{100}$

**b.** $\Pr(\mathrm{S} \mid \mathrm{F}) = \dfrac{12 \text{ Science Females}}{49 \text{ Females}} = \dfrac{12}{49}$

## *Exercises 7.5*

**1.** ${}_6P_4 = \dfrac{6!}{(6-4)!} = \dfrac{6\cdot5\cdot4\cdot3\cdot2!}{2!} = 360$

**3.** ${}_{10}P_6 = \dfrac{10!}{(10-6)!} = \dfrac{10\cdot9\cdot8\cdot7\cdot6\cdot5\cdot4!}{4!} = 151{,}200$

**5.** ${}_5P_0 = \dfrac{5!}{(5-0)!} = \dfrac{5!}{5!} = 1$

**7. a.** By Fundamental Counting Principle:
$6\cdot5\cdot4\cdot3 = 360$

**b.** By Fundamental Counting Principle:
$6\cdot6\cdot6\cdot6 = 1296$

**9.** ${}_nP_n = \dfrac{n!}{(n-n)!} = \dfrac{n!}{1} = n!$

**11.** $\dfrac{(n+1)!}{n!} = \dfrac{(n+1)n!}{n!} = n+1$

**13.** $(n+1)! = 17n!$
$(n+1)n! = 17n!$
$n+1 = 17$
$n = 16$

**15.** ${}_{100}C_{98} = \frac{100!}{98!2!} = \frac{100 \cdot 99}{2} = 4950$

**17.** ${}_{4}C_{4} = \frac{4!}{0!4!} = \frac{4!}{1 \cdot 4!} = 1$

**19.** $\binom{5}{0} = \frac{5!}{0!5!} = 1$

**21.**
$$\begin{aligned} {}_nC_6 &= {}_nC_4 \\ \frac{n!}{6!(n-6)!} &= \frac{n!}{4!(n-4)!} \\ 4!(n-4)! &= 6!(n-6)! \\ 4!(n-4)(n-5)(n-6)! &= 6 \cdot 5 \cdot 4!(n-6)! \\ n^2 - 9n + 20 &= 30 \\ n^2 - 9n - 10 &= 0 \\ (n-10)(n+1) &= 0 \end{aligned}$$
$n = 10$ is the only possible solution.

**23.** **a.** By FCP: $2 \cdot 4 = 8$
**b.** By FCP: $2 \cdot 4 \cdot 5 \cdot 6 = 240$

**25.** By FCP: $10 \cdot 9 \cdot 8 \cdot 7 \cdot 6 \cdot 5 \cdot 4 = 604{,}800$

**27.** By FCP: $6 \cdot 5 \cdot 4 = 120$

**29.** By FCP: $4 \cdot 3 \cdot 2 \cdot 1 = 24$

**31.** By FCP: $4 \cdot 4 \cdot 4 = 64$

**33.** By FCP: $6 \cdot 5 \cdot 4 \cdot 3 \cdot 2 \cdot 1 = 720$

**35.** By FCP: $2 \cdot 2 \cdot 2 \cdots 2 = 2^{10} = 1024$
Question | | | |
1 2 3 10

**37.** The choice for the first card does not matter in the calculation: $4 \cdot {}_{13}C_5 = 4 \cdot \frac{13!}{8!5!} = 5148$

**39.**

Old: $26 \cdot 26 \cdot 10 \cdot 10 \cdot 10 \cdot 10 = 6{,}760{,}000$
New: $26 \cdot 26 \cdot 26 \cdot 10 \cdot 10 \cdot 10 = 17{,}576{,}000$
Difference: $17{,}576{,}000 - 6{,}760{,}000 = 10{,}816{,}000$

**41.** ${}_{30}C_{20} = \frac{30!}{20!10!} = 30{,}045{,}015$

**43.** ${}_{12}C_5 = \frac{12!}{5!7!} = \frac{12 \cdot 11 \cdot 10 \cdot 9 \cdot 8 \cdot 7!}{5 \cdot 4 \cdot 3 \cdot 2 \cdot 1 \cdot 7!} = 792$

**45.** ${}_{10}C_4 = \frac{10!}{4!6!} = \frac{10 \cdot 9 \cdot 8 \cdot 7 \cdot 6!}{4 \cdot 3 \cdot 2 \cdot 1 \cdot 6!} = 210$

**47.** "AND" means multiply.
${}_{20}C_6 \cdot {}_{22}C_6 = \frac{20!}{6!14!} \cdot \frac{22!}{6!16!} = 2{,}891{,}999{,}880$

**49.** By FCP: $10 \cdot 370{,}000 = 3{,}700{,}000$

## *Exercises 7.6*

**1.** $n(E) = 1$ DOG is the only way for a success.
$n(S) = \underline{6} \cdot \underline{5} \cdot \underline{4} = 120$
So, $\Pr(E) = \frac{1}{120}$. Also, $\Pr(E) = \frac{1}{{}_6P_3} = \frac{1}{120}$

**3.** **a.** By FCP: $5 \cdot 4 \cdot 3 \cdot 2 \cdot 1 = 120$
**b.** $\Pr(\text{order on survey}) = \frac{1}{120}$

**5.** Total License Plates:
$26 \cdot 26 \cdot 26 \cdot 10 \cdot 10 \cdot 10 = 17{,}576{,}000$
Letters and numbers different:
$26 \cdot 25 \cdot 24 \cdot 10 \cdot 9 \cdot 8 = 11{,}232{,}000$
$\Pr(\text{Success}) = \frac{11{,}232{,}000}{17{,}576{,}000} \approx 0.639$

**7.** **a.** $\Pr(\text{Success}) = \frac{1}{10 \cdot 10 \cdot 10 \cdot 10} = \frac{1}{10{,}000}$
**b.** $\Pr(\text{Success}) = \frac{1}{10 \cdot 9 \cdot 8 \cdot 7} = \frac{1}{5040}$

**9.** Total Phone Numbers:
$8\cdot10\cdot10\cdot10\cdot10\cdot10\cdot10 = 8{,}000{,}000$
There are 8 phone numbers that have the same digits.
$$\Pr(\text{Success}) = \frac{8}{8{,}000{,}000} = \frac{1}{1{,}000{,}000}$$

**11.** $$\Pr(\text{John and Jill}) = \frac{\binom{4}{2}\cdot\binom{496}{2}}{\binom{500}{4}} \approx 2.8\times10^{-4}$$

**13.** Total ways to answer test $= \underline{10}\cdot\underline{9}\cdot\underline{8}\cdots\underline{1} = 10!$
Question: 1 2 3⋯10
Also, order is important, thus ${}_{10}P_{10} = 10!$ is another way to obtain the total.
$$\Pr(\text{All correct}) = \frac{1}{10!} = \frac{1}{3{,}628{,}800}$$

**15. a.** $$\Pr(\text{DD}) = \frac{\binom{3}{2}}{\binom{12}{2}} = \frac{3}{66} = \frac{1}{22} \approx 0.0455$$

**b.** $$\Pr(\text{GG}) = \frac{\binom{9}{2}}{\binom{12}{2}} = \frac{36}{66} = \frac{6}{11} \approx 0.5455$$

**c.** $$\Pr(1\text{D}) = 1 - [\Pr(\text{DD}) + \Pr(\text{GG})]$$
$$\approx 1 - 0.5910 = 0.409$$
Also,
$$\Pr(1\text{G}1\text{D}) = \frac{\binom{9}{1}\binom{3}{1}}{\binom{12}{2}} = \frac{9\cdot3}{66} = \frac{9}{22} \approx 0.409$$

**17.** Method 1:
Def Good Def Good
$$\Pr(E) = \frac{\binom{2}{1}\cdot\binom{98}{4} + \binom{2}{2}\cdot\binom{98}{3}}{\binom{100}{5}}$$
$$= \frac{7{,}224{,}560 + 152{,}096}{75{,}287{,}520} \approx 0.098$$
Method 2:
$$n(1 \text{ or } 2 \text{ Def}) = \text{Total} - \text{All Good} = \binom{100}{5} - \binom{98}{5}$$
$$= 75{,}287{,}520 - 67{,}910{,}864$$
$$= 7{,}376{,}656$$
$$\Pr(E) = \frac{7{,}376{,}656}{\binom{100}{5}} \approx 0.098$$

**19.** In the sample of 30, there are 2 minority loans. Thus, ${}_{10}C_2$ is the numbers of ways to obtain 2 minority loans from the 10 minority loans. Also, ${}_{90}C_{28}$ is the number of ways of getting 28 non minority loans from the total of non minority loans.
$$\Pr(E) = \frac{\binom{90}{28}\cdot\binom{10}{2}}{\binom{100}{30}}$$

**21. a.** $$\Pr(\text{No minority}) = \frac{\binom{6}{4}}{\binom{9}{4}} = \frac{15}{126} = \frac{5}{42} \approx 0.119$$

**b.** $$\Pr(\text{All minority}) = \frac{\binom{6}{1}\cdot\binom{3}{3}}{\binom{9}{4}} = \frac{6}{126}$$
$$= \frac{1}{21} \approx 0.0476$$

**c.** $$\Pr(1 \text{ minority}) = \frac{\binom{6}{3}\cdot\binom{3}{1}}{\binom{9}{4}} = \frac{20\cdot3}{126} = \frac{10}{21}$$
$$\approx 0.476$$

**23.** $$\Pr(E) = \frac{\binom{6}{5}}{\binom{10}{5}} = \frac{6}{252} = \frac{1}{42} \approx 0.0238$$

**25. a.** $$\Pr(2\text{M}) = \frac{\binom{23}{2}}{\binom{27}{2}} = \frac{253}{351} \approx 0.721$$

**b.** $$\Pr(\text{MW}) = \frac{\binom{23}{1}\cdot\binom{4}{1}}{\binom{27}{2}} = \frac{23\cdot4}{351} = \frac{92}{351} \approx 0.262$$

**c.** $$\Pr(\text{at least 1W}) = 1 - \Pr(2\text{M})$$
$$\approx 1 - 0.721 = 0.279$$
$$\text{Also, } \Pr(\text{at least 1W}) = \frac{\binom{23}{1}\binom{4}{1} + \binom{4}{2}}{\binom{27}{2}}$$
$$= \frac{92 + 6}{351} = \frac{98}{351} \approx 0.279$$

**27. a.** $$\Pr(E) = \frac{1 \text{ most able}}{3 \text{ men}} = \frac{1}{3}$$

**b.** $$\Pr(E) = \frac{1}{{}_3P_3} = \frac{1}{3\cdot2\cdot1} = \frac{1}{6}$$

**29.** $\Pr(E) = \dfrac{\text{ways to get 10 winning numbers}}{\text{total ways to get 10 numbers}}$

$= \dfrac{{}_{20}C_{10}}{{}_{80}C_{10}} \approx 1.12\times10^{-7}$

**31.** **a.** $\Pr(E) = \dfrac{\binom{2}{2}\binom{23}{3}}{\binom{25}{5}} = \dfrac{1\cdot 1771}{53{,}130} \approx 0.033$

**b.** $\Pr(E) = \dfrac{\binom{23}{5}}{\binom{25}{5}} = \dfrac{33{,}649}{53{,}130} \approx 0.633$

**33.** Order is not important.

**a.** $\Pr(5 \text{ spades}) = \dfrac{\binom{13}{5}}{\binom{52}{5}} = \dfrac{1287}{2{,}598{,}960} \approx 0.0005$

**b.** $\Pr(5 \text{ of same suit})$

$= \Pr(\text{All S or All C or All D or All H})$

$\approx 4(0.0005) = 0.002$

**35.** There are $52\cdot51\cdot50\cdot49\cdot48 = 311{,}875{,}200$ possible poker hands and $52\cdot12\cdot11\cdot10\cdot9 = 617{,}760$ ways to get a flush.

$\Pr(\text{flush}) = \dfrac{617{,}760}{311{,}875{,}200} \approx 0.00198$

**37.** **a.** $\dfrac{1}{10}\cdot\dfrac{1}{10}\cdot\dfrac{1}{10} = \dfrac{1}{1000} = 0.001$

**b.** 444, 446, 464, 466, 644, 646, 664, 666

**c.** $\dfrac{1}{2}\cdot\dfrac{1}{2}\cdot\dfrac{1}{2} = \dfrac{1}{8} = 0.125$

## *Exercises 7.7*

**1.** Can be since $\dfrac{1}{4}+\dfrac{3}{4}=1.$

**3.** Cannot be since $\dfrac{1}{4}+\dfrac{3}{5}+\dfrac{1}{6}\neq 1.$

**5.** Cannot be since matrix is not square.

**7.** Can be since the sum of entries in each row is 1, and matrix is square.

**9.** $\begin{bmatrix}0.2 & 0.8\end{bmatrix}\begin{bmatrix}0.1 & 0.9\\ 0.3 & 0.7\end{bmatrix} = \begin{bmatrix}0.26 & 0.74\end{bmatrix};$

$\begin{bmatrix}0.26 & 0.74\end{bmatrix}\begin{bmatrix}0.1 & 0.9\\ 0.3 & 0.7\end{bmatrix} = \begin{bmatrix}0.248 & 0.752\end{bmatrix}$

**11.** $\begin{bmatrix}0.1 & 0.3 & 0.6\end{bmatrix}\begin{bmatrix}0.5 & 0.3 & 0.2\\ 0.3 & 0.5 & 0.2\\ 0.1 & 0.1 & 0.8\end{bmatrix} = \begin{bmatrix}0.20 & 0.24 & 0.56\end{bmatrix};$

$\begin{bmatrix}0.20 & 0.24 & 0.56\end{bmatrix}\begin{bmatrix}0.5 & 0.3 & 0.2\\ 0.3 & 0.5 & 0.2\\ 0.1 & 0.1 & 0.8\end{bmatrix} = \begin{bmatrix}0.228 & 0.236 & 0.536\end{bmatrix}$

**13.** $A_{10} = A_0P^{10} = \begin{bmatrix}0.2 & 0.8\end{bmatrix}\begin{bmatrix}0.1 & 0.9\\ 0.3 & 0.7\end{bmatrix}^{10} \approx \begin{bmatrix}0.2 & 0.8\end{bmatrix}\begin{bmatrix}0.25 & 0.75\\ 0.25 & 0.75\end{bmatrix} \approx \begin{bmatrix}0.25 & 0.75\end{bmatrix}$

**15.** $A_8 = A_0P^8 = \begin{bmatrix} 0.1 & 0.3 & 0.6 \end{bmatrix} \begin{bmatrix} 0.5 & 0.3 & 0.2 \\ 0.3 & 0.5 & 0.2 \\ 0.1 & 0.1 & 0.8 \end{bmatrix}^8$

$$\approx \begin{bmatrix} 0.1 & 0.3 & 0.6 \end{bmatrix} \begin{bmatrix} 0.254 & 0.254 & 0.492 \\ 0.254 & 0.254 & 0.492 \\ 0.246 & 0.246 & 0.508 \end{bmatrix} \approx \begin{bmatrix} 0.249 & 0.249 & 0.502 \end{bmatrix}$$

**17.** $\begin{bmatrix} V_1 & V_2 \end{bmatrix} \begin{bmatrix} 0.1 & 0.9 \\ 0.3 & 0.7 \end{bmatrix} = \begin{bmatrix} V_1 & V_2 \end{bmatrix}$ or $\begin{matrix} 0.1V_1 + 0.3V_2 = V_1 \\ 0.9V_1 + 0.7V_2 = V_2 \end{matrix}$ or $\begin{matrix} -0.9V_1 + 0.3V_2 = 0 \\ 0.9V_1 - 0.3V_2 = 0 \end{matrix}$

Thus, $V_1 = \frac{0.3}{0.9}V_2 = \frac{1}{3}V_2$. $V_1 + V_2 = 1$ gives $\frac{1}{3}V_2 + V_2 = 1$ or $V_2 = \frac{3}{4}$.

Then, $V_1 = \frac{1}{4}$. The steady-state vector is $\begin{bmatrix} \frac{1}{4} & \frac{3}{4} \end{bmatrix}$.

**19.** $\begin{bmatrix} V_1 & V_2 & V_3 \end{bmatrix} \begin{bmatrix} 0.5 & 0.3 & 0.2 \\ 0.3 & 0.5 & 0.2 \\ 0.1 & 0.1 & 0.8 \end{bmatrix} = \begin{bmatrix} V_1 & V_2 & V_3 \end{bmatrix}$

$$\begin{matrix} 0.5V_1 + 0.3V_2 + 0.1V_3 = V_1 \\ 0.3V_1 + 0.5V_2 + 0.1V_3 = V_2 \\ 0.2V_1 + 0.2V_2 + 0.8V_3 = V_3 \end{matrix} \quad \text{or} \quad \begin{matrix} -0.5V_1 + 0.3V_2 + 0.1V_3 = 0 \\ 0.3V_1 - 0.5V_2 + 0.1V_3 = 0 \\ 0.2V_1 + 0.2V_2 - 0.2V_3 = 0 \end{matrix}$$

$$\left[\begin{array}{ccc|c} -0.5 & 0.3 & 0.1 & 0 \\ 0.3 & -0.5 & 0.1 & 0 \\ 0.2 & 0.2 & -0.2 & 0 \end{array}\right] \to \left[\begin{array}{ccc|c} 1 & -0.6 & -0.2 & 0 \\ 0.3 & -0.5 & 0.1 & 0 \\ 0.2 & 0.2 & -0.2 & 0 \end{array}\right] \to \left[\begin{array}{ccc|c} 1 & -0.6 & -0.2 & 0 \\ 0 & -0.32 & 0.16 & 0 \\ 0 & 0.32 & -0.16 & 0 \end{array}\right] \to \left[\begin{array}{ccc|c} 1 & -0.6 & -0.2 & 0 \\ 0 & -.32 & 0.16 & 0 \\ 0 & 0 & 0 & 0 \end{array}\right]$$

$$\to \left[\begin{array}{ccc|c} 1 & -0.6 & -0.2 & 0 \\ 0 & 1 & -0.5 & 0 \\ 0 & 0 & 0 & 0 \end{array}\right] \to \left[\begin{array}{ccc|c} 1 & 0 & -0.5 & 0 \\ 0 & 1 & -0.5 & 0 \\ 0 & 0 & 0 & 0 \end{array}\right]$$

So, $V_1 = 0.5V_3$

$V_2 = 0.5V_3$

$V_1 + V_2 + V_3 = 0.5V_3 + 0.5V_3 + V_3 = 1$

Thus, $V_3 = \frac{1}{2}$ and the steady-state vector is $\begin{bmatrix} \frac{1}{4} & \frac{1}{4} & \frac{1}{2} \end{bmatrix}$.

$V_3 = \frac{2}{11}, V_1 = \frac{6}{11}, V_2 = \frac{3}{11}$ Steady-state vector is $\begin{bmatrix} \frac{6}{11} & \frac{3}{11} & \frac{2}{11} \end{bmatrix}$.

**21.** $\begin{bmatrix}1 & 0 & 0\end{bmatrix}\begin{bmatrix}0.5 & 0.4 & 0.1\\0.4 & 0.5 & 0.1\\0.3 & 0.3 & 0.4\end{bmatrix}=\begin{bmatrix}0.5 & 0.4 & 0.1\end{bmatrix}$; $\begin{bmatrix}0.5 & 0.4 & 0.1\end{bmatrix}\begin{bmatrix}0.5 & 0.4 & 0.1\\0.4 & 0.5 & 0.1\\0.3 & 0.3 & 0.4\end{bmatrix}=\begin{bmatrix}0.44 & 0.43 & 0.13\end{bmatrix}$

$$\begin{bmatrix}0.44 & 0.43 & 0.13\end{bmatrix}\begin{bmatrix}0.5 & 0.4 & 0.1\\0.4 & 0.5 & 0.1\\0.3 & 0.3 & 0.4\end{bmatrix}=\begin{bmatrix}0.431 & 0.430 & 0.139\end{bmatrix}$$

$$\begin{bmatrix}0.431 & 0.430 & 0.139\end{bmatrix}\begin{bmatrix}0.5 & 0.4 & 0.1\\0.4 & 0.5 & 0.1\\0.3 & 0.3 & 0.4\end{bmatrix}=\begin{bmatrix}0.4292 & 0.4291 & 0.1417\end{bmatrix}$$

**23.**

$$\text{Mother}\;\begin{array}{c} \\ R \\ N\end{array}\!\!\begin{array}{c}\text{Daughter}\\ \begin{array}{cc}R & N\end{array}\\ \begin{bmatrix}0.8 & 0.2\\0.3 & 0.7\end{bmatrix}\end{array}$$

**25.** $\begin{bmatrix}0 & 1\end{bmatrix}\begin{bmatrix}0.8 & 0.2\\0.3 & 0.7\end{bmatrix}=\begin{bmatrix}0.3 & 0.7\end{bmatrix}$

$$\begin{bmatrix}0.3 & 0.7\end{bmatrix}\begin{bmatrix}0.8 & 0.2\\0.3 & 0.7\end{bmatrix}=\begin{bmatrix}0.45 & 0.55\end{bmatrix}$$

Pr(Granddaughter regular) = 0.45

**27.**

$$\begin{array}{c|ccc} & A & F & VW\\ \hline A & 0 & 0.7 & 0.3\\ F & 0.6 & 0 & 0.4\\ VW & 0.8 & 0.2 & 0\end{array}$$

**29.** $\begin{bmatrix}0 & 1 & 0\end{bmatrix}\begin{bmatrix}0 & 0.7 & 0.3\\0.6 & 0 & 0.4\\0.8 & 0.2 & 0\end{bmatrix}=\begin{bmatrix}0.6 & 0 & 0.4\end{bmatrix}$

$$\begin{bmatrix}0.6 & 0 & 0.4\end{bmatrix}\begin{bmatrix}0 & 0.7 & 0.3\\0.6 & 0 & 0.4\\0.8 & 0.2 & 0\end{bmatrix}=\begin{bmatrix}0.32 & 0.50 & 0.18\end{bmatrix};$$

$$\begin{bmatrix}0.32 & 0.50 & 0.18\end{bmatrix}\begin{bmatrix}0 & 0.7 & 0.3\\0.6 & 0 & 0.4\\0.8 & 0.2 & 0\end{bmatrix}=\begin{bmatrix}0.444 & 0.260 & 0.296\end{bmatrix}$$

$$\begin{bmatrix}0.444 & 0.260 & 0.296\end{bmatrix}\begin{bmatrix}0 & 0.7 & 0.3\\0.6 & 0 & 0.4\\0.8 & 0.2 & 0\end{bmatrix}=\begin{array}{ccc}[0.3928 & 0.37 & 0.2372]\\ A & F & VW\end{array}$$

**31.** $\begin{bmatrix} V_1 & V_2 & V_3 \end{bmatrix}\begin{bmatrix} 0 & 0.7 & 0.3 \\ 0.6 & 0 & 0.4 \\ 0.8 & 0.2 & 0 \end{bmatrix} = \begin{bmatrix} V_1 & V_2 & V_3 \end{bmatrix}$ $\begin{aligned} 0.6V_2 + 0.8V_3 &= V_1 \\ 0.7V_1 + 0.2V_3 &= V_2 \\ 0.3V_1 + 0.4V_2 &= V_3 \end{aligned}$ or $\begin{aligned} -V_1 + 0.6V_2 + 0.8V_3 &= 0 \\ 0.7V_1 - V_2 + 0.2V_3 &= 0 \\ 0.3V_1 + 0.4V_2 - V_3 &= 0 \end{aligned}$

$$\left[\begin{array}{ccc|c} 1 & -0.6 & -0.8 & 0 \\ 0.7 & -1 & 0.2 & 0 \\ 0.3 & 0.4 & -1 & 0 \end{array}\right] \to \left[\begin{array}{ccc|c} 1 & -0.60 & -0.80 & 0 \\ 0 & -0.58 & 0.76 & 0 \\ 0 & 0.58 & -0.76 & 0 \end{array}\right] \to \left[\begin{array}{ccc|c} 1 & -0.60 & -0.80 & 0 \\ 0 & -0.58 & 0.76 & 0 \\ 0 & 0 & 0 & 0 \end{array}\right] \to \left[\begin{array}{ccc|c} 1 & -0.60 & -0.80 & 0 \\ 0 & 1 & -1.31 & 0 \\ 0 & 0 & 0 & 0 \end{array}\right]$$

$$\to \left[\begin{array}{ccc|c} 1 & 0 & -1.586 & 0 \\ 0 & 1 & -1.310 & 0 \\ 0 & 0 & 0 & 0 \end{array}\right] \quad \begin{aligned} V_1 &= 1.586V_3 \\ V_2 &= 1.31V_3 \\ V_1 + V_2 + V_3 &= 1 \end{aligned}$$

So, $1.586V_3 + 1.31V_3 + V_3 = 1$ or $V_3 = \frac{1}{3.896} \approx 0.257$. $V_1 \approx 1.586(0.257) \approx 0.407$ and $V_2 \approx 1.31(0.257) \approx 0.336$

Steady-state vector $\approx \begin{bmatrix} 0.407 & 0.336 & 0.257 \end{bmatrix}$. Actual steady-state vector is $\begin{bmatrix} \frac{46}{113} & \frac{38}{113} & \frac{29}{113} \end{bmatrix}$.

**33.**
$$\begin{array}{c} \\ \text{R} \\ \text{U} \end{array}\begin{array}{c} \text{R} \quad \text{U} \\ \begin{bmatrix} 0.7 & 0.3 \\ 0.1 & 0.9 \end{bmatrix} \end{array}$$

$$\begin{bmatrix} V_1 & V_2 \end{bmatrix}\begin{bmatrix} 0.7 & 0.3 \\ 0.1 & 0.9 \end{bmatrix} = \begin{bmatrix} V_1 & V_2 \end{bmatrix}$$

$\begin{aligned} 0.7V_1 + 0.1V_2 &= V_1 \\ 0.3V_1 + 0.9V_2 &= V_2 \end{aligned}$ or $\begin{aligned} -0.3V_1 + 0.1V_2 &= 0 \\ 0.3V_1 - 0.1V_2 &= 0 \end{aligned}$;

$V_1 = \frac{1}{3}V_2$, $V_1 + V_2 = 1$ gives $V_1 = \frac{1}{4}$, $V_2 = \frac{3}{4}$.

Steady-state vector is $\begin{bmatrix} \frac{1}{4} & \frac{3}{4} \end{bmatrix}$.

**35.** $\begin{bmatrix} V_1 & V_2 & V_3 \end{bmatrix}\begin{bmatrix} 0.7 & 0.2 & 0.1 \\ 0.1 & 0.6 & 0.3 \\ 0 & 0.1 & 0.9 \end{bmatrix} = \begin{bmatrix} V_1 & V_2 & V_3 \end{bmatrix}$ or $\begin{aligned} -0.3V_1 + 0.1V_2 &= 0 \\ 0.2V_1 - 0.4V_2 + 0.1V_3 &= 0 \\ 0.1V_1 + 0.3V_2 - 0.1V_3 &= 0 \end{aligned}$

$$\left[\begin{array}{ccc|c} 1 & 3 & -1 & 0 \\ -0.3 & 0.1 & 0 & 0 \\ 0.2 & -0.4 & 0.1 & 0 \end{array}\right] \to \left[\begin{array}{ccc|c} 1 & 3 & -1 & 0 \\ 0 & 1 & -0.3 & 0 \\ 0 & -1 & 0.3 & 0 \end{array}\right] \to \left[\begin{array}{ccc|c} 1 & 0 & -0.1 & 0 \\ 0 & 1 & -0.3 & 0 \\ 0 & 0 & 0 & 0 \end{array}\right] \quad \begin{aligned} V_1 &= \frac{1}{10}V_3 \\ V_2 &= \frac{3}{10}V_3 \end{aligned}$$

$V_1 + V_2 + V_3 = 1$ So, $V_3 = \frac{10}{14} = \frac{5}{7}$.

This gives $V_1 = \frac{1}{14}$ and $V_2 = \frac{3}{14}$.

Thus, the steady-state vector is $\begin{bmatrix} \frac{1}{14} & \frac{3}{14} & \frac{5}{7} \end{bmatrix}$.

**37.** $\begin{bmatrix} V_1 & V_2 & V_3 \end{bmatrix}\begin{bmatrix} 0.7 & 0.2 & 0.1 \\ 0.4 & 0.4 & 0.2 \\ 0.4 & 0.4 & 0.2 \end{bmatrix} = \begin{bmatrix} V_1 & V_2 & V_3 \end{bmatrix}$ or $\begin{array}{r} -0.3V_1 + 0.4V_2 + 0.4V_3 = 0 \\ 0.2V_1 - 0.6V_2 + 0.4V_3 = 0 \\ 0.1V_1 + 0.2V_2 - 0.8V_3 = 0 \end{array}$

$$\left[\begin{array}{ccc|c} 1 & 2 & -8 & 0 \\ -0.3 & 0.4 & 0.4 & 0 \\ 0.2 & -0.6 & 0.4 & 0 \end{array}\right] \to \left[\begin{array}{ccc|c} 1 & 2 & -8 & 0 \\ 0 & 1 & -2 & 0 \\ 0 & -1 & 2 & 0 \end{array}\right] \to \left[\begin{array}{ccc|c} 1 & 0 & -4 & 0 \\ 0 & 1 & -2 & 0 \\ 0 & 0 & 0 & 0 \end{array}\right] \quad \begin{array}{l} V_1 = 4V_3 \\ V_2 = 2V_3 \end{array}$$

$4V_3 + 2V_3 + V_3 = 1$ or $V_3 = \frac{1}{7}$, $V_1 = \frac{4}{7}$, $V_2 = \frac{2}{7}$

Steady-state vector is $\begin{bmatrix} \frac{4}{7} & \frac{2}{7} & \frac{1}{7} \end{bmatrix}$.

**39.** From exercise 38, it is clear that the vector $\begin{bmatrix} 0.49 & 0.42 & 0.09 \end{bmatrix}$ is the steady-state vector for the given transition matrix.

# Chapter 7: Introduction to Probability

## *Chapter 7 Review Exercises*

**1.** **a.** $\Pr(\text{odd}) = \frac{5}{9}$

**b.** $\Pr(3, 6, 9) = \frac{3}{9} = \frac{1}{3}$

**c.** $\Pr(3, 9) = \frac{2}{9}$

**2.** **a.** $S_1$ is equiprobable with each outcome having probability $\frac{1}{12}$.

**b.** $\Pr(\text{R}) = \frac{9}{12} = \frac{3}{4}$

**c.** $\Pr(\text{odd}) = \frac{6}{12} = \frac{1}{2}$

**d.** $\Pr(10, 12) = \frac{1}{6}$

**e.** $\Pr(\text{W or } 1, 3, 5, 7, 9) = \frac{3+5}{12} = \frac{2}{3}$

**3.** $\Pr(\text{E}) = \frac{3}{7}$ means Win 3, Lose 4.

**a.** Odds for E: 3:4
**b.** Odds against E: 4:3

**4.** Equiprobable sample space:
$\{(\text{HH}), (\text{HT}), (\text{TH}), (\text{TT})\}$

**a.** $\Pr(0 \text{ heads}) = \frac{1}{4}$

**b.** $\Pr(1 \text{ head}) = \frac{2}{4} = \frac{1}{2}$

**c.** $\Pr(2 \text{ heads}) = \frac{1}{4}$

**5.** {HHH, HHT, HTH, HTT, TTT, TTH, THT, THH}

**a.** $\Pr(2\text{H}) = \frac{3}{8}$

**b.** $\Pr(3\text{H}) = \frac{1}{8}$

**c.** $\Pr(1\text{H}) = \frac{3}{8}$

**6.** $\Pr(\text{Queen or Jack}) = \frac{8}{52} = \frac{2}{13}$

**7.** $\Pr(\text{Success}) = \frac{16}{52} \cdot \frac{16}{52} = \frac{16}{169}$

**8.** $\Pr(\text{L}) = 1 - \Pr(\text{W}) = \frac{3}{4}$

**9.** $\Pr(\text{A or } 10) = \frac{8}{52} = \frac{2}{13}$

**10.** $\Pr(\text{King or Red}) = \Pr(\text{King}) + \Pr(\text{Red}) - \Pr(\text{King and Red}) = \frac{4}{52} + \frac{26}{52} - \frac{2}{52} = \frac{28}{52} = \frac{7}{13}$

**11.** **a.** There are 2 even numbered R. $\Pr(\text{Even and R}) = \frac{2}{9}$

**b.** $\Pr(\text{R or E}) = \Pr(\text{R}) + \Pr(\text{E}) - \Pr(\text{R and E}) = \frac{4}{9} + \frac{4}{9} - \frac{2}{9} = \frac{2}{3}$

**c.** $\Pr(\text{W or O}) = \Pr(\text{W}) + \Pr(\text{O}) - \Pr(\text{W and O}) = \frac{5}{9} + \frac{5}{9} - \frac{3}{9} = \frac{7}{9}$

**12.** $\Pr(\text{both balls red}) = \Pr(\text{1st ball red}) \cdot \Pr(\text{2nd ball red} \mid \text{1st ball red}) = \left(\frac{4}{7}\right)\left(\frac{3}{6}\right) = \frac{12}{42} = \frac{2}{7}$

**13.** There are 4 balls in box, given that the first ball drawn is black.
$S = \{\text{R, R, B, B}\}$ Thus, $\Pr(\text{R} \mid \text{B}) = \frac{2}{4} = \frac{1}{2}$.

# Chapter 7: Introduction to Probability

**14.** $\Pr(\text{R,R,B}) = \Pr(\text{R})\cdot\Pr(\text{R})\cdot\Pr(\text{B}) = \frac{5}{20}\cdot\frac{5}{20}\cdot\frac{7}{20} = \frac{175}{8000} = \frac{7}{320}$

**15.** $\Pr(\text{R,R,B}) = \Pr(\text{R})\cdot\Pr(\text{2nd R}\,|\,\text{1st R})\cdot\Pr(\text{3rd B}\,|\,\text{1st R and 2nd R}) = \frac{5}{20}\cdot\frac{4}{19}\cdot\frac{7}{18} = \frac{140}{6840} = \frac{7}{342}$

**16.** There are $16\cdot15\cdot14 = 3360$ ways to choose three balls if order matters.
There are six orders that yield the desired results: $\{\text{WRB, WBR, BWR, BRW, RWB, RBW}\}$. For each of these outcomes, there are 8 ways to choose white, 3 ways to choose red, and 5 ways to choose black. So $8\cdot3\cdot5 = 120$ ways to get the particular order. Thus, $(6\text{ orders})(120\text{ ways to get each order}) = 720$ outcomes that are desired.

$\Pr(\text{one of each color}) = \frac{720}{3360} = \frac{3}{14}.$

**17.** $\Pr(\text{W and R}) = \Pr(\text{WR or RW}) = \Pr(\text{W})\cdot\Pr(\text{R}\mid\text{W}) + \Pr(\text{R})\cdot\Pr(\text{W}\mid\text{R}) = \left(\frac{6}{10}\right)\cdot\left(\frac{4}{9}\right) + \left(\frac{4}{10}\right)\cdot\left(\frac{6}{9}\right) = \frac{48}{90} = \frac{8}{15}$

**18. a.** $\Pr(\text{U1 and R}) = \Pr(\text{U1})\cdot\Pr(\text{R}\mid\text{U1}) = \frac{1}{2}\cdot\frac{3}{7} = \frac{3}{14}$

**b.** $\Pr(\text{R}) = \Pr((\text{U1 and R}) \text{ or } (\text{U2 and R}))$

$= \Pr(\text{U1})\cdot\Pr(\text{R}\mid\text{U1}) + \Pr(\text{U2})\cdot\Pr(\text{R}\mid\text{U2}) = \frac{1}{2}\cdot\frac{3}{7} + \frac{1}{2}\cdot\frac{5}{7} = \frac{8}{14} = \frac{4}{7}$

**c.** Using Bayes' formula, we have

$$\Pr(\text{U1}\mid\text{R}) = \frac{\Pr(\text{U1})\cdot\Pr(\text{R}\mid\text{U1})}{\Pr(\text{U1})\cdot\Pr(\text{R}\mid\text{U1}) + \Pr(\text{U2})\cdot\Pr(\text{R}\mid\text{U2})} = \frac{\frac{1}{2}\cdot\frac{3}{7}}{\frac{1}{2}\cdot\frac{3}{7} + \frac{1}{2}\cdot\frac{5}{7}} = \frac{3}{3+5} = \frac{3}{8}$$

**19.** $$\Pr(\text{Bag B}\mid\text{white ball}) = \frac{\Pr(\text{Bag B})\cdot\Pr(\text{White}\mid\text{Bag B})}{\Pr(\text{Bag B})\cdot\Pr(\text{White}\mid\text{Bag B}) + \Pr(\text{Bag A})\cdot\Pr(\text{White}\mid\text{Bag A})} = \frac{\frac{1}{2}\cdot\frac{7}{16}}{\frac{1}{2}\cdot\frac{7}{16} + \frac{1}{2}\cdot\frac{5}{14}} = \frac{49}{89}$$

**20.** ${}_6P_2 = \frac{6!}{4!} = 30$

**21.** ${}_7C_3 = \frac{7!}{3!4!} = \frac{7\cdot6\cdot5\cdot4!}{3!4!} = \frac{7\cdot6\cdot5}{3\cdot2} = 35$

**22.** $26^3 = 17{,}576$

**23.** ${}_8C_5 = \frac{8!}{5!3!} = \frac{8\cdot7\cdot6\cdot5!}{5!3!} = \frac{8\cdot7\cdot6}{3\cdot2} = 56$

**24. a.** It is not a square matrix.
**b.** The row sums are not equal to 1.

**25.** $\begin{bmatrix}0.76 & 0.24\end{bmatrix}\begin{bmatrix}0.4 & 0.6\\ 0.8 & 0.2\end{bmatrix} = \begin{bmatrix}0.496 & 0.504\end{bmatrix}$

$\begin{bmatrix}0.1 & 0.9\end{bmatrix}\begin{bmatrix}0.4 & 0.6\\ 0.8 & 0.2\end{bmatrix} = \begin{bmatrix}0.76 & 0.24\end{bmatrix}$

**26.** $\begin{bmatrix}V_1 & V_2\end{bmatrix}\begin{bmatrix}0.6 & 0.4\\ 0.1 & 0.9\end{bmatrix} = \begin{bmatrix}V_1 & V_2\end{bmatrix}$

$\begin{aligned}0.6V_1 + 0.1V_2 &= V_1\\ 0.4V_1 + 0.9V_2 &= V_2\end{aligned}$ or $\begin{aligned}-0.4V_1 + 0.1V_2 &= 0\\ 0.4V_1 - 0.1V_2 &= 0\end{aligned}$

Thus, $V_1 = \frac{-0.1}{-0.4}V_2 = \frac{1}{4}V_2$ and $V_1 + V_2 = 1$.

$\frac{1}{4}V_2 + V_2 = 1 \rightarrow \frac{5}{4}V_2 = 1$

$V_2 = \frac{4}{5} = 0.8,\ V_1 = 0.2$

The steady-state vector is $\begin{bmatrix}0.2 & 0.8\end{bmatrix}$.

**27.** $\Pr(\le 50) = \frac{50{,}000}{80{,}000} = \frac{5}{8}$

**28.** $\Pr(\text{read newspaper regularly}) = \frac{1}{3+1} = \frac{1}{4}$

**29.** $\Pr(\text{French or German})$
$= \Pr(\text{Fr}) + \Pr(\text{Ger}) - \Pr(\text{Fr and Ger})$
$= \frac{30}{100} + \frac{40}{100} - \frac{12}{100} = \frac{29}{50} = 0.58$

**30.** $\Pr(\text{Demo and Med. Pro}) = \frac{25}{280} = \frac{5}{56}$

H + A − HA

**31.** $\Pr(\text{HP or A}) = \frac{75}{280} + \frac{120}{280} - \frac{30}{280} = \frac{165}{280} = \frac{33}{56}$

**32.** No formulas are needed. Read table.
$\Pr(\text{Auth} \mid \text{Med Pro}) = \frac{75}{110} = \frac{15}{22}$

**33.**

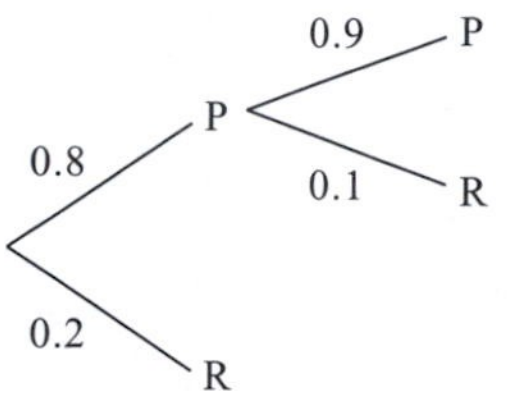


$\Pr(\text{pass both}) = (0.8)(0.9) = 0.72$

**34. a.** $\Pr(\text{Color Blind}) = \frac{63}{2000}$

**b.** $\Pr(\text{M} \mid \text{CB}) = \frac{60}{63}$

**35.**

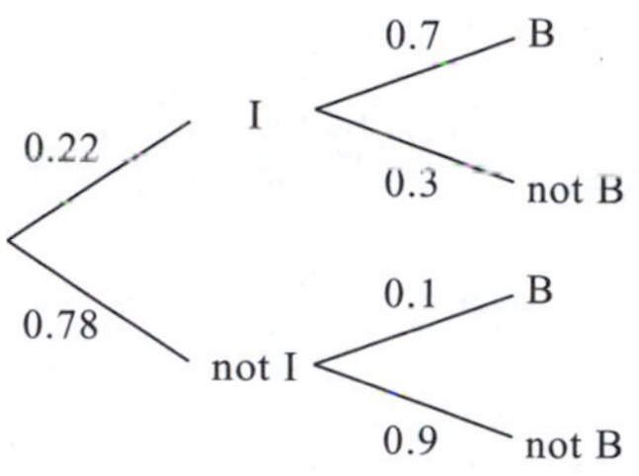


Using Bayes' formula and the probability tree, we have
$\Pr(\text{No intention} \mid \text{Bought})$
$= \frac{(0.78)(0.10)}{(0.78)(0.10) + (0.22)(0.7)} = \frac{39}{116}$

**36.** By FCP: $4 \cdot 3 \cdot 2 \cdot 1 = 24$

**37.** ${}_8P_4 = \frac{8!}{4!} = 1680$ 8·7·6·5 (4 positions)

**38.** $\binom{12}{4} = \frac{12!}{4!8!} = 495$

**39.** ${}_8C_4 = \frac{8!}{4!4!} = 70$

**40. a.** ${}_{12}C_2 = \binom{12}{2} = \frac{12!}{2!10!} = 66$

**b.** ${}_{12}C_3 = \binom{12}{3} = \frac{12!}{3!9!} = 220$

**41.** There are 35 choices for each position on the plate and repetitions are allowed.
Total plates $= 35 + 35 \cdot 35 + 35 \cdot 35 \cdot 35 + 35^4$
$= 1{,}544{,}760$

26 letters
9 digits = 35

**42.** $(\text{\# of blood groups}) \cdot (\text{pos or neg}) \cdot (\text{\# blood types}) \cdot (\text{\#Rh types})$
$= 4 \cdot 2 \cdot 4 \cdot 8$
$= 256$
So, the conclusion that there are 288 unique groups is incorrect if the given information is correct.

**43.** Number of ways to enter courses = 4! = 24.
$\Pr(\text{alphabetical}) = \frac{1}{24}$.

**44.** Number of three digit possibilities $= 10^3 = 1000$.
Number of permutations of 3 different digits = 6.
$\Pr(\text{winner}) = \frac{6}{1000} = \frac{3}{500}$

**45.** There are 24 ways to win. Assume 0 can be in any position and digits can be repeated.

$$\Pr(\text{Win}) = \frac{24}{10\cdot 10\cdot 10\cdot 10} = \frac{3}{1250}$$

**46. a.** $\Pr(0\text{ def}) = \dfrac{\binom{180}{10}}{\binom{200}{10}} \approx \dfrac{7.6283\times 10^{15}}{2.2451\times 10^{16}} \approx 0.3398$

**b.** $\Pr(2\text{ def}) = \dfrac{\binom{20}{2}\cdot\binom{180}{8}}{\binom{200}{10}} \approx \dfrac{(190)(2.3342\times 10^{13})}{2.2451\times 10^{16}}$

$\approx 0.1975$

**47.** $\Pr(3\text{ best}) = \dfrac{1}{{}_5C_3} = \dfrac{1}{\binom{5}{3}} = \dfrac{1}{10}$

**48. a.** $\Pr(1\text{ def}) = \dfrac{\binom{2}{1}\cdot\binom{10}{5}}{\binom{12}{6}} = \dfrac{2(252)}{924} = \dfrac{6}{11} \approx 0.545$

**b.** $\Pr(\text{At least 1 def}) = 1 - \Pr(\text{No def}) = 1 - \dfrac{\binom{10}{6}}{\binom{12}{6}} = 1 - \dfrac{210}{924} = 1 - \dfrac{5}{22} = \dfrac{17}{22} \approx 0.773$

Alternate Method: $\Pr(E) = \dfrac{\binom{2}{1}\cdot\binom{10}{5} + \binom{2}{2}\cdot\binom{10}{4}}{\binom{12}{6}} = \dfrac{504+210}{924} \approx 0.773$

**49.** $$\begin{bmatrix}0.7 & 0.2 & 0.1\end{bmatrix}\begin{bmatrix}0.15 & 0.60 & 0.25\\ 0.15 & 0.35 & 0.50\\ 0 & 0.20 & 0.80\end{bmatrix} = \begin{bmatrix}0.135 & 0.51 & 0.355\end{bmatrix}$$

$$\begin{bmatrix}0.135 & 0.51 & 0.355\end{bmatrix}\begin{bmatrix}0.15 & 0.60 & 0.25\\ 0.15 & 0.35 & 0.50\\ 0 & 0.20 & 0.80\end{bmatrix} = \begin{bmatrix}0.09675 & 0.3305 & 0.57275\end{bmatrix}$$

$$\begin{bmatrix}0.09675 & 0.3305 & 0.57275\end{bmatrix}\begin{bmatrix}0.15 & 0.60 & 0.25\\ 0.15 & 0.35 & 0.50\\ 0 & 0.20 & 0.80\end{bmatrix} = \begin{bmatrix}0.0640875 & 0.288275 & 0.6476375\end{bmatrix}$$

**50.** $$\begin{bmatrix}V_1 & V_2 & V_3\end{bmatrix}\begin{bmatrix}0.15 & 0.60 & 0.25\\ 0.15 & 0.35 & 0.50\\ 0 & 0.20 & 0.80\end{bmatrix} = \begin{bmatrix}V_1 & V_2 & V_3\end{bmatrix}$$

$$\begin{array}{lcl} 0.15V_1 + 0.15V_2 = V_1 & & -0.85V_1 + 0.15V_2 = 0 \\ 0.60V_1 + 0.35V_2 + 0.20V_3 = V_2 & \text{or} & 0.60V_1 - 0.65V_2 + 0.20V_3 = 0 \\ 0.25V_1 + 0.50V_2 + 0.80V_3 = V_3 & & 0.25V_1 + 0.50V_2 - 0.20V_3 = 0 \end{array}$$

$$\left[\begin{array}{ccc|c}1 & 2 & -0.80 & 0\\ -0.85 & 0.15 & 0 & 0\\ 0.60 & -0.65 & 0.20 & 0\end{array}\right] \to \left[\begin{array}{ccc|c}1 & 2 & -0.80 & 0\\ 0 & 1.85 & -0.68 & 0\\ 0 & -1.85 & 0.68 & 0\end{array}\right] \to \left[\begin{array}{ccc|c}1 & 2 & -.80 & 0\\ 0 & 1 & -\frac{68}{185} & 0\\ 0 & 0 & 0 & 0\end{array}\right] \to \left[\begin{array}{ccc|c}1 & 0 & -\frac{12}{185} & 0\\ 0 & 1 & -\frac{68}{185} & 0\\ 0 & 0 & 0 & 0\end{array}\right]$$

$$V_1 = \frac{12}{185}V_3,\ V_2 = \frac{68}{185}V_3;\ \text{So } \frac{12}{185}V_3 + \frac{68}{185}V_3 + V_3 = 1 \text{ gives } V_3 = \frac{37}{53}.$$

$$V_1 = \frac{12}{185}\cdot\frac{37}{53} = \frac{12}{265} \qquad V_2 = \frac{68}{185}\cdot\frac{37}{53} = \frac{68}{265}$$

The steady-state vector is $\begin{bmatrix}\frac{12}{265} & \frac{68}{265} & \frac{37}{53}\end{bmatrix}$.

## *Chapter 7 Test*

**1. a.** $\Pr(\text{odd}) = \frac{4}{7}$

**b.** $\Pr(\text{white}) = \frac{3}{7}$

**2. a.** $\Pr(\text{black and even}) = \frac{2}{7}$

**b.** $\Pr(\text{white or even}) = \frac{5}{7}$

**3. a.** $\Pr(\text{red}) = 0$

**b.** $\Pr(<8) = 1$

**4.** $\Pr(\text{sum} = 7)$

$= \Pr((1, 6), (2, 5), (3, 4))$

$= \frac{3}{\binom{7}{2}} = \frac{3}{21} = \frac{1}{7}$

**5.** $\Pr(\text{W and W}) = \frac{3}{7}\cdot\frac{2}{6} = \frac{1}{7}$

**6. a.** $\Pr(\text{W1 and B2}) = \frac{3}{7}\cdot\frac{4}{6} = \frac{2}{7}$

**b.** $\Pr(\text{W1B2 or B1W2}) = \frac{3}{7}\cdot\frac{4}{6} + \frac{4}{7}\cdot\frac{3}{6} = \frac{4}{7}$

**7.** $\Pr(\text{both odd}) = \frac{4}{7}\cdot\frac{3}{6} = \frac{2}{7}$

**8.**

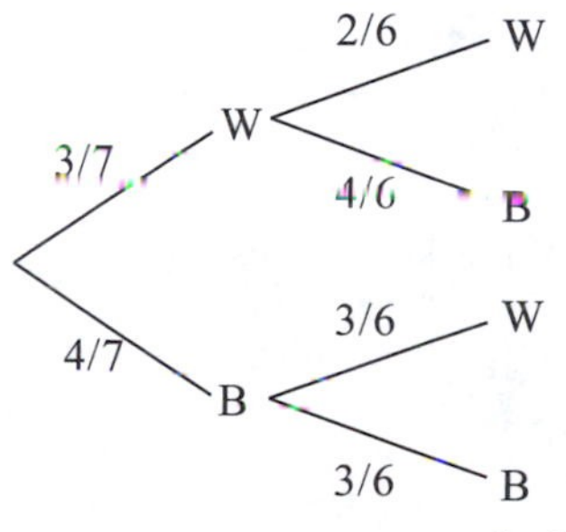


$\Pr(\text{W1W2 or B1W2}) = \frac{3}{7}\cdot\frac{2}{6} + \frac{4}{7}\cdot\frac{3}{6} = \frac{3}{7}$

**9.** Using Bayes' formula and the probability tree in exercise 8, we have

$$\Pr(\text{B1} \mid \text{W2}) = \frac{\frac{4}{7}\cdot\frac{3}{6}}{\frac{4}{7}\cdot\frac{3}{6} + \frac{3}{7}\cdot\frac{2}{6}} = \frac{12}{18} = \frac{2}{3}$$

**10.** Total outcomes: $26\cdot 26\cdot 26$

$\Pr(\text{RAT}) = \frac{1}{26^3} = \frac{1}{17{,}576}$

**11.** Assume a large sample space and events are independent.

$\Pr(2 \text{ Am in } 3 \text{ draws})$

$= \binom{3}{2}(0.35)^2(0.65) = 0.238875$

**12. a.** $\Pr(\text{left handed}) = \frac{4}{20} = \frac{1}{5}$

**b.** $\Pr(\text{ambidextrous}) = \frac{1}{20}$

**13. a.** $\Pr(\text{LL}) = \frac{4}{20}\cdot\frac{3}{19} = \frac{3}{95}$

**b.** There are other ways to solve this problem.

$$\Pr(\text{1R and 1L}) = \frac{\binom{15}{1}\binom{4}{1}}{\binom{20}{2}} = \frac{6}{19}$$

**c.** $\Pr(2\text{R}) = \frac{15}{20}\cdot\frac{14}{19} = \frac{21}{38}$

**d.** $\Pr(2 \text{ ambidextrous}) = 0$

**14. a.** $\binom{42}{6} = 5{,}245{,}786$

**b.** $\Pr(\text{Win}) = \frac{1}{5{,}245{,}786}$

**15. a.** $\binom{50}{5} = 2{,}118{,}760$

**b.** $\Pr(\text{Win}) = \frac{1}{2{,}118{,}760}$

**16.** $\Pr(\text{Def}) = \Pr(\#1 \text{ and Def or } \#2 \text{ and Def})$
$= (0.8)(0.06) + (0.2)(0.08) = 0.064$

**17.**

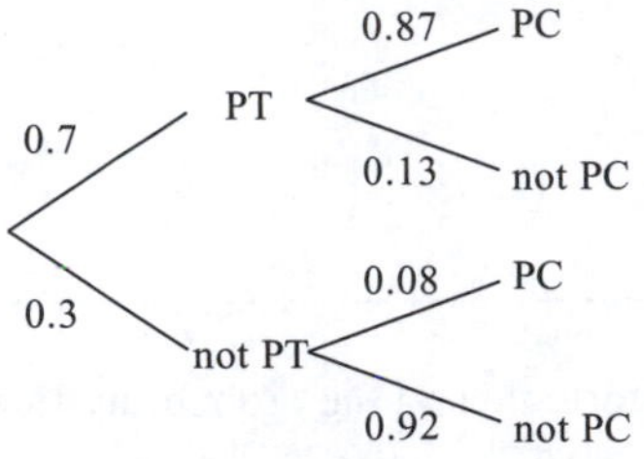


**a.** $\Pr(\text{Pass T and Pass C or Fail T and Pass C})$
$= (0.7)(0.87) + (0.3)(0.08) = 0.633$

**b.** $\Pr(\text{Pass T} \mid \text{Pass C})$
$$= \frac{(0.7)(0.87)}{(0.7)(0.87) + (0.3)(0.08)}$$
$\approx 0.96$

**18.**

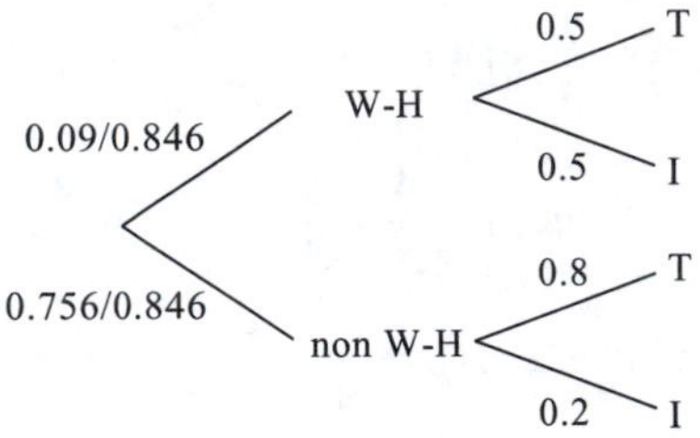


Using Bayes' formula and the probability tree, we have

$$\Pr(\text{H} \mid \text{I}) = \frac{(\frac{0.09}{0.846})(0.5)}{(\frac{0.09}{0.846})(0.5) + (\frac{0.756}{0.846})(0.2)} \approx 0.229$$

**19. a.** $\Pr(\text{Def WW}) = \frac{70}{350} = \frac{1}{5}$

**b.** $\Pr(\text{Def TL}) = \frac{25}{350} = \frac{1}{14}$

**c.** $\Pr(\text{No Def TL}) = 1 - \Pr(\text{Def TL}) = \frac{13}{14}$

**20.** $\Pr(\text{Paint}) = 0.14$
$\Pr(\text{BW and Paint}) = 0.03$
$$\Pr(\text{BW} \mid \text{Paint}) = \frac{\Pr(\text{BW and Paint})}{\Pr(\text{Paint})} = \frac{0.03}{0.14} = \frac{3}{14}$$

**21. a.** off/on $2 \cdot 2 \cdot 2 \cdot \ldots \cdot 2 = 2^{10} = 1024$
Switch 1 2 3 … 10

**b.** $\Pr(\text{open}) = \frac{1}{1024}$

**c.** $\Pr(\text{open}) = \frac{1}{3}$

**d.** Change the code (sequence).

**22. a.**
$$A = \begin{bmatrix} 0.80 & 0.20 \\ 0.07 & 0.93 \end{bmatrix} \begin{matrix} \text{Text} \\ \text{Other} \end{matrix}$$
(columns T, O)

**b.** $[0.25 \quad 0.75] \cdot A = [0.2525 \quad 0.7475]$
$[0.2525 \quad 0.7475] \cdot A$
$= [0.254325 \quad 0.745675]$
$[0.2543 \quad 0.7457] \cdot A$
$= [0.25565725 \quad 0.74434275]$
Percent of market share 3 editions later is $\approx 25.6\%$.

**c.** $[V_1 \quad V_2] \begin{bmatrix} 0.80 & 0.20 \\ 0.07 & 0.93 \end{bmatrix} = [V_1 \quad V_2]$

$$\begin{cases} 0.8V_1 + 0.07V_2 = V_1 \\ -0.2V_1 + 0.07V_2 = 0 \end{cases}$$

$$\begin{cases} 0.2V_1 + 0.93V_2 = V_2 \\ 0.2V_1 - 0.07V_2 = 0 \end{cases}$$

$V_1 = \frac{7}{20} V_2$, $V_1 + V_2 = 1$ gives

$V_1 = \frac{7}{27}$, $V_2 = \frac{20}{27}$.

Steady-state vector is $\left[\frac{7}{27}, \frac{20}{27}\right]$. Hence the percent of market this text will have is $\frac{7}{27} \approx 25.9\%$.

# Chapter 8: Further Topics in Probability; Data Description

## *Exercises 8.1*

**1.** $p=0.3,\ q=0.7,\ n=6,\ x=4$

$$\binom{6}{4}(0.3)^4(0.7)^2=15(0.0081)(0.49)\approx 0.0595$$

**3.** **a.** $p=\dfrac{1}{6}$ **b.** $q=\dfrac{5}{6}$ **c.** $n=18$

**d.** $\Pr(6\ 4\text{'s})=\binom{18}{6}\left(\dfrac{1}{6}\right)^6\left(\dfrac{5}{6}\right)^{12}=18564\left(\dfrac{1}{46656}\right)\left(\dfrac{244140625}{2176782336}\right)\approx 0.04463$

**5.** $p=\dfrac{1}{2},\ q=\dfrac{1}{2},\ n=6$

**a.** $x=6$ $\binom{6}{6}\left(\dfrac{1}{2}\right)^6\left(\dfrac{1}{2}\right)^0=1\left(\dfrac{1}{64}\right)(1)=\dfrac{1}{64}$

**b.** $x=3$ $\binom{6}{3}\left(\dfrac{1}{2}\right)^3\left(\dfrac{1}{2}\right)^3=20\left(\dfrac{1}{8}\right)\left(\dfrac{1}{8}\right)=\dfrac{5}{16}$

**c.** $x=2$ $\binom{6}{2}\left(\dfrac{1}{2}\right)^2\left(\dfrac{1}{2}\right)^4=15\left(\dfrac{1}{4}\right)\left(\dfrac{1}{16}\right)=\dfrac{15}{64}$

**7.** $p=\dfrac{1}{6},\ q=\dfrac{5}{6},\ n=12,\ x=5$

$$\binom{12}{5}\left(\frac{1}{6}\right)^5\left(\frac{5}{6}\right)^7=792\left(\frac{1}{7776}\right)\left(\frac{78{,}125}{279{,}936}\right)\approx 0.0284$$

**9.** **a.** $p=\dfrac{3}{5},\ q=\dfrac{2}{5},\ n=5,\ x=2$

$$\binom{5}{2}\left(\frac{3}{5}\right)^2\left(\frac{2}{5}\right)^3=10\left(\frac{9}{25}\right)\left(\frac{8}{125}\right)=0.2304$$

**b.** $p=\dfrac{2}{5},\ q=\dfrac{3}{5},\ n=5,\ x=5$

$$\binom{5}{5}\left(\frac{2}{5}\right)^5\left(\frac{3}{5}\right)^0=1\left(\frac{32}{3125}\right)(1)=0.01024$$

**c.** $\Pr(\text{At least 3B})=\binom{5}{3}\left(\dfrac{2}{5}\right)^3\left(\dfrac{3}{5}\right)^2+\binom{5}{4}\left(\dfrac{2}{5}\right)^4\left(\dfrac{3}{5}\right)+\binom{5}{5}\left(\dfrac{2}{5}\right)^5\left(\dfrac{3}{5}\right)^0$

$$=10\left(\frac{8}{125}\right)\left(\frac{9}{25}\right)+5\left(\frac{16}{625}\right)\left(\frac{3}{5}\right)+0.01024=0.2304+0.0768+0.01024=0.31744$$

**11.** $p=\dfrac{4}{36}=\dfrac{1}{9},\ q=\dfrac{8}{9},\ n=4,\ x=2;$

$$\binom{4}{2}\left(\frac{1}{9}\right)^2\left(\frac{8}{9}\right)^2=6\left(\frac{1}{81}\right)\left(\frac{64}{81}\right)\approx 0.0585$$

**13.** $p=0.85$, $q=0.15$, $n=10$, $x=8$

$$\binom{10}{8}(0.85)^8(0.15)^2 \approx 45(0.2725)(0.0225) \approx 0.2759$$

**15.** $p=\frac{1}{2}$, $q=\frac{1}{2}$, $n=4$

**a.** $x=2$ $\quad \binom{4}{2}\left(\frac{1}{2}\right)^2\left(\frac{1}{2}\right)^2 = 6\cdot\frac{1}{4}\cdot\frac{1}{4}=\frac{3}{8}=0.375$

**b.** $x=4$ $\quad \binom{4}{4}\left(\frac{1}{2}\right)^4\left(\frac{1}{2}\right)^0 = 1\cdot\frac{1}{16}\cdot 1 = 0.0625$

**17.** Pr(Def) $=\frac{1}{12}$, $n=4$

**a.** Pr(2Def) $=\binom{4}{2}\left(\frac{1}{12}\right)^2\left(\frac{11}{12}\right)^2 = 6\cdot\frac{1}{144}\cdot\frac{121}{144}=\frac{121}{3456}\approx 0.035$

**b.** Pr(0Def) $=\binom{4}{0}\left(\frac{1}{12}\right)^0\left(\frac{11}{12}\right)^4 = \frac{14641}{20736}\approx 0.706$

**19.** Pr(Blue) $=\frac{1}{4}$, $n=4$

**a.** Pr(1 Blue) $=\binom{4}{1}\left(\frac{1}{4}\right)^1\left(\frac{3}{4}\right)^3 = 4\cdot\frac{1}{4}\cdot\frac{27}{64}=\frac{27}{64}$

**b.** Pr(2 Blue) $=\binom{4}{2}\left(\frac{1}{4}\right)^2\left(\frac{3}{4}\right)^2 = 6\cdot\frac{1}{16}\cdot\frac{9}{16}=\frac{27}{128}$

**c.** Pr(0 Blue) $=\binom{4}{0}\left(\frac{1}{4}\right)^0\left(\frac{3}{4}\right)^4 = 1\cdot 1\cdot\frac{81}{256}=\frac{81}{256}$

**21.** Pr(Death) = 0.1, $n=5$

**a.** Pr(2 Deaths) $=\binom{5}{2}(.1)^2(.9)^3 = 10(.01)(.729) = 0.0729$

**b.** Pr(0 Deaths) $=\binom{5}{0}(.1)^0(.9)^5 = 1(1)(.59049) = 0.59049$

**c.** Pr(0 or 1 or 2 Deaths) $= 0.59049+\binom{5}{1}(.1)^1(.9)^4+0.0729 = .59049+.32805+.0729 = 0.9914$

**23.** Pr(Boy) $=\frac{105}{205}\approx 0.5122$, $n=6$

Pr(4 Boys) $=\binom{6}{4}(.5122)^4(.4878)^2 \approx 15(.06883)(.2379)\approx 0.2457$

**25.** Pr(Fire) = 0.004, $n=10$

Pr(2 Fires) $=\binom{10}{2}(.004)^2(.996)^8 \approx 45(.000016)(.96844)\approx 0.0007$

**27.** Pr(Hit) = 0.3, $n = 5$

**a.** $\Pr(3 \text{ Hits}) = \binom{5}{3}(.3)^3(.7)^2 = 10(.027)(.49) = 0.1323$

**b.** $\Pr(4 \text{ or } 5 \text{ Hits}) = \binom{5}{4}(.3)^4(.7)^1 + \binom{5}{5}(.3)^5(.7)^0 = 5(.0081)(.7) + 1(.00243)(1)$

$= 0.02835 + 0.00243 = 0.03078$

**29.** Pr(Def) = 0.01, $n = 10$

**a.** $\Pr(0 \text{ Def}) = \binom{10}{0}(.01)^0(.99)^{10} = (.99)^{10} \approx 0.9044$

**b.** $\Pr(1 \text{ Def}) = \binom{10}{1}(.01)^1(.99)^9 \approx 10(.01)(.9135) = 0.09135$

**c.** $\Pr(>1 \text{ Def}) = 1 - \Pr(0 \text{ or } 1 \text{ Def}) \approx 1 - (.9044 + .09135) = 0.00425$

**31.** $\Pr(\text{correct}) = \frac{1}{3}$

**a.** To get 60%, student needs to get at least 1 of the last 5 correct.

$\Pr(\geq 60\%) = 1 - \Pr(0 \text{ correct}) = 1 - \binom{5}{0}\left(\frac{1}{3}\right)^0\left(\frac{2}{3}\right)^5 = 1 - \frac{32}{243} = \frac{211}{243} \approx 0.8683$

**b.** To get 80%, student needs to get at least 3 of the last 5 correct.

$\Pr(\geq 80\%) = \Pr(3 \text{ or } 4 \text{ or } 5 \text{ correct}) = \binom{5}{3}\left(\frac{1}{3}\right)^3\left(\frac{2}{3}\right)^2 + \binom{5}{4}\left(\frac{1}{3}\right)^4\left(\frac{2}{3}\right) + \binom{5}{5}\left(\frac{1}{3}\right)^5\left(\frac{2}{3}\right)^0$

$= 10 \cdot \frac{1}{27} \cdot \frac{4}{9} + 5 \cdot \frac{1}{81} \cdot \frac{2}{3} + \frac{1}{243} \approx 0.1646 + 0.0412 + 0.0041 = 0.2099$

**33.** Pr(No suicide) = 0.997, $n = 100$

$\Pr(\text{No suicide}) = \binom{100}{100}(.997)^{100}(.003)^0 \approx 0.7405$

## *Exercises 8.2*

**1.**

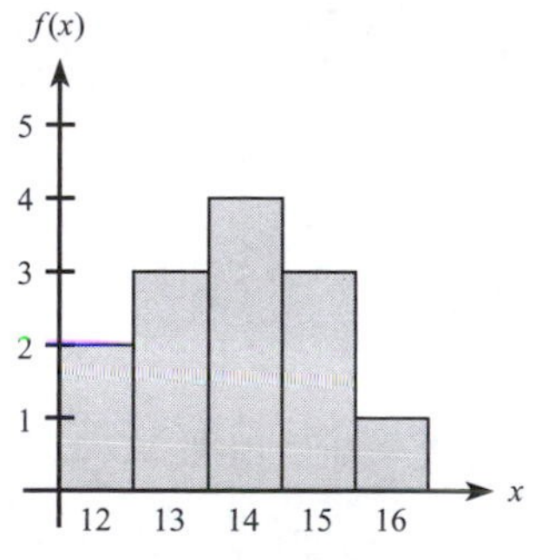


**3.**

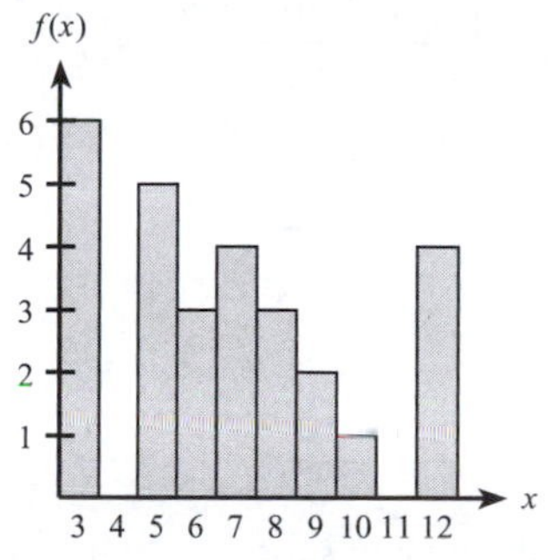

**5.**

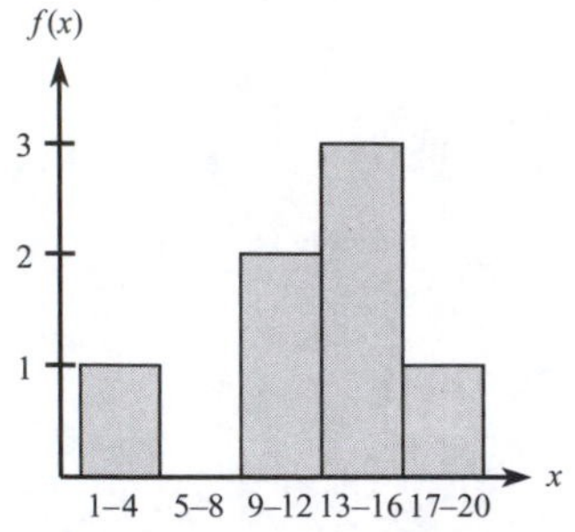


**7.**

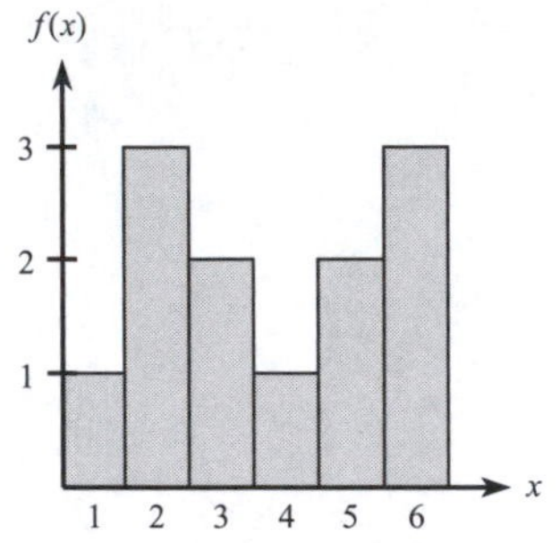


**9.** The mode is the score that occurs most frequently. 3 (4 times)

**11.** 13 (3 times)

**13.** Arranged in order: 1, 3, **5**, 6, 7.
The median is 5.

**15.** Arranged in order: 4, 7, 9, **12**, 14, 18, 36.
The median is 12.

**17.** Arranged in order: 1, 2, 2, **3**, **6**, 8, 12, 14
The mode is 2. The median is $(3+6)\div 2 = 4.5$.
The mean is $\overline{x} = (1+2+2+3+6+8+12+14)\div 8 = 6$.

**19.** Arranged in order: 14, 17, **17**, **20**, 31, 42
The mode is 17 since 17 occurs most often.
The median is $(17+20)\div 2 = 18.5$.
The mean is $\overline{x} = (14+17+17+20+31+42)\div 6 = 141\div 6 = 23.5$.

**21.** Arranged in order: 2.8, 5.3, **5.3**, 6.4, 6.8
The mode is 5.3. The median is 5.3.
The mean is $\overline{x} = (2.8+5.3+5.3+6.4+6.8)\div 5 = (26.6)\div 5 = 5.32$.

**23.**

| Scores | Class marks | Frequencies |
|---|---|---|
| 1 – 4 | 2.5 | 1 |
| 5 – 8 | 6.5 | 0 |
| 9 – 12 | 10.5 | 2 |
| 13 – 16 | 14.5 | 3 |
| 17 – 20 | 18.5 | 1 |

mean: $\overline{x} = \dfrac{2.5(1)+6.5(0)+10.5(2)+14.5(3)+18.5(1)}{1+0+2+3+1} \approx 12.21$

mode : 14.5 (most frequent score)
median: 14.5 (middle (4th) score)

**25.** Range: $11 - 2 = 9$

**27.** Range: $11 - (-3) = 14$

**29.** $\bar{x} = (5+7+1+3+0+8+6+2) \div 8 = 4$

$s^2 = \begin{bmatrix} 1^2 + 3^2 + (-3)^2 + (-1)^2 + \\ (-4)^2 + 4^2 + 2^2 + (-2)^2 \end{bmatrix} \div 7 = \frac{60}{7} \approx 8.57$

$s = \sqrt{8.57} \approx 2.93$

**31.** $\bar{x} = (11+12+13+14+15+16+17) \div 7 = 14$

$s^2 = \left[(-3)^2 + (-2)^2 + (-1)^2 + 0^2 + 1^2 + 2^2 + 3^2\right] \div 6$

$= \frac{28}{6} \approx 4.67$

$s = \sqrt{4.67} \approx 2.16$

**33.**

$\bar{x} = (3 \cdot 1 + 1 \cdot 2 + 4 \cdot 3 + 2 \cdot 4 + 1 \cdot 5) \div 11 = \frac{30}{11} \approx 2.73$

$s^2 = \left[3(-1.73)^2 + 1(-0.73)^2 + 4(0.27)^2 + 2(1.27)^2 + 1(2.27)^2\right] \div 10 \approx 1.82$

$s = \sqrt{1.82} \approx 1.35$

**35.** $\bar{x} = (6 \cdot 3 + 0 \cdot 4 + 5 \cdot 5 + 3 \cdot 6 + 4 \cdot 7 + 3 \cdot 8 + 2 \cdot 9 + 1 \cdot 10 + 0 \cdot 11 + 4 \cdot 12) \div 28 = \frac{189}{28} = 6.75$

$s^2 = \frac{6(-3.75)^2 + 5(-1.75)^2 + 3(-.75)^2 + 4(.25)^2 + 3(1.25)^2 + 2(2.25)^2 + 1(3.25)^2 + 4(5.25)^2}{27}$

$= \frac{237.25}{27} \approx 8.787$

$s = \sqrt{8.787} \approx 2.96$

**f37.** **a.**

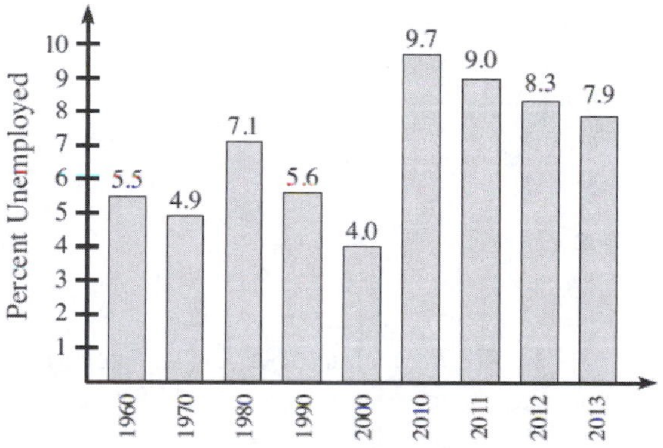


**b.** $\bar{x} = 6.9,\ s = 1.98$

**39.**

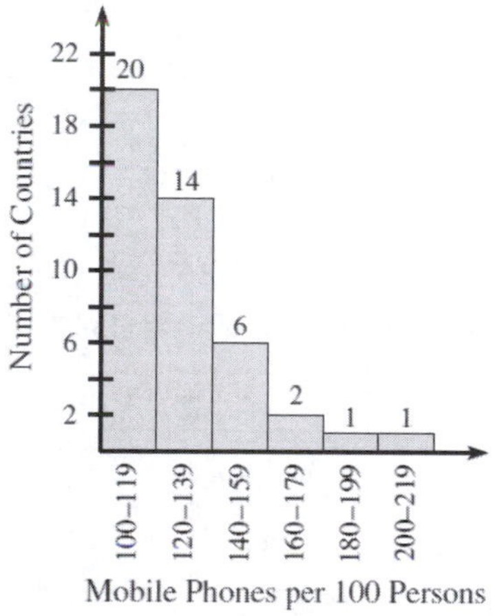


**41.** Using the mean will give the highest measure.

**43.** Using the median would give the most representative average.

**45.** Using technology:

**a.** $\bar{x} = 63.4\%,\ s = 19.26$

**b.** yes

**47.**

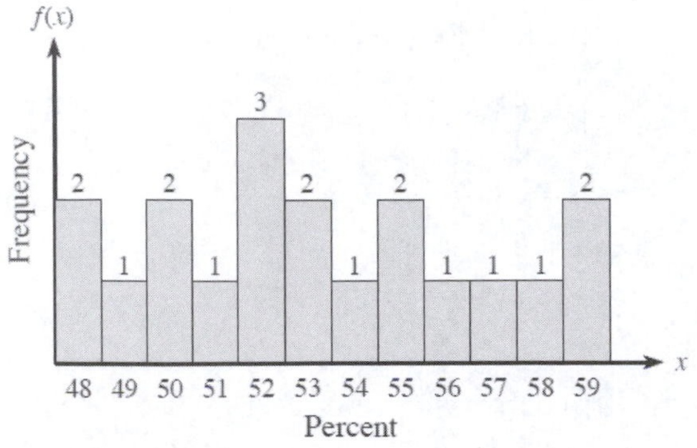


**49.** $\overline{x} = \dfrac{\Sigma xf}{\Sigma f} = \dfrac{531.8}{160} = 3.32375$ $\quad s^2 = \dfrac{72.970}{159} \approx 0.4589$ $\quad s = 0.6774$

**51.** **a.** $\overline{x} = \dfrac{160,000(1)+120,000(1)+60,000(2)+40,000(1)+32,000(5)}{10} = \$60,000$

**b.** 32,000, 32,000, 32,000, 32,000, <u>32,000</u>, <u>40,000</u>, 60,000, 60,000, 12,000, 160,000

Median $= \dfrac{32,000+40,000}{2} = \$36,000$

**c.** \$32,000

**53.** **a.** $\overline{x} = \dfrac{17.32+17.67+18.02+\ldots+30.80+32.55}{15} \approx \$23.33$

**b.** $s^2 = \left[(17.32-23.33)^2+(17.67-23.33)^2+\ldots+(32.55-23.33)^2\right] \div 15 \approx 26.4$

$s \approx \$5.14$

**55.** Using technology: **a.** $\overline{x} = 35,434,000$ **b.** $s = 13,312,000$

**57.** **a.**

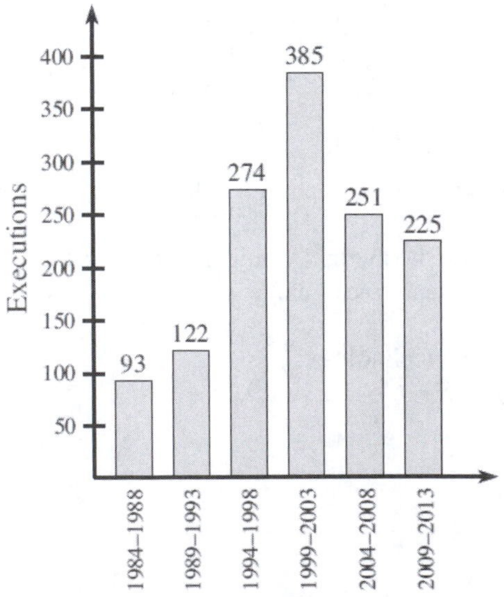


**b.** Using technology: $\overline{x} = 45$ executions

**c.** Using technology: $s = 22.3$ executions

**d.** No

# Chapter 8: Further Topics in Probability; Data Description

## *Exercises 8.3*

**1.** No. $\Pr(x)$ must be $\geq 0$.

**3.** Yes. $0 \leq \Pr(x) \leq 1$ and $\sum \Pr(x) = 1$.

**5.** Yes. $0 \leq \Pr(x) \leq 1$ and $\sum \Pr(x) = 1$.

**7.** No. $\sum \Pr(x) \neq 1$.

**9.** $E(x) = \sum x \Pr(x) = 0 \cdot \frac{1}{8} + 1 \cdot \frac{1}{4} + 2 \cdot \frac{1}{4} + 3 \cdot \frac{3}{8} = \frac{15}{8}$

**11.** $E(x) = \sum x \Pr(x) = 4 \cdot \frac{1}{3} + 5 \cdot \frac{1}{3} + 6 \cdot \frac{1}{3} + 7 \cdot 0 = 5$

**13.** $\mu = \sum x \Pr(x) = 0 \cdot \frac{1}{4} + 1 \cdot \frac{1}{4} + 2 \cdot \frac{1}{8} + 3 \cdot \frac{3}{8} = \frac{13}{8}$

$$\sigma^2 = \sum (x-\mu)^2 \Pr(x) = \left(0 - \frac{13}{8}\right)^2 \cdot \frac{1}{4} + \left(1 - \frac{13}{8}\right)^2 \cdot \frac{1}{4} + \left(2 - \frac{13}{8}\right)^2 \cdot \frac{1}{8} + \left(3 - \frac{13}{8}\right)^2 \cdot \frac{3}{8}$$

$$= \frac{169}{256} + \frac{25}{256} + \frac{9}{512} + \frac{363}{512} = \frac{760}{512} = \frac{95}{64} \approx 1.48$$

$\sigma \approx \sqrt{1.48} \approx 1.22$

**15.** $\mu = 0 \cdot \frac{0}{21} + 1 \cdot \frac{1}{21} + 2 \cdot \frac{2}{21} + 3 \cdot \frac{3}{21} + 4 \cdot \frac{4}{21} + 5 \cdot \frac{5}{21} + 6 \cdot \frac{6}{21} = \frac{91}{21} = \frac{13}{3}$

$$\sigma^2 = \left(-\frac{13}{3}\right)^2 \cdot 0 + \left(-\frac{10}{3}\right)^2 \cdot \frac{1}{21} + \left(-\frac{7}{3}\right)^2 \cdot \frac{2}{21} + \left(-\frac{4}{3}\right)^2 \cdot \frac{3}{21} + \left(-\frac{1}{3}\right)^2 \cdot \frac{4}{21} + \left(\frac{2}{3}\right)^2 \cdot \frac{5}{21} + \left(\frac{5}{3}\right)^2 \cdot \frac{6}{21}$$

$$= \frac{1}{21}\left(\frac{100}{9} + \frac{98}{9} + \frac{48}{9} + \frac{4}{9} + \frac{20}{9} + \frac{150}{9}\right) = \frac{420}{9 \cdot 21} = \frac{20}{9} \approx 2.22$$

$\sigma = \sqrt{2.22} \approx 1.49$

**17.** $E(x) = \sum x \Pr(x) = 0 \cdot \frac{0}{10} + 1 \cdot \frac{1}{10} + 2 \cdot \frac{2}{10} + 3 \cdot \frac{3}{10} + 4 \cdot \frac{4}{10} = 3$

**19.** $E(x) = 0 \cdot \frac{1}{5} + 1 \cdot \frac{1}{5} + 2 \cdot \frac{1}{5} + 3 \cdot \frac{1}{5} + 4 \cdot \frac{1}{5} = \frac{10}{5} = 2$

**21. a.** Let $x$ = number of fives

| $x$ | $\Pr(x)$ |
|---|---|
| 0 | $\binom{3}{0}\left(\frac{1}{6}\right)^0\left(\frac{5}{6}\right)^3 = \frac{125}{216}$ |
| 1 | $\binom{3}{1}\left(\frac{1}{6}\right)^1\left(\frac{5}{6}\right)^2 = \frac{75}{216} = \frac{25}{72}$ |
| 2 | $\binom{3}{2}\left(\frac{1}{6}\right)^2\left(\frac{5}{6}\right)^1 = \frac{15}{216} = \frac{5}{72}$ |
| 3 | $\binom{3}{3}\left(\frac{1}{6}\right)^3\left(\frac{5}{6}\right)^0 = \frac{1}{216}$ |

**b.** $\mu = 3\left(\frac{1}{6}\right) = \frac{1}{2}$

**c.** $\sigma = \sqrt{3\left(\frac{1}{6}\right)\left(\frac{5}{6}\right)} = \sqrt{\frac{15}{36}} = \frac{1}{6}\sqrt{15} \approx 0.645$

**23. a.** $\mu = 60(0.7) = 42$

**b.** $\sigma = \sqrt{60(0.7)(0.3)} \approx 3.55$

**25. a.** $\mu = 50\left(\frac{3}{5}\right) = 30$

**b.** $\sigma = \sqrt{40(0.8)(0.2)} \approx 2.53$

**b.** $\sigma = \sqrt{50\left(\frac{3}{5}\right)\left(\frac{2}{5}\right)} = \sqrt{\frac{300}{25}} = \frac{2\sqrt{3}}{5} \approx 0.693$

**27.** $\Pr(\text{sum of } 6) = \frac{5}{36}$.

$E(x) = 900\left(\frac{5}{36}\right) = 125.$

**29.** $(a+b)^6 = \binom{6}{0}a^6 + \binom{6}{1}a^5b + \binom{6}{2}a^4b^2 + \binom{6}{3}a^3b^3 + \binom{6}{4}a^2b^4 + \binom{6}{5}ab^5 + \binom{6}{6}b^6$

$= a^6 + 6a^5b + 15a^4b^2 + 20a^3b^3 + 15a^2b^4 + 6ab^5 + b^6$

**31.** $(x+h)^4 = \binom{4}{0}x^4 + \binom{4}{1}x^3h + \binom{4}{2}x^2h^2 + \binom{4}{3}xh^3 + \binom{4}{4}h^4 = x^4 + 4x^3h + 6x^2h^2 + 4xh^3 + h^4$

**33.** $E(x) = 0(.04) + 1(.35) + 2(.38) + 3(.18) + 4(.05) = 1.85$

**35.** TV: $E(x) = 100{,}000(0.01) + 50{,}000(0.47) + 25{,}000(0.52) = 37{,}500$

PA: $E(x) = 80{,}000(0.02) + 50{,}000(0.47) + 20{,}000(0.51) = 35{,}300$

**37.** $E(x) = 39{,}900\left(\frac{1}{1500}\right) + 4900\left(\frac{1}{1500}\right) + 2400\left(\frac{1}{1500}\right) + 1400\left(\frac{1}{1500}\right) - 100\left(\frac{1496}{1500}\right)$

$= 26.60 + 3.27 + 1.60 + 0.93 - 99.73 = -67.33$

Expected loss \$67.33

**39.** $E(x) = 15 \cdot \frac{1}{13} + 10 \cdot \frac{1}{13} + 1 \cdot \frac{1}{13} - 4(1) = -\$2.00$

**41.** Buy 0: There is no profit.

Buy 100: $E(x) = 4(100)(0.25) + 4(50)(0.20) + 4(10)(0.55) - 2(50)(0.20) - 2(90)(0.55) = \$43$

Buy 200: $E(x) = 4(180)(.25) + 4(50)(.20) + 4(10)(.55) - 2(20)(.25) - 2(150)(.20) - 2(190)(.55) = -\$37$

Buy 100 for the best profit.

**43.** E(cost with policy) = \$100 + 0(0.92) + 100(0.08) = \$108; E(cost without policy) = 0(0.92) + 1000(0.08) = \$80

Save \$28 per year by "taking the chance."

**45.** No. To be usable the diameter must be within 0.01 inches of 2 inches. Although the average is 2 inches, some of the pipes may fall outside this range.

**47.** $n = 100,\ p = 0.1$

**a.** $\mu = np = 100(0.1) = 10$

**b.** $\sigma = \sqrt{npq} = \sqrt{100(0.1)(0.9)} = 3$

**49.** $n = 100,000$, $p = 0.6$

**a.** $\mu = np = 100,000(0.6) = 60,000$

**b.** $\sigma = \sqrt{npq}$
$= \sqrt{100,000(0.6)(0.4)} = \sqrt{24,000} \approx 155$

**51.** In problem 49 we have $\sigma \approx 155$. Thus, 2 standard deviations is 2(155) = 310 votes. The candidate actually received 60,000 – 310 = 59,690 votes.

**53.** $n = 20,\ p = \frac{1}{5},\ q = \frac{4}{5}$

**a.** $\mu = np = 20 \cdot \frac{1}{5} = 4$

**b.** $\sigma = \sqrt{npq} = \sqrt{20 \cdot \frac{1}{5} \cdot \frac{4}{5}} = \sqrt{\frac{16}{5}} \approx 1.79$

**55.** $n = 200$, $\Pr(\text{Def}) = 0.01 = p$, $q = 0.99$
$\mu = np = 200(0.01) = 2;$
$\sigma = \sqrt{200(0.01)(0.99)} \approx 1.41$

**57.** Pr(Germinate) = 0.85,
Pr(Fail to germinate) = 0.15
For this question $n = 2000$, $p = 0.15$
$\mu = 2000(0.15) = 300$

## *Exercises 8.4*

**1.** $\Pr(0 \le z \le 1.8) = 0.4641$

**3.** $\Pr(-0.6 \le z \le 0) = 0.2258$

**5.** $\Pr(-1.5 \le z \le 0) = 0.4332$
$\Pr(0 \le z \le 2.1) = 0.4821$
$\Pr(-1.5 \le z \le 2.1) = 0.4332 + 0.4821 = 0.9153$

**7.** $\Pr(-1.9 \le z \le 0) = 0.4713$
$\Pr(-1.1 \le z \le 0) = 0.3643$
$\Pr(-1.9 \le z \le -1.1) = 0.4713 - 0.3643 = 0.1070$

**9.** $\Pr(0 \le z \le 3) = 0.4987$
$\Pr(0 \le z \le 2.1) = 0.4821$
$\Pr(2.1 \le z \le 3) = 0.4987 - 0.4821 = 0.0166$

**11.** $\Pr(0 < z \le 2) = 0.4773$
$\Pr(z > 2) = 0.5 - 0.4773 = 0.0227$

**13.** $\Pr(0 \le x \le 1.2) = 0.3849$
$\Pr(z < 1.2) = 0.5 + 0.3849 = 0.8849$

**15.** 20: $z = \frac{20-20}{5} = 0$ 22.5: $z = \frac{22.5-20}{5} = 0.5$
$\Pr(20 \le x \le 22.5) = \Pr(0 \le z \le 0.5) = 0.1915$

**17.** 13.75: $z = \frac{13.75-20}{5} = -1.25$ 20: $z = \frac{20-20}{5} = 0$
$\Pr(13.75 \le x \le 20) = \Pr(0 \le z \le 1.25) = 0.3944$

**19.** 45: $z = \frac{45-50}{10} = -0.5$ 55: $z = \frac{55-50}{10} = 0.5$
$\Pr(45 \le x \le 55) = 2\Pr(0 \le z \le 0.5) = 2(0.1915) = 0.3830$

**21.** 35: $z = \frac{35-50}{10} = -1.5$ 60: $z = \frac{60-50}{10} = 1$
$\Pr(35 \le x \le 60) = \Pr(0 \le z \le 1.5) + \Pr(0 \le z \le 1) = 0.4332 + 0.3413 = 0.7745$

**23.** 134: $z=\dfrac{134-110}{12}=2$

$\Pr(x<134)=0.5+\Pr(0\le z\le 2)=0.5+0.4773=0.9773$

**25.** 128: $z=\dfrac{128-110}{12}=1.5$

$\Pr(x>128)=0.5-\Pr(0\le z\le 1.5)=0.5-0.4332=0.0668$

**27.** Since $\Pr(x>A)>0.5$, $A$ is negative. We have $\Pr(A\le x\le 0)=0.95-.5=0.45$, so $\Pr(0\le x\le -A)=0.45$. Using Appendix C we average the $z$-scores corresponding to area values 0.4495 and 0.4505 to get $-A=\dfrac{1.64+1.65}{2}=1.645$. Thus $A=-1.645$.

**29.** Since $\Pr(x<B)=0.75$, $\Pr(0<x<B)=0.75-0.5=0.25$. Appendix C gives a corresponding $z$-score of approximately 0.675. Solving $0.675=\dfrac{B-(-3)}{4}$ gives $B=-0.3$.

**31.** **a.** 15: $z=0$ 19: $z=\dfrac{19-15}{4}=1$ $\Pr(0\le z\le 1)=0.3413$

**b.** 10: $z=\dfrac{10-15}{4}=-1.25$ $\Pr(-1.25\le z\le 0)=0.3944$

**33.** 10: $z=\dfrac{10-20}{4}=-2.5$ $z=\dfrac{30-20}{4}=2.5$

$\Pr(10\le x\le 30)=\Pr(-2.5\le z\le 0)+\Pr(0\le z\le 2.5)=0.4938+0.4938=0.9876$

**35.** **a.** 160: $z=0$ 181: $z=\dfrac{181-160}{15}=1.4$ $\Pr(0\le z\le 1.4)=0.4192$

**b.** 190: $z=\dfrac{190-160}{15}=2$ $\Pr(z\ge 2)=0.5000-0.4773=0.0227$

**c.** $\Pr(1.4\le z\le 2)=0.4773-0.4192=0.0581$

**d.** 130: $z=\dfrac{130-160}{15}=-2$

$\Pr(-2\le z\le 1.4)=\Pr(-2\le z\le 0)+\Pr(0\le z\le 1.4)=0.4773+0.4192=0.8965$

**37.** **a.** 22: $z=\dfrac{22-28}{4}=-1.5$

$\Pr(z\le -1.5)=0.5000-0.4332=0.0668$

**b.** 30: $z=\dfrac{30-28}{4}=0.5$

$\Pr(z\ge 0.5)=0.5000-0.1915=0.3085$

**c.** 26: $z=\dfrac{26-28}{4}=-0.5$

$\Pr(-0.5\le z\le 0.5)=\Pr(-0.5\le z\le 0)+\Pr(0\le z\le 0.5)=0.1915+0.1915=0.3830$

**39.** **a.** 140: $z = \frac{140-120}{12} = 1.67$ $\Pr(z \geq 1.67) = 0.5000 - 0.4525 = 0.0475$

**b.** 110: $z = \frac{110-120}{12} = -0.83$ $\Pr(z \leq -0.83) = 0.5000 - 0.2967 = 0.2033$

**c.** 130: $z = 0.83$ $\Pr(-0.83 \leq z \leq 0.83) = 0.2967 + 0.2967 = 0.5934$

**41.** **a.** 0.9: $z = \frac{0.9-0.7}{0.1} = 2$ $\Pr(z \geq 2) = 0.5000 - 0.4773 = 0.0227$

**b.** 0.6: $z = \frac{0.6-0.7}{0.1} = -1$ $\Pr(z \leq -1) = 0.5000 - 0.3413 = 0.1587$

**c.** $\Pr(-1 \leq z \leq 2) = 0.3413 + 0.4773 = 0.8186$

**43.** Let $A$ be the score above which a student earns an A. $\Pr(78 < x < A) = 0.4$, and the $z$-score yielding this probability is about 1.28. Solving $1.28 = \frac{A-78}{6}$ gives $A \approx 85.7$. Score 86 or higher for an A.

**45.** We want to find the fill-amount setting $\mu$. Using Appendix C with $A = 0.50 - 0.01 = 0.49$ to find $z = -2.33$, we solve $-2.33 = \frac{128-\mu}{0.3}$ to get $\mu \approx 128.7$ oz.

**47.** We want to find the academic-suspension threshold $G$. Using Appendix C with $A = 0.5 - 0.15 = 0.35$ to find that $z \approx -1.035$, we solve $-1.035 = \frac{G-2.48}{0.9}$ to get $G \approx 1.55$.

## *Exercises 8.5*

**1.** $np = 28(0.3) = 8.4$
$nq = 28(0.7) = 19.6$
Yes, the normal approximation can be used.

**3.** $np = 12(0.3) = 3.6$
$nq = 12(0.7) = 8.4$
No, the normal approximation cannot be used.

**5.** $\mu = 150(0.6) = 90$
$\sigma = \sqrt{150(0.6)(0.4)} = 6$
Find $\Pr(90.5 \leq x \leq 91.5)$.
$z_1 = \frac{90.5-90}{6} \approx 0.08$
$z_2 = \frac{91.5-90}{6} \approx 0.25$
$\Pr(90.5 \leq x \leq 91.5)$
$= \Pr(0.08 \leq z \leq 0.25) = 0.0668$

**7.** $\mu = 180(0.4) = 72$
$\sigma = \sqrt{180(0.4)(0.6)} = 6.57$
Find $\Pr(95.5 \leq x \leq 96.5)$.
$z_1 = \frac{95.5-72}{6.57} \approx 3.58$
$z_2 = \frac{96.5-72}{6.57} \approx 3.73$
$\Pr(95.5 \leq x \leq 96.5)$
$= \Pr(3.58 \leq z \leq 3.73) = 0.00008$

**9.** $\mu = 150(0.6) = 90$
$\sigma = \sqrt{150(0.6)(0.4)} = 6$
Find $\Pr(74.5 < x < 80.5)$.
$z_1 = \frac{74.5-90}{6} \approx -2.58$
$z_2 = \frac{80.5-90}{6} \approx -1.58$
$\Pr(74.5 \leq x \leq 80.5)$
$= \Pr(-2.58 \leq z \leq -1.58) = 0.0521$

**11.** $\mu = 100(0.7) = 70$

$\sigma = \sqrt{100(0.7)(0.3)} = 4.58$

Find $\Pr(x \geq 80.5)$.

$z = \dfrac{80.5 - 70}{4.58} \approx 2.29$

$\Pr(x \geq 80.5) = \Pr(z \geq 2.29) = 0.0110$

**13.** $\mu = 100(0.7) = 70$

$\sigma = \sqrt{100(0.7)(0.3)} = 4.58$

Find $\Pr(x \leq 80.5)$.

$z = \dfrac{80.5 - 70}{4.58} \approx 2.29$

$\Pr(x \geq 80.5) = \Pr(z \leq 2.29) = 0.9890$

**15.** $\mu = 500(0.2) = 100$

$\sigma = \sqrt{500(0.2)(0.8)} = 8.94$

Find $\Pr(x \geq 94.5)$.

$z = \dfrac{94.5 - 100}{8.94} \approx -0.62$

$\Pr(x \geq 94.5) = \Pr(z \geq -0.62) = 0.7324$

**17.** $\mu = 180(0.4) = 72$

$\sigma = \sqrt{180(0.4)(0.6)} = 6.57$

Find $\Pr(x \leq 69.5)$.

$z = \dfrac{69.5 - 72}{6.57} \approx -0.38$

$\Pr(x \leq 69.5) = \Pr(z \leq -0.38) = 0.3520$

**19.** $\mu = 300\left(\dfrac{1}{6}\right) = 50$

$\sigma = \sqrt{300\left(\dfrac{1}{6}\right)\left(\dfrac{5}{6}\right)} = 6.46$

Find $\Pr(54.5 \leq x \leq 55.5)$.

$z_1 = \dfrac{54.5 - 50}{6.46} \approx 0.70$

$z_2 = \dfrac{55.5 - 50}{6.46} \approx 0.85$

$\Pr(54.5 \leq x \leq 55.5)$

$= \Pr(0.70 \leq z \leq 0.85) = 0.0443$

**21.** $\mu = 100(0.5) = 50$

$\sigma = \sqrt{100(0.5)(0.5)} = 5$

Find $\Pr(x \leq 50.5)$.

$z = \dfrac{50.5 - 50}{5} = 0.10$

$\Pr(x \leq 50.5) = \Pr(z \leq 0.10) = 0.5398$

**23.** $\mu = 1000(0.01) = 10$

$\sigma = \sqrt{1000(0.01)(0.99)} = 3.15$

Find $\Pr(x \geq 7.5)$.

$z = \dfrac{7.5 - 10}{3.15} \approx -0.79$

$\Pr(x \geq 7.5) = \Pr(z \geq -0.79) = 0.7852$

**25.** $\mu = 10800(0.25) = 2700$

$\sigma = \sqrt{10800(0.25)(0.75)} = 45$

Find $\Pr(x \geq 2800.5)$.

$z = \dfrac{2800.5 - 2700}{45} \approx 2.23$

$\Pr(x \geq 2800.5) = \Pr(z \geq 2.23) = 0.0129$

**27.** $\mu = 200(.35) = 70$

$\sigma = \sqrt{200(.35)(.65)} = 6.75$

Find $\Pr(x \geq 74.5)$.

$z = \dfrac{74.5 - 70}{6.75} \approx 0.67$

$\Pr(x \geq 74.5) = \Pr(z \geq 0.67) = 0.2514$

**29.** $\mu = 2000(0.488) = 976$

$\sigma = \sqrt{2000(0.488)(0.512)} = 22.35$

Find $\Pr(1029.5 \leq x \leq 1030.5)$.

$z_1 = \dfrac{1029.5 - 976}{22.35} \approx 2.39$

$z_2 = \dfrac{1030.5 - 976}{22.35} \approx 2.44$

$\Pr(1029.5 \leq x \leq 1030.5)$

$= \Pr(2.39 \leq z \leq 2.44) = 0.0011$

**31.** $\mu = 12000(0.06) = 720$

$\sigma = \sqrt{12000(0.06)(0.94)} = 26.02$

Find $\Pr(x \le 800.5)$.

$z = \dfrac{800.5 - 720}{26.02} \approx 3.09$

$\Pr(x \le 800.5) = \Pr(z \le 3.09) = 0.9990$

**33. a.** $\mu = 60\left(\dfrac{1}{5}\right) = 12$

$\sigma = \sqrt{60\left(\dfrac{1}{5}\right)\left(\dfrac{4}{5}\right)} = 3.10$

Find $\Pr(11.5 \le x \le 12.5)$.

$z_1 = \dfrac{11.5 - 12}{3.10} \approx -0.16$

$z_2 = \dfrac{12.5 - 12}{3.10} \approx 0.16$

$\Pr(11.5 \le x \le 12.5)$

$= \Pr(-0.16 \le z \le 0.16) = 0.1271$

**b.** Find $\Pr(x \ge 12.5)$.

$z = \dfrac{12.5 - 12}{3.10} \approx 0.16$

Find $\Pr(x \ge 12.5) = \Pr(z \ge 0.16) = 0.4364$

**35. a.** $\mu = 2000(0.53) = 1060$

$\sigma = \sqrt{2000(0.53)(0.47)} = 22.32$

Find $\Pr(x \ge 1119.5)$.

$z = \dfrac{1119.5 - 1060}{22.32} \approx 2.67$

$\Pr(x \ge 1119.5) = \Pr(z \ge 2.67) = 0.0038$

**c.** Yes, it would be unusual for 1120 out of 2000 students to pass; maybe they participated in a test prep seminar.

**37.** $\mu = 1300(0.025) = 32.5$

$\sigma = \sqrt{1300(0.025)(0.975)} = 5.63$

Find $\Pr(x \ge 44.5)$.

$z = \dfrac{44.5 - 32.5}{5.63} \approx 2.13$

$\Pr(x \ge 44.5) = \Pr(z \ge 2.13) = 0.0166$

# Chapter 8: Further Topics in Probability; Data Description

## *Chapter 8 Review Exercises*

**1.** $\Pr(5 \text{ successes}) = \binom{7}{5}(0.4)^5(0.6)^2 = 21(0.01024)(0.36) = 0.07741$

**2.** **a.** $\Pr(2 \text{ black out of } 4) = \binom{4}{2}\left(\frac{5}{12}\right)^2\left(\frac{7}{12}\right)^2 \approx 0.354$

**b.** $\Pr(\text{at least } 2) = \Pr(2)+\Pr(3)+\Pr(4) = \binom{4}{2}\left(\frac{5}{12}\right)^2\left(\frac{7}{12}\right)^2 + \binom{4}{3}\left(\frac{5}{12}\right)^3\left(\frac{7}{12}\right)^1 + \binom{4}{4}\left(\frac{5}{12}\right)^4\left(\frac{7}{12}\right)^0$

$= 0.3545 + 0.1688 + 0.03014 = 0.5534$

**3.** Pr(At least 2 rolls are greater than 4) = Pr(2 rolls) + Pr(3 rolls) + Pr(4 rolls)

$= \binom{4}{2}\left(\frac{1}{3}\right)^2\left(\frac{2}{3}\right)^2 + \binom{4}{3}\left(\frac{1}{3}\right)^3\left(\frac{2}{3}\right) + \binom{4}{4}\left(\frac{1}{3}\right)^4\left(\frac{2}{3}\right)^0 = 6\cdot\frac{1}{9}\cdot\frac{4}{9} + 4\cdot\frac{1}{27}\cdot\frac{2}{3} + 1\cdot\frac{1}{81}\cdot 1 = \frac{11}{27} \approx 0.407$

**4.**

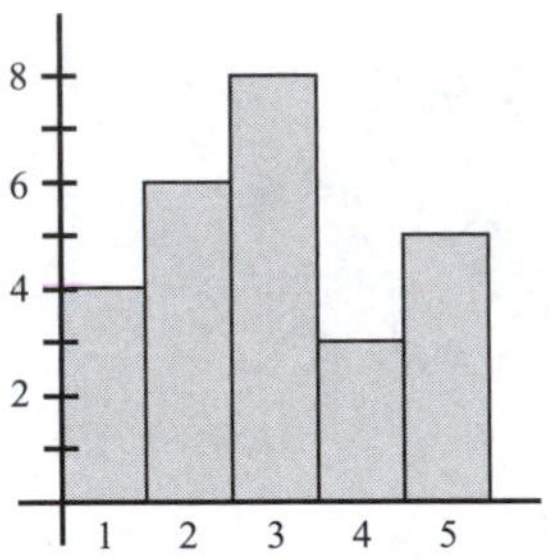


**5.** mode = 3

**6.** $\bar{x} = \dfrac{4\cdot 1 + 6\cdot 2 + 8\cdot 3 + 3\cdot 4 + 5\cdot 5}{4+6+8+3+5} \approx 2.96$

**7.** Median is between the 13th and 14th score.
It is $\dfrac{3+3}{2} = 3.$

**8.**

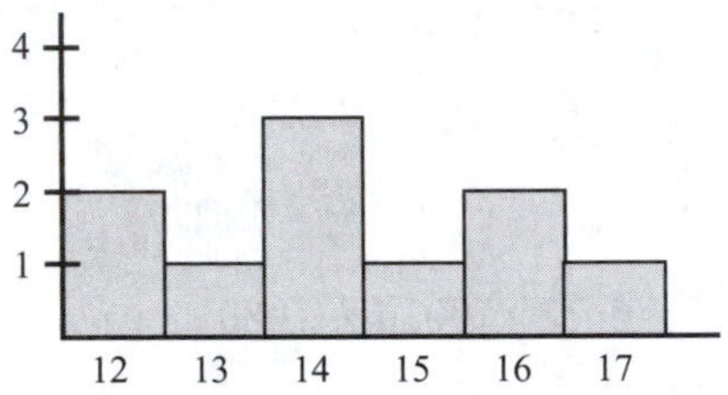


**9.** Median is between the 5th and 6th score. It is 14.

**10.** Mode is 14 (3 times).

**11.** $\text{mean} = \dfrac{\text{sum of scores}}{10} = \dfrac{143}{10} = 14.3$

**12.** Mean: $\bar{x} = \dfrac{4+3+4+6+8+0+2}{7} = \dfrac{27}{7} \approx 3.86$

Variance: $s^2 = \dfrac{2(0.14)^2 + (-0.86)^2 + (2.14)^2 + (4.14)^2 + (-3.86)^2 + (-1.86)^2}{6}$

$\approx 6.81$

Standard deviation: $s = \sqrt{6.81} \approx 2.61$

**13.** $\bar{x} = \dfrac{\Sigma x}{n} = \dfrac{20}{10} = 2$

$s^2 = \dfrac{\Sigma(x-\bar{x})^2}{n-1} = \dfrac{1+0+1+9+1+4+4+0+1+1}{9} = \dfrac{22}{9} = 2.4\overline{4}$

$s \approx 1.56$

# Chapter 8: Further Topics in Probability; Data Description

**14.** $E(x) = 1(0.2) + 2(0.3) + 3(0.4) + 4(0.1) = 2.4$

**15.** $\sum \Pr(x) = \frac{1}{15} + \frac{2}{15} + \frac{3}{15} + \frac{4}{15} + \frac{5}{15} = 1$
Yes.

**16.** No, because $\sum \Pr(x) \neq 1$.

**17.** $\sum \Pr(x) = \frac{1}{6} + \frac{1}{3} + \frac{1}{3} + \frac{1}{6} = 1$ Yes.

**18.** No, because $\Pr(x)$ cannot be negative.

**19.** $E(x) = 1(0.4) + 2(0.3) + 3(0.2) + 4(0.1) = 2.0$

**20. a.** $\mu = \sum x \Pr(x) = 1\left(\frac{1}{16}\right) + 2\left(\frac{2}{16}\right) + 3\left(\frac{3}{16}\right) + 4\left(\frac{4}{16}\right) + 6\left(\frac{6}{16}\right) = \frac{66}{16} = 4.125$

**b.** $\sigma^2 = \sum (x-\mu)^2 \Pr(x)$

$$= \left(1 - \frac{33}{8}\right)^2 \cdot \frac{1}{16} + \left(2 - \frac{33}{8}\right)^2 \cdot \frac{2}{16} + \left(3 - \frac{33}{8}\right)^2 \cdot \frac{3}{16} + \left(4 - \frac{33}{8}\right)^2 \cdot \frac{4}{16} + \left(6 - \frac{33}{8}\right)^2 \cdot \frac{6}{16} = \frac{1}{16} \cdot \frac{175}{4} = 2.734375$$

**c.** $\sigma = \sqrt{2.734375} \approx 1.6536$

**21. a.** $\mu = \sum x \Pr(x) = 1 \cdot \frac{1}{12} + 2 \cdot \frac{1}{6} + 3 \cdot \frac{1}{3} + 4 \cdot \frac{5}{12} = \frac{37}{12}$

**b.** $\sigma^2 = \sum (x-\mu)^2 \cdot \Pr(x) = \left(-\frac{25}{12}\right)^2 \cdot \frac{1}{12} + \left(-\frac{13}{12}\right)^2 \cdot \frac{1}{6} + \left(-\frac{1}{12}\right)^2 \cdot \frac{1}{3} + \left(\frac{11}{12}\right)^2 \cdot \frac{5}{12} = \frac{131}{144} \approx 0.9097$

**c.** $\sigma = \sqrt{0.9097} \approx 0.9538$

**22.** $\mu = np = 6\left(\frac{2}{3}\right) = 4$

$\sigma = \sqrt{npq} = \sqrt{6\left(\frac{2}{3}\right)\left(\frac{1}{3}\right)} = \frac{\sqrt{12}}{3} \approx 1.15$

**23.** $\mu = np = 18\left(\frac{1}{6}\right) = 3$

**24.** $(x+y)^5 = x^5 + 5x^4y + 10x^3y^2 + 10x^2y^3 + 5xy^4 + y^5$

**25.** $\Pr(-1.6 \le z \le 1.9) = 0.4452 + 0.4713 = 0.9165$

**26.** $\Pr(-1 \le z \le -0.5) = 0.3413 - 0.1915 = 0.1498$

**27.** $\Pr(1.23 \le z \le 2.55) = 0.4946 - 0.3907 = 0.1039$

**28.** $\Pr(25 \le x \le 30) = \Pr\left(\frac{25-25}{5} \le \frac{x-\mu}{\sigma} \le \frac{30-25}{5}\right) = \Pr(0 \le z \le 1) = 0.3413$

**29.** $z$ scores are $-1$ and 1 respectively.
$\Pr(-1 \le z \le 1) = 0.3413 + 0.3413 = 0.6826$

**30.** $\Pr(30 \le x \le 35) = \Pr\left(\frac{30-25}{5} \le \frac{x-\mu}{\sigma} \le \frac{35-25}{5}\right) = \Pr(1 \le z \le 2) = 0.4773 - 0.3413 = 0.1360$

**31.** $np = 25(0.15) = 3.75$
$nq = 25(0.85) = 21.25$
No, the normal approximation cannot be used.

**32.** $np = 15(0.4) = 6$
$nq = 15(0.6) = 9$
Yes, the normal approximation can be used.

**33.** $\mu = 200(0.2) = 40$
$\sigma = \sqrt{200(0.2)(0.8)} = 5.657$
Find $\Pr(49.5 \le x \le 50.5)$.
$z_1 = \frac{49.5 - 40}{5.657} \approx 1.68$ $z_1 = \frac{50.5 - 40}{5.657} \approx 1.86$
$\Pr(49.5 \le x \le 50.5) = \Pr(1.68 \le z \le 1.86) = 0.0151$

**34.** $\mu = 60(0.8) = 48$

$\sigma = \sqrt{60(0.8)(0.2)} = 3.098$

Find $\Pr(x \geq 42.5)$.

$z = \dfrac{42.5 - 48}{3.098} \approx -1.78$

$\Pr(x \geq 42.5) = \Pr(z \geq -1.78) = 0.9625$

**35.** $\mu = 80(0.4) = 32$

$\sigma = \sqrt{80(0.4)(0.6)} = 4.382$

Find $\Pr(18.5 \leq x \leq 36.5)$.

$z_1 = \dfrac{18.5 - 32}{4.382} \approx -3.08 \qquad z_1 = \dfrac{36.5 - 32}{4.382} \approx 1.03$

$\Pr(18.5 \leq x \leq 36.5) = \Pr(-3.08 \leq z \leq 1.03) = 0.8475$

**36.** $\mu = 250(0.6) = 150$

$\sigma = \sqrt{250(0.6)(0.4)} = 7.746$

Find $\Pr(x \leq 132.5)$.

$z = \dfrac{132.5 - 150}{7.746} \approx -2.26$

$\Pr(x \geq 132.5) = \Pr(z \leq -2.26) = 0.0119$

**37.** $\Pr(2 \text{ blond}) = \binom{6}{2}\left(\frac{1}{4}\right)^2\left(\frac{3}{4}\right)^4 = 15 \cdot \frac{1}{16} \cdot \frac{81}{256} \approx 0.297$

**38.** $\Pr(x \geq 3) = \binom{5}{3}(0.3)^3(0.7)^2 + \binom{5}{4}(0.3)^4(0.7)^1 + \binom{5}{5}(0.3)^5(0.7)^0 = 0.1323 + 0.02835 + 0.00243 = 0.16308$

**39.** $\Pr(\text{victim}) = \dfrac{1}{5} \qquad \Pr(\text{not a victim}) = \dfrac{4}{5}$

$\Pr(2 \text{ victims}) = \binom{5}{2}\left(\frac{1}{5}\right)^2\left(\frac{4}{5}\right)^3 = 10\left(\frac{1}{25}\right)\left(\frac{64}{125}\right) = \frac{640}{3125} = 0.2048$

**40.** **a.** Pr(exactly 1) $= \binom{100,000}{1}\left(\dfrac{1}{100,000}\right)^1\left(\dfrac{99,999}{100,000}\right)^{99,999} \approx 100,000\left(\dfrac{1}{100,000}\right)(0.36788) \approx 0.37$

**b.** Pr(at least 1) $= 1 - \Pr(0 \text{ get disease})$

$= 1 - \binom{100,000}{0}\left(\dfrac{1}{100,000}\right)^0\left(\dfrac{99,999}{100,000}\right)^{100,000} \approx 1 - 0.3679 \approx 0.63$

**41.**

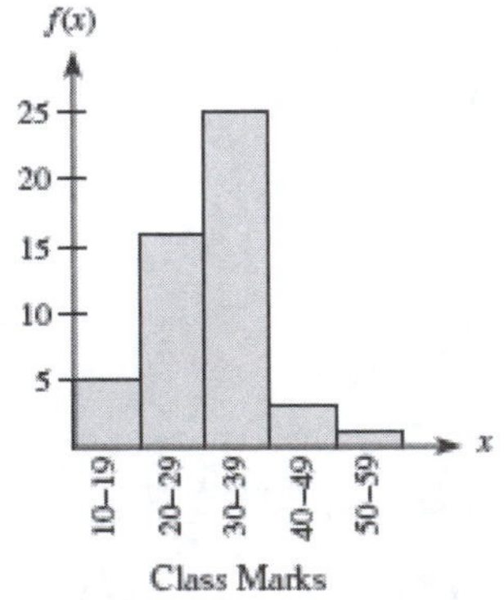

**42.**

| Percentages | Class Marks | Frequencies |
|---|---|---|
| 10–19 | 14.5 | 5 |
| 20–29 | 24.5 | 16 |
| 30–39 | 34.5 | 25 |
| 40–49 | 44.5 | 3 |
| 50–59 | 54.5 | 1 |

$$\bar{x} = \frac{5(14.5)+16(24.5)+25(34.5)+3(44.5)+1(54.5)}{50} = \frac{1515}{50} = 30.3\%$$

**43.** $$s^2 = \frac{5(14.5-30.3)^2+16(24.5-30.3)^2+25(34.5-30.3)^2+3(44.5-30.3)^2+1(54.5-30.3)^2}{49} \approx 69.76$$

$s = \sqrt{69.76} \approx 8.35\%$

**44.** $\Pr(\text{testing positive}) = 0.91$

Thus, $500(0.91) = 455$ of the 500 are expected to test positively.

**45.** Each capsule is an independent event.
Expected number of empty capsules

$$= np = 100\left(\frac{60,000}{2,000,000}\right) = 3.$$

**46.** $\Pr(\text{Win}) = \frac{3}{3+12} = \frac{1}{5}$

Expected value $= 98\left(\frac{1}{5}\right) + \frac{4}{5}(-2) = \$18.00$

**47.** $E(x) = (-1)(0.999) + 499(0.001) = -\$0.50$

**48.**

| $x$ | 0 | 1 | 2 | 3 | 4 | 5 |
|---|---|---|---|---|---|---|
| $\Pr(x)$ | $\frac{1024}{5^5}$ | $\frac{1280}{5^5}$ | $\frac{640}{5^5}$ | $\frac{160}{5^5}$ | $\frac{20}{5^5}$ | $\frac{1}{5^5}$ |

**a.** $$E(x) = 0\cdot\frac{1024}{5^5} + 1\cdot\frac{1280}{5^5} + 2\cdot\frac{640}{5^5} + 3\cdot\frac{160}{5^5} + 4\cdot\frac{20}{5^5} + 5\cdot\frac{1}{5^5} = \frac{3125}{5^5} = 1$$

**b.** $$\Pr(1 \text{ victim}) = \binom{5}{1}\left(\frac{1}{5}\right)\left(\frac{4}{5}\right)^4 = \left(\frac{4}{5}\right)^4 = 0.4096$$

**49. a.** $$\Pr(1500 \le x \le 2100) = \Pr\left(\frac{1500-1500}{300} \le \frac{x-\mu}{\sigma} \le \frac{2100-1500}{300}\right) = \Pr(0 \le z \le 2) = 0.4773$$

**b.** $\Pr(1800 \le x \le 2100) = \Pr(1 \le z \le 2) = 0.4773 - 0.3413 = 0.1360$

**c.** $\Pr(x > 2100) = \Pr(z > 2) = 0.5000 - 0.4773 = 0.0227$

**50.** $\sigma = 96,000,\ \mu = 611,000$

$$z_1 = \frac{700,000-611,000}{96,000} \approx 0.93 \qquad z_2 = \frac{800,000-611,000}{96,000} \approx 1.97$$

$\Pr(0.93 \le z \le 1.97) = 0.4756 - 0.3238 = 0.1518 \approx 15\%$

**51.** $p=0.51$

$np=100(0.51)=51;\ nq=100(0.49)=49$

Yes, the normal approximation can be used.

$\mu=100(0.51)=51$

$\sigma=\sqrt{100(0.51)(0.49)}=4.999$

Find $\Pr(x\geq 52.5)$.

$z=\dfrac{52.5-51}{4.999}\approx 0.30$

$\Pr(x\geq 52.5)=\Pr(z\geq 0.30)=0.3821$

**52.** $p=0.05$

$np=100(0.05)=5;\ nq=100(0.98)=98$

Yes, the normal approximation can be used.

$\mu=100(0.05)=5$

$\sigma=\sqrt{100(0.05)(0.95)}=2.179$

Find $\Pr(x\geq 5.5)$.

$z=\dfrac{5.5-5}{2.179}\approx 0.23$

$\Pr(x\geq 5.5)=\Pr(z\geq 0.23)=0.4090$

**53.** $p=0.10$

$np=7500(0.10)=750;\ nq=7500(0.90)=6750$

Yes, the normal approximation can be used.

$\mu=7500(0.10)=750$

$\sigma=\sqrt{7500(0.10)(0.90)}=25.981$

Find $\Pr(x\geq 800.5)$.

$z=\dfrac{800.5-750}{25.981}\approx 1.94$

$\Pr(x\geq 800.5)=\Pr(z\geq 1.94)=0.0262$

**54.** $p=0.05$

$np=500(0.05)=25;\ nq=500(0.95)=475$

Yes, the normal approximation can be used.

$\mu=500(0.05)=25$

$\sigma=\sqrt{500(0.05)(0.95)}=4.873$

Find $\Pr(x\leq 20.5)$.

$z=\dfrac{20.5-25}{4.873}\approx -0.92$

$\Pr(x\leq 20.5)=\Pr(z\leq -0.92)=0.1788$

### *Chapter 8 Test*

**1. a.** $\Pr(3\text{H})=\binom{5}{3}\left(\frac{1}{3}\right)^3\left(\frac{2}{3}\right)^2=\frac{40}{243}$

**b.** $\Pr(3\text{H or 4H or 5H})=\frac{40}{243}+\binom{5}{4}\left(\frac{1}{3}\right)^4\left(\frac{2}{3}\right)+\binom{5}{5}\left(\frac{1}{3}\right)^5\left(\frac{2}{3}\right)^0=\frac{51}{243}=\frac{17}{81}$

**2. a.** For a binomial distribution, $\text{E}(x)=np,$ so the expected number of heads is $12\left(\frac{1}{3}\right)=4$.

**b.** $\mu=np=12\left(\frac{1}{3}\right)=4$

$$\sigma=\sqrt{npq}=\sqrt{12\left(\frac{1}{3}\right)\left(\frac{2}{3}\right)}=\sqrt{\frac{8}{3}}=\sqrt{\frac{24}{9}}=\frac{2}{3}\sqrt{6}\approx 1.6$$

The variance $\sigma^2=\left(\sqrt{\frac{8}{3}}\right)^2=\frac{8}{3}$.

**3.** For each $x$, $\Pr(x)\geq 0$ and $\sum\Pr(x)=1$.

**4.** $\text{E}(x)=\sum x\Pr(x)$

$=0.6+1.2+0.5+0.6+1.4+0.8$

$=5.1$

**5.** $\mu = E(x) = \sum x \Pr(x) = 1 + 3.6 + 1.5 + 3.6 + 2 + 5 = 16.7$

$\sigma^2 = \sum (x - \mu)^2 \Pr(x) = 26.61$

$\sigma = \sqrt{26.61} \approx 5.16$

**6.** mean $\mu = \dfrac{100 + 147 + 66 + 92 + 48}{5 + 7 + 3 + 4 + 2} \approx 21.57$; median = 21 (11th score), mode = 21

**7.** $z = \dfrac{x - \mu}{\sigma}$

**a.** $z_1 = \dfrac{14 - 16}{6} \approx -.33$, $z_2 = \dfrac{22 - 16}{6} = 1$

$\Pr(14 \le x \le 22) = \Pr(-0.33 \le z \le 1)$

$= 0.1293 + 0.3413 = 0.4706$

**b.** $\Pr(x \le 22) = \Pr(z \le 1) = 0.5000 + 0.3413 = 0.8413$

**c.** $\Pr(22 \le x \le 24) = \Pr(1 \le z \le 1.33)$

$= 0.4082 - 0.3413 = 0.0669$

**8.** $z = \dfrac{x - \mu}{\sigma}$

**a.** $\Pr(73 \le x \le 97) = \Pr(0.25 \le z \le 2.25)$

$= 0.4878 - 0.0987 = 0.3891$

**b.** $\Pr(65 \le x \le 84) = \Pr(-0.42 \le z \le 1.17)$

$= 0.1628 + 0.3790 = 0.5418$

**c.** $\Pr(x \ge 84) = \Pr(z \ge 1.17) = 0.5000 - 0.3790 = 0.1210$

**9.** Using Appendix C to find that the $z$-score corresponding to $A = 0.5 - 0.02 = 0.48$ is about 2.055, solve $2.055 = \dfrac{B - 22}{8}$ to get $B \approx 38.4$.

**10.** $\mu = 50(0.3) = 15$

$\sigma = \sqrt{50(0.3)(0.7)} = 3.240$

Find $\Pr(x \ge 9.5)$.

$z = \dfrac{9.5 - 15}{3.240} \approx -1.70$

$\Pr(x \ge 9.5) = \Pr(z \ge -1.70) = 0.9554$

**11.** $\mu = 120(0.20) = 24$

$\sigma = \sqrt{120(0.20)(0.80)} = 4.382$

Find $\Pr(x \le 25.5)$.

$z = \dfrac{25.5 - 24}{4.382} \approx 0.34$

$\Pr(x \le 25.5) = \Pr(z \le 0.34) = 0.6331$

**12.**

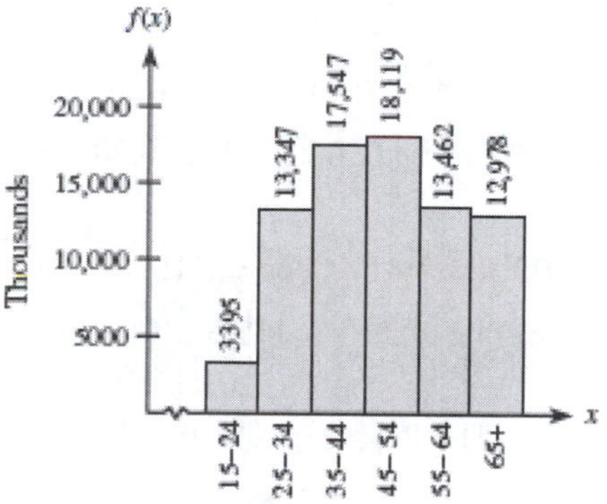


**13.** Class marks: 19.5, 29.5, …, 59.5, 74

$\bar{x} = \dfrac{\Sigma \text{ (class mark)} \cdot \text{number}}{\Sigma \text{ number}}$

$= \dfrac{3811297}{78848} \approx 48.33$ years

$s^2 = \dfrac{\Sigma (x - \bar{x})^2 \cdot \text{number}}{n-1} = \dfrac{191447798.08}{78848}$

$\approx 243.24$

$s = \sqrt{243.24} \approx 15.59$

**14.** **a.** $\mu = \dfrac{\Sigma\text{ (class mark)} \cdot \text{number}}{\Sigma\text{number}}$

$$= \frac{7(68.1)+22(67.3)+37(66.8)+52(63.3)+67(53.3)+82(22.5)}{341.3}$$

$$= \frac{13136.6}{341.3} \approx 38.5$$

**b.** Under 30 would likely be the largest group, which would make the mean age lower.

**15.** **a.** $\mu = \dfrac{2(9.5)+1(29.5)+\ldots+1(209.5)}{64} = 109.5$

**b.** We would expect this mean to increase unless a new technology replaces mobile phones.

**16.** **a.** $\Pr(\text{10 of 100 are defective}) = \dbinom{100}{10}(0.02)^{10}(0.98)^{90} \approx 0.00003$

**b.** If 1500 chips are shipped, the expected number of defective chips is (0.02)(1500) = 30.

**17.** In this group of 30, 0.06(30) = 1.8 or 2 could expect to become pregnant.

**18.** In this group of 30, 0.18(30) = 5.4 or 5 could expect to become pregnant.

**19.** In this group of 30, 0.06(0.03)(30) = 0.054 or 0 could expect to become pregnant.

**20.** Use $z = \dfrac{x-\mu}{\sigma}$, $\mu = 18620$, and $\sigma = 4012$

**a.** $z = \dfrac{10000-18620}{4012} \approx -2.15$

$\Pr(x < 10000) = \Pr(z < -2.15) = 0.5000 - 0.4842 = 0.0158$

**b.** $z = \dfrac{24000-18620}{4012} \approx 1.34$

$\Pr(x > 24000) = \Pr(z > 1.34) = 0.5000 - 0.4099 = 0.0901$

**c.** $\Pr(15000 < x < 21000) = \Pr(-0.90 < z < 0) + \Pr(0 < z < 0.59) = 0.3159 + 0.2224 = 0.5383$

**21.** $\mu = 11000(0.01) = 110;\ \ \sigma = \sqrt{11000(0.01)(0.99)} = 10.436$

Find $\Pr(x \ge 119.5)$: $z = \dfrac{119.5-110}{10.436} \approx 0.91$

$\Pr(x \ge 119.5) = \Pr(z \ge 0.91) = 0.1814$

# Chapter 9: Derivatives

## *Exercises 9.1*

**1.** **a.** $\lim_{x\to c} f(x) = -8$

**b.** $f(c) = -8$

**3.** **a.** $\lim_{x\to c} f(x) = 10$

**b.** $f(x)$ is not defined at $x = c$.

**5.** **a.** $\lim_{x\to c} f(x) = 0$

**b.** $f(c) = -6$

**7.** **a.** $\lim_{x\to c^-} f(x) = +\infty$

**b.** $\lim_{x\to c^+} f(x) = +\infty$

**c.** $\lim_{x\to c} f(x) = +\infty$

**d.** $f(c)$ is not defined.

**9.** **a.** $\lim_{x\to c^-} f(x) = 3$

**b.** $\lim_{x\to c^+} f(x) = -6$

**c.** $\lim_{x\to c} f(x)$ does not exist.

**d.** $f(c) = -6$

**11.**

| $x$ | $f(x)$ |
|---|---|
| 0.9 | −2.9 |
| 0.99 | −2.99 |
| 0.999 | −2.999 |
| 1.001 | −3.001 |
| 1.01 | −3.01 |
| 1.1 | −3.1 |

$\lim_{x\to 1} f(x) = -3$

**13.**

| $x$ | $f(x)$ |
|---|---|
| 0.9 | 3.5 |
| 0.99 | 3.95 |
| 0.999 | 3.995 |
| 1.001 | 4.995999 |
| 1.01 | 4.9599 |
| 1.1 | 4.59 |

$\lim_{x\to 1^-} f(x) = 4$ and $\lim_{x\to 1^+} f(x) = 5$. These limits differ so $\lim_{x\to 1} f(x)$ does not exist.

**15.** $\lim_{x\to -35} (34 + x) = 34 + (-35) = -1$

**17.** $\lim_{x\to -1} (4x^3 - 2x^2 + 2) = 4(-1) - 2(1) + 2 = -4$

**19.** $\lim_{x\to -\frac{1}{2}} \dfrac{4x-2}{4x^2+1} = \dfrac{-2-2}{1+1} = -2$

**21.** $\lim_{x\to 3} \dfrac{x^2-9}{x-3} = \lim_{x\to 3} \dfrac{(x-3)(x+3)}{x-3} = \lim_{x\to 3}(x+3) = 6$

**23.** $\lim_{x\to 7} \dfrac{(x^2-8x+7)}{(x^2-6x-7)} = \lim_{x\to 7} \dfrac{(x-7)(x-1)}{(x-7)(x+1)} = \dfrac{7-1}{7+1} = \dfrac{3}{4}$

**25.** $\lim_{x\to 10} \dfrac{3x^2-30x}{x^2-100} = \lim_{x\to 10} \dfrac{3x(x-10)}{(x-10)(x+10)} = \dfrac{30}{20} = \dfrac{3}{2}$

**27.** $\lim_{x\to -2} \dfrac{x^2+4x+4}{x^2+3x+2} = \lim_{x\to -2} \dfrac{(x+2)(x+2)}{(x+2)(x+1)} = \dfrac{0}{-1} = 0$

**29.** $\lim_{x\to 3^-} f(x) = 4$ $\quad\lim_{x\to 3^+} f(x) = 6$ $\quad\lim_{x\to 3} f(x)$ does not exist.

**31.** $\lim_{x\to(-1)^-} f(x) = -3 \quad \lim_{x\to(-1)^+} f(x) = -3 \quad \lim_{x\to-1} f(x) = -3$

**33.** $\lim_{x\to2} \frac{x^2+6x+9}{x-2} = \frac{\lim_{x\to2}(x^2+6x+9)}{\lim_{x\to2}(x-2)} = \frac{25}{0}$ = does not exist (unbounded near $x = 2$)

**35.** $\lim_{x\to-1} \frac{x^2+5x+6}{x+1} = \frac{\lim_{x\to-1}(x^2+5x+6)}{\lim_{x\to-1}(x+1)} = \frac{2}{0}$ = does not exist (unbounded near $x = -1$)

**37.** $\lim_{h\to0} \frac{(x+h)^3 - x^3}{h} = \lim_{h\to0} \frac{x^3+3x^2h+3xh^2+h^3-x^3}{h} = \lim_{h\to0}(3x^2+3xh+h^2) = 3x^2$

**39.** $\lim_{x\to10} \frac{x^2-19x+90}{3x^2-30x} = \lim_{x\to10} \frac{(x-10)(x-9)}{3x(x-10)} = \lim_{x\to10} \frac{x-9}{3x} = \frac{1}{30}$

**41.** $\lim_{x\to-1} \frac{x^3-x}{x^2+2x+1} = \lim_{x\to-1} \frac{x(x+1)(x-1)}{(x+1)(x+1)} = \lim_{x\to-1} \frac{x(x-1)}{x+1} = \frac{2}{0}$ Limit does not exist.

**43.** $\lim_{x\to-2} \frac{x^4-4x^2}{x^2+8x+12} = \lim_{x\to-2} \frac{x^2(x-2)(x+2)}{(x+6)(x+2)} = \frac{4(-4)}{4} = -4$

**45.** $\lim_{x\to4^-} f(x) = 9 \quad \lim_{x\to4^+} f(x) = 9 \quad \lim_{x\to4} f(x) = 9$

**47.**

| $a$ | 0.1 | 0.01 | 0.001 | 0.0001 | 0.00001 | $\to 0$ |
|---|---|---|---|---|---|---|
| $(1+a)^{1/a}$ | 2.5937 | 2.7048 | 2.7169 | 2.7181 | 2.71827 | $e \approx 2.71828$ |

**49. a.** $\lim_{x\to3}[f(x)+g(x)] = 4+(-2) = 2$

**b.** $\lim_{x\to3}[f(x)-g(x)] = 4-(-2) = 6$

**c.** $\lim_{x\to3}[f(x)\cdot g(x)] = 4(-2) = -8$

**d.** $\lim_{x\to3}\left[\frac{g(x)}{f(x)}\right] = -\frac{2}{4} = -\frac{1}{2}$

**51. a.** $\lim_{x\to2} f(x) = \lim_{x\to2}[f(x)+g(x)] - \lim_{x\to2} g(x) = 5-11 = -6$

**b.** $\lim_{x\to2}\{[f(x)]^2 - [g(x)]^2\} = \left[\lim_{x\to2} f(x)\right]^2 - \left[\lim_{x\to2} g(x)\right]^2 = (-6)^2 - 11^2 = -85$

**c.** $\lim_{x\to2} \frac{3g(x)}{f(x)-g(x)} = \frac{3\cdot\lim_{x\to2} g(x)}{\lim_{x\to2} f(x) - \lim_{x\to2} g(x)} = \frac{3(11)}{(-6)-11} = -\frac{33}{17}$

**53.** $\lim_{x\to100}(1600x - x^2) = 160,000 - 10,000 = \$150,000$

**55. a.** $\lim_{x\to4^+}\left(\frac{4}{x} + 30 + \frac{x}{4}\right) = 1+30+1 = \$32$ (thousands)

**b.** $\lim\limits_{x\to 100^-}\left(\dfrac{4}{x}+30+\dfrac{x}{4}\right)=0.04+30+25=\$55.04$ (thousands)

**57.** **a.** $S(0)=400+\dfrac{2400}{1}=\$2800$

**b.** $\lim\limits_{t\to 7}\left(400+\dfrac{2400}{t+1}\right)=400+\dfrac{2400}{8}=\$700$

**c.** $\lim\limits_{t\to 14}\left(400+\dfrac{2400}{t+1}\right)=400+\dfrac{2400}{15}=\$560$

**59.** **a.** $\lim\limits_{t\to 4}\dfrac{128t(t+6)}{(t^2+6t+18)^2}=\dfrac{128\cdot 4\cdot 10}{(16+24+18)^2}\approx 1.52$ units/hr

**b.** $\lim\limits_{t\to 8^-}\dfrac{128t(t+6)}{(t^2+6t+18)^2}=\dfrac{128\cdot 8\cdot 14}{(64+48+18)^2}\approx 0.85$ units/hr

**c.** lunch time

**61.** **a.** $\lim\limits_{p\to 100^-}C(p)=0$ This is basically untreated water.

**b.** $\lim\limits_{p\to 0^+}C(p)=\infty$

**c.** No. The cost would be extremely large since $C(0)$ is undefined.

**63.** **a.** $\lim\limits_{x\to 36,900^-}T(x)=\$907.50+0.15(36,900-9075)=\$5081.25$

**b.** $\lim\limits_{x\to 36,900^+}T(x)=\$5081.25+0.25(36,900-36,900)=\$5081.25$

**c.** $\lim\limits_{x\to 36,900}T(x)=\$5081.25=\$5081.25$

**65.** $C(x)=\begin{cases}12.76+15.96x & 0\le x\le 10\\ 172.36+13.56(x-10) & 10<x\le 120\\ 1675 & x>120\end{cases}$

$\lim\limits_{x\to 10}C(x)=172.36$

**67.** $\lim\limits_{t\to 9:30\text{AM}^+}D(t)\approx 11,228.00$

This corresponds to the Dow Jones opening average.

**69.** **a.** $\lim\limits_{x\to 60}S(x)=\dfrac{0.264(60)^2+10.7(60)-66.9}{-0.00850(60)^2+1.25(60)+0.854}\approx 33.7$

**b.** This predicts the percent of obese Americans who are severely obese as the year approaches 2040.

**c.** Yes. The table shows increases of 3% to 6% per decade, and severe obesity is expected to get worse.

## *Exercises 9.2*

**1.** **a.** Continuous

**b.** Discontinuous; $f(1)$ does not exist.

**c.** Discontinuous; $\lim\limits_{x\to 3}f(x)$ does not exist.

**d.** Discontinuous; $\lim\limits_{x\to 0}f(x)$ does not exist. $f(0)$ does not exist.

**3.** $\lim\limits_{x\to -2}f(x)=f(-2)=0$, so $f$ is continuous at $x=-2$.

**5.** Because $y=\dfrac{x^2-x-12}{x^2+3x}$ is not defined at $x=-3$, it is discontinuous at $x=-3$.

# Chapter 9: Derivatives

**7.** Because $\lim_{x\to 2^-} f(x) = -1$ and $\lim_{x\to 2^+} f(x) = 1$, $\lim_{x\to 2} f(x)$ does not exist. The function is not continuous at $x = 2$.

**9.** $f$ is continuous everywhere.

**11.** $g$ is discontinuous at $x = -2$ since $g(-2)$ is not defined and $\lim_{x\to -2} g(x)$ does not exist.

**13.** $f$ is continuous everywhere. Note that $x^2 + 1$ is always positive (never zero).

**15.** $f$ is continuous everywhere because $\lim_{x\to 1^-} f(x) = \lim_{x\to 1^+} f(x) = f(1) = 3$.

**17.** $f(-1)$ is not defined. $f$ is discontinuous at $x = -1$.

**19.** $f$ is discontinuous at $x = 3$ since $\lim_{x\to 3} f(x)$ does not exist.

**21. a.** From the graph, VA: $x = -2$, $\lim_{x\to +\infty} f(x) = 0$, $\lim_{x\to -\infty} f(x) = 0$, and HA: $y = 0$.

**b.** The denominator is 0 and the numerator is not 0 at $x = -2$, so $f$ has a vertical asymptote at $x = -2$.

$$\lim_{x\to +\infty} \frac{8}{x+2} = \lim_{x\to +\infty} \frac{\frac{8}{x}}{1+\frac{2}{x}} = \frac{0}{1} = 0$$

$$\lim_{x\to -\infty} \frac{8}{x+2} = \lim_{x\to -\infty} \frac{\frac{8}{x}}{1+\frac{2}{x}} = \frac{0}{1} = 0$$

Since $\lim_{x\to \pm\infty} f(x) = 0$, there is a horizontal asymptote at $y = 0$.

**23. a.** From the graph, VA: $x = -2$ and $x = 3$, $\lim_{x\to -\infty} f(x) = 2$, $\lim_{x\to +\infty} f(x) = 2$, and HA: $y = 2$.

**b.** The denominator is 0 and the numerator is not 0 at $x = -2$ and $x = 3$, so $f$ has vertical asymptotes at $x = -2$ and $x = 3$.

$$\lim_{x\to +\infty} \frac{2(x+1)^3(x+5)}{(x-3)^2(x+2)^2} = \lim_{x\to +\infty} \frac{2(x+1)^3(x+5)\cdot\frac{1}{x^4}}{(x-3)^2(x+2)^2\cdot\frac{1}{x^4}}$$

$$= \lim_{x\to +\infty} \frac{2\left(1+\frac{1}{x}\right)^3\left(1+\frac{5}{x}\right)}{\left(1-\frac{3}{x}\right)^2\left(1+\frac{2}{x}\right)^2} = \frac{2(1)(1)}{(1)(1)} = 2$$

Similarly, $\lim_{x\to -\infty} f(x) = 2$.

Since $\lim_{x\to \pm\infty} f(x) = 2$, there is a horizontal asymptote at $y = 2$.

**25. a.** $\lim_{x\to +\infty} \frac{3}{x+1} = \lim_{x\to +\infty} \frac{\frac{3}{x}}{1+\frac{1}{x}} = \frac{0}{1+0} = 0$

**b.** $y = 0$ is a horizontal asymptote.

**27. a.** $\lim_{x\to +\infty} \frac{x^3-1}{x^3+4} = \lim_{x\to +\infty} \frac{1-\frac{1}{x^3}}{1+\frac{4}{x^3}} = \frac{1-0}{1+0} = 1$

**b.** $y = 1$ is a horizontal asymptote.

**29. a.** $\lim_{x\to -\infty} \frac{5x^3-4x}{3x^3-2} = \lim_{x\to -\infty} = \frac{5-\frac{4}{x^2}}{3-\frac{2}{x^3}} = \frac{5-0}{3-0} = \frac{5}{3}$

**b.** $y = \frac{5}{3}$ is a horizontal asymptote.

**31. a.** $\lim_{x\to +\infty} \frac{3x^2+5x}{6x^2+1} = \lim_{x\to +\infty} \frac{3+\frac{5}{x}}{6+\frac{1}{x^2}} = \frac{3+0}{6+0} = \frac{1}{2}$

**b.** $y = \frac{1}{2}$ is a horizontal asymptote.

**33.**

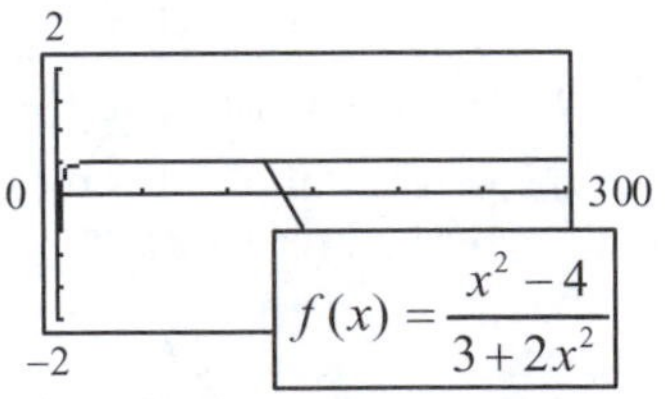


**a.** $\lim_{x\to +\infty} f(x) = 0.5$

**b.** $f(100{,}000) = 0.49999$

$f(1{,}000{,}000) = 0.5$

The table supports the conclusion.

**35.** $f(x)=\frac{1000(x-1)}{x+1000}$

**a.** Discontinuous at $x=-1000$

**b.** Since $f(x)=\frac{1000\left(1-\frac{1}{x}\right)}{1+\frac{1000}{x}}$, $y=1000$ is a horizontal asymptote.

**c.** For large values, an attempt to find the appropriate window is difficult. The asymptote may never be located.

**37.** $\lim_{x\to\infty}\frac{a_nx^n+a_{n-1}x^{n-1}+\ldots+a_1x+a_0}{b_mx^m+b_{m-1}x^{m-1}+\ldots+b_1x+b_0}\cdot\frac{\left(\frac{1}{x^m}\right)}{\left(\frac{1}{x^m}\right)}$

$=\lim_{x\to\infty}\frac{a_n+a_{n-1}\left(\frac{1}{x}\right)+\ldots+a_1\left(\frac{1}{x^{m-1}}\right)+a_0\left(\frac{1}{x^m}\right)}{b_m+b_{m-1}\left(\frac{1}{x}\right)+\ldots+b_1\left(\frac{1}{x^{m-1}}\right)+b_0\left(\frac{1}{x^m}\right)}=\frac{a_n}{b_m}$

Similarly, $\lim_{x\to-\infty}f(x)=\frac{a_n}{b_n}$. Thus, there is a horizontal asymptote at $y=\frac{a_n}{b_n}$.

**39.** $y=\frac{32}{(p+8)^{2/5}}$

**a.** If all values are allowed, then $y$ is not continuous at $p=-8$.

**b.** Yes

**c.** Yes

**d.** $p>0$

**41. a.** Discontinuous at $q=-1$.

**b.** Yes

**43. a.** $\lim_{n\to\infty}A_n=\lim_{n\to\infty}R\left(\frac{1-\frac{1}{(1+i)^n}}{i}\right)=R\left[\frac{1-0}{i}\right]=\frac{R}{i}$

**b.** $i=0.12$, $R=100$; $A=\frac{100}{0.12}\approx\$833.33$

**45.** $p\le 100$ for problem to have meaning. Thus, the function is continuous for the domain of the function.

**47.** $p=\frac{100C}{7300+C}=\frac{100}{\frac{7300}{C}+1}$ $\lim_{C\to\infty}p=\frac{100}{0+1}=100\%$

100% of pollution cannot be removed. Cost would be impossible to afford.

**49.** $R$ is discontinuous at each value where the rate changes: $x=18{,}150$; $x=73{,}800$; $x=148{,}850$; $x=226{,}850$; $x=405{,}100$; and $x=457{,}600$.

**51. a.** $C(1100)=98.80+0.10(1100-500)$
$=\$158.80$

**b.** $\lim_{x\to100^-}C(x)=20+0.188(100)=38.80$

$\lim_{x\to100^+}C(x)=38.80+0.15(100-100)$
$=38.80$

Thus, $\lim_{x\to100}C(x)=38.80$.

Similarly, $\lim_{x\to500}C(x)=98.80$

**c.** Yes, since in each case $\lim_{x\to a}C(x)=C(a)$.

**53. a.** $m(x)=0.591x+43.2; w(x)=0.787x+20.9$

**b.** $r(x)=\frac{0.591x+43.2}{0.787x+20.9}$

**c.** $\lim_{x\to0}r(x)\approx2.07$ means that for years approaching 1950 there were about 2.07 men per woman in the U.S. workforce.
$\lim_{x\to100}r(x)\approx1.03$ means that for years approaching 2050 it is projected that there will be 1.03 men per woman in the U.S. workforce.

**d.** $\lim_{x\to\infty}r(x)\approx0.751\approx\frac{3}{4}$ means that the long-term projection is for 3 men per 4 women in the U.S. workforce.

## *Exercises 9.3*

1. **a.** average rate of change $=\frac{f(5)-f(0)}{5-0}=\frac{18-(-12)}{5}=\frac{30}{5}=6$

**b.** average rate of change $=\frac{f(10)-f(-3)}{10-(-3)}=\frac{98-(-6)}{13}=\frac{104}{13}=8$

# Chapter 9: Derivatives

**3.** **a.** average rate of change $= \dfrac{f(5)-f(2)}{5-2} = \dfrac{30-20}{3} = \dfrac{10}{3} = 3.\overline{3}$

**b.** average rate of change $= \dfrac{f(4)-f(3.8)}{4-3.8} = \dfrac{16-17}{0.2} = \dfrac{-1}{0.2} = -5$

**5.** **a.** average rate of change over $[2.9,3] = \dfrac{f(3)-f(2.9)}{3-2.9} = \dfrac{-3-(-2.61)}{3-2.9} = \dfrac{-0.39}{0.1} = -3.9$

average rate of change over $[2.99,3] = \dfrac{f(3)-f(2.99)}{3-2.99} = \dfrac{-3-(-2.96)}{3-2.99} = \dfrac{-0.04}{0.01} = -4$

**b.** average rate of change over $[3,3.1] = \dfrac{f(3.1)-f(3)}{3.1-3} = \dfrac{-3.41-(-3)}{3.1-3} = \dfrac{-0.41}{0.1} = -4.1$

average rate of change over $[3,3.01] = \dfrac{f(3.01)-f(3)}{3.01-3} = \dfrac{-3.04-(-3)}{3.01-3} = \dfrac{-0.04}{0.01} = 4$

**c.** The calculations suggest that the instantaneous rate of change of $f(x)$ at $x=3$ might be $-4$.

**7.** **a.** Instantaneous rate of change means $f'(x)$. $f'(4) = 8\cdot 4 = 32.$

**b.** Slope of tangent to the graph means $f'(x)$. $f'(4) = 8\cdot 4 = 32.$

**c.** $f(4) = 4(4)^2 = 64$ Point on graph: (4, 64)

**9.** $f(x) = 3x^2 - 2x$

**a.**
$$\begin{aligned} f(x+h)-f(x) &= \left[3(x+h)^2 - 2(x+h)\right] - [3x^2 - 2x] \\ &= 3\left(x^2+2xh+h^2\right) - 2x - 2h - 3x^2 + 2x \\ &= 3x^2 + 6xh + 3h^2 - 2x - 2h - 3x^2 + 2x \\ &= 6xh + 3h^2 - 2h \end{aligned}$$

$$\frac{f(x+h)-f(x)}{h} = 6x + 3h - 2$$

$$f'(x) = \lim_{h\to 0}\frac{f(x+h)-f(x)}{h} = \lim_{h\to 0}(6x+3h-2) = 6x-2$$

**b.** The instantaneous rate of change is given by $f'(-1) = 6(-1) - 2 = -8.$

**c.** The slope of the tangent to the graph at $x=-1$ is given by $f'(-1) = 6(-1)-2 = -8.$

**d.** $f(-1) = 3(-1)^2 - 2(-1) = 5$ Point on graph: $(-1,5)$.

**11.** **a.** $P(x,y) = P(1,1)$ $A(x,y) = A(3,0)$

**b.** $m_{\tan} = \dfrac{y_2-y_1}{x_2-x_1} = \dfrac{0-1}{3-1} = -\dfrac{1}{2}$

**c.** $f'(1) = -\dfrac{1}{2}$

**d.** $f'(1) = -\dfrac{1}{2}$

**13.** **a.** $P(x, y) = P(1, 3)$ $A(x, y) = A(0, 3)$

**b.** $m_{\tan} = \dfrac{y_2-y_1}{x_2-x_1} = \dfrac{3-3}{0-1} = 0$

**c.** $f'(1) = 0$

**d.** $f'(1) = 0$

**15.** $f(x)=5x^2+6x-11$

**a.** $$f(x+h)-f(x)=\left[5(x+h)^2+6(x+h)-11\right]-\left(5x^2+6x-11\right)$$
$$=\left[5x^2+10xh+5h^2+6x+6h-11\right]-5x^2-6x+11$$
$$=10xh+5h^2+6h$$
$$\frac{f(x+h)-f(x)}{h}=\frac{10xh+5h^2+6h}{h}=10x+5h+6$$
$$f'(x)=\lim_{h\to 0}\frac{f(x+h)-f(x)}{h}=\lim_{h\to 0}(10x+5h+6)=10x+6$$

**b.** $f'(x)=10x+6$; $f'(-2)=10(-2)+6=-14$

**c.** $f'(-2)=-14$

**17. a.** $$p(q+h)-p(q)=\left[2(q+h)^2+(q+h)+5\right]-\left(2q^2+q+5\right)$$
$$=\left[2q^2+4qh+2h^2+q+h+5\right]-2q^2-q-5$$
$$=4qh+2h^2+h$$
$$\frac{p(q+h)-p(q)}{h}=\frac{4qh+2h^2+h}{h}=4q+2h+1$$
$$p'(q)=\lim_{h\to 0}\frac{p(q+h)-p(q)}{h}=\lim_{h\to 0}(4q+2h+1)=4q+1$$

**b.** $p'(q)=4q+1$; $p'(10)=4(10)+1=41$

**c.** $p'(10)=41$

**19. a.** $nDeriv(3x^4-7x-5,\ x,\ 2)=89.000024$

**b.** $\frac{f(2.0001)-f(2)}{0.0001}\approx\frac{29.0089-29}{0.0001}\approx 89.0072$

**c.** $f'(2)\approx 89$

**21. a.** $nDeriv((2x-1)^3,\ x,\ 4)=294.000008$

**b.** $\frac{f(4.0001)-f(4)}{0.0001}\approx\frac{343.0294-343}{0.0001}\approx 294.0084$

**c.** $f'(4)\approx 294$

**23.** $f'(13)=\frac{f(12.99)-f(13)}{-0.01}=\frac{17.42-17.11}{-0.01}=-31$

**25.**

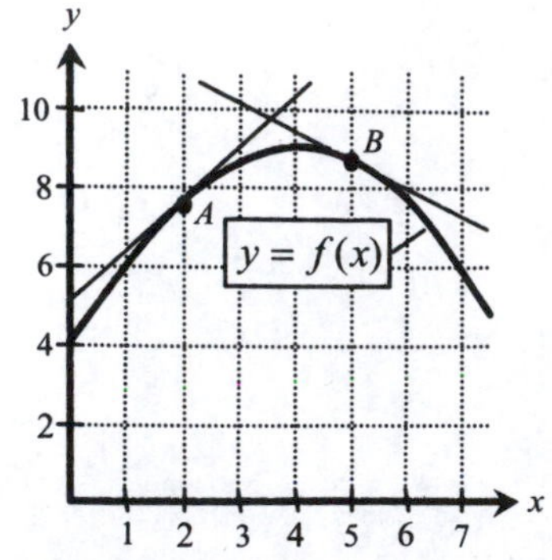


**a.** $f'(x)$ is greater at point *A*. The slope of the tangent line is positive and rate of change is greater.

**b.** Answers may vary. $f'(x)$ at $B \approx -\frac{1}{3}$.

**27.** $(4,-11) = (a, f(a))$, thus $f(4) = -11$.

Solve $7x - 3y = 61$ for *y* to find the slope:

$$7x - 3y = 61$$
$$-3y = -7x + 61$$
$$y = \frac{7}{3}x - \frac{61}{3}$$

The slope of the tangent line, $f'(4)$ is $\frac{7}{3}$.

**29.** $f'(2) = 5$

$$y - y_1 = m(x - x_1)$$
$$y - (-4) = 5(x - 2)$$
$$y + 4 = 5x - 10$$
$$y = 5x - 14$$

**31. a.** $f'(x) > 0$ means $f(x)$ is increasing.
Answer: *a, b, d*

**b.** $f'(x) < 0$ means $f(x)$ is decreasing.
Answer: *c*

**c.** $f'(x) = 0$ at *A, C,* and *E.*

**33. a.** Function is continuous at *A, B, C,* and *D.*

**b.** Function is differentiable at *A* and *D.*
($f'(B)$ and $f'(C)$ are undefined.)

**35. a.**

$$f(x+h) - f(x)$$
$$= [(x+h)^2 + (x+h)] - (x^2 + x)$$
$$= 2xh + h^2 + h$$
$$\frac{f(x+h) - f(x)}{h} = \frac{2xh + h^2 + h}{h} = 2x + h + 1$$
$$f'(x) = \lim_{h \to 0}(2x + h + 1) = 2x + 1$$

**b.** $f'(2) = 2(2) + 1 = 5$

**c.** $y - y_1 = m(x - x_1)$
$y - 6 = 5(x - 2)$ yields $y = 5x - 4$

**d.**

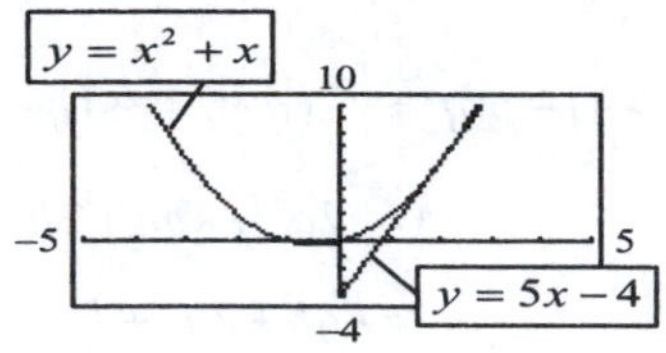


**37. a.** $f(x+h) - f(x)$

$$= [(x+h)^3 + 3] - (x^3 + 3)$$
$$= 3x^2h + 3xh^2 + h^3$$
$$\frac{f(x+h) - f(x)}{h} = \frac{3x^2h + 3xh^2 + h^3}{h}$$
$$= 3x^2 + 3xh + h^2$$
$$f'(x) = \lim_{h \to 0}(3x^2 + 3xh + h^2) = 3x^2$$

**b.** $f'(1) = 3(1)^2 = 3$

**c.** $y - y_1 = m(x - x_1)$
$y - 4 = 3(x - 1)$ yields $y = 3x + 1$

**d.**

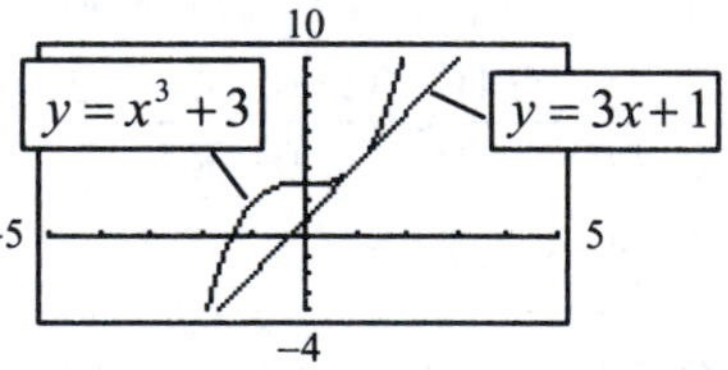


**39. a.**

$$\frac{f(300) - f(100)}{300 - 100} = \frac{14550 - 5950}{200}$$
$$= \frac{8600}{200}$$
$$= 43$$

**b.**

$$\frac{f(600) - f(300)}{600 - 300} = \frac{43200 - 14550}{300}$$
$$= \frac{28650}{300}$$
$$= 95.50$$

**c.** The average rate of change of total cost of production is greater when 300 to 600 printers are produced.

**b.** $\frac{D(100)-D(25)}{100-25} = \frac{(100-1)-(199)}{75} = -\frac{4}{3}$

**41. a.** $\frac{D(25)-D(1)}{25-1} = \frac{\left(\frac{1000}{\sqrt{25}}-1\right)-\left(\frac{1000}{\sqrt{1}}-1\right)}{24}$

$= \frac{199-999}{24} = -\frac{100}{3}$

**43.** Smallest to largest: $A$ to $B$, $A$ to $C$, $B$ to $C$. The slope of the line from $A$ to $B$ was less than the slope from $A$ to $C$ and the slope of the line from $A$ to $C$ was less than the slope of the line from $B$ to $C$.

**45.** $R(x) = 300x - x^2$

**a.** $R'(x) = \lim_{h\to 0}\frac{[300(x+h)-(x+h)^2]-(300x-x^2)}{h} = 300-2x$

**b.** $R'(50) = 300-2(50) = 200$ A positive value for $R'(x)$ means that $R(x)$ is increasing.

**c.** $R'(200) = 300-2(200) = -100$ A negative value for $R'(x)$ means that $R(x)$ is decreasing.

**d.** $R'(150) = 300-2(150) = 0$ A zero value for $R'(x)$ means that $R(x)$ is stationary.

**e.** The revenue changes from increasing to decreasing at $x = 150$.

**47.** $Q(x) = 15{,}000 + 2x^2$

$Q'(x) = \lim_{h\to 0}\frac{[15{,}000+2(x+h)^2]-(15{,}000+2x^2)}{h} = \lim_{h\to 0}(4x+2h) = 4x$

$Q'(50) = 4(50) = 200$

**49.** $P(x) = 500x - x^2 - 100$

**a.** $nDeriv(500x - x^2 - 100,\ x,\ 200) = 100$ Profit is increasing.

**b.** $nDeriv(500x - x^2 - 100,\ x,\ 300) = -100$ Profit is decreasing.

**51.** $f(x) = 0.009x^2 + 0.139x + 91.875$

**a.** $nDeriv(0.009x^2 + 0.139x + 91.875,\ x,\ 50) = 1.039$

**b.** If humidity changes 1%, the heat index will change by about 1.039°F.

**53. a.** The $\overline{MR}$ is greater at 300. The slope of the tangent line to the curve is greater at 300 than it is at 700.

**b.** The sale of the 301st cell phone brings in more revenue since the rate of change in revenue is greater at 300 and it indicates the predicted increase in total revenue when one more unit is sold.

## *Exercises 9.4*

**1.** $y = 4$

$y'(x) = 0$

Derivative of a constant is zero.

**3.** $f(t) = t$

$f'(t) = 1\cdot t^{1-1} = 1$

**5.** $y = 6 - 8x + 2x^2$

$y' = -8 + 2\cdot 2x^{2-1} = 4x - 8$

**7.** $f(x) = 3x^4 - x^6$

$f'(x) = 4\cdot 3x^3 - 6\cdot x^5$

$= 12x^3 - 6x^5$

**9.** $y = 10x^5 - 3x^3 + 5x - 11$

$y' = 10 \cdot 5x^4 - 3 \cdot 3x^2 + 5 \cdot 1x^0 - 0$

$= 50x^4 - 9x^2 + 5$

**11.** $w = z^7 - 3z^6 + 13$

$w' = 7z^{7-1} - 6 \cdot 3z^{6-1}$

$= 7z^6 - 18z^5$

**13.** $g(x) = 2x^{12} - 5x^6 + 9x^4 + x - 5$

$g'(x) = 2 \cdot 12x^{11} - 5 \cdot 6x^5 + 9 \cdot 4x^3 + 1x^0 - 0$

$= 24x^{11} - 30x^5 + 36x^3 + 1$

**15.** $y = 7x^2 + 2x + 1$

**a.** $y' = 14x + 2$

At $x = 2$ we have $y' = 14(2) + 2 = 30$.

**b.** At $x = 2$ we have $y' = 30$.

**17.** $P(x) = x^3 - 6x$

**a.** $P'(x) = 3x^2 - 6$; $P'(2) = 3(2)^2 - 6 = 6$

**b.** $P'(2) = 6$

**19.** $y = x^{-5} + x^{-8} - 3$

$y' = -5x^{-5-1} + (-8)x^{-8-1} - 0$

$= -5x^{-6} - 8x^{-9} = -\dfrac{5}{x^6} - \dfrac{8}{x^9}$

**21.** $z = 3t^{11/3} - 2t^{7/4} - t^{1/2} + 8$

$z' = 3\left(\dfrac{11}{3}t^{8/3}\right) - 2\left(\dfrac{7}{4}t^{3/4}\right) - \dfrac{1}{2}t^{-1/2} + 0$

$= 11t^{8/3} - \dfrac{7}{2}t^{3/4} - \dfrac{1}{2}t^{-1/2}$

$= 11\sqrt[3]{t^8} - \dfrac{7}{2}\sqrt[4]{t^3} - \dfrac{1}{2\sqrt{t}}$

**23.** $f(x) = 5x^{-4/5} + 2x^{-4/3}$

$f'(x) = 5\left(-\dfrac{4}{5}x^{-9/5}\right) + 2\left(-\dfrac{4}{3}x^{-7/3}\right)$

$= -4x^{-9/5} - \dfrac{8}{3}x^{-7/3} = -\dfrac{4}{\sqrt[5]{x^9}} - \dfrac{8}{3\sqrt[3]{x^7}}$

**25.** $g(x) = 3x^{-5} + 2x^{-4} + 6x^{1/3}$

$g'(x) = 3\left(-5x^{-5-1}\right) + 2\left(-4x^{-4-1}\right) + 6\left(\dfrac{1}{3}x^{1/3-1}\right)$

$= -15x^{-6} - 8x^{-5} + 2x^{-2/3}$

$= -\dfrac{15}{x^6} - \dfrac{8}{x^5} + \dfrac{2}{\sqrt[3]{x^2}}$

**27.** $y = x^3 - 5x^2 + 7$; $y' = 3x^2 - 10x$

At $x = 1$, $y = (1)^3 - 5(1)^2 + 7 = 1 - 5 + 7 = 3$.

At $x = 1$, $y' = 3(1)^2 - 10(1) = -7$.

$y - y_1 = m(x - x_1)$

$y - 3 = -7(x - 1)$

$y - 3 = -7x + 7$

$y = -7x + 10$

**29.** $f(x) = 4x^2 - x^{-1}$

$f'(x) = 8x + x^{-2} = 8x + \dfrac{1}{x^2}$

$f\left(-\dfrac{1}{2}\right) = 4\left(-\dfrac{1}{2}\right)^2 - \dfrac{1}{-\frac{1}{2}} = 1 + 2 = 3$

$f'\left(-\dfrac{1}{2}\right) = 8\left(-\dfrac{1}{2}\right) + \dfrac{1}{(-\frac{1}{2})^2} = -4 + 4 = 0$

$y - y_1 = m(x - x_1)$

$y - 3 = 0\left(x - \dfrac{1}{2}\right)$

$y = 3$

**31–33 A horizontal tangent means $f'(x) = 0$**

**31.** $f(x) = -x^3 + 9x^2 - 15x + 6$

$f'(x) = -3x^2 + 18x - 15$

$= -3(x^2 - 6x + 5)$

$= -3(x - 5)(x - 1)$

$f'(x) = 0$ if $x = 1$ or $5$

Answer: (1, –1), (5, 31)

**33.** $f(x) = x^4 - 4x^3 + 9$

$f'(x) = 4x^3 - 12x^2$

$= 4x^2(x - 3)$

$f'(x) = 0$ if $x = 0$ or $3$

Answer: (0, 9), (3, –18)

**35.** $y = 5 - 2x^{1/2}$

**a.** $y' = -x^{-1/2} = \dfrac{-1}{\sqrt{x}}$

At $x = 4$, $y' = \dfrac{-1}{\sqrt{4}} = -\dfrac{1}{2}$

**b.** $nDeriv\left(5 - 2x^{1/2},\ x,\ 4\right) = -0.500000004$

**37.** $f(x) = 2x^3 + 5x - \pi^4 + 8$

**a.** $f'(x) = 6x^2 + 5$

**b.**

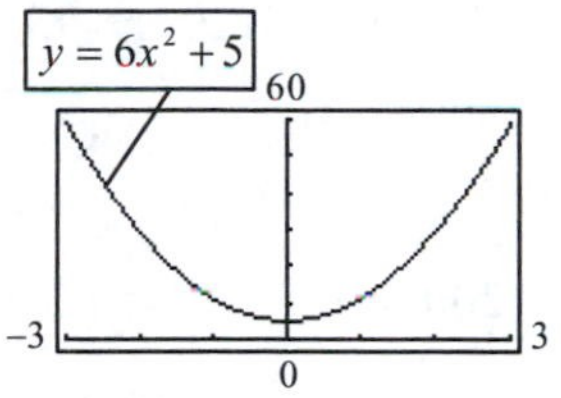


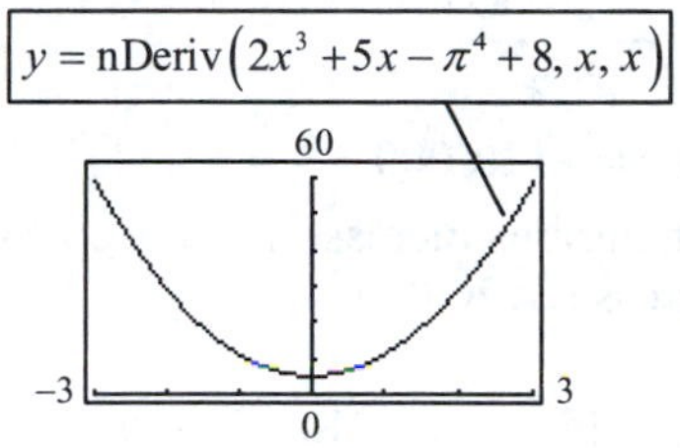


**39.** $h(x) = 10x^{-3} - 10x^{-2/5} + x^2 + 1$

**a.** $h'(x) = -30x^{-4} + 4x^{-7/5} + 2x$

$= \dfrac{-30}{x^4} + \dfrac{4}{\sqrt[5]{x^7}} + 2x$

**b.**

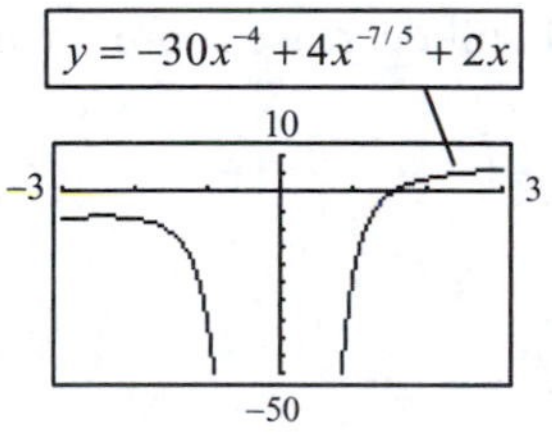


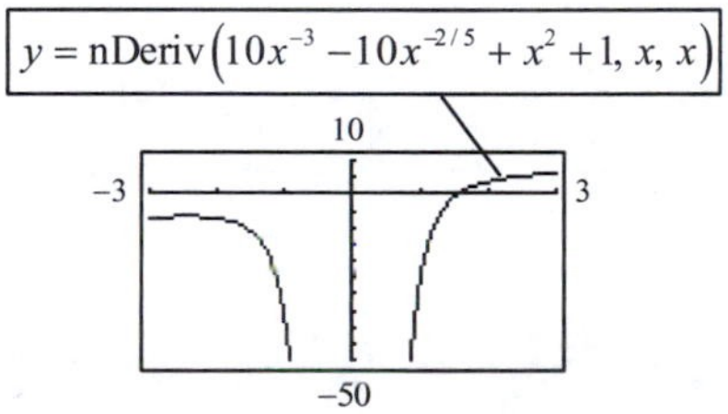


**41. a.** $f(x) = 3x^2 + 2x \qquad y - y_1 = m(x - x_1)$

$f'(x) = 6x + 2 \qquad y - 5 = 8(x - 1)$

$f'(1) = 6 + 2 = 8 \qquad y = 8x - 3$

**b.**

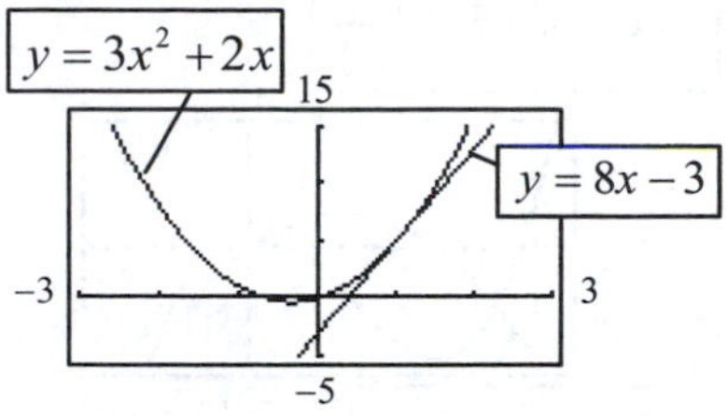


**c.** Answers may vary.
Starting with the window $[0, 2] \times [0, 10]$ and zooming in four times, we find that, in the window
$[0.99609375, 1.00390625] \times$
$[4.98046875, 5.01953125]$,
the function and tangent line cannot be distinguished.

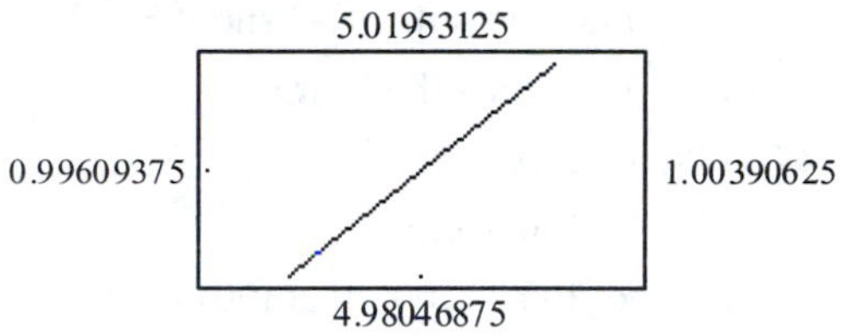


**43.** $f(x) = 8 - 2x - x^2$

**a.** $f'(x) = -2 - 2x$

**b.**

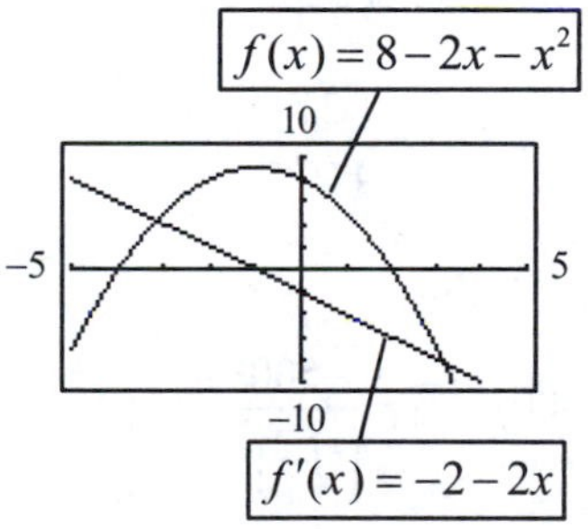


**c.** $f'(x) = 0$ at $x = -1$.
$f'(x) > 0$ if $x < -1$.
$f'(x) < 0$ if $x > -1$.

**d.** $f(x)$ has a max if $x = -1$
$f(x)$ rises if $x < -1$.
$f(x)$ falls if $x > -1$.

**45.** $f(x) = x^3 - 12x - 5$

**a.** $f'(x) = 3x^2 - 12$

**b.**

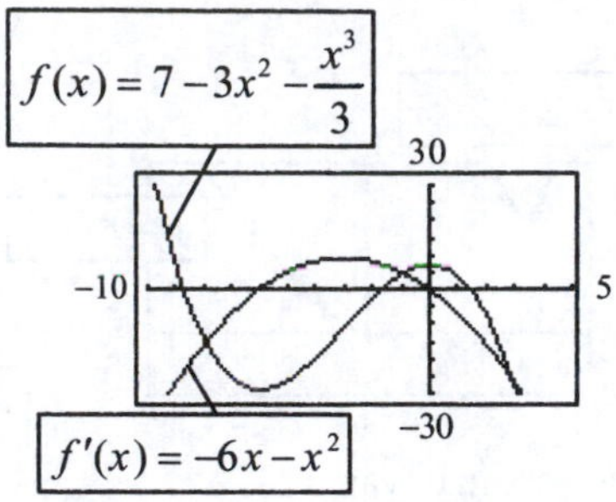


**c.** $f'(x) = 0$ if $x = \pm 2$.

$f'(x) > 0$ if $x < -2$ and if $x > 2$.

$f'(x) < 0$ if $-2 < x < 2$.

**d.** $f(x)$ has a max if $x = -2$.

$f(x)$ has a min if $x = 2$.

$f(x)$ rises if $x < -2$ and if $x > 2$.

$f(x)$ falls if $-2 < x < 2$.

**47.** $R(x) = 100x - 0.1x^2$

$R'(x) = 100 - 0.2x$

**a.** $R'(300) = 100 - 0.2(300) = 40$

The additional revenue from the 301st unit is about \$40.

**b.** $R'(600) = 100 - 0.2(600) = -20$

The sale of the 601st unit results in a loss of revenue of about \$20.

**49.** $\dfrac{dq}{dM} = \dfrac{3}{4}\left(kM^{-1/4}\right) = \dfrac{3k}{4M^{1/4}}$

**51.** $q = 1000p^{-1/2} - 1$

$q'(p) = -500p^{-3/2} = \dfrac{-500}{\left(\sqrt{p}\right)^3}$

**a.** $q'(25) = \dfrac{-500}{\left(\sqrt{25}\right)^3} = \dfrac{-500}{125} = -4$

If price increases by \$1, the demand will drop approximately 4 units.

**b.** $q'(100) = \dfrac{-500}{\left(\sqrt{100}\right)^3} = \dfrac{-500}{1000} = -\dfrac{1}{2}$

If price increases to \$101, the demand will drop approximately $\frac{1}{2}$ unit.

**53.** $\overline{C}(x) = \dfrac{4000}{x} + 55 + 0.1x$

**a.** $\overline{C}'(x) = \dfrac{-4000}{x^2} + 0.1$

**b.** $\dfrac{-4000}{x^2} + 0.1 = 0$

$0.1x^2 = 4000$

$x^2 = 40{,}000$

$x = 200$

**c.** $C'(x) = 55 + 0.2x$. So $C'(200) = 95$.

$\overline{C}(200) = \frac{4000}{200} + 55 + 0.1(200) = 95$.

So $C'(200) = \overline{C}(200)$.

**55.** $C = \dfrac{120{,}000}{p} - 1200$

$C'(p) = -\dfrac{120{,}000}{p^2}$

**a.** $C'(1) = -120{,}000$

**b.** If impurities increase 1%, the cost will decrease \$120,000.

**57. a.** $WC = 35.74 + 0.6215(15) - 35.75s^{0.16} + 0.4275ts^{0.16}$

$= 45.0625 - 29.3375s^{0.16}$

**b.** $WC'(s) = 0 - 4.694s^{-0.84}$

$WC'(25) = -4.694(25)^{-0.84} \approx -0.31$

**c.** If the wind speed increases 1 mph, then the wind-chill will decrease approximately $-31^{\circ}$F .

**NOTE: Answers to 59 and 61 may vary slightly depending on calculator.**

**59. a.** $y = 0.0682x^2 + 1.76x + 96.0$

**b.** $\frac{120.56 - 100}{2020 - 2012} = 2.57$; \$2.57 per year

**c.** $\frac{dy}{dx} = 0.1364x + 1.76$

**d.** $y'(10) = 3.124$; about \$3.12 per year

**e.** $120.56 + 2(3.124) = 126.808$; \$126.81

**61. a.** $P(t) = -0.0000738t^3 + 0.0102t^2 + 2.20t + 163$

**b.** $P'(t) = 3\left(-0.0000738t^{3-1}\right) + 2\left(0.0102t^{2-1}\right) + 2.20$

$= -0.0002214t^2 + 0.0204t + 2.20$

**c.** The year 2000 corresponds to $t = 50$ and the year 2025 corresponds to $t = 75$.

$P'(50) \approx 2.67$ means that for 2001, the U.S. population will rise by about 2.67 million people.

$P'(75) \approx 2.48$ means that for 2026, the U.S. population is expected to rise by about 2.48 million people.

## Exercises 9.5

**1.** $y=(5x+3)(x^2-2x)$

$y'=(5x+3)(2x-2)+(x^2-2x)\cdot 5$

$=10x^2-10x+6x-6+5x^2-10x$

$=15x^2-14x-6$

**3.** $f(x)=(x^{12}+3x^4+4)(2x^3-1)$

$f'(x)=(x^{12}+3x^4+4)(6x^2)+(2x^3-1)(12x^{11}+12x^3)$

$=6x^{14}+18x^6+24x^2+24x^{14}+24x^6-12x^{11}-12x^3$

$=30x^{14}-12x^{11}+42x^6-12x^3+24x^2$

**5.** $y=(7x^6-5x^4+2x^2-1)(4x^9+3x^7-5x^2+3x)$

$$\frac{dy}{dx}=(7x^6-5x^4+2x^2-1)(36x^8+21x^6-10x+3)+(4x^9+3x^7-5x^2+3x)(42x^5-20x^3+4x)$$

**7.** $y=(x^2+x+1)(x^{1/3}-2x^{1/2}+5)$

$$\frac{dy}{dx}=(x^2+x+1)\left(\frac{1}{3}x^{-2/3}-x^{-1/2}\right)+(x^{1/3}-2x^{1/2}+5)(2x+1)$$

**9.** $f(x)=(x^2+1)(x^3-4x)$

$f'(x)=(x^2+1)(3x^2-4)+(x^3-4x)(2x)$

$=5x^4-9x^2-4$

$f'(-2)=5(-2)^4-9(-2)^2-4=40$

**a.** Slope of tangent line at (–2, 0) is $f'(-2)=40$.

**b.** Instantaneous rate of change at (–2, 0) is $f'(-2)=40$.

**11.** $p=\frac{q^2+3}{2q-1}$ $\quad$ $$\frac{dp}{dq}=\frac{(2q-1)(2q)-(q^2+3)(2)}{(2q-1)^2}=\frac{4q^2-2q-2q^2-6}{(2q-1)^2}=\frac{2q^2-2q-6}{(2q-1)^2}$$

**13.** $y=\frac{1-2x^2}{x^4-2x^2+5}$ $\quad$ $$\frac{dy}{dx}=\frac{(x^4-2x^2+5)(-4x)-(1-2x^2)(4x^3-4x)}{(x^4-2x^2+5)^2}=\frac{4x^5-4x^3-16x}{(x^4-2x^2+5)^2}\text{ or }\frac{4x(x^4-x^2-4)}{(x^4-2x^2+5)^2}$$

**15.** $z=x^2+\frac{x^2}{1-x-2x^2}$ $\quad$ $$\frac{dz}{dx}=2x+\frac{(1-x-2x^2)(2x)-(x^2)(-1-4x)}{(1-x-2x^2)^2}=2x+\frac{-x^2+2x}{(1-x-2x^2)^2}$$

**17.** $p=\frac{3\sqrt[3]{q}}{1-q}=\frac{3q^{1/3}}{1-q}$ $\quad$ $$\frac{dp}{dq}=\frac{(1-q)\left(\frac{1}{q^{2/3}}\right)-3q^{1/3}(-1)}{(1-q)^2}\cdot\frac{q^{2/3}}{q^{2/3}}=\frac{(1-q)-3q(-1)}{q^{2/3}(1-q)^2}=\frac{1+2q}{q^{2/3}(1-q)^2}$$

**19.** $y = \frac{x(x^2+4)}{x-2} = \frac{x^3+4x}{x-2}$ $\quad y' = \frac{(x-2)(3x^2+4)-(x^3+4x)(1)}{(x-2)^2} = \frac{2x^3-6x^2-8}{(x-2)^2}$

**21.** $f(x) = \frac{x^2+1}{x+3}$ $\quad f'(x) = \frac{(x+3)(2x)-(x^2+1)(1)}{(x+3)^2} = \frac{x^2+6x-1}{(x+3)^2}$

$f'(2) = \frac{4+12-1}{25} = \frac{15}{25} = \frac{3}{5}$

**a.** The slope of the tangent line at (2, 1) is $\frac{3}{5}$.

**b.** The instantaneous rate of change is also $\frac{3}{5}$ at this point.

**23.** $y = (9x^2-6x+1)(1+2x)$ $\quad \frac{dy}{dx} = (9x^2-6x+1)(2)+(1+2x)(18x-6)$

At $x = 1$, $\frac{dy}{dx} = 4(2)+3(12) = 44$. At $x = 1, y = 4(3) = 12$.

Tangent line: $y - y_1 = m(x - x_1) \rightarrow y - 12 = 44(x-1) \rightarrow y = 44x - 32$

**25.** $f(x) = \frac{3x^4-2x-1}{4-x^2}$ $\quad f'(x) = \frac{(4-x^2)(12x^3-2)-(3x^4-2x-1)(-2x)}{(4-x^2)^2} = \frac{-6x^5+48x^3-2x^2-2x-8}{(4-x^2)^2}$

At $x = 1$ we have $f(x) = \frac{3-2-1}{3} = 0$. At $x = 1$ we have $f'(x) = \frac{-6+48-2-2-8}{9} = \frac{30}{9} = \frac{10}{3}$.

Tangent line: $y - y_1 = m(x - x_1) \rightarrow y - 0 = \frac{10}{3}(x-1) \rightarrow y = \frac{10}{3}x - \frac{10}{3}$

**27.** $nDeriv\left(\left(4\sqrt{x}+3x^{-1}\right)\cdot\left(3x^{1/3}-5x^{-2}-25\right),\ x,\ 1\right) = 104.0002584$

**29.** $nDeriv\left((4x-4)/\left(3x^{2/3}\right),\ x,\ 1\right) = 1.333334074$

**31.** $f(x) = (x^2+4x+4)(x-7)$

**a.** $f'(x) = (x^2+4x+4)(1)+(x-7)(2x+4)$

$= 3x^2 - 6x - 24$

$= 3(x^2 - 2x - 8)$

$= 3(x-4)(x+2)$

**b.**

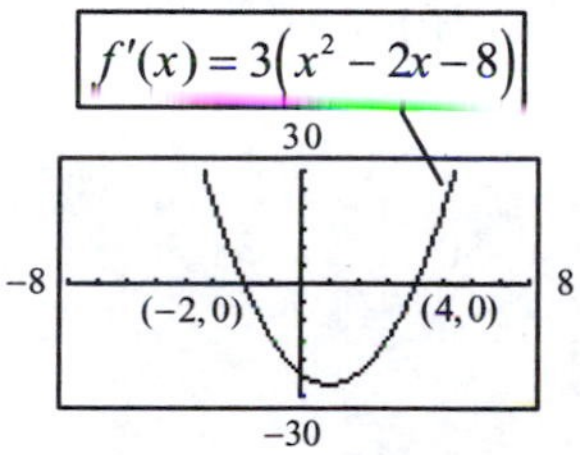


The slope of the tangent is 0 at $x = -2$ and at $x = 4$.

**c.**

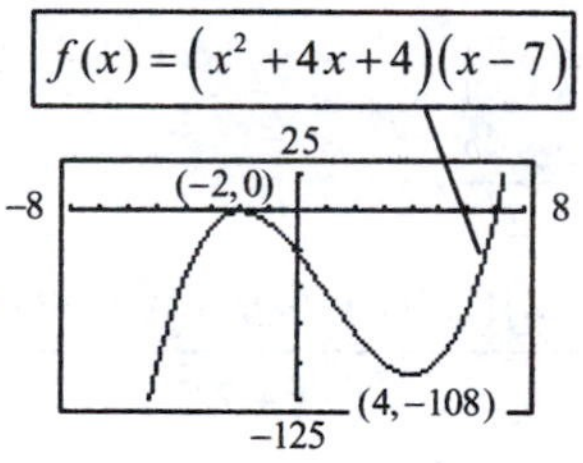

**33.** $y = \dfrac{x^2}{x-2}$

**a.** $\dfrac{dy}{dx} = \dfrac{(x-2)(2x) - x^2(1)}{(x-2)^2}$

$= \dfrac{x^2 - 4x}{(x-2)^2} = \dfrac{x(x-4)}{(x-2)^2}$

**b.**

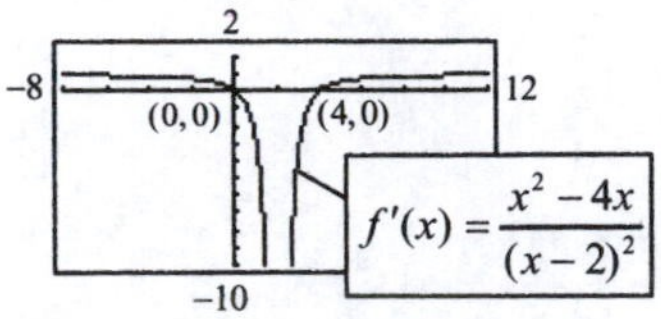


The slope of the tangent is 0 at $x = 0$ and at $x = 4$.

**c.**

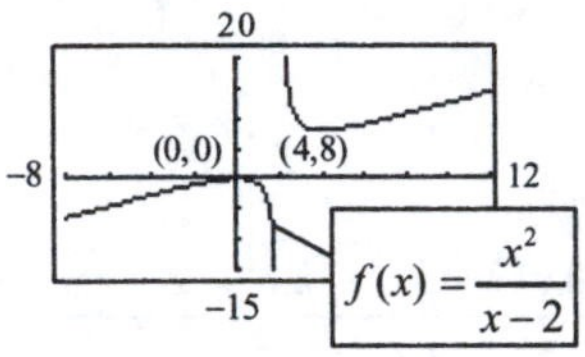


**35.** $f(x) = \dfrac{10x^2}{x^2+1}$

**a.** $f'(x) = \dfrac{(x^2+1)(20x) - 10x^2(2x)}{(x^2+1)^2} = \dfrac{20x}{(x^2+1)^2}$

**b.**

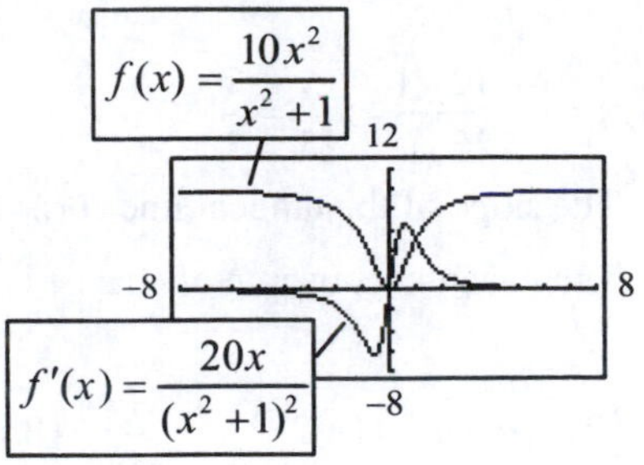


**c.** $f'(x) = 0$ if $x = 0$.
$f'(x) > 0$ if $x > 0$.
$f'(x) < 0$ if $x < 0$.
Read graph of $f'(x)$.

**d.** $f$ has a min at $x = 0$.
$f$ is increasing for $x > 0$.
$f$ is decreasing for $x < 0$.

**37.** $f'(x) = \lim\limits_{h\to 0} \dfrac{\frac{u(x+h)}{v(x+h)} - \frac{u(x)}{v(x)}}{h} = \lim\limits_{h\to 0} \dfrac{u(x+h)v(x) - u(x)v(x+h)}{h \cdot v(x)v(x+h)}$

$= \lim\limits_{h\to 0} \dfrac{u(x+h)v(x) - u(x)v(x) + u(x)v(x) - u(x)v(x+h)}{h \cdot v(x)v(x+h)}$

$= \lim\limits_{h\to 0} \dfrac{v(x)\left[\frac{u(x+h)-u(x)}{h}\right] - u(x)\left[\frac{v(x+h)-v(x)}{h}\right]}{v(x)v(x+h)} = \dfrac{v(x)u'(x) - u(x)v'(x)}{[v(x)]^2}$

**39.** $C(p) = \dfrac{8100p}{100-p}$ $\quad C'(p) = \dfrac{(100-p)(8100) - (8100p)(-1)}{(100-p)^2} = \dfrac{810{,}000}{(100-p)^2}$

**41.** $R(x) = \dfrac{60x^2 + 74x}{2x+2}$

$R'(x) = \dfrac{(2x+2)(120x+74) - (60x^2+74x)(2)}{(2x+2)^2} = \dfrac{120x^2 + 240x + 148}{(2x+2)^2}$ $\quad R'(49) = \dfrac{300{,}028}{(100)^2} \approx \$30$

The revenue from the sale of the next unit is approximately \$30.

**43.** $R(x) = (25+x)(300-10x) = 7500 + 50x - 10x^2$ $\quad R'(x) = 50 - 20x$ $\quad R'(5) = -50$

The revenue will decrease approximately \$50 if the group adds one person.

**45.** $R(x) = 500x^2 - \dfrac{1}{3}x^3$ $\quad R'(x) = 1000x - x^2$

**47.** $R(n) = \dfrac{nr}{1+nr-r}$, $r$ is a constant. $R'(n) = \dfrac{(1+nr-r)r - nr(r)}{(1+nr-r)^2} = \dfrac{r(1-r)}{[1+(n-1)r]^2}$

**49.** $P(t) = \dfrac{13t}{t^2+100} + 0.18$ $P'(t) = \dfrac{(t^2+100)(13)-(13t)(2t)}{(t^2+100)^2} = \dfrac{-13t^2+1300}{(t^2+100)^2}$

**a.** $P'(6) \approx 0.045$ In the next month the recognition will increase about 4.5%.

**b.** $P'(12) \approx -0.010$ In the next month the recognition will decrease about 1%.

**c.** Positive means increasing recognition.

**51.** $f(s) = \dfrac{289.173 - 58.5731s}{s+1}$

$f'(s) = \dfrac{(s+1)(-58.5731)-(289.173-58.5731s)(1)}{(s+1)^2} = \dfrac{-347.7461}{(s+1)^2}$

**a.** $f'(20) \approx -0.79$

**b.** At 0°F, if the wind speed increases 1 mph, the wind-chill will decrease about 0.79°F.

**53. a.** $B'(t) = (0.01t+3)(0.0476t-9.79)+(0.01)(0.0238t^2-9.79t+3100)$

$= 0.000476t^2 + 0.0449t - 29.37 + 0.000238t^2 - 0.0979t + 31$

$= 0.000714t^2 - 0.053t + 1.63$

**b.** Instantaneous rate of change in 2020 $= B'(70) \approx 1.42$

This means that in 2020 the number of beneficiaries will be changing at the rate of about 1.42 million per year.

**c.** From 2010 to 2020 $= \dfrac{68.8-53.3}{10} = 1.55$ From 2020 to 2030 $= \dfrac{82.7-68.8}{10} = 1.39$

From 2010 to 2030 $= \dfrac{82.7-53.3}{20} = 1.47$

The average rate of change over 2020 to 2030 best but is still off by almost 0.03 million per year.

**55. a.** $p(t) = \dfrac{78.6t+2090}{1.38t+64.1}$

$p'(t) = \dfrac{(1.38t+64.1)(78.6)-(78.6t+2090)(1.38)}{(1.38t+64.1)^2} = \dfrac{2154.06}{(1.38t+64.1)^2}$

**b.** The year 2005 corresponds to $t = 55$. $p'(55) \approx 0.1099$.

The year 2020 corresponds to $t = 70$, $p'(70) \approx 0.0834$.

**c.** $p'(55)$ means that in 2005 the percent females in the work force was changing about 0.1099% per year.

$p'(70)$ means that in 2020 the percent females in the work force was changing about 0.0834% per year.

# Chapter 9: Derivatives

## *Exercises 9.6*

**Unless otherwise specified the Power Rule will be used in calculating the derivatives.**

**1.** $y = u^3$ and $u = x^2 + 1$

$\frac{dy}{du} = 3u^2; \ \frac{du}{dx} = 2x;$

$\frac{dy}{dx} = \frac{dy}{du} \cdot \frac{du}{dx} = 3u^2 \cdot 2x = 6x\left(x^2+1\right)^2$

**3.** $y = u^4$ and $u = 4x^2 - x + 8$

$\frac{dy}{du} = 4x^3; \ \frac{du}{dx} = 8x - 1;$

$\frac{dy}{dx} = \frac{dy}{du} \cdot \frac{du}{dx} = 4u^3(8x-1)$

$= 4(8x-1)\left(4x^2 - x + 8\right)^3$

**5.** $f(x) = \left(3x^5 - 2\right)^{20}$

$f'(x) = 20\left(3x^5-2\right)^{19}\left(15x^4\right)$

$= 300x^4\left(3x^5-2\right)^{19}$

**7.** $h(x) = \frac{3}{4}\left(x^5 - 2x^3 + 5\right)^8$

$h'(x) = \frac{3}{4} \cdot 8\left(x^5 - 2x^3 + 5\right)^7\left(5x^4 - 6x^2\right)$

$= 6\left(5x^4 - 6x^2\right)\left(x^5 - 2x^3 + 5\right)^7$

$= 6x^2\left(5x^2 - 6\right)\left(x^5 - 2x^3 + 5\right)^7$

**9.** $s(t) = 5t - 3\left(2t^4 + 7\right)^3$

$s'(t) = 5 - 9\left(2t^4+7\right)^2\left(8t^3\right)$

$= 5 - 72t^3\left(2t^4+7\right)^2$

**11.** $g(x) = (x^4 - 5x)^{-2}$

$g'(x) = -2(x^4 - 5x)^{-3}(4x^3 - 5) = \frac{-2\left(4x^3-5\right)}{\left(x^4-5x\right)^3}$

**13.** $f(s) = \frac{3}{\left(2s^5+1\right)^4} = 3\left(2s^5+1\right)^{-4}$

$f'(s) = 3(-4)\left(2s^5+1\right)^{-5}\left(10s^4\right) = \frac{-120s^4}{\left(2s^5+1\right)^5}$

**15.** $g(x) = \left(2x^3 + 3x + 5\right)^{-3/4}$

$g'(x) = -\frac{3}{4}\left(2x^3+3x+5\right)^{-7/4}\left(6x^2+3\right)$

$= \frac{-18x^2 - 9}{4\left(2x^3+3x+5\right)^{7/4}}$

**17.** $y = \sqrt{3x^2+4x+9} = \left(3x^2+4x+9\right)^{1/2}$

$y' = \frac{1}{2}\left(3x^2+4x+9\right)^{-1/2}(6x+4)$

$= \frac{3x+2}{\sqrt{3x^2+4x+9}}$

**19.** $y = \frac{11\left(x^3-7\right)^6}{9} = \frac{11}{9}\left(x^3-7\right)^6$

$y' = \frac{11}{9} \cdot 6\left(x^3-7\right)^5\left(3x^2\right) = 22x^2\left(x^3-7\right)^5$

**21.** $y = \frac{(3w+1)^5 - 3w}{7} = \frac{1}{7}\left[(3w+1)^5 - 3w\right]$

$y' = \frac{1}{7}\left[5(3w+1)^4(3) - 3\right] = \frac{3}{7}\left[5(3w+1)^4 - 1\right]$

**23.** $y = f(x) = \left(x^3 + 2x\right)^4$

$f'(x) = 4\left(x^3+2x\right)^3\left(3x^2+2\right)$

$f'(2) = 4(8+4)^3(12+2) = 96,768$

**a.** The slope of the tangent line at $x = 2$ is 96,768.

**b.** The instantaneous rate of change of the function at $x = 2$ is also 96,768.

$nDeriv\left(\left(x^3+2x\right)^4, x, 2\right) = 96,768.28378$

**25.** $f(x) = \left(x^3+1\right)^{1/2}$

$f'(x) = \frac{1}{2}\left(x^3+1\right)^{-1/2}\left(3x^2\right) = \frac{3x^2}{2\sqrt{x^3+1}}$

$f'(2) = \frac{12}{2\sqrt{8+1}} = \frac{6}{\sqrt{9}} = 2$

**a.** The slope of the tangent line at (2, 3) is 2.

**b.** The instantaneous rate of change of the function at (2, 3) is also 2.

$nDeriv\left(\left(x^3+1\right)^{1/2}, x, 2\right) = 2$

**27.** $y = \left(x^2-3x+3\right)^3$

$y' = 3\left(x^2-3x+3\right)^2\left(2x-3\right)$

At $x = 2,\ y' = 3\left(2^2-3\cdot 2+3\right)^2\left(2\cdot 2-3\right) = 3$

and $y = \left(2^2-3\cdot 2+3\right)^3 = 1.$

$y - y_1 = m(x - x_1)$

$y - 1 = 3(x-2)$

$y = 3x - 5$

**29.** $y = \left(3x^2-2\right)^{1/2}$

$y' = \frac{1}{2}\left(3x^2-2\right)^{-1/2}\left(6x\right) = \frac{3x}{\sqrt{3x^2-2}}$

At $x = 3,\ y = \sqrt{27-2} = 5$ and $y' = \frac{9}{\sqrt{27-2}} = \frac{9}{5}.$

$y - y_1 = m(x - x_1)$

$y - 5 = \frac{9}{5}(x-3)$

$9x - 5y = 2$

**31.** $f(x) = \left(x^2-4\right)^3 + 12$

**a.** $f'(x) = 3\left(x^2-4\right)^2\left(2x\right) = 6x\left(x-2\right)^2\left(x+2\right)^2$

**b.**

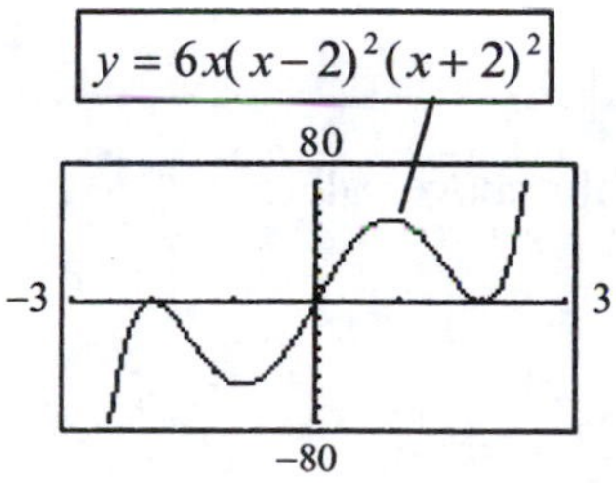


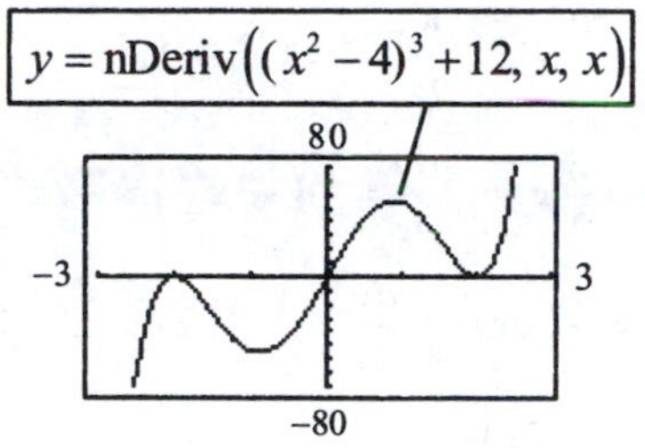


**c.** $f'(x) = 0$ at $x = 0, 2, -2$

**d.** At

| | | | |
|---|---|---|---|
| $x =$ | 0 | 2 | −2 |
| $y =$ | −52 | 12 | 12 |

Points: (0, –52), (2, 12), (–2, 12)

**e.**

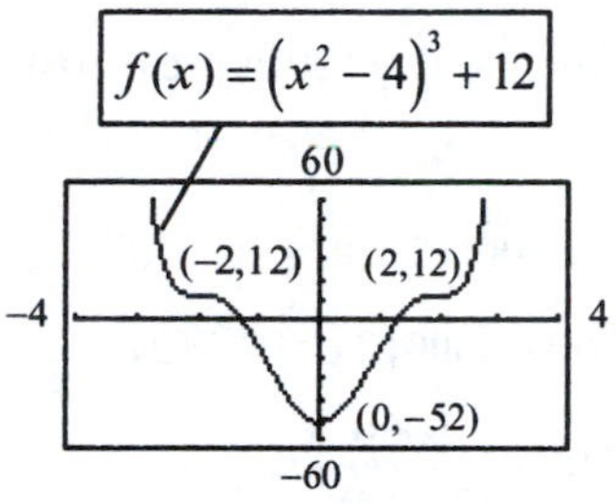


**33.** $f(x) = 12 - 3\left(1-x^2\right)^{4/3}$

**a.** $f'(x) = -4\left(1-x^2\right)^{1/3}\left(-2x\right) = 8x\left(1-x^2\right)^{1/3}$

**b.**

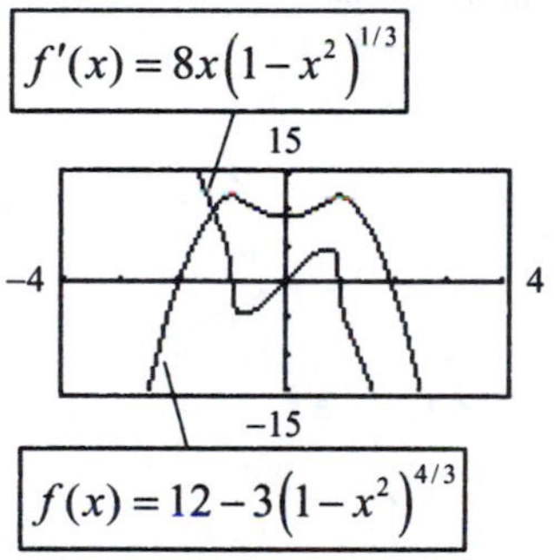


**c.**

| $x$ | | −1 | | 0 | | 1 | |
|---|---|---|---|---|---|---|---|
| $f'(x)$ | + | 0 | − | 0 | + | 0 | − |

**d.** Max at $x = -1$ and $x = 1$. Min at $x = 0$.
$f(x)$ increasing for $x < -1$ and $0 < x < 1$.
$f(x)$ decreasing for $x > 1$ and $-1 < x < 0$.

**35.** **a.** $y=\frac{2}{3}x^3 \quad \frac{dy}{dx}=\frac{2}{3}\cdot 3x^2=2x^2$

**b.** $y=\frac{2}{3}x^{-3} \quad \frac{dy}{dx}=\frac{2}{3}\left(-3x^{-4}\right)=-2x^{-4}=-\frac{2}{x^4}$

**c.** $y=\frac{1}{3}(2x)^3 \quad \frac{dy}{dx}=\frac{1}{3}\cdot 3(2x)^2\cdot 2=2(2x)^2$

**d.** $y=2(3x)^{-3}$

$\frac{dy}{dx}=2(-3)(3x)^{-4}(3)=-\frac{18}{(3x)^4}$

**37.** $s=27-(3-10t)^3$

$s'(t)=0-3(3-10t)^2(0-10)=30(3-10t)^2$

$s'\left(\frac{1}{10}\right)=30(3-1)^2=120\text{ in/sec}=10\text{ ft/sec}$

**39.** $R=1500x+3000(2x+3)^{-1}-1000$

$R'(x)=1500-3000(2x+3)^{-2}(2)$

$=1500-\frac{6000}{(2x+3)^2}$

$R'(100)=1500-\frac{6000}{203^2}=\$1499.85$

So if the sales go from 100 units sold to 101 units sold, the revenue will increase by about \$1499.85.

**41.** $y=32(3p+1)^{-2/5}$

$y'(p)=32\left(-\frac{2}{5}\right)(3p+1)^{-7/5}(3)$

$=\frac{-192}{5(3p+1)^{7/5}}$

**a.** $y'(21)=\frac{-192}{5(64)^{7/5}}\approx\frac{-192}{5(337.79)}\approx -0.114$

**b.** If the price increases \$1, the sales volume will decrease by 114 units.

**43.** $p=200{,}000(q+1)^{-2}$

$p'(q)=-400{,}000(q+1)^{-3}(1)$

**a.** $p'(49)=\frac{-400{,}000}{50^3}=-\$3.20$

**b.** If the quantity demanded increases one unit, the price will decrease about \$3.20.

**45.** $y=k(x-x_0)^{8/5}$ $k$ and $x_0$ are constants.

$\frac{dy}{dx}=k\cdot\frac{8}{5}(x-x_0)^{3/5}=\frac{8k}{5}(x-x_0)^{3/5}$

**47.** $p=100(2q+1)^{-1/2}$

$p'(q)=-50(2q+1)^{-3/2}(2)=\frac{-100}{(2q+1)^{3/2}}$

**49.** $K_c=4\sqrt{4v+1}$

$K_c'=4\cdot\frac{1}{2}(4v+1)^{-1/2}\cdot 4=\frac{8}{\sqrt{4v+1}}$

**51.** $S=1000\left[1+\frac{0.01r}{12}\right]^{240}$

$S'(r)=240{,}000\left[1+\frac{0.01r}{12}\right]^{239}\left(\frac{0.01}{12}\right)=200\left[1+\frac{0.01r}{12}\right]^{239}$

**a.** $S'(6)=200[1+0.005]^{239}=200(3.2937)=\$658.75$

**b.** $S'(12)=200[1+0.01]^{239}=200(10.7847)=\$2156.94$

In each case the future value of the investment would have an increase of the above values.

**53. a.** $A(t) = 445(0.1t+1)^3 - 2120(0.1t+1)^2 + 4570(0.1t+1) - 1600$

$A'(t) = 1335(0.1t+1)^2(0.1) - 4240(0.1t+1)(0.1) + 4570(0.1) + 0$

$= 133.5(0.1t+1)^2 - 424(0.1t+1) + 457$

The year 2008 corresponds to $t = 8$, so $A'(8) \approx 126.3$.

The year 2015 corresponds to $t = 15$, so $A'(15) \approx 231.375$

These mean that the total national expenditures for health were predicted to change by about \$126.3 billion from 2000 to 2008 and about \$231.4 billion from 20015 to 2016.

**b.** The average rate of change for 2014 to 2015 is $\frac{3541-3313}{2015-2014} = \$228$ billion per year, which is close.

**55. a.** $G(t) = 212.9(0.2t+5)^3 - 5016(0.2t+5)^2 + 8810.4t + 104{,}072$

$G'(t) = 212.9 \cdot 3(0.2t+5)^2(0.2) - 5016 \cdot 2(0.2t+5)(0.2) + 8810.4$

$= 127.74(0.2t+5)^2 - 2006.4(0.2t+5) + 8810.4$

The year 2005 corresponds to $t = 5$, so $G'(5) \approx 1370.64$.

The year 2015 corresponds to $t = 15$, so $G'(15) \approx 934.56$.

These mean that the GDP was changing at the rate of \$1370.64 billion per year in 2005 and \$934.5 billion per year in 2015.

**b.** The average rate of change in the GDP between '05 and '15 is $\frac{21{,}270-12{,}145}{2015-2005} = \$912.5$ billion per year.

**c.** In 2010, $G'(10) \approx 1024.86$. The answer from part (b) is not a good approximation to $G'(10)$.

## *Exercises 9.7*

**1.** $f(x) = \pi^4 \quad f'(x) = 0$

Derivative of a constant.

**3.** $g(x) = 4x^{-4} \quad g'(x) = 4(-4)x^{-5} = \frac{-16}{x^5}$

**5.** $g(x) = 5x^3 + 4x^{-1}$

$g'(x) = 15x^2 - 4x^{-2} = 15x^2 - \frac{4}{x^2}$

**7.** $y = (x^2-2)(x+4)$

$y' = (x^2-2)(1) + (x+4)(2x) = 3x^2 + 8x - 2$

**9.** $f(x) = \frac{x^3+1}{x^2} = \frac{x^3}{x^2} + \frac{1}{x^2} = x + x^{-2}$

Preferred: $f'(x) = 1 - 2x^{-3} = 1 - \frac{2}{x^3} = \frac{x^3-2}{x^3}$

Acceptable: $f'(x) = \frac{x^2(3x^2) - (x^3+1)(2x)}{x^4}$

$= \frac{x^4-2x}{x^4} = \frac{x^3-2}{x^3}$

**11.** $y = \frac{1}{10}(x^3-4x)^{10}$

$y' = \frac{1}{10} \cdot 10(x^3-4x)^9(3x^2-4)$

$= (3x^2-4)(x^3-4x)^9$

**13.** $y = \frac{5}{3}x^3\left(4x^5-5\right)^3$

$$y' = \frac{5}{3}x^3 \cdot 3\left(4x^5-5\right)^2\left(20x^4\right)+\left(4x^5-5\right)^3 \cdot \frac{5}{3}\cdot 3x^2$$
$$= \left(4x^5-5\right)^2\left[\left(5x^3\right)\left(20x^4\right)+5x^2\left(4x^5-5\right)\right]$$
$$= 5x^2\left(4x^5-5\right)^2\left[20x^5+\left(4x^5-5\right)\right]$$
$$= 5x^2\left(4x^5-5\right)^2\left(24x^5-5\right)$$

**15.** $f(x) = (x-1)^2\left(x^2+1\right)$

$$f'(x) = (x-1)^2(2x)+\left(x^2+1\right)(2)(x-1)(1)$$
$$= (x-1)\left[(x-1)2x+2\left(x^2+1\right)\right]$$
$$= (x-1)\left(4x^2-2x+2\right)$$
$$= 2(x-1)\left(2x^2-x+1\right)$$

**17.** $y = \frac{\left(x^2-4\right)^3}{x^2+1}$

$$y' = \frac{\left(x^2+1\right)\cdot 3\left(x^2-4\right)^2(2x)-\left(x^2-4\right)^3(2x)}{\left(x^2+1\right)^2} = \frac{2x\left(x^2-4\right)^2\left[3\left(x^2+1\right)-\left(x^2-4\right)\right]}{\left(x^2+1\right)^2} = \frac{2x\left(x^2-4\right)^2\left(2x^2+7\right)}{\left(x^2+1\right)^2}$$

**19.** $p = (q+1)^3\left(q^3-3\right)^3$

$$p' = (q+1)^3\cdot 3\left(q^3-3\right)^2\left(3q^2\right)+\left(q^3-3\right)^3\cdot 3(q+1)^2(1) = 3\left(q^3-3\right)^2\left[(q+1)^3\left(3q^2\right)+\left(q^3-3\right)(q+1)^2\right]$$
$$= 3\left(q^3-3\right)^2(q+1)^2\left[(q+1)\left(3q^2\right)+\left(q^3-3\right)\right] = 3\left(q^3-3\right)^2(q+1)^2\left(4q^3+3q^2-3\right)$$

**21.** $R(x) = x^8(x^2+3x)^4$

$$R'(x) = x^8\cdot 4\left(x^2+3x\right)^3(2x+3)+\left(x^2+3x\right)^4\cdot 8x^7 = 4\left(x^2+3x\right)^3\left[x^8(2x+3)+\left(x^2+3x\right)2x^7\right]$$
$$= 4x^7\left(x^2+3x\right)^3\left[x(2x+3)+2\left(x^2+3x\right)\right] = 4x^7\left(x^2+3x\right)^3\left[4x^2+9x\right] = 4x^8\left(x^2+3x\right)^3(4x+9)$$

**23.** $y = \frac{(2x-1)^4}{\left(x^2+x\right)^4}$

$$y' = \frac{\left(x^2+x\right)^4\cdot 4(2x-1)^3(2)-(2x-1)^4\cdot 4\left(x^2+x\right)^3(2x+1)}{\left(x^2+x\right)^8}$$
$$= \frac{4\left(x^2+x\right)^3\left[\left(x^2+x\right)(2x-1)^3(2)-(2x-1)^4(2x+1)\right]}{\left(x^2+x\right)^8} = \frac{4(2x-1)^3\left(-2x^2+2x+1\right)}{\left(x^2+x\right)^5}$$

**25.** $g(x) = (8x^4+3)^2(x^3-4x)^3$

$$g'(x) = (8x^4+3)^2\cdot 3(x^3-4x)^2(3x^2-4)+(x^3-4x)^3\cdot 2(8x^4+3)(32x^3)$$
$$= (x^3-4x)^2[3(8x^4+3)^2(3x^2-4)+2(x^3-4x)(8x^4+3)(32x^3)]$$
$$= (x^3-4x)^2(8x^4+3)[3(8x^4+3)(3x^2-4)+64x^3(x^3-4x)]$$
$$= (x^3-4x)^2(8x^4+3)(136x^6-352x^4+27x^2-36)$$

**27.** $f(x)=\dfrac{(x^2+5)^{1/3}}{4-x^2}$

$$f'(x)=\frac{(4-x^2)\cdot\frac{1}{3}(x^2+5)^{-2/3}(2x)-(x^2+5)^{1/3}(-2x)}{(4-x^2)^2}$$

$$=\frac{2[x(4-x^2)-(x^2+5)(-3x)]}{3(x^2+5)^{2/3}(4-x^2)^2}=\frac{2(2x^3+19x)}{3(x^2+5)^{2/3}(4-x^2)^2}=\frac{2x(2x^2+19)}{3(x^2+5)^{2/3}(4-x^2)^2}$$

**29.** $y=x^2(4x-3)^{1/4}$

$$y'=x^2\cdot\frac{1}{4}(4x-3)^{-3/4}(4)+(4x-3)^{1/4}(2x)=\frac{x^2}{(4x-3)^{3/4}}+\frac{2x(4x-3)^{1/4}}{1}=\frac{x^2+2x(4x-3)}{(4x-3)^{3/4}}=\frac{9x^2-6x}{(4x-3)^{3/4}}$$

**31.** $c(x)=2x(x^3+1)^{1/2}$

$$c'(x)=(2x)\cdot\frac{1}{2}(x^3+1)^{-1/2}(3x^2)+(x^3+1)^{1/2}\cdot(2)$$

$$=\frac{3x^3}{(x^3+1)^{1/2}}+\frac{2(x^3+1)^{1/2}}{1}=\frac{3x^3+2(x^3+1)}{(x^3+1)^{1/2}}=\frac{5x^3+2}{(x^3+1)^{1/2}}$$

**33.** **a.** $F_1(x)=\dfrac{3}{5}(x^4+1)^5\quad F_1'(x)=\dfrac{3}{5}\cdot5(x^4+1)^4(4x^3)=12x^3(x^4+1)^4$

**b.** $F_2(x)=\dfrac{3}{5}(x^4+1)^{-5}\quad F_2'(x)=\dfrac{3}{5}\cdot(-5)(x^4+1)^{-6}(4x^3)=\dfrac{-12x^3}{(x^4+1)^6}$

**c.** $F_3(x)=\dfrac{1}{5}(3x^4+1)^5\quad F_3'(x)=\dfrac{1}{5}\cdot5(3x^4+1)^4(12x^3)=12x^3(3x^4+1)^4$

**d.** $F_4(x)=3\left(5x^4+1\right)^{-5}\quad F_4'(x)=3\cdot(-5)\left(5x^4+1\right)^{-6}\left(20x^3\right)=\dfrac{-300x^3}{\left(5x^4+1\right)^6}$

**35.** $P=10(3x+1)^3-10\quad \dfrac{dP}{dx}=30(3x+1)^2(3)=90(3x+1)^2$

**37.** $R(x)=60,000x+40,000(10+x)^{-1}-4000$

$R'(x)=60,000-40,000(10+x)^{-2}(1)$

**a.** $R'(10)=60,000-\dfrac{40,000}{20^2}=\$59,900$

**b.** The revenue is increasing.

**39.** $C(y)=2(y+1)^{1/2}+0.4y+4\quad C'(y)=2\cdot\dfrac{1}{2}(y+1)^{-1/2}(1)+0.4=\dfrac{1}{(y+1)^{1/2}}+0.4$

**41.** $V=x(12-2x)^2$

$$V'(x)=x\cdot2(12-2x)(0-2)+(12-2x)^2\cdot1=(12-2x)[-4x+(12-2x)]$$

$$=(12-2x)(12-6x)=144-96x+12x^2$$

**43.** $S(t) = \dfrac{200t}{(t+1)^2}$

$S'(t) = \dfrac{(t+1)^2(200) - 200t \cdot 2(t+1) \cdot 1}{(t+1)^4} = \dfrac{200(t+1)[(t+1) - 2t]}{(t+1)^4} = \dfrac{200(1-t)}{(t+1)^3}$ $\quad S'(9) = \dfrac{200(-8)}{10^3} = -1.6$

From the 9th to the 10th week, sales will decrease approximately \$1.60 thousand or \$1,600.

**45. a.** $y(x) = \dfrac{4.38(x-10)^2 + 78(x-10) + 1430}{0.0029x + 0.25}$

Let $d(x)$ be the denominator of $y(x)$.

Let $n(x)$ be the numerator of $y(x)$.

$d'(x) = 0.0029$

$n'(x) = 8.76(x-10) + 78$

$y'(x) = \dfrac{d(x) \cdot n'(x) - n(x)d'(x)}{[d(x)]^2}$

$= \dfrac{(0.0029x + 0.25)[8.76(x-10) + 78] - [4.38(x-10)^2 + 78(x-10) + 1430](0.0029)}{(0.0029x + 0.25)^2}$

The year 2005 corresponds to $x = 15$, $y'(15) \approx 350$.

The year 2015 corresponds to $x = 25$, $y'(25) \approx 549$.

**b.** In 2015, the per capita health care expenditures are predicted to be changing at the rate of \$549 per year.

**c.** Avereage rate $= \dfrac{7091 - 6331}{2} = 380$. This approximates the instantaneous rate in 2005 quite well.

## *Exercises 9.8*

**1.** $f(x) = 2x^{10} - 18x^5 - 12x^3 + 4$

$f'(x) = 20x^9 - 90x^4 - 36x^2$

$f''(x) = 180x^8 - 360x^3 - 72x$

**3.** $g(x) = x^3 - x^{-1}$

$g'(x) = 3x^2 + x^{-2}$

$g''(x) = 6x - 2x^{-3} = 6x - \dfrac{2}{x^3}$

**5.** $y = x^3 - x^{1/2}$

$y' = 3x^2 - \dfrac{1}{2}x^{-1/2}$

$y'' = 6x + \dfrac{1}{4}x^{-3/2} = 6x + \dfrac{1}{4x^{3/2}}$

**7.** $y = x^5 - 16x^3 + 12$

$y' = 5x^4 - 48x^2$

$y'' = 20x^3 - 96x$

$y''' = 60x^2 - 96$

**9.** $f(x) = 2x^9 - 6x^6$

$f'(x) = 18x^8 - 36x^5$

$f''(x) = 144x^7 - 180x^4$

$f'''(x) = 1008x^6 - 720x^3$

**11.** $y = x^{-1}$

$y' = -1x^{-2}$

$y'' = 2x^{-3}$

$y''' = -6x^{-4}$

**13.** $y = x^5 - x^{1/2}$

$\dfrac{dy}{dx} = 5x^4 - \dfrac{1}{2}x^{-1/2}$

$\dfrac{d^2y}{dx^2} = 20x^3 + \dfrac{1}{4}x^{-3/2}$

**15.** $f(x) = (x+1)^{1/2}$

$f'(x) = \dfrac{1}{2}(x+1)^{-1/2}(1)$

$f''(x) = -\dfrac{1}{4}(x+1)^{-3/2}(1)$

$f'''(x) = \dfrac{3}{8}(x+1)^{-5/2}(1) = \dfrac{3}{8}(x+1)^{-5/2}$

**17.** $y = 4x^3 - 16x$

$y' = 12x^2 - 16$

$y'' = 24x$

$y''' = 24$

$y^{(4)} = 0$

**19.** $f(x) = x^{1/2}$

From problem 15 we have $f^{(3)}(x) = \dfrac{3}{8}x^{-5/2}$.

$f^{(4)}(x) = -\dfrac{15}{16}x^{-7/2}$

**21.** $y' = (4x-1)^{1/2}$

$y'' = \dfrac{1}{2}(4x-1)^{-1/2}(4) = 2(4x-1)^{-1/2}$

$y''' = -1(x-1)^{-3/2}(4) = -4(4x-1)^{-3/2}$

$y^{(4)} = \dfrac{12}{2}(4x-1)^{-5/2}(4) = 24(4x-1)^{-5/2}$

**23.** $f^{(4)}(x) = \dfrac{x}{x+1}$

$f^{(5)}(x) = \dfrac{(x+1)(1) - x(1)}{(x+1)^2} = (x+1)^{-2}$

$f^{(6)}(x) = -2(x+1)^{-3}(1) = -\dfrac{2}{(x+1)^3}$

**25.** $f(x) = 16x^2 - x^3$

$f'(x) = 32x - 3x^2$

$f''(x) = 32 - 6x$

At $x = 1$, $f''(1) = 32 - 6 = 26$.

**27.** $f''(3) \approx nDeriv(nDeriv(x^3 - 27/x, x, x,), x, 3)$

$= 15.99999925$

**29.**

$f''(21) \approx nDeriv(nDeriv(\sqrt{\ }(x^2+4), x, x,), x, 21)$

$= 0.000426$

**31.** $f(x) = x^3 - 3x^2 + 5$

**a.** $f'(x) = 3x^2 - 6x$

$f''(x) = 6x - 6$

**b.**

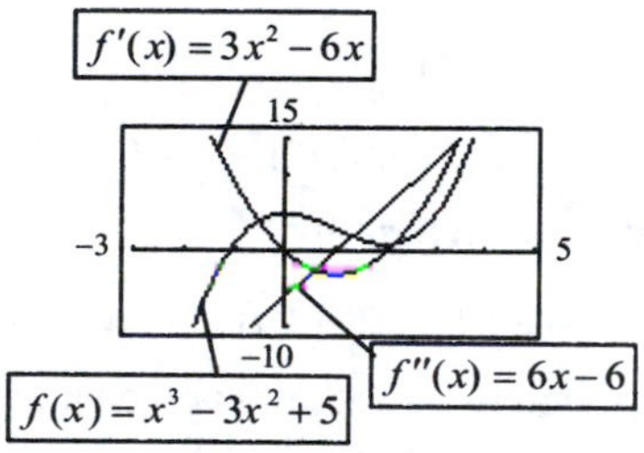


**c.** $f''(x) = 0$ at $x = 1$.

$f''(x) > 0$ if $x > 1$;

$f''(x) < 0$ if $x < 1$.

**d.** $f'(x)$ [NOT $f(x)$] min at $x = 1$.

$f'(x)$ increasing if $x > 1$. ($f''(x) > 0$)

$f'(x)$ decreasing if $x < 1$. ($f''(x) < 0$)

**e.** $f''(x) < 0$

**f.** $f''(x) > 0$ [For now look at graphs.]

**33.** $s(t) = 100 + 10t + 0.01t^3$ distance

$s'(t) = 10 + 0.03t^2$ velocity

$s''(t) = 0.06t$ acceleration

$s''(2) = 0.12$

**35.** $R(x) = 100x - 0.01x^2$

$R'(x) = 100 - 0.02x$

$R''(x) = -0.02$

**37.** $R = m^2\left(\frac{c}{2} - \frac{m}{3}\right) = \frac{c}{2}m^2 - \frac{1}{3}m^3$

**a.** $R' = cm - m^2$

**b.** $R'' = c - 2m$

**c.** Second derivative

**39.** $R(x) = 15x + 30(4x+1)^{-1} - 30$

$\overline{MR} = R'(x) = 15 - 30(4x+1)^{-2}(4)$

$= 15 - 120(4x+1)^{-2}$

$\overline{MR}' = R''(x) = 0 + 240(4x+1)^{-3}(4) = \frac{960}{(4x+1)^3}$

**a.** $R''(25) = \frac{960}{101^3} \approx 0.0009$

**b.** When the next unit is sold the marginal revenue will increase about $0.90 per unit.

**41.** $S(t) = 1 + 3(t+3)^{-1} - 18(t+3)^{-2}$

**a.** $S'(t) = -3(t+3)^{-2} + 36(t+3)^{-3}$

$= \frac{-3}{(t+3)^2} + \frac{36}{(t+3)^3}$

**b.** $S''(t) = 6(t+3)^{-3} - 108(t+3)^{-4}$

$= \frac{6}{(t+3)^3} - \frac{108}{(t+3)^4}$

$S''(15) = \frac{6}{18^3} - \frac{108}{18^4} = 0$

**c.** The rate of change of the rate of change in sales (after 15 weeks) is zero.

**43. a.** $W(t) = 0.0212t^{2.11}$

$W'(t) = 2.11\left(0.0212t^{1.11}\right) \approx 0.0447t^{1.11}$

**b.** $W''(t) \approx 0.0497t^{0.11}$

**c.** $W(50) \approx 81.5$, $W'(50) \approx 3.44$, and $W''(50) \approx 0.0764$. These mean that in 2025 the average annual wage is expected to be $81,500 and changing at a rate of about $3440 per year. Also that rate is expected to be changing at about $76.40 per year per year.

**45. a.** $R(x) = -0.000190x^3 + 0.0519x^2 - 4.06x + 192$

**b.** $R'(x) = -0.000570x^2 + 0.1038x - 4.06$

**c.** $R''(x) = -0.00114x + 0.1038$

**d.** $R'(90) = -0.000570(90)^2 + 0.1038(90) - 4.06 \approx 0.665$ and $R''(90) = -0.00114(90) + 0.1038 \approx 0.0012$.

In 2040, the economic dependency ratio is expected to be changing at a rate of about 0.665 per year, and this rate is expected to be changing at about 0.0012 per year per year.

**47. a.** $f'(x) = 0.000864\left(3x^2\right) - 0.128(2x) + 6.61 = 0.002592x^2 - 0.256x + 6.61$

**b.** $f'(25) = 1.83$ and $f'(55) = 0.3708$. At age 25, the rate of change is about $1830 per year, and at age find the rate of change is about $371 per year.

**c.** $f''(x) = 0.005184x - 0.256$

**d.** $f''(25) = -0.1264$ and $f''(55) = 0.02912$. At age 25, the rate of change changes at about $-\$126$ per year per year of age, and age 55 the rate of change changes at about $29 per year per year of age.

**e.** The rate of change in the median income of 55-year-old workers is about $371 per year of age, and it is increasing at a rate of about $29 per year per year of age.

# Chapter 9: Derivatives

## *Exercises 9.9*

**1.** **a.** $\overline{MR} = R'(x) = 4$

**b.** Each unit sold brings in \$4 revenue.

**3.** $R(x) = 36x - 0.01x^2$

**a.** $R(100) = 3600 - 0.01(10{,}000)$

$R(100) = \$3500$

100 units produce \$3500 of revenue.

**b.** $\overline{MR} = R'(x) = 36 - 0.02x$

**c.** $R'(100) = 36 - 2 = 34$

The sale of the next unit will increase the revenue by about \$34. The sale of the next 3 units will increase the revenue by about \$102.

**d.** $R(101) - R(100) = 3533.99 - 3500 = \$33.99$

(Actual revenue from 101st unit.)

**5.** **a.** $R(x) = px = (80 - 0.4x)x = 80x - 0.4x^2$

(in hundreds of dollars)

**b.** $50 = 80 - 0.4x$

$0.4x = 80 - 50 = 30$

$x = \dfrac{30}{0.4} = 75$ hundreds

(or 7500 subscribers)

$R = 50 \cdot 7500 = \$375{,}000$

**c.** More subscribers are attracted by lower prices.

**d.** $R'(x) = 80 - 0.8x$

$p = 50 \;\rightarrow\; x = 75$

$R'(75) = 80 - 0.8(75) = \$20$

If the number of subscribers increases from 75 to 76 (hundred) the revenue increases $\$20(100) = \$2{,}000$. Increasing the number of subscribers will occur by lowering the monthly charges.

**7.** **a.** $\overline{MR} = 36 - 0.02x$

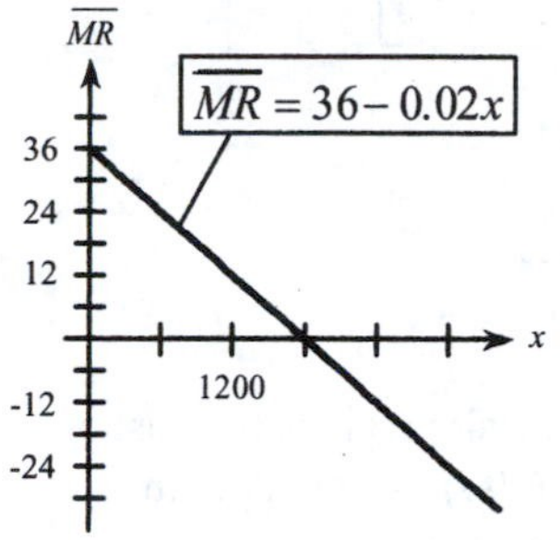


**b.** Maximum revenue at $x = 1800$ $(\overline{MR} = 0)$

**c.** $R(1800) = \$32{,}400$

**9.** $C(x) = 40 + 8x;\ \overline{MC} = C'(x) = 8$

**11.** $C(x) = 500 + 13x + x^2;\ \overline{MC} = C'(x) = 13 + 2x$

**13.** $C = x^3 - 6x^2 + 24x + 10$

$\overline{MC} = C' = 3x^2 - 12x + 24$

**15.** $C = 400 + 27x + x^3;\ \overline{MC} = C' = 27 + 3x^2$

**17.** **a.** $C(x) = 40 + x^2$

$\overline{MC} = C'(x) = 2x$

$C'(5) = 2(5) = 10$

The cost to produce the 6th unit is predicted to be \$10.

**b.** $C(6) - C(5) = 76 - 65 = \$11$

**19.** $C(x) = x^3 - 4x^2 + 30x + 20$

$\overline{MC} = C'(x) = 3x^2 - 8x + 30$

$C'(4) = 48 - 32 + 30 = 46$

The cost of producing 1 additional unit will increase by \$46. The cost of producing 3 additional units will increase by $3 \times \$46 = \$138$.

**21.** $C(x) = 300 + 4x + x^2 \quad \overline{MC} = C'(x) = 4 + 2x$

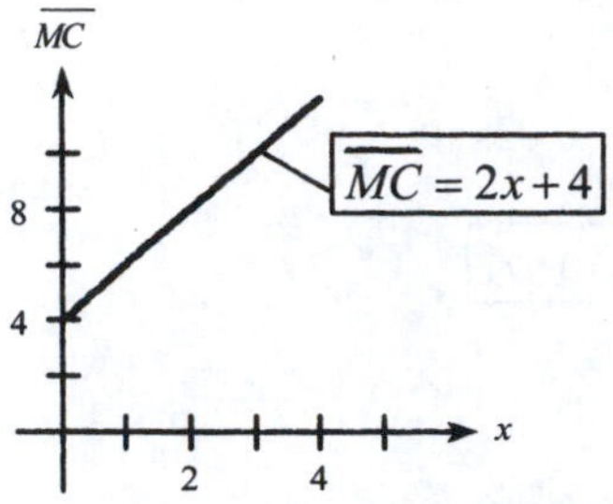


**23.** **a.** Remember that the slope gives the cost of the next unit. $C'(100)$ is greater than $C'(500)$. Thus the 101st unit will cost more than the 501st unit.

**b.** Since the slope is decreasing, the cost of each additional unit is less than the previous unit. The manufacturing process is more efficient.

**25.** $P(x) = 5x - 25 \qquad \overline{MP} = P'(x) = 5$

The next unit sold earns a \$5 profit.

**27.** $P(x) = R(x) - C(x)$

$$= 50x - \left(1900 + 30x + 0.01x^2\right)$$
$$= 50x - 1900 - 30x - 0.01x^2$$
$$= -0.01x^2 + 20x - 1900$$

**a.** $P(500) = \$5,600$

**b.** $P'(x) = -0.02x + 20$

$\overline{MP} = 20 - 0.02x$

**c.** $\overline{MP}(500) = P'(500) = -0.02(500) + 20 = 10$

The total profit will increase by approximately \$10 on the sale of the next (501st) unit.

**d.**

$P(501) - P(500) = (5609.99) - (5600) = \$9.99$

This is the actual profit on the sale of the 501st unit.

**29.** **a.** From $P = R - C$ we obtain the following profits in order from smallest to largest: 100, 700, and 400 units. There is a loss at 100 units since the cost curve is above the revenue curve.

**b.** $MP = MR - MC$ Look at the slopes of each curve at $x = 100$, 400, and 700. Ranking from high to low:

$\overline{MP}(100) > \overline{MP}(400) > \overline{MP}(700)$.

$\overline{MP}(700) < 0$ since $R' - C' < 0$.

**31.** **a.** From low to high: $A < B < C$.

The graph shows a loss at $A$.

**b.** $P' > 0$ at each point.

Evaluate the slope at each point to obtain $P'(C) < P'(B) < P'(A)$.

**33.** $\overline{MP} = 30 - 2x$

**a.**

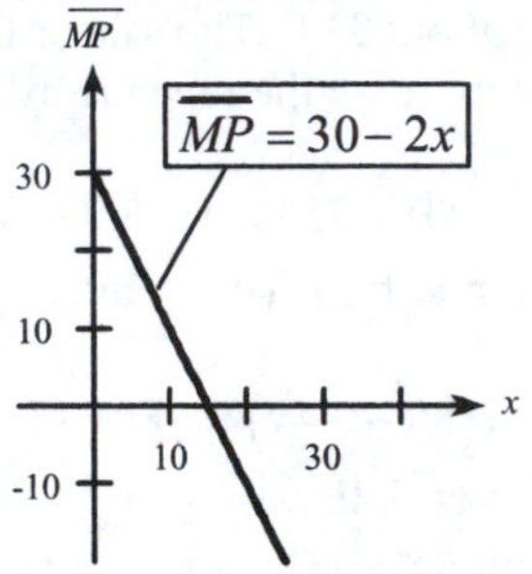


**b.** $\overline{MP} = 0$ if $30 - 2x = 0$. Thus, $x = 15$.

**c.** Profit is a maximum at the vertex of the parabola $P(x) = 30x - x^2 - 200$.

Vertex is at $x = \dfrac{-b}{2a} = \dfrac{-30}{2(-1)} = 15$.

**d.** $P(15) = 450 - 225 - 200 = \$25$

**35.** Note that cost is per unit. Total cost is the cost per unit times the number of units.

$$P(x) = R(x) - C(x) = 300x - (160 + 0.1x)x$$
$$= 300x - 160x - 0.1x^2$$
$$= 140x - 0.1x^2$$

$P'(x) = 140 - 0.2x$

$P'(x) = 0$ at $140 - 0.2x = 0$ or $x = 700$.

Production of 700 units will maximize profit.

**37.** $P(x) = R(x) - C(x)$

$$= 70x - (10 + 0.1x)x$$
$$= 70x - 10x - 0.1x^2$$
$$= 60x - 0.1x^2$$

$P'(x) = 60 - 0.2x$

$P'(x) = 0$ when $60 - 0.2x = 0$ or $x = 300$.

Maximum daily profit = $P(300)$ = \$9,000

# Chapter 9: Derivatives

## *Chapter 9 Review Exercises*

1. **a.** $f(-2) = 2$

   **b.** $\lim_{x\to -2} f(x) = 2$

2. **a.** From the graph, $f(-1) = 0$.

   **b.** $\lim_{x\to -1} f(x) = 0$

3. **a.** $f(4) = 2$

   **b.** $\lim_{x\to 4^-} f(x) = 1$

4. **a.** From the graph, $\lim_{x\to 4^+} f(x) = 2$.

   **b.** $\lim_{x\to 4} f(x)$ does not exist.

5. **a.** $f(1)$ does not exist

   **b.** $\lim_{x\to 1} f(x) = 2$

6. **a.** From the graph, $f(2)$ does not exist.

   **b.** $\lim_{x\to 2} f(x)$ does not exist.

7. $\lim_{x\to 4}(3x^2 + x + 3) = 48 + 4 + 3 = 55$

8. $\lim_{x\to 4}\dfrac{x^2-16}{x+4} = \dfrac{4^2-16}{4+4} = 0$

9. $\lim_{x\to -1}\dfrac{x^2-1}{x+1} = \lim_{x\to -1}\dfrac{(x+1)(x-1)}{x+1}$

   $= \lim_{x\to -1}(x-1) = -2$

10. $\lim_{x\to 5}\dfrac{x^2-6x+5}{x^2-5x} = \lim_{x\to 5}\dfrac{(x-5)(x-1)}{x(x-5)}$

    $= \lim_{x\to 5}\dfrac{x-1}{x} = \dfrac{5-1}{5} = \dfrac{4}{5}$

11. $\lim_{x\to 2}\dfrac{4x^3-8x^2}{4x^3-16x} = \lim_{x\to 2}\dfrac{4x^2(x-2)}{4x(x-2)(x+2)}$

    $= \lim_{x\to 2}\dfrac{x}{x+2} = \dfrac{2}{4} = \dfrac{1}{2}$

12. $\lim_{x\to -\frac{1}{2}}\dfrac{x^2-\frac{1}{4}}{6x^2+x-1} = \lim_{x\to -\frac{1}{2}}\dfrac{\left(x+\frac{1}{2}\right)\left(x-\frac{1}{2}\right)}{(3x-1)(2)\left(x+\frac{1}{2}\right)}$

    $= \lim_{x\to -\frac{1}{2}}\dfrac{x-\frac{1}{2}}{2(3x-1)} = \dfrac{-1}{-5} = \dfrac{1}{5}$

13. $\lim_{x\to 3}\dfrac{x^2-16}{x-3} = \dfrac{-7}{0}$

    Thus, the limit does not exist.

14. $\lim_{x\to -3}\dfrac{x^2-x-12}{2x-6} = \dfrac{(-3)^2-(-3)-12}{2(-3)-6} = 0$

15. $\lim_{x\to 1}\dfrac{x^2-9}{x-3} = \dfrac{-8}{-2} = 4$

16. $\lim_{x\to 2}\dfrac{x^2-8}{x-2}$ does not exist.

17. $\lim_{x\to 1^-} f(x) = 4 - 1 = 3$

    $\lim_{x\to 1^+} f(x) = 2(1) + 1 = 3$

    Thus, $\lim_{x\to 1} f(x) = 3$

18. $\lim_{x\to -2^-} f(x) = -8 + 2 = -6$

    $\lim_{x\to -2^+} f(x) = 2 - 4 = -2$

    Since these two limits are not equal, $\lim_{x\to -2} f(x)$ does not exist.

19. $\lim_{h\to 0}\dfrac{3(x+h)^2-3x^2}{h}$

    $= \lim_{h\to 0}\dfrac{3x^2+6xh+3h^2-3x^2}{h}$

    $= \lim_{h\to 0}\dfrac{6xh+3h^2}{h} = \lim_{h\to 0}(6x+3h) = 6x$

20. $\lim_{h\to 0}\dfrac{[(x+h)-2(x+h)^2]-(x-2x^2)}{h}$

    $= \lim_{h\to 0}\dfrac{x+h-2x^2-4xh-2h^2-x+2x^2}{h}$

    $= \lim_{h\to 0}\dfrac{h-4xh-2h^2}{h}$

    $= \lim_{h\to 0}(1-4x-2h) = 1-4x$

21. $\lim_{x\to 2}\dfrac{(x+12)(x-2)}{(x-3)(x-2)}$

    $= \lim_{x\to 2}\dfrac{x+12}{x-3} = \dfrac{14}{-1} = -14$

22. $\lim_{x\to -\frac{1}{2}}\dfrac{\left(x+\frac{1}{2}\right)\left(x-\frac{1}{3}\right)}{\left(x+\frac{1}{2}\right)\left(x+\frac{1}{3}\right)}$

    $= \lim_{x\to -\frac{1}{2}}\dfrac{x-\frac{1}{3}}{x+\frac{1}{3}} = \dfrac{-\frac{1}{2}-\frac{1}{3}}{-\frac{1}{2}+\frac{1}{3}} = 5$

23. **a.** From the graph, $f(-1) = 0 = \lim_{x\to -1} f(x)$, so $f(x)$ is continuous at $x = -1$.

    **b.** From the graph, $f(1)$ does not exist, so $f(x)$ is not continuous at $x = 1$.

**24. a.** From the graph, $f(-2) = 2 = \lim_{x\to-2} f(x)$, so $f(x)$ is continuous at $x = -2$.

**b.** From the graph, $f(2)$ and $\lim_{x\to2} f(x)$ do not exist, so $f(x)$ is not continuous at $x = 2$.

**25.** $\lim_{x\to-1^+} f(x) = (-1)^2 + 1 = 2$

$\lim_{x\to-1^-} f(x) = (-1)^2 + 1 = 2$

Thus, $\lim_{x\to-1} f(x) = 2$

**26.** $\lim_{x\to0^-} f(x) = 0^2 + 1 = 1$

$\lim_{x\to0^+} f(x) = 0$

Thus, $\lim_{x\to0} f(x)$ does not exist.

**27.** $\lim_{x\to1^+} f(x) = 2(1)^2 - 1 = 1$

$\lim_{x\to1^-} f(x) = 1$

Thus, $\lim_{x\to1} f(x) = 1$

**28.** $f(x)$ is not continuous at $x = 0$ since the $\lim_{x\to0} f(x)$ does not exist.

**29.** $f(1) = 2(1)^2 - 1 = 1 = \lim_{x\to1} f(x)$

Yes, $f(x)$ is continuous at $x = 1$.

**30.** Since $f(-1) = 2 = \lim_{x\to-1} f(x)$, $f(x)$ is continuous at $x = -1$.

**31.** Function is discontinuous at $x = 5$. (Function is not defined.)

**32.** $f(x)$ is not continuous at $x = 2$ since $f(2)$ does not exist.

**33.** $f(2) = 4$

$\lim_{x\to2} f(x) = 4$

Function is continuous everywhere.

**34.** $f(x)$ is not continuous at $x = 1$ since $\lim_{x\to1} f(x)$ does not exist.

**35. a.** $f(x)$ is discontinuous at $x = 0$ and $x = 1$.

**b.** $\lim_{x\to+\infty} f(x) = 0$

**c.** $\lim_{x\to-\infty} f(x) = 0$

**36. a.** From the graph, $f(x)$ is discontinuous at $x = 0$ and $x = -1$.

**b.** $\lim_{x\to+\infty} f(x) = \frac{1}{2}$

**c.** $\lim_{x\to-\infty} f(x) = \frac{1}{2}$

**37.** $\lim_{x\to-\infty} \frac{2x^2}{1-x^2} = \lim_{x\to-\infty} \frac{2}{\frac{1}{x^2}-1} = \frac{2}{0-1} = -2$

There is a horizontal asymptote at $y = -2$.

**38.** $\lim_{x\to+\infty} \frac{3x^{2/3}}{x+1} = \lim_{x\to+\infty} \frac{\frac{3}{x^{1/3}}}{1+\frac{1}{x}} = \frac{0}{1} = 0$

There is a horizontal asymptote at $y = 0$.

**39.** $\text{Avg} = \frac{f(2)-f(-1)}{2-(-1)} = \frac{33-12}{3} = 7$

**40.** True.

**41.** False. The given expression gives the slope of the tangent line at $x = c$.

**42.** $f(x) = 3x^2 + 2x - 1$

$$f'(x) = \lim_{h\to0} \frac{\left[3(x+h)^2 + 2(x+h) - 1\right] - \left(3x^2 + 2x - 1\right)}{h} = \lim_{h\to0} \frac{6xh + 3h^2 + 2h}{h} = \lim_{h\to0}(6x + 3h + 2) = 6x + 2$$

**43.** $$f'(x) = \lim_{h\to0} \frac{\left[x + h - (x+h)^2\right] - \left(x - x^2\right)}{h} = \lim_{h\to0} \frac{h - 2xh - h^2}{h} = \lim_{h\to0}(1 - 2x - h) = 1 - 2x$$

**44.** $[-3, 0]$: $\text{Avg} = \frac{f(0)-f(-3)}{0-(-3)} = \frac{1-0}{3} = \frac{1}{3}$

$[-1, 0]$: $\text{Avg} = \frac{f(0)-f(-1)}{0-(-1)} = \frac{1-0}{1} = 1$

From $[-1, 0]$ the change is greater.

**45. a.** No **b.** No

**46. a.** From the graph, $f(x)$ is differentiable at $x = -2$.

**b.** $f(x)$ is not differentiable at $x = 2$.

**47. a.** $nDeriv\left((4x)^{1/3}/(3x^2-10)^2,\ x,\ 2\right)$

$=-5.917062766$

**b.** $\dfrac{f(2+0.0001)-f(2)}{0.0001}$

$=\dfrac{0.49941-0.5}{0.0001}=-5.9$

**48. a.** $\text{Avg}=\dfrac{g(5)-g(2)}{5-2}=\dfrac{18.1-13.2}{3}=1.633$

**b.** $g'(4)=\dfrac{g(4.3)-g(4)}{4.3-4}=\dfrac{2.1}{0.3}=7$

**49.** Approximate the tangent through $(0,2)$ and $(8,0)$. $f'(4)\approx\dfrac{0-2}{8-0}=-\dfrac{1}{4}$

**50.** $A: f'(2)\approx 0$

$B: f'(6)\approx\dfrac{-1-1}{8-4}=-\dfrac{1}{2}$

$C: \text{Avg}=\dfrac{-1-1.2}{10-2}=-0.275$

Rank: $B<C<A$

$B<0$ and $C<0$; the tangent line at $x=6$ falls more steeply than the segment over $[2,10]$.

**51.** $c=4x^5-6x^3$

$c'=4(5x^4)-6(3x^2)=20x^4-18x^2$

**52.** $f(x)=10x^9-5x^6+4x-2^7+19$

$f'(x)=10\left(9x^8\right)-5\left(6x^5\right)+4\left(1x^0\right)+0$

$=90x^8-30x^5+4$

**53.** $p=q+\sqrt{7}\qquad \dfrac{dp}{dq}=1+0=1$

**54.** $y=\sqrt{x}=x^{1/2}\qquad y'=\dfrac{1}{2}x^{-1/2}=\dfrac{1}{2\sqrt{x}}$

**55.** $f(z)=2^{4/3}\qquad f'(z)=0$

**56.** $v(x)=\dfrac{4}{\sqrt[3]{x}}=4x^{-1/3}$

$v'(x)=-\dfrac{4}{3}x^{-4/3}=-\dfrac{4}{3\sqrt[3]{x^4}}$

**57.** $y=x^{-1}-x^{-1/2}$

$y'=-1(x^{-2})-\left(-\dfrac{1}{2}x^{-3/2}\right)=-\dfrac{1}{x^2}+\dfrac{1}{2x^{3/2}}$

**58.** $f(x)=\dfrac{3}{2x^2}-\sqrt[3]{x}+4^5=\dfrac{3}{2}x^{-2}-x^{1/3}+4^5$

$f'(x)=-3x^{-3}-\dfrac{1}{3}x^{-2/3}=-\dfrac{3}{x^3}-\dfrac{1}{3\sqrt[3]{x^2}}$

**59.** $f(x)=3x^5-6\qquad f'(x)=15x^4$

$f(1)=3-6=-3\qquad f'(1)=15$

So, $y-(-3)=15(x-1)$ or $y=15x-18$.

**60.** $f(x)=3x^3-2x\qquad f(2)=20$

$f'(x)=9x^2-2\qquad f'(2)=34$

So, $y-20=34(x-2)$ or $y=34x-48$.

**61.** $f'(x)=3x^2-6x=3x(x-2)$

**a.** $f'(x)=0$ if $x=0,\ 2$.

**b.** $f(0)=0-0+1=1$ Point: (0, 1)

$f(2)=8-12+1=-3$ Point: (2, −3)

**c.**

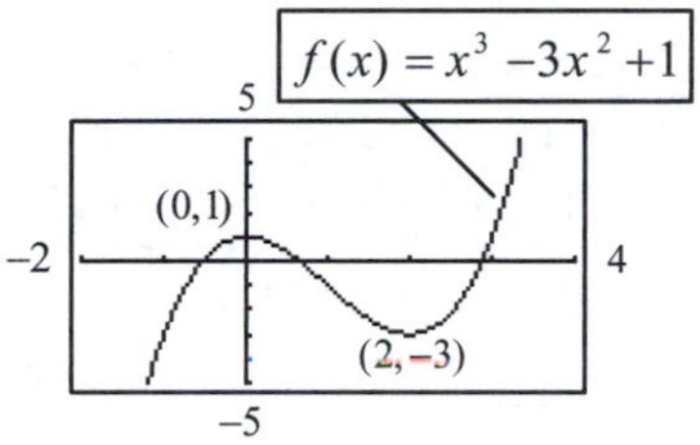


**62.** $f(x)=x^6-6x^4+8$

$f'(x)=6x^5-24x^3=6x^3(x^2-4)$

$=6x^3(x+2)(x-2)$

**a.** $f'(x)=0$ when $x=0$, $x=-2$, and when $x=2$.

**b.** The associated points where the slope equals zero are (0, 8), (−2, −24) and (2, −24).

**c.**

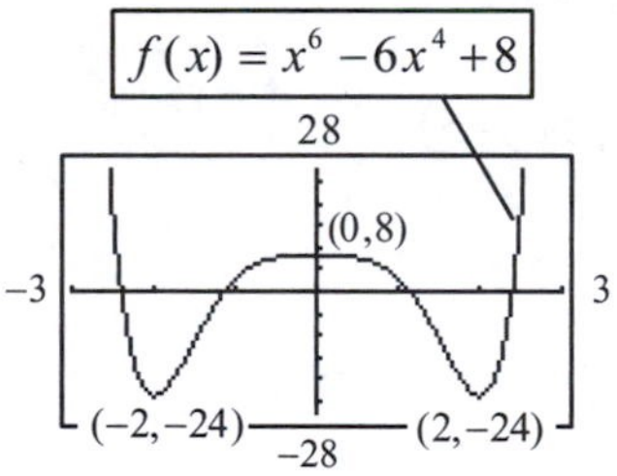


**63.** $f(x)=(3x-1)(x^2-4x)$

$f'(x)=(3x-1)(2x-4)+(x^2-4x)(3)$

$=9x^2-26x+4$

**64.** $y=\left(x^4+3\right)\left(3x^3+1\right)$

$y'=\left(x^4+3\right)\left(9x^2\right)+\left(3x^3+1\right)\left(4x^3\right)$

$=9x^6+27x^2+12x^6+4x^3$

$=21x^6+4x^3+27x^2$

**65.** $p=\dfrac{5q^3}{2q^3+1}$

$$\frac{dp}{dq}=\frac{\left(2q^3+1\right)\left(15q^2\right)-\left(5q^3\right)\left(6q^2\right)}{\left(2q^3+1\right)^2}=\frac{30q^5+15q^2-30q^5}{\left(2q^3+1\right)^2}=\frac{15q^2}{\left(2q^3+1\right)^2}$$

**66.** $s=\dfrac{\sqrt{t}}{3t+1}=\dfrac{t^{1/2}}{3t+1}$ $\quad \dfrac{ds}{dt}=\dfrac{(3t+1)\cdot\frac{1}{2}t^{-1/2}-t^{1/2}\cdot 3}{(3t+1)^2}=\dfrac{\frac{1}{2}t^{-1/2}(3t+1-6t)}{(3t+1)^2}=\dfrac{1-3t}{2\sqrt{t}(3t+1)^2}$

**67.** $y=\sqrt{x}(3x+2)=3x^{3/2}+2x^{1/2}$ $\quad \dfrac{dy}{dx}=\dfrac{9}{2}x^{1/2}+x^{-1/2}=\dfrac{9x+2}{2\sqrt{x}}$

**68.** $C=\dfrac{5x^4-2x^2+1}{x^3+1}$

$$\frac{dC}{dx}=\frac{(x^3+1)(20x^3-4x)-(5x^4-2x^2+1)(3x^2)}{(x^3+1)^2}$$

$$=\frac{20x^6-4x^4+20x^3-4x-15x^6+6x^4-3x^2}{(x^3+1)^2}=\frac{5x^6+2x^4+20x^3-3x^2-4x}{(x^3+1)^2}$$

**69.** $y=\left(x^3-4x^2\right)^3$ $\quad y'=3\left(x^3-4x^2\right)^2\left(3x^2-8x\right)=\left(x^3-4x^2\right)^2\left(9x^2-24x\right)$

**70.** $y=\left(5x^6+6x^4+5\right)^6$ $\quad y'=6\left(5x^6+6x^4+5\right)^5\left(30x^5+24x^3\right)$

**71.** $y=\left(2x^4-9\right)^9$ $\quad \dfrac{dy}{dx}=9\left(2x^4-9\right)^8\cdot 8x^3=72x^3\left(2x^4-9\right)^8$

**72.** $g(x)=\dfrac{1}{\sqrt{x^3-4x}}=\left(x^3-4x\right)^{-1/2}$ $\quad g'(x)=-\dfrac{1}{2}\left(x^3-4x\right)^{-3/2}\left(3x^2-4\right)=-\dfrac{3x^2-4}{2\sqrt{(x^3-4x)^3}}$

**73.** $f(x)=x^2\left(2x^4+5\right)^8$

$$f'(x)=x^2\cdot 8\left(2x^4+5\right)^7\left(8x^3\right)+\left(2x^4+5\right)^8(2x)=2x\left(2x^4+5\right)^7\left[32x^4+\left(2x^4+5\right)\right]=2x\left(2x^4+5\right)^7\left(34x^4+5\right)$$

**74.** $S=\dfrac{(3x+1)^2}{x^2-4}$

$$S'=\frac{(x^2-4)(2)(3x+1)(3)-(3x+1)^2(2x)}{(x^2-4)^2}=\frac{(3x+1)[(6x^2-24)-(6x^2+2x)]}{(x^2-4)^2}=-\frac{2(3x+1)(x+12)}{(x^2-4)^2}$$

**75.** $y = (3x+1)^{12}(2x^3-1)^{12}$

$$\frac{dy}{dx} = (3x+1)^{12}\cdot 12(2x^3-1)^{11}(6x^2) + (2x^3-1)^{12}\cdot 12(3x+1)^{11}(3)$$
$$= 12(2x^3-1)^{11}\left[6x^2(3x+1)^{12} + 3(2x^3-1)(3x+1)^{11}\right]$$
$$= 36(2x^3-1)^{11}(3x+1)^{11}\left[2x^2(3x+1) + (2x^3-1)\right]$$
$$= 36\left[(2x^3-1)(3x+1)\right]^{11}(8x^3+2x^2-1)$$

**76.** $y = \left(\frac{x+1}{1-x^2}\right)^3 = \left(\frac{1}{1-x}\right)^3 = (1-x)^{-3}$ $\quad y' = -3(1-x)^{-4}(-1) = \frac{3}{(1-x)^4}$

**77.** $y = x(x^2-4)^{1/2}$

$$y' = x\cdot\frac{1}{2}(x^2-4)^{-1/2}(2x) + (x^2-4)^{1/2}\cdot 1 = \frac{x^2}{\sqrt{x^2-4}} + \frac{\sqrt{x^2-4}}{1} = \frac{x^2+(x^2-4)}{\sqrt{x^2-4}} = \frac{2x^2-4}{\sqrt{x^2-4}}$$

**78.** $y = \frac{x}{\sqrt[3]{3x-1}} = \frac{x}{(3x-1)^{1/3}}$ $\quad \frac{dy}{dx} = \frac{(3x-1)^{1/3}\cdot 1 - x\cdot\frac{1}{3}(3x-1)^{-2/3}\cdot 3}{(3x-1)^{2/3}} = \frac{(3x-1)^{-2/3}\left[(3x-1)-x\right]}{(3x-1)^{2/3}} = \frac{2x-1}{(3x-1)^{4/3}}$

**79.** $y = x^{1/2} - x^2$ $\quad y' = \frac{1}{2}x^{-1/2} - 2x$ $\quad y'' = -\frac{1}{4}x^{-3/2} - 2$

**80.** $y = x^4 - x^{-1}$ $\quad y' = 4x^3 + x^{-2}$ $\quad y'' = 12x^2 - 2x^{-3} = 12x^2 - \frac{2}{x^3}$

**81.** $y = (2x+1)^4$ Since the largest power in any term is 4, the fifth derivative equals 0.

**82.** $y = \frac{1}{24}(1-x)^6$ $\quad y' = -\frac{1}{4}(1-x)^5$ $\quad y'' = \frac{5}{4}(1-x)^4$

$y''' = -5(1-x)^3$ $\quad y^{(4)} = 15(1-x)^2$ $\quad y^{(5)} = -30(1-x)$

**83.** $\frac{dy}{dx} = (x^2-4)^{1/2}$

$$\frac{d^2y}{dx^2} = \frac{1}{2}(x^2-4)^{-1/2}(2x) = \frac{x}{(x^2-4)^{1/2}}$$

$$\frac{d^3y}{dx^3} = \frac{(x^2-4)^{1/2}(1) - x\left[\frac{x}{(x^2-4)^{1/2}}\right]}{(x^2-4)}\cdot\frac{(x^2-4)^{1/2}}{(x^2-4)^{1/2}} = \frac{x^2-4-x^2}{(x^2-4)^{3/2}} = \frac{-4}{(x^2-4)^{3/2}}$$

**84.** $\frac{d^2y}{dx^2} = \frac{x}{x^2+1}$

$$\frac{d^3y}{dx^3} = \frac{x^2+1-x(2x)}{(x^2+1)^2} = \frac{-x^2+1}{(x^2+1)^2}$$

$$\frac{d^4y}{dx^4} = \frac{(x^2+1)^2(-2x) - (-x^2+1)(2)(x^2+1)(2x)}{(x^2+1)^4} = \frac{(x^2+1)[-2x(x^2+1) - 4x(-x^2+1)]}{(x^2+1)^4} = \frac{2x^3-6x}{(x^2+1)^3} = \frac{2x(x^2-3)}{(x^2+1)^3}$$

**85. a.** $\lim_{x\to 4000} R(x) = 140(4000) - 0.01(4000)^2 = 400{,}000$

**b.** $\lim_{x\to 4000} C(x) = 60(4000) + 70{,}000 = 310{,}000$

**c.** $\lim_{x\to 4000} P(x) = \lim_{x\to 4000} R(x) - \lim_{x\to 4000} C(x) = 90{,}000$

**86. a.** $\lim_{x\to 0^+} C(x) = 60(0) + 70{,}000 = 70{,}000$ The fixed costs of the company (at zero units) are $70,000.

**b.** $\lim_{x\to 1000} P(x) = \lim_{x\to 1000} [R(x) - C(x)] = \lim_{x\to 1000} \left[(140x - 0.01x^2) - (60x + 70000)\right]$

$= \left(140(1000) - 0.01(1000)^2\right) - \left(60(1000) + 70000\right) = 0$

1000 units is the break-even point for the company.

**87. a.** $\lim_{x\to 0^+} \bar{R}(x) = \lim_{x\to 0^+} \frac{140x - 0.01x^2}{x} = \lim_{x\to 0^+} (140 - 0.01x) = 140 - 0.01(0) = 140$

$140 per unit

**b.** $\lim_{x\to 0^+} \bar{C}(x) = \lim_{x\to 0^+} \frac{60x + 70000}{x} = \lim_{x\to 0^+} \left(60 + \frac{70000}{x}\right) = \infty$ ; the limit does not exist.

**88. a.** $\lim_{x\to \infty} C(x) = \lim_{x\to \infty} (60x + 70000) = \infty$ ; the limit does not exist. As the number of units produced increases without bound, so does the total cost.

**b.** $\lim_{x\to \infty} \bar{C}(x) = \lim_{x\to \infty} \frac{60x + 70000}{x} = \lim_{x\to \infty} \left(60 + \frac{70000}{x}\right) = 60$ . The average cost per unit approaches $60 as the number of units increases.

**89. a.** Men: $\frac{19.6 - 16.3}{40} = 0.0285\%$ per year

Between 1990 and 2030 the average annual percent change of elderly men in the workforce is 0.0825 percentage points per year.

**b.** Women: $\frac{11.7 - 8.6}{40} = 0.0775\%$ per year

Between 1990 and 2030 the average annual percent change of elderly women in the workforce is 0.0775 percentage points per year.

**90. a.** 1970-1980 for elderly men: $\frac{19.0 - 26.8}{10} = -0.78$ percentage points per year

2030-2040 for elderly men: $\frac{17.3 - 19.6}{10} = -0.23$ percentage points per year

**b.** 1970-1980 for elderly women: $\frac{8.1 - 9.7}{10} = -0.16$ percentage points per year

2030-2040 for elderly women: $\frac{10.1 - 11.7}{10} = -0.16$ percentage points per year

**91.** $x(p) = \frac{100}{p} - 1 \qquad x'(p) = \frac{-100}{p^2}$

**a.** $x'(10) = \frac{-100}{10^2} = -1$

As the price per unit increases from $10 to $11, the demand is expected to drop by 1 unit.

**b.** $x'(20) = \dfrac{-100}{20^2} = -\dfrac{1}{4}$

As the price per unit increases from \$20 to \$21, the demand is expected to drop by $\dfrac{1}{4}$ units.

**92.** $h = 0.000595s^{1.922}$ $\quad h(100) = 0.000595(100)^{1.922} \approx 4.1545$

$h' = 0.00114359s^{0.922}$ $\quad h'(100) = 0.00114359(100)^{0.922} \approx 0.08$

When the speed of an updraft is 100 mph, the diameter of the hail is 4.15 inches, and the size of the hail is changing at a rate of 0.08 inches per mph of updraft.

**93.** The $(A+1)$st item will produce more revenue. Reason: $R'(A) > R'(B)$.

**94.** $R(x) = (830+30x)(100-x)$

$R'(x) = (830+30x)(-1)+(100-x)(30)$

$= -830-30x+3000-30x = -60x+2170$

$R'(10) = -60(10)+2170 = 1570$

This tells us that rent should be raised. An 11th rent increase of \$30 (and hence an 11th vacancy) would change revenue by about \$1570.

**95.** $P(x) = 3+\dfrac{70x^2}{x^2+1000}$

$P(20) = 3+\dfrac{70(20)^2}{(20)^2+1000} = 23$ means productivity is 23 units per hour after 20 hours of training and experience.

$P'(x) = 0+\dfrac{(x^2+1000)(140x)-70x^2(2x)}{(x^2+1000)^2} = \dfrac{140x^3+140000x-140x^3}{(x^2+1000)^2} = \dfrac{140000x}{(x^2+1000)^2}$

$P'(20) = \dfrac{140000(20)}{(20^2+1000)^2} \approx 1.4$ means that the 21st hour of training or experience will change productivity by about 1.4 units per hour.

**96.** $q = 10{,}000-50(0.02p^2+500)^{1/2}$ $\quad q' = -25(0.02p^2+500)^{-1/2}(0.04p) = -\dfrac{p}{\sqrt{0.02p^2+500}}$

**97.** $x(p) = \sqrt{p-1}$ $\quad x'(p) = \dfrac{1}{2\sqrt{p-1}}$ $\quad x'(10) = \dfrac{1}{2\sqrt{9}} = \dfrac{1}{6}$

If the price increases \$1, the number of units supplied will increase about 1/6.

**98.** $s(t) = 16+140t+8t^{1/2}$

$v(t) = s'(t) = 140+4t^{-1/2}$

$a(t) = v'(t) = -2t^{-3/2}$; $a(4) = -2(4)^{-3/2} = -\dfrac{2}{8} = -\dfrac{1}{4}$

The acceleration at 4 seconds is $-0.25$ ft/s$^2$.

**99.** $P(x) = 70x-0.1x^2-5500$

$P'(x) = 70-0.2x$; $P'(300) = 70-0.2(300) = 10$

$P''(x) = -0.2$; $P''(300) = -0.2$

The marginal profit is \$10 per unit. The profit on the 301st unit is \$10. The marginal profit decreases at a constant rate of \$0.20 per unit per unit.

**100.** $C(x) = 3x^2 + 6x + 600$

**a.** $C'(x) = 6x + 6 = \overline{MC}$

**b.** When $x = 30$, $C'(30) = 186$.

**c.** If a 31st unit is produced, costs will increase by about \$186.

**101.** $C(x) = 400 + 5x + x^3 \quad \overline{MC} = C'(x) = 5 + 3x^2 \quad C'(4) = 5 + 48 = 53$

It will cost about \$53 for the 5th unit. Also, the 5th unit produced will increase total costs by \$53.

**102.** $R = 40x - 0.02x^2$

**a.** $R' = 40 - 0.04x$

**b.** $\overline{MR} = 0$ when $40 - 0.04x = 0$ or when $x = 1000$.

**103.** $P(x) = 60x - (200 + 10x + 0.1x^2) \quad P'(x) = 50 - 0.2x \quad P'(10) = 50 - 2 = 48$

The next unit sold will have a profit of about \$48.

**104.** $R = 80x - 0.04x^2$

**a.** $R' = 80 - 0.08x$

**b.** At $x = 100$, $R' = 80 - 0.08(100) = 72$.

**c.** Selling the 101st unit is expected to result in \$72 extra revenue.

**105.** $R(x) = \dfrac{60x^2}{2x+1} \quad R'(x) = \dfrac{(2x+1)(120x) - 60x^2(2)}{(2x+1)^2} = \dfrac{120x^2 + 120x}{(2x+1)^2} = \dfrac{120x(x+1)}{(2x+1)^2}$

**106.** $C(x) = 45{,}000 + 100x + x^3 \quad R(x) = 4600x$

$P(x) = R(x) - C(x) = 4600x - (45{,}000 + 100x + x^3) = -x^3 + 4500x - 45{,}000$

$P'(x) = \overline{MP} = -3x^2 + 4500$

**107.** $P(x) = 46x - \left(100 + 30x + \dfrac{x^2}{10}\right) = 16x - 100 - \dfrac{x^2}{10} \quad P'(x) = 16 - \dfrac{x}{5} = 16 - 0.2x$

**108.** Answers are decided from $C'$, $R'$, and $R' - C' = P'$

**a.** At *C*: Tangent line to $R(x)$ has smallest slope at *C*, so $\overline{MR}(C)$ is smallest and the next item at *C* will earn the least revenue.

**b.** At *B*: $R(x) > C(x)$ at both *B* and *C*. Distance between $R(x)$ and $C(x)$ gives the amount of profit and is greatest at *B*.

**c.** At *A*: $\overline{MR}$ greatest at *A* and $\overline{MC}$ least at *A*, as seen from the slopes of the tangents. Hence $\overline{MP}(A)$ is greatest, so the next item at *A* will give the greatest profit.

**d.** At *C*: $\overline{MC}(C) > \overline{MR}(C)$, as seen from the slopes of the tangents. Hence $\overline{MP}(C) < 0$, so the next unit sold reduces profit.

# Chapter 9: Derivatives

## *Chapter 9 Test*

**1.** **a.** $\lim_{x\to -2}\frac{4x-x^2}{4x-8}=\lim_{x\to -2}\frac{x(4-x)}{4(x-2)}=\frac{(-2)(6)}{4(-4)}=\frac{3}{4}$

**b.** $\lim_{x\to\infty}\frac{8x^2-4x+1}{2+x-5x^2}=\lim_{x\to\infty}\frac{8-\frac{4}{x}+\frac{1}{x^2}}{-5+\frac{1}{x}+\frac{2}{x^2}}=-\frac{8}{5}$

**c.** $\lim_{x\to 7}\frac{x^2-5x-14}{x^2-6x-7}=\lim_{x\to 7}\frac{(x-7)(x+2)}{(x-7)(x+1)}=\frac{7+2}{7+1}=\frac{9}{8}$

**d.** $\lim_{x\to -5}\frac{5x-25}{x+5}$ does not exist. (division by zero)

**2.** **a.** $f'(x)=\lim_{h\to 0}\frac{f(x+h)-f(x)}{h}$

**b.** $f'(x)=\lim_{h\to 0}\frac{3(x+h)^2-(x+h)+9-(3x^2-x+9)}{h}=\lim_{h\to 0}\frac{6xh+3h^2-h}{h}=\lim_{h\to 0}(6x+3h-1)=6x-1$

**3.** $f(x)=\frac{4x}{x(x-8)}$ $f(x)$ is not continuous at $x=0$ and $x=8$ as there is division by zero.

**4.** **a.** $B=0.523W-5176;\ B'=0.523$

**b.** $p=9t^{10}-6t^7-17t+23;\ p'=90t^9-42t^6-17$

**c.** $y=\frac{3x^3}{2x^7+11};\ \frac{dy}{dx}=\frac{(2x^7+11)(9x^2)-3x^3(14x^6)}{(2x^7+11)^2}=\frac{-24x^9+99x^2}{(2x^7+11)^2}$

**d.** $f(x)=(3x^5-2x+3)(4x^{10}+10x^4-17)$

$f'(x)=(3x^5-2x+3)(40x^9+40x^3)+(4x^{10}+10x^4-17)(15x^4-2)$

**e.** $g(x)=\frac{3}{4}(2x^5+7x^3-5)^{12};\ g'(x)=9(2x^5+7x^3-5)^{11}(10x^4+21x^2)$

**f.** $y=(x^2+3)(2x+5)^6$

$y'=(x^2+3)\cdot 6(2x+5)^5(2)+(2x+5)^6\cdot 2x$

$=2(2x+5)^5[6(x^2+3)+x(2x+5)]$

$=2(2x+5)^5(8x^2+5x+18)$

**g.** $f(x)=12x^{1/2}-10x^{-2}+17;\ f'(x)=6x^{-1/2}+20x^{-3}=\frac{6}{\sqrt{x}}+\frac{20}{x^3}$

**5.** $y=x^3-x^{-3};\ y'=3x^2+3x^{-4};\ y''=6x-12x^{-5};\ y'''=\frac{d^3y}{dx^3}=6+60x^{-6}$

**6.** $f(x)=x^3-3x^2-24x-10$

**a.** $f'(x)=3x^2-6x-24=3(x-4)(x+2)$

$f(-1)=(-1)^3-3(-1)^2-24(-1)-10=10$

$f'(-1)=3(-1)^2-6(-1)-24=-15$

$y-y_1=m(x-x_1)$

$\to\ y-10=-15(x+1)$

$\to\ y=-15x-5$

**b.** $f'(x)=0$ at $x=4$ and $x=-2$ (from **a**).

$f(4)=-90,\ f(-2)=18$

Points: $(4, -90), (-2, 18)$

**7.** $f(x)=4-x-2x^2$ over $[1,6]$.

Avg rate $=\frac{f(6)-f(1)}{6-1}=\frac{-74-1}{5}=-15$

# Chapter 9: Derivatives

**8.** **a.** $\lim_{x\to 5} f(x) = 2$

**b.** $\lim_{x\to 5} g(x) =$ DNE

**c.** $\lim_{x\to 5^-} g(x) = -4$

**9.** $g(x) = \begin{cases} 6-x \text{ if } x \le -2 \\ x^3 \text{ if } x > -2 \end{cases}$

$g(-2) = 6-(-2) = 8$

$\lim_{x\to -2^-} g(x) = 8 \qquad \lim_{x\to -2^+} g(x) = -8$

$g(x)$ is not continuous at $x = -2$ since the left hand limit does not equal the right hand limit.

**10.** **a.** $R'(x) = 250 - 0.02x$

**b.** $R(72) = 17948.16$; the revenue for 72 units is \$17,948.16. $R'(72) = 248.56$; when 72 units have been produced, the 73rd unit will increase the revenue \$248.56.

**11.** $C(x) = 200x + 10000 \qquad R(x) = 250x - 0.01x^2$

**a.** $P(x) = R(x) - C(x) = -0.01x^2 + 50x - 10,000$

**b.** $\overline{MP} = P'(x) = -0.02x + 50$

**c.** $P'(1000) = 30$. The company will make approximately \$30 on the sale of the next unit.

**12.** $f'(3) \approx \dfrac{f(3) - f(2.999)}{3 - 2.999} = \dfrac{0.104}{0.001} = 104$

**13.** **a.** $f(1) = -5$

**b.** $\lim_{x\to 6} f(x) = -1$

**c.** $\lim_{x\to 3^-} f(x) = 4$

**d.** $\lim_{x\to -4} f(x) =$ DNE

**e.** $\lim_{x\to -\infty} f(x) = 2$

**f.** $f'(4) \approx \dfrac{3}{2}$

**g.** $f'(x) =$ DNE at $x = -4$, 1, 3, and 6.

**h.** $f(x)$ is not continuous at $x = -4$, 3, and 6.

**i.** $f'(-2) <$ average rate $< f'(2)$

**14.** $y = \dfrac{2}{3}x - 8$ is tangent to $f(x)$ at $x = 6$.

**a.** $f'(6) = \dfrac{2}{3}$

**b.** $f(6) = \dfrac{2}{3}(6) - 8 = -4$

**c.** At $x = 6$, the instantaneous rate of change is

**d.** $\dfrac{2}{3}$.

**15.** **a.** *B*; $R(x) - C(x)$ is greatest at this point. $R(x) > C(x)$ at *B*, so there is a profit. Distance between $R(x)$ and $C(x)$ gives the amount of profit.

**b.** $A$; $C(x) - R(x) > 0$

**c.** *A* and *B*; $R'(x) - C'(x) > 0$ at each of these points. Slope of $R(x)$ is greater than the slope of $C(x)$. Hence, $\overline{MR} > \overline{MC}$ and $\overline{MP} > 0$.

**d.** C; $R'(x) - C'(x) < 0$ at this point. Slope of $C(x)$ is greater than the slope of $R(x)$. Hence, $\overline{MC} > \overline{MR}$ and $\overline{MP} < 0$.

# Chapter 10: Applications of Derivatives

## *Exercises 10.1*

**1.** **a.** (1, 5)
**b.** (4, 1)
**c.** (–1, 2)

**3.** **a.** (1, 5)
**b.** (4, 1)
**c.** (–1, 2)

**5.** **a.** Critical values: $x = 3$ and $x = 7$
**b.** Increasing: $3 < x < 7$
**c.** Decreasing: $x < 3$, $x > 7$
**d.** Max at $x = 7$
**e.** Min at $x = 3$

**7.** **a.** $y = 2x^3 - 12x^2 + 6$

$y' = 6x^2 - 24x = 6x(x-4)$

$y' = 0$ if $x = 0$ or 4

Critical values: 0, 4

**b.** From sign chart below:

Relative maximum at 0

Relative minimum at 4

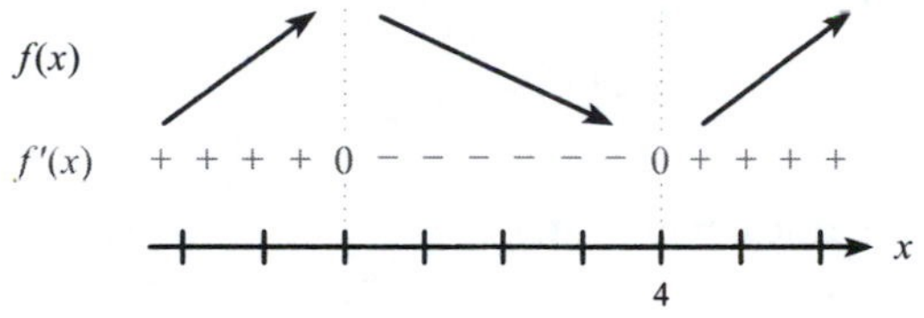


**9.** **a.** $y = 2x^5 + 5x^4 - 11$

$y' = 10x^4 + 20x^3$

$y' = 0$ if $x = 0$ or $x = -2$

Critical values: 0, –2

**b.**

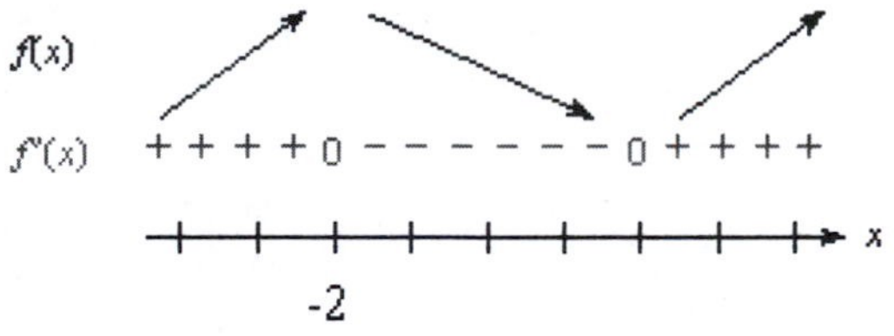


**11.** $y = x^3 - 3x + 4$

**a.** Relative max: $(-1, 6)$

Relative min: $(1, 2)$

**b.** $f'(x) = 3x^2 - 3 = 3(x^2 - 1)$

$f'(x) = 0$ if $3(x+1)(x-1) = 0$ or

$x = -1, 1$.

**c.** At $x = -1$ we have

$y = (-1)^3 - 3(-1) + 4 = 6$.

At $x = 1$ we have $y = 1^3 - 3 \cdot 1 + 4 = 2$

Critical points (–1, 6), (1, 2)

**d.** Yes.

**13.** **a.** HPI: $(-1, -3)$

**b.** $y = x^3 + 3x^2 + 3x - 2$

$f'(x) = 3x^2 + 6x + 3$

Critical values: $3(x^2 + 2x + 1) = 0$ or $x = -1$

**c.** Critical point: (–1, –3)

**d.** Yes. From the graph, (–1, –3) is a horizontal point of inflection.

**15.** $y = \frac{1}{2}x^2 - x$

**a.** $f'(x) = x - 1$

**b.** $f'(x) = 0$ if $x = 1$

**c.** At $x = 1$ we have $y = \frac{1}{2}(1)^2 - 1 = -\frac{1}{2}$.

Critical point: $\left(1, -\frac{1}{2}\right)$

**d.**

| $x$ | 0 | 1 | 2 |
|---|---|---|---|
| $f'(x)$ | – | 0 | + |

Decreasing if $x < 1$

Increasing if $x > 1$

**e.**

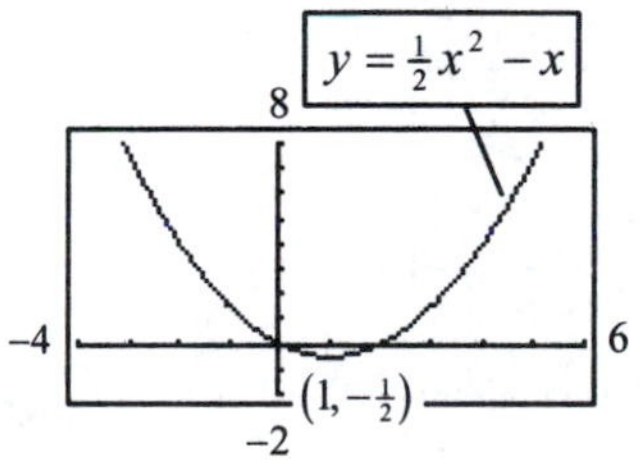


**17.** $y = \frac{1}{3}x^3 + \frac{1}{2}x^2 - 2x + 1$

**a.** $f'(x) = x^2 + x - 2$

**b.** $x^2 + x - 2 = 0$

$(x+2)(x-1) = 0$

Critical values are $x = -2, 1$.

**c.** $f(-2) = -\frac{8}{3} + 2 + 4 + 1 = \frac{13}{3}$

$f(1) = \frac{1}{3} + \frac{1}{2} - 2 + 1 = -\frac{1}{6}$

Critical points: $\left(-2, \frac{13}{3}\right), \left(1, -\frac{1}{6}\right)$

**d.**

| $x$ | | $-2$ | | $1$ | |
|---|---|---|---|---|---|
| $f'(x)$ | + | 0 | − | 0 | + |

Increasing: $x < -2$ and $x > 1$
Decreasing: $-2 < x < 1$

**e.**

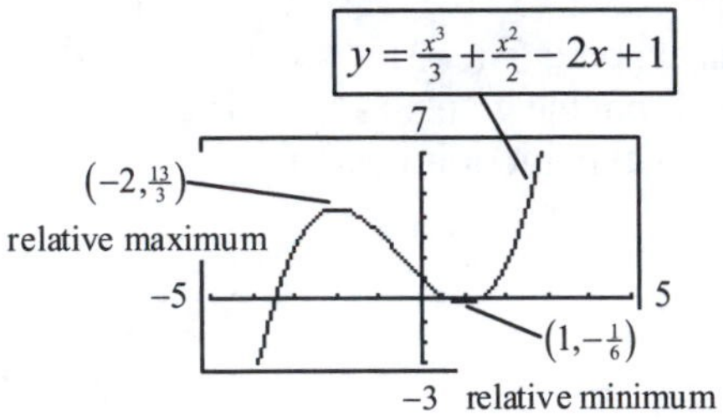


**19.** $y = x^{2/3}$

**a.** $f'(x) = \frac{2}{3x^{1/3}}$

**b.** Critical value is $x = 0$ since $f'(x)$ does not exist at $x = 0$.

**c.** Critical point: (0, 0)

**d.** Decreasing: $x < 0$ ( $f'(x) < 0$ )
Increasing: $x > 0$

**e.**

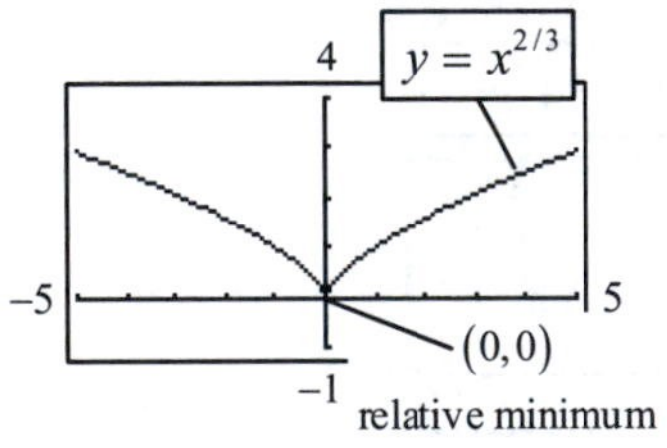


**21. a.** $y' = 0$ if $x = -\frac{1}{2}$; $y'(x) > 0$ if $x < -\frac{1}{2}$;

$y'(x) < 0$ if $x > -\frac{1}{2}$

**b.** $y'(x) = -1 - 2x$

$-1 - 2x = 0$ if $x = -\frac{1}{2}$

Substitute −1 and +1 to verify conclusion.

**23. a.** $y'(x) = 0$ at $x = 0, 3, -3$
$y'(x) > 0$ for $-3 < x < 3,\ x \neq 0$
$y'(x) < 0$ for $x < -3$ and for $x > 3$

**b.** $y'(x) = 3x^2 - \frac{1}{3}x^4 = 3x^2\left(1 - \frac{1}{9}x^2\right)$

Substitute −4, −3, −1, 0, 1, 3, and 4 to verify conclusion.

**25.**

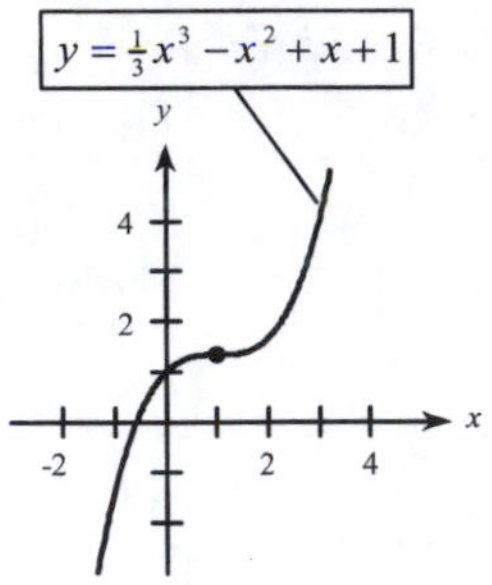


$y = \frac{1}{3}x^3 - x^2 + x + 1$

$\frac{dy}{dx} = x^2 - 2x + 1 = (x-1)^2$

Critical value is $x = 1$.

$f(1) = \frac{1}{3} - 1 + 1 + 1 = \frac{4}{3}$

Critical point: $\left(1, \frac{4}{3}\right)$

| $x$ | | $1$ | |
|---|---|---|---|
| $f'(x)$ | + | 0 | + |

No relative maximum or minimum.
Horizontal point of inflection at $x = 1$.

**27.**

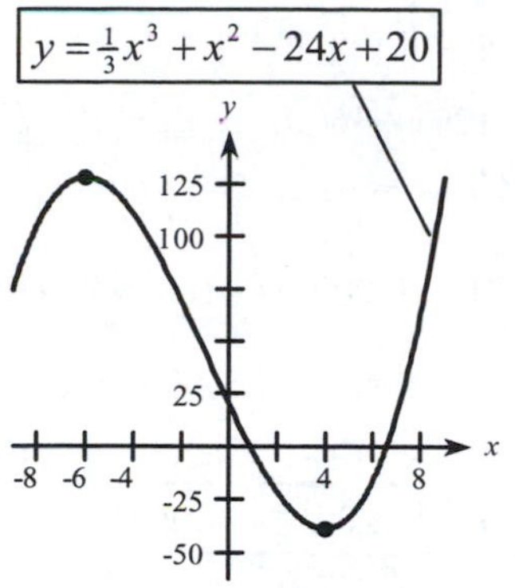


$$y = \frac{1}{3}x^3 + x^2 - 24x + 20$$

$$\frac{dy}{dx} = x^2 + 2x - 24 = (x+6)(x-4)$$

Critical values are $x = -6, 4$.

$$f(-6) = 128;\ f(4) = -\frac{116}{3}$$

Critical points: $(-6, 128)$, $\left(4, -\frac{116}{3}\right)$

| $x$ | | $-6$ | | $4$ | |
|---|---|---|---|---|---|
| $f'(x)$ | + | 0 | − | 0 | + |

Relative maximum at $x = -6$.
Relative minimum at $x = 4$.

**29.**

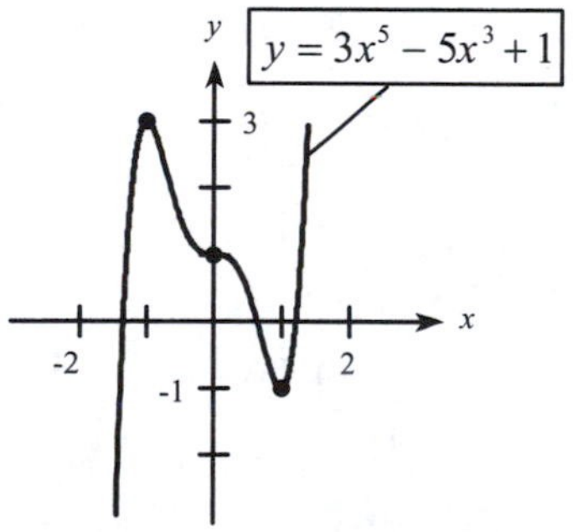


$$y = 3x^5 - 5x^3 + 1$$

$$y' = 15x^4 - 15x^2 = 15x^2(x^2 - 1)$$

Critical values are $x = 0, 1, -1$.
Critical points are $(0, 1)$, $(1, -1)$, $(-1, 3)$

| $x$ | | $-1$ | | $0$ | | $1$ | |
|---|---|---|---|---|---|---|---|
| $f'(x)$ | + | 0 | − | 0 | − | 0 | + |

HPI at $x = 0$.
Relative maximum at $x = -1$.
Relative minimum at $x = 1$.

**31.**

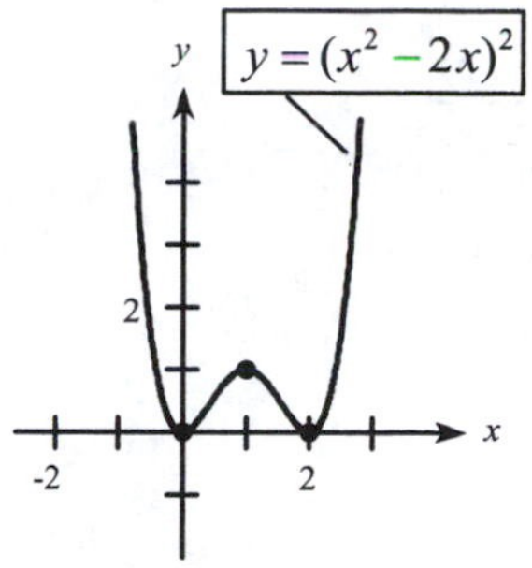


$$y = (x^2 - 2x)^2$$

$$y' = 4x(x-1)(x-2)$$

Critical values at $x = 0, 1, 2$.
$f(0) = 0;\ f(1) = 1;\ f(2) = 0$
Critical points: $(0, 0)$, $(1, 1)$, $(2, 0)$

| $x$ | | $0$ | | $1$ | | $2$ | |
|---|---|---|---|---|---|---|---|
| $f'(x)$ | − | 0 | + | 0 | − | 0 | + |

Relative minimum at $x = 0$, 2.Relative maximum at $x = 1$.

**33.**

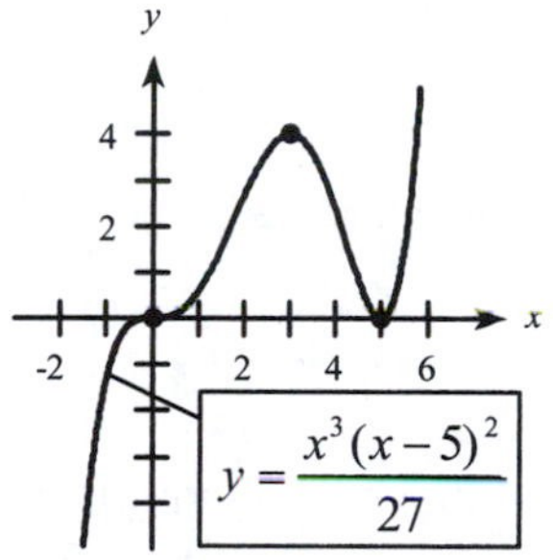


$$y = \frac{1}{27}x^3(x-5)^2$$

$$y' = \frac{5x^2(x-3)(x-5)}{27}$$

Critical values are $x = 0, 5, 3$.
Critical points are $(0, 0)$, $(5, 0)$, $(3, 4)$.

| $x$ | | $0$ | | $3$ | | $5$ | |
|---|---|---|---|---|---|---|---|
| $f'(x)$ | + | 0 | + | 0 | − | 0 | + |

Relative maximum at $x = 3$.
Relative minimum at $x = 5$.
HPI at $x = 0$.

**35.**

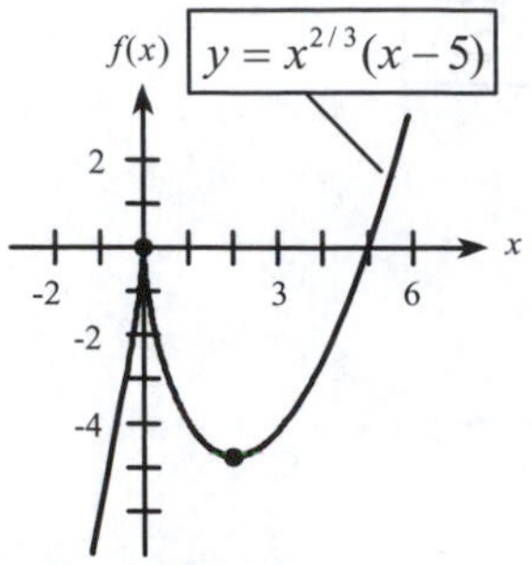


$f(x) = x^{2/3}(x-5)$

$f'(x) = \dfrac{5(x-2)}{3x^{1/3}}$

Critical values are $x = 0, 2$.

Critical points are (0, 0) and $\left(2, -3\sqrt[3]{4}\right)$.

Note that $-3\sqrt[3]{4} \approx -4.8$.

| $x$ | −1 | 0 | 1 | 2 | 3 |
|---|---|---|---|---|---|
| $f'(x)$ | + | * | − | 0 | + |

* means $f'(0)$ is undefined.

Relative maximum at $x = 0$.
Relative minimum at $x = 2$.

**37.** $f(x) = x^3 - 225x^2 + 15000x - 12000$

$f'(x) = 3x^2 - 450x + 15000$

$= 3(x^2 - 150x + 5000)$

$= 3(x-100)(x-50)$

Critical values: $x = 50$ and $x = 100$
Critical points: (50, 300,500), (100, 238,000)
Viewing window: $0 \le x \le 150$, $0 \le y \le 400,000$

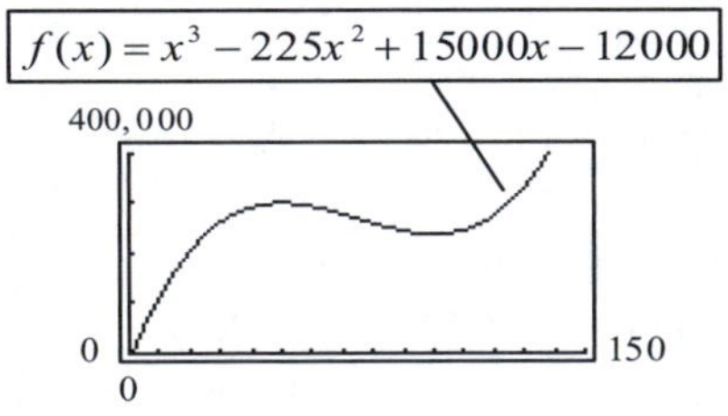


**39.** $f(x) = x^4 - 160x^3 + 7200x^2 - 40000$

$f'(x) = 4x^3 - 480x^2 + 14400x$

$= 4x(x^2 - 120x + 3600)$

$= 4x(x-60)(x-60)$

Critical values: $x = 0$ and $x = 60$
Critical points: (0, –40,000) and (60, 4,280,000)
Viewing window:
$-30 \le x \le 90$, $-500,000 \le y \le 5,000,000$
(There are other windows.)

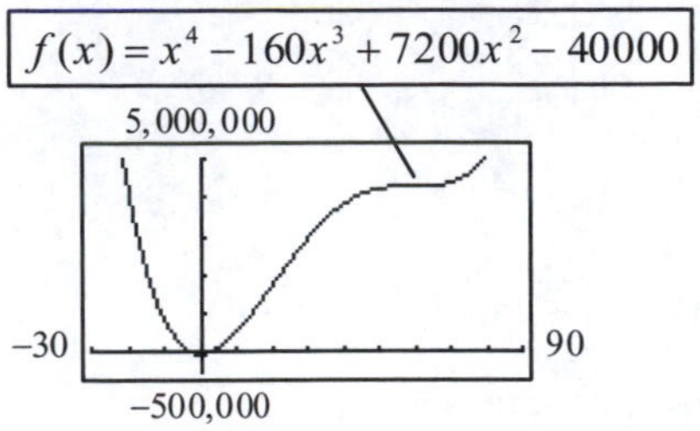


**41.** $y = 7.5x^4 - x^3 + 2$

$y' = 30x^3 - 3x^2 = 3x^2(10x-1)$

Critical values: $x = 0, 0.1$
Critical points: (0, 2) and (0.1, 1.99975)
Viewing window:
$-0.1 \le x \le 0.2$, $1.9997 \le y \le 2.0007$
Viewing windows can vary.

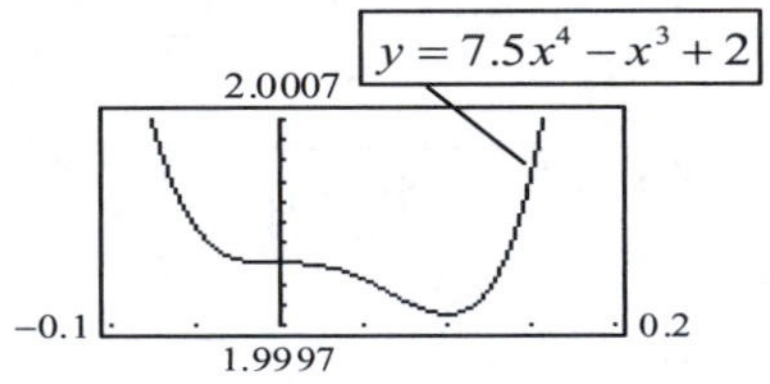


**43.** Possible graph for $f(x)$.

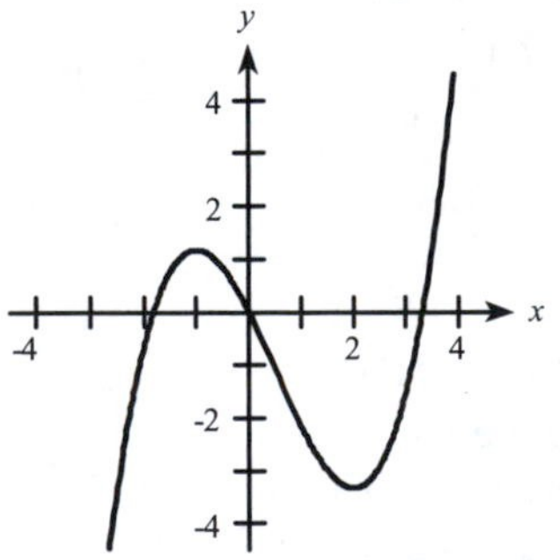


$f'(x) = x^2 - x - 2 = (x-2)(x+1)$

Note: The given graph is $f'(x)$ and NOT $f(x)$.

Critical values: $x = -1$, $x = 2$
$f(x)$ is increasing if $x < -1$ and $x > 2$.
$f(x)$ is decreasing if $-1 < x < 2$.
Thus, rel max at $x = -1$; rel min at $x = 2$.

**45.** Possible graph of $f(x)$.

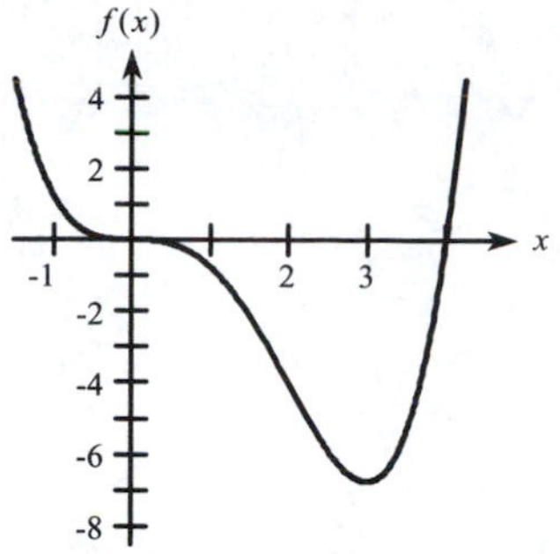


$f'(x) = x^3 - 3x^2 = x^2(x-3)$

Note: The given graph is $f'(x)$ and NOT $f(x)$.

Critical values: $x = 0$, $x = 3$

$f(x)$ is increasing if $x > 3$.

$f(x)$ is decreasing if $x < 3$, $x \neq 0$.

Thus, relative min at $x = 3$; HPI at $x = 0$.

**47.** Graph on the left is $f(x)$.
Graph on the right is $f'(x)$.
If $f'(x) > 0$, then $f(x)$ is increasing (rising).
If $f'(x) < 0$, then $f(x)$ is decreasing (falling).

**49.** $S(t) = 1000 + 400(t+1)^{-1}$ $\quad S'(t) = \dfrac{-400}{(t+1)^2}$

$S'(t) < 0$ for all $t$. Thus, $S(t)$ is always decreasing for $t \geq 0$.

**51.** $P(t) = 27t + 6t^2 - t^3$

$P'(t) = 27 + 12t - 3t^2 = 3(9 + 4t - t^2)$

**a.** $t = \dfrac{-4 \pm \sqrt{16+36}}{-2} = \dfrac{4 \pm 2\sqrt{13}}{2} = 2 \pm \sqrt{13}$

**b.** $t = 2 + \sqrt{13}$ is the only value in the domain of the function.

**c.**

| $t$ | $2+\sqrt{13}$ |
|---|---|
| $P'(t)$ | $+ \quad 0 \quad -$ |

$P(t)$ increasing if $0 \leq t < 2 + \sqrt{13}$.

**d.**

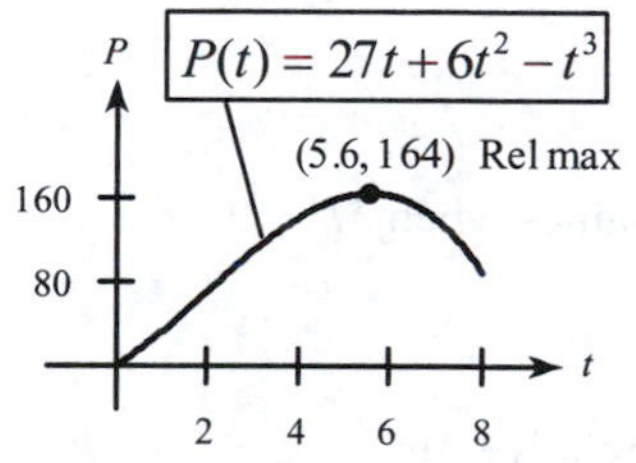


**53.** $\overline{C}(x) = 5000x + 125{,}000x^{-1}$

$\overline{C}'(x) = 5000 - 125{,}000x^{-2}$

**a.** $\overline{C}'(x) = 0$ if $x^2 - 25 = 0$

Critical value: $x = 5$

**b.** $\overline{C}'(x) < 0$ if $x < 5$.

**c.** $\overline{C}'(x) > 0$ if $x > 5$.

**55. a.** Increasing at $x = 150$, decreasing at $x = 350$, changing from increasing to decreasing at $x = 250$.

**b.** $R'(x) > 0$ if $x < 250$

**c.** $x = 250$ units

**57.** $R(t) = \dfrac{50t}{t^2 + 36}$

$R'(t) = \dfrac{(t^2+36)(50) - 50t(2t)}{(t^2+36)^2} = \dfrac{1800 - 50t^2}{(t^2+36)^2}$

**a.** $R'(t) = 0$ if $50(36 - t^2) = 0$ or $t = 6$.
(–6 is not in the domain.)

**b.**

| $t$ | $6$ |
|---|---|
| $R'(t)$ | $+ \quad 0 \quad -$ |

Revenue will increase for 6 weeks.

**59.** $P(t) = \dfrac{13t}{t^2+100} + 0.18$

$P'(t) = \dfrac{(t^2+100)(13) - 13t(2t)}{(t^2+100)^2} = \dfrac{13(100-t^2)}{(t^2+100)^2}$

Domain = $\{t: t \ge 0\}$. Critical value is $t = 10$.

| $t$ | | 10 | |
|---|---|---|---|
| $P'(t)$ | + | 0 | − |

**a.** Ten months after the campaign starts the recognition is a maximum.
**b.** Begins on Jan. 1.

**61.** $C(t) = 0.000286t^3 - 0.0443t^2 + 1.49t - 5.36$

$C'(t) = 0.000858t^2 - 0.0886t + 1.49$

Using the quadratic formula or a graphing utility, we find the critical values when $C'(t) = 0$.

Critical values $\approx 21.1$ or $\approx 82.1$.

The derivative changes sign from positive to negative at $t \approx 21.1$, and for that value of t $C(t) \approx 9.04$. In 2022 the model will achieve a maximum of 9.04 billion subscriberships.

**63. a.** $y = -0.000487x^3 - 0.162x^2 + 19.8x + 388$

**b.** $y' = -0.001461x^2 - 0.324x + 19.8$

Using the quadratic formula or a graphing utility, we find the critical values when $y' = 0$.

Critical value ($x > 0$): $\approx 49.9$

$y'$ changes from + to − at $x \approx 49.9$ (relative max)

$y(49.9) \approx 912$

The maximum size of China's labor pool will be about 912 million in 2020.

**65. a.** $M(t) = 0.0000889t^3 - 0.0111t^2 + 0.237t + 11.5$

**b.** $y' = 0.2832t^2 - 51.88t + 2273.3$

$M' = 0.0002667t^2 - 0.0222t + 0.237$

Using the quadratic formula or a graphing utility, we find the critical values when $M' = 0$.

Critical values: $\approx 12.6, 70.7$

$t \approx 12.6$ gives a maximum and $t \approx 70.7$ gives a minimum

The critical points are $(12.6, 12.9)$, a relative max, and $(70.7, 4.2)$, a relative min.

**c.** No. The years corresponding to the critical points are 2023 and 2081. Despite the fact that the given model is an excellent fit to the data given, it is unrealistic for the model to be accurate for more than 40 years beyond the data set.

## *Exercises 10.2*

**1.** $f(x) = x^3 - 3x^2 + 1$

$f'(x) = 3x^2 - 6x$

$f''(x) = 6x - 6 = 6(x-1)$

**a.** $f''(-2) < 0$ concave downward

**b.** $f''(3) > 0$ concave upward

**3.** $(a, c)$ and $(d, e)$

**5.** $(c, d)$ and $(e, f)$

**7.** $c$, $d$ and $e$

**9.** $f(x) = x^3 - 6x^2 + 5x + 6$

$f'(x) = 3x^2 - 12x + 5$

$f''(x) = 6x - 12$

$f''(x) = 0$ if $x = 2$. We have $f(2) = 0$.

| $x$ | 2 |
|---|---|
| $f''(x)$ | − 0 + |

The point (2, 0) is a point of inflection.
$f(x)$ is concave up if $x > 2$. $f(x)$ is concave down if $x < 2$.

**11.** $y = f(x) = \frac{1}{4}x^4 + \frac{1}{2}x^3 - 3x^2 + 3$

$y' = x^3 + \frac{3}{2}x^2 - 6x$

$y'' = 3x^2 + 3x - 6 = 3(x+2)(x-1)$

$f(-2) = -9 \qquad f(1) = \frac{3}{4}$

| $x$ | | −2 | | 1 | |
|---|---|---|---|---|---|
| $y''$ | + | 0 | − | 0 | + |

The points (–2, –9) and $\left(1, \frac{3}{4}\right)$ are points of inflection.
$f(x)$ is concave up if $x < -2$ and if $x > 1$.
$f(x)$ is concave down if $-2 < x < 1$.

**13.** $y = x^2 - 4x + 2$

$y' = 2x - 4$

$y'' = 2$

Critical value: $x = 2 \quad y'' > 0$ at $x = 2$.
There is no point of inflection.
There is a relative minimum at (2, –2).

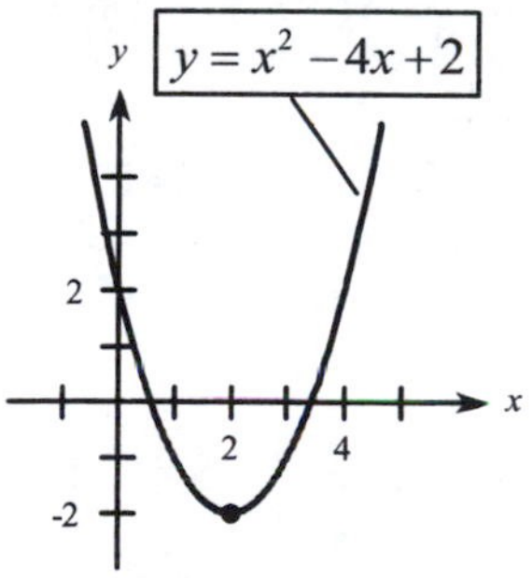


**15.** $y = \frac{1}{3}x^3 - 2x^2 + 3x + 2$

$y' = x^2 - 4x + 3 = (x-3)(x-1)$

$y'' = 2x - 4$

Point of inflection at $x = 2$. (concavity changes)
Relative minimum at $x = 3$ ($y'' > 0$).
Relative maximum at $x = 1$ ($y'' < 0$).

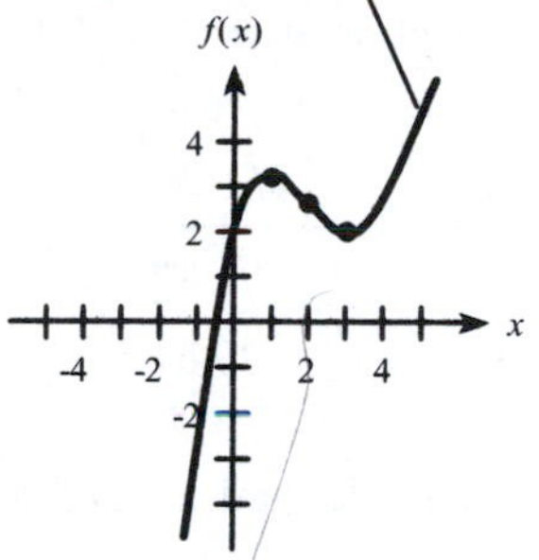


**17.** $y = x^4 - 16x^2$

$y' = 4x^3 - 32x = 4x(x^2 - 8)$

$y'' = 12x^2 - 32 = 4(3x^2 - 8)$

Critical values are $x = 0, \pm 2\sqrt{2}$.

Possible inflection points at $x = \pm\frac{2\sqrt{6}}{3}$

| $x$ | | $-\frac{2\sqrt{6}}{3}$ | | $\frac{2\sqrt{6}}{3}$ | |
|---|---|---|---|---|---|
| $y''$ | + | 0 | − | 0 | + |

IPs: $\left(\frac{-2\sqrt{6}}{3}, -\frac{320}{9}\right)$ and $\left(\frac{2\sqrt{6}}{3}, -\frac{320}{9}\right)$

| $x$ | | $-2\sqrt{2}$ | | 0 | | $2\sqrt{2}$ | |
|---|---|---|---|---|---|---|---|
| $y'$ | − | 0 | + | 0 | − | 0 | + |

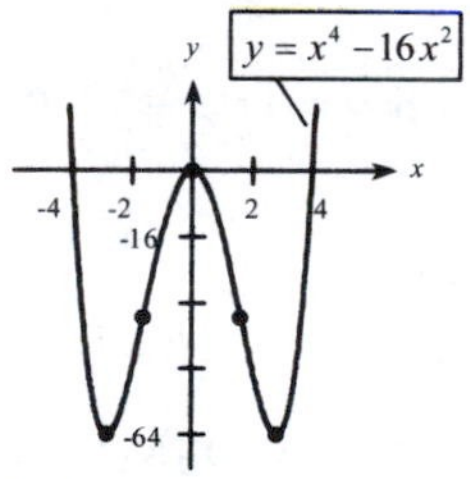


Relative max: (0, 0)
Relative min: $(-2\sqrt{2}, -64)$ and $(2\sqrt{2}, -64)$

**19.**

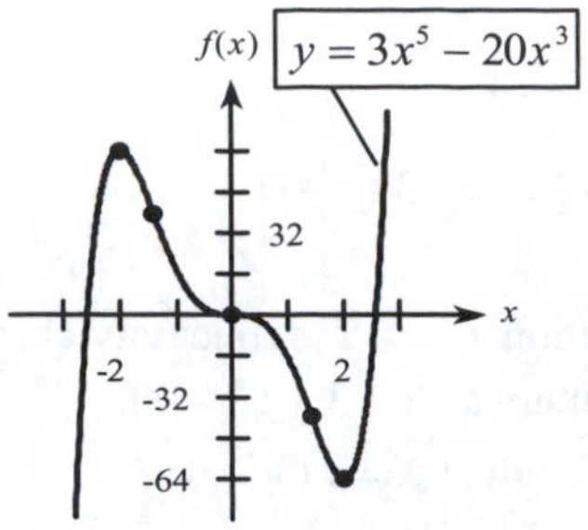


Critical values are $x = -2, 0, 2$.

Possible points of inflection at $x = -\sqrt{2}$, $x = 0$, and $x = \sqrt{2}$.

| $x$ | | $-\sqrt{2}$ | | $0$ | | $\sqrt{2}$ | |
|---|---|---|---|---|---|---|---|
| $f''(x)$ | $-$ | $0$ | $+$ | $0$ | $-$ | $0$ | $+$ |

$(-\sqrt{2}, 39.6)$, $(0, 0)$ and $(\sqrt{2}, -39.6)$ are points of inflection.

$f''(0) = 0$ means second derivative test fails.

$f'(x) < 0$ for values near 0, both to the left and right of 0 means HPI at (0, 0).

$f''(-2) < 0$ means relative maximum at (–2, 64).

$f''(2) > 0$ means relative minimum at (2, –64).

**21.**

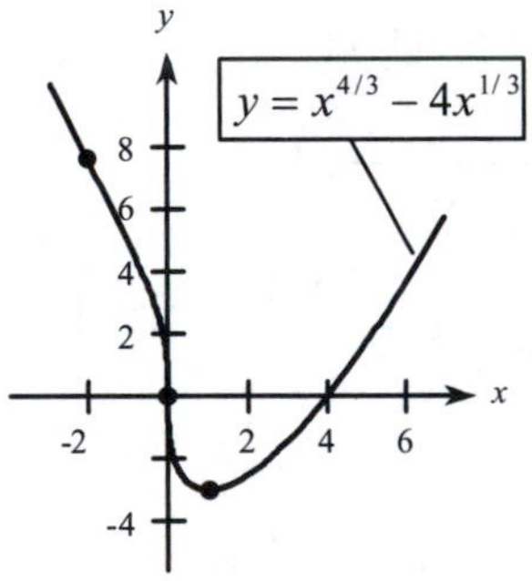


Critical values are $x = 0, 1$.

Possible points of inflection are at $x = -2, 0$.

| $x$ | | $-2$ | | $0$ | |
|---|---|---|---|---|---|
| $f''(x)$ | $+$ | $0$ | $-$ | * | $+$ |

* means $f''(0)$ is undefined.

$(0,0)$ and $(-2, 7.6)$ are points of inflection.

| $x$ | | $0$ | | $1$ | |
|---|---|---|---|---|---|
| $f'(x)$ | $-$ | * | $-$ | $0$ | $+$ |

* means $f'(0)$ is undefined

(1, –3) is a relative minimum.

**23.** **a.** $f''(x) > 0$ if $x < 1$

$f''(x) < 0$ if $x > 1$

$f''(x) = 0$ if $x = 1$

**b.** $f'(x)$ [NOT $f(x)$] has rel max at $x = 1$.

**c.** $f(x) = -\frac{1}{3}x^3 + x^2 + 8x - 12$

$f'(x) = -x^2 + 2x + 8$

$f''(x) = -2x + 2$

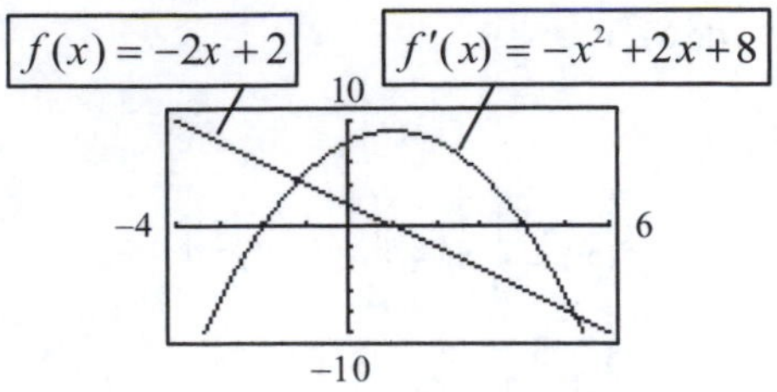


**25.** $f'(x) = 4x - x^2$

**a.** If $x < 2$, $f'(x)$ is increasing.

Thus, $f''(x) > 0$ and $f$ is concave up.

If $x > 2$, $f'(x)$ is decreasing.

Thus, $f''(x) < 0$ and $f$ is concave down.

**b.** Point of inflection at $x = 2$.

**c.** $f''(x) = 4 - 2x$

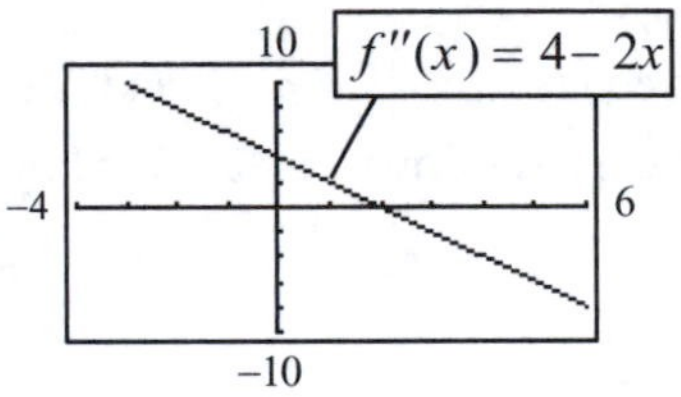


**d.** All graphs have the same basic form.

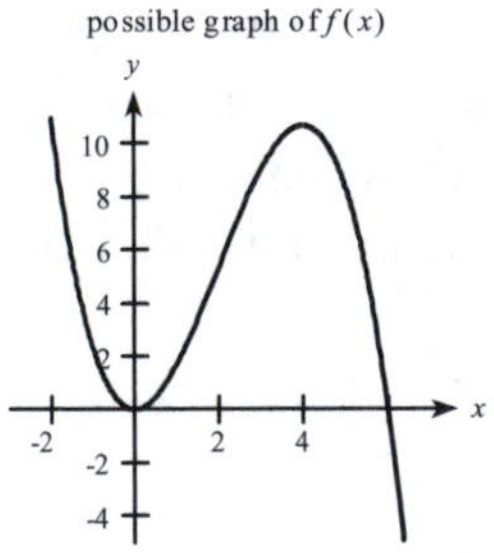


**27.** **a.** G

**b.** C

**c.** F

**d.** H

**e.** I

**29. a.** Concave up if $x < 0$; concave down if $x > 0$; point of inflection at $x = 0$.

**b.** Concave up if $-1 < x < 1$; concave down if $x < -1$ and if $x > 1$; points of inflection at $x = \pm 1$.

**c.** Concave up if $x > 0$; concave down if $x < 0$; point of inflection at $x = 0$.

**31. a.** $P'(t)$ or $\dfrac{dP}{dt}$

**b.** $P''(t) = 0$ at $B$.

**c.** $C$

**33. a.** Concavity changes at $C$.

**b.** $S''(t) > 0$ means concave up or to the right of $C$.

**c.** Yes. $S'(t)$ is the rate of change of sales. $S''(t)$ changes from – to + at $C$.

**35.** $P(t) = 27t + 12t^2 - t^3$

$P'(t) = 27 + 24t - 3t^2 = 3(9-t)(1+t)$

$P''(t) = 24 - 6t = 6(4-t)$

**a.**

| $t$ | | 9 | |
|---|---|---|---|
| $P'(t)$ | + | 0 | – |

Production is maximized at $t = 9$. But the domain is $0 \le t \le 8$. Thus, production is maximized when $t = 8$.

**b.** $P''(t)$ changes signs at $t = 4$. Point of diminishing returns begins at $t = 4$.

**37.** $S(t) = 3(t+3)^{-1} - 18(t+3)^{-2} + 1$

**a.** $S'(t) = \dfrac{-3}{(t+3)^2} + \dfrac{36}{(t+3)^3}$

$= \dfrac{-3(t+3)+36}{(t+3)^3} = \dfrac{27-3t}{(t+3)^3}$

Critical value is $t = 9$.

| $t$ | | 9 | |
|---|---|---|---|
| $S'(t)$ | + | 0 | – |

Sales volume is maximized after 9 days.

**b.** $S''(t) = \dfrac{6}{(t+3)^3} - \dfrac{108}{(t+3)^4}$

$= \dfrac{6(t+3)-108}{(t+3)^4} = \dfrac{6t-90}{(t+3)^4}$

Critical value is $t = 15$.

| $t$ | | 15 | |
|---|---|---|---|
| $S''(t)$ | – | 0 | + |

Point of diminishing returns is $t = 15$ days.

**39. a.** $E(t) = 0.000772t^3 - 0.0760t^2 + 1.83t + 87.3$

$E'(t) = 0.002316t^2 - 0.152t + 1.83$

$E''(t) = 0.004632t - 0.152$

$E'(t) = 0$ at $t \approx 15.9$ or $t \approx 49.7$ using the quadratic formula.

The critical points are $(15.9, 100.3)$ and $(49.7, 85.3)$.

**b.** $E''(15.9) \approx -0.08 < 0$, so there is a relative maximum at $(15.9, 100.3)$.

$E''(49.7) \approx 0.08 > 0$, so there is a relative minimum at $(49.7, 85.3)$.

These mean that over the years 1985 to 2035, the energy use per capita reached a maximum of 100.3% of the 1995 use during 1996 and is expected to reach a minimum of 85.3% of the 1995 use during 2030.

**41. a.** $y = -0.000102x^3 + 0.00884x^2 + 1.43x + 57.9$

**b.** $y' = -0.000306x^2 + 0.01768x + 1.43$

$y'' = -0.000612x + 0.01768$

Setting $y'' = 0$ gives $x \approx 28.9$.

The second derivative predicts that the rate of change of the civilian labor force begins to decrease in 1979.

**43. a.** $C(t) = 0.342t^3 - 6.83t^2 + 105t + 1330$

**b.** $C'(t) = 1.026t^2 - 13.66t + 105$

$C''(t) = 2.052t - 13.66$; $C''(t) = 0$ at $t \approx 6.66$. $C(6.66) \approx 1827$.

These costs increase at a decreasing rate until $t = 6.66$ (2012) and increase at an increasing rate afterward.

## *Exercises 10.3*

**1.** $f(x) = x^3 - 2x^2 - 4x + 2$, $[-1, 3]$

$f'(x) = 3x^2 - 4x - 4 = (3x+2)(x-2)$

$f'(x) = 0$ if $x = 2$ or $x = -\frac{2}{3}$

| $x$ | $-1$ | $-\frac{2}{3}$ | $2$ | $3$ |
|---|---|---|---|---|
| $f(x)$ | $3$ | $\frac{94}{27}$ | $-6$ | $-1$ |

Maximum is $\frac{94}{27}$ at $x = -\frac{2}{3}$.

Minimum is $-6$ at $x = 2$.

**3.** $f(x) = x^3 + x^2 - x + 1$, $[-2, 0]$

$f'(x) = 3x^2 + 2x - 1 = (3x-1)(x+1)$

$f'(x) = 0$ if $x = -1$ or $x = \frac{1}{3}$

| $x$ | $-2$ | $-1$ | $0$ | ($\frac{1}{3}$ not in domain) |
|---|---|---|---|---|
| $f(x)$ | $-1$ | $2$ | $1$ | |

Maximum is 2 at $x = -1$.

Minimum is $-1$ at $x = -2$.

**5. a.** $R(x) = 36x - 0.01x^2$  $R'(x) = 36 - 0.02x$

$R'(x) = 0$ if $36 - 0.02x = 0$ or $x = 1800$ units

$R'(1000) > 0$; $R'(1800) = 0$; $R'(2000) < 0$

Total revenue is maximized at 1800 units.

$R(1800) = \$32,400$ is the maximum revenue.

**b.** $R(1500) = \$31,500$

**7.** $R(x) = 2000x - 20x^2 - x^3$

$R'(x) = 2000 - 40x - 3x^2$

$= (100 + 3x)(20 - x)$

$R'(x) = 0$ if $x = 20$ units  (Domain $\geq 0$)

$R'(10) > 0$; $R'(20) = 0$; $R'(30) < 0$

Total revenue is maximized at 20 units.

$R(20) = 40,000 - 8,000 - 8,000 = \$24,000$

**9.** Let $x$ = number of people above 70.

$R$ = (number of people)(price per person)

$= (70 + x)(100 - 1.00x)$

$= 7000 + 30x - x^2$

$R'(x) = 30 - 2x$

$R'(x) = 0$ if $x = 15$

| $x$ | $0$ | $15$ | $20$ |
|---|---|---|---|
| $R(x)$ | $7000$ | $7225$ | $7200$ |

Therefore, $70 + 15 = 85$ people will maximize revenue.

**11.** Let $x$ = new customers

$R(x) = (4000 + 50x)(110 - x)$

$= 440,000 + 1500x - 50x^2$

Since the price must not be negative, we have $0 \leq x \leq 110$.

$R'(x) = 1500 - 100x$

$R'(x) = 0$ if $x = 15$

| $x$ | $0$ | $15$ | $110$ |
|---|---|---|---|
| $R(x)$ | $440,000$ | $451,250$ | $0$ |

We maximize revenue with a price of \$95.00 (750 new customers). The maximum revenue is $R(15) = \$451,250$.

**13.** $R(x) = 2000x + 20x^2 - x^3$

$\overline{R}(x) = \frac{R(x)}{x} = 2000 + 20x - x^2$

**a.** $\overline{R}'(x) = 20 - 2x$

$\overline{R}'(x) = 0$ if $x = 10$.

$\overline{R}'(0) > 0;\ \overline{R}'(10) = 0;\ \overline{R}'(20) < 0$

$\overline{R}(10) = 2000 + 200 - 100 = \$2100$ is the maximum average revenue.

**b.** $\overline{R} = \frac{R(x)}{(x)}$ $\quad \overline{R}'(x) = \frac{x \cdot R'(x) - R(x) \cdot 1}{x^2}$

$\overline{R}'(x) = 0$ if $x R'(x) - R(x) = 0$ or

$R'(x) = \frac{R(x)}{x}$.

Note: $\frac{R(x)}{x} = \overline{R}(x)$ and $R'(x) = \overline{MR}$.

$\overline{R}(x)$ is a maximum where $\overline{R}(x) = \overline{MR}$.

$2000 + 20x - x^2 = 2000 + 40x - 3x^2$

$2x^2 - 20x = 0$

$2x(x - 10) = 0$

$x = 10$

**15.** $C(x) = 250 + 33x + 0.1x^2$

$\overline{C}(x) = \frac{250}{x} + 33 + 0.1x$

$\overline{C}'(x) = -\frac{250}{x^2} + 0.1$

$-\frac{250}{x^2} + 0.1 = 0$ (multiply both sides by $x^2$)

$-250 + 0.1x^2 = 0$

$0.1x^2 = 250$

$x^2 = 2500$

$x = 50$ (positive only)

$\overline{C}'(x) = 0$ if $x = 50$.

$\overline{C}'(40) < 0;\ \overline{C}'(50) = 0;\ \overline{C}'(60) > 0$

$\overline{C}(50) = \frac{250}{50} + 33 + 0.1(50) = \$43$

$\overline{C} = \$43$ at 50 units of production.

**17.** $C(x) = 810 + 0.1x^2$

$\overline{C}(x) = \frac{810}{x} + 0.1x$

$\overline{C}'(x) = -\frac{810}{x^2} + 0.1$

$-\frac{810}{x^2} + 0.1 = 0$ (multiply both sides by $x^2$)

$-810 + 0.1x^2 = 0$

$0.1x^2 = 810$

$x^2 = 8100$

$x = 90$ (positive only)

$\overline{C}'(x) = 0$ if $x = 90$.

$\overline{C}'(80) < 0;\ \overline{C}'(90) = 0;\ \overline{C}'(100) > 0$

$\overline{C}(90) = \frac{810}{90} + 0.1(90) = \$18$

$\overline{C} = \$18$ at 90 units of production.

**19.** $C(x) = 100(0.02x + 4)^3$

$\overline{C}(x) = \frac{100(0.02x + 4)^3}{x}$

$\overline{C}'(x) = \frac{6x(0.02x + 4)^2 - 100(0.02x + 4)^3}{x^2}$

$= \frac{(0.02x + 4)^2(6x - 100(0.02x + 4))}{x^2}$

$= \frac{(0.02x + 4)^2(4x - 400)}{x^2}$

$\overline{C}'(x) = 0$ if $4x - 400 = 0$ or $x = 100$.

Total number of units is $100(100) = 10{,}000$.

$\overline{C}'(50) < 0;\ \overline{C}'(100) = 0;\ \overline{C}'(150) > 0$

$\overline{C}(100) = \$216$ per 100 units (or \$2.16 per unit).

**21.** $\overline{C}(x) = \frac{C(x)}{x}$ $\quad \overline{C}'(x) = \frac{x \cdot C'(x) - C(x) \cdot 1}{x^2}$

$\overline{C}'(x) = 0$ if $x \cdot C'(x) - C(x) = 0$ or $C'(x) = \frac{C(x)}{x}$

But, $\frac{C(x)}{x} = \overline{C}(x)$ and $C'(x) = \overline{MC}$. Thus, the minimum average cost occurs where $\overline{C}(x) = \overline{MC}$. Note that this is true for all cost functions. Using the $C(x)$ of problem 15:

$\overline{MC} = 13 + 2x = \frac{25}{x} + 13 + x = \overline{C}(x)$

$x^2 = 25$ or $x = 5$

**23.** **a.** A line from $(0,0)$ to $(x, C(x))$ has slope $\frac{C(x)}{x} = \overline{C}(x)$. This is minimized when the line has the least rise. This occurs when the line is tangent to $C(x)$.

**b.** The level is approximately 600 units.

**25.** $P(x) = 5600x + 85x^2 - x^3 - 200,000$

$P'(x) = 5600 + 170x - 3x^2 = (80 - x)(70 + 3x)$

$P'(x) = 0$ if $x = 80$ units, (domain: $x \ge 0$)

$P'(70) > 0$; $P'(80) = 0$; $P'(90) < 0$

$P(80) = \$280,000$ is the maximum profit.

**27.** $P(x) = 4600x - (45,000 + 100x + x^3)$

$= 4500x - x^3 - 45,000$

$P'(x) = 4500 - 3x^2$

$P'(x) = 0$ if $3x^2 = 4500$ or if $x^2 = 1500$ or $x = 10\sqrt{15} \approx 39$ units.

$P'(30) > 0$; $P'(39) \approx 0$; $P'(40) < 0$

$P(39) = \$71,181$ is the maximum profit.

**29.** $P(x) = 250x - 0.01x^2 - (300 + 200x)$

$= -0.01x^2 + 50x - 300$

$P'(x) = -0.02x + 50 \qquad P'(x) = 0$ if $x = 2500$.

Since $P'(x) > 0$ for all $x$ in the domain ($0 \le x \le 1000$), produce the maximum of 1000 units and obtain the maximum profit. That maximum profit is $P(1000) = \$39,700$.

**31.** **a.** B

**b.** B

**c.** B ($P' = 0$ if $R' - C' = 0$)

**d.** $P$ is a max when $P' = 0$ or $\overline{MR} = \overline{MC}$

**33.** Let $x$ = number of vacant units.

$P(x) = (720 + 20x)(50 - x) - 12(50 - x)$

$= 35,400 + 292x - 20x^2$

$P'(x) = 292 - 40x$

$P'(x) = 0$ if $x = 7.3$ units

| $x$ | 0 | 7 | 8 | 50 |
|---|---|---|---|---|
| $P(x)$ | 35,400 | 36,464 | 36,456 | 0 |

Using $x = 7$ units, maximum profit is obtained if rent is $720 + 20(7) = \$860$.

**35.** $C(x) = 45,000 + 100x + x^2$

$R(x) = 1600x$

$P(x) = R(x) - C(x) = -x^2 + 1500x - 45,000$

$P'(x) = -2x + 1500$

$P'(x) = 0$ if $x = 750$, but this is not in the domain.

| $x$ | 0 | 600 |
|---|---|---|
| $P(x)$ | −45,000 | 495,000 |

Maximum profit is at 600 units due to limited production. $P(600) = \$495,000$

**37.** $R(x) = p \cdot x = 600x - \frac{1}{2}x^2$

$C(x) = \overline{C}(x) \cdot x = 300x + 2x^2$

$P(x) = \left(600x - \frac{1}{2}x^2\right) - (300x + 2x^2)$

$= 300x - \frac{5}{2}x^2$

$P'(x) = 300 - 5x$

$P''(x) = -5 < 0$ for all $x$.

$P'(x) = 0$ if $x = 60$.

**a.** Maximum profit is at $x = 60$.

**b.** Selling price is $p = 600 - \frac{1}{2} \cdot 60 = \$570$.

**c.** Maximum profit is $P(60) = \$9000$.

**39.** $R(x) = p \cdot x = 1960x - \frac{1}{3}x^3$

$C(x) = \overline{C}(x) \cdot x = 1000x + 2x^2 + x^3$

$P(x) = 960x - 2x^2 - \frac{4}{3}x^3, 0 \le x \le 10$

1000 units = 10 hundreds

$P'(x) = 960 - 4x - 4x^2$

$= 4(16 + x)(15 - x)$

$P'(x) = 0$ if $x = 15$, but this is not in the domain.

**a.** Maximum profit at 1000 units ($x = 10$) due to limited production.

**b.** $P(10) = \$8066.67$

**41.** $R(x) = p \cdot x = 120x - 0.015x^2$

$C(x) = \overline{C}(x) \cdot x$

$= 10,000 + 60x - 0.03x^2 + 0.00001x^3$

$P(x) = -10,000 + 60x + 0.015x^2 - 0.00001x^3$

$P'(x) = 60 + 0.03x - 0.00003x^2$

$P'(x) = 0$ if

$$x = \frac{-0.03 \pm \sqrt{0.0009 + 0.0072}}{-0.00006}$$

$$= \frac{-0.03 \pm 0.09}{-0.00006}$$

$= 2000$ (positive value only)

$P'(1090) > 0;\ P'(2000) = 0;\ P'(2010) < 0$

Maximum profit is at $x = 2000$ units.

Price is $p = 120 - 0.015(2000) = \$90$ per unit.

Maximum profit is $P(2000) = \$90,000$.

**43. a.** $R(x) = 2x - 0.0004x^2$

$P(x) = 1.8x - 0.0005x^2 - 800$

**b.** $p = \$1.28, x = 1800, p(1800) = \$820$

**c.** $p = \$1.25, x = 1875, p(1875) = \$817.19$

Coastal would still provide sodas; profits almost the same.

**45. a.** $y = 0.000252x^3 - 0.0279x^2 + 1.63x + 2.16$

**b.** $y' = 0.000756x^2 - 0.0558x + 1.63$

$y'' = 0.001512x - 0.0558$ $y''$ changes signs (equals zero) at

$$x = \frac{0.0558}{0.001512} \approx 36.90.$$

The point of inflection is $(36.9, 37.0)$.

**c.**

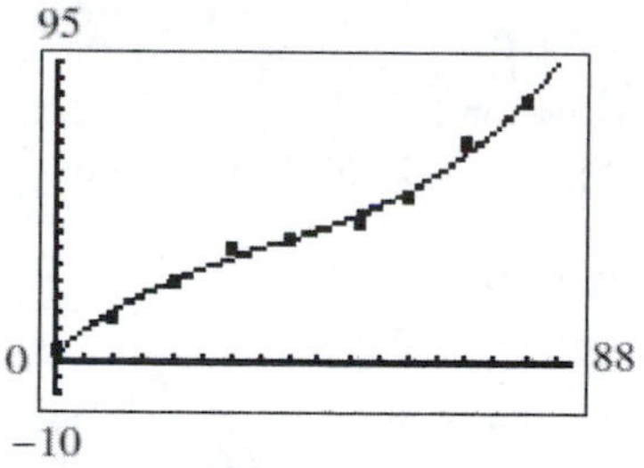


The rate of change of the number of beneficiaries was descreasing until 1987, after which the rate has been increasing. Hence, since 1987 the number of beneficiaries has been increasing at an increasing rate.

**47. a.** Absolute max was reached in May.

**b.** Absolute min was reached in September. This was triggered by the 9-11 terrorist attack.

**49. a.** From the graph, the absolute maximum for $f(t)$ is 16.5.

**b.** From the graph, the absolute minimum for $f(t)$ is 1.9.

**c.** The graph suggests that Social Security taxes might rise in the early twenty-first century in order that fewer people can pay in enough to cover those already retired.

## *Exercises 10.4*

**1.** $S_1 = 30 + 20x_1 - 0.4x_1^2$

$S_2 = 20 + 36x_2 - 1.3x_2^2$

**a.** $S_1' = 20 - 0.8x_1$

$S_1' = 0$ if $x_1 = \$25$ million

$S_2' = 36 - 2.6x_2$

$S_2' = 0$ if $x_2 \approx \$13.846$ million

**b.** Total money needed is $25 + 13.846 = \$38.846$ million.

**3.** $P(x) = 200x - x^2$

$P'(x) = 200 - 2x$

$P'(x) = 0$ if $x = 100$

$P'(90) > 0;\ P'(100) = 0;\ P'(110) < 0$

Thus, 100 trees per acre will maximize the profit.

**5.** $y = 70t + \frac{1}{2}t^2 - t^3$

$y' = 70 + t - 3t^2 = (5 - t)(14 + 3t)$

**a.** $y' = 0$ if $t = 5$.

$y'(4) > 0$

$y'(5) = 0$

$y'(6) < 0$

Production is maximized after 5 hours.

**b.** $y(5) = 350 + 12.5 - 125 = 237.5$

**7.** $E(p) = 10{,}000p - 100p^2$

$E'(p) = 10{,}000 - 200p$

$E'(p) = 0$ if $p = 50$

$E'(40) > 0,\ E'(50) = 0,\ E'(60) < 0$

Expenditure is greatest at $p = \$50$.

**9.** $R = \frac{c}{2}m^2 - \frac{m^3}{3}$

$R' = cm - m^2$

$R' = 0$ if $m = 0$ or $m = c$.

$R'(0) = 0; R'\left(\frac{c}{2}\right) > 0; R'(c) = 0; R'(2c) < 0$

Reaction is a max when $m = c$.

**11.** $S(t) = \frac{200t}{(t+1)^2}$

$S'(t) = \frac{(t+1)^2(200) - 200t \cdot 2 \cdot (t+1)}{(t+1)^4}$

$= \frac{200(1-t)}{(t+1)^3}$

$S'(t) = 0$ if $t = 1$

$S'(0) > 0,\ S'(1) = 0,\ S'(2) < 0$

Sales are maximized after one week.

**13.** $p(t) = \frac{6.4t}{t^2 + 64} + 0.05$

$p'(t) = \frac{(t^2+64)(6.4) - (6.4t)(2t)}{(t^2+64)^2}$

$= \frac{6.4(64 - t^2)}{(t^2+64)^2}$

$p'(t) = 0$ if $t = 8$.

$p'(7) > 0;\ p'(8) = 0;\ p'(9) < 0$

Awareness is maximized after 8 days.
Maximum percentage that identify 2 defendants is $p(8) = 45\%$.

**15.** Quantity being minimized: $F = 4x + 3y$

Another equation: $A = xy = 1200$ or $y = \frac{1200}{x}$

Substituting: $F(x) = 4x + \frac{3600}{x}$

$F'(x) = 4 - \frac{3600}{x^2}$

$F'(x) = 0$ if $x = 30$.

$F'(29) < 0;\ F'(30) = 0;\ F'(31) > 0$

Fence needed is $F(30) = 120 + 120 = 240$ feet.

**17.** Quantity being minimized: $C = 20x + 5(2y)$

Another equation: $A = xy = 45{,}000$ or

$y = \frac{45{,}000}{x}$

Substituting: $C(x) = 20x + \frac{450{,}000}{x}$

$C'(x) = 20 - \frac{450{,}000}{x^2}$

$C'(x) = 0$ if $x^2 = 22{,}500$ or $x = 150$

$C'(140) < 0; C'(150) = 0; C'(160) > 0$

Then $y = \frac{45{,}000}{150} = 300$.

Dimensions are 150 ft for fence parallel to the river and 300 ft for each of the other sides.

**19.** Quantity being maximized: $A = xy$

Another equation:

$800 = 10(2x) + 10(2y) + 20(2y)$

$800 = 20x + 60y$ or $x = 40 - 3y$

Substituting: $A(y) = y(40 - 3y) = 40y - 3y^2$

$A'(y) = 40 - 6y = 0$ if $y = \frac{20}{3}$

$A'(6) > 0; A'\left(\frac{20}{3}\right) = 0; A'(7) < 0$

Then $x = 40 - 3\left(\frac{20}{3}\right) = 20$.

Dimensions are $\frac{20}{3}$ ft for the side parallel to the dividers and 20 ft for the other side.

**21.**

$$256 = x(2x)h$$
$$h = \frac{128}{x^2}$$
$$C = 10(x)(2x) + 5(x)(2x) + 5(2)(h)2x + 5(2x)(h)$$
$$= 30x^2 + 30xh$$
$$C(x) = 30x^2 + 30x\left(\frac{128}{x^2}\right) = 30x^2 + \frac{3840}{x}$$
$$C'(x) = 60x - \frac{3840}{x^2}$$
$C'(x) = 0$ if $60x^3 = 3840$ or
$x^3 = 64$ or $x = 4$
$C'(3) < 0;\ C'(4) = 0;\ C'(5) > 0$

Minimum cost is at $x = 4$. Then $h = \frac{128}{16} = 8$.

Dimensions are 8" by 4" by 8" (high).

**23.** $C(x) = \frac{1,500,000}{x}(600) + 1,500,000(15) + \frac{x}{2}\cdot 2$

$$C'(x) = \frac{-900\times 10^6}{x^2} + 1$$
$C'(x) = 0$ if $x^2 = 900\times 10^6$ or $x = 30,000$.
$C'(1) < 0$, $C'(30,000) = 0, C'(40,000) > 0$
30,000 items should be produced in each run.

**25.** $C(x) = \frac{150,000}{x}(360) + 150,000(7) + \frac{x}{2}\cdot\frac{3}{4}$

$$C'(x) = \frac{-54,000,000}{x^2} + \frac{3}{8}$$
$C'(x) = 0$ if $x^2 = \frac{8}{3}(54,000,000)$ or $x = 12,000$
$C'(11,000) < 0;\ C'(12,000) = 0;\ C'(13,000) > 0$
12,000 items should be produced in each run.

**27.** $V(x) = x(12 - 2x)^2$

$$V'(x) = x(2)(12 - 2x)(-2) + (12 - 2x)^2(1)$$
$$= (12 - 2x)(12 - 6x)$$
$V'(x) = 0$ if $x = 2$
Domain: $0 < x < 6$
$V'(1) > 0;\ V'(2) = 0;\ V'(3) < 0$
Volume is a maximum at $x = 2$.

**29.** Let $x$ = time in weeks to pick oranges.
Quantity being maximized:

$$R(x) = (24 - 1.5x)(5 + 0.5x), 0 \le x \le 5$$
$$R'(x) = (24 - 1.5x)(0.5) + (5 + 0.5x)(-1.5)$$
$$= 4.5 - 1.50x$$
$R'(x) = 0$ if $4.5 - 1.5x = 0$ or $x = \frac{4.5}{1.5} = 3$

| $x$ | 0 | 3 | 5 |
|---|---|---|---|
| $R(x)$ | 120 | 126.75 | 123.75 |

Thus, 3 weeks from now oranges should be picked.

**31.** Let $x$ = number of plates and $y$ = number of impressions.
Quantity being minimized:
$$C = 20x + 125\left(\frac{y}{1000}\right) = 20x + 0.125y$$
Another equation: $xy = 100,000$ or $y = \frac{100,000}{x}$

Substituting: $C(x) = 20x + \frac{12,500}{x}$
$$C'(x) = 20 - \frac{12,500}{x^2}$$
$C'(x) = 0$ if $x = 25$.
$C'(24) < 0;\ C'(25) = 0;\ C'(26) > 0$
Thus, 25 plates will minimize the cost.

**33.** $R(t) = -0.00044t^3 + 0.0042t^2 + 0.52t + 19$
$R'(t) = -0.00132t^2 + 0.0084t + 0.52$

**a.** Using the quadratic formula or a graphing utility, we find (for $t > 0$) that $R'(t) = 0$ at $t \approx 23.28$. $R'$ changes from $+$ to $-$ at 23.28, so the model predicts that the onshore oil reserves in the lower 48 states will reach a maximum in 2034.

| $t$ | 1 | 23.3 | 30 |
|---|---|---|---|
| $R(t)$ | 19.5 | 27.8 | 26.5 |

The predicted maximum is $R = 27.8$ billion barrels. This is the absolute maximum.

**b.** The absolute min is $(1, 19.5)$.

**c.** $R''(t) = -0.00264t + 0.0084$
$R''(t) = 0$ at $t \approx 3.2$. The rate reaches its maximum at $t \approx 3.2$ (in 2014); $R''(t)$ changes from positive to negative at $t \approx 3.2$. After this $R(t)$ increases at a decreasing rate until its maximum

# Chapter 10: Applications of Derivatives

## *Exercises 10.5*

**1.** **a.** $x = 2$

**b.** $\lim\limits_{x\to\infty} f(x) = 1$

**c.** $\lim\limits_{x\to-\infty} f(x) = 1$

The denominator of *f* is 0 when $x = 2$, but the numerator is not 0 when $x = 2$, so $x = 2$ is a vertical asymptote.

$$\lim_{x\to\infty}\frac{x-4}{x-2} = \lim_{x\to\infty}\frac{1-\frac{4}{x}}{1-\frac{2}{x}} = \frac{1-0}{1-0} = 1$$

$$\lim_{x\to-\infty}\frac{x-4}{x-2} = \lim_{x\to-\infty}\frac{1-\frac{4}{x}}{1-\frac{2}{x}} = \frac{1-0}{1-0} = 1$$

**d.** Horizontal asymptote: $y = 1$

**3.** $f(x) = \dfrac{3(x^4 + 2x^3 + 6x^2 + 2x + 5)}{(x^2-4)^2}$

**a.** $x = \pm 2$

**b.** $\lim\limits_{x\to\infty} f(x) = 3$

**c.** $\lim\limits_{x\to-\infty} f(x) = 3$

When $x = \pm 2$, the denominator of *f* is 0, but the numerator is not, so $x = \pm 2$ are equations of vertical asymptotes.

$$\lim_{x\to\infty}\frac{3(x^4 + 2x^3 + 6x^2 + 2x + 5)}{(x^2-4)^2}$$

$$= 3\lim_{x\to\infty}\frac{1+\frac{2}{x}+\frac{6}{x^2}+\frac{2}{x^3}+\frac{5}{x^4}}{1-\frac{8}{x}+\frac{16}{x^2}} = 3$$

Likewise, $\lim\limits_{x\to-\infty} f(x) = 3$.

**d.** Horizontal asymptote: $y = 3$

**5.** $y = \dfrac{2x}{x-3}$

$$\lim_{x\to\infty}\frac{2x}{x-3} = \lim_{x\to\infty}\frac{2}{1-\frac{3}{x}} = 2$$

Horizontal asymptote: $y = 2$

Vertical asymptote: $x = 3$

**7.** $y = \dfrac{x+1}{(x+2)(x-2)}$

$$\lim_{x\to\infty}\frac{x+1}{x^2-4} = \lim_{x\to\infty}\frac{\frac{1}{x}-\frac{1}{x^2}}{1-\frac{4}{x^2}} = \frac{0}{1} = 0$$

Horizontal asymptote: $y = 0$

Vertical asymptotes: $x = \pm 2$

**9.** $y = \dfrac{3x^3-6}{x^2+4}$

Vertical asymptote: None since $x^2 + 4 \neq 0$

Horizontal asymptote: None since degree of numerator > degree of denominator.

**11.** $f(x) = \dfrac{2x+2}{x-3}$

$$f'(x) = \frac{(x-3)2-(2x+2)}{(x-3)^2} = \frac{-8}{(x-3)^2}$$

VA; $x = 3$

$$\lim_{x\to\infty} f(x) = \lim_{x\to\infty}\frac{2+\frac{2}{x}}{1-\frac{3}{x}} = 2$$

HA : $y = 2$

No critical values. Thus, there can be no maximum or minimum points.

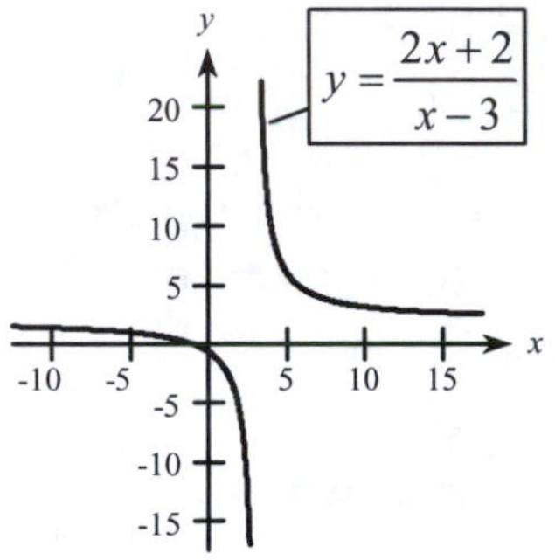

**13.** $y = \dfrac{x^2+4}{x}$

$$\frac{dy}{dx} = \frac{x(2x)-(x^2+4)}{x^2} = \frac{(x+2)(x-2)}{x^2}$$

Critical values: $x = -2, 2$

| $x$ | | $-2$ | | $0$ | | $2$ | |
|---|---|---|---|---|---|---|---|
| $y'$ | $+$ | $0$ | $-$ | $*$ | $-$ | $0$ | $+$ |

* means $y'$ is undefined at $x = 0$. In fact, 0 is not in the domain.

Relative maximum point: $(-2, -4)$

Relative minimum point: $(2, 4)$

$y > 0$ if $x > 0$ and $y < 0$ if $x < 0$.

VA: $x = 0$

HA: None (Deg $x^2+4$ > Deg $x$)

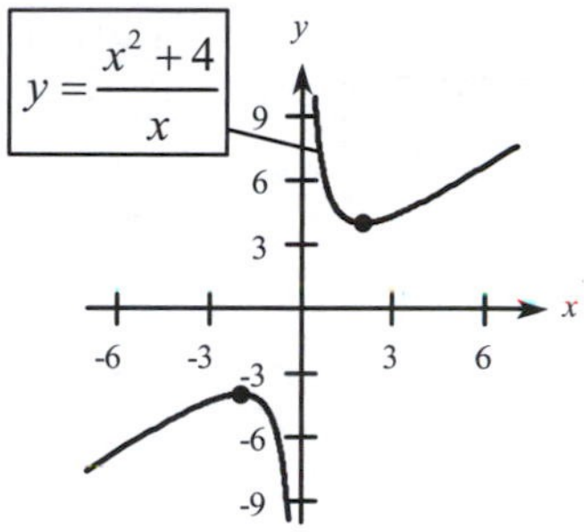


**15.** $y = \dfrac{27x^2}{(x+1)^3}$

$$y' = \frac{(x+1)^3 \cdot 54x - 27x^2 \cdot 3(x+1)^2}{(x+1)^6}$$

$$= \frac{27x(x+1)^2[2(x+1)-3x]}{(x+1)^6} = \frac{27x(2-x)}{(x+1)^4}$$

VA: $x = -1$

$$\lim_{x\to\infty} y = \lim_{x\to\infty} \frac{27x^2}{x^3+3x^2+3x+1}$$

$$= \lim_{x\to\infty} \frac{\frac{27}{x}}{1+\frac{3}{x}+\frac{3}{x^2}+\frac{1}{x^3}} = \frac{0}{1} = 0$$

HA: $y = 0$

Critical values: $x = 0, 2$

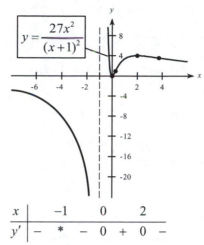

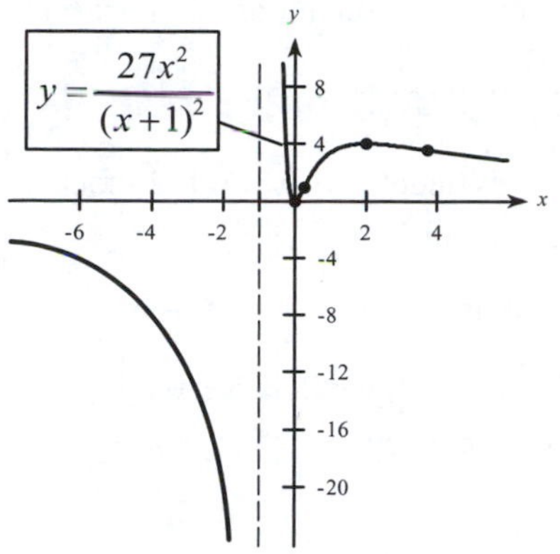


| $x$ | | $-1$ | | $0$ | | $2$ | |
|---|---|---|---|---|---|---|---|
| $y'$ | $-$ | $*$ | $-$ | $0$ | $+$ | $0$ | $-$ |

*means $y'$ is undefined at $x = -1$. In fact, $-1$ is not in the domain.

Relative minimum point: (0, 0)

Relative maximum point: (2, 4)

**17.** $f(x) = \dfrac{16x}{x^2+1}$

$$f'(x) = \frac{(x^2+1)16 - 16x(2x)}{(x^2+1)^2} = \frac{16-16x^2}{(x^2+1)^2}$$

Critical values: $x = -1, 1$

| $x$ | | $-1$ | | $1$ | |
|---|---|---|---|---|---|
| $f'(x)$ | $-$ | $0$ | $+$ | $0$ | $-$ |

Relative maximum point: (1, 8)

Relative minimum point: (–1, –8)

VA: None since $x^2+1 \neq 0$. HA: $y = 0$ since degree $(x^2+1)$ > degree $(16x)$.

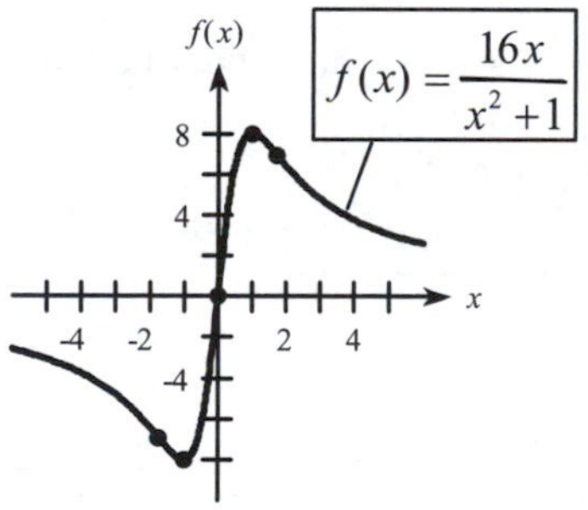

**19.** Critical value is $x = -1$.
($x = 1$ is not in the domain of the function.)

| $x$ | | $-1$ | | $1$ | |
|---|---|---|---|---|---|
| $y'$ | $-$ | $0$ | $+$ | $*$ | $-$ |

$*$ means $y'$ is undefined at $x = 1$. In fact, 1 is not in the domain.

Relative minimum point: $\left(-1, -\frac{1}{4}\right)$

$y > 0$ if $x > 0$ and $y < 0$ if $x < 0$.
(0, 0) is an $x$-intercept.

$\lim_{x\to\infty} \frac{x}{(x-1)^2} = 0$

HA: $y = 0$, VA: $x = 1$

Point of inflection at $x = -2$ ($y''$ changes signs)

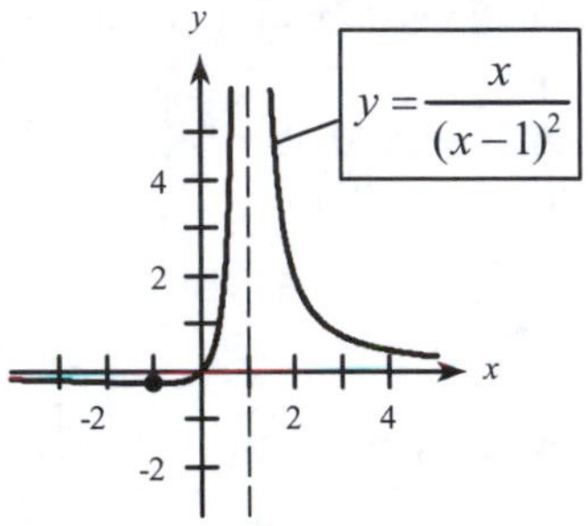


**21.** VA: $x = 3$, Critical values: $x = 2, 4$

| $x$ | | $2$ | | $3$ | | $4$ | |
|---|---|---|---|---|---|---|---|
| $f'(x)$ | $+$ | $0$ | $-$ | $*$ | $-$ | $0$ | $+$ |

$*$ means $f'(3)$ is undefined. In fact, 3 is not in the domain.
Relative maximum at $x = 2$.
Relative minimum at $x = 4$.
$y'' < 0$ when $x < 3$, and
$y'' > 0$ when $x > 3$ but $x = 3$ is not in the domain of the function, so there are no inflection points.

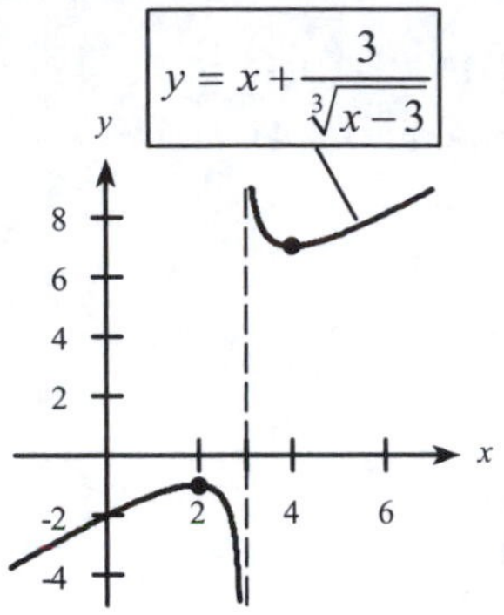


**23.** VA: $x = 0$

$$\lim_{x\to\infty} f(x) = \lim_{x\to\infty} \frac{9(x-2)^{2/3}}{(x^3)^{2/3}} = \lim_{x\to\infty} 9\left(\frac{x-2}{x^3}\right)^{2/3} = \lim_{x\to\infty} 9\left(\frac{\frac{1}{x^2}-\frac{2}{x^3}}{1}\right)^{2/3} = 9\cdot\left(\frac{0}{1}\right)^{2/3} = 0$$

HA: $y = 0$, Critical values: $x = 3$ and $x = 2$

| $x$ | | $0$ | | $2$ | | $3$ | |
|---|---|---|---|---|---|---|---|
| $f'(x)$ | $+$ | $*$ | $-$ | $*$ | $+$ | $0$ | $-$ |

$*$ means $f'(0)$ and $f'(2)$ are undefined. In fact, 0 is not in the domain.
Relative minimum: (2, 0)
Relative maximum: (3, 1)
Possible inflection points: $x = 2$ and

$$x = \frac{42 \pm \sqrt{42^2 - 4(7)(54)}}{2(7)} = \frac{42 \pm 6\sqrt{7}}{14} = 3 \pm \frac{3\sqrt{7}}{7} \approx 1.87 \text{ and } 4.13$$

| $x$ | | $0$ | | $3-\frac{3\sqrt{7}}{7}$ | | $2$ | | $3+\frac{3\sqrt{7}}{7}$ | |
|---|---|---|---|---|---|---|---|---|---|
| $f''(x)$ | $+$ | $*$ | $+$ | $0$ | $-$ | $*$ | $-$ | $0$ | $+$ |

* means $f''(0)$ and $f''(2)$ are undefined.

Points of inflection at approximately (1.87, 0.68) and (4.13, 0.87)

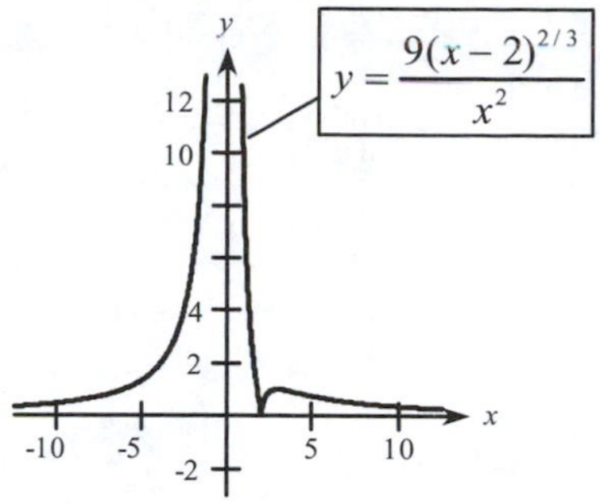


**25. a.** HA: approximately $y = -2$; VA: approximately $x = 4$

**b.** $f(x) = \dfrac{9x}{17-4x} = \dfrac{9}{\frac{17}{x}-4}$, HA: $y = -\dfrac{9}{4}$; VA: $x = \dfrac{17}{4}$

**27. a.** HA: approximately $y = 2$; VA: approximately $x = \pm 2.5$

**b.** $f(x) = \dfrac{20x^2+98}{9x^2-49} = \dfrac{20+\frac{98}{x^2}}{9-\frac{49}{x^2}}$, HA: $y = \dfrac{20}{9}$; VA: $x = \pm\dfrac{7}{3}$

**29. a.** Standard Viewing Window

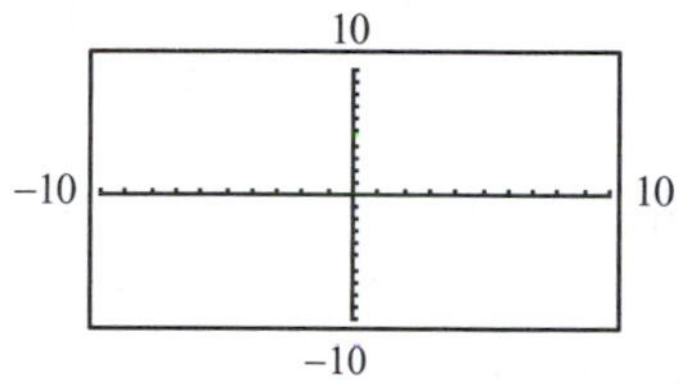


**b.** $f(x) = \dfrac{x+25}{x^2+1400}$

HA: $y = 0$

degree $(x^2+1400)$ > degree $(x+25)$

$$f'(x) = \frac{(x^2+1400)\cdot 1-(x+25)2x}{(x^2+1400)^2}$$

$$= \frac{(70+x)(20-x)}{(x^2+1400)^2}$$

| $x$ | | −70 | | 20 | |
|---|---|---|---|---|---|
| $f'(x)$ | − | 0 | + | 0 | − |

Relative minimum at $x = -70$. Relative maximum at $x = 20$.

**c.**

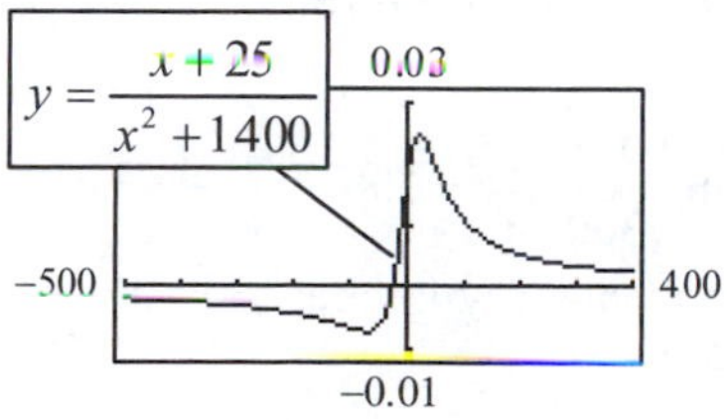


**31. a.** Standard Viewing Window

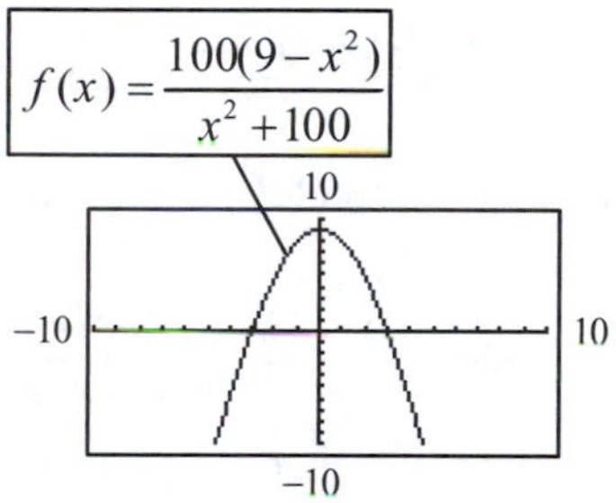


**b.** $f(x) = \dfrac{100(9-x^2)}{x^2+100}$

$$\lim_{x\to\infty} f(x) = \lim_{x\to\infty} \frac{100\left(\frac{9}{x^2}-1\right)}{1+\frac{100}{x^2}} = -100$$

HA: $y = -100$

$$f'(x) = 100\left[\frac{(x^2+100)(-2x)-(9-x^2)2x}{(x^2+100)^2}\right]$$

$$= \frac{100(-218x)}{(x^2+100)^2}$$

| $x$ | | 0 | |
|---|---|---|---|
| $f'(x)$ | + | 0 | − |

Relative maximum: (0, 9)

For viewing window try: $x$: –75 to 75

$y$: –120 to 20.

**c.**

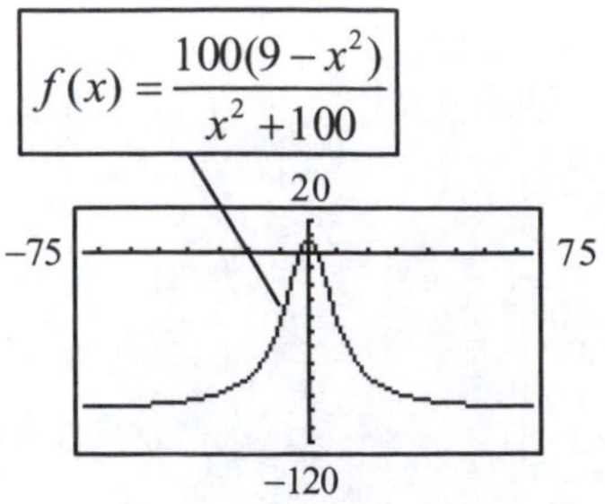


**33. a.** Standard viewing window: See graphing section below problem.

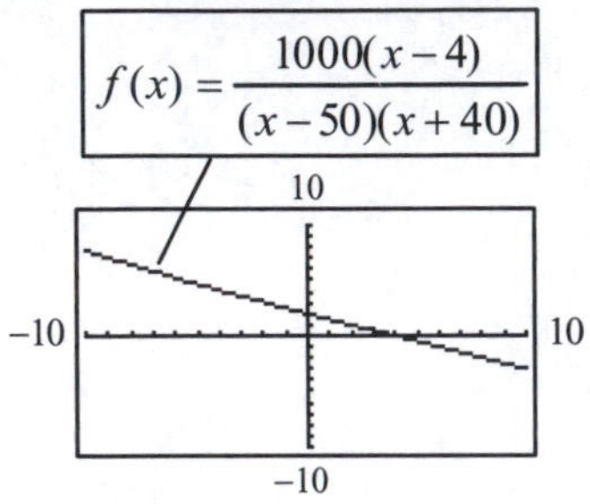


**b.** $f(x) = \frac{1000(x-4)}{(x-50)(x+40)}$ $\qquad \lim_{x\to\infty} f(x) = \lim_{x\to\infty} \frac{\frac{1000}{x} - \frac{4000}{x^2}}{1 - \frac{10}{x} - \frac{2000}{x^2}} = \frac{0}{1} = 0$

Horizontal asymptote: $y = 0$, Vertical asymptotes: $x = -40$, $x = 50$

$$f'(x) = 1000 \cdot \frac{(x^2 - 10x - 2000)\cdot 1 - (x-4)(2x-10)}{[(x-50)(x+40)]^2} = \frac{1000(-x^2 + 8x - 2040)}{[(x-50)(x+40)]^2}$$

Numerator is never zero. No relative maximum or relative minimum.

**c.** For viewing window, try $x$: –200 to 200, $y$: –200 to 200

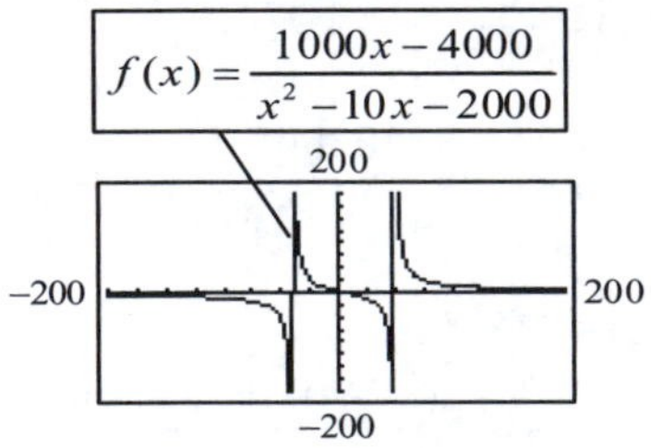


**35.** $p(C) = \frac{100C}{7300 + C}$

$$p'(C) = \frac{(7300 + C)100 - 100C(1)}{(7300 + C)^2}$$

$$= \frac{730000}{(7300 + C)^2}$$

**a.** Domain = $\{C : C \geq 0\}$. Thus $p(C)$ exists for all $C$ in the domain.

**b.** $p'(C)$ is always positive. Thus, $p(C)$ is increasing if $C \geq 0$.

**c.** $\lim_{C\to\infty} \frac{100C}{7300 + C} = 100$

Horizontal asymptote: $p = 100$

**d.** No

**37.** $R(t) = \dfrac{50t}{t^2+36}$

$R'(t) = \dfrac{(t^2+36)50 - 50t(2t)}{(t^2+36)^2} = \dfrac{50(36-t^2)}{(t^2+36)^2}$

Critical value: $t = 6$

$R'(5) > 0,\ R'(6) = 0,\ R'(7) < 0$

Horizontal asymptote: $y = 0$ (for $t > 0$)

**a.**

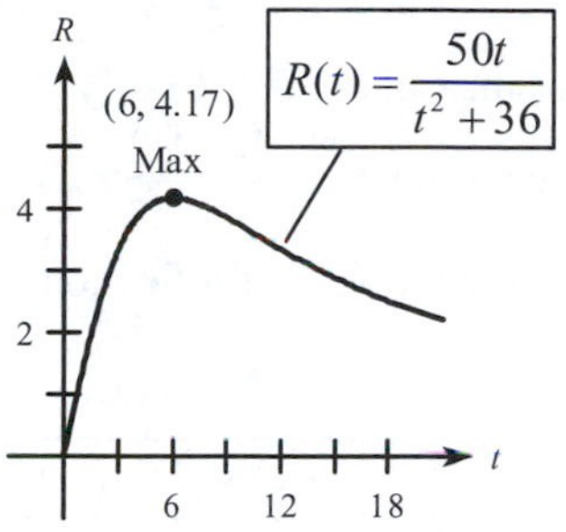


**b.** Revenue is maximized at $t = 6$.

**c.** Revenue is decreasing for $t > 6$. The video will be released in $6+4+12 = 22$ weeks from date of original release.

**39.** $f(x) = \dfrac{289.173 - 58.5731x}{x+1}$

**a.** Yes; $x = -1$

**b.** No; Domain: $x \geq 5$

**c.** Yes; $y = -58.5731$

**d.** At high wind speeds any additional wind will have little effect on the wind-chill. Interpretation is meaningful.

**41. a.** $P = C$

**b.** $C$

**c.** 0

**d.** 0

**43. a.** Since the degree of the denominator and the numerator are the same:

$$\lim_{t\to\infty} \frac{78.6t + 2090}{1.38t + 64.1} = \frac{78.6}{1.38} \approx 57.0$$

**b.** The model predicts that in the long run, 57% of the workers will be female.

**c.** No. The only vertical asymptote occurs at $t \approx -46.4$.

**d.** $p(t) > 0$ for $t > 0$ and $p(t)$ never exceeds 100, so the model is never inappropriate.

**45. a.** No. Barometric pressure can drop off the scale (as shown), but it cannot decrease without bound. In fact, it must always be positive.

**b.** A blizzard struck the East Coast, causing \$3-6 billion in damages and approximately 270 deaths.

## Chapter 10 Review Exercises

**1.** $y = -x^2$

$y' = -2x$

Critical value is at $x = 0$. Critical point is (0, 0).

$y'(-1) > 0,\ y'(0) = 0,\ y'(1) < 0$

(0, 0) is a maximum point.

**2.** $p = q^2 - 4q - 5$

$p' = 2q - 4$

Since $p' = 0$ for $q = 2$, the point (2, –9) is a critical point. To the left of (2, –9) we have $p' < 0$ and to the right of (2, –9) we have $p' > 0$. The point (2, –9) is a minimum point.

**3.** $f(x) = 1 - 3x + 3x^2 - x^3$

$$f'(x) = -3 + 6x - 3x^2 = -3(x^2 - 2x + 1) = -3(x-1)^2$$

$f'(x) = 0$ if $x = 1$

$f(1) = 1 - 3 + 3 - 1 = 0$

Critical point is (1, 0).

| $x$ | | 1 | |
|---|---|---|---|
| $f'(x)$ | – | 0 | – |

(1, 0) is a horizontal point of inflection.

**4.** $f(x) = \dfrac{3x}{x^2+1}$

$$f'(x) = \frac{(x^2+1)(3) - (3x)(2x)}{(x^2+1)^2} = \frac{-3x^2+3}{(x^2+1)^2} = \frac{-3(x+1)(x-1)}{(x^2+1)^2}$$

$f'(x) = 0$ if $x = -1$ or $x = 1$. Critical points are $\left(-1, -\frac{3}{2}\right)$ and $\left(1, \frac{3}{2}\right)$.

$f'(-2) < 0$; $f'(-1) = 0$; $f'(0) > 0$; $f'(1) = 0$; $f'(2) < 0$ means that we have a relative minimum at $\left(-1, -\frac{3}{2}\right)$ and a relative maximum at $\left(1, \frac{3}{2}\right)$.

**5.** $f(x) = x^3 + x^2 - x - 1$

$f'(x) = 3x^2 + 2x - 1 = (3x-1)(x+1)$

**a.** Critical points are (–1, 0) and $\left(\frac{1}{3}, -\frac{32}{27}\right)$.

| $x$ | | –1 | | $\frac{1}{3}$ | |
|---|---|---|---|---|---|
| $f'(x)$ | + | 0 | – | 0 | + |

**b.** Relative maximum: (–1, 0)

Relative minimum: $\left(\frac{1}{3}, -\frac{32}{27}\right)$

**c.** No horizontal points of inflection.

**d.**

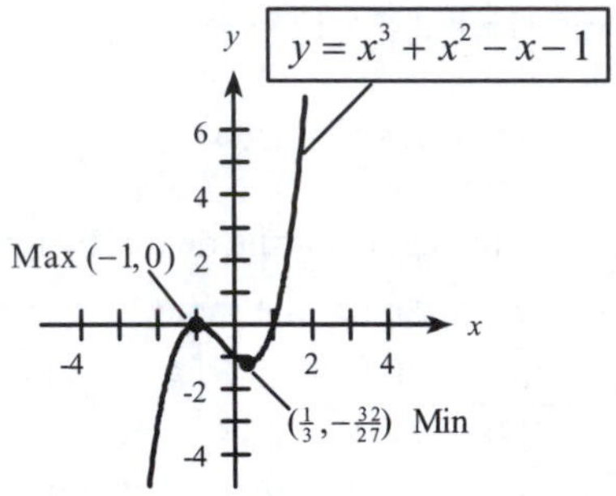


**6.** $f(x) = 4x^3 - x^4$

**a.** $f'(x) = 12x^2 - 4x^3 = 4x^2(3-x)$

Critical values: 0, 3

**b.** $f'(x) > 0$ when $x < 3$, $x \neq 0$.

$f'(x) < 0$ when $x > 3$.

Relative maximum at (3, 27).

**c.** Horizontal point of inflection: (0, 0)

**d.**

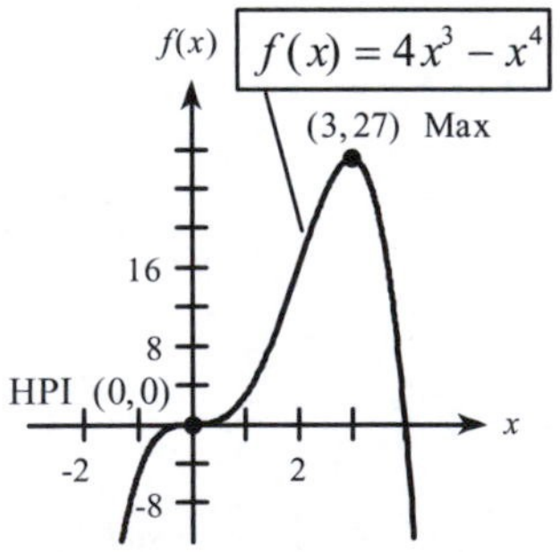


**7.** $f(x) = x^3 - \dfrac{15}{2}x^2 - 18x + \dfrac{3}{2}$

$f'(x) = 3x^2 - 15x - 18 = 3(x-6)(x+1)$

$f'(x) = 0$ at $x = -1, 6$.

**a.** Critical points: (–1, 11) and (6, –160.5).

| $x$ | | $-1$ | | $6$ | |
|---|---|---|---|---|---|
| $f'(x)$ | $+$ | $0$ | $-$ | $0$ | $+$ |

**b.** Relative maximum: (–1, 11)
Relative minimum: (6, –160.5)

**c.** No horizontal points of inflection.

**d.**

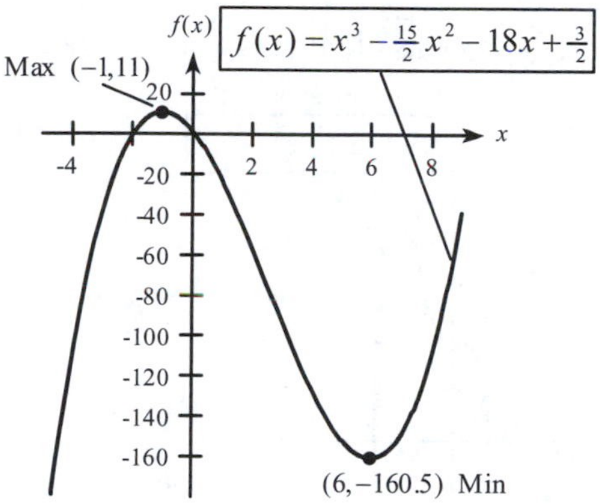


**8.** $y = f(x) = 5x^7 - 7x^5 - 1$

**a.** $y' = f'(x) = 35x^6 - 35x^4$

$= 35x^4(x+1)(x-1)$

Critical values: 0, –1, 1

**b.** $f'(x) > 0$ when $x < -1$ and when $x > 1$.

$f'(x) < 0$ when $-1 < x < 1$, $x \neq 0$.

Relative maximum: (–1, 1)
Relative minimum: (1, –3)

**c.** Horizontal point of inflection: (0, –1)

**d.**

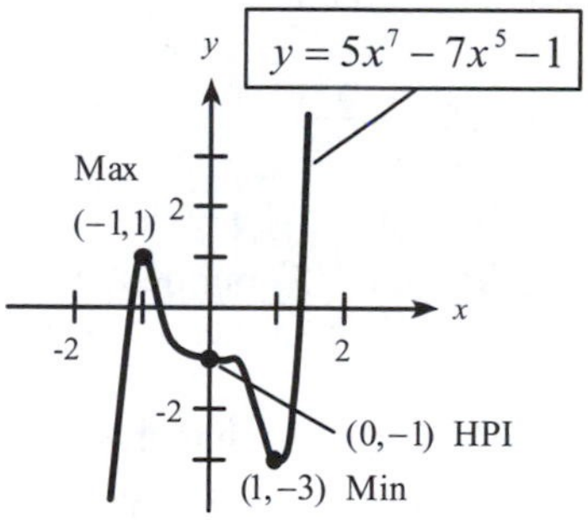


**9.** $y = x^{2/3} - 1$

$y'(x) = \dfrac{2}{3x^{1/3}}$

**a.** Critical value is at $x = 0$.

$y'(-1) < 0$, $y'(0)$ is undefined, $y'(1) > 0$

**b.** (0, –1) is a minimum point.

**c.** No horizontal points of inflection.

**d.**

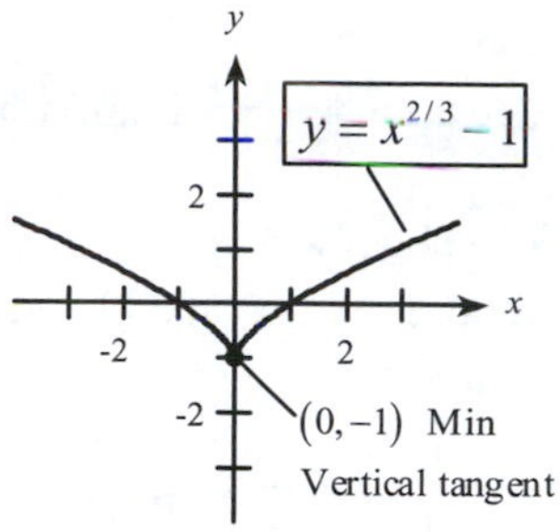


**10.** $y = f(x) = x^{2/3}(x-4)^2$

**a.** $f'(x) = x^{2/3} \cdot 2(x-4) + (x-4)^2 \cdot \dfrac{2}{3}x^{-1/3}$

$= \dfrac{2}{3}(x-4)x^{-1/3}(4x-4)$

Critical values: 0, 1, 4

**b.** $f'(x) > 0$ when $0 < x < 1$ and when $x > 4$.

$f'(x) < 0$ when $x < 0$ and when $1 < x < 4$.

Relative minimum: (0, 0); Relative minimum: (4, 0); Relative maximum: (1, 9)

**c.** No horizontal points of inflection.

**d.**

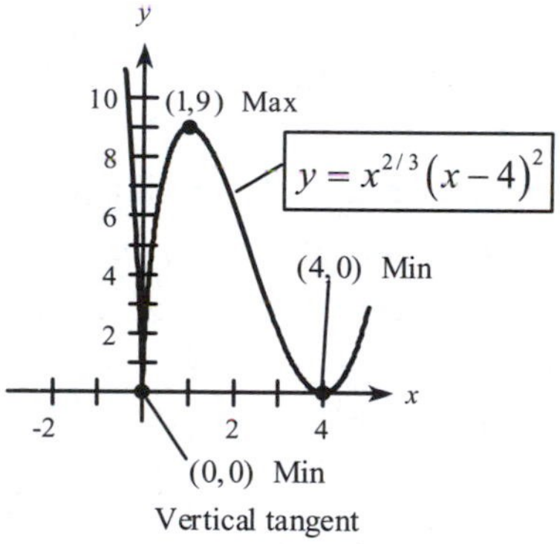


**11.** $y = x^4 - 3x^3 + 2x - 1$

$y' = 4x^3 - 9x^2 + 2$

$y'' = 12x^2 - 18x$

At $x = 2$ we have $y'' = 12$. Thus, the curve is concave up.

**12.** $y = f(x) = x^4 - 2x^3 - 12x^2 + 6$

$f'(x) = 4x^3 - 6x^2 - 24x$

$f''(x) = 12x^2 - 12x - 24$

$= 12(x^2 - x - 2)$

$= 12(x-2)(x+1)$

$f''(x) > 0$ when $x > 2$ and when $x < -1$.
$f''(x) < 0$ when $-1 < x < 2$.
$f(x)$ is concave up when $x > 2$ and when $x < -1$;
$f(x)$ is concave down when $-1 < x < 2$.
Points of inflection: (–1, –3) and (2, –42).

**13.** $y = x^3 - 3x^2 - 9x + 10$

$y' = 3x^2 - 6x - 9 = 3(x-3)(x+1)$
Critical values are $x = -1$ and $x = 3$.
$y'' = 6x - 6$. So, possible point of inflection is at $x = 1$.

| $x$ | –1 | 1 | 3 |
|---|---|---|---|
| $y''$ | – | 0 | + |

There is a point of inflection at (1, –1).
$f''(-1) < 0$ means there is a relative maximum at (–1, 15).
$f''(3) > 0$ means there is a relative minimum at (3, –17).

**14.** $y = x^3 - 12x$

$y' = 3x^2 - 12$
$y' = 0$ when $x = 2$ and when $x = -2$.
Critical points: (2, –16) and (–2, 16)
$y'' = f''(x) = 6x$
$f''(2) > 0$ means that (2, –16) is a relative minimum. $f''(-2) < 0$ means that (–2, 16) is a relative maximum. $y'' = 0$ when $x = 0$. $y'' < 0$ when $x < 0$ and $y'' > 0$ when $x > 0$. So, (0, 0) is a point of inflection.

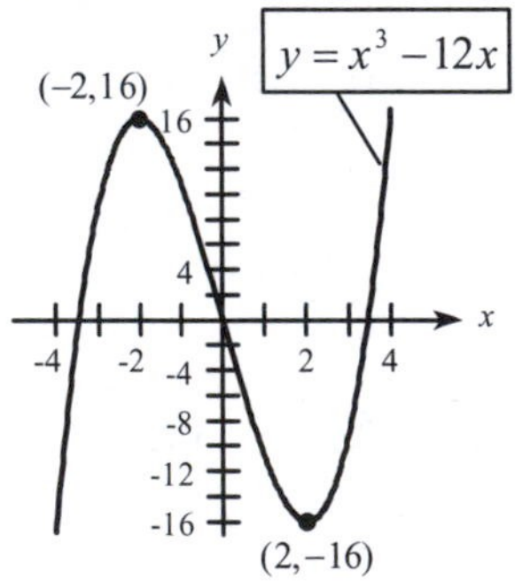


**15.** $y = 2 + 5x^3 - 3x^5$

$y' = 15x^2 - 15x^4 = 15x^2(1-x^2)$
Critical values: $x = 0, 1, -1$

| $x$ | | –1 | | 0 | | 1 | |
|---|---|---|---|---|---|---|---|
| $y'$ | – | 0 | + | 0 | + | 0 | – |

Relative minimum: (–1, 0)
Relative maximum: (1, 4)
$y'' = 30x - 60x^3 = 30x(1-2x^2)$

Possible points of inflection at $x = 0, \frac{1}{\sqrt{2}}, -\frac{1}{\sqrt{2}}$.

| $x$ | | $-\frac{1}{\sqrt{2}}$ | | 0 | | $\frac{1}{\sqrt{2}}$ | |
|---|---|---|---|---|---|---|---|
| $y''$ | + | 0 | – | 0 | + | 0 | – |

There are points of inflection at $x = 0, \pm\frac{1}{\sqrt{2}}$.

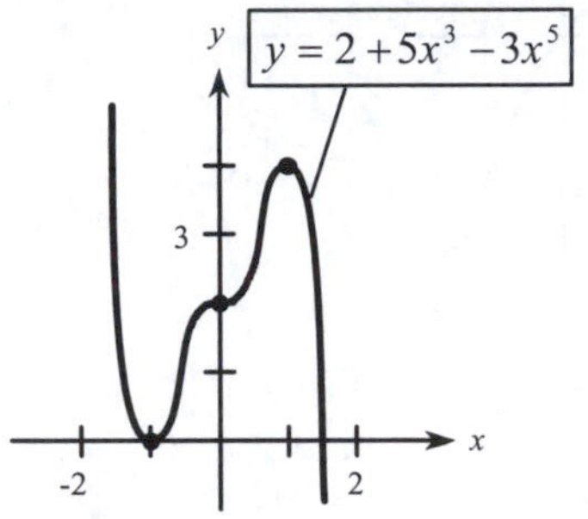


**16.** $R(x) = 280x - x^2$

**a.** $0 \le x \le 200$
$R'(x) = 280 - 2x$
$R'(x) = 0$ when $x = 140$
$R(0) = 0$, $R(140) = 19{,}600$,
$R(200) = 16{,}000$
Absolute maximum: (140, 19,600)
Absolute minimum: (0, 0)

**b.** $0 \le x \le 100$
Absolute maximum: (100, 18,000)
Absolute minimum: (0, 0)

**17.** $y = 6400x - 18x^2 - \frac{x^3}{3}$

$y' = 6400 - 36x - x^2 = (100 + x)(64 - x)$
Absolute maximum and minimum occur at critical values or at endpoints of the domain.

**a.** No critical values.
$y(50) \approx 233{,}333$ absolute maximum
$y(0) = 0$ absolute minimum

**b.** Critical value is at $x = 64$.
$y(64) \approx 248{,}491$ absolute maximum
$y(0) = 0$ absolute minimum
$y(100) \approx 126{,}667$

**18. a.** Vertical asymptote: $x = 1$

**b.** Horizontal asymptote: $y = 0$

**c.** $\lim_{x\to\infty} f(x) = 0$

**d.** $\lim_{x\to-\infty} f(x) = 0$

**19. a.** $x=-1$

**b.** $y=\frac{1}{2}$

**c.** $\frac{1}{2}$

**d.** $\frac{1}{2}$

**20.** $y=\frac{3x+2}{2x-4}$

Vertical asymptote: $x=2$

Horizontal asymptote:

$$y=\lim_{x\to\infty}\frac{3x+2}{2x-4}=\lim_{x\to\infty}\frac{3+\frac{2}{x}}{2-\frac{4}{x}}=\frac{3}{2}$$

**21.** $y=\frac{x^2}{1-x^2}$

$$\lim_{x\to\infty}\frac{x^2}{1-x^2}=\lim_{x\to\infty}\frac{1}{\frac{1}{x^2}-1}=-1$$

Horizontal asymptote: $y=-1$

Vertical asymptotes: $x=\pm1$

**22.** $y=\frac{3x}{x+2}$

**a.** Vertical asymptote: $x=-2$

Horizontal asymptote: $y=\lim_{x\to\infty}\frac{3x}{x+2}=3$

**b.** $y'=\frac{(x+2)(3)-(3x)}{(x+2)^2}=\frac{6}{(x+2)^2}$

$y'$ is never zero, no maximum nor minimum.

**c.**

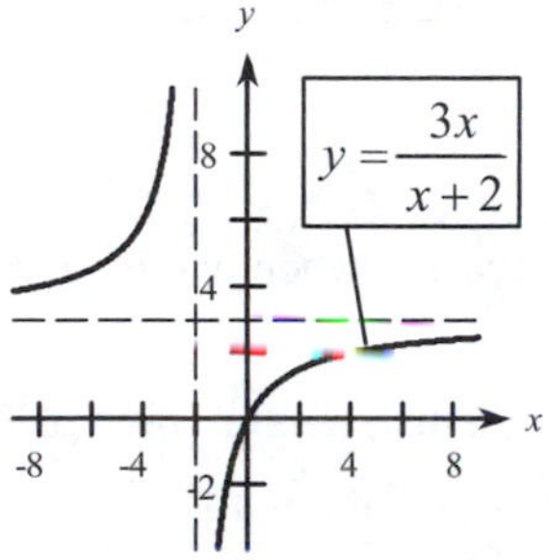


**23.** $y=\frac{8x-16}{x^2}$

$$\frac{dy}{dx}=\frac{x^2\cdot 8-(8x-16)2x}{x^4}=\frac{8(4-x)}{x^3}$$

Critical value at $x=4$.

$y'(3)>0,\ y'(4)=0,\ y'(5)<0$

**a.** Horizontal asymptote: $y=0$
Vertical asymptote: $x=0$

**b.** Relative maximum: (4, 1)

**c.**

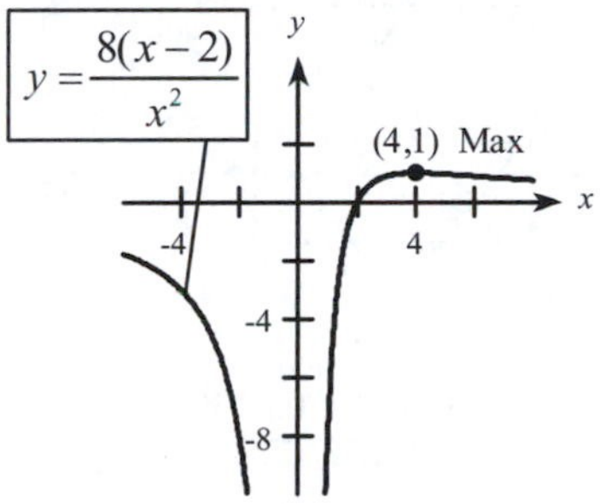


**24.** $y=\frac{x^2}{x-1}$

**a.** Vertical asymptote: $x=1$
Horizontal asymptote: none

**b.** $y'=\frac{(x-1)(2x)-x^2}{(x-1)^2}=\frac{x^2-2x}{(x-1)^2}$

$y'=0$ when $x=0$ and when $x=2$.

Critical points: (0, 0), (2, 4)

$y'=f'(-1)>0$, $f'(0.5)<0$, $f'(1.5)<0$, $f'(3)>0$

Relative maximum: (0, 0)
Relative minimum: (2, 4)

**c.**

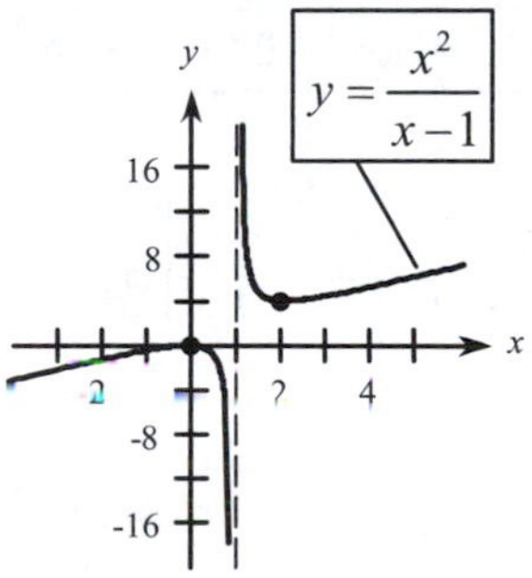

**25. a.** $f'(x) > 0$ if $x < \frac{2}{3}$ and if $x > 2$.

$f'(x) < 0$ if $\frac{2}{3} < x < 2$.

$f'(x) = 0$ if $x = \frac{2}{3}$ and if $x = 2$.

**b.** $f''(x) > 0$ if $x > \frac{4}{3}$.

$f''(x) < 0$ if $x < \frac{4}{3}$.

$f''(x) = 0$ if $x = \frac{4}{3}$.

**c.**

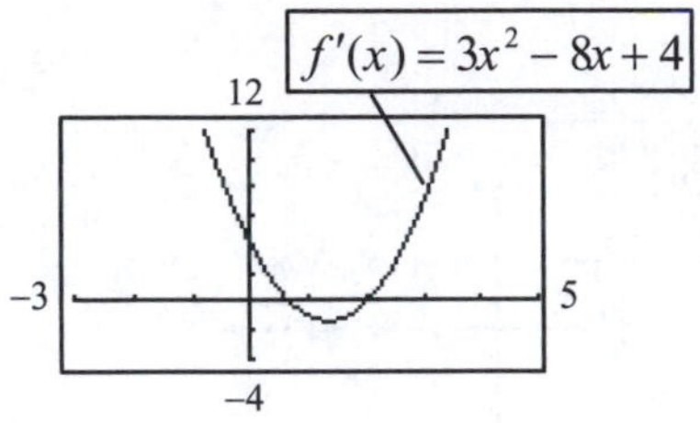


**d.**

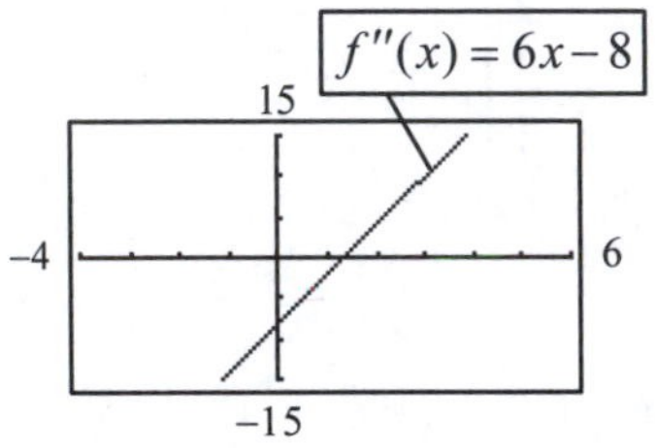


**26.** $f(x) = 0.0025x^4 + 0.02x^3 - 0.48x^2 + 0.08x + 4$

**a.** From the graph, the estimated values are

$f'(x) > 0$ when $-13 < x < 0$ and $x > 7$.

$f'(x) < 0$ when $x < -13$ and $0 < x < 7$.

$f'(x) = 0$ when $x = -13$, $x = 0$ and $x = 7$.

**b.** Estimates from the graph:

$f''(x) > 0$ when $x < -8$ and $x > 4$.

$f''(x) < 0$ when $-8 < x < 4$.

$f''(x) = 0$ when $x = -8$ and $x = 4$.

**c.**

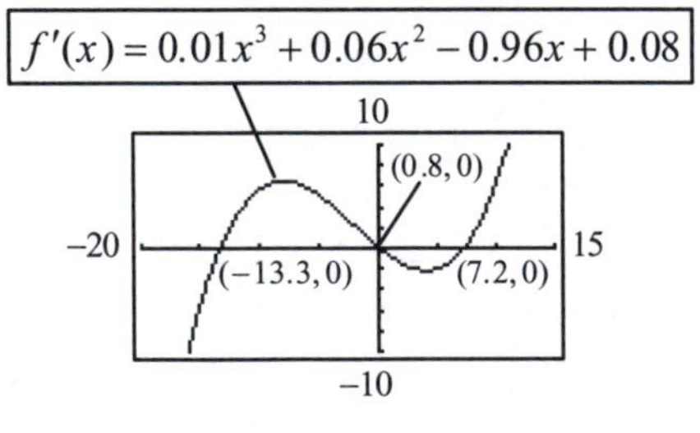


**d.**

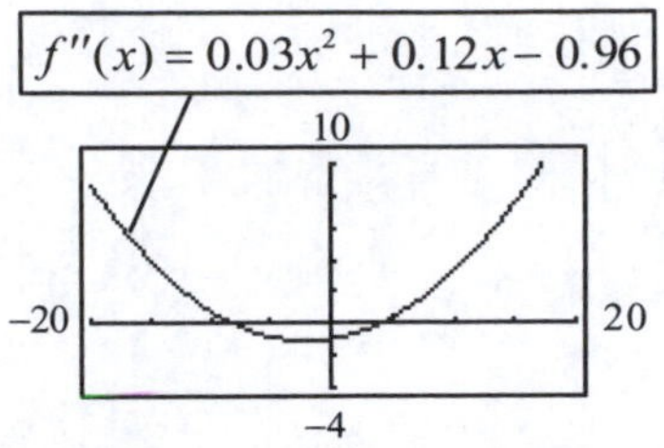


**27. a.** $f(x)$ is increasing when $f'(x) > 0$.

$f(x)$ is decreasing when $f'(x) < 0$.

Increasing if $x < -5$ and if $x > 1$.

Decreasing if $-5 < x < 1$.

Relative maximum at $x = -5$ (+ to –)

Relative minimum at $x = 1$ (– to +)

**b.** If $f'(x)$ is increasing, then $f''(x) > 0$.

If $f'(x)$ is decreasing, then $f''(x) < 0$.

$f''(x) > 0$ if $x > -2$, $f''(x) < 0$ if $x < -2$

$f''(x) = 0$ if $x = -2$.

**c.** $f(x) = \dfrac{x^3}{3} + 2x^2 - 5x \quad f'(x) = x^2 + 4x - 5$

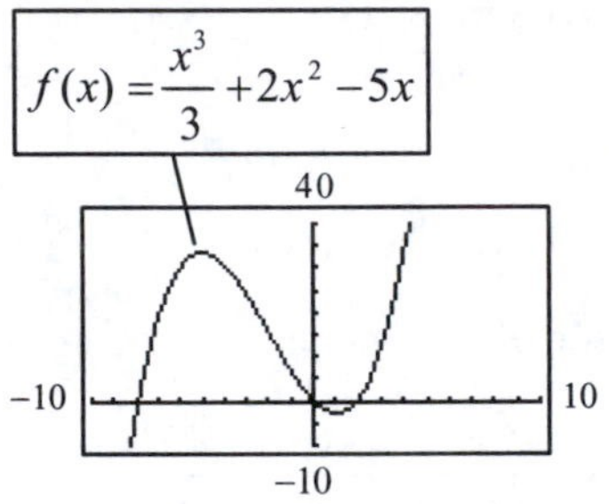


**d.**

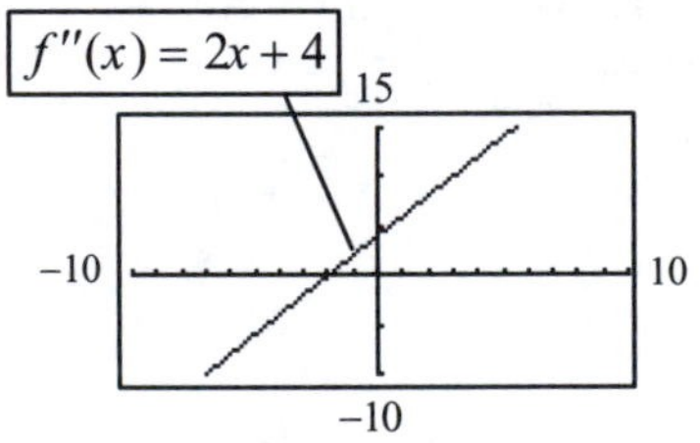


**28.** $f'(x) = 6x^2 - x^3$

**a.** From the graph,

$f(x)$ is increasing when $x < 6$, $x \neq 0$.

$f(x)$ is decreasing when $x > 6$.

$f(x)$ has a relative maximum at $x = 6$.

$f(x)$ has a horizontal point of inflection at $x = 0$.

**b.** From the graph,
$f''(x) > 0$ when $0 < x < 4$.
$f''(x) < 0$ when $x < 0$ and when $x > 4$.
$f''(x) = 0$ when $x = 0$ and $x = 4$.

**c.** $f(x) = 2x^3 - \frac{x^4}{4}$, $f'(x) = 6x^2 - x^3$

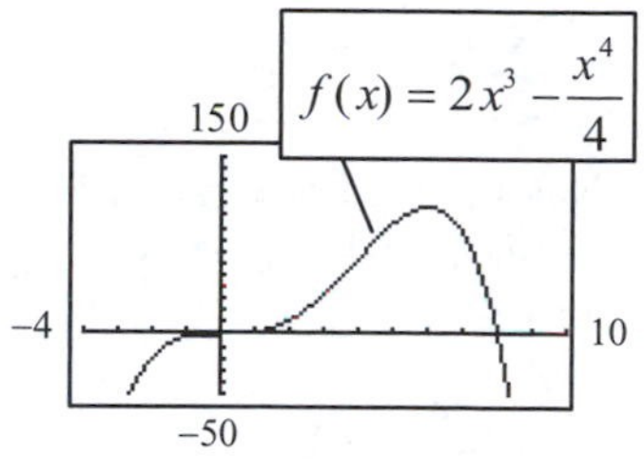


**d.** $f''(x) = 12x - 3x^2 = 3x(4 - x)$
Graph of

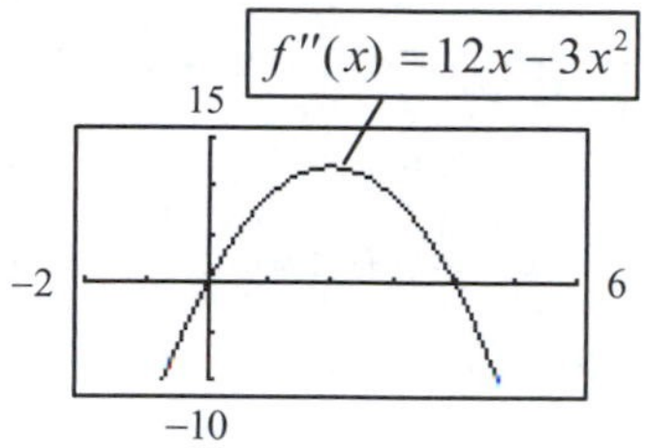


$f''$

**29. a.** $f(x)$ is concave up if $f''(x) > 0$.
$f(x)$ is concave down if $f''(x) < 0$.
Concave up if $x < 4$.
Concave down if $x > 4$.
There is a point of inflection at $x = 4$.

**b.** $f(x) = 2x^2 - \frac{x^3}{6}$, $f'(x) = 4x - \frac{1}{2}x^2$,
$f''(x) = 4 - x$

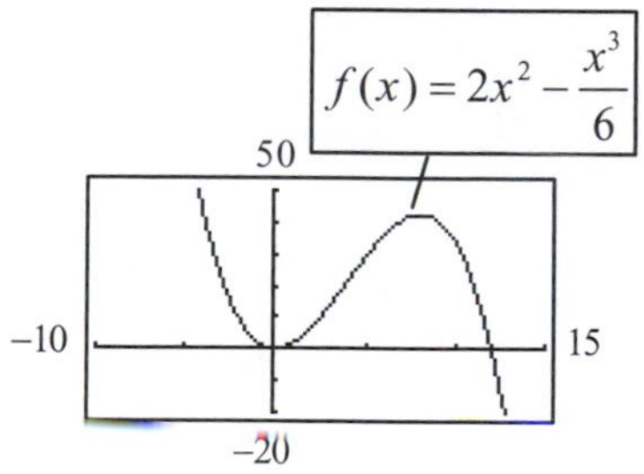


**30.** $f''(x) = 6 - x - x^2$

**a.** From the given graph,
$f(x)$ is concave up when $-3 < x < 2$.
$f(x)$ is concave down when $x < -3$, and when $x > 2$. $f(x)$ has a point of inflection at $x = -3$ and at $x = 2$.

**b.** $f(x) = 3x^2 - \frac{x^3}{6} - \frac{x^4}{12}$
$f'(x) = 6x - \frac{1}{2}x^2 - \frac{1}{3}x^3$
$f''(x) = 6 - x - x^2$

Graph of $f$:

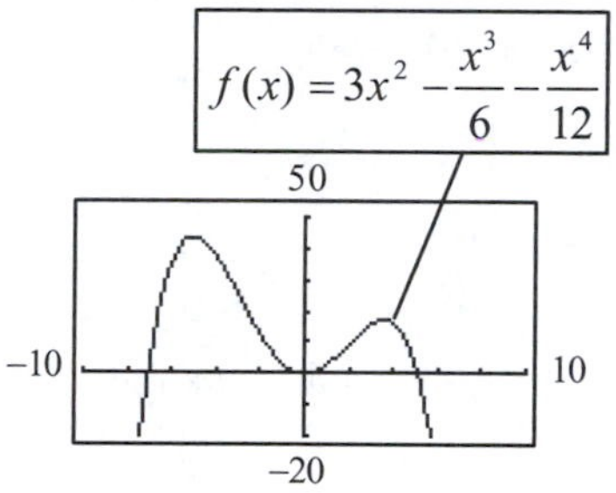


**31.** $\overline{C}(x) = 3x + 15 + \frac{75}{x}$

$\overline{C}'(x) = 3 - \frac{75}{x^2}$

$\overline{C}'(x) = 0$ if $x^2 = 25$ or $x = 5$.

| $x$ | 5 |
|---|---|
| $\overline{C}'(x)$ | − 0 + |

$\overline{C}(5) = 15 + 15 + 15 = 45$
Minimum average cost is $45 at 5 units.

**a.** $R(x) = 32x - 0.01x^2$
$R'(x) = 32 - 0.02x$
$R'(x) = 0$ when $x = 1600$.

| $x$ | 1600 |
|---|---|
| $R'(x)$ | + 0 − |

$R(1600) = \$25,600$
The maximum revenue is $25,600 when $x = 1600$ units are produced.

**b.** If production is limited to 1500, then the maximum revenue occurs when $x = 1500$. So, the maximum revenue would then be $25,500.

**32.** $R(x) = 32x - 0.01x^2$

**a.** $R'(x) = 32 - 0.02x$

$R'(x) = 0 \rightarrow x = 1600$

1600 units maximize total revenue.

$R(1600) = \$25,600$

**b.** Since $R'(x) > 0$ for $x < 1600$, if production is limited to 1500 units, then 1500 units will maximize revenue. The maximum revenue is $R(1500) = \$25,500$.

**33.** $P(x) = 1080x + 9.6x^2 - 0.1x^3 - 50,000$

$P'(x) = 1080 + 19.2x - 0.3x^2$

$P'(x) = 0$ if $-0.3(x^2 - 64x - 3600)$ or if $(x-100)(x+36) = 0$.

| $x$ | | 100 | |
|---|---|---|---|
| $P'(x)$ | + | 0 | − |

Maximum profit is at $x = 100$.

$P(100) = \$54,000$ is the maximum profit.

**34.** $R(x) = 46x - 0.01x^2$

$C(x) = 0.05x^2 + 10x + 1100$

$P(x) = R(x) - C(x)$

$= \left(46x - 0.01x^2\right) - \left(0.05x^2 + 10x + 1100\right)$

$= -0.06x^2 + 36x - 1100$

$P'(x) = -0.12x + 36$

$P'(x) = 0$ when $x = 300$

Since $P''(x) = -0.12 < 0$ for all *x*, the profit is maximized when $x = 300$.

**35.** $P(x) = 80x - \dfrac{1}{4}x^2 - (800 + 4x)$

$= -\dfrac{1}{4}x^2 + 76x - 800$

$P'(x) = -\dfrac{1}{2}x + 76$

$P'(x) = 0$ if $x = 152$.

| $x$ | | 152 | |
|---|---|---|---|
| $P'(x)$ | + | 0 | − |

Since 152 is not in the domain and *P* is increasing for $0 \le x \le 150$, the maximum profit is at 150 units.

**36.** $C = 2x^2 + 54x + 98$

$\overline{C} = 2x + 54 + \dfrac{98}{x}$

$\overline{C}' = 2 - \dfrac{98}{x^2} = \dfrac{2x^2 - 98}{x^2}$

$\overline{C}' = 0$ when $x = 7$.

Since $\overline{C}'' = \dfrac{196}{x^3} > 0$ for all $x > 0$, the average cost is minimized when $x = 7$.

**37.** Profit is maximized when $\overline{MP}$ changes from + to −. This occurs at $x = 500$ units.

**38.** $P'(t)$ is to be maximized (not $P(t)$).

$$P(t) = \frac{95t^2}{t^2 + 2700} + 5$$

$$P'(t) = \frac{(t^2 + 2700)\cdot 190t - 95t^2 \cdot 2t}{(t^2 + 2700)^2} = \frac{513,000t}{(t^2 + 2700)^2}$$

$$P''(t) = \frac{(t^2 + 2700)^2 \cdot 513,000 - 513,000t \cdot 2(t^2 + 2700)\cdot 2t}{(t^2 + 2700)^4}$$

$$= \frac{513,000(t^2 + 2700)\left[(t^2 + 2700) - 4t^2\right]}{(t^2 + 2700)^4}$$

$$= \frac{513,000(2700 - 3t^2)}{(t^2 + 2700)^3}$$

$P''(t)$ changes from + to − at $t = 30$. Point of diminishing returns is at $t = 30$ hours.

**39. a.** Diminishing returns or point of inflection occurs at $x = 60$.

**b.** $m = \frac{f(I)-0}{I-0} = \frac{f(I)}{I} = \text{Average output}$

**c.** Maximum average output occurs when slope of average output is closest to $f'(I)$. This occurs at $x = 70$.

**40.** $R = (54+x)(1540-10x) = 83{,}160 + 1000x - 10x^2$

$R' = 1000 - 20x$

$R'$ changes from + to − at $x = 50$. Revenue is maximized with selling price of $1540-10(50) = \$1040$ per bike.

**41.** $P = 83{,}160 + 1000x - 10x^2 - 680(54+x) = 46{,}440 + 320x - 10x^2$

$P' = 320 - 20x$

Profit is maximized when $P' = 0$ or $x = 16$. The selling price will be $1540-10(16) = \$1380$ per bike.

**42.** Equilibrium price means supply = demand.

$1200 - 2x = 200 + 2x$ gives equilibrium quantity $x = 250$. Equilibrium price is $p = 200 + 2(250) = 700$.

Revenue = $700x$. Cost $= 12{,}000 + 50x + x^2$.

$P(x) = 700x - (12{,}000 + 50x + x^2) = -12{,}000 + 650x - x^2$

$P'(x) = 650 - 2x$

$P'(x) = 0$ if $x = 325$.

| $x$ | | 325 | |
|---|---|---|---|
| $P'(x)$ | + | 0 | − |

$P(325) = -12{,}000 + 211{,}250 - 105{,}625 = \$93{,}625$

**43.** $\overline{C} = 200 + x,\ C = 200x + x^2,\ p = 800 - x,\ R = 800x - x^2$

**a.** $P(x) = 800x - x^2 - 200x - x^2 = 600x - 2x^2$

$P'(x) = 600 - 4x$

$P'(x) = 0$ when $x = 150$.

$P''(x) = -4 < 0$ for all $x$ means that $x = 150$ maximizes profit.

**b.** The selling price is $p = 800 - 150 = \$650$.

**44.** $R(x) = 7000x - 10x^2 - \frac{x^3}{3}$ $\qquad C(x) = 40{,}000 + 600x + 8x^2$

$P(x) = -\frac{x^3}{3} - 18x^2 + 6400x - 40{,}000$

$P'(x) = -x^2 - 36x + 6400 = (64-x)(100+x)$

Profit is a maximum at $x = 64$. $(P'(64) = 0)$

| $x$ | | 64 | |
|---|---|---|---|
| $P'(x)$ | + | 0 | |

$P(64) = -87{,}381.33 - 73{,}728 + 409{,}600 - 40{,}000 = \$208{,}490.67$

**45.** $R(x) = x^2\left(500 - \frac{x}{3}\right) = 500x^2 - \frac{x^3}{3}$ $\qquad R'(x) = 1000x - x^2$

$R'(x) = 0$ when $x = 0$ and when $x = 1000$.

$R''(x) = 1000 - 2x$ and $R''(1000) < 0$. So, the maximum reaction occurs at $x = 1000$.

**46.** $N(t) = 4 + 3t^2 - t^3 \quad N'(t) = 6t - 3t^2 = 3t(2-t)$

Critical values: $t = 0, 2$

$N''(t) = 6 - 6t, N''(2) = -6 < 0$

Thus, maximum at $t = 2$. Maximum production occurs at $8:00 + 2$ hrs =10:00 a.m.

**47.** $P = 300 + 10t - t^2$, $t = 0$ is 2015 $0 \le t \le 10$

$P' = 10 - 2t = 0$ when $t = 5$.

| $t$ | 5 |
|---|---|
| $P'(t)$ | + 0 − |

When $t = 5$, $P = 325$. The largest graduating class is the class of 2020 with 325 graduates.

**48.**

$$b(x) = \frac{8k}{x^2} + \frac{k}{(30-x)^2}$$

$$b'(x) = \frac{-16k}{x^3} - \frac{2k}{(30-x)^3}(-1)$$

Set $b'(x) = 0$ and simplify. $x^3 = 8(30-x)^3$

$x = 2(30-x)$

$3x = 60$ or $x = 20$

| $x$ | 20 |
|---|---|
| $b'(x)$ | − 0 + |

Build the observatory 20 miles from $A$ (10 miles from $B$).

**49.** Maximize $A = xy$ where $2x + 2y = 16$.

$A = x(8-x) = 8x - x^2 \qquad A' = 8 - 2x$

$A' = 0$ when $x = 4$.

Since $A'' < 0$ for all $x$, a $4 \times 4$ playpen gives a maximum area.

**50.** Quantity to be minimized: $A = (x+2)\left(y + \frac{7}{4}\right)$

Another equation: $xy = 56$ or $y = \frac{56}{x}$

Substituting: $A = (x+2)\left(\frac{56}{x} + \frac{7}{4}\right)$

$$A = 56 + \frac{7}{4}x + \frac{112}{x} + \frac{7}{2}$$

$$A' = \frac{7}{4} - \frac{112}{x^2}$$

Set $A' = 0$ and simplify. $7x^2 = 448$

$x^2 = 64$ or $x = 8$

| $x$ | 8 |
|---|---|
| $A'$ | − 0 + |

So, $x = 8$ and $y = 7$ minimizes $A$.

Page dimensions are (8 + 2) by $\left(7 + 1\frac{3}{4}\right)$ or 10″ by $8\frac{3}{4}''$.

**51.** $R(x) = x^2\left(500 - \frac{x}{3}\right) = 500x^2 - \frac{x^3}{3}$

$R'(x) = 1000x - x^2$

$R''(x) = 1000 - 2x$

$R''(x) = 0$ when $x = 500$.

| $x$ | 500 |
|---|---|
| $R''(x)$ | + 0 − |

So, the dosage $x = 500$ maximizes sensitivity.

**52. a.** $C(t) = 0.118t^3 - 2.51t^2 + 40.2t + 677$

$C'(t) = 0.354t^2 - 5.02t + 40.2$

$C''(t) = 0.708t - 5.02$

$C''(t) = 0$ when $x \approx 23.3$

$C''$ changes from + to − at $\approx 7.09$, thus this is maximum for $C'$.

The rate of change of federal tax per capita reaches its maximum in the year 1994.

**b.** The graph of $C$ has a point of inflection at $t \approx 7.09$.

**53.** Total production cost: $\left(\dfrac{288{,}000}{x}\right)(1500)+(288{,}000)(30)$ Total storage cost: $\left(\dfrac{x}{2}\right)(1.5)$

$$C(x)=\left(\frac{288{,}000}{x}\right)(1500)+(288{,}000)(30)+\frac{1.5x}{2}=\frac{432{,}000{,}000}{x}+8{,}640{,}000+0.75x$$

$C'(x)=-\dfrac{432{,}000{,}000}{x^2}+0.75$ $C'(x)=0$ when $x^2=576{,}000{,}000$ or when $x=24{,}000$.

$C''(x)=\dfrac{864{,}000{,}000}{x^3}>0$ for $x>0$.

So, the minimum cost occurs for a run of $x=24{,}000$.

**54. a.** $\overline{C}(x)=\dfrac{4500+120x+0.05x^2}{x}$

Vertical asymptote at $x=0$.

Horizontal asymptote: $\lim\limits_{x\to\infty}\dfrac{4500+120x+0.05x^2}{x}=\lim\limits_{x\to\infty}\left(\dfrac{4500}{x}+120+0.05x\right)=\infty$ (so none).

**b.**

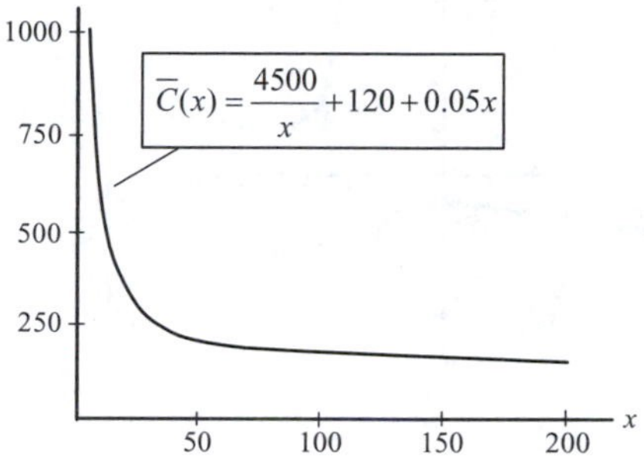


**55. a.** $M(0)=\dfrac{3.8(0)^2+3}{0.1(0)^2+1}=3\%$

**b.** Horizontal asymptote: $\lim\limits_{t\to\infty}\dfrac{3.8t^2+3}{0.1t^2+1}=\lim\limits_{t\to\infty}\dfrac{3.8+\frac{3}{t^2}}{0.1+\frac{1}{t^2}}=\dfrac{3.8}{0.1}=38\%$

The limit for the company's percent share of the market is 38%.

## *Chapter 10 Test*

**1.** $f(x) = x^3 + 6x^2 + 9x + 3$

$f'(x) = 3x^2 + 12x + 9 = 3(x+3)(x+1)$

| $x$ | | $-3$ | | $-1$ | |
|---|---|---|---|---|---|
| $f'(x)$ | + | 0 | − | 0 | + |

Relative maximum at $x = -3$ $(-3, 3)$
Relative minimum at $x = -1$ $(-1, -1)$
Polynomials have no asymptotes.

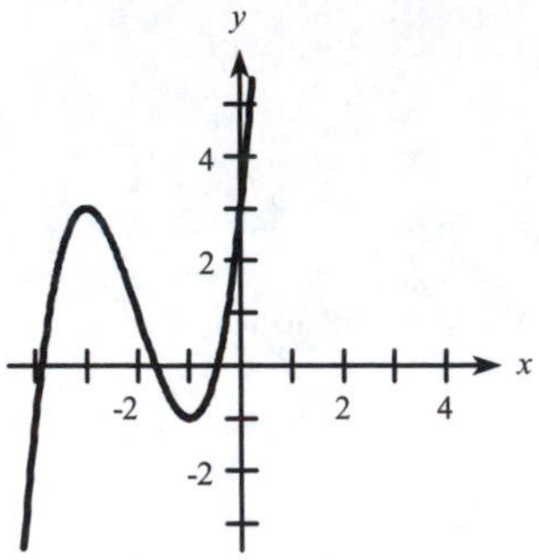


**2.** $y = 4x^3 - x^4 - 10$

$y' = 12x^2 - 4x^3 = 4x^2(3 - x)$

| $x$ | | $0$ | | $3$ | |
|---|---|---|---|---|---|
| $f'(x)$ | + | 0 | + | 0 | − |

Maximum at $(3, 17)$.

$y'' = 24x - 12x^2 = 12x(2 - x)$

| $x$ | | $0$ | | $2$ | |
|---|---|---|---|---|---|
| $f''(x)$ | − | 0 | + | 0 | − |

Inflection points at $(0, -2)$ and $(2, 6)$.

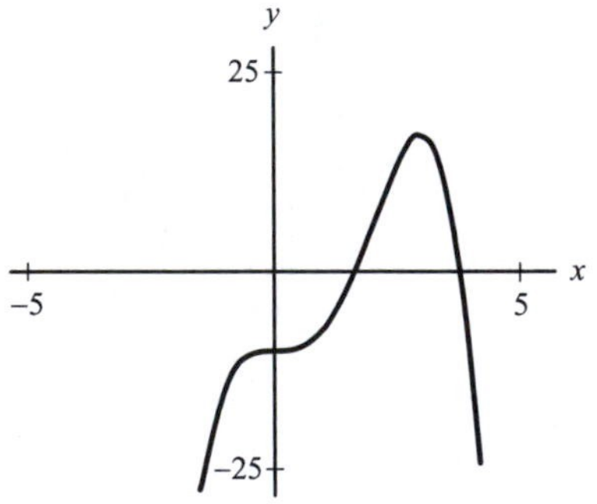


**3.** $y = \dfrac{x^2 - 3x + 6}{x - 2}$

$y' = \dfrac{(x-2)(2x-3) - (x^2 - 3x + 6)(1)}{(x-2)^2} = \dfrac{x(x-4)}{(x-2)^2}$

| $x$ | | $0$ | | $2$ | | $4$ | |
|---|---|---|---|---|---|---|---|
| $y'$ | + | 0 | − | * | − | 0 | + |

* means $y'$ is undefined when $x = 2$.
In fact, $x = 2$ is not in the domain.
Relative maximum at $x = 0$ $(0, -3)$
Relative minimum at $x = 4$ $(4, 5)$
Vertical asymptote: $x = 2$

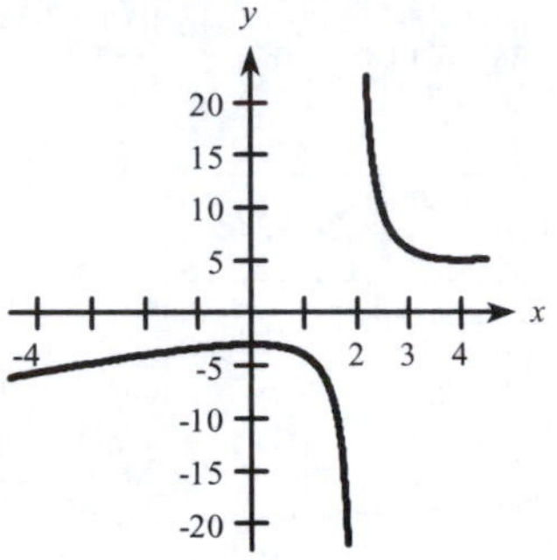


**4–6.** $y = 3x^5 - 5x^3 + 2$

$y' = 15x^4 - 15x^2 = 15x^2(x+1)(x-1)$

$y'' = 60x^3 - 30x$

$= 30x(2x^2 - 1)$

$= 30x\left[2\left(x + \dfrac{1}{\sqrt{2}}\right)\left(x - \dfrac{1}{\sqrt{2}}\right)\right]$

**4.** Concave up means $y'' > 0$.

$\left(\dfrac{1}{\sqrt{2}} \approx 0.707\right)$ $\quad -\dfrac{1}{\sqrt{2}} < x < 0, \quad x > \dfrac{1}{\sqrt{2}}$

**5.** $y''$ changes signs for inflection points.

$x = 0, \pm\dfrac{1}{\sqrt{2}} \approx \pm 0.707$

**6.** Relative maximum and minimum occur at $x = \pm 1$.

$y'' > 0$ at $x = 1$ gives a relative minimum (1, 0).

$y'' < 0$ at $x = -1$ gives a relative maximum (−1, 4).

**7.** $f(x) = 2x^3 - 15x^2 + 3$; [–2, 8]

$f'(x) = 6x^2 - 30x = 6x(x-5)$

| $x$ | –2 | 0 | 5 | 8 |
|---|---|---|---|---|
| $f(x)$ | –73 | 3 | –122 | 67 |

Absolute maximum of 67 at $x = 8$.
Absolute minimum of –122 at $x = 5$.

**8.** $f(x) = \frac{200x - 500}{x + 300} = \frac{200 - \frac{500}{x}}{1 + \frac{300}{x}}$

Horizontal asymptote: $y = 200$

Vertical asymptote: $x = -300$

**9.**

| Point | $f$ | $f'$ | $f''$ |
|---|---|---|---|
| $A$ | – | + | – |
| $B$ | + | – | 0 |
| $C$ | + | 0 | + |

$f(x) < 0$ if below $x$-axis

$f'(x)$ is positive if $f(x)$ is increasing.

$f''(x)$ is positive if $f(x)$ is concave up.

**10.** **a.** $\lim_{x \to -\infty} f(x) = 2$

**b.** $x = -3$ is the vertical asymptote

**c.** Since $\lim_{x \to -\infty} f(x) = 2$ and $\lim_{x \to \infty} f(x) = 2$, the horizontal asymptote is $y = 2$.

**11.** $f(6) = 10, f'(6) = 0, f''(6) = -3$

By the 2nd Derivative Test there is a local maximum at (6, 10).

**12.** $A(t) = -0.000497t^3 + 0.0449t^2 - 0.669t + 22.3$

$A'(t) = -0.001491t^2 + 0.0898t - 0.669$

**a.** Using the quadratic formula or a graphing utility, we find where $A'(t) = 0$:

$t \approx 8.7$ or $t \approx 51.5$

$A'$ changes from – to + at 8.7, which tells us that there is a relative minimum at $(8.7, 19.6)$.

$A'$ changes from + to – at 51.5, which tells us that there is a relative maximum at $(51.5, 39.0)$.

**b.** The point $(8.7, 19.6)$ means that when $t = 8.7$ (during 1999), the aged dependency ratio reached a minimum of 19.6 aged individuals per 100 individuals ages 20 – 64. The point $(51.5, 39.0)$ means that when $t = 51.5$ (during 2042), the aged dependence ratio is expected to reach a maximum of 39.0 aged individuals per 100 individuals ages 20-64.

**13.** $R(x) = 164x \qquad C(x) = 0.01x^2 + 20x + 300$

$P(x) = 164x - (0.01x^2 + 20x + 300)$

$= 144x - 0.01x^2 - 300$

**a.** $P'(x) = 144 - 0.02x$

Critical value at $x = \frac{144}{0.02} = 7200$.

| $x$ | | 7200 | |
|---|---|---|---|
| $P'(x)$ | + | 0 | – |

$x = 7200$ gives maximum profit

**b.** $P(7200) = \$518{,}100$

**14.** $C(x) = 100 + 20x + 0.01x^2$

$\overline{C}(x) = \frac{C(x)}{x} = \frac{100}{x} + 20 + 0.01x$

$\overline{C}'(x) = -\frac{100}{x^2} + 0.01$ Critical value at $x = 100$.

| $x$ | | 100 | |
|---|---|---|---|
| $\overline{C}'(x)$ | – | 0 | + |

Minimum occurs when $\overline{C}'(x) = 0$, or $x = 100$.

**15.** Let $x$ = number of additional units sold.

$R(x) = (100 + x)(300 - 2x) = 30{,}000 + 100x - 2x^2$

$R'(x) = 100 - 4x$

$R'(x) = 0$ if $x = 25$

| $x$ | | 25 | |
|---|---|---|---|
| $R'(x)$ | + | 0 | – |

Price to get maximum revenue
$= 300 - 2(25) = \$250$.

**16.**

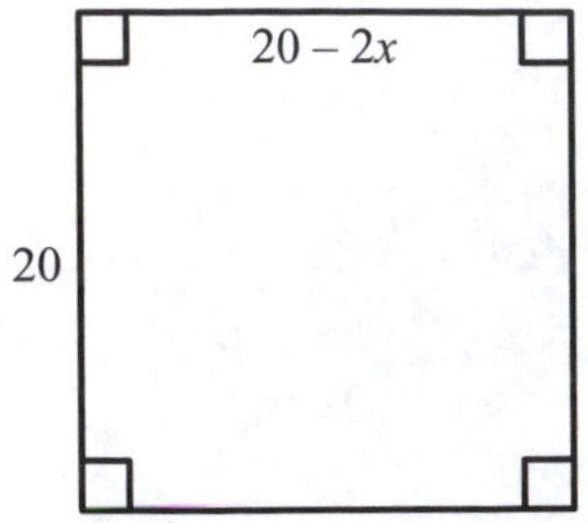


$x$ = length of removed square.

$$V = x(20-2x)^2$$
$$V'(x) = x(2)(20-2x)(-2) + (20-2x)^2(1)$$
$$= (20-2x)[-4x + (20-2x)]$$
$$= (20-2x)(20-6x)$$

Critical value occurs if

$V'(x) = 0$ and $x = \frac{20}{6} = \frac{10}{3}$ cm.

($x = 10$ is not in domain)

| $x$ | $\frac{10}{3}$ |
|---|---|
| $V'(x)$ | $+ \quad 0 \quad -$ |

$x = \frac{10}{3}$ gives maximum volume.

**17.** Inventory problem

$x$ = number in each production run

$$C = \frac{784000}{x}(2500) + 784000(420) + \frac{x}{2}(5)$$

Production cost    Storage cost

$$C' = \frac{-784000(2500)}{x^2} + \frac{5}{2}$$

$$C' = 0 \text{ if } x^2 = \frac{2(2500)(784000)}{5}$$

Critical value occurs when $x$ = 28,000 units.

| $x$ | 28000 |
|---|---|
| $C'$ | $- \quad 0 \quad +$ |

$x = 28{,}000$ gives minimum costs.

**18.** **a.** $y = -0.0000700x^3 + 0.00567x^2 + 0.863x + 16.0$

**b.** $y' = -0.0002100x^2 + 0.01134x + 0.863$

$y'' = -0.0004200x + 0.01134$

$y'' = 0$ when $x \approx 27$

$y''$ changes from $-$ to $+$ at $\approx 27$, thus this is minimum for $y'$. The rate of change of the national debt reaches its minimum in the year 1977.

**c.** This represents the $x$-coordinate of a point of inflection on the graph of $y$.

# Chapter 11: Derivatives Continued

## *Exercises 11.1*

**1.** $f(x) = 4\ \ln x$

$f'(x) = 4 \cdot \frac{1}{x} = \frac{4}{x}$

**3.** $y = \ln 8x$

$\frac{dy}{dx} = \frac{1}{8x}(8) = \frac{1}{x}$

**5.** $y = \ln x^4$

$\frac{dy}{dx} = \frac{1}{x^4} \cdot 4x^3 = \frac{4}{x}$

**7.** $f(x) = \ln(4x+9)$

$f'(x) = \frac{1}{4x+9}(4) = \frac{4}{4x+9}$

**9.** $y = \ln(2x^2 - x) + 3x$

$\frac{dy}{dx} = \frac{1}{2x^2 - x}(4x-1) + 3$

$= \frac{4x-1}{2x^2-x} + 3$

**11.** $p = \ln(q^2+1)$

$\frac{dp}{dq} = \frac{1}{q^2+1} \cdot 2q = \frac{2q}{q^2+1}$

**13. a.** $y = \ln x - \ln(x-1)$

$\frac{dy}{dx} = \frac{1}{x} - \frac{1}{x-1} = \frac{x-1-x}{x(x-1)} = \frac{-1}{x(x-1)}$

**b.** $y = \ln \frac{x}{x-1}$

$\frac{dy}{dx} = \frac{1}{\frac{x}{x-1}}\left[\frac{(x-1)(1) - x(1)}{(x-1)^2}\right]$

$= \frac{x-1}{x}\left[\frac{-1}{(x-1)^2}\right] = \frac{-1}{x(x-1)}$

**15. a** $y = \frac{1}{3}\ln(x^2-1)$

$\frac{dy}{dx} = \frac{1}{3}\left[\frac{1}{x^2-1} \cdot 2x\right] = \frac{2x}{3(x^2-1)}$

**b.** Using properties of logs.

$y = \ln\sqrt[3]{x^2-1} = \ln(x^2-1)^{1/3} = \frac{1}{3}\ln(x^2-1)$

Thus, $\frac{dy}{dx} = \frac{2x}{3(x^2-1)}$.

**17. a.** $y = \ln(4x-1) - 3\ln x$

$\frac{dy}{dx} = \frac{1}{4x-1} \cdot 4 - 3 \cdot \frac{1}{x}$

$= \frac{4x - 3(4x-1)}{x(4x-1)} = \frac{3-8x}{x(4x-1)}$

**b.** $y = \ln\left(\frac{4x-1}{x^3}\right) = \ln(4x-1) - 3\ln x$

$\frac{dy}{dx} = \frac{1}{4x-1} \cdot 4 - 3 \cdot \frac{1}{x}$

$= \frac{4}{4x-1} - \frac{3}{x} = \frac{3-8x}{x(4x-1)}$

**19.** $p = \ln\left(\frac{q^2-1}{q}\right)$

$= \ln(q^2-1) - \ln q$

$\frac{dp}{dq} = \frac{2q}{q^2-1} - \frac{1}{q} = \frac{2q^2 - q^2 + 1}{q(q^2-1)} = \frac{q^2+1}{q(q^2-1)}$

**21.** $y = \ln\left(\frac{t^2+3}{\sqrt{1-t}}\right) = \ln(t^2+3) - \frac{1}{2}\ln(1-t)$

$\frac{dy}{dt} = \frac{2t}{t^2+3} - \frac{-1}{2(1-t)} = \frac{4t - 4t^2 + t^2 + 3}{2(1-t)(t^2+3)}$

$= \frac{3 + 4t - 3t^2}{2(1-t)(t^2+3)} = \frac{-(3t^2 - 4t - 3)}{2(1-t)(t^2+3)}$

**23.** $y = \ln\left(x^3\sqrt{x+1}\right) = 3\ln x + \frac{1}{2}\ln(x+1)$

$\frac{dy}{dx} = 3 \cdot \frac{1}{x} + \frac{1}{2} \cdot \frac{1}{x+1} \cdot 1 = \frac{3}{x} + \frac{1}{2(x+1)}$ or

$= \frac{3(2)(x+1) + x \cdot 1}{2x(x+1)} = \frac{7x+6}{2x(x+1)}$

**25.** $y = x - \ln x$

$\frac{dy}{dx} = 1 - \frac{1}{x}$

**27.** $y = \frac{\ln x}{x}$

$\frac{dy}{dx} = \frac{x \cdot \frac{1}{x} - (\ln x)(1)}{x^2} = \frac{1 - \ln x}{x^2}$

**29.** $y = \ln(x^4+3)^2 = 2\ln(x^4+3)$

$\frac{dy}{dx} = 2 \cdot \frac{1}{x^4+3} \cdot 4x^3 = \frac{8x^3}{x^4+3}$

**31.** $y=(\ln x)^4$

$$\frac{dy}{dx}=4(\ln x)^3\cdot\frac{1}{x}=\frac{4(\ln x)^3}{x}$$

**33.** $y=\left[\ln(x^4+3)\right]^2$

$$\frac{dy}{dx}=2\left[\ln(x^4+3)\right]\cdot\frac{1}{x^4+3}\cdot 4x^3$$
$$=\frac{8x^3\ln(x^4+3)}{x^4+3}$$

**35.** $y=\log_4 x=\frac{\ln x}{\ln 4}=\frac{1}{\ln 4}(\ln x)$

$$\frac{dy}{dx}=\frac{1}{\ln 4}\cdot\frac{1}{x}=\frac{1}{x\ln 4}$$

**37.** $y=\log_6(x^4-4x^3+1)=\frac{1}{\ln 6}\ln(x^4-4x^3+1)$

$$\frac{dy}{dx}=\frac{1}{\ln 6}\cdot\frac{1}{x^4-4x^3+1}\cdot(4x^3-12x^2)$$
$$=\frac{4x^3-12x^2}{(x^4-4x^3+1)\ln 6}$$

**39.** $y=x\ln x$

$y'=x\cdot\frac{1}{x}+(\ln x)(1)=1+\ln x$

Set $y'=0$.

$1+\ln x=0$ gives $\ln x=-1$ or $x=e^{-1}$

$y=e^{-1}\ln e^{-1}=e^{-1}(-1)=-e^{-1}$

Rel min at $(e^{-1}, -e^{-1})$.

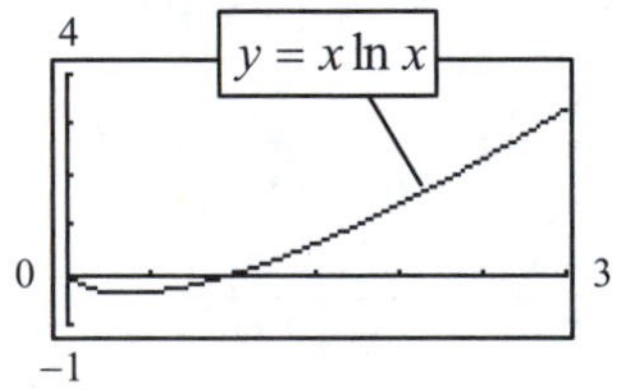


**41.** $y=x^2-8\ln x$

$y'=2x-8\cdot\frac{1}{x}$

$2x^2-8=0$ or $x^2-4=0$ gives $x=\pm 2$.

Note: $\ln(-2)$ is not defined.

$x=2$ gives $y=2^2-8\ln 2=4-8\ln 2$

Rel min at $(2,\ 4-8\ln 2)$.

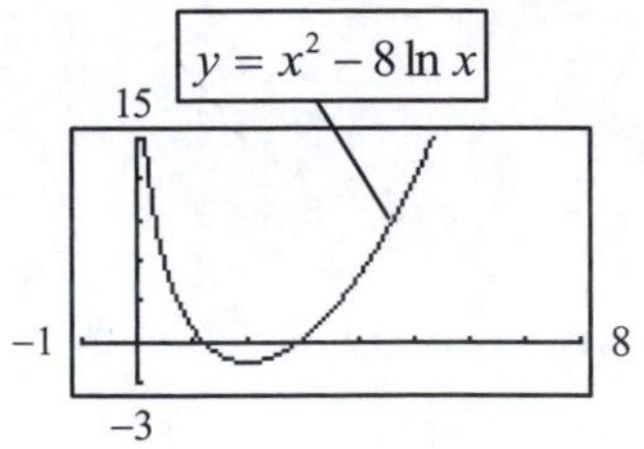


**43.** $C(x)=1500+200\ln(2x+1)$

**a.** $C'(x)=200\cdot\frac{1}{2x+1}\cdot 2=\frac{400}{2x+1}$

**b.** $C(200)=\frac{400}{401}\approx \$1.00$

It will cost approximately \$1.00 to make the next unit.

**c.** $\overline{MC}>0$. Yes.

**45.** $R(x)=\frac{2500x}{\ln(10x+10)}$

**a.** $R'(x)=\frac{2500\ln(10x+10)-2500x\left(\frac{1}{10x+10}\cdot 10\right)}{(\ln(10x+10))^2}=\frac{2500\ln(10x+10)-\frac{2500x}{x+1}}{(\ln(10x+10))^2}$

$$=\frac{2500(x+1)\ln(10x+10)-2500x}{(x+1)\left[\ln(10x+10)\right]^2}=\frac{2500\left[(x+1)\ln(10x+10)-x\right]}{(x+1)\left[\ln(10x+10)\right]^2}$$

**b.** $R'(100)=\frac{2500(101)\ln 1010-250{,}000}{101(\ln 1010)^2}=309.67$

Selling one additional unit yields \$309.67.

**47.** $p(x) = \dfrac{4000}{\ln(x+10)}$ $\quad \dfrac{dp}{dx} = \dfrac{(\ln(x+10))(0) - \frac{4000}{x+10}}{(\ln(x+10))^2} = -\dfrac{4000}{(x+10)(\ln(x+10))^2}$

**a.** When $x = 40$, $\dfrac{dp}{dx} = -\dfrac{4000}{(40+10)(\ln(40+10))^2} = -5.23$

**b.** When $x = 90$, $\dfrac{dp}{dx} = -\dfrac{4000}{(90+10)(\ln(90+10))^2} = -1.89$

**c.** $p''(x) = \dfrac{0 + 4000\left[(x+10)2\ln(x+10)\cdot\frac{1}{x+10} + (\ln(x+10))^2(1)\right]}{\left[(x+10)(\ln(x+10))^2\right]^2} = \dfrac{4000\left[2\ln(x+10) + (\ln(x+10))^2\right]}{\left[(x+10)(\ln(x+10))^2\right]^2}$

When $x = 40$, $p''(x) > 0$. Thus, $p'(x)$ is increasing at 40 units.

**49.**

$y = A\ln t - Bt + C$

$y' = \dfrac{A}{t} - B$

Solving $0 = \dfrac{A}{t} - B$ for $t$ gives $t = \dfrac{A}{B}$.

$y'' = -\dfrac{A}{t^2}$, which is $< 0$ for all $t$.

Thus, $t = \dfrac{A}{B}$ is a maximum.

**51.** $R = \dfrac{1}{\ln 10}(\ln I - \ln I_0)$

$\dfrac{dR}{dI} = \dfrac{1}{\ln 10}\left(\dfrac{1}{I} - 0\right) = \dfrac{1}{I\ln 10}$

**53. a.** $y = -31.7 + 18.7\ln x$

**b.** $y' = \dfrac{18.7}{x}$

**c.** The year 2020 corresponds to $x = 40$.

$y'(40) \approx 0.47$

In 2020, obesity is predicted to increase by 0.47 percentage points per year.

## *Exercises 11.2*

**1.** $y = 5e^x - x$

$y' = 5e^x - 1$

**3.** $f(x) = e^x - x^e$

$f'(x) = e^x - ex^{e-1}$

**5.** $g(x) = 500\left(1 - e^{-0.1x}\right)$

$\dfrac{dy}{dx} = -0.1\cdot 500\left(1 - e^{-0.1x}\right) = -50\left(1 - e^{-0.1x}\right)$

**7.** $y = e^{x^3}$

$\dfrac{dy}{dx} = e^{x^3}\cdot 3x^2 = 3x^2e^{x^3}$

**9.** $y = 6e^{3x^2}$

$\dfrac{dy}{dx} = 6e^{3x^2}\cdot 6x = 36xe^{3x^2}$

**11.** $y = 2e^{(x^2+1)^3}$

$\dfrac{dy}{dx} = 2e^{(x^2+1)^3}\cdot 3(x^2+1)^2(2x)$

$= 12x(x^2+1)^2e^{(x^2+1)^3}$

**13.** $y = e^{\ln x^3} = x^3$

$y' = 3x^2$

**15.** $y = e^{-1/x} = e^{-(x^{-1})}$

$y' = e^{-(x^{-1})}\cdot(1x^{-2}) = \dfrac{e^{-1/x}}{x^2}$

**17.** $y = e^{-1/x^2} + e^{-x^2} = e^{-(x^{-2})} + e^{-x^2}$

$y' = e^{-(x^{-2})}(2x^{-3}) + e^{-x^2}(-2x) = \dfrac{2e^{-1/x^2}}{x^3} - 2xe^{-x^2}$

**19.** $s = t^2e^t$

$s' = t^2(e^t) + e^t(2t) = te^t(t+2)$

**21.**

$y = e^{x^4} - \left(e^x\right)^4$

$y' = e^{x^4} \cdot 4x^3 - 4\left(e^x\right)^3 \cdot e^x = 4x^3 e^{x^4} - 4e^{4x}$

**23.** $y = \ln(e^{4x} + 2)$

$y' = \dfrac{1}{e^{4x}+2} \cdot e^{4x}(4) = \dfrac{4e^{4x}}{e^{4x}+2}$

**25.** $y = e^{-3x}\ln(2x)$

$y' = e^{-3x} \cdot \dfrac{1}{2x} \cdot 2 + \ln(2x) \cdot e^{-3x}(-3)$

$= \dfrac{e^{-3x}}{x} - 3e^{-3x}\ln(2x)$

**27.** $y = \dfrac{1+e^{5x}}{e^{3x}} = \dfrac{1}{e^{3x}} + e^{5x-3x} = e^{-3x} + e^{2x}$

$y' = e^{-3x}(-3) + e^{2x}(2) = 2e^{2x} - 3e^{-3x}$

**29.** $y = (e^{3x} + 4)^{10}$

$y' = 10(e^{3x}+4)^9(e^{3x} \cdot 3) = 30e^{3x}(e^{3x}+4)^9$

**31.** $y = 6^x$

$y' = 6^x \cdot \ln 6$

**33.** $y = 4^{x^2}$

$y' = 4^{x^2}(2x \ln 4)$

**35.** **a.** $y = xe^{-x}$

$y' = xe^{-x}(-1) + e^{-x}(1) = e^{-x} - xe^{-x}$

$y'(1) = e^{-1} - 1e^{-1} = 0$

**b.** If $x = 1$, then $y = 1e^{-1} = e^{-1}$.

$y - e^{-1} = 0(x-1)$

$y = e^{-1}$

**37.** **a.** $y = \dfrac{1}{\sqrt{2\pi}} e^{-z^2/2}$

$y' = \dfrac{1}{\sqrt{2\pi}} e^{-z^2/2}(-z)$

$y' = 0$ at $z = 0$.

Maximum occurs at $z = 0$.

**b.**

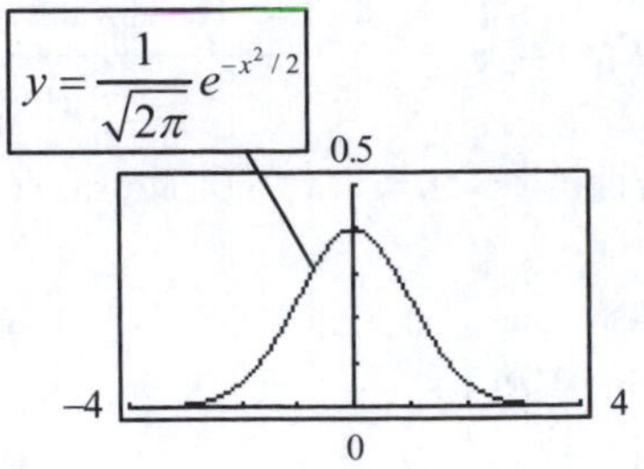


**39.** $y = \dfrac{e^x}{x}$

$y' = \dfrac{xe^x - e^x(1)}{x^2} = \dfrac{e^x(x-1)}{x^2}$

$y' = 0$ if $e^x(x-1) = 0$.

Since $e^x \neq 0$ for any $x$, the critical value is $x = 1$.

Relative minimum at $(1, e)$.

**41.** $y = x - e^x$

$y' = 1 - e^x$

$1 - e^x = 0$ gives $x = 0$ as the critical value.

Relative maximum at $(0, -1)$.

**43.** $S = Pe^{0.1n}$

**a.** $\dfrac{dS}{dn} = Pe^{0.1n}(0.1)$

**b.** When $n = 1$, $\dfrac{dS}{dn} = 0.1Pe^{0.1}$

**c.** Yes, since $e^{0.1} > 1$.

**45.** $S(t) = 100{,}000e^{-0.5t}$

**a.** $S'(t) = 100{,}000e^{-0.5t}(-0.5) = -50{,}000e^{-0.5t}$

**b.** The function is decay exponential. The derivative is always negative.

**47.** $C(x) = 10{,}000 + 20xe^{x/600}$

$C'(x) = 20x \cdot e^{x/600} \cdot \dfrac{1}{600} + e^{x/600}(20)$

$= e^{x/600}\left(\dfrac{x}{30} + 20\right)$

$C'(600) = e(20+20) = 40e \approx \$108.73/\text{unit}$

**49.** $y = 100(1 - e^{-0.462t}) = 100 - 100e^{-0.462t}$

**a.** $y' = -100e^{-0.462t}(-0.462) = 46.2e^{-0.462t}$

**b.** $y'(1) = 46.2e^{-0.462(1)} = 29.107$

**51.** $x(t) = 0.05 + 0.18e^{-0.38t}$

$x'(t) = (0.18)e^{-0.38t}(-0.38) = -0.068e^{-0.38t}$

**53.** $H = 624e^{0.07t}$

$N' = 624e^{0.07t} \cdot 0.07 = 43.68e^{0.07t}$

$N'(20) \approx 177.1$ ($billion/year)

**55.** $\dfrac{I}{I_0} = 10^R$

Multiplying both sides by $I_0$ to be in the form $y = a^u$, we obtain $I = I_0 10^R$, so

$I' = I_0 10^R \ln 10.$

If $I_0 = 1$, $\dfrac{dI}{dR} = 10^R \ln 10.$

**57. a.** $d(t) = 1.60e^{0.083t}$

$d'(t) = 1.60e^{0.083t} \cdot 0.083 = 0.1328e^{0.083t}$

**b.** $d'(50) \approx \$8.42$ billion per year

$d'(125) \approx \$4256$ billion per year

**59.** $y = \dfrac{10,000}{1 + 9999e^{-0.99t}}$

$$y' = \frac{-10,000 \cdot (-0.99 \cdot 9999)e^{-0.99t}}{\left(1 + 9999e^{-0.99t}\right)^2} = \frac{98,990,100e^{-0.99t}}{\left(1 + 9999e^{-0.99t}\right)^2}$$

**61. a.** $P(t) = \dfrac{250}{1 + 1.91e^{-0.0280t}}$

$$P'(t) = \frac{0 - \left(-0.0280 \cdot 1.91e^{-0.0280t}\right)(250)}{\left(1 + 1.91e^{-0.0280t}\right)^2}$$

$$= \frac{13.37e^{-0.0280t}}{\left(1 + 1.19e^{-0.0280t}\right)^2}$$

$P'(45) \approx 1.60$; The rate of increase in this population was approximately 1.60 million per year in 1995.

**b.** $P'(90) \approx 0.808$; The rate of increase in this population is predicted to be 0.808 million (808 thousand) per year in 2040.

**c.** The rate of increase in 2040 is approximately half of the rate of increase in 1995.

**63. a.** $y = 0.544\left(1.07^x\right)$

**b.** $y' = 0.544 \ln(1.07) \cdot \left(1.07^x\right) = 0.0368\left(1.07^x\right)$

**c.** $y'(40) \approx 0.55$; Obesity is predicted to increase at the rate of 0.55 percentage points per year in 2020.

**65. a.** $y = 165.550\left(1.055^x\right)$

$y = 165.550\left(1.055^x\right)$

$y' = 165.550\left(1.055^x\right) \cdot \ln 1.055 = 8.864\left(1.055^x\right)$

**b.** $y'(35) = \$57.73$ billion

**67. a.** $y = 42.1\left(1.04^x\right)$

**b.** $y' = 42.1\left(1.04^x\right)(\ln 1.04)$

$y' \approx 1.65\left(1.04^x\right)$

**c.** The year 2040 corresponds to $x = 30$.

$y'(30) \approx 5.35$

The rate of growth of the average annual wage in 2040 is predicted to be 5.35 thousand dollars per year.

# Chapter 11: Derivatives Continued

## Exercises 11.3

**1.** $x^2 - 4y - 17 = 0$

$2x - 4\frac{dy}{dx} = 0$

At $(1, -4)$, $\frac{dy}{dx} = \frac{1}{2}$.

**3.** $xy^2 = 8$

$x \cdot 2y\frac{dy}{dx} + y^2 = 0$

At (2, 2), $8\frac{dy}{dx} + 4 = 0.$

$8\frac{dy}{dx} = -4$

$\frac{dy}{dx} = -\frac{1}{2}$

**5.** $x^2 + 3xy - 4 = 0$

$2x + 3\left[x\frac{dy}{dx} + y(1)\right] = 0$

At (1, 1), $2(1)+3\left[(1)\frac{dy}{dx} + (1)(1)\right] = 0.$

$2 + 3\frac{dy}{dx} + 3 = 0$

$3\frac{dy}{dx} = -5$

$\frac{dy}{dx} = -\frac{5}{3}$

**7.** $x^2 + 2y^2 - 4 = 0$

$2x + 4y\frac{dy}{dx} - 0 = 0$

$\frac{dy}{dx} = -\frac{2x}{4y} = -\frac{x}{2y}$

**9.** $x^2 + 4x + y^2 - 3y + 1 = 0$

$2x + 4 + 2y\frac{dy}{dx} - 3\frac{dy}{dx} + 0 = 0$

$\frac{dy}{dx} = -\frac{2x+4}{2y-3}$

**11.** $x^2 + y^2 = 4$

$2x + 2yy' = 0$

$y' = -\frac{x}{y}$

**13.** $xy^2 - y^3 = 1$

$x \cdot 2y\frac{dy}{dx} + y^2(1) - 3y^2\frac{dy}{dx} = 0$

$\frac{dy}{dx}(2xy - 3y^2) = -y^2$

$\frac{dy}{dx} = \frac{-y^2}{2xy - 3y^2}$

$\frac{dy}{dx} = \frac{-y}{2x - 3y}$

**15.** $p^2q = 4p - 2$

$p^2(1) + q \cdot 2p\frac{dp}{dq} = 4\frac{dp}{dq}$

$(2qp - 4)\frac{dp}{dq} = -p^2$

$\frac{dp}{dq} = \frac{p^2}{4 - 2qp}$

**17.** $3x^5 - 5y^3 = 5x^2 + 3y^5$

$15x^4 - 15y^2\frac{dy}{dx} = 10x + 15y^4\frac{dy}{dx}$

$\frac{dy}{dx}(-15y^2 - 15y^4) = 10x - 15x^4$

$y' = -\frac{5\left(2x - 3x^4\right)}{15\left(y^2 + y^4\right)} = -\frac{2x - 3x^4}{3\left(y^2 + y^4\right)}$

**19.** $x^4 + 2x^3y^2 = x - y^3$

$4x^3 + 2\left[x^3 \cdot 2y\frac{dy}{dx} + y^2 \cdot 3x^2\right] = 1 - 3y^2\frac{dy}{dx}$

$3y^2\frac{dy}{dx} + 2x^3 \cdot 2y\frac{dy}{dx} = 1 - 4x^3 - 6x^2y^2$

$\frac{dy}{dx} = \frac{1 - 4x^3 - 6x^2y^2}{3y^2 + 4x^3y}$

**21.**
$$x^4+3x^3y^2-2y^5=(2x+3y)^2$$
$$4x^3+3x^3\cdot 2y\frac{dy}{dx}+y^2\cdot 9x^2-10y^4\frac{dy}{dx}=2(2x+3y)\left(2+3\frac{dy}{dx}\right)$$
$$(6x^3y-10y^4)\frac{dy}{dx}=-4x^3-9x^2y^2+8x+12x\frac{dy}{dx}+12y+18y\frac{dy}{dx}$$
$$(6x^3y-10y^4-12x-18y)\frac{dy}{dx}=-4x^3-9x^2y^2+8x+12y$$
$$\frac{dy}{dx}=\frac{-4x^3-9x^2y^2+8x+12y}{6x^3y-10y^4-12x-18y}=\frac{4x^3+9x^2y^2-8x-12y}{10y^4+12x+18y-6x^3y}$$

**23.** $x^2+4x+y^2+2y-4=0$

$2x+4+2yy'+2y'=0$

$2(y+1)y'=-2(x+2)$

$$y'=-\frac{x+2}{y+1}$$

At $(1, -1)$, $y'$ is undefined.

**25.** $x^2+2xy+3=0$

$2x+2xy'+y(2)=0$

$2xy'=-2x-2y$

$$y'=\frac{-2x-2y}{2x}=-\frac{x+y}{x}$$

At $(-1, 2)$, $y'=1$.

**27.** $x^2-2y^2+4=0$

$2x-4yy'=0$

$$y'=\frac{x}{2y}$$

At $(2, 2)$ we have $y'=\frac{2}{4}=\frac{1}{2}$.

The equation of the tangent line is

$y-2=\frac{1}{2}(x-2)$ or $y=\frac{1}{2}x+1$.

**29.** $4x^2+3y^2-4y-3=0$

$8x+6yy'-4y'=0$

At $(-1, 1)$, we have $8(-1)+6(1)y'-4y'=0$

or $2y'=8$ or $y'=4$. So, $m=4$.

The equation of the tangent line

$y-1=4(x-(-1))$

$y-1=4x+4$

$y=4x+5$.

**31.** $\ln x=y^2$

$$\frac{1}{x}=2y\frac{dy}{dx}$$
$$\frac{dy}{dx}=\frac{1}{2xy}$$

**33.** $y^2\ln x=4$

$$y^2\left(\frac{1}{x}\right)+\ln x(2yy')=0$$
$$y'=\frac{-y^2}{2xy\ln x}=\frac{-y}{2x\ln x}$$

**35.** $y^2\ln x+x^2y=3$

$$y^2\cdot\frac{1}{x}+\ln x\cdot 2y\frac{dy}{dx}+x^2\frac{dy}{dx}+y\cdot 2x=0$$
$$\frac{dy}{dx}\left(2y\ln x+x^2\right)=-\frac{y^2}{x}-2xy$$
$$\frac{dy}{dx}\left(2y\ln x+x^2\right)=-\frac{y^2+2x^2y}{x}$$
$$\frac{dy}{dx}=-\frac{y^2+2x^2y}{2xy\ln x+x^3}$$

At $(1, 3)$, $\frac{dy}{dx}=-\frac{3^2+2(1)^2(3)}{2(1)(3)\ln 1+1^3}=-15.$

**37.**
$$xe^y=6$$
$$xe^y\cdot\frac{dy}{dx}+e^y(1)=0$$
$$\frac{dy}{dx}=\frac{-e^y}{xe^y}=-\frac{1}{x}$$

**39.** $xe^{xy} = 4$

$$xe^{xy}\left(x \cdot \frac{dy}{dx} + y \cdot 1\right) + e^{xy} \cdot 1 = 0$$

$$x^2e^{xy}\frac{dy}{dx} + xye^{xy} + e^{xy} = 0$$

$$x^2e^{xy}\frac{dy}{dx} = -xye^{xy} - e^{xy}$$

$$\frac{dy}{dx} = \frac{-xye^{xy} - e^{xy}}{x^2e^{xy}} = \frac{-xy-1}{x^2}$$

**41.** $ye^x - y = 3$

$$y \cdot e^x + e^x \cdot \frac{dy}{dx} - \frac{dy}{dx} = 0$$

$$\frac{dy}{dx} = \frac{ye^x}{1-e^x}$$

**43.** $ye^x = y^2 + x - 2$

$$ye^x + e^x\frac{dy}{dx} = 2y\frac{dy}{dx} + 1$$

$$ye^x - 1 = 2y\frac{dy}{dx} - e^x\frac{dy}{dx}$$

$$ye^x - 1 = \frac{dy}{dx}\left(2y - e^x\right)$$

$$\frac{dy}{dx} = \frac{ye^x - 1}{2y - e^x}$$

$$\left.\frac{dy}{dx}\right|_{(0,2)} = \frac{2e^0 - 1}{2(2) - e^0} = \frac{2-1}{4-1} = \frac{1}{3}$$

**45.** $xe^y = 2y + 3$

$$xe^y\frac{dy}{dx} + e^y = 2\frac{dy}{dx}$$

$$e^y = 2\frac{dy}{dx} - xe^y\frac{dy}{dx}$$

$$e^y = \frac{dy}{dx}\left(2 - xe^y\right)$$

$$\frac{dy}{dx} = \frac{e^y}{2 - xe^y}$$

$$\left.\frac{dy}{dx}\right|_{(3,0)} = \frac{e^0}{2 - 3e^0} = \frac{1}{2-3} = -1$$

Line: $y - 0 = -1(x-3)$

$y = -x + 3$

**47.** $x^2 + 4y^2 - 4x - 4 = 0$

$$2x + 8yy' - 4 = 0$$

$$y' = \frac{4-2x}{8y} = \frac{2-x}{4y}$$

**a.** There is a horizontal tangent if $y' = 0$ or $x = 2$. Then,

$$4 + 4y^2 - 8 - 4 = 0$$

$$4y^2 = 8$$

$y^2 = 2$ or $y = \pm\sqrt{2}$.

Horizontal tangents at $(2, \sqrt{2})$ and at $(2, -\sqrt{2})$.

**b.** There is a vertical tangent when $y = 0$. Then, $x^2 - 4x - 4 = 0$ gives $x = 2 \pm 2\sqrt{2}.$ Vertical tangents at $(2+2\sqrt{2}, 0)$ and at $(2-2\sqrt{2}, 0)$.

**49.** $y' = -\frac{x}{y} = \frac{-x}{y}$

**a.** $y'' = \frac{y(-1)-(-x)y'}{y^2} = \frac{-y+xy'}{y^2}$

**b.** $y'' = \frac{-y + x\left(-\frac{x}{y}\right)}{y^2} \cdot \frac{y}{y}$

$$= \frac{-y^2 - x^2}{y^3} = \frac{-\left(x^2 + y^2\right)}{y^3}$$

**c.** Yes. $x^2 + y^2 = 4$

**51.** $\sqrt{x} + \sqrt{y} = 1$

$$\frac{1}{2}x^{-1/2} + \frac{1}{2}y^{-1/2}y' = 0$$

$$y' = -\frac{x^{-1/2}}{y^{-1/2}} = -\frac{y^{1/2}}{x^{1/2}}$$

$$y'' = \frac{-x^{1/2} \cdot \frac{1}{2}y^{-1/2}y' - (-y^{1/2}) \cdot \frac{1}{2}x^{-1/2}}{x}$$

$$= \frac{\frac{y^{1/2}}{x^{1/2}} - \frac{x^{1/2}}{y^{1/2}}y'}{2x} = \frac{\frac{y^{1/2}}{x^{1/2}} - \frac{x^{1/2}}{y^{1/2}} \cdot \frac{-y^{1/2}}{x^{1/2}}}{2x}$$

$$= \frac{\frac{y^{1/2}}{x^{1/2}} + 1}{2x} \cdot \frac{x^{1/2}}{x^{1/2}} = \frac{y^{1/2} + x^{1/2}}{2x^{3/2}}$$

Since $y^{1/2} + x^{1/2} = 1$ we have

$$y'' = \frac{1}{2x^{3/2}} = \frac{1}{2x\sqrt{x}}.$$

**53.** $x^2 + y^2 - 9 = 0$

$$2x + 2yy' = 0$$

$$y' = \frac{-2x}{2y} = -\frac{x}{y}$$

$-\frac{x}{y} = 0$ gives $x = 0$.

Max at $x = 0$, $y = 3$, and min at $x = 0$, $y = -3$.

**55.** $xy - 20x + 10y = 0$

$xy' + y(1) - 20 + 10y' = 0$

$y' = \frac{20 - y}{x + 10}$

At $x = 10$, we have $y = 10$.

Thus $y' = \frac{20 - 10}{10 + 10} = \frac{1}{2} = 0.5.$

**57.**

$(x+1)^{3/4}(y+2)^{1/3} = 384$

$(x+1)^{3/4} \cdot \frac{1}{3}(y+2)^{-2/3} \cdot y' + (y+2)^{1/3} \cdot \frac{3}{4}(x+1)^{-1/4} \cdot 1 = 0$

Multiplying both sides by $\left(12(x+1)^{1/4}(y+2)^{2/3}\right)$ and simplifying and solving for $y'$ we have $y' = \frac{-9(y+2)}{4(x+1)}$.

At

$(255, 214)$ we have $y' = \frac{-9(216)}{4(256)} = -\frac{243}{128} \approx -1.898.$

**59.** $p(q+1)^2 = 200,000$

$p \cdot 2(q+1)\frac{dq}{dp} + (q+1)^2 \cdot 1 = 0$

$\frac{dq}{dp} = -\frac{q+1}{2p}$

At $p = 80$, we have $q = 49$.

Thus, $\frac{dq}{dp} = \frac{-50}{160} = \frac{-5}{16}$.

If the price is increased by \$1.00, the demand will decrease by $\frac{5}{16}$ unit.

**61.** $-0.000436t = \ln y - \ln(100)$

$-0.000436 = \frac{1}{y} \cdot \frac{dy}{dt}$

So, $\frac{dy}{dt} = -0.000436y$

**63.** $\text{THI} = t - 0.55(1-h)(t-58)$

$0 = 1 - 0.55\left[(1-h)(1) + (t-58)\left(-\frac{dh}{dt}\right)\right]$

$0.55(t-58)\frac{dh}{dt} = -1 + 0.55(1-h)$

$\frac{dh}{dt} = \frac{-0.55h - 0.45}{0.55(t-58)}$

At $t = 70$, $\frac{dh}{dt} = \frac{-0.55h - 0.45}{0.55(12)} = \frac{-h}{12} - \frac{3}{44}$

## *Exercises 11.4*

**1.** $y = x^3 - 3x$

$\frac{dy}{dt} = 3x^2\frac{dx}{dt} - 3\frac{dx}{dt}$

If $x = 2$ and $\frac{dx}{dt} = 4$,

$\frac{dy}{dt} = 3(2)^2(4) - 3(4) = 36$

**3.** $xy = 4$ gives $y = \frac{4}{x}$

$\frac{dy}{dt} = -\frac{4}{x^2} \cdot \frac{dx}{dt}$

If $x = 8$ and $\frac{dx}{dt} = -2$,

$\frac{dy}{dt} = -\frac{4}{8^2}(-2) = \frac{1}{8}$

**5.** $x^2 + y^2 = 169$

$\frac{d}{dt}(x^2 + y^2) = \frac{d}{dt}(169)$

$2x \cdot \frac{dx}{dt} + 2y \cdot \frac{dy}{dt} = 0$

$\frac{dx}{dt} = -\frac{y}{x} \cdot \frac{dy}{dt}$

$\frac{dx}{dt} = -\frac{12}{5} \cdot 2 = -\frac{24}{5}$

**7.** $y^2 = 2xy + 24$

$\frac{d}{dt}(y^2) = \frac{d}{dt}(2xy + 24)$

$2y \cdot \frac{dy}{dt} = 2x \cdot \frac{dy}{dt} + 2y \cdot \frac{dx}{dt} + 0$

$\frac{dx}{dt} = \frac{y-x}{y} \cdot \frac{dy}{dt}$

$\frac{dx}{dt} = \frac{12-5}{12} \cdot 2 = \frac{7}{6}$

**9.** $x^2 + y^2 = z^2$

$\frac{d}{dt}(x^2 + y^2) = \frac{d}{dt}(z^2)$

$2x \cdot \frac{dx}{dt} + 2y \cdot \frac{dy}{dt} = 2z \cdot \frac{dz}{dt}$

$\frac{dy}{dt} = \frac{z \cdot \frac{dz}{dt} - x \cdot \frac{dx}{dt}}{y}$

$x^2 + y^2 = z^2$ yields $3^2 + 4^2 = z^2$.

So, $z^2 = 25$ or $z = \pm 5$.

When $z = 5$, $\frac{dy}{dt} = \frac{5 \cdot 2 - 3 \cdot 10}{4} = -5$.

When $z = -5$, $\frac{dy}{dt} = \frac{-5 \cdot 2 - 3 \cdot 10}{4} = -10$.

**11.** $y = -4x^2$

$\frac{dy}{dt} = -8x \cdot \frac{dx}{dt} = -8(5)(2) = -80$ units/sec.

**13.** $A = \pi r^2$

$\frac{dA}{dt} = A'(t) = 2\pi r \cdot \frac{dr}{dt}$

$A'(t) = 2\pi \cdot 3 \cdot 2 = 12\pi$ sq ft/min

**15.** $V = x^3$

$V'(t) = 3x^2 \cdot \frac{dx}{dt}$

$64 = 3 \cdot 36 \cdot \frac{dx}{dt}$

$\frac{dx}{dt} = \frac{64}{108} = \frac{16}{27}$ in./sec

**17.** $P = 180x - \frac{1}{1000}x^2 - 2000$

$P'(t) = 180 \cdot \frac{dx}{dt} - \frac{1}{500}x \cdot \frac{dx}{dt}$

If $x = 100$ and $\frac{dx}{dt} = 10$, then

$P'(t) = 180(10) - \frac{1}{500}(100)(10)$

$= \$1798/\text{day}$

**19.** $p = \frac{1000 - 10x}{400 - x}$

$\frac{dp}{dt} = \frac{(400-x)(-10) - (1000-10x)(-1)}{(400-x)^2} \cdot \frac{dx}{dt}$

If $\frac{dx}{dt} = -20$ and $x = 20$,

then $\frac{dp}{dt} = \frac{380(-10) - 800(-1)}{(380)^2}(-20)$

$= \frac{(-3000)(-20)}{380 \cdot 380}$

$= 0.42$ dollars/day.

**21.** $x = 30y + 20y^2$

$\frac{dx}{dt} = 30\frac{dy}{dt} + 40y\frac{dy}{dt}$

If $y = 10$ and $\frac{dy}{dt} = 1$, then

$\frac{dx}{dt} = 30(1) + 40(10)(1) = 430$ units/month

Note: $y$ is the number of thousands.

**23.** $V = \frac{4}{3}\pi r^3$

$\frac{dV}{dt} = 3 \cdot \frac{4}{3}\pi r^2 \cdot \frac{dr}{dt}$

Substituting $\frac{dr}{dt} = -1$ and $r = 3$, we have

$\frac{dV}{dt} = 4\pi \cdot 3^2(-1)$

$= -36\pi \text{ mm}^3/\text{month}$.

$V$ is decreasing at the rate of $36\pi$ mm$^3$/month.

**25.** $W = kL^3$

$$\frac{dW}{dt} = 3kL^2 \frac{dL}{dt}$$

Percentage rate of change:

$$\frac{\frac{dW}{dt}}{W} = \frac{3kL^2 \frac{dL}{dt}}{W}$$

$$\frac{\frac{dW}{dt}}{W} = \frac{3kL^2 \frac{dL}{dt}}{kL^3} = \frac{3\frac{dL}{dt}}{L} = 3\left(\frac{\frac{dL}{dt}}{L}\right)$$

**27.** $C = 0.11W^{1.54}$

$$\frac{dC}{dt} = 1.54\left(0.11W^{0.54}\right) \cdot \frac{dW}{dt}$$

$$= 0.1694W^{0.54} \cdot \frac{dW}{dt}$$

$$\frac{\frac{dC}{dt}}{C} = \frac{0.1694W^{0.54} \cdot \frac{dW}{dt}}{C}$$

$$= \frac{0.1694W^{0.54} \frac{dW}{dt}}{0.11W^{1.54}} = 1.54\left(\frac{\frac{dW}{dt}}{W}\right)$$

**29.** $V = \frac{4}{3}\pi r^3$

$$\frac{dV}{dt} = \frac{4}{3}\pi \cdot 3r^2 \cdot \frac{dr}{dt}$$

Substituting $\frac{dV}{dt} = 4,\ r = 2$ , we have

$4 = 4\pi \cdot 2^2 \cdot \frac{dr}{dt}$ or $\frac{dr}{dt} = \frac{1}{4\pi}$ micrometers/day.

**31.** $V = \frac{4}{3}\pi r^3$

$$\frac{dV}{dt} = \frac{4}{3}\pi \cdot 3r^2 \cdot \frac{dr}{dt}$$

Substituting $\frac{dV}{dt} = 5 \text{ in}^3/\text{min}$, $r = 5$ gives

$5 = 4\pi \cdot 5^2 \cdot \frac{dr}{dt}$ or $\frac{dr}{dt} = \frac{5}{100\pi} = \frac{1}{20\pi}$ in./min.

**33.** Let $y$ be the height of the ladder on the wall and let $x$ be the distance from the wall to the bottom of the ladder.

$$\frac{dx}{dt} = 1 \text{ ft/sec}$$

$$30^2 = x^2 + y^2$$

$$2x\frac{dx}{dt} + 2y\frac{dy}{dt} = 0$$

When $x = 18,\ y = \sqrt{576} = 24$.

$$2(18)(1) + 2(24)\frac{dy}{dt} = 0$$

$48\frac{dy}{dt} = -36$ or $\frac{dy}{dt} = -\frac{36}{48} = -0.75$ ft/sec

The ladder is sliding down at the rate of $\frac{3}{4}$ ft/sec.

**35.** Let $x$ be the horizontal distance from the plane to the observer and let $y$ be the direct distance.

$$\frac{dx}{dt} = -300 \text{ mph}$$

$$y^2 = x^2 + 1^2$$

$$2y\frac{dy}{dt} = 2x\frac{dx}{dt}$$

When $y = 5,\ x = \sqrt{24}$.

$2(5)\frac{dy}{dt} = 2\sqrt{24}(-300)$ or

$\frac{dy}{dt} = -\sqrt{24}(60) \approx -293.94$ mph.

The plane is approaching the observer at the rate of 293.94 mph.

**37.** Let $x$ be the distance of car A from the intersection, let $y$ be the distance of car B from the intersection, and let $z$ be the distance between them.

$$\frac{dx}{dt} = -40,\ \frac{dy}{dt} = -55$$

$$z^2 = x^2 + y^2$$

$$2z\frac{dz}{dt} = 2x\frac{dx}{dt} + 2y\frac{dy}{dt}$$

When $x = 15$ and $y = 8$, $z = 17$.

$2(17)\frac{dz}{dt} = 2(15)(-40) + 2(8)(-55)$ or

$\frac{dz}{dt} \approx -61.18$ mph.

The distance between the cars is decreasing at the rate of 61.18 mph.

**39.** $V = 10(25)(r) = 250r$, $\frac{dV}{dt} = 10$

Find $\frac{dr}{dt}$ when $r = 4$.

$\frac{dV}{dt} = 250\frac{dr}{dt}$; $10 = 250\frac{dr}{dt}$ or $\frac{dr}{dt} = \frac{1}{25}$ ft/hr.

## *Exercises 11.5*

**1.** $p + 4q = 80$

$1 + 4\frac{dq}{dp} = 0$ gives $\frac{dq}{dp} = -\frac{1}{4}$

**a.** At (10, 40) we have $\eta = -\frac{40}{10}\left(-\frac{1}{4}\right) = 1$.

**b.** Since the demand is unitary, there will be no change in the revenue with a price increase.

**3.** $p^2 + 2p + q = 49$

$2p + 2 + \frac{dq}{dp} = 0$ gives $\frac{dq}{dp} = -2p - 2$

**a.** At (1, 6) we have $\eta = -\frac{6}{1}(-14) = 84$.

**b.** Since the demand is elastic, an increase in price will decrease the total revenue.

**5.** $pq + p = 5000$

**a.** $p \cdot \frac{dq}{dp} + q \cdot 1 + 1 = 0$ gives $\frac{dq}{dp} = \frac{-(q+1)}{p}$.

At (99, 50) the elasticity is

$\eta = -\frac{50}{99}\left(-\frac{100}{50}\right) = \frac{100}{99}$.

**b.** $\eta > 1$, so demand is elastic.

**c.** A price increase will decrease the total revenue.

**7.** $pq + p + 100q = 50,000$

**a.** $p \cdot \frac{dq}{dp} + q \cdot 1 + 1 + 100\frac{dq}{dp} = 0$ gives

$\frac{dq}{dp} = \frac{-(q+1)}{p+100}$. At $p = 401$ we have $q = 99$.

Thus, elasticity is $\eta = -\frac{401}{99}\left(-\frac{100}{501}\right) \approx 0.81$

at $(99, 401)$.

**b.** $\eta < 1$, so demand is inelastic.

**c.** An increase in price will result in an increase in total revenue.

**9.** $p = \frac{1}{2}[\ln(5000 - q) - \ln(q+1)]$

**a.** $1 = \frac{1}{2}\left[\frac{1}{5000-q}(-1) \cdot \frac{dq}{dp} - \frac{1}{q+1} \cdot \frac{dq}{dp}\right]$

$2 = -\left[\frac{q+1+5000-q}{(5000-q)(q+1)}\right] \cdot \frac{dq}{dp}$

$\frac{dq}{dp} = \frac{-2(5000-q)(q+1)}{5001}$

At $p = 3.71$, we were given that $q = 2$.

Elasticity is $\eta = -\frac{3.71}{2}\left(\frac{-29,988}{5001}\right) \approx 11.1$.

**b.** Demand is elastic.

**11.** $p = 120\sqrt[3]{125 - q}$

**a.** $1 = 40(125-q)^{-2/3}(-1) \cdot \frac{dq}{dp}$ gives

$\frac{dq}{dp} = \frac{-(125-q)^{2/3}}{40}$

Elasticity is

$\eta = -\frac{p}{q} \cdot \frac{dq}{dp}$

$= -\frac{120(125-q)^{1/3}}{q} \cdot \frac{-(125-q)^{2/3}}{40}$

$= \frac{3(125-q)}{q}$

**b.** $\eta = 1$ gives $1 = \frac{375 - 3q}{q}$

$4q = 375$ or $q = 93.75$ for unitary elasticity.

Inelastic means $\frac{375-3q}{q} < 1$ or $q > 93.75$.

Elastic means $\frac{375-3q}{q} > 1$ or $0 < q < 93.75$.

**c.** Revenue increases when $\eta > 1$ or $0 < q < 93.75$.

Revenue decreases when $\eta < 1$ or $q > 93.75$.

Revenue is maximized at $q = 93.75$.

**d.** $R = pq = 120q\sqrt[3]{125-q}$

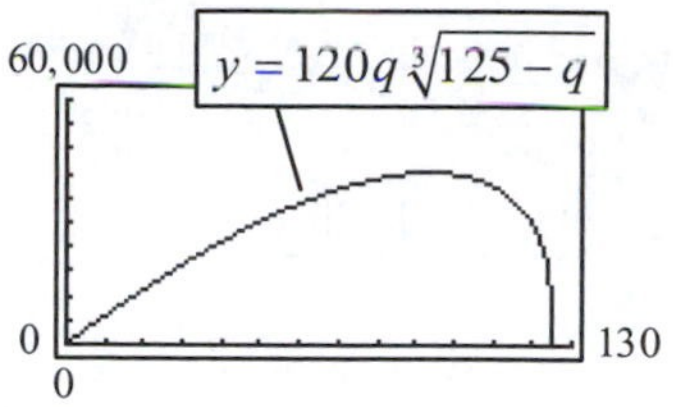


Yes.

**13.** a. $p = 250 - 0.125q$

**b.** $\eta = \dfrac{2000}{q} - 1$

**c.** $\eta \approx 2.33$; elastic. no

**d.** $q = 1000; p = \$125; \max R = \$125,000$

**15.** $\$12$ / item

**17.** After taxation the supply function is $p = 100 + 0.5q + t$.
Then, $100 + 0.5q + t = 800 - 2q$ yields $t = 700 - 2.5q$.
Total tax revenue: $T = t \cdot q = 700q - 2.5q^2$.
$T'(q) = 700 - 5q = 0$ if $q = 140$.
Since $T''(q) < 0$, the tax revenue is maximized at $q = 140$ and $t = 700 - 2.5(140) = \$350$.

**19.** After taxation the supply function is $p = 20 + 3q + t$. Then, $20 + 3q + t = 200 - 2q^2$ yields $t = 180 - 3q - 2q^2$.
Total tax revenue: $T = t \cdot q = 180q - 3q^2 - 2q^3$.
$T'(q) = 180 - 6q - 6q^2$.
$T'(q) = 0$ if $6(6+q)(5-q) = 0$ and $T''(q) < 0$.
Total tax revenue is maximized at $q = 5$ and $t = 180 - 15 - 50 = \$115$ / item.

**21.** After taxation the supply function is $p = 0.02q^2 + 0.55q + 7.4 + t$.
Then, $0.02q^2 + 0.55q + 7.4 + t = 840 - 2q$ yields $t = 832.6 - 2.55q - 0.02q^2$.
$T(q) = 832.6q - 2.55q^2 - 0.02q^3$ and
$T'(q) = 832.6 - 5.1q - 0.06q^2$.
$T'(q) = 0$ if

$$q = \frac{5.1 \pm \sqrt{26.01 + 199.824}}{-0.12} = \frac{5.1 \pm 15.03}{-0.12} \approx 83.$$

Tax per item is
$t = 832.6 - 2.55(83) - 0.02(83)^2 = \$483.17$ / item
.
Total tax revenue (to nearest dollar) = $40,100.

**23.** After taxation the supply function is $p = 300 + 5q + 0.5q^2 + t$.
Then, $300 + 5q + 0.5q^2 + t = 2100 - 10q - 0.5q^2$ gives $t = 1800 - 15q - q^2$.
Total tax revenue $T = t \cdot q = 1800q - 15q^2 - q^3$.
$T'(q) = 1800 - 30q - 3q^2 = 3(30+q)(20-q) = 0$ if $q = 20$. Note $T''(q) < 0$.
Tax per unit that will maximize total tax revenue is $t = 1800 - 300 - 400 = \$1100$.

# Chapter 11: Derivatives Continued

## *Chapter 11 Review Exercises*

**1.** $y = 10e^{3x^2-x}$

$\frac{dy}{dx} = 10e^{3x^2-x}(6x-1)$

**2.** $y = 3\ln(4x+11)$

$y' = 3\cdot\frac{1}{4x+11}\cdot 4 = \frac{12}{4x+11}$

**3.** $p = \ln q - \ln(q^2-1)$

$\frac{dp}{dq} = \frac{1}{q} - \frac{2q}{q^2-1} = \frac{q^2-1-2q^2}{q(q^2-1)} = \frac{-1-q^2}{q(q^2-1)}$

**4.** $y = xe^{x^2}$

$\frac{dy}{dx} = x\left(e^{x^2}\cdot 2x\right) + e^{x^2}(1) = e^{x^2}(2x^2+1)$

**5.** $f(x) = 5e^{2x} - 40e^{-0.1x} + 11$

$f'(x) = 5e^{2x}\cdot 2 - 40\left(e^{-0.1x}\right)(-0.1)$

$= 10e^{2x} + 4e^{-0.1x}$

**6.** $g(x) = \left(2e^{3x+1}-5\right)^3$

$g'(x) = 3\left(2e^{3x+1}-5\right)^2\left(2e^{3x+1}\cdot 3\right)$

$= 18e^{3x+1}\left(2e^{3x+1}-5\right)^2$

**7.** $s = \frac{3}{4}\ln\left(x^{12}-2x^4+5\right)$

$\frac{ds}{dx} = \frac{3}{4}\cdot\frac{1}{x^{12}-2x^4+5}\left(12x^{11}-8x^3\right)$

$= \frac{3\left(3x^{11}-2x^3\right)}{x^{12}-2x^4+5}$

**8.** $\frac{dw}{dt} = \left(t^2+1\right)\cdot\frac{1}{t^2+1}\cdot 2t + \ln\left(t^2+1\right)(2t) - 2t$

$\frac{dw}{dt} = 2t\ln\left(t^2+1\right)$

**9.** $y = 3^{3x-4}$

$\frac{dy}{dx} = 3^{3x-4}(3)\ln 3 = 3^{3x-3}(\ln 3)$

**10.** $y = 1 + \log_8\left(x^{10}\right) = 1 + \frac{10}{\ln 8}\ln x$

$\frac{dy}{dx} = \frac{10}{\ln 8}\cdot\frac{1}{x} = \frac{10}{x\ln 8}$

**11.** $y = \frac{\ln x}{x}$

$\frac{dy}{dx} = \frac{x\cdot\frac{1}{x} - \ln x\cdot 1}{x^2} = \frac{1-\ln x}{x^2}$

**12.** $y = \frac{1+e^{-x}}{1-e^{-x}}$

$\frac{dy}{dx} = \frac{(1-e^{-x})(-e^{-x}) - (1+e^{-x})(e^{-x})}{(1-e^{-x})^2}$

$= \frac{-2e^{-x}}{(1-e^{-x})^2}$

**13.** $y = 4e^{x^3}$

$y' = 4e^{x^3}(3x^2)$

At $x = 1$, $y' = 12e$

Tangent line: $y - 4e = 12e(x-1)$ or

$y = 12ex - 8e$.

**14.** $y = 8 + 3x^2\ln x$ (If $x = 1$, $y = 8$)

$\frac{dy}{dx} = 3x^2\cdot\frac{1}{x} + (\ln x)(6x^2) = 3x + 6x\ln x$

At $x = 1$, $\frac{dy}{dx} = 3$.

Thus, $y - 8 = 3(x-1)$ or $y = 3x + 5$.

**15.** $y\ln x = 5y^2 + 11$

$y\cdot\frac{1}{x} + \ln x\cdot\frac{dy}{dx} = 10\frac{dy}{dx}$ gives

$\frac{dy}{dx} = \frac{y}{x}\div(10-\ln x) = \frac{y}{x(10-\ln x)}$.

**16.**

$e^{xy} = y$

$e^{xy}\left(x\frac{dy}{dx} + y\right) = \frac{dy}{dx}$

$e^{xy}y = \frac{dy}{dx}(1-xe^{xy})$

$\frac{dy}{dx} = \frac{ye^{xy}}{1-xe^{xy}}$

**17.** $y^2 = 4x - 1$

$2y\frac{dy}{dx} = 4$ gives $\frac{dy}{dx} = \frac{2}{y}$

**18.** $x^2+3y^2+2x-3y+2=0$

$$2x+6y\frac{dy}{dx}+2-3\frac{dy}{dx}=0$$

$$\frac{dy}{dx}(6y-3)=-(2x+2)$$

$$\frac{dy}{dx}=-\frac{2x+2}{6y-3}=\frac{2(x+1)}{3(1-2y)}$$

**19.** $3x^2+2x^3y^2-y^5=7$

$$6x+2x^3\cdot 2y\cdot\frac{dy}{dx}+y^2(6x^2)-5y^4\cdot\frac{dy}{dx}=0$$

gives $\dfrac{dy}{dx}=\dfrac{6x+6x^2y^2}{5y^4-4x^3y}=\dfrac{6x(1+xy^2)}{y(5y^3-4x^3)}$

**20.** $x^2+y^2=1$

$$2x+2y\frac{dy}{dx}=0$$

$$\frac{dy}{dx}=\frac{-x}{y}$$

$$y''=\frac{y(-1)-(-x)y'}{y^2}$$

$$=\frac{-y-\frac{x^2}{y}}{y^2}=-\frac{y^2+x^2}{y^3}=-\frac{1}{y^3}$$

**21.** $x^2+4x-3y^2+6=0$

$2x+4-6y\dfrac{dy}{dx}=0$ gives $\dfrac{dy}{dx}=\dfrac{x+2}{3y}$. At $(3,3)$, we have $\dfrac{dy}{dx}=\dfrac{5}{9}$.

**22.** $x^2+4x-3y^2+6=0$

$$2x+4-6y\frac{dy}{dx}=0$$

$$\frac{dy}{dx}=\frac{-2x-4}{-6y}=\frac{x+2}{3y}$$

$\dfrac{dy}{dx}=0$ when $x=-2$.

When $x=-2$, $y=\pm\sqrt{\dfrac{2}{3}}$.

So, horizontal tangents at $\left(-2,\pm\sqrt{\dfrac{2}{3}}\right)$.

**23.** $3x^2-2y^3=10y$

$$6x\frac{dx}{dt}-6y^2\frac{dy}{dt}=10\frac{dy}{dt}$$

$$\frac{dy}{dt}=\frac{6x\cdot\frac{dx}{dt}}{10+6y^2}$$

At $(10,5)$, we have $\dfrac{dy}{dt}=\dfrac{6(10)2}{10+6(25)}=\dfrac{120}{160}=\dfrac{3}{4}$.

**24.** $A=\dfrac{1}{2}xy$

$$\frac{dA}{dt}=\frac{1}{2}x\frac{dy}{dt}+y\left(\frac{1}{2}\frac{dx}{dt}\right)$$

Use $\dfrac{dx}{dt}=2$, $\dfrac{dy}{dt}=5$, $x=4$, and $y=1$.

$$\frac{dA}{dt}=\frac{1}{2}\cdot4\cdot5+1\cdot\frac{1}{2}\cdot2=11\text{ units}^2/\text{min}$$

**25.** $P(t)=96.1+17.4\ln t$

**a.** $P'(t)=\dfrac{17.4}{t}$

**b.** The year 2028 corresponds to $x=58$. $P(58)\approx166.8$; $P'(58)\approx0.3$. These mean that in 2028 this population is projected to be about 166.8 million and growing at about 0.3 million (300,000) per year.

**26.** $L(y)=-14.6+7.10\ln y$

**a.** $L'(y)=\dfrac{7.10}{y}$

**b.** $L(50)\approx13.2$ and $L(80)\approx16.5$. These mean that the expected number of additional years of life expectancy at age 65 is about 13.2 in 2000 and 16.5 in 2030.

**c.** $L'(50)\approx0.14$ and $L'(80)\approx0.09$. These mean that the number of additional years of life expectancy at age 65 is expected to change at the rate of about 0.14 years of life per year in 2000 and 0.09 years of life per year in 2030.

**27. a.** $S=1000e^{0.12n}$

$$\frac{dS}{dn}=1000e^{0.12n}(0.12)$$

$$=120e^{0.12n}$$

**b.** At $n=1$, $\dfrac{dS}{dn}=120e^{0.12(1)}\approx\$135.30/\text{yr}$

At $n=10$, $\dfrac{dS}{dn}=120e^{0.12(10)}\approx\$398.41/\text{yr}$

**28.** $D(t)=10,020e^{0.02292t}$

**a.** $D'(t)=10,020e^{0.02292t}(0.02292)$

$\approx 229.7e^{0.02292t}$

**b.** $D(20)\approx 15,850$ and $D'(20)\approx 363$. These mean than in 2030 total U.S. disposable income is expected to beat about \$15,850 billion and changing at the rate of about \$363 billion per year.

**29.** $A(t)=A_0e^{-0.00002876t}$

$A'(t)=A_0e^{-0.00002876t}(-0.00002876)$

**a.** $A'(24,101)=-0.00002876A_0e^{-0.00002876(24,101)}$

$\approx -0.00001438A_0$

**b.** $A'(1)=-0.00002876A_0$

Note that $e^{-0.00002876}\approx e^0=1$.

**c.** The rate of decay when $t=24,101$ is approximately $-1.44\times10^{-5}$ and the rate when $t=1$ is approximately $-2.88\times10^{-5}$, so the rate of decay at its half-life is less than its rate of decay after 1 year.

**30.** $\overline{C}=600e^{x/600}$

$C=600xe^{x/600}$

$\overline{MC}=C'=600xe^{x/600}\cdot\frac{1}{600}+600e^{x/600}$

$=e^{x/600}(x+600)$

When $x=600$, $\overline{MC}=1200e$.

**31.** $P(t)=60,000e^{-0.0488t}$

$P'(t)=-2970^{-0.0488t}$

$P'(10)=-2970e^{-0.0488(10)}$

$\approx -\$1797.36/\text{yr}$

**32.** $V=\frac{4}{3}\pi r^3$

$\frac{dV}{dt}=4\pi r^2\frac{dr}{dt}$

Using $r=2.5$ and $\frac{dV}{dt}=-1$ gives

$-1=4\pi(2.5)^2\frac{dr}{dt}\ \ \frac{dr}{dt}=-\frac{1}{25\pi}$ mm/min

**33.** Let $y$ be the height of the sign above the workers' hands and let $z$ be the length of the guide line.

$\frac{dy}{dt}=2\text{ft}/\text{min}$

$z^2=y^2+7^2$

$2z\frac{dz}{dt}=\frac{dy}{dt}(2y)+0$

If $z=25$, $y=24$.

$2(25)\frac{dz}{dt}=2(24)(2)$ gives $\frac{dz}{dt}=\frac{48}{25}$ ft/min.

**34.** $S=kA^{1/3}$

$\frac{dS}{dt}=\frac{1}{3}\cdot k\cdot\frac{1}{A^{2/3}}\cdot\frac{dA}{dt}$

$\frac{\frac{dS}{dt}}{S}=\frac{1}{S}\cdot\frac{k}{3}\cdot\frac{1}{A^{2/3}}\cdot\frac{dA}{dt}$

$=\frac{1}{kA^{1/3}}\cdot\frac{k}{3}\cdot\frac{1}{A^{2/3}}\cdot\frac{dA}{dt}=\frac{1}{3}\left(\frac{\frac{dA}{dt}}{A}\right)$

**35.** Yes.

**36.** After taxation, the supply function is $p=400+2q+t$.

$400+2q+t=2800-8q-\frac{1}{3}q^2$ gives

$t=2400-10q-\frac{1}{3}q^2$.

$T(q)=tq=2400q-10q^2-\frac{1}{3}q^3$

$T'(q)=2400-20q-q^2=(60+q)(40-q)$

$T(q)$ is maximized at $q=40$ and

$t=2400-400-\frac{1600}{3}=\$1466.67$ /unit. The maximum tax revenue is $T(40)\approx 58,666.80$.

**37.** After taxation, the supply function is $p=40+20q+t$.

$40+20q+t=\frac{5000}{q+1}$ gives $t=\frac{5000}{q+1}-40-20q$

$T=tq=\frac{5000q}{q+1}-40q-20q^2$ and

$T'(q)=\frac{(q+1)5000-5000q(1)}{(q+1)^2}-40-40q$

Set $T'(q)=0$. Then, $5000-40(q+1)^3=0$ or $q+1=5$ or $q=4$.

So, $t = \frac{5000}{5} - 40 - 80 = \$880/\text{unit}$. The maximum tax revenue is $T(4) \approx \$3520$.

**38. a.** $pq = 27$

$$p \cdot \frac{dq}{dp} + q \cdot 1 = 0 \text{ or } \frac{dq}{dp} = -\frac{q}{p}$$

$$\eta = -\frac{p}{q} \cdot \frac{dq}{dp} = -\frac{p}{q} \cdot \left(-\frac{q}{p}\right) = 1$$

So at $(9, 3)$ we have $\eta = 1$.

**b.** Since $\eta = 1$, there is no change in total revenue with a price increase.

**39.** $p^2(2q+1) = 10{,}000$

$$p^2 \cdot 2 \cdot \frac{dq}{dp} + (2q+1)2p = 0$$

$$\frac{dq}{dp} = \frac{-(2q+1)}{p}$$

$$\eta = -\frac{p}{q} \cdot \frac{dq}{dp} = -\frac{20}{12}\left(-\frac{25}{20}\right) = \frac{25}{12}$$

**40. a.** $p = 100e^{-0.1q}$

$$1 = 100e^{-0.1q}(-0.1)\frac{dq}{dp} \text{ or } \frac{dq}{dp} = \frac{-e^{0.1q}}{10}$$

At $p = 36.79$, $q = 10$ we have

$$\eta = -\frac{36.79}{10}\left(-\frac{e}{10}\right) \approx 1.$$

**b.** revenue decreases

**41. a.**

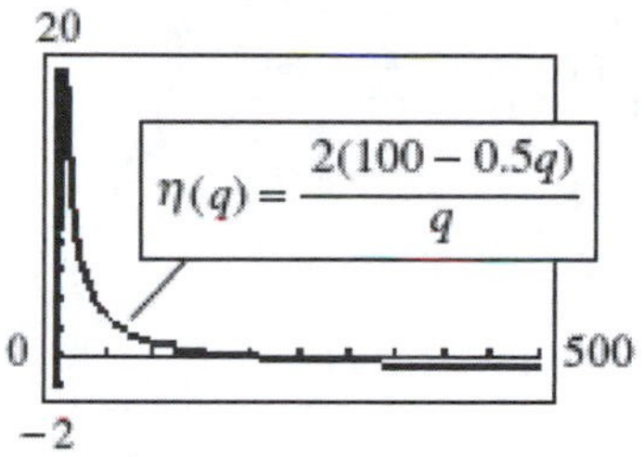


**b.** $q = 100$

**c.** max revenue at $q = 100$

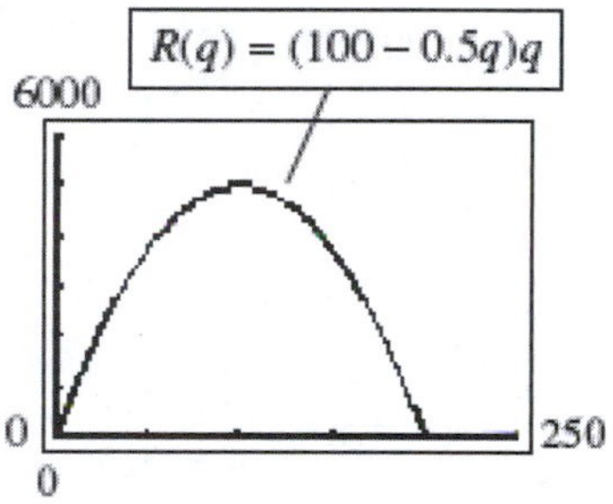


**d.** Revenue is maximized where elasticity is unitary.

## *Chapter 11 Test*

**1.** $y = 5e^{x^3} + x^2$

$$y' = 5 \cdot e^{x^3}(3x^2) + 2x = 15x^2e^{x^3} + 2x$$

**2.** $y = 4\ln(x^3+1)$

$$y' = 4 \cdot \frac{1}{x^3+1} \cdot 3x^2 = \frac{12x^2}{x^3+1}$$

**3.** $y = \ln(x^4+1)^3 = 3\ln(x^4+1)$

$$y' = 3 \cdot \frac{1}{x^4+1} \cdot 4x^3 = \frac{12x^3}{x^4+1}$$

**4.** $f(x) = 10(3^{2x})$

$$f'(x) = 10(3^{2x} \cdot 2\ln 3) = 20(3^{2x})\ln 3$$

**5.** $S = te^{t^4}$

$$S'(t) = t\left(4t^3e^{t^4}\right) + e^{t^4}(1) = e^{t^4}(4t^4+1)$$

**6.** $y = \frac{e^{x^3+1}}{x}$

$$y' = \frac{x(3x^2e^{x^3+1}) - e^{x^3+1}(1)}{x^2} = \frac{e^{x^3+1}(3x^3-1)}{x^2}$$

**7.** $y = \frac{3\ln x}{x^4}$

$$y' = \frac{x^4 \cdot \frac{3}{x} - 3\ln x(4x^3)}{\left(x^4\right)^2} = \frac{3 - 12\ln x}{x^5}$$

**8.** $g(x) = 2\log_5(4x+7)$

$$g'(x) = 2 \cdot \frac{1}{(4x+7)} \cdot \frac{4}{\ln 5} = \frac{8}{(4x+7)\ln 5}$$

**9.** $3x^4+2y^2+10=0$

$12x^3+4yy'=0$

$y'=\frac{-3x^3}{y}$

**10.** $x^2+y^2=100,\ \frac{dx}{dt}=2,\ x=6,\ y=8$

$2x\cdot\frac{dx}{dt}+2y\frac{dy}{dt}=0$

$2(6)\cdot 2+2(8)\frac{dy}{dt}=0$

$\frac{dy}{dt}=-\frac{3}{2}$

**11.** $xe^y=10y$

$x(e^y y')+e^y(1)=10y'$

$y'=\frac{e^y}{10-xe^y}$

**12.** $R(x)=300x-0.001x^2$

$C(x)=4000+30x$

$P(x)=-0.001x^2+270x-4000$

$\frac{dP}{dt}=-0.002x\frac{dx}{dt}+270\frac{dx}{dt}$

If $x=50,\ \frac{dx}{dt}=5$ then

$\frac{dP}{dt}=(-0.002)(50)(5)+270(5)=\$1349.50$ /week

**13.** $p^2+3p+q=1500$

$p=30$ gives $q=510$.

$q=1500-p^2-3p$

$\frac{dq}{dp}=-2p-3$ If $p=30,\ \frac{dq}{dp}=-63.$

$\eta=-\frac{p}{q}\cdot\frac{dq}{dp}=\frac{-30}{510}(-63)\approx 3.71$

Revenue decreases.

**14.** $(p+1)q^2=10{,}000$

Find $\frac{dq}{dp}$ if $p=99$ $(q=10)$.

$(p+1)\left(2q\frac{dq}{dt}\right)+q^2(1)=0$

$\frac{dq}{dt}=\frac{-q}{2(p+1)}$

$\frac{dq}{dt}=\frac{-10}{2(99+1)}=-\frac{1}{20}$ units/dollar

**15.** $S=80{,}000e^{-0.4t}$

$S'(t)=80{,}000(e^{-0.4t}\cdot(-0.4))=-32{,}000e^{-0.4t}$

$S'(10)=-32{,}000e^{-4}=-586$ sales/day

**16. a.** $y=1319e^{0.062t}$

$y'=81.778e^{0.062t}$

**b.** $y'(5)\approx 111.5$ (billion dollars per year)

$y'(20)\approx 282.6$ (billion dollars per year)

**17.** $D: p=1100-5q \quad S: p=20+0.4q$

New supply: $p=20+0.4q+t$ ($t$ = tax/unit)

Break-even: $1100-5q=20+0.4q+t$

$t=1080-5.4q$

Total tax revenue: $T=tq=1080q-5.4q^2$

$T'(q)=1080-10.8q$

Tax revenue is maximized when $T'=0$ or $q=100$.

Then $t=1080-5.4(100)=\$540$ / unit .

**18. a.** $y=11.851\ln(x)-12.975$

**b.** $y'=\frac{11.851}{x}$

**c.** $f(25)=11.851\ln(25)-12.975=25.172$

$f'(25)=y'=\frac{11.851}{25}=0.474$

**19. a.** $P'(t)=1.078(0.9732^t)\cdot\ln(0.9732)$

$\approx -0.02928(0.9732^t)$

**b.** $P(18)\approx 0.661$ and $P'(18)\approx -0.018$. These mean that in 2028 the purchasing power of \$1 is about 66.1% of what it was in 2012 and is expected to be decreasing about 1.8% per year.

# Chapter 12: Indefinite Integrals

## *Exercises 12.1*

**1.** $f'(x) = 4x^3$

$f(x) = \frac{4x^{3+1}}{4} + C = x^4 + C$

**3.** $f'(x) = x^6$

$f(x) = \frac{x^{6+1}}{7} + C = \frac{x^7}{7} + C$

**5.** $\int x^7 dx = \frac{x^8}{8} + C$

**7.** $8\int x^5 dx = 8\left(\frac{x^6}{6}\right) + C = \frac{4}{3}x^6 + C$

**9.** $\int (3^3 + x^{13}) dx = 27x + \frac{x^{14}}{14} + C$

**11.** $\int (3 - x^{3/2}) dx = 3x - \frac{x^{5/2}}{\frac{5}{2}} = 3x - \frac{2}{5}x^{5/2} + C$

**13.** $\int (x^4 - 9x^2 + 3) dx = \frac{x^5}{5} - \frac{9x^3}{3} + 3x + C$

$= \frac{1}{5}x^5 - 3x^3 + 3x + C$

**15.** $\int (13 - 6x + 21x^6) dx = 13x - 6\frac{x^2}{2} + 21\frac{x^7}{7} + C$

$= 13x - 3x^2 + 3x^7 + C$

**17.** $\int (2 + 2\sqrt{x}) dx = \int (2 + 2x^{1/2}) dx$

$= 2x + 2\cdot\frac{x^{3/2}}{\frac{3}{2}} + C = 2x + \frac{4}{3}x\sqrt{x} + C$

**19.** $\int 6\sqrt[4]{x}\, dx = \int 6x^{1/4} dx = 6\cdot\frac{x^{5/4}}{\frac{5}{4}} + C$

$= \frac{24}{5}x^{5/4} + C = \frac{24}{5}x\sqrt[4]{x} + C$

**21.** $\int \frac{5}{x^4} dx = \int 5x^{-4} dx = 5\cdot\frac{x^{-3}}{-3} + C$

$= -\frac{5}{3}x^{-3} + C = -\frac{5}{3x^3} + C$

**23.** $\int \frac{dx}{2\sqrt[3]{x^2}} = \frac{1}{2}\int x^{-2/3} dx = \frac{1}{2}\cdot\frac{x^{1/3}}{\frac{1}{3}} + C$

$= \frac{3}{2}x^{1/3} + C = \frac{3}{2}\sqrt[3]{x} + C$

**25.** $\int \left(x^3 - 4 + \frac{5}{x^6}\right) dx = \frac{x^4}{4} - 4x + 5\cdot\frac{x^{-5}}{-5} + C$

$= \frac{1}{4}x^4 - 4x - \frac{1}{x^5} + C$

**27.** $\int \left(x^9 - \frac{1}{x^3} + \frac{2}{\sqrt[3]{x}}\right) dx = \frac{x^{10}}{10} - \frac{x^{-2}}{-2} + 2\cdot\frac{x^{2/3}}{\frac{2}{3}} + C$

$= \frac{1}{10}x^{10} + \frac{1}{2x^2} + 3x^{2/3} + C$

**29.** $\int (4x^2 - 1)^2 x^3 dx = \int (16x^4 - 8x^2 + 1)(x^3) dx$

$= \int (16x^7 - 8x^5 + x^3) dx$

$= 16\cdot\frac{x^8}{8} - 8\cdot\frac{x^6}{6} + \frac{x^4}{4} + C$

$= 2x^8 - \frac{4}{3}x^6 + \frac{1}{4}x^4 + C$

**31.** $\int \frac{x+1}{x^3} dx = \int \left(\frac{1}{x^2} + \frac{1}{x^3}\right) dx$

$= \int (x^{-2} + x^{-3}) dx = \frac{x^{-1}}{-1} + \frac{x^{-2}}{-2} + C$

$= -\frac{1}{x} - \frac{1}{2x^2} + C$

**33.** $\int (2x + 3) dx = 2\cdot\frac{x^2}{2} + 3x + C = x^2 + 3x + C$

$f(x) = 4x - \frac{x^2}{2} + C$
$(C = -8, -4, 0, 4, 8)$

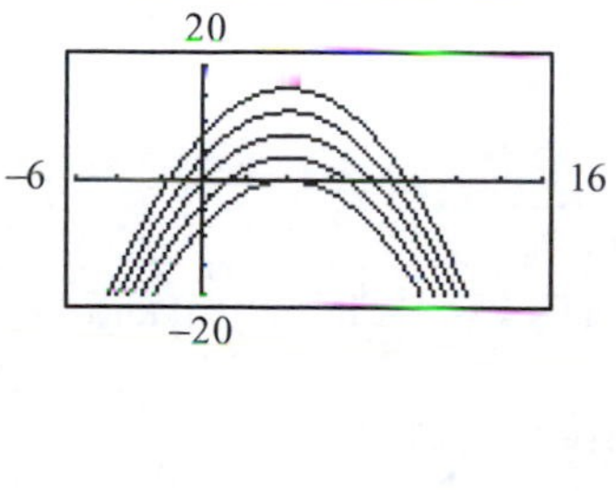

**35.** $\int f(x)dx = 2x^9 - 7x^5 + C$

$$f(x) = \frac{d}{dx}\left(2x^9 - 7x^5 + C\right) = 18x^8 - 35x^4$$

**37.** $F(x) = 5x - \frac{x^2}{4} + C$

$$F'(x) = 5 - \frac{1}{4}\cdot 2x = 5 - \frac{1}{2}x$$

$$\int\left(5 - \frac{1}{2}x\right)dx$$

**39.** $F(x) = x^3 - 3x^2 + C$

$$F'(x) = 3x^2 - 6x$$

$$\int(3x^2 - 6x)dx$$

**41.** $\overline{MR} = -0.4x + 30$

$$R(x) = \int\left(-0.4x + 30\right)dx = -0.4\cdot\frac{x^2}{2} + 30x + C$$

$C = 0$ since $R(0) = 0$.

$$R(x) = -0.2x^2 + 30x$$

**43.** $\overline{MR} = -0.3x + 450$

$$R(x) = \int\left(-0.3x + 450\right)dx = -0.3\cdot\frac{x^2}{2} + 450x + C$$

$C = 0$ since $R(0) = 0$.

$$R(50) = \frac{-0.3(50)^2}{2} + 450(50) = \$22{,}125$$

**45.** $\int(t^3 + 4t^2 + 6)dt = \frac{t^4}{4} + 4\cdot\frac{t^3}{3} + 6t + C$

When $t = 0$, $C = 0$. So, $P(t) = \frac{1}{4}t^4 + \frac{4}{3}t^3 + 6t$.

**47.** $\frac{dx}{dt} = \frac{1}{600}t^{3/4}$

**a.** $x = \frac{1}{600}\cdot\frac{t^{7/4}}{\frac{7}{4}} + C$

$t = 0$ means $x = 0$. Thus $C = 0$ and

$$x(t) = \frac{1}{1050}t^{7/4}.$$

**b.** $x(52) = \frac{1}{1050}(52)^{7/4} \approx \frac{1}{1050}(1006.95) \approx 0.96$ tons

**49.** $\overline{C}'(x) = \frac{1}{4} - \frac{100}{x^2} = \frac{1}{4} - 100x^{-2}$

**a.** $\overline{C}(x) = \frac{1}{4}x - \frac{100x^{-1}}{-1} + K = \frac{1}{4}x + \frac{100}{x} + K$

Since $\overline{C}(20) = 40$, we have

$40 = \frac{1}{4}(20) + \frac{100}{20} + K$. So, $K = 30$. Thus,

$$\overline{C}(x) = \frac{1}{4}x + \frac{100}{x} + 30.$$

**b.** $\overline{C}(100) = \frac{1}{4}(100) + \frac{100}{100} + 30 = \$56$

**51.** $\frac{dE}{dt} = 41.22t - 116.4$

**a.** $E(t) = \int\left(41.22t - 116.4\right)dt$

$$E(t) = 41.22\cdot\frac{t^2}{2} - 116.4t + C$$

$$E(t) = 20.61t^2 - 116.4t + C$$

Using $E(14) = 9808$:

$$9808 = 20.61(14)^2 - 116.4(14)t + C$$

$$C \approx 7398$$

$$E(t) = 20.61t^2 - 116.4t + 7398$$

**b.** $E(20) = 20.61(20)^2 - 116.4(20) + 7398 \approx \$13{,}314$ per person

**53.** $\frac{dt}{dw} = -4.352w^{-0.84}$

**a.** The number increases because $\frac{dt}{dw} < 0$ for $w > 0$. The rate increases because $\frac{d^2t}{dw^2} > 0$ for $w > 0$.

**b.** $t(w) = \int -4.352w^{-0.84}dt = -4.352\left(\frac{w^{0.16}}{0.16}\right) + C$

$t(w) = -27.2w^{0.16} + C$

Using $t(31) = 1$:

$1 = -27.2(31)^{0.16} + C$

$C \approx 48.12$

$t(w) = -27.2w^{0.16} + 48.12$

**55.** $\frac{dP}{dt} = -0.0002187t^2 + 0.0276t + 1.98$

**a.** Using a graphing utility, we graph $\frac{dP}{dt}$ and find that the maximum occurs at $\approx (63.1, 2.85)$. Rounding up to the nearest year, this corresponds to the year 2024.

**b.** $P(t) = \int\left(-0.0002187t^2 + 0.0276t + 1.98\right)dt$

$P(t) = -0.0000729t^3 + 0.0138t^2 + 1.98t + C$

Using $P(0) = 181$, we find $C = 181$.

$P(t) = -0.0000729t^3 + 0.0138t^2 + 1.98t + 181$

**c.** $P(65) \approx 348$ million

## *Exercises 12.2*

**1.** $u = 2x^5 + 9;\ du = 10x^4dx$

**3. a.** Let $u = 3x^4 - 7$. Then $du = 12x^3\,dx$. Since there is not a factor of $x^3$ in $12x\,dx$, the power rule cannot be used.

**b.** Let $u = 5x^3 + 11$. Then $du = 15x^2\,dx$.

$\int\left(5x^3 + 11\right)^7 15x^2\,dx = \int u^7\,du = \frac{u^8}{8} + C$

$= \frac{1}{8}\left(5x^3 + 11\right)^8 + C$

**5.** Let $u = x^2 + 3$. Then, $du = 2xdx$

$\int (x^2 + 3)^3 2x\,dx = \int u^3du = \frac{u^4}{4} + C$

$= \frac{1}{4}\left(x^2 + 3\right)^4 + C$

**7.** Let $u = 5x^3 + 11$. Then, $du = 15x^2dx$.

$\int\left(5x^3 + 11\right)^4 15x^2dx = \int u^4du = \frac{u^5}{5} + C$

$= \frac{1}{5}(5x^3 + 11)^5 + C$

**9.** Let $u = 3x - x^3$. Then, $du = (3 - 3x^2)dx$.

$\int\left(3x - x^3\right)^2\left(3 - 3x^2\right)dx = \int u^2du = \frac{u^3}{3} + C$

$= \frac{1}{3}(3x - x^3)^3 + C$

**11.** Let $u = 7x^4 + 12$. Then, $du = 28x^3\,dx$.

$\int 4x^3\left(7x^4 + 12\right)^3 dx = \frac{1}{7}\int 28x^3\left(7x^4 + 12\right)^3 dx$

$= \frac{1}{7}\int u^3du = \frac{1}{7}\cdot\frac{u^4}{4} + C$

$= \frac{1}{28}\left(7x^4 + 12\right)^4 + C$

**13.** $\int 7\left(4x - 1\right)^6 dx = 7\cdot\frac{1}{4}\int\left(4x - 1\right)^6\left(4\,dx\right)$

$= \frac{7}{4}\int u^6du = \frac{7}{4}\cdot\frac{u^7}{7} + C$

$= \frac{1}{4}(4x - 1)^7 + C$

**15.** Let $u = 4x^6 + 15$. Then, $du = 24x^5 dx$.

$$\int 8x^5\left(4x^6+15\right)^{-3} dx = \frac{1}{3}\int\left(4x^6+15\right)^{-3}\left(24x^5\,dx\right)$$
$$= \frac{1}{3}\int u^{-3}du = \frac{1}{3}\cdot\frac{u^{-2}}{-2} + C$$
$$= -\frac{1}{6}\left(4x^6+15\right)^{-2} + C$$

**17.** Let $u = x^2 - 2x + 5$. Then $du = (2x-2)dx$.

$$\int (x-1)\left(x^2-2x+5\right)^4 dx$$
$$= \frac{1}{2}\int(2x-2)\left(x^2-2x+5\right)^4 dx$$
$$= \frac{1}{2}\int u^4 du = \frac{1}{2}\cdot\frac{u^5}{5} + C$$
$$= \frac{1}{10}\left(x^2-2x+5\right)^5 + C$$

**19.** Let $u = x^4 - 4x + 3$. Then $du = \left(4x^3 - 4\right)dx$.

$$\int 2\left(x^3-1\right)\left(x^4-4x+3\right)^{-5} dx$$
$$= \frac{1}{2}\int\left(4x^3-4\right)\left(x^4-4x+3\right)^{-5} dx$$
$$= \frac{1}{2}\int u^{-5}du = \frac{1}{2}\cdot\frac{u^{-4}}{-4} + C$$
$$= -\frac{1}{8}(x^4-4x+3)^{-4} + C$$

**21.** Let $u = x^4 + 6$. Then $du = 4x^3 dx$.

$$\int 7x^3\sqrt{x^4+6}\,dx = 7\cdot\frac{1}{4}\int 4x^3\left(x^4+6\right)^{1/2} dx$$
$$= \frac{7}{4}\int u^{1/2}du = \frac{7}{4}\cdot\frac{u^{3/2}}{3/2} + C$$
$$= \frac{7}{4}\cdot\frac{2}{3}\left(x^4+6\right)^{3/2} + C$$
$$= \frac{7}{6}\left(x^4+6\right)^{3/2} + C$$

**23.** If we let $u = x^3 + 1$, then $du = 3x^2 dx$. This substitution does not work, so we use algebra to simplify the expression first

$$\int\left(x^3+1\right)^2(3x\,dx) = \int\left(x^6+2x^3+1\right)(3x)\,dx$$
$$= \int\left(3x^7+6x^4+3x\right)dx$$
$$= \frac{3}{8}x^8 + \frac{6}{5}x^5 + \frac{3}{2}x^2 + C$$

**25.** If we let $u = 3x^4 - 1$, then $du = 12x^3 dx$. This substitution does not work, so we use algebra to simplify the expression first.

$$\int\left(3x^4-1\right)^2 12x\,dx = \int\left(9x^8-6x^4+1\right)(12x)\,dx$$
$$= \int\left(108x^9-72x^5+12x\right)dx$$
$$= \frac{108}{10}x^{10} - \frac{72}{6}x^6 + \frac{12}{2}x^2 + C$$
$$= 10.8x^{10} - 12x^6 + 6x^2 + C$$

**27.** Let $u = x^3 - 3x$. Then $du = \left(3x^2 - 3\right)dx$.

$$\int\sqrt{x^3-3x}\,(x^2-1)dx = \frac{1}{3}\int\left(x^3-3x\right)^{1/2}\left(3x^2-3\right)dx$$
$$= \frac{1}{3}\int u^{1/2}du = \frac{1}{3}\cdot\frac{u^{3/2}}{3/2} + C$$
$$= \frac{1}{3}\cdot\frac{2}{3}\left(x^3-3x\right)^{3/2} + C$$
$$= \frac{2}{9}\left(x^3-3x\right)^{3/2} + C$$

**29.** Let $u = 2x^5 - 5$. Then $du = 10x^4 dx$.

$$\int\frac{3x^4dx}{\left(2x^5-5\right)^4} = 3\cdot\frac{1}{10}\int\left(2x^5-5\right)^{-4}\left(10x^4\right)dx$$
$$= \frac{3}{10}\int u^{-4}du = \frac{3}{10}\cdot\frac{u^{-3}}{-3} + C$$
$$= -\frac{1}{10}\left(2x^5-5\right)^{-3} + C$$

**31.** Let $u = x^4 - 4x$. Then $du = \left(4x^3 - 4\right)dx$.

$$\int\frac{x^3-1}{\left(x^4-4x\right)^3}dx = \frac{1}{4}\int\left(x^4-4x\right)^{-3}\left(4x^3-4\right)dx$$
$$= \frac{1}{4}\int u^{-3}du = \frac{1}{4}\cdot\frac{u^{-2}}{-2} + C$$
$$= -\frac{1}{8}\left(x^4-4x\right)^{-2} + C$$

**33.** Let $u = x^3 - 6x^2 + 2$. Then $du = \left(3x^2 - 12x\right)dx$.

$$\int \frac{x^2 - 4x}{\sqrt{x^3 - 6x^2 + 2}}\,dx$$
$$= \frac{1}{3}\int \left(x^3 - 6x^2 + 2\right)^{-1/2}\left(3x^2 - 12x\right)dx$$
$$= \frac{1}{3}\int u^{-1/2}du = \frac{1}{3}\cdot\frac{u^{1/2}}{1/2} + C$$
$$= \frac{1}{3}\cdot 2\cdot\left(x^3 - 6x^2 + 2\right)^{1/2} + C$$
$$= \frac{2}{3}\sqrt{x^3 - 6x^2 + 2} + C$$

**35.** $\int f(x)dx = (7x - 13)^{10} + C$

$$f(x) = \frac{d}{dx}\left[(7x - 13)^{10} + C\right]$$
$$= 10(7x - 13)^9(7)$$
$$= 70(7x - 13)^9$$

**37. a.** $\int x\left(x^2 - 1\right)^3 dx = \frac{1}{2}\int 2x\left(x^2 - 1\right)^3 dx$

$$= \frac{1}{2}\cdot\frac{\left(x^2 - 1\right)^4}{4} + C = \frac{1}{8}\left(x^2 - 1\right)^4 + C$$

**b.**

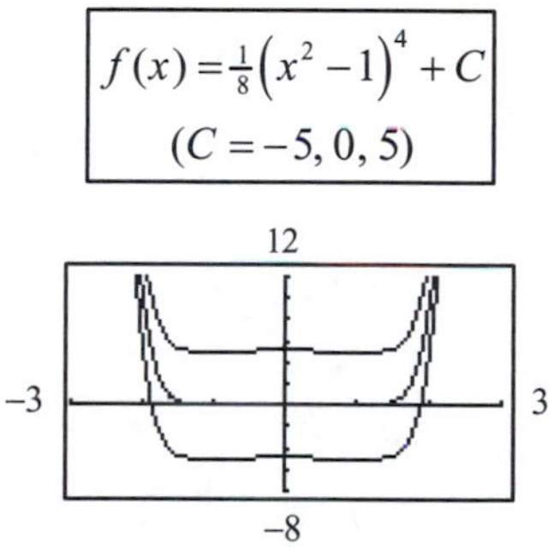


**39. a.** $\int \frac{3\,dx}{(2x - 1)^{3/5}} = 3\cdot\frac{1}{2}\int (2x - 1)^{-3/5}\cdot 2\,dx$

$$= \frac{3}{2}\cdot\frac{(2x - 1)^{2/5}}{2/5} + C = \frac{15}{4}(2x - 1)^{2/5} + C$$

**b.** When $x = 0$, $y = 2$, so we have

$$2 = \frac{15}{4}(-1)^{2/5} + C$$
$$2 = \frac{15}{4} + C \quad\rightarrow\quad -\frac{7}{4} = C$$
$$F(x) = \frac{15}{4}(2x - 1)^{2/5} - \frac{7}{4}$$

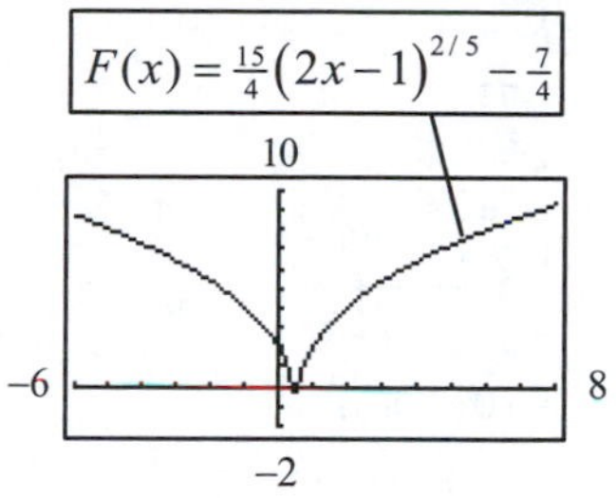


**c.** $f(x)$ is not defined at $x = \frac{1}{2}$.

**d.** The tangent line to $F(x)$ at $x = \frac{1}{2}$ is vertical.

**41.** $F(x) = \left(x^2 - 1\right)^{4/3} + C$

$$F'(x) = \frac{4}{3}\left(x^2 - 1\right)^{1/3}(2x) = \frac{8}{3}x\left(x^2 - 1\right)^{1/3}$$

The indefinite integral that gives the family is

$$\int \frac{8x\left(x^2 - 1\right)^{1/3}}{3}dx.$$

**43.** Only (b) can be integrated using methods we have studied:

**b.** $\int 7x^2\left(x^3 + 4\right)^{-2} dx = \frac{7}{3}\frac{\left(x^3 + 4\right)^{-1}}{-1} + C$

$$= \frac{-7}{3\left(x^3 + 4\right)} + C$$

**d.** One form is $\int x\left(x^3 + 4\right)^{-p} dx$. ($p > 0$)

**45.** $\overline{MR} = \frac{-30}{(2x + 1)^2} + 30$

$$\int\left(-30(2x + 1)^{-2} + 30\right)dx = \frac{-30}{2}\cdot\frac{(2x + 1)^{-1}}{-1} + 30x + C$$
$$= \frac{15}{2x + 1} + 30x + C$$

$R(0) = 0$ gives $C = -15$.

Thus, $R(x) = \frac{15}{2x + 1} + 30x - 15$.

**47.** $P(x)=\int 90(x+1)^2\,dx$

$$=\frac{90(x+1)^3}{3}+C$$

$$=30(x+1)^3+C$$

$P(0)=0$ or $0=30(1)^3+C$ or $C=-30$ and

$P(x)=30(x+1)^3-30.$

$P(4)=30(5)^3-30=3720$

**49. a.** $\int 5(x+1)^{-1/2}\,dx=5\cdot\frac{(x+1)^{1/2}}{1/2}+C$

$$=10(x+1)^{1/2}+C$$

$10=10(0+1)^{1/2}+C$ gives $C=0$.

Thus, $s(x)=10\sqrt{x+1}$.

**b.** $s(24)=10\sqrt{24+1}=50$

**51.** $A(t)=\int\left[-100(t+10)^{-2}+2000(t+10)^{-3}\right]dt$

$$=\frac{-100(t+10)^{-1}}{-1}+\frac{2000(t+10)^{-2}}{-2}+C$$

$$=\frac{100}{t+10}-\frac{1000}{(t+10)^2}+C$$

$A(0)=0$ or $0=10-10+C$ or $C=0$.

**a.** $A(t)=\frac{100}{t+10}-\frac{1000}{(t+10)^2}$

**b.** $A(10)=\frac{100}{20}-\frac{1000}{400}=2.5$ (million)

**53.** $\frac{dN}{dx}=\frac{-300}{\sqrt{x+9}}$

$N(x)=\int -300(x+9)^{-1/2}\,dx$

$$=-300\cdot\frac{(x+9)^{1/2}}{1/2}+C$$

$$=-600(x+9)^{1/2}+C$$

Using $N(0)=8000$:

$8000=-600(0+9)^{1/2}+C \rightarrow C=9800$

$N(x)=-600(x+9)^{1/2}+9800$

$N(7)=-600(16)^{1/2}+9800=7400$

The population in 7 years is expected to be 7400.

**55.** $f'(x)=\frac{375}{(0.743x+6.97)^2}=375(0.743x+6.97)^{-2}$

**a.** $f(x)=\int 375(0.743x+6.97)^{-2}\,dx$

Let $u=0.743x+6.97$. Then $du=0.743\,dx$.

$$f(x)=\frac{1}{0.743}\int 375u^{-2}\,du=\frac{375u^{-1}}{-0.743}+C=\frac{375}{-0.743u}+C=\frac{375}{-0.743(0.743x+6.97)}+C$$

Using $f(15)=8.3$:

$$8.3=\frac{375}{-0.743(0.743(15)+6.97)}+C$$

$C\approx 36.2$

$$f(x)=\frac{-505}{0.743x+6.97}+36.2$$

**b.** $f(45)\approx 23.7\%$

**57.** $\dfrac{dp}{dt} = \dfrac{2154.18}{(1.38t + 64.1)^2}$

**a.** $p(t) = \displaystyle\int \frac{2154.18}{(1.38t + 64.1)^2}\,dt$ Let $u = 1.38t + 64.1$, then $du = 1.38dt$.

$$p(t) = 2154.18\int (1.38t + 64.1)^{-2}\,dt = 1561\int u^{-2}du = 1561\cdot\frac{u^{-1}}{-1} + C = -\frac{1561}{1.38t + 64.1} + C$$

Using $p(18) = 39.8$:

$$39.8 = -\frac{1561}{1.38\times 18 + 64.1} + C$$

$$39.8 \approx -17.55 + C \;\rightarrow\; 57.35 \approx C$$

$$p(t) = -\frac{1561}{1.38\times 18 + 64.1} + 57.35 \text{ or } p(t) = 57.35 - \frac{1561}{1.38\times 18 + 64.1}$$

**b.**

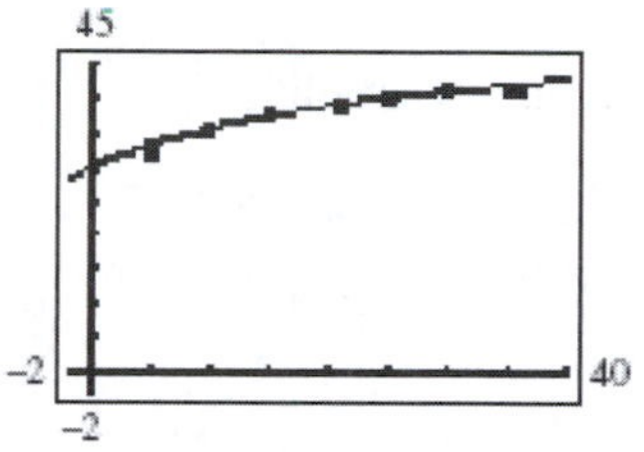


**c.** The model is a good fit to the data.

## *Exercises 12.3*

**1.** Let $u = 3x$. Then $du = 3dx$.

$$\int e^{3x}(3dx) = \int e^u du = e^u + C = e^{3x} + C$$

**3.** Let $u = -x$. Then $du = (-1)dx$.

$$\int e^{-x}dx = (-1)\int e^{-x}(-1dx)$$
$$= -1\int e^u du = (-1)e^u + C$$
$$= -e^{-x} + C$$

**5.** Let $u = 0.1x$. Then $du = 0.1dx$.

$$\int 1000e^{0.1x}dx = \frac{1000}{0.1}\int e^{0.1x}(0.1dx)$$
$$= 10{,}000\int e^u du = 10{,}000e^u + C$$
$$= 10{,}000e^{0.1x} + C$$

**7.** Let $u = -0.7x$. Then $du = -0.7dx$.

$$\int 840e^{-0.7x}dx = \frac{840}{-0.7}\int e^{-0.7x}(-0.7dx)$$
$$= -1200\int e^u du = -1200e^u + C$$
$$= -1200e^{-0.7x} + C$$

**9.** Let $u = 3x^4$. Then $du = 12x^3dx$.

$$\int x^3e^{3x^4}dx = \frac{1}{12}\int e^{3x^4}(12x^3dx)$$
$$= \frac{1}{12}\int e^u du = \frac{1}{12}e^u + C = \frac{1}{12}e^{3x^4} + C$$

**11.** Let $u = -2x$. Then $du = -2dx$.

$$\int 3e^{-2x}dx = \frac{3}{-2}\int e^{-2x}(-2dx) = -\frac{3}{2}\int e^u du$$
$$= -\frac{3}{2}e^u + C = -\frac{3}{2}e^{-2x} + C$$

**13.** Let $u = 3x^6 - 2$. Then $du = 18x^5 dx$.

$$\int \frac{x^5}{e^{2-3x^6}}dx = \int e^{-(2-3x^6)}x^5 dx$$
$$= \frac{1}{18}\int e^{3x^6-2}\left(18x^5 dx\right) = \frac{1}{18}\int e^u du$$
$$= \frac{1}{18}e^u + C = \frac{1}{18}e^{3x^6-2} + C$$

**15.** There are two substitutions in this problem:

$u = 4x,\ du = 4dx$

$v = -\frac{x}{2},\ dv = -\frac{1}{2}dx$

$$\int\left(e^{4x} - \frac{3}{e^{x/2}}\right)dx$$
$$= \int\left(e^{4x} - 3e^{-x/2}\right)dx = \int e^{4x}dx - 3\int e^{-x/2}dx$$
$$= \frac{1}{4}\int e^{4x}(4dx) - 3(-2)\int e^{-x/2}\left(-\frac{1}{2}dx\right)$$
$$= \frac{1}{4}\int e^u du + 6\int e^v dv$$
$$= \frac{1}{4}e^u + 6e^v + C = \frac{1}{4}e^{4x} + 6e^{-x/2} + C$$

**17.** Let $u = x^3 + 4$. Then $du = 3x^2 dx$.

$$\int \frac{3x^2}{x^3+4}dx = \int \frac{du}{u} = \ln|u| + C = \ln\left|x^3+4\right| + C$$

**19.** Let $u = 4z + 1$. Then $du = 4dz$.

$$\int \frac{dz}{4z+1} = \frac{1}{4}\int \frac{4dz}{4z+1} = \frac{1}{4}\int \frac{du}{u}$$
$$= \frac{1}{4}\ln|u| + C = \frac{1}{4}\ln|4z+1| + C$$

**21.** Let $u = 2x^4 + 1$. Then $du = 8x^3 dx$.

$$\int \frac{6x^3}{2x^4+1}dx = \frac{6}{8}\int \frac{8x^3}{2x^4+1}dx$$
$$= \frac{3}{4}\int \frac{du}{u} = \frac{3}{4}\ln|u| + C = \frac{3}{4}\ln\left|2x^4+1\right| + C$$

Because $2x^4 + 1 > 0$ for all $x$:

$$= \frac{3}{4}\ln\left(2x^4+1\right) + C$$

**23.** Let $u = 5x^2 - 4$. Then, $du = 10x\,dx$.

$$\int \frac{4x}{5x^2-4}dx = \frac{4}{10}\int \frac{10xdx}{5x^2-4}$$
$$= \frac{2}{5}\int \frac{du}{u} = \frac{2}{5}\ln|u| + C = \frac{2}{5}\ln\left|5x^2-4\right| + C$$

**25.** Let $u = x^3 - 2x$. Then, $du = \left(3x^2 - 2\right)dx$.

$$\int \frac{3x^2-2}{x^3-2x}dx = \int \frac{du}{u} = \ln|u| + C = \ln\left|x^3-2x\right| + C$$

**27.** $$\int \frac{z^2+1}{z^3+3z+17}dz = \frac{1}{3}\int \frac{\left(3z^2+3\right)dz}{z^3+3z+17} = \frac{1}{3}\int \frac{du}{u}$$
$$= \frac{1}{3}\ln|u| + C = \frac{1}{3}\ln\left|z^3+3z+17\right| + C$$

**29.** Use long division to simplify:

$$\begin{array}{r} x^2 \\ x-1\overline{)x^3 - x^2 + 1} \\ \underline{x^3 - x^2} \\ \text{Rem } 1 \end{array}$$

$$\int \frac{x^3-x^2+1}{x-1}dx = \int\left(x^2 + \frac{1}{x-1}\right)dx$$
$$= \int x^2 dx + \int \frac{dx}{x-1} = \frac{1}{3}x^3 + \ln|x-1| + C$$

**31.** $$\int \frac{x^2+x+3}{x^2+3}dx = \int\left(\frac{x^2+3}{x^2+3} + \frac{x}{x^2+3}\right)dx$$
$$= \int\left(1 + \frac{x}{x^2+3}\right)dx = \int 1dx + \frac{1}{2}\int \frac{2xdx}{x^2+3}$$
$$= x + \frac{1}{2}\ln\left|x^2+3\right| + C = x + \frac{1}{2}\ln\left(x^2+3\right) + C$$

**33.** $h(x) = f(x) = -e^{-(1/2)x}$

$$\int f(x)dx = \int\left(-e^{-(1/2)x}\right)dx$$
$$= 2\int e^{-(1/2)x}\left(-\tfrac{1}{2}\right)dx$$
$$= 2e^{-(1/2)x} + C$$
$$= g(x) + C$$

**35.** $\int \frac{1}{3-x}dx = -\ln|3-x| + C = F(x)$

$0 = -\ln|3-0| + C$ gives $C = \ln 3$.

$F(x) = \ln 3 - \ln|3-x|$

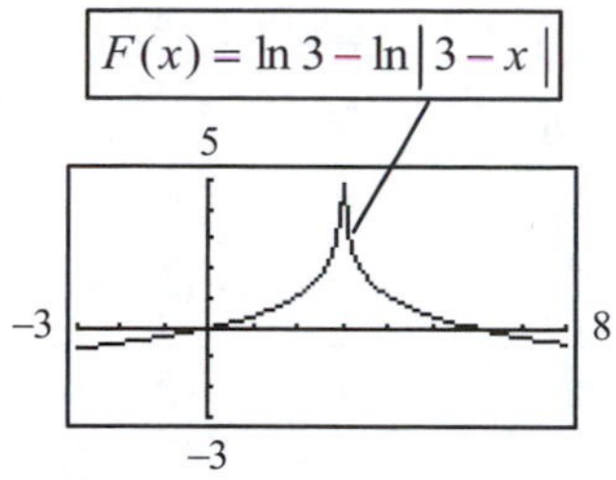


**37.** $F'(x) = f(x);$

$f(x) = 1 + \frac{1}{x};$

$F(x) = \int \left(1 + \frac{1}{x}\right)dx$

**39.** $F'(x) = f(x);$

$f(x) = 5x(-e^{-x}) + e^{-x}(5);$

$F(x) = \int \left(5e^{-x} - 5xe^{-x}\right)dx$

**41.** Only (c) and (d) can be integrated using methods we have studied:

**c.** $\frac{1}{3}\int \frac{3\left(x^2+2x\right)}{x^3+3x^2+7}dx = \frac{1}{3}\ln\left(x^3+3x^2+7\right) + C$

**d.** $\frac{5}{8}\int e^{2x^4}\left(8x^3\right)dx = \frac{5}{8}e^{2x^4} + C$

**43.** $R(x) = 6\int e^{0.01x}dx = \frac{6}{0.01}\int e^{0.01x}\left(0.01dx\right)$

$= 600e^{0.01x} + K$

$R(0) = 0$ gives $K = -600$. Thus,

$R(x) = 600e^{0.01x} - 600.$

$R(100) = 600e^{0.01(100)} - 600 = 600e - 600$

$= \$1030.97$

**45.** $n = \int n_0(-K)e^{-Kt}dt = n_0\int e^{-Kt}\left(-K\,dt\right) = n_0e^{-Kt}$

(Given $C = 0$.)

**47.** $v(t) = \int \frac{40}{t+1}dt = 40\ln(t+1) + C$

$v(0) = 0$ gives $C = 0$. Recall $\ln 1 = 0$.

$v(3) = 40\ln 4 \approx 55.45$ or 55 words.

**49. a.** $S(n) = P\int e^{0.1n}\left(0.1dn\right) = Pe^{0.1n} \quad (C = 0)$

**b.** $S = 2P$

$2P = Pe^{0.1n}$

$0.1n = \ln 2$

$n = \frac{\ln 2}{0.1} \approx 6.9$ years

**51. a.** $p = 46.645\int e^{-0.491t}\left(-1dt\right)$

$= \frac{46.645}{0.491}\int e^{-0.491t}\left(-0.491dt\right)$

$= 95e^{-0.491t} + C$

$p = 95$ when $t = 0$ gives $C = 0$.

**b.** $p(0.1) = 95e^{-0.0491} \approx 90.45$

**53.** $\frac{dl}{dt} = \frac{14.304}{t+20}$

**a.** $l(t) = \int \frac{14.304}{t+20}dt = 14.304\int \frac{dt}{t+20}$

$l(t) = 14.304\ln|t+20| + C$

Using $l(80) = 76.9$:

$76.9 = 14.304\ln|80+20| + C$

$76.9 \approx 65.87 + C \rightarrow C \approx 11.03$

$l(t) = 14.304\ln|t+20| + 11.03$

**b.**

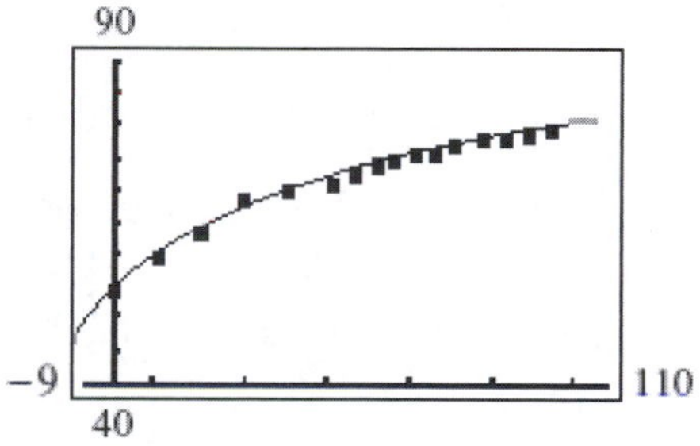


**c.** The model is a very good fit to the data.

**55.** $\frac{dC}{dt} = 3.087e^{0.0384t}$

**a.** Yes. The rate is an exponential that is always positive. Hence the function is always increasing.

**b.** $C(t) = \int \left(3.087e^{0.0384t}\right) dt$

Let $u = 0.0384t$; $du = 0.0384dt$.

$$C(t) = \frac{3.087}{0.0384}\int \left(e^{0.0384t}\right)(0.0384dt)$$
$$= 80.39\int e^{u}\,du = 80.39e^{u} + C$$
$$C(t) = 80.39e^{0.0384t} + C$$

## *Exercises 12.4*

**1.** $C(x) = \int (2x+100)\,dx = x^2 + 100x + K$

From the fact that $C(0) = 200$ we have $200 = 0 + 0 + K$ or $K = 200$.

$C(x) = x^2 + 100x + 200$

**3.** $C(x) = \int (4x+2)\,dx = 2x^2 + 2x + K$

$C(10) = 300 = 200 + 20 + K$ or $K = 80$.

$C(x) = 2x^2 + 2x + 80$

**5.** $C(x) = \int (4x+40)\,dx = 2x^2 + 40x + K$

$C(25) = 3000 = 1250 + 1000 + K$ gives $K = 750$.

$C(30) = 1800 + 1200 + 750 = \$3750$

**7. a.** Optimal level is when $\overline{MR} = \overline{MC}$.

Thus, $44 - 5x = 3x + 20$

$24 = 8x$

So, $x = 3$ units is the optimal level.

**b.** $C(x) = \int (3x+20)\,dx = \frac{3}{2}x^2 + 20x + K$

Now, $C(80) = 11{,}400 = 9600 + 1600 + K$ gives $K = 200$.

$$R(x) = \int (44 - 5x)dx = 44x - \frac{5}{2}x^2$$
$$P(x) = R(x) - C(x)$$
$$= 44x - \frac{5}{2}x^2 - \left(\frac{3}{2}x^2 + 20x + 200\right)$$
$$= 24x - 4x^2 - 200$$

**c.** $P(3) = 72 - 36 - 200 = -\$164$ (a loss)

**9.** $C(x) = \int 30(x+4)^{1/2}\,dx = \frac{30(x+4)^{3/2}}{3/2} + K$

$= 20(x+4)^{3/2} + K$

$C(0) = 1000$ gives $1000 = 20(8) + K$ or $K = 840$.

$R(x) = \int 900\,dx = 900x$

$P(x) = 900x - 20(x+4)^{3/2} - 840$

**a.** $P(5) = 4500 - 540 - 840 = \$3120$

**b.** $30\sqrt{x+4} = 900$

$\sqrt{x+4} = 30$

$x + 4 = (30)^2 = 900$

Thus, $x = 896$ units will yield maximum profit.

**11.** $\overline{C}(x) = \int \left(-6x^{-2} + \frac{1}{6}\right) dx$

$$= \frac{-6x^{-1}}{-1} + \frac{1}{6}x + K$$
$$= \frac{6}{x} + \frac{x}{6} + K$$

$\overline{C}(6) = 10 = \frac{6}{6} + \frac{6}{6} + K$ gives $K = 8$.

**a.** $\overline{C}(x) = \frac{6}{x} + \frac{x}{6} + 8$

**b.** $\overline{C}(12) = \frac{6}{12} + \frac{12}{6} + 8 = \$10.50$

**13. a.** $C(x)=\int 1.05(x+180)^{0.05}\,dx$

$$=1.05\cdot\frac{(x+180)^{1.05}}{1.05}+K$$

$$=(x+180)^{1.05}+K$$

$C(0)=180^{1.05}+K=200$ or $K\approx -33.365$.

$C(x)=(x+180)^{1.05}-33.365$

$$R(x)=\int\left(\frac{1}{\sqrt{0.5x+4}}+2.8\right)dx$$

$$=\int\left[(0.5x+4)^{-1/2}+2.8\right]dx$$

$$=2\cdot\frac{(0.5x+4)^{1/2}}{1/2}+2.8x+C$$

$$=4(0.5x+4)^{1/2}+2.8x+C$$

$R(0)=0=4(0+4)^{1/2}+0+C$ or $C=-8$

$R(x)=4(0.5x+4)^{1/2}+2.8x-8$

**b.**

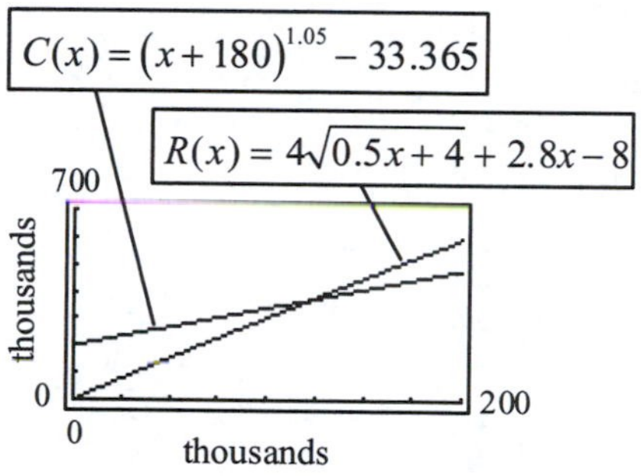


**c.** Using the graph, the maximum profit is at $x=200$ thousand units.

$P(x)=4\sqrt{0.5x+4}+2.8x-8$

$\quad -(x+180)^{1.05}+33.365$

$P(200)\approx \$114.743$ thousand

**15.** $\frac{dC}{dy}=0.80 \quad C(y)=\int 0.80dy=0.80y+K$

$C(0)=7=0.80(0)+K$ gives $K=7$.

So, $C(y)=0.80y+7$.

**17.** $C(y)=\int\left[0.3+0.2\left(y^{-1/2}\right)\right]dy$

$$=0.3y+0.2\cdot\frac{y^{1/2}}{1/2}+K$$

$C(0)=8$ gives $K=8$. Thus,

$C(y)=0.3y+0.4\sqrt{y}+8$.

**19.** $C(y)=\int\left[(y+1)^{-1/2}+0.4\right]dy$

$$=2(y+1)^{1/2}+0.4y+K$$

$C(0)=6$ gives $K=4$.

Thus, $C(y)=2(y+1)^{1/2}+0.4y+4$.

**21.** $C(y)=\int\left(0.7-e^{-2y}\right)dy=0.7y+\frac{1}{2}e^{-2y}+K$

$C(0)=5.65$ gives $5.65=0+0.5+K$ or $K=5.15$.

Thus, $C(y)=0.7y+\frac{1}{2}e^{-2y}+5.15$.

**23.** $\frac{dS}{dy}=1-\frac{dC}{dy}$

So, $0.15=1-\frac{dC}{dy}$ or $C'(y)=0.85$.

Then, $C(y)=\int 0.85\,dy=0.85y+K$

$C(0)=5.15$ means $5.15=0+K$ or $K=5.15$.

Thus, $C(y)=0.85y+5.15$.

**25.** $0.2-\frac{1}{\sqrt{3y+7}}=1-C'(y)$ or

$C'(y)=0.8+(3y+7)^{-1/2}$

$C(y)=\int\left[0.8+(3y+7)^{-1/2}\right]dy$

$$=0.8y+\frac{2}{3}(3y+7)^{1/2}+K$$

$C(0)=6=0+\frac{2}{3}(0+7)^{1/2}+K$ gives $K\approx 4.24$.

Thus, $C(y)=0.8y+\frac{2}{3}(3y+7)^{1/2}+4.24$.

# Chapter 12: Indefinite Integrals

## *Exercises 12.5*

**1.** $4y - 2xy' = 0 \qquad y = x^2$ gives $y' = 2x$

$4(x^2) - 2x(2x) = 0$

$4x^2 - 4x^2 = 0$ ✓

**3.** $y = 3x^2 + 1 \qquad y' = 6x$

$2y\,dx - x\,dy = 2dx$ or $(2y - 2dx) = x\,dy$

Then, $\dfrac{dy}{dx} = \dfrac{2y-2}{x}$

$6x = \dfrac{2(3x^2+1)-2}{x} = \dfrac{6x^2+2-2}{x} = 6x$ ✓

**5.** $dy = xe^{x^2+1}dx$

$y = \int xe^{x^2+1}dx = \dfrac{1}{2}\int e^{x^2+1}(2x\,dx) = \dfrac{1}{2}e^{x^2+1} + C$

**7.** $2y\,dy = 4x\,dx$

$\int 2y\,dy = \int 4x\,dx$

$2 \cdot \dfrac{y^2}{2} = 4 \cdot \dfrac{x^2}{2} + C$

$y^2 = 2x^2 + C$

**9.** $3y^2dy = (2x-1)dx$

$\int 3y^2dy = \int (2x-1)dx$

$3 \cdot \dfrac{y^3}{3} = 2 \cdot \dfrac{x^2}{2} - x + C$

$y^3 = x^2 - x + C$

**11.** $y' = e^{x-3}$

$y = \int e^{x-3}dx$

$y = e^{x-3} + C$

$y(0) = e^{0-3} + C = 2$ gives $C = 2 - e^{-3}$.

So, $y = e^{x-3} + 2 - e^{-3}$

**13.** $dy = \left(\dfrac{1}{x} - x\right)dx$

$y = \int\left(\dfrac{1}{x} - x\right)dx = \int\left(x^{-1} - x\right)dx$

$y = \ln|x| - \dfrac{x^2}{2} + C$

$y(1) = \ln 1 - \dfrac{(1)^2}{2} + C = 0$ or $0 - \dfrac{1}{2} + C = 0$ or

$C = \dfrac{1}{2}$. So, $y = \ln|x| - \dfrac{x^2}{2} + \dfrac{1}{2}$.

**15.** $\dfrac{dy}{dx} = \dfrac{x^2}{y}$

$y\,dy = x^2dx$

$\int y\,dy = \int x^2dx$

$\dfrac{y^2}{2} = \dfrac{x^3}{3} + C$

**17.** $dx = x^3y\,dy$

$x^{-3}dx = y\,dy$

$\int x^{-3}dx = \int y\,dy$

$\dfrac{x^{-2}}{-2} = \dfrac{y^2}{2} + \bar{C}$

or $\dfrac{1}{2x^2} + \dfrac{y^2}{2} = C \quad (C = -\bar{C})$

**19.** $dx = \left(x^2y^2 + x^2\right)dy$

$dx = x^2\left(y^2+1\right)dy$

$x^{-2}dx = \left(y^2+1\right)dy$

$\int x^{-2}dx = \int\left(y^2+1\right)dy$

$\dfrac{x^{-1}}{-1} = \dfrac{y^3}{3} + y + \bar{C}$

or $C = \dfrac{1}{x} + y + \dfrac{1}{3}y^3$

**21.**
$$y^2 dx = x\,dy$$
$$x^{-1}dx = y^{-2}dy$$
$$\int x^{-1}dx = \int y^{-2}dy$$
$$\ln|x| = \frac{y^{-1}}{-1} + C$$
$$C = \frac{1}{y} + \ln|x|$$

**23.**
$$\frac{dy}{dx} = \frac{x}{y}$$
$$x\,dx = y\,dy$$
$$\int x\,dx = \int y\,dy$$
$$\frac{x^2}{2} = \frac{y^2}{2} + \bar{C}$$
$$x^2 - y^2 = C$$

**25.** $(x+1)\dfrac{dy}{dx} = y$
$$y^{-1}dy = (x+1)^{-1}\,dx$$
$$\int y^{-1}dy = \int (x+1)^{-1}\,dx$$
$$\ln|y| = \ln|x+1| + C$$
$$\ln|y| = \ln|x+1| + \ln\bar{C}$$
$$\ln|y| = \ln|\bar{C}(x+1)|$$
$$y = \bar{C}(x+1)$$
$$(\ln\bar{C} = C)$$

**27.** $(x^2+2)e^{y^2}dx = xydy$
$$\frac{(x^2+2)}{x}dx = \frac{y}{e^{y^2}}dy$$
$$(x+2x^{-1})dx = ye^{-y^2}dy$$
$$\int (x+2x^{-1})dx = \int ye^{-y^2}dy$$
$$\frac{x^2}{2} + 2\ln|x| = -\frac{1}{2}\int e^{-y^2}(-2ydy)$$
$$\frac{x^2}{2} + 2\ln|x| = -\frac{1}{2}e^{-y^2} + C$$
Multiply both sides by 2 and collect terms.
$$x^2 + 4\ln|x| + e^{-y^2} = C$$

**29.** $y^3dy = x^2dx$ gives $\dfrac{1}{4}y^4 = \dfrac{1}{3}x^3 + C.$

At $x = 1$, $y = 1$, we have

$\dfrac{1}{4} = \dfrac{1}{3} + C$ or $C = -\dfrac{1}{12}.$

Using this value for $C$ and clearing fractions, we have $3y^4 - 4x^3 + 1 = 0.$

**31.** $\dfrac{2\,dx}{x^2} = \dfrac{3\,dy}{y^2}$ gives $\dfrac{2x^{-1}}{-1} = \dfrac{3y^{-1}}{-1} + C$ or

$-\dfrac{2}{x} + \dfrac{3}{y} = C.$ At $x = 2$, $y = -1$, we have

$-1 + (-3) = C$ or $C = -4$.

Thus, $\dfrac{3}{y} - \dfrac{2}{x} + 4 = 0$ or $y = \dfrac{3x}{2-4x}.$

**33.**
$$e^{2y}dy = \frac{x^3+1}{x^2}dx$$
$$\frac{1}{2}(e^{2y}2\,dy) = (x + x^{-2})dx$$
$$\frac{1}{2}e^{2y} = \frac{x^2}{2} + \frac{x^{-1}}{-1} + C$$

At $(1, 0)$ we have $\dfrac{1}{2}e^0 = \dfrac{1}{2} - \dfrac{1}{1} + C$ which gives

$C = 1$. $(e^0 = 1)$

Thus, $\dfrac{1}{2}e^{2y} = \dfrac{1}{2}x^2 - \dfrac{1}{x} + 1$ or $xe^{2y} = x^3 - 2 + 2x.$

**35.** $\dfrac{2y}{y^2+1}dy = \dfrac{dx}{x}$ gives

$\ln(y^2+1) = \ln|x| + \ln C = \ln(C|x|)$ or

$y^2 + 1 = C|x|$

At $(1, 2)$, we have $4 + 1 = C\cdot 1$ or $C = 5$.

Thus, $y^2 + 1 = 5|x|.$

**37.** $\dfrac{dy}{y} = k\dfrac{dx}{x}$ gives $\ln|y| = k\ln|x| + \ln|C| = \ln|Cx^k|$

or $y = |Cx^k|.$

# Chapter 12: Indefinite Integrals

**39. a.** $\frac{dx}{dt} = rx;\ r = 0.06$

$\frac{dx}{x} = r\,dt$

$\ln x = rt + C$

$x = e^{rt+C} = e^{rt} \cdot e^{C} = Ke^{rt}$

At $t = 0$, $x = 10{,}000$ and therefore,

$x = 10{,}000e^{rt}$.

**b.** $x = 10{,}000e^{0.06} = \$10{,}618.37$

$x = 10{,}000e^{0.30} = \$13{,}498.59$

**c.** $20{,}000 = 10{,}000e^{0.06t}$

$2 = e^{0.06t}$

$\ln 2 = 0.06t$

$t = \frac{\ln 2}{0.06} \approx 11.55$ years

**41.** Rearrange $\frac{dP}{dt} = kP$ to get $\frac{dP}{P} = kdt.$

Then integrate: $\int \frac{dP}{P} = \int kdt$

$\ln P = kt + C$

$P = e^{kt+C} = e^{kt} \cdot e^{C} = Ke^{kt}$

At $t = 0$, $P = 100{,}000$:

$100{,}000 = Ke^{k(0)} \Rightarrow K = 100{,}000$

At $t = 15$, $P = 211{,}700$:

$211{,}700 = 100{,}000e^{15k}$

$2.117 = e^{15k}$

$\ln 2.117 = 15k$

$k = \frac{\ln 2.117}{15} \approx 0.05$

So, $P = 100{,}000e^{0.05t}$.

**43.** $\frac{dy}{y} = k\,dt$ gives $\ln|y| = kt + C$ or $y = Ae^{kt}$

At $t = 0$, we have $y = 10{,}000$ and, therefore, $A = 10{,}000$. Thus, $y = 10{,}000e^{kt}$. At $t = 2$, we have $y = 30{,}000$ and, therefore, $3 = e^{2k}$ or $2k = \ln 3$ or $k \approx 0.55$. Then, if $y = 1{,}000{,}000$, we have $100 = e^{0.55t}$ or $\ln 100 = 0.55t$ or $t \approx 8.4$ hr.

**45.** $\frac{dy}{dp} = -\frac{2}{5}\left(\frac{y}{p+8}\right)$

$\int \frac{dy}{y} = -\frac{2}{5}\int \frac{dp}{p+8}$

$\ln y = -\frac{2}{5}\ln|p+8| + \ln C = \ln\left[(p+8)^{-2/5}C\right]$

$y = C(p+8)^{-2/5}$

At $p = 24$, $y = 8$, we have $8 = C(24+8)^{-2/5}$ or $C = 2^3(32)^{2/5} = 2^5 = 32$. Thus, $y = 32(p+8)^{-2/5}$

$\ln 18 = \ln\frac{1}{5} + C$

$\ln 18 = \ln 1 - \ln 5 + C$

$\ln 18 + \ln 5 = C$

$\ln 90 = C$

So, $\ln y = \ln\frac{1}{\sqrt{p+5}} + \ln 90$ or

$\ln y = \ln\frac{90}{\sqrt{p+5}}$ or $y = \frac{90}{\sqrt{p+5}}$.

**47.** $\frac{dy}{dt} = -kt$ gives, after several steps, $y = Ae^{-kt}$.

At $t = 0$, $y = 100$ and, thus, $A = 100$. So, $y = 100e^{-kt}$. At $t = 10$, $0.9997(100) = 99.97$ pounds remain. Solving $99.97 = 100e^{-10k}$ for $k$ gives $k = 0.00003$. With $y = 50$ (half the original 100), we have $50 = 100e^{-0.00003t}$. Solving, we have $-0.00003t = \ln(0.5)$ or $t \approx 23{,}100$ years.

**49.** $\frac{dx}{dt} = 5(0.06) - 5 \cdot \frac{x}{100} = \frac{3}{10} - \frac{x}{20} = \frac{6-x}{20}$

Thus, $\frac{dx}{6-x} = \frac{dt}{20}$ gives $-\ln|6-x| = \frac{t}{20} + C$ or $x = 6 - ke^{-t/20}$. At $t = 0$, we have $x = 0$ and, thus, $k = 6$. Then, $x = 6 - 6e^{-t/20}$.

**51.** $\frac{dx}{dt} = 5(0.1) - 5 \cdot \frac{x}{200} = \frac{20-x}{40}$

Thus, $\frac{dx}{20-x} = \frac{dt}{40}$ gives $-\ln|20-x| = \frac{t}{40} + C$ or

$\ln|20-x| = -\frac{t}{40} - C$ So, $20 - x = ke^{-t/40}$.

At $t = 0$, we have $x = 10$ and, thus, $k = 10$.

Then, $x = 20 - 10e^{-t/40}$. $\left(\frac{t}{40} = 0.025t\right)$

**53.** $\frac{dV}{V} = 0.2e^{-0.1t}dt$ gives $\ln|V| = -2e^{-0.1t} + C$.

At $t = 0$, $V = 1.86$ and $\ln 1.86 = -2(1) + C$ or

$C = 2 + \ln 1.86$. So, $\ln|V| = -2e^{-0.1t} + 2 + \ln 1.86$

or $\ln\left(\frac{|V|}{1.86}\right) = 2 - 2e^{-0.1t}$. Thus, $V = 1.86e^{2-2e^{-0.1t}}$.

**55.** $\frac{dV}{dt} = kV^{2/3}$ or $V^{-2/3}dV = k\,dt$ gives

$\frac{V^{1/3}}{1/3} = kt + C$ or $3V^{1/3} = kt + C$. At $t = 0$,

$V = 0$ and $3(0)^{1/3} = k(0) + C$ gives $C = 0$. So,

$3V^{1/3} = kt$ or $V^{1/3} = \frac{kt}{3}$ or $V = \frac{k^3t^3}{27}$.

**57.** $\frac{du}{u-T} = k\,dt$ gives $\ln|u-T| = kt + C$ and after several steps $u - T = Ae^{kt}$.

At $t = 0$, $u - T = 0 - 20 = A \cdot 1$ gives $A = -20$.

Thus, $u = 20 - 20e^{kt}$ $(T = 20)$

At $t = 1$, $u = 8$, we have

$$8 = 20 - 20e^{k(1)}$$
$$-12 = -20e^{k}$$
$$e^{k} = 0.6$$

$\ln 0.6 = k \ln e$ or $k \approx -0.51$

Thus, $u = 20 - 20e^{-0.51t}$. If $u = 18$, then

$18 = 20 - 20e^{-0.51t}$ gives $t \approx 4.5$ hours.

**59. a.** $\frac{dE}{dt} = 0.0164E$ or $\frac{dE}{E} = 0.0164dt$ gives

$\ln E = 0.0164t + C$ or $E = ke^{0.0164t}$.

Then, $18.5 = ke^{0}$ gives $k = 18.5$.

Thus, $E(t) = 18.5e^{0.0164t}$

**b.**

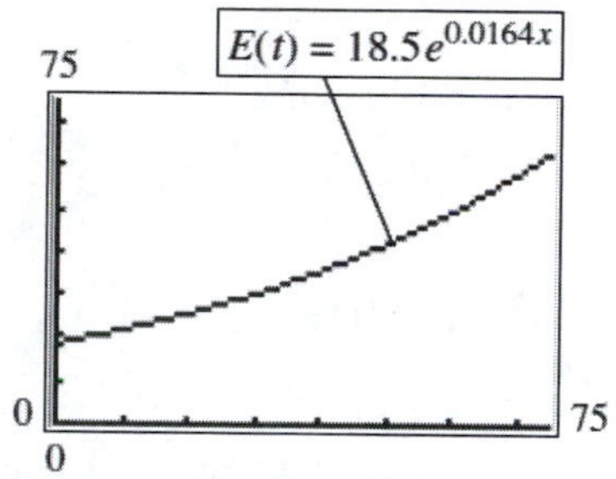


The graph of the model looks similar except at the right end, where it rises more sharply than the graph of the data.

**61. a.** $\frac{dP}{dt} = -0.05P$

$\int \frac{dP}{P} = -\int 0.05dt$

$\ln P = -0.05t + C$

$P = ke^{-0.05t}$

Using $P(0) = 80{,}000$:

$80{,}000 = ke^{0} \Rightarrow k = 80{,}000$

$P(t) = 80{,}000e^{-0.05t}$

**b.** $P(15) \approx 37{,}789.32$

# Chapter 12: Indefinite Integrals

## *Chapter 12 Review Exercises*

**1.** $\int x^6 dx = \frac{1}{7}x^7 + C$

**2.** $\int x^{1/2} dx = \frac{2}{3}x^{3/2} + C$

**3.** $\int (12x^3 - 3x^2 + 4x + 5)dx$
$= 3x^4 - x^3 + 2x^2 + 5x + C$

**4.** $\int 7(x^2-1)^2 dx = 7\int (x^4 - 2x^2 + 1)dx$
$= 7\left(\frac{1}{5}x^5 - \frac{2}{3}x^3 + x\right) + C$
$= \frac{7}{5}x^5 - \frac{14}{3}x^3 + 7x + C$

**5.** $\int 7x(x^2-1)^2 dx$
Let $u = x^2 - 1$; $du = 2xdx$.
$\int \frac{7}{2}u^2 du = \frac{7}{2}\cdot\frac{1}{3}u^3 + C = \frac{7}{6}(x^2-1)^3 + C$

**6.** $\int (x^3 - 3x^2)^5 (x^2 - 2x)dx$
$= \frac{1}{3}\int (x^3 - 3x^2)^5 (3x^2 - 6x)dx$
$= \frac{1}{18}(x^3 - 3x^2)^6 + C$

**7.** $\int (x^3+4)^2 3xdx = \int (x^6 + 8x^3 + 16)3xdx$
$= \int (3x^7 + 24x^4 + 48x)dx$
$= \frac{3}{8}x^8 + \frac{24}{5}x^5 + 24x^2 + C$

**8.** $\int 5x^2(3x^3+7)^6 dx$
Let $u = 3x^3 + 7$; $du = 9x^2 dx$.
$= \frac{5}{9}\int u^6 du = \frac{5}{9}\cdot\frac{u^7}{7} + C = \frac{5}{63}(3x^3+7)^7 + C$

**9.** $\int \frac{x^2}{x^3+1}dx = \frac{1}{3}\int \frac{3x^2 dx}{x^3+1}$
$= \frac{1}{3}\ln|x^3+1| + C$

**10.** $\int \frac{x^2}{(x^3+1)^2}dx = \frac{1}{3}\int \frac{3x^2 dx}{(x^3+1)^2}$
$= \frac{-1}{3(x^3+1)} + C$

**11.** $\int (x^3-4)^{-1/3} x^2 dx = \frac{1}{3}\int (x^3-4)^{-1/3}(3x^2 dx)$
$= \frac{1}{3}\cdot\frac{(x^3-4)^{2/3}}{2/3} + C$
$= \frac{1}{2}(x^3-4)^{2/3} + C$

**12.** $\int \frac{x^2}{x^3-4}dx = \frac{1}{3}\int \frac{3x^2 dx}{x^3-4} = \frac{1}{3}\ln|x^3-4| + C$

**13.** $\int \frac{x^3+1}{x^2}dx = \int (x + x^{-2})dx$
$= \frac{x^2}{2} + \frac{x^{-1}}{-1} + C$
$= \frac{1}{2}x^2 - \frac{1}{x} + C$

**14.** $\int \frac{x^3 - 3x + 1}{x-1}dx = \int \left(x^2 + x - 2 - \frac{1}{x-1}\right)dx$
$= \frac{1}{3}x^3 + \frac{1}{2}x^2 - 2x - \ln|x-1| + C$

**15.** $\int y^2 e^{y^3} dy = \frac{1}{3}\int e^{y^3}(3y^2 dy) = \frac{1}{3}e^{y^3} + C$

**16.** $\int (3x-1)^{12} dx = \frac{1}{3}\int (3x-1)^{12}(3dx)$
$= \frac{1}{3}\cdot\frac{1}{13}(3x-1)^{13} + C$
$= \frac{1}{39}(3x-1)^{13} + C$

**17.** $\int \frac{3x^2}{2x^3-7}dx = \frac{1}{2}\int 2(3x^2)(2x^3-7)^{-1} dx$
$= \frac{1}{2}\ln|2x^3-7| + C$

# Chapter 12: Indefinite Integrals

**18.** $\int \frac{5\,dx}{e^{4x}} = -\frac{5}{4}\int e^{-4x}(-4\,dx)$

$= -\frac{5}{4}e^{-4x} + C$

$= -\frac{5}{4e^{4x}} + C$

**19.** $\int (x^3 - e^{3x})\,dx = \int x^3 dx - \frac{1}{3}\int e^{3x}(3\,dx)$

$= \frac{1}{4}x^4 - \frac{1}{3}e^{3x} + C$

**20.** $\int xe^{1+x^2}\,dx = \frac{1}{2}\int e^{1+x^2}(2x\,dx)$

$= \frac{1}{2}e^{1+x^2} + C$

**21.** $\int \frac{6x^7}{(5x^8+7)^3}\,dx = 6\cdot\frac{1}{40}\int (5x^8+7)^{-3}(40x^7 dx)$

$= \frac{6}{40}\cdot\frac{(5x^8+7)^{-2}}{-2}$

$= \frac{-3}{40(5x^8+7)^2} + C$

**22.** $\int \frac{7x^3}{(1-x^4)^{1/2}}\,dx = -\frac{7}{4}\int \frac{-4x^3 dx}{(1-x^4)^{1/2}}$

$= -\frac{7}{4}\left[2(1-x^4)^{1/2}\right] + C$

$= -\frac{7}{2}\sqrt{1-x^4} + C$

**23.** $\int \left(\frac{e^{2x}}{2} + \frac{2}{e^{2x}}\right)dx$

$= \frac{1}{2}\cdot\frac{1}{2}\int e^{2x}(2\,dx) + 2\left(-\frac{1}{2}\right)\int e^{-2x}(-2\,dx)$

$= \frac{1}{4}e^{2x} - e^{-2x} + C$

**24.** $\int \left(x - \frac{1}{(x+1)^2}\right)dx = \frac{1}{2}x^2 + \frac{1}{x+1} + C$

**25. a.** $\int (x^2-1)^4 (x\,dx) = \frac{1}{2}\int (x^2-1)^4 (2x\,dx)$

$= \frac{1}{10}(x^2-1)^5 + C$

**b.** $\int (x^2-1)^{10}(x\,dx) = \frac{1}{2}\int (x^2-1)^{10}(2x\,dx)$

$= \frac{1}{22}(x^2-1)^{11} + C$

**c.** $\int (x^2-1)^7 (3x\,dx) = \frac{3}{2}\int (x^2-1)^7 (2x\,dx)$

$= \frac{3}{16}(x^2-1)^8 + C$

**d.** $\int (x^2-1)^{-2/3}(x\,dx) = \frac{1}{2}\int (x^2-1)^{-2/3}(2x\,dx)$

$= \frac{3}{2}(x^2-1)^{1/3} + C$

**26. a.** $\int \frac{2x}{x^2-1}\,dx = \ln|x^2-1| + C$

**b.** $\int \frac{2x}{(x^2-1)^2}\,dx = -\frac{1}{x^2-1} + C$

**c.** $\int \frac{3x\,dx}{(x^2-1)^{1/2}} = \frac{3}{2}\int \frac{2x\,dx}{(x^2-1)^{1/2}}$

$= 3(x^2-1)^{1/2} + C$

$= 3\sqrt{x^2-1} + C$

**d.** $\int \frac{3x\,dx}{x^2-1} = \frac{3}{2}\int \frac{2x\,dx}{x^2-1} = \frac{3}{2}\ln|x^2-1| + C$

**27.** $\frac{dy}{dt} = 4.6e^{-0.05t}$

$dy = 4.6e^{-0.05t}dt$

$\int dy = -\frac{1}{0.05}\int 4.6e^{-0.05t}(-0.05\,dt)$

$y = -20(4.6)e^{-0.05t} + C$

$y = C - 92e^{-0.05t}$

**28.** $dy = (64 + 76x - 36x^2)\,dx$

$\int dy = \int (64 + 76x - 36x^2)\,dx$

$y = 64x + 38x^2 - 12x^3 + C$

**29.** $\frac{dy}{dx} = \frac{4x}{y-3}$ or $(y-3)\,dy = 4x\,dx$ gives

$\frac{y^2}{2} - 3y = 2x^2 + C.$

Also: $\frac{(y-3)^2}{2} = 2x^2 + C_1$ or $(y-3)^2 = 4x^2 + C$

**30.** $$t\,dy = \frac{dt}{y+1}$$
$$\int (y+1)\,dy = \int \frac{dt}{t}$$
$$\frac{(y+1)^2}{2} = \ln|t| + C_1$$
$$(y+1)^2 = 2\ln|t| + C$$

**31.** $\frac{dy}{dx} = \frac{x}{e^y}$ or $e^y dy = x\,dx$ gives $e^y = \frac{x^2}{2} + C.$

**32.** $$\frac{dy}{dt} = \frac{4y}{t}$$
Write in separated form and integrate.
$$\int \frac{dy}{y} = 4\int \frac{dt}{t}$$
$$\ln|y| = 4\ln|t| + C$$
Assume $y > 0$, $\ln y = \ln t^4 + \ln C_1$ where $C = \ln C_1$.
$$\ln y = \ln C_1 t^4$$
$$y = C_1 t^4$$

**33.** $y' = \frac{x^2}{y+1}$ or $x^2 dx = (y+1)dy$ gives
$\frac{x^3}{3} = \frac{(y+1)^2}{2} + C.$ $y(0) = 4$ or $\frac{0}{3} = \frac{(4+1)^2}{2} + C$
gives $C = -\frac{25}{2}$. Then, $3(y+1)^2 = 2x^3 + 75.$

**34.** $(1+2y)dy = 2x\,dx$ gives $y + y^2 = x^2 + C.$
At $(2,0)$ we have $0+0 = 4+C$ or $C = -4$.
So, $x^2 = y + y^2 + 4.$

**35.** $\overline{MR} = 0.06x + 12$
$$\int (0.06x+12)\,dx = 0.06\cdot\frac{x^2}{2} + 12x + C$$
$$R(x) = 0.03x^2 + 12x + C$$
$R(0) = 0$, so $C = 0$.
$$R(x) = 0.03x^2 + 12x$$
$$R(800) = 0.03(800)^2 + 12(800) = \$28{,}800$$

**36.** $$\frac{dp}{dt} = 27 + 24t - 3t^2$$
$$p(t) = 27t + 12t^2 - t^3 + C$$
At $t = 0$, $p = 0$ and this yields $C = 0$.
Then $p(8) = 216 + 768 - 512 = 472$.

**37.** $$P'(t) = 400\left[\frac{5}{(t+5)^2} - \frac{50}{(t+5)^3}\right]$$
$$P(t) = 400\int\left[5(t+5)^{-2} - 50(t+5)^{-3}\right]dt$$
$$= 400\left[\frac{5(t+5)^{-1}}{-1} - \frac{50(t+5)^{-2}}{-2}\right] + C$$
$$= 400\left[\frac{-5}{t+5} + \frac{25}{(t+5)^2}\right] + C$$
Since $P(0) = 400$, $400 = 400(-1+1) + C$ gives $C = 400$.
$$P(t) = 400\left[1 - \frac{5}{t+5} + \frac{25}{(t+5)^2}\right]$$

**38.** $$\frac{dp}{dt} = \frac{100{,}000}{(t+100)^2}$$
$$p(t) = 100{,}000\cdot\frac{(t+100)^{-1}}{-1} + C$$
$p(1) = 1000$ gives $1000 = \frac{-100{,}000}{101} + C$ or
$C \approx 1990.099$. Thus,
$$p(t) = 1990.099 - \frac{100{,}000}{t+100}$$

**39. a.** $\frac{dy}{dt} = 2.4e^{-0.04t}$ or $dy = 2.4e^{-0.04t}dt$ gives
$$y = -60e^{-0.04t} + C$$
$$0 = -60(1) + C \text{ or } C = 60$$
$$y = 60 - 60e^{-0.04t}$$

**b.** $y = 60 - 60e^{-0.04(12)} \approx 23\%$

**40.** $$R'(x) = \frac{800}{x+2}$$
$$R(x) = 800\ln(x+2) + C$$
Using $R(0) = 0$:
$$0 = 800\ln(0+2) + C$$
$$C = -800\ln 2 \approx -554.5$$
$$R(x) = 800\ln(x+2) - 554.5$$

**41.** $\overline{MC} = 6x + 4$

**a.** $C(x) = \int (6x+4)\,dx = 3x^2 + 4x + K$

$3(100)^2 + 4(100) + K = 31,400$ gives

$K = 1000$ fixed costs.

**b.** $C(x) = 3x^2 + 4x + 1000$

**42.** $\overline{MR} = \overline{MC}$

$46 = 30 + \frac{1}{5}x$

$16 = \frac{1}{5}x$

$x = 80$

So, 80 units are needed to maximize profit.

$C(x) = 30x + \frac{1}{10}x^2 + K$. $C(0) = 200$ gives $K = 200$.

Since $R(x) = 46x$ we have

$P(x) = 16x - \frac{1}{10}x^2 - 200.$

$P(80) = 1280 - 640 - 200 = \$440$

**43.** $\frac{dC}{dy} = (2y+16)^{-1/2} + 0.6$ or

$dC = \frac{1}{2}(2y+16)^{-1/2}(2\,dy) + 0.6\,dy$ gives

$C = \frac{1}{2} \cdot \frac{(2y+16)^{1/2}}{1/2} + 0.6y + K = \sqrt{2y+16} + 0.6y + K.$

$8.5 = \sqrt{0+16} + 0 + K$ gives $K = 4.5$. So,

$C = \sqrt{2y+16} + 0.6y + 4.5.$

**44.** $\frac{dS}{dy} = 0.2 - 0.1e^{-2y}$

$\frac{dC}{dy} = 1 - \frac{dS}{dy}$

$\frac{dC}{dy} = 0.80 + 0.1e^{-2y}$

$C(y) = \int (0.80 + 0.1e^{-2y})\,dy$

$C(y) = 0.8y - \frac{0.1}{2}e^{-2y} + K$

Using $C(0) = 7.8$ to find $K$ gives $K = 7.85$.

So, $C(y) = 0.8y - 0.05e^{-2y} + 7.85$.

**45.** $\frac{dW}{dL} = \frac{3W}{L}$ or $\frac{dW}{W} = 3\frac{dL}{L}$ gives

$\ln W = 3\ln L + \ln C = \ln(L^3 \cdot C)$.

Thus, $W = CL^3$

**46. a.** $\frac{dP}{dt} = kP \Rightarrow \frac{dP}{P} = kdt$

Integrating: $\ln|P| = kt + C$

**b.** $P = e^{kt+C} = e^{kt} \cdot e^C = Ke^{kt}$

**c.** Using $P(0) = 50,000$:

$50000 = Ke^{k(0)} \Rightarrow K = 50,000$

$P(t) = 50,000e^{kt}$

Using $P(10) = 135,914$:

$135914 = 50000e^{k(10)}$

$2.7183 = e^{10k}$

$\ln 2.7183 = 10k$

$k = \frac{\ln 2.7183}{10} \approx 0.1$

$P(t) = 50,000e^{0.1t}$

**d.** $k$ represents the annual interest rate, which is 10%.

**47.** $\frac{dx}{dt} = kx$ As before $x = Ce^{kt}$.

When $t = 0$, $x = 10$ and from this information we have $C = 10$. So, $x = 10e^{kt}$. Use the half-life, $t = 4.6$ million years and $x = 5$, to determine $k$. The result is $k \approx -0.15$. So, $x = 10e^{-0.15t}$.

With 20% left, $x = 2$. Now determine the corresponding $t$.

$2 = 10e^{-0.15t}$

$\ln 0.2 = -0.15t$

$t \approx 10.73$ million years

**48.** $\frac{dx}{dt} = 4 \cdot 3 - 4 \cdot \frac{x}{120} = \frac{360 - x}{30}$

Thus, $\int \frac{dx}{360 - x} = \int \frac{dt}{30}$ gives

$-\ln(360 - x) = \frac{1}{30}t + C_1$ or

$\ln(360 - x) = -\frac{t}{30} + C$. After algebraic

manipulation, we have $360-x=Ae^{-t/30}$. Then, $360-0=A\cdot 1$ or $A=360$ and $x=360-360e^{-t/30}$.

**49.** $\frac{dx}{dt}=3\cdot 2-3\cdot\frac{x}{300}$

$$\frac{dx}{dt}=6-\frac{x}{100}$$

$$\frac{dx}{dt}=\frac{600-x}{100}$$

$$\int\frac{dx}{600-x}=\frac{1}{100}\int dt$$

$$-\ln(600-x)=\frac{1}{100}t+C$$

$x=100$ when $t=0$.

Using $x=100$ when $t=0$ gives $C=-\ln 500$.

So, $-\ln(600-x)=0.01t-\ln 500$

$$\ln\left(\frac{600-x}{500}\right)=\ln e^{-0.01t}$$

$$600-x=500e^{-0.01t}$$

$$x=600-500e^{-0.01t}$$

When $x=500$ we have $500=600-500e^{-0.01t}$

$$0.2=e^{-0.01t}$$

$$\ln 0.2=-0.01t$$

$$t\approx 161 \text{ minutes}$$

## *Chapter 12 Test*

**1.** $\int(6x^2+8x-7)\,dx=\frac{6x^3}{3}+\frac{8x^2}{2}-7x+C$

$$=2x^3+4x^2-7x+C$$

**2.** $\int(11-2x^3)\,dx=11x-\frac{2x^4}{4}+C$

$$=11x-\frac{x^4}{2}+C$$

**3.** $\int 5(x^2-1)\,dx=5\left(\frac{x^3}{3}-x\right)+C$

$$=\frac{5x^3}{3}-5x+C$$

**4.** $\int(4+x^{1/2}-x^{-2})\,dx=4x+\frac{x^{3/2}}{\frac{3}{2}}-\frac{x^{-1}}{-1}+C$

$$=4x+\frac{2}{3}x\sqrt{x}+\frac{1}{x}+C$$

**5.** Let $u=7+2x^3$. Then $du=6x^2\,dx$.

$$\int 6x^2(7+2x^3)^9\,dx=\int u^9\,du$$

$$=\frac{u^{10}}{10}+C$$

$$=\frac{(7+2x^3)^{10}}{10}+C$$

**6.** $5\int(4x^3-7)^9(x^2\,dx)=\frac{5}{12}\int(4x^3-7)^9(12x^2dx)$

$$=\frac{5}{12}\frac{(4x^3-7)^{10}}{10}+C$$

$$=\frac{1}{24}(4x^3-7)^{10}+C$$

**7.** $\int (3x^2 - 6x + 1)^{-3}(2x - 2)\,dx$

Let $u = 3x^2 - 6x$;

$du = (6x - 6)\,dx = 3(2x - 2)\,dx.$

$= \frac{1}{3}\int u^{-3}\,du = \frac{1}{3} \cdot \frac{u^{-2}}{-2} + C$

$= -\frac{1}{6}(3x^2 - 6x + 1)^{-2} + C$

**8.** $\int \left(e^x + \frac{5}{x} - 1\right)dx = e^x + 5\ln|x| - x + C$

**9.** $\int \frac{s^3}{2s^4 - 5}\,ds = \frac{1}{8}\int \frac{8s^3\,ds}{2s^4 - 5} = \frac{1}{8}\ln|2s^4 - 5| + C$

**10.** $100\int e^{-0.01x}\,dx = \frac{100}{-0.01}\int e^{-0.01x}(-0.01\,dx)$

$= -10{,}000e^{-0.01x} + C$

**11.** $5\int e^{2y^4 - 1}(y^3\,dy) = \frac{5}{8}\int e^{2y^4 - 1}(8y^3\,dy)$

$= \frac{5}{8}e^{2y^4 - 1} + C$

**12.** Simplify using long division:

$$\begin{array}{r} 2x - 1 \\ 2x + 1 \overline{)\,4x^2 \phantom{+2x}} \\ \underline{4x^2 + 2x} \\ -2x \phantom{-1} \\ \underline{-2x - 1} \end{array}$$

Remainder: 1

$\int \frac{4x^2}{2x + 1}\,dx = \int \left(2x - 1 + \frac{1}{2x + 1}\right)dx$

$= x^2 - x + \frac{1}{2}\ln|2x + 1| + C$

(Let $u = 2x + 1$ to integrate $\frac{1}{2x + 1}$.)

**13.** $\int f(x)\,dx = 2x^3 - x + 5e^x + C$. $f(x)$ is the derivative of the right side. $f(x) = 6x^2 - 1 + 5e^x$

**14.** $f(x) = \int \left(\frac{x^2}{3} - \frac{5}{8}\right)dx = \frac{1}{3}\left(\frac{x^3}{3}\right) - \frac{5}{8}x + C$

$f(x) = \frac{x^3}{9} - \frac{5}{8}x + C$

**15.** $y' = 4x^3 + 3x^2$; $y(0) = 4$

$y = x^4 + x^3 + C$

$4 = 0^4 + 0^3 + C$

$y = x^4 + x^3 + 4$

**16.** $\frac{dy}{dx} = e^{4x}$; $y(0) = 2$

$y = \frac{1}{4}e^{4x} + C$

$2 = \frac{1}{4}(1) + C$ or $C = \frac{7}{4}$

$y = \frac{1}{4}e^{4x} + \frac{7}{4}$

**17.** $\frac{dy}{dx} = x^3y^2$

$\int \frac{dy}{y^2} = \int x^3\,dx$

$\frac{y^{-1}}{-1} = \frac{x^4}{4} + \overline{C}$ or $\frac{x^4}{4} + \frac{1}{y} = K$

$\frac{1}{y} = \frac{4K - x^4}{4}$ or $y = \frac{4}{C - x^4}$

**18.** $p'(t) = 2000t^{1.04}$; $p(0) = 50{,}000$; $p(10) = ?$

$p(t) = 2000\frac{t^{2.04}}{2.04} + C$

$50{,}000 = 0 + C = C$

$p(10) = \frac{2000}{2.04}10^{2.04} + 50{,}000 = 157{,}498$

**19.** $\overline{MC} = 4x + 50$; $\overline{MR} = 500$; $C(10) = 1000$

$C = 2x^2 + 50x + K$; $R = 500x$

$1000 = 2(10)^2 + 50(10) + K$ gives $K = 300$

$P(x) = 500x - (2x^2 + 50x + 300)$

$= 450x - 2x^2 - 300$

**20.** $\frac{dS}{dy} = 0.22 - \frac{0.25}{\sqrt{0.5y + 1}}$ Since $\frac{dC}{dy} = 1 - \frac{dS}{dy}$, we have $\frac{dC}{dy} = 0.78 + \frac{0.25}{\sqrt{0.5y + 1}}$.

$C = 0.78y + \sqrt{0.5y + 1} + K$

$6.6 = 0 + \sqrt{0 + 1} + K$ gives $K = 5.6$

$C(y) = 0.78y + \sqrt{0.5y + 1} + 5.6$

**21.** $\frac{dx}{dt} = kx$ ; $\frac{dx}{x} = k\,dt$ ; $\ln x = kt + C$ gives $x = e^{kt+C} = e^{kt} \cdot e^{C} = Ae^{kt}$. At $t = 0$, $x = A$. Then at $t = 100$, $x = \frac{1}{2}A$.

$\frac{1}{2}A = Ae^{100k}$ gives $k = \frac{\ln\left(\frac{1}{2}\right)}{100} \approx -0.00693$

$x = Ae^{-0.00693t}$; $x = 0.1A$ gives $0.1A = Ae^{-0.00693t}$

gives $t = \frac{\ln(0.1)}{-.00693} \approx 332.3$ days

**22.** Let $x$ be the amount of drug (in g) in the organ at time $t$. Note also that $x(0) = 0$.

Rate of change = Rate in – Rate out

$$\text{Rate in} = \frac{4\text{ cc}}{\text{sec}} \cdot \frac{0.1\text{ g}}{\text{cc}} = 0.4\frac{\text{g}}{\text{sec}}$$

$$\text{Rate out} = \left(\frac{x}{160}\frac{\text{g}}{\text{cc}}\right)\left(4\frac{\text{cc}}{\text{sec}}\right) = \frac{x}{40}\,\frac{\text{g}}{\text{sec}}$$

$$\frac{dx}{dt} = 0.4 - \frac{x}{40}$$

$$\frac{dx}{dt} = \frac{16 - x}{40}$$

$$-\int \frac{-dx}{16 - x} = \int \frac{1}{40}\,dt$$

$$-\ln(16 - x) = \frac{1}{40}t + C$$

$$\ln(16 - x) = -\frac{1}{40}t + C$$

$$16 - x = ke^{-t/40}$$

$$16 - 0 = ke^{0} \rightarrow k = 16$$

$$16 - x = 16e^{-t/40}$$

$$x = 16 - 16e^{-t/40}$$

# Chapter 13: Definite Integrals: Techniques of Integration

## *Exercises 13.1*

**1.** $f(x) = 4x - x^2$

Width of rectangles $= \dfrac{2-0}{2} = 1$.

Height of each rectangle: $f(1) = 4(1) - 1^2 = 3$

$f(2) = 4(2) - 2^2 = 4$

$A \approx 1(3+4) = 7$

**3.** $f(x) = 9 - x^2$

Width of rectangles $= \dfrac{3-1}{4} = \dfrac{1}{2}$

Height of rectangles:

$f(1.5) = 9 - 2.25 = 6.75$ $\quad f(2) = 9 - 4 = 5$

$f(2.5) = 9 - 6.25 = 2.75$ $\quad f(3) = 9 - 9 = 0$

$A = 0.5(6.75 + 5 + 2.75 + 0) = 7.25$

**5.** $f(x) = 4x - x^2$

Width of rectangles $= \dfrac{2-0}{2} = 1$

Height of each rectangle:

$f(0) = 4(0) - 0^2 = 0$

$f(1) = 4(1) - 1^2 = 3$

$A \approx 1(0+3) = 3$

**7.** $f(x) = 9 - x^2$

Width of rectangles $= \dfrac{3-1}{4} = \dfrac{1}{2}$

Height of rectangles:

$f(1) = 9 - 1 = 8$ $\quad f\left(\frac{3}{2}\right) = 9 - 2.25 = 6.75$

$f(2) = 9 - 4 = 5$ $\quad f\left(\frac{5}{2}\right) = 9 - 6.25 = 2.75$

$\frac{1}{2}(8 + 6.75 + 5 + 2.75) = 11.25$

**9.** $S_L(10) = \dfrac{14}{3} - \dfrac{6}{10} + \dfrac{4}{300} = \dfrac{1224}{300} = 4.08$

$S_R(10) = \dfrac{14(100) + 180 + 4}{300} = \dfrac{1584}{300} = 5.28$

$S_R(100) = \dfrac{14(100)^2 + 18(100) + 4}{3(100)^2} = 4.7268$

**11.** $\lim\limits_{n\to\infty} S_L = \lim\limits_{n\to\infty}\left(\dfrac{14}{3} - \dfrac{6}{n} + \dfrac{4}{3n^2}\right)$

$= \dfrac{14}{3} - 0 + 0 = \dfrac{14}{3}$

$\lim\limits_{n\to\infty} S_R = \lim\limits_{n\to\infty}\left(\dfrac{14}{3} + \dfrac{6}{n} + \dfrac{4}{3n^2}\right)$

$= \dfrac{14}{3} + 0 + 0 = \dfrac{14}{3}$

**13.** If a point within each subinterval is selected and the area of each rectangle is determined, then the total area $A$ would satisfy $S_L \le A \le S_R$.

Since $\lim\limits_{n\to\infty} S_L = \lim\limits_{n\to\infty} S_R = \dfrac{14}{3}$, it follows that

$\lim\limits_{n\to\infty} A = \dfrac{14}{3}$.

**15.** $\sum\limits_{i=1}^{4} x_i = 3 + (-1) + 3 + (-2) = 3$

**17.** $\sum\limits_{j=2}^{5} (j^2 - 3)$

$= (4-3) + (9-3) + (16-3) + (25-3)$

$= 42$

No formulas are used because $j$ begins at 2 and the formulas require that $j$ begin at 1.

**19.** $\sum\limits_{j=0}^{4} (j^2 - 4j + 1)$

$= (1) + (-2) + (-3) + (-2) + (1) = -5$

**21.** $\sum\limits_{j=1}^{60} 3 = 3(60) = 180$

**23.** $\sum\limits_{k=1}^{30} (k^2 + 4k) = \dfrac{30(31)(61)}{6} + \dfrac{4(30)(31)}{2}$

$= 11{,}315$

**25.** $\sum_{i=1}^{n}\left(1-\frac{2i}{n}+\frac{i^2}{n^2}\right)\left(\frac{3}{n}\right)$

$= \frac{3}{n}\left[n - \frac{2}{n}\cdot\frac{n(n+1)}{2} + \frac{1}{n^2}\cdot\frac{n(n+1)(2n+1)}{6}\right]$

$= 3 - \frac{3}{n}(n+1) + \frac{(n+1)(2n+1)}{2n^2}$

$= \frac{2n^2 - 3n + 1}{2n^2}$

**27.** **a.** $f(x) = 2x$

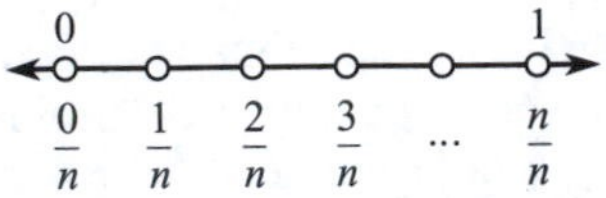


Area 1st rectangle: $\frac{1}{n}\cdot f\left(\frac{0}{n}\right) = \frac{1}{n}\cdot 0$

Area 2nd rectangle: $\frac{1}{n}\cdot f\left(\frac{1}{n}\right) = \frac{1}{n}\cdot\frac{2}{n}$

$\vdots$

Area $i$th rectangle:

$S(n) = \frac{1}{n}\left[0 + \frac{2}{n} + \frac{4}{n} + \cdots + \frac{2n-2}{n}\right]$

$= \frac{1}{n}\cdot\frac{2}{n}(0+1+2+\cdots+(n-1)]$

$= \frac{2}{n^2}\cdot\frac{n}{2}(0+n-1) = \frac{(n-1)}{n}$

**b.** $S(10) = \frac{10-1}{10} = \frac{9}{10}$

**c.** $S(100) = \frac{100-1}{100} = \frac{99}{100}$

**d.** $S(1000) = \frac{1000-1}{1000} = \frac{999}{1000}$

**e.** $\lim_{n\to\infty} S(n) = \lim_{n\to\infty}\left(1-\frac{1}{n}\right) = 1$

**29.** Area of 1st rectangle: $\frac{1}{n}\cdot f(1/n) = \frac{1}{n}\cdot\frac{1}{n^2}$

Area of 2nd rectangle: $\frac{1}{n}\cdot f(2/n) = \frac{1}{n}\cdot\frac{4}{n^2}$

Area of $i$th rectangle: $\frac{1}{n}\cdot f(i/n) = \frac{1}{n}\cdot\frac{i^2}{n^2}$

$S(n) = \frac{1}{n}\left[\frac{1}{n^2} + \frac{4}{n^2} + \frac{9}{n^2} + \cdots + \frac{n^2}{n^2}\right]$

$$= \frac{1}{n^3}\left[\frac{n(n+1)(2n+1)}{6}\right] = \frac{(n+1)(2n+1)}{6n^2}$$

**a.** $S(10) = \dfrac{11 \cdot 21}{600} = \dfrac{77}{200}$

**b.** $S(100) = \dfrac{101 \cdot 201}{60{,}000} = \dfrac{6767}{20{,}000}$

**c.** $S(1000) = \dfrac{1001 \cdot 2001}{6{,}000{,}000} = \dfrac{667{,}667}{2{,}000{,}000}$

**d.** $\lim\limits_{n\to\infty} S(n) = \lim\limits_{n\to\infty} \dfrac{2n^2+3n+1}{6n^2}$

$$= \lim_{n\to\infty}\left(\frac{2}{6} + \frac{1}{2n} + \frac{1}{6n^2}\right)$$

$$= \frac{2}{6} + 0 + 0 = \frac{1}{3}$$

**31.** $f(x) = x^2 - 6x + 8$

Use right hand endpoints.

Area of 1st rectangle: $\dfrac{2}{n} \cdot f(2/n)$

Area of 2nd rectangle: $\dfrac{2}{n} \cdot f(4/n)$

$\vdots$

Area of ith rectangle: $\dfrac{2}{n} \cdot f(2i/n)$

$$S(n) = \frac{2}{n}\left[\left\{\left(\frac{2}{n}\right)^2 - 6\left(\frac{2}{n}\right) + 8\right\} + \left\{\left(\frac{4}{n}\right)^2 - 6\left(\frac{4}{n}\right) + 8\right\} + \cdots + \left\{\left(\frac{2n}{n}\right)^2 - 6\left(\frac{2n}{n}\right) + 8\right\}\right]$$

We are going to group like terms.

$$S(n) = \frac{2}{n}\left[\left\{\left(\frac{2}{n}\right)^2 + \left(\frac{4}{n}\right)^2 + \cdots + \left(\frac{2n}{n}\right)^2\right\} - 6\left\{\frac{2}{n} + \frac{4}{n} + \cdots + \frac{2n}{n}\right\} + \{8n\}\right]$$

$$= \frac{2}{n}\left[\left\{4\left(\frac{1}{n}\right)^2 + 4\left(\frac{2}{n}\right)^2 + \cdots + 4\left(\frac{n}{n}\right)^2\right\} - 12\left\{\frac{1}{n} + \frac{2}{n} + \cdots + \frac{n}{n}\right\} + 8n\right]$$

$$= \frac{2}{n}\left[\frac{4}{n^2}(1 + 2^2 + \cdots + n^2) - \frac{12}{n}(1 + 2 + \cdots + n) + 8n\right]$$

$$= \frac{2}{n}\left[\frac{4}{n^2} \cdot \frac{n(n+1)(2n+1)}{6} - \frac{12}{n} \cdot \frac{n(n+1)}{2} + 8n\right] = \frac{20n^2 - 24n + 4}{3n^2}$$

$$A = \lim_{n\to\infty}\left(\frac{20}{3} - \frac{24}{3n} + \frac{4}{3n^2}\right) = \frac{20}{3}$$

**33. a.** $A \approx [(1020 \times 2) + (1022 \times 2) + (1097 \times 2) + (1209 \times 2)] = 8696$ square units

**b.** This represents the total per capita out-of-pocket expenses for health care between 2013 and 2021.

**35.** There are approximately 90 squares under the curve.

Each square represents $10\times\frac{1\text{ hr}}{3600}\times\frac{1\text{ miles}}{\text{hr}}=\frac{1}{360}$ miles.

Area represents $\approx 90\cdot\frac{1}{360}=\frac{1}{4}$ miles.

**37.** $A\approx 10(0+15+18+18+30+27+24+23)=1550$ sq ft

**39.** Width of rectangles: $\frac{15-10}{10}=\frac{5}{10}=\frac{1}{2}$

$$\begin{aligned}\text{Area} &\approx 0.5(T(10.5)+T(11)+T(11.5)+\ldots+T(14.5)+T(15))\\ &\approx 0.5(19.5608+19.9872+20.4192+20.8568+21.3+21.7488+22.2032+22.6632+23.1288+23.6)\\ &\approx 107.734 \text{ square units}\end{aligned}$$

This represents the total sulphur dioxide emissions (in millions of short tons) from electricity generation from 2010 to 2015.

## *Exercises 13.2*

**1.** $\int_0^3 4x\,dx=4\cdot\frac{x^2}{2}\Big|_0^3=2\left[3^2-0^2\right]=18$

**3.** $\int_2^4 dx=x\big|_2^4=4-2=2$

**5.** $\int_2^4 x^3dx=\frac{1}{4}x^4\Big|_2^4=\frac{256}{4}-\frac{16}{4}=60$

**7.** $$\int_0^5 4x^{2/3}dx=\frac{12}{5}x^{5/3}\Big|_0^5=\frac{12}{5}(5^{5/3}-0)=\frac{12}{5}\left[5(5)^{2/3}\right]=12\sqrt[3]{25}$$

**9.** $$\int_1^4(10-4x)dx=\left(10x-2x^2\right)\Big|_1^4=(40-32)-(10-2)=0$$

**11.** $$\int_2^4(4x^3-6x^2-5x)dx=\left(x^4-2x^3-\frac{5}{2}x^2\right)\Big|_2^4=(256-128-40)-(16-16-10)=98$$

**13.** $$\int_3^4(x-4)^9\,dx=\frac{1}{10}(x-4)^{10}\Big|_3^4=\frac{1}{10}\left[0^{10}-(-1)^{10}\right]=-\frac{1}{10}$$

**15.** $$\int_2^4(x^2+2)^3\,xdx=\frac{1}{2}\int_2^4(x^2+2)^3(2x)dx=\frac{1}{2}\cdot\frac{1}{4}(x^2+2)^4\Big|_2^4=\frac{1}{8}(104{,}976-1296)=12{,}960$$

**17.** $$\int_0^3(2x-x^2)^4(1-x)dx=\frac{1}{2}\int_0^3(2x-x^2)^4(2-2x)dx=\frac{1}{10}(2x-x^2)^5\Big|_0^3=-\frac{243}{10}$$

**19.** $$\int_{-2}^2 15x^3(x^4-6)^6\,dx=15\left(\frac{x^{28}}{28}-\frac{3x^{24}}{2}+27x^{20}-270x^{16}+1620x^{12}-5832x^8+11664x^4\right)\Bigg|_0^1=0$$

**21.** $\int_0^4 \sqrt{4x+9}\,dx$

$= \frac{1}{4}\int_0^4 (4x+9)^{1/2}(4x\,dx)$

$= \frac{1}{4}\cdot\frac{2}{3}(4x+9)^{3/2}\Big|_0^4 = \frac{1}{6}\left[25^{3/2} - 9^{3/2}\right]$

$= \frac{1}{6}[125-27] = \frac{1}{6}(98) = \frac{49}{3} = 16\frac{1}{3}$

**23.** $\int_1^3 \frac{3}{y^2}dy = \int_1^3 3y^{-2}dy$

$= 3\cdot\frac{y^{-1}}{-1}\Big|_1^3 = -3\left[3^{-1} - 1^{-1}\right]$

$= -3\left[\frac{1}{3} - 1\right] = 2$

**25.** $\int_0^1 e^{3x}dx = \frac{1}{3}\int_0^1 e^{3x}(3\,dx)$

$= \frac{1}{3}e^{3x}\Big|_0^1 = \frac{1}{3}\left[e^3 - e^0\right] = \frac{1}{3}\left(e^3 - 1\right)$

**27.** $\int_1^e \frac{4}{z}dz = 4\int_1^e z^{-1}dz$

$= 4\ln|z|\,\Big|_1^e = 4[\ln e - \ln 1] = 4$

**29.** $\int_0^2 8x^2e^{-x^3}dx$

$= 8\left(-\frac{1}{3}\right)\int_0^2 e^{-x^3}\left(-3x^2dx\right) = -\frac{8}{3}e^{-x^3}\Big|_0^2$

$= -\frac{8}{3}\left(e^{-8} - e^0\right) = -\frac{8}{3}\left(\frac{1}{e^8} - 1\right) = \frac{8}{3}\left(1 - \frac{1}{e^8}\right)$

**31. a.** $\int_3^6 \frac{x}{3x^2+4}dx = \frac{1}{6}\int_3^6 \frac{6x}{3x^2+4}dx$

$= \frac{1}{6}\ln\left(3x^2+4\right)\Big|_3^6 = \frac{1}{6}[\ln 112 - \ln 31]$

$= \frac{1}{6}\ln\left(\frac{112}{31}\right)$

**b.** Graphing utility gives 0.2140853.

**33. a.** $\int_1^2 \frac{x^2+3}{x}dx = \int_1^2\left(x + 3x^{-1}\right)dx$

$= \left(\frac{x^2}{2} + 3\ln|x|\right)\Big|_1^2 = 2 + 3\ln 2 - \left(\frac{1}{2} + 3\ln 1\right)$

$= \frac{3}{2} + 3\ln 2 = 3.5794415$

**b.** Graphing utility gives 3.5794415.

**35. a.** *A*, *C*

**b.** *B*

**37. a.** $\int_0^4\left(2x - \frac{1}{2}x^2\right)dx$

**b.** $A = \left[2\cdot\frac{x^2}{2} - \frac{1}{2}\cdot\frac{x^3}{3}\right]\Big|_0^4$

$= \left(x^2 - \frac{1}{6}x^3\right)\Big|_0^4 = 16 - \frac{1}{6}(64) - 0$

$= \frac{96-64}{6} = \frac{16}{3}$

**39. a.** $A = \int_{-1}^0\left(x^3 + 1\right)dx$

**b.** $A = \left(\frac{x^4}{4} + x\right)\Big|_{-1}^0 = 0 - \left(\frac{1}{4} - 1\right) = \frac{3}{4}$

**41.** $\int_1^2\left(-x^2 + 3x - 2\right)dx$

$= \left(-\frac{1}{3}x^3 + \frac{3}{2}x^2 - 2x\right)\Big|_1^2$

$= \left(-\frac{8}{3} + 6 - 4\right) - \left(-\frac{1}{3} + \frac{3}{2} - 2\right) = \frac{1}{6}$

**43.** $\int_1^3 e^{x^2}x\,dx = \frac{1}{2}\int_1^3 e^{x^2}(2x\,dx)$

$= \frac{1}{2}e^{x^2}\Big|_1^3 = \frac{1}{2}\left(e^9 - e\right)$

**45.** $\int_0^a g(x)\,dx > \int_0^a f(x)\,dx$ because there is more positive area under $g(x)$ than under $f(x)$ on this interval.

**47.** The integrals differ only in sign.

**49.** $\int_1^2 (2x - x^2)\,dx + \int_2^4 (2x - x^2)\,dx = (-1)\int_1^2 (x^2 - 2x)\,dx + (-1)\int_2^4 (x^2 - 2x)\,dx = (-1)\int_1^4 (x^2 - 2x)\,dx$

Thus, we have $\int_1^2 (2x - x^2)\,dx + \int_2^4 (2x - x^2)\,dx = \frac{2}{3} + \left(-\frac{20}{3}\right) = -6.$ So, $\int_1^4 (x^2 - 2x)\,dx = (-1)(-6) = 6.$

**51.** $\int_4^4 \sqrt{x^2 - 2}\,dx = 0$

**53.** **a.** $D = 50{,}000 \cdot 5 + 30{,}000 \cdot 5 = \$400{,}000$ (RHP)
or $60{,}000 \cdot 5 + 50{,}000 \cdot 5 = \$550{,}000$ (LHP)
or $\frac{1}{2}(10)(60{,}000 + 30{,}000) = \$450{,}000$ (trapezoid)

**b.** $D = \int_0^{10} 3000(20 - t)\,dt = 3000\left(20t - \frac{t^2}{2}\right)\Big|_0^{10} = 3000(200 - 50) - 0 = \$450{,}000$

**55.** **a.** $\int_0^7 (-30t^2 + 360t)\,dt = (-10t^3 + 180t^2)\Big|_0^7 = -3430 + 8820 = \$5390$

**b.** $\int_7^{14} (-30t^2 + 360t)\,dt = (-10t^3 + 180t^2)\Big|_7^{14} = (-27{,}440 + 35{,}280) - (-3430 + 8820) = \$2450$

**57.** $\int_0^3 120e^{0.01t}\,dt = \frac{120}{0.01}\int_0^3 e^{0.01t}(0.01dt) = 12{,}000e^{0.01t}\Big|_0^3 = 12{,}000(e^{0.03} - e^0)$
$= 12{,}000(1.03045 - 1) = 365.40$ thousands $= \$365{,}400$

**59.** $\int_0^{10} (19.12t + 319.0)\,dt = (9.56t^2 + 319t)\Big|_0^{10}$
$= (956 + 3190) - (0)$
$= 4146$ represents the total million metric tons of $CO_2$ emissions from 2010 to 2020.

**61.** $\int_0^{0.3} (0.3 - 3.33r^2)2\pi r\,dr = \int_0^{0.3} (0.6\pi r - 6.66\pi r^3)\,dr = \left(0.6\pi \cdot \frac{r^2}{2} - 6.66\pi \cdot \frac{r^4}{4}\right)\Big|_0^{0.3}$
$= (0.3\pi r^2 - 1.665\pi r^4)\Big|_0^{0.3} = \left[0.3\pi(0.3)^2 - 1.665\pi(0.3)^4\right] - (0 - 0) = \pi(0.027 - 0.0134865) = 0.04\text{ cm}^3$

**63.** $x = \int_0^5 200\left(1 + \frac{400}{(t+40)^2}\right)dt = 200\left(t + \frac{400(t+40)^{-1}}{-1}\right)\Big|_0^5 = 200\left[\left(5 - \frac{400}{45}\right) - (0 - 10)\right] = 1222.22$

**65.** $\int_8^{10} (0.012t^2 - 0.0012t^3)\,dt = (0.004t^3 - 0.003t^4)\Big|_8^{10}$
$= (0.004(10)^3 - 0.003(10)^4) - (0.004(8)^3 - 0.003(8)^4) = 0.1808$

**67.** **a.** $\int_0^3 0.3e^{-0.3t}\,dt = -\int_0^3 e^{-0.3t}(-0.3dt) = -e^{-0.3t}\Big|_0^3 = -e^{-0.3(3)} - \left(-e^{-0.3(0)}\right) = -e^{-0.9} + 1 \approx 0.5934$

**b.** $\int_5^{10} 0.3e^{-0.3t}\,dt = -\int_5^{10} e^{-0.3t}(-0.3dt) = -e^{-0.3t}\Big|_5^{10} = -e^{-0.3(10)} - \left(-e^{-0.3(5)}\right) = -e^{-3} + e^{-1.5} \approx 0.1733$

**69.** **a.** Using technology, $G(t)=-0.157t^2-0.196t+133$.

**b.** $\int_0^{10} G(t)\,dt=\int_0^{10}\left(-0.157t^2-0.196t+133\right)dt$

$\approx 1267.9$

The total amount of gasoline used by motor vehicles in the United States from 2014 to 2024 is predicted to be 1267.9 billion gallons.

## *Exercises 13.3*

**1.** **.** $\int_0^2\left(4-x^2\right)dx$

**b.** $\int_0^2\left(4-x^2\right)dx=\left(4x-\frac{x^3}{3}\right)\Big|_0^2=4(2)-\frac{2^3}{3}=8-\frac{8}{3}=\frac{24-8}{3}=\frac{16}{3}$

**3.** **a.** $A=\int_1^8\left(x^{1/3}-2+x\right)dx$

**b.** $A=\left|\left(\frac{3}{4}x^{4/3}-2x+\frac{x^2}{2}\right)\right|_1^8=28-\left(-\frac{3}{4}\right)=28\frac{3}{4}$

**5.** **a.** $\int_1^2\left[\left(4-x^2\right)-\left(\frac{1}{4}x^3-2\right)\right]dx$

**b.** $\int_1^2\left(6-x^2-\frac{1}{4}x^3\right)dx=\left(6x-\frac{1}{3}x^3-\frac{1}{16}x^4\right)\Big|_1^2=\left(12-\frac{8}{3}-1\right)-\left(6-\frac{1}{3}-\frac{1}{16}\right)=\frac{131}{48}$

**7.** **a.**

$x+2=x^2$

$x^2-x-2=0$

$(x-2)(x+1)=0$

$x=2$ or $x=-1$

If $x=2$, $y=4$. If $x=-1$, $y=1$.

**b., c.** $\int_{-1}^2\left[(x+2)-x^2\right]dx=\left(\frac{x^2}{2}+2x-\frac{x^3}{3}\right)\Big|_{-1}^2=\left(2+4-\frac{8}{3}\right)-\left(\frac{1}{2}-2+\frac{1}{3}\right)=\frac{18-8}{3}-\frac{3-12+2}{6}=\frac{20}{6}+\frac{7}{6}=\frac{9}{2}$

**9.** **a.**

$x^2-4x=x-x^2$

$2x^2-5x=0$

$x(2x-5)=0$

$x=0$ or $x=\frac{5}{2}$ If $x=0$, $y=0$. If $x=\frac{5}{2}$, $y=-\frac{15}{4}$

**b. c.** $\int_0^{5/2}\left[\left(x-x^2\right)-\left(x^2-4x\right)\right]dx=\int_0^{5/2}\left(5x-2x^2\right)dx=\left(\frac{5}{2}x^2-\frac{2}{3}x^3\right)\Big|_0^{5/2}=\left(\frac{125}{8}-\frac{250}{24}\right)-0=\frac{125}{24}$

**11. a.**
$$x^3 - 2x = 2x$$
$$x^3 - 4x = 0$$
$$x(x^2 - 4) = 0$$
$$x(x+2)(x-2) = 0$$
$$x = 0, x = -2, x = 2$$

If $x = 0$, then $y = 0$. If $x = -2$, then $y = -4$. If $x = 2$, then $y = 4$.

**b., c.** $\int_{-2}^{0}(x^3 - 2x - 2x)\,dx + \int_{0}^{2}\left[2x - (x^3 - 2x)\right]dx = \int_{-2}^{0}(x^3 - 4x)\,dx + \int_{0}^{2}(-x^3 + 4x)\,dx$

$$= \left[\frac{x^4}{4} - 4\cdot\frac{x^2}{2}\right]\Bigg|_{-2}^{0} + \left[-\frac{x^4}{4} + 4\cdot\frac{x^2}{2}\right]\Bigg|_{0}^{2} = 0 - \left(\frac{16}{4} - 2(-2)^2\right) + \left(-\frac{16}{4} + 2(2)^2\right) - 0 = 8$$

**13.** $f(1) = 3,\ g(1) = -1,\ f(x) \ge g(x)$

$$\int_0^2\left[(x^2+2) - (-x^2)\right]dx$$
$$= \left(\frac{2}{3}x^3 + 2x\right)\Bigg|_0^2 = \frac{16}{3} + 4 = \frac{28}{3}$$

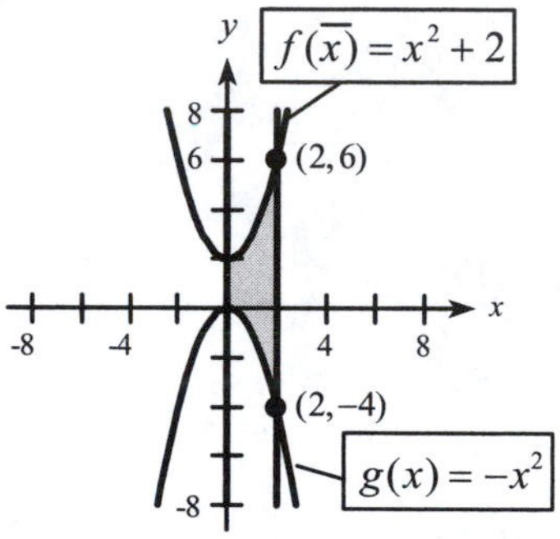


**15.** $x^3 - 1 = x - 1$

$$x^3 - x = x(x-1)(x+1) = 0$$
$$x = 0, x = 1$$
$$f\left(\frac{1}{2}\right) = \frac{1}{8} - 1 = -\frac{7}{8}$$
$$g\left(\frac{1}{2}\right) = \frac{1}{2} - 1 = -\frac{1}{2}$$
$$g(x) \ge f(x)$$
$$\int_0^1\left[(x-1) - (x^3 - 1)\right]dx$$
$$= \int_0^1 (x - x^3)\,dx$$
$$= \left(\frac{x^2}{2} - \frac{x^4}{4}\right)\Bigg|_0^1 = \frac{1}{2} - \frac{1}{4} = \frac{1}{4}$$

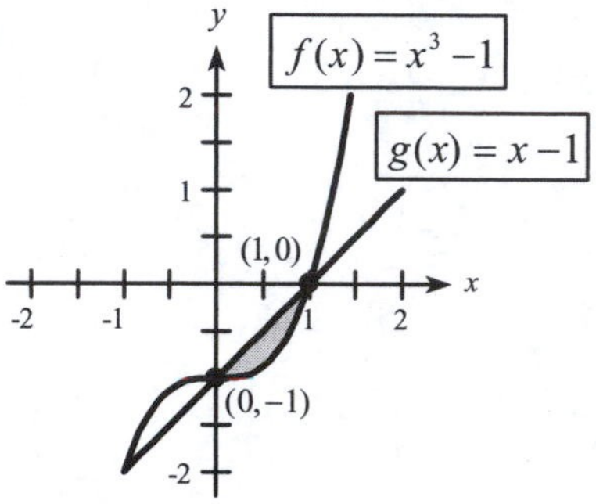


**17.** $x^2 - 2x = \frac{1}{2}x^2$

$$\frac{1}{2}x^2 - 2x = \frac{1}{2}x(x-4)$$
$$= 0$$
$$x = 0, x = 4$$
$$f(1) = \frac{1}{2} \quad g(1) = 1 - 2 = -1,\ f(x) \ge g(x)$$
$$\int_0^4\left[\frac{1}{2}x^2 - (x^2 - 2x)\right]dx$$
$$= \int_0^4\left(2x - \frac{1}{2}x^2\right)dx$$
$$= \left(x^2 - \frac{1}{6}x^3\right)\Bigg|_0^4$$
$$= \left(16 - \frac{64}{6}\right) = \frac{16}{3}$$

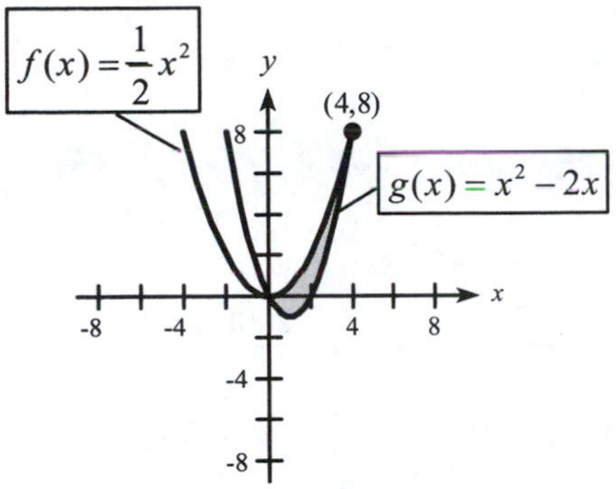


**19.** $x^2=\sqrt{x}$

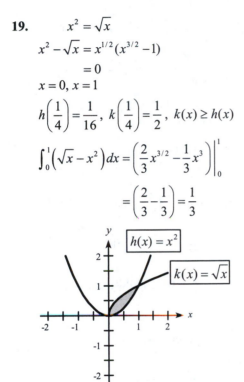

$x^2-\sqrt{x}=x^{1/2}(x^{3/2}-1)$
$=0$
$x=0,\ x=1$

$h\left(\frac{1}{4}\right)=\frac{1}{16},\ k\left(\frac{1}{4}\right)=\frac{1}{2},\ k(x)\ge h(x)$

$$\int_0^1\left(\sqrt{x}-x^2\right)dx=\left(\frac{2}{3}x^{3/2}-\frac{1}{3}x^3\right)\Big|_0^1$$
$$=\left(\frac{2}{3}-\frac{1}{3}\right)=\frac{1}{3}$$

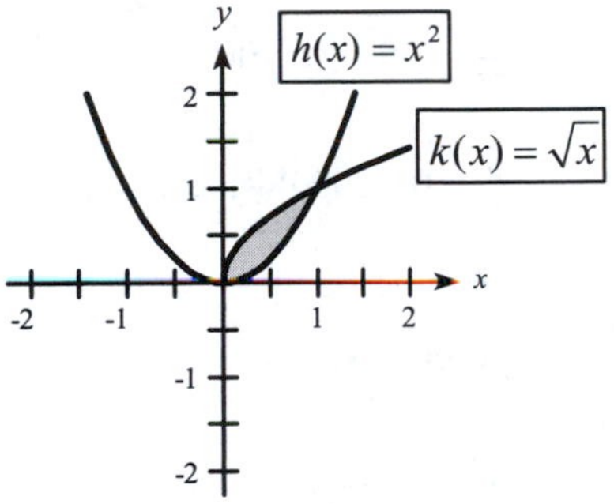


**21.** $x^3=x^2+2x$

$x^3-x^2-2x=x(x-2)(x+1)$
$=0$
$x=-1,\ x=0,\ x=2$
$f(1)=1,\ \ g(1)=3,$
$g(x)\ge f(x)$ over $[0,2]$

$$\int_{-1}^0\left[x^3-\left(x^2+2x\right)\right]dx+\int_0^2\left[\left(x^2+2x\right)-x^3\right]dx$$
$$=\left(\frac{x^4}{4}-\frac{x^3}{3}-x^2\right)\Big|_{-1}^0+\left(\frac{x^3}{3}+x^2-\frac{x^4}{4}\right)\Big|_0^2$$
$$=\left[0-\left(\frac{1}{4}+\frac{1}{3}-1\right)\right]+\left[\left(\frac{8}{3}+4-4\right)-0\right]$$
$$=\frac{5}{12}+\frac{8}{3}=\frac{37}{12}$$

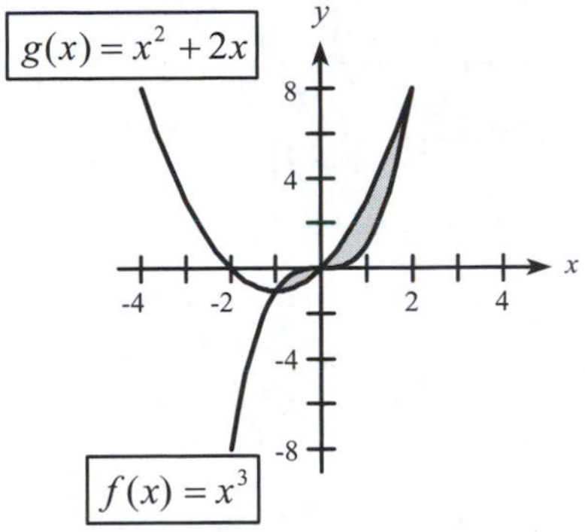


**23.** $\frac{3}{x}=4-x$

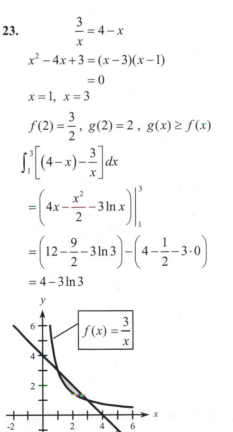

$x^2-4x+3=(x-3)(x-1)$
$=0$
$x=1,\ \ x=3$

$f(2)=\frac{3}{2},\ g(2)=2,\ g(x)\ge f(x)$

$$\int_1^3\left[(4-x)-\frac{3}{x}\right]dx$$
$$=\left(4x-\frac{x^2}{2}-3\ln x\right)\Big|_1^3$$
$$=\left(12-\frac{9}{2}-3\ln 3\right)-\left(4-\frac{1}{2}-3\cdot 0\right)$$
$$=4-3\ln 3$$

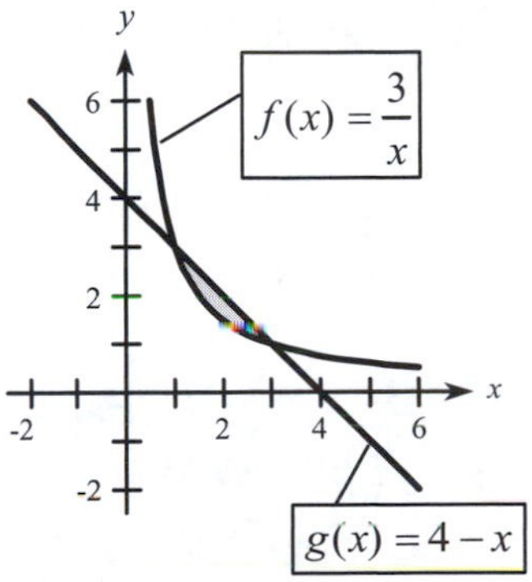

**25.** $\sqrt{x+3}=2$

$x+3=4$ or $x=1$, $x=-3$

$f(0)=\sqrt{3}$, $g(0)=2$, $g(x)\ge f(x)$

$$\int_{-3}^{1}(2-\sqrt{x+3})dx=\left(2x-\frac{2}{3}(x+3)^{3/2}\right)\Bigg|_{-3}^{1}$$

$$=\left(2-\frac{2}{3}\cdot 8\right)-(-6-0)=\frac{8}{3}$$

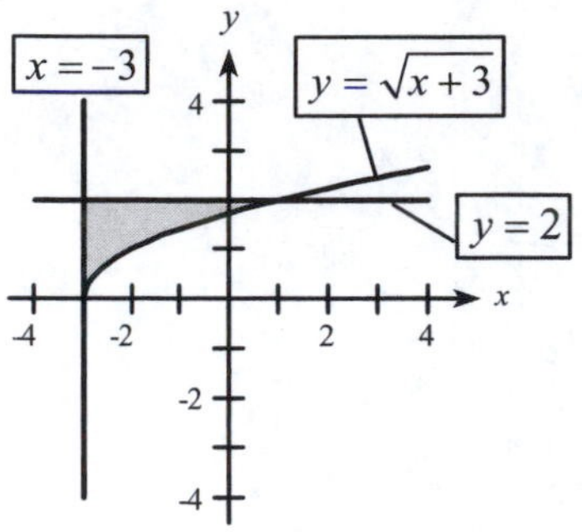


**27.** $$\text{Avg value}=\frac{1}{b-a}\int_a^b f(x)dx$$

$$=\frac{1}{3-0}\int_0^3\left(9-x^2\right)dx$$

$$=\frac{1}{3}\left(9x-\frac{x^3}{3}\right)\Bigg|_0^3=\frac{1}{3}(27-9)=6$$

**29.** $$\text{Avg value}=\frac{1}{b-a}\int_a^b f(x)dx$$

$$=\frac{1}{1-(-1)}\int_{-1}^{1}\left(x^3-x\right)dx$$

$$=\frac{1}{2}\left(\frac{x^4}{4}-\frac{x^2}{2}\right)\Bigg|_{-1}^{1}=\frac{1}{2}\left(\frac{1}{4}-\frac{1}{4}\right)=0$$

**31.** $$\text{Avg value}=\frac{1}{4-1}\int_1^4\left(\sqrt{x}-2\right)dx$$

$$=\frac{1}{3}\left(\frac{2}{3}x^{3/2}-2x\right)\Bigg|_1^4$$

$$=\frac{1}{3}\left[-\frac{8}{3}-\left(-\frac{4}{3}\right)\right]=-\frac{4}{9}$$

**33.** Use appropriate technology. Answer: 11.83

**35.** $$AP=\frac{1}{x_1-x_0}\int_{x_0}^{x_1}\left(R(x)-C(x)\right)dx$$

**37.** $C(x)=x^2+400x+2000$

**a.** $$\overline{C}(x)=\frac{C(x)}{x}=x+400+\frac{2000}{x}$$

$\overline{C}(1000)=1000+400+2=\$1402$

**b.**

$$AV=\frac{1}{1000-0}\int_0^{1000}\left(x^2+400x+2000\right)dx$$

$$=\frac{1}{1000}\left(\frac{1}{3}x^3+200x^2+2000x\right)\Bigg|_0^{1000}$$

$$=\frac{1000}{1000}\left(\frac{1{,}000{,}000}{3}+200(1000)+2000\right)$$

$$=\$535{,}333\frac{1}{3}$$

**39. a.** $$\text{Average}=\frac{1}{20-0}\int_0^{20}\left[100e^{-x^2}\cdot x+100\right]dx$$

$$=\frac{1}{20}\left(-50e^{-x^2}+100x\right)\Bigg|_0^{20}$$

$$=\frac{1}{20}[(0+2000)-(-50+0)]$$

$$=102.5$$

**b.** $$\frac{1}{30-20}\int_{20}^{30}\left(100xe^{-x^2}+100\right)dx$$

$$=\frac{1}{10}\left(-50e^{-x^2}+100x\right)\Bigg|_{20}^{30}$$

$$\approx\frac{1}{10}(3000-2000)=100$$

**41. a.** 801.067 million per year

**b.** 1398.67 million per year

**43.** Average $= \frac{1}{4-0}\int_0^4 \left(30x^{18/7} - 240x^{11/7} + 480^{4/7}\right)dx$

$= \frac{1}{4}\left(\frac{7(30)}{25}\cdot x^{25/7} - \frac{7(240)}{18}\cdot x^{18/7} + \frac{7(480)}{11}\cdot x^{11/7}\right)\Bigg|_0^4$

$= \frac{1}{4}(1187.12 - 3297.55 + 2697.99) = 146.89 = 147$ mg

**45.** The Gini coefficient of income for Black households is

$2\int_0^1 \left[x - \left(2.292x^4 - 3.127x^3 + 2.006x^2 - 0.1710x + 0.0005556\right)\right]dx \approx 0.479.$

The Gini coefficient of income for Hispanic households is

$2\int_0^1 \left[x - \left(1.830x^4 - 2.564x^3 + 1.806x^2 - 0.1137x + 0.0003512\right)\right]dx \approx 0.459.$

The income distribution for Black households is more unequal than it is for Hispanic households.

**47.** The Gini coefficient of income for 2012 is $2\int_0^1 \left(x - x^{2.760}\right)dx \approx 0.468.$

The Gini coefficient of income for 2004 if $2\int_0^1 \left(x - x^{2.671}\right)dx \approx 0.455.$

2012 shows more income inequality than 2004.

**49.** $G = \frac{\int_0^1 \left(x - x^p\right)dx}{1/2} = 2\int_0^1 \left(x - x^p\right)dx = 2\left(\frac{1}{2}x^2 - \frac{1}{p+1}x^{p+1}\right)\Bigg|_0^1 = 2\left(\frac{1}{2} - \frac{1}{p+1}\right) = 1 - \frac{2}{p+1} = \frac{p+1}{p+1} - \frac{2}{p+1} = \frac{p-1}{p+1}$

### *Exercises 13.4*

**1.** Total income $= \int_0^{10} 12,000e^{0.01t}\,dt$

$= 1,200,000\int_0^{10} 0.01e^{0.01t}\,dt$

$= 1,200,000e^{0.01t}\Big|_0^{10}$

$= 1,200,000e^{0.1} - 1,200,000$

$= \$126,205.10$

**3.** Total income $= \int_0^{12} 24,000e^{0.03t}\,dt$

$= 800,000e^{0.03t}\Big|_0^{12}$

$= 800,000(1.4333 - 1)$

$= \$346,664$

**5.** Total income $= \int_0^{10} 80e^{-0.1t}\,dt$

$= -800e^{-0.1t}\Big|_0^{10}$

$= -800(0.3678 - 1)$

$= \$505.70$ thousand

$=\$505,700$

**7.** Total Income $= \int_6^{12} 3000e^{0.004t}\,dt$

$= 750,000e^{0.004t}\Big|_6^{12}$

$= 750,000(0.0249)$

$= \$18,675$

**9.** Present Value $= \int_0^8 12,000e^{0.04t}\cdot e^{-0.08t}\,dt$

$= \int_0^8 12,000e^{-0.04t}\,dt$

$= -300,000e^{-0.04t}\Big|_0^8$

$= -300,000(0.72615 - 1)$

$= \$82,155$

**11.** Present Value $= \int_0^5 630{,}000e^{-0.07t}\,dt$

$= -9{,}000{,}000e^{-0.07t}\Big|_0^5$

$= -9{,}000{,}000(0.70469-1)$

$= \$2{,}657{,}807$

$\text{FV} = e^{.35}\int_0^5 630{,}000e^{-0.07t}\,dt$

$= e^{.35}(-9{,}000{,}000e^{-0.07t})\Big|_0^5$

$= \$3{,}771{,}608$

**13.**

Present Value $= \int_0^{10} 97.5e^{-0.2(t+3)}e^{-0.06t}\,dt$

$= 97.5\int_0^{10} e^{-0.26t-0.6}\,dt$

$= -\dfrac{97.5}{0.26}e^{-0.26t-0.6}\Big|_0^{10}$

$= -\dfrac{97.5}{0.26}\left[e^{-3.2} - e^{-0.6}\right]$

$= 190.519$ thousand dollars

$= \$190{,}519$

By pattern in 11, $FV = e^{.6}PV = \$347{,}148$

**15.** $\text{PV} = \int_0^7 30{,}000e^{-0.1t}$

$= -300{,}000e^{-0.1t}\Big|_0^7$

$= -300{,}000(0.4966-1)$

$= \$151{,}024$

Present value of gift shop is \$151,024

$\text{PV} = \int_0^7 21{,}600e^{0.08t}\left(e^{-0.1t}\right)dt$

$= \int_0^7 21{,}600e^{-0.02t}\,dt$

$= -1{,}080{,}000e^{-0.02t}\Big|_0^7$

$= -1{,}080{,}000(0.8694-1)$

$= \$141{,}048$

Present value of video store is \$141,048. The gift shop is the better buy.

**17.** $9 = 34 - x^2$ gives $x^2 = 25$ or $x = 5$.

Equilibrium point is $(5,9)$.

$CS = \int_0^5 (34 - x^2)dx - 5\cdot 9$

$= \left(34x - \dfrac{1}{3}x^3\right)\Big|_0^5 - 45$

$= 170 - \dfrac{125}{3} - 45 = \dfrac{250}{3} = \$83.33$

**19.** $p = \dfrac{200}{8+2} = 20$.

Equilibrium point is $(8,20)$.

$CS = \int_0^8 \dfrac{200}{x+2}\,dx - 8\cdot 20$

$= 200\ln(x+2)\Big|_0^8 - 160$

$= 200(\ln 10 - \ln 2) - 160$

$= \$161.89$

**21.** $x^2 + 4x + 11 = 81 - x^2$

$2x^2 + 4x - 70 = 0$

$2(x+7)(x-5) = 0$

Equilibrium point is $(5,56)$, since $p(5) = 81 - 25 = 56$.

$CS = \int_0^5 (81 - x^2)\,dx - 5\cdot 56$

$= \left(81x - \dfrac{1}{3}x^3\right)\Big|_0^5 - 280$

$= 405 - \dfrac{125}{3} - 280 = \$83.33$

**23.** $\dfrac{12}{x+1} = 1 + 0.2x$

$12 = 1 + 1.2x + 0.2x^2$

$2x^2 + 12x - 110 = 0$

$2(x+11)(x-5) = 0$

Equilibrium point is $(5,2)$.

$CS = \int_0^5 \dfrac{12}{x+1}\,dx - 5\cdot 2$

$= 12\ln(x+1)\Big|_0^5 - 10$

$= 12(\ln 6 - \ln 1) - 10 = \$11.50$

**25.** $R = px = 360x - 3x^2 - 2x^3$

$P = R - C = -2x^3 - 9x^2 + 240x - 1000$

$P'(x) = -6x^2 - 18x + 240$

$= -6(x^2 + 3x - 40)$

$= -6(x+8)(x-5)$

Maximum profit is at $x = 5$ with

$p(5) = 360 - 15 - 50 = 295$

$CS = \int_0^5 (360 - 3x - 2x^2)dx - 5(295)$

$= \left(360x - \frac{3}{2}x^2 - \frac{2}{3}x^3\right)\Big|_0^5 - 1475$

$= 1800 - 37.50 - 83.33 - 1475$

$= \$204.17$

**27.** $422 = 4x^2 + 2x + 2$

$4x^2 + 2x - 420 = 0$

$2(2x+21)(x-10) = 0$

Equilibrium point is $(10, 422)$.

$PS = 10 \cdot 422 - \int_0^{10} (4x^2 + 2x + 2)dx$

$= 4220 - \left(\frac{4}{3}x^3 + x^2 + 2x\right)\Big|_0^{10}$

$= 4220 - (1333.33 + 100 + 20)$

$= \$2766.67$

**29.** $p(x) = 10e^{x/3}$

$p(15) = 10e^5 = 1484.13$

$PS = 15(1484.13) - \int_0^{15} 10e^{x/3}dx$

$= 22{,}261.95 - (30e^{x/3})\Big|_0^{15}$

$= \$17{,}839.58$

**31.** $x^2 + 4x + 11 = 81 - x^2$

$2x^2 + 4x - 70 = 2(x+7)(x-5) = 0$

At $x = 5$, we have $p = 81 - 25 = 56$.

$PS = 5(56) - \int_0^5 (x^2 + 4x + 11)dx$

$= 280 - \left(\frac{1}{3}x^3 + 2x^2 + 11x\right)\Big|_0^5$

$= 280 - (41.67 + 50 + 55)$

$= \$133.33$

**33.** See problem 23 to find equilibrium point $(5, 2)$.

$PS = 5 \cdot 2 - \int_0^5 (1 + 0.2x)dx$

$= 10 - (x + 0.1x^2)\Big|_0^5$

$= 10 - (5 + 2.5)$

$= \$2.50$

**35.** $x^2 + 33x + 48 = 144 - 2x^2$

$3x^2 + 33x - 96 = 3(x^2 + 11x - 32) = 0$

Using the quadratic formula to solve for $x$ and the supply function to find the equilibrium price, we have that the equilibrium point is $(2.39, 132.58)$.

$PS = 2.39(132.58) - \int_0^{2.39} (x^2 + 33x + 48)dx$

$= 316.87 - \left(\frac{1}{3}x^3 + \frac{33}{2}x^2 + 48x\right)\Big|_0^{2.39}$

$= 316.87 - 213.52 = \$103.35$

## *Exercises 13.5*

**1.** formula 5: $\int \frac{dx}{16 - x^2} = \int \frac{dx}{4^2 - x^2} = \frac{1}{2(4)} \ln\left|\frac{4+x}{4-x}\right| + C$

**3.** formula 11: $\int_1^4 \frac{dx}{x\sqrt{9+x^2}} = \int_1^4 \frac{dx}{x\sqrt{3^2+x^2}} = -\frac{1}{3}\ln\left(\frac{3+\sqrt{9+x^2}}{x}\right)\Bigg|_1^4 = -\frac{1}{3}\left[\ln 2 - \ln\left(3+\sqrt{10}\right)\right]$

$= \frac{1}{3}\left[\ln\left(3+\sqrt{10}\right) - \ln 2\right] = \frac{1}{3}\ln\left[\frac{3+\sqrt{10}}{2}\right]$

**5.** formula 14: $\int \ln w\,dw = w(\ln w - 1) + C$

# Chapter 13: Definite Integrals: Techniques of Integration

**7.** formula 12: $\int_0^2 \frac{q\,dq}{6q+9} = \left[\frac{q}{6} - \frac{9}{36}\ln(6q+9)\right]\Big|_0^2 = \left(\frac{2}{6} - \frac{1}{4}\ln 21\right) - \left(0 - \frac{1}{4}\ln 9\right) = \frac{1}{3} + \frac{1}{4}(\ln 9 - \ln 21) = \frac{1}{3} + \frac{1}{4}\ln\frac{3}{7}$

**9.** formula 13: $\int \frac{dv}{v(3v+8)} = \frac{1}{8}\ln\left|\frac{v}{3v+8}\right| + C$

**11.** formula 7: $\int_5^7 \sqrt{x^2-25}\,dx = \int_5^7 \sqrt{x^2-5^2}\,dx = \frac{1}{2}\left[x\sqrt{x^2-25} - 25\ln\left(x+\sqrt{x^2-25}\right)\right]\Big|_5^7$

$= \frac{1}{2}\left[7\sqrt{24} - 25\ln\left(7+\sqrt{24}\right) - (0 - 25\ln 5)\right] = \frac{1}{2}\left[7\sqrt{24} - 25\ln\left(7+\sqrt{24}\right) + 25\ln 5\right]$

**13.** formula 16: $\int w\sqrt{4w+5}\,dw = \frac{2(12w-10)(4w+5)^{3/2}}{15(16)} + C = \frac{(6w-5)(4w+5)^{3/2}}{60} + C$

**15.** formula 3: $\int x5^{x^2}\,dx = \frac{1}{2}\int 5^{x^2}(2x\,dx) = \frac{1}{2}\cdot\frac{5^{x^2}}{\ln 5} + C$ or $\frac{1}{2}\left(5^{x^2}\right)\log_5 e + C$

**17.** formula 1: $\int_0^3 x\sqrt{x^2+4}\,dx = \frac{1}{2}\int_0^3 \left(x^2+4\right)^{1/2}(2x\,dx) = \frac{1}{2}\cdot\frac{\left(x^2+4\right)^{3/2}}{\frac{3}{2}}\Bigg|_0^3 = \frac{1}{3}\left(x^2+4\right)^{3/2}\Big|_0^3 = \frac{1}{3}\left(13^{3/2}-8\right)$

**19.** formula 9: $5\int \frac{dx}{x\sqrt{4-9x^2}} = 5\int \frac{3\,dx}{3x\sqrt{2^2-(3x)^2}} = 5\cdot\left(-\frac{1}{2}\right)\ln\left|\frac{2+\sqrt{4-9x^2}}{3x}\right| + C = -\frac{5}{2}\ln\left|\frac{2+\sqrt{4-9x^2}}{3x}\right| + C$

**21.** formula 10: $\int \frac{dx}{\sqrt{9x^2-4}} = \frac{1}{3}\int \frac{3\,dx}{\sqrt{(3x)^2-2^2}} = \frac{1}{3}\ln\left|3x+\sqrt{9x^2-4}\right| + C$

**23.** formula 15: $\int \frac{3x\,dx}{(2x-5)^2} = 3\cdot\frac{1}{2^2}\left(\ln|2x-5| + \frac{-5}{2x-5}\right) + C = \frac{3}{4}\left(\ln|2x-5| - \frac{5}{2x-5}\right) + C$

**25.** formula 8: $\int \frac{dx}{\sqrt{(3x+1)^2+1}} = \frac{1}{3}\int \frac{(3\,dx)}{\sqrt{(3x+1)^2+1}} = \frac{1}{3}\ln\left|3x+1+\sqrt{(3x+1)^2+1}\right| + C$

**27.** formula 6: $\int_0^3 x\sqrt{(x^2+1)^2+9}\,dx = \frac{1}{2}\int_0^3 \sqrt{(x^2+1)^2+3^2}\,(2x\,dx)$

$= \frac{1}{4}\left[\left(x^2+1\right)\sqrt{(x^2+1)^2+9} + 9\ln\left(\left(x^2+1\right)+\sqrt{(x^2+1)^2+9}\right)\right]\Big|_0^3$

$= \frac{1}{4}\left[10\sqrt{109} + 9\ln\left(10+\sqrt{109}\right) - \sqrt{10} - 9\ln\left(1+\sqrt{10}\right)\right]$

**29.** formula 2: $\int \frac{x\,dx}{7-3x^2} = -\frac{1}{6}\int \frac{1}{7-3x^2}(-6x\,dx) = -\frac{1}{6}\ln\left|7-3x^2\right| + C$

**31.** formula 8: $\int \frac{dx}{\sqrt{4x^2+7}} = \frac{1}{2}\int \frac{(2\,dx)}{\sqrt{(2x)^2+7}} = \frac{1}{2}\ln\left|2x+\sqrt{4x^2+7}\right|+C$

**33.** Using technology: $\int_2^3 \frac{e^{\sqrt{x-1}}}{\sqrt{x-1}}dx \approx 2.7899$

**35.** Using technology: $\int_0^1 \frac{x^3 dx}{\left(4x^2+5\right)^2} \approx 0.004479$

**37.** At $x = 20$, we have $p = 40 + 200\ln 21 = 648.90$

$$PS = 20(648.90) - \int_0^{20}\left[40+200\ln(x+1)\right]dx = 12{,}978 - \left[40x + 200(x+1)(\ln(x+1)-1)\right]\Big|_0^{20}$$

$$= 12{,}978 - \left[800 + 4200(\ln 21 - 1) - (0 + 200(-1))\right] = 12{,}978 - 800 - 12{,}787 + 4200 - 200 = \$3391$$

**39.** **a.** $C(x) = \int\sqrt{x^2+9}dx = \frac{1}{2}\left(x\sqrt{x^2+9} + 9\ln\left(x+\sqrt{x^2+9}\right)\right) + K$

(By formula 6)

With the interpretation that $C(0) = 300$, we get a $K$ value of 295.06.

**b.** $C(4) = 19.888 + 295.06 = 314.95$

**41.** $TI = \int_0^{120}\left(10\ln(t+1) - 0.1t\right)dt = \left[10(t+1)(\ln(t+1)-1) - \frac{.1t^2}{2}\right]\Big|_0^{120}$

$= 10\cdot 121(\ln 121 - 1) - \frac{1440}{2} - [10(-1) - 0]$ (by formula 14)

$\approx \$3882.9$ thousands

## *Exercises 13.6*

**1.** $\int xe^{2x}dx = uv - \int v\,du$

$u = x \quad dv = e^{2x}dx$

$du = dx \quad v = \frac{1}{2}e^{2x}$

$\int xe^{2x}dx = \frac{1}{2}xe^{2x} - \frac{1}{2}\int e^{2x}dx$

$= \frac{1}{2}xe^{2x} - \frac{1}{4}e^{2x} + C$

**3.** $\int x^2\ln x\,dx = uv - \int v\,du$

$u = \ln x \quad dv = x^2 dx$

$du = \frac{1}{x}dx \quad v = \frac{1}{3}x^3$

$\int x^2 \ln x\,dx = \frac{1}{3}x^3\ln x - \frac{1}{3}\int x^3\cdot\frac{1}{x}dx$

$= \frac{1}{3}x^3\ln x - \frac{1}{9}x^3 + C$

**5.** $\int_4^6 q\sqrt{q-4}dq = \left[uv - \int v\,du\right]\Big|_4^6$

$u = q \quad dv = (q-4)^{1/2}dq$

$du = dq \quad v = \frac{2}{3}(q-4)^{3/2}$

$\int_4^6 q\sqrt{q-4}dq$

$$= \left[ \frac{2q}{3}(q-4)^{3/2} - \frac{2}{3}\int (q-4)^{3/2}\,dq \right] \Bigg|_4^6$$

$$= \left[ \frac{2q}{3}(q-4)^{3/2} - \frac{4}{15}(q-4)^{5/2} \right] \Bigg|_4^6$$

$$= 4\left(2^{3/2}\right) - \frac{4}{15}\left(2^{5/2}\right) - 0$$

$$= 4\left(2^{3/2}\right) - \frac{8}{15}\left(2^{3/2}\right)$$

$$= \frac{52}{15}\left(2^{3/2}\right) = \frac{104}{15}\sqrt{2}$$

**7.** $\int \frac{\ln x}{x^2}\,dx = uv - \int v\,du$

$u = \ln x \quad dv = x^{-2}dx$

$du = \frac{dx}{x} \qquad v = -\frac{1}{x}$

$$\int \frac{\ln x}{x^2}dx - \frac{1}{x}\ln x + \int \frac{dx}{x^2} - \frac{1}{x}\ln x - \frac{1}{x} + C$$

**9.** $\int_1^e \ln x\,dx = \left[ uv - \int v\,du \right] \Big|_1^e$

$u = \ln x \quad dv = dx$

$du = \frac{dx}{x} \qquad v = x$

$$\int_1^e \ln x dx = \left[ x\ln x - \int dx \right] \Big|_1^e$$

$$= \left[ x\ln x - x \right] \Big|_1^e$$

$$= (e\ln e - e) - (0 - 1)$$

$$= e - e + 1 = 1$$

**11.** $\int x\ln(2x-3)\,dx = uv - \int v\,du$

$u = \ln(2x-3) \quad dv = xdx$

$du = \frac{2\,dx}{2x-3} \qquad v = \frac{x^2}{2}$

$$\int x\ln(2x-3)\,dx$$

$$= \frac{1}{2}x^2\ln(2x-3) - \int \frac{x^2}{2x-3}dx$$

$$= \frac{1}{2}x^2\ln(2x-3) - \int\left( \frac{1}{2}x + \frac{3}{4} + \frac{9/4}{2x-3} \right)dx$$

$$= \frac{1}{2}x^2\ln(2x-3) - \frac{1}{4}x^2 - \frac{3}{4}x - \frac{9}{8}\ln(2x-3) + C$$

**13.** $\frac{1}{2}\int q^2\left(\sqrt{q^2-3}\cdot 2q\,dq\right) = uv - \int v\,du$

$u = q^2 \qquad dv = \sqrt{q^2-3}\cdot 2q\,dq$

$du = 2qdq \qquad v = \frac{2}{3}(q^2-3)^{3/2}$

$$\frac{1}{2}\int q^2\left(\sqrt{q^2-3}\cdot 2q\,dq\right)$$

$$= \frac{1}{2}\left[ \frac{2}{3}q^2\left(q^2-3\right)^{3/2} - \frac{2}{3}\int\left(q^2-3\right)^{3/2}(2q\,dq) \right]$$

$$= \frac{1}{3}q^2\left(q^2-3\right)^{3/2} - \frac{1}{3}\cdot\frac{\left(q^2-3\right)^{5/2}}{5/2} + C$$

$$= \left(q^2-3\right)^{3/2}\left[ \frac{1}{3}q^2 - \frac{2}{15}\left(q^2-3\right) \right] + C$$

$$= \left(q^2-3\right)^{3/2}\cdot\frac{3q^2+6}{15} + C$$

$$= \frac{\left(q^2-3\right)^{3/2}\left(q^2+2\right)}{5} + C$$

**15.**

$$\frac{1}{2}\int_0^4 x^2\left(x^2+9\right)^{1/2}2x\,dx = \frac{1}{2}\left[ uv - \int v\,du \right] \Bigg|_0^4$$

$u = x^2 \qquad dv = \left(x^2+9\right)^{1/2}(2x\,dx)$

$du = 2xdx \qquad v = \frac{2}{3}\left(x^2+9\right)^{3/2}$

$$\frac{1}{2}\int_0^4 x^2\left(x^2+9\right)^{1/2}2xdx$$

$$= \frac{1}{2}\left[ \frac{2}{3}x^2\left(x^2+9\right)^{3/2} - \frac{2}{3}\int\left(x^2+9\right)^{3/2}(2x\,dx) \right] \Bigg|_0^4$$

$$= \left[ \frac{1}{3}x^2\left(x^2+9\right)^{3/2} - \frac{1}{3}\cdot\frac{\left(x^2+9\right)^{5/2}}{5/2} \right] \Bigg|_0^4$$

$$= \left[ \frac{16}{3}(125) - \frac{2}{15}(3125) \right] - \left[ 0 - \frac{2}{15}(243) \right]$$

$$= 666.67 - 416.67 + 32.4 = 282.4$$

**17.** $\int x^2e^{-x}dx$

$u = x^2 \qquad dv = e^{-x}dx$

$du = 2xdx \qquad v = -e^{-x}$

$\int x^2e^{-x}dx = -e^{-x}x^2 - \int(-e^{-x})2x\,dx$

$= -x^2e^{-x} + 2\int xe^{-x}dx$

$u = x \qquad dv = e^{-x}dx$

$du = dx \qquad v = -e^{-x}$

$\int x^2e^{-x}dx$

$= -x^2e^{-x} + 2\left(-xe^{-x} + \int e^{-x}dx\right)$

$= -x^2e^{-x} - 2xe^{-x} - 2e^{-x} + C$

$= -e^{-x}\left(x^2 + 2x + 2\right) + C$

**19.** $\int_0^2 x^3e^{x^2}dx = \int_0^2 x^2e^{x^2}x\,dx$

$u = x^2 \qquad dv = e^{x^2}x\,dx$

$du = 2xdx \qquad v = \frac{1}{2}e^{x^2}$

$\int_0^2 x^3e^{x^2}dx = \frac{1}{2}x^2e^{x^2}\Big|_0^2 - \frac{1}{2}\int_0^2 e^{x^2}2x\,dx$

$= \left(\frac{1}{2}x^2e^{x^2} - \frac{1}{2}e^{x^2}\right)\Big|_0^2$

$= \frac{1}{2}(4)e^4 - \frac{1}{2}e^4 - \left(0 - \frac{1}{2}\right)$

$= 2e^4 - \frac{1}{2}e^4 + \frac{1}{2} = \frac{3e^4+1}{2}$

**21.** $\int x^3(\ln x)^2\,dx$

$u = (\ln x)^2 \qquad dv = x^3dx$

$du = \frac{2\ln x}{x}dx \qquad v = \frac{1}{4}x^4$

$\int x^3(\ln x)^2\,dx = \frac{1}{4}x^4(\ln x)^2 - \frac{1}{2}\int x^3\ln x\,dx$

$u = \ln x \quad dv = x^3dx$

$du = \frac{dx}{x} \qquad v = \frac{1}{4}x^4$

$\int x^3(\ln x)^2\,dx$

$= \frac{1}{4}x^4(\ln x)^2 - \frac{1}{2}\left[\frac{1}{4}x^4\ln x - \int\frac{1}{4}x^3dx\right]$

$= \frac{1}{4}x^4(\ln x)^2 - \frac{1}{8}x^4\ln x + \frac{1}{32}x^4 + C$

**23.**

$\int e^{2x}\sqrt{e^x+1}dx = \int\left(e^x+1\right)^{1/2}e^xe^xdx$

$u = e^x \qquad dv = \left(e^x+1\right)^{1/2}e^xdx$

$du = e^xdx \qquad v = \frac{2}{3}\left(e^x+1\right)^{3/2}$

$\int e^{2x}\sqrt{e^x+1}dx$

$= \frac{2}{3}e^x\left(e^x+1\right)^{3/2} - \frac{2}{3}\int\left(e^x+1\right)^{3/2}e^xdx$

$= \frac{2}{3}e^x\left(e^x+1\right)^{3/2} - \frac{2}{3}\cdot\frac{2}{5}\left(e^x+1\right)^{5/2} + C$

$= \left(e^x+1\right)^{3/2}\left[\frac{2e^x}{3} - \frac{4}{15}\left(e^x+1\right)\right] + C$

$= (e^x+1)^{3/2}\left[\frac{6e^x}{15} - \frac{4}{15}\right] + C$

$= \frac{2}{15}\left(e^x+1\right)^{3/2}\left(3e^x-2\right) + C$

**25.** $\int e^{x^2}x\,dx = \frac{1}{2}\int e^u du$ where $u = x^2$ and $du = 2xdx$.

$\int e^{x^2}x\,dx = \frac{1}{2}e^{x^2} + C$

**27.** $\int\sqrt{e^x+1}\,e^xdx = \int\left(e^x+1\right)^{1/2}e^xdx = \int u^{1/2}du$ where $u = e^x+1$ and $du = e^xdx$.

$\int\sqrt{e^x+1}\,e^xdx = \frac{2}{3}\left(e^x+1\right)^{3/2} + C$

**29.** I. $\int_0^4\frac{t}{e^t}dt = \int_0^4 e^{-t}t\,dt$

$u = t \qquad dv = e^{-t}dt$

$du = dt \qquad v = -e^{-t}$

$\int_0^4\frac{t}{e^t}dt = -te^{-t}\Big|_0^4 + \int_0^4 e^{-t}dt$

$= -te^{-t}\Big|_0^4 + \left(-e^{-t}\right)\Big|_0^4$

$= -4e^{-4} + \left(-e^{-4} + e^0\right)$

$= -5e^{-4} + 1$

**31.** At $x = 30$, we have $p = 30 + 100\ln 61 = 441.09$.

$$PS = 30(441.09) - \int_0^{30}\left[30 + 100\ln(2x+1)\right]dx = 13{,}232.62 - \left(30x + \frac{100}{2}(2x+1)\left[\ln(2x+1) - 1\right]\right)\Bigg|_0^{30}$$

$$= 13{,}232.62 - \left\{900 + 50(61)\left[\ln 61 - 1\right] - \left[0 + 50(\ln 1 - 1)\right]\right\}$$

$$= 13{,}232.62 - 900 - 3050(3.11087) - 50 = \$2794.46$$

(by formula 14)

**33.** Present Value

$$= \int_0^5 (10{,}000 - 500t)e^{-0.1t}dt = \left[-100{,}000e^{-0.1t} - \int 500te^{-0.1t}dt\right]\Bigg|_0^5$$

$u = t \qquad dv = e^{-0.1t}dt$

$du = dt \qquad v = -10e^{-0.1t}$

(So, $\int te^{-0.1t}dt = -10te^{-0.1t} - 100e^{-0.1t}$.)

$$= \left[-100{,}000e^{-0.1t} + 5000te^{-0.1t} + 50{,}000e^{-0.1t}\right]\Bigg|_0^5 = \left[5000te^{-0.1t} - 50{,}000e^{-0.1t}\right]\Bigg|_0^5$$

$$= \left[e^{-0.1t}(5000t - 50{,}000)\right]\Bigg|_0^5 = 0.6065[-25{,}000] - 1[0 - 50{,}000] = \$34{,}837$$

**35.** Gini coeff $= \dfrac{\int_0^1 \left(x - xe^{x-1}\right)dx}{1/2}$

For $xe^{x-1}$, let $u = x \qquad dv = e^{x-1}dx$

$du = dx \qquad v = e^{x-1}$

$$\int xe^{x-1}dx = xe^{x-1} - \int e^{x-1}dx = xe^{x-1} - e^{x-1}$$

$$\text{Gini coeff} = 2\left(\frac{x^2}{2} - xe^{x-1} + e^{x-1}\right)\Bigg|_0^1 = 2\left(\frac{1}{2} - 1 + 1 - \left(e^{-1}\right)\right) = 0.264$$

**37.** Average population $= \dfrac{1}{10}\int_{50}^{60}(96.1 + 17.4\ln x)dx = \dfrac{96.1}{10}\int_{50}^{60}dx + \dfrac{17.4}{10}\int_{50}^{60}\ln x\,dx$

First part: $9.61\int_{50}^{60}dx = 9.61x\Big|_{20}^{30} = 9.61(30 - 20) = 96.1$

Second part uses integration by parts: $u = \ln x,\ du = \dfrac{1}{x}dx,\ dv = dx,\ v = x$

$$1.74\int_{50}^{60}\ln x\,dt = 1.74\left(x\ln x\Big|_{50}^{60} - \int_{50}^{60}x\cdot\frac{1}{x}dx\right) = 1.74\left[60\ln 60 - 50\ln 50 - \int_{50}^{60}dx\right]$$

$$\approx 1.74\left[50.0595 - \left(x\Big|_{50}^{60}\right)\right] = 1.74\left[50.0595 - (60 - 50)\right] \approx 69.7$$

Together: $96.1 + 69.7 \approx 166$ million

# Chapter 13: Definite Integrals: Techniques of Integration

## *Exercises 13.7*

**1.** $\int_1^{\infty} \frac{dx}{x^6} = \lim_{a\to\infty} \int_1^a x^{-6} dx$

$= \frac{1}{5} \lim_{a\to\infty} \left(-x^{-5}\right)\Big|_1^a$

$= \frac{1}{5} \lim_{a\to\infty} \left(-\frac{1}{a^5} - (-1)\right) = \frac{1}{5}$

**3.** $\int_1^{\infty} \frac{dt}{t^{3/2}} = \lim_{a\to\infty} \int_1^a t^{-3/2} dt$

$= \lim_{a\to\infty} -2t^{-1/2}\Big|_1^a$

$= \lim_{a\to\infty} \left(-2\frac{1}{a^{1/2}} - (-2(1))\right) = 2$

**5.** $\int_1^{\infty} e^{-x} dx = (-1) \lim_{b\to\infty} \int_1^b e^{-x}(-1)dx$

$= (-1) \lim_{b\to\infty} e^{-x}\Big|_1^b$

$= (-1) \lim_{b\to\infty} \left(\frac{1}{e^b} - \frac{1}{e}\right)$

$= (-1)\left(-\frac{1}{e}\right) = \frac{1}{e}$

**7.** $\int_1^{\infty} \frac{dt}{t^{1/3}} = \lim_{a\to\infty} \int_1^a t^{-1/3} dt$

$= \lim_{a\to\infty} \frac{3}{2} t^{2/3}\Big|_1^a$

$= \lim_{a\to\infty} \left(\frac{3}{2}a^{2/3} - \frac{3}{2}\right)$

Thus, the integral diverges.

**9.** $\int_0^{\infty} e^{3x} dx = \lim_{a\to\infty} \frac{1}{3} \int_0^a e^{3x} \cdot 3dx$

$= \lim_{a\to\infty} \frac{1}{3} e^{3x}\Big|_0^a$

$= \lim_{a\to\infty} \frac{1}{3}\left(e^{3a} - 1\right)$

Thus, the integral diverges.

**11.** $\int_{-\infty}^{-1} 10x^{-2} dx = \lim_{a\to\infty} \int_{-a}^{-1} 10x^{-2} dx$

$= \lim_{a\to\infty} \frac{10x^{-1}}{-1}\Big|_{-a}^{-1}$

$= \lim_{a\to\infty} \left(\frac{-10}{-1} - \frac{-10}{-a}\right) = 10$

**13.** $\int_{-\infty}^{0} x^2 e^{-x^3} dx = -\frac{1}{3} \lim_{a\to\infty} \int_{-a}^{0} e^{-x^3}\left(-3x^2 dx\right)$

$= -\frac{1}{3} \lim_{a\to\infty} e^{-x^3}\Big|_{-a}^{0}$

$= -\frac{1}{3} \lim_{a\to\infty} \left(1 - e^{a^3}\right) = \infty$

Thus, the integral diverges.

$[-(-a)^3 = -(-a^3) = a^3]$

**15.** $\int_{-\infty}^{-1} \frac{6dx}{x} = \lim_{a\to\infty} \int_{-a}^{-1} \frac{6dx}{x}$

$= \lim_{a\to\infty} 6\ln|x|\Big|_{-a}^{-1}$

$= \lim_{a\to\infty} \left(6\ln 1 - 6\ln|a|\right) = -\infty$

Thus, the integral diverges.

**17.** $\int_{-\infty}^{\infty} \frac{2x}{\left(x^2+1\right)^2} dx = \lim_{a\to\infty} \int_{-a}^{0} \frac{2x}{\left(x^2+1\right)^2} dx + \lim_{b\to\infty} \int_{0}^{b} \frac{2x}{\left(x^2+1\right)^2} dx = \lim_{a\to\infty} \int_{-a}^{0} 2x\left(x^2+1\right)^{-2} dx + \lim_{b\to\infty} \int_{0}^{b} 2x\left(x^2+1\right)^{-2} dx$

$= \lim_{a\to\infty} -\left(x^2+1\right)^{-1}\Big|_{-a}^{0} + \lim_{b\to\infty} -\left(x^2+1\right)^{-1}\Big|_{0}^{b} = \lim_{a\to\infty}\left[-(1)^{-1} + \left(a^2+1\right)^{-1}\right] + \lim_{b\to\infty}\left[-\left(b^2+1\right)^{-1} + (1)^{-1}\right] = -1+0-0+1 = 0$

**19.** $\int_{-\infty}^{\infty} x^3 e^{-x^4} dx = -\frac{1}{4}\left[\lim_{a\to\infty} \int_{-a}^{0} e^{-x^4}\left(-4x^3 dx\right) + \lim_{b\to\infty} \int_{0}^{b} e^{-x^4}\left(-4x^3 dx\right)\right]$

$= -\frac{1}{4}\left[\lim_{a\to\infty} e^{-x^4}\Big|_{-a}^{0} + \lim_{b\to\infty} e^{-x^4}\Big|_{0}^{b}\right] = -\frac{1}{4}\left[\lim_{a\to\infty}\left(e^0 - 1/e^{a^4}\right) + \lim_{b\to\infty}\left(1/e^{b^4} - e^0\right)\right] = -\frac{1}{4}(1-0+0-1) = 0$

**21.** $\int_0^{\infty} \frac{c}{e^{0.5t}}\,dt = 1$

$\lim_{a\to\infty} \int_0^a ce^{-0.5t}\,dt = c\lim_{a\to\infty} -2\int_0^a e^{-0.5t}(-0.5)\,dt = c\lim_{a\to\infty}(-2)e^{-0.5t}\Big|_0^a = -2c\lim_{a\to\infty}\left(e^{-0.5a}-1\right) = 2c$

So, $2c = 1$ gives $c = \frac{1}{2}$.

**23.** $\int_1^{\infty} \frac{x}{e^{x^2}}\,dx = \int_1^{\infty} e^{-x^2}x\,dx$

$$= \lim_{a\to\infty} -\frac{1}{2}\int_1^a e^{-x^2}(-2)x\,dx$$

$$= \lim_{a\to\infty} -\frac{1}{2}\left(e^{-x^2}\right)\Big|_1^a$$

$$= \lim_{a\to\infty} -\frac{1}{2}\left[e^{-a^2} - e^{-1}\right] = \frac{1}{2e}$$

**25.** $\int_1^{\infty} \frac{1}{\sqrt[3]{x^5}}\,dx = \lim_{a\to\infty}\int_1^a x^{-5/3}\,dx$

$$= \lim_{a\to\infty} -\frac{3}{2}x^{-2/3}\Big|_1^a$$

$$= \lim_{a\to\infty} -\frac{3}{2}\left[a^{-2/3} - 1\right] = \frac{3}{2}$$

**27.** $\int_{-\infty}^{\infty} f(x)\,dx = \int_{-\infty}^{10} f(x)\,dx + \int_{10}^{\infty} f(x)\,dx$

$$= \int_{-\infty}^{10} 0\cdot dx + \int_{10}^{\infty} 200x^{-3}\,dx$$

$$= 0 + \lim_{b\to\infty} \frac{200}{-2}x^{-2}\Big|_{10}^b$$

$$= \lim_{b\to\infty}\left(-\frac{100}{b^2} - \frac{-100}{100}\right) = 1$$

**29.** $\int_{-\infty}^{\infty} f(x)\,dx = \int_{-\infty}^{1} f(x)\,dx + \int_1^{\infty} f(x)\,dx$

$$= \int_{-\infty}^{1} 0\cdot dx + \int_1^{\infty} cx^{-2}\,dx$$

$$= 0 + \lim_{b\to\infty} c\cdot\frac{x^{-1}}{-1}\Big|_1^b$$

$$= (-c)\lim_{b\to\infty}\left(\frac{1}{b} - \frac{1}{1}\right) = c$$

Thus, $f(x)$ is a probability density function if $c = 1$.

**31.** $\int_{-\infty}^{\infty} f(x)\,dx = \int_{-\infty}^{0} f(x)\,dx + \int_0^{\infty} f(x)\,dx$

$$= \int_{-\infty}^{0} 0\cdot dx + \int_0^{\infty} ce^{-x/4}\,dx$$

Therefore, $\int_{-\infty}^{\infty} f(x)\,dx = 0 + \lim_{b\to\infty}\frac{c}{-.25}e^{-x/4}\Big|_0^b$

$$= -4c\lim_{b\to\infty}\left(\frac{1}{e^{b/4}} - e^0\right)$$

Thus, $-4c(0-1) = 1$ or $c = \frac{1}{4}$.

**33.** Mean $= \int_{-\infty}^{\infty} x\,f(x)\,dx$

$$= \int_{-\infty}^{10} x\,f(x)\,dx + \int_{10}^{\infty} x\,f(x)\,dx$$

$$= 0 + \int_{10}^{\infty} x\left(200x^{-3}\right)dx$$

$$= \lim_{b\to\infty} 200\cdot\frac{x^{-1}}{-1}\Big|_{10}^b$$

$$= -200\lim_{b\to\infty}\left(\frac{1}{b} - \frac{1}{10}\right) = 20$$

**35.** $A = 8\int_0^{\infty} xe^{-3x}(3\,dx)$ (Alternate form used)

$u = x \qquad dv = e^{-3x}\cdot 3\,dx$

$du = dx \qquad v = -e^{-3x}$

$$A = 8\left[-xe^{-3x}\Big|_0^{\infty} + \int_0^{\infty} e^{-3x}\,dx\right]$$

$$= 8\left[\lim_{b\to\infty}\left(-\frac{b}{e^{3b}} + 0\right) + \lim_{b\to\infty}\left(-\frac{1}{3e^{3b}} + \frac{1}{3}\right)\right]$$

$$= \frac{8}{3}$$

If necessary ask your instructor why $\lim_{b\to\infty}\frac{b}{e^{3b}} = 0$.

**37.** $\int_0^\infty Ae^{-rt}\,dt = \lim_{b\to\infty}\int_0^b Ae^{-rt}\,dt$

$= \lim_{b\to\infty} -\frac{A}{r}e^{-rt}\Big|_0^b = -\frac{A}{r}\lim_{b\to\infty}\left(\frac{1}{e^{br}} - e^0\right)$

$= -\frac{A}{r}(0-1) = \frac{A}{r}$

**39.** $CV = \int_0^\infty 120e^{0.04t}e^{-0.09t}\,dt$

$= \lim_{b\to\infty}\int_0^b 120e^{-0.05t}\,dt$

$= \lim_{b\to\infty}\left(-2400e^{-0.05t}\right)\Big|_0^b$

$= \lim_{b\to\infty}\left(-\frac{2400}{e^{0.05b}} - \left(-2400e^0\right)\right)$

= \$2400 thousands

=\$2,400,000

**41.** $\int_0^\infty 56,000e^{0.02t}e^{-0.1t}\,dt = \lim_{a\to\infty}\int_0^a 56,000e^{-0.08t}\,dt = \lim_{a\to\infty} 56,000\cdot\frac{1}{-0.08}\int_0^a e^{-0.08t}(-0.08)\,dt$

$= \lim_{a\to\infty}\left(-700,000e^{-0.08t}\right)\Big|_0^a = -700,000\lim_{a\to\infty}\left[e^{-0.08a} - 1\right] = \$700,000$

**43.** **a.** $\int_2^\infty 0.5e^{-0.5t}\,dt = -\int_2^\infty e^{-0.5t}(-0.5t\,dt) = -\lim_{b\to\infty}\int_2^b e^{-0.5t}(-0.5t\,dt) = -\lim_{b\to\infty} e^{-0.5t}\Big|_2^b = -\lim_{b\to\infty}\left(e^{-0.5b} - e^{-1}\right) = \frac{1}{e} \approx 0.368$

**b.** Similarly, $\int_8^\infty 0.5e^{-0.5t}\,dt = -\lim_{b\to\infty}\int_8^b e^{-0.5t}(-0.5t\,dt) = -\lim_{b\to\infty} e^{-0.5t}\Big|_8^b = -\lim_{b\to\infty}\left(e^{-0.5b} - e^{-4}\right) = \frac{1}{e^4} \approx 0.0183$

**45.** $P(>24) = \int_{24}^\infty 0.08e^{-0.08t}\,dt = \lim_{b\to\infty}\int_{24}^b e^{-0.08t}(-0.08)(-1)\,dt$

$= (-1)\lim_{b\to\infty} e^{-0.08t}\Big|_{24}^b = (-1)\lim_{b\to\infty}\left(\frac{1}{e^{0.08b}} - \frac{1}{e^{1.92}}\right) = (-1)\left(0 - \frac{1}{6.821}\right) = 0.1466$

**47.** **a** . $500\int_0^b te^{-0.03(b-t)}\,dt$

Using integration by parts: $u = t \qquad dv = e^{-0.03(b-t)}\,dt$

$du = dt \qquad v = \frac{1}{0.03}e^{-0.03(b-t)}$

$500\int_0^b te^{-0.03(b-t)}\,dt = \left[\frac{500te^{-0.03(b-t)}}{0.03} - \frac{500}{0.03}\int e^{-0.03(b-t)}\,dt\right]\Bigg|_0^b = \left[\frac{500te^{-0.03(b-t)}}{0.03} - \frac{500e^{-0.03(b-t)}}{(0.03)(0.03)}\right]\Bigg|_0^b$

$= \frac{500e^{-0.03(b-t)}}{0.03}\left(t - \frac{1}{0.03}\right)\Bigg|_0^b = \frac{500e^0}{0.03}\left(\frac{0.03b-1}{0.03}\right) - \frac{500e^{-0.03b}}{0.03}\left(0 - \frac{1}{0.03}\right)$

$= \frac{500}{0.0009}\left(0.03b - 1 + e^{-0.03b}\right)$

**b.** $\lim_{b\to\infty}\int_0^b f(t)dt = \lim_{b\to\infty}\frac{500}{0.0009}\left(0.03b-1+\frac{1}{e^{0.03b}}\right)=\infty$

Waste is produced more rapidly than existing waste decays.

## *Exercises 13.8*

**1.** $[0,2]\quad n=4\quad h=\frac{2-0}{4}=\frac{1}{2}$

$x_0=0,\ x_1=\frac{1}{2},\ x_2=1,\ x_3=\frac{3}{2},\ x_4=2$

**3.** $[1,4]\quad n=6\quad h=\frac{4-1}{6}=\frac{1}{2}$

$x_0=1,\ x_1=\frac{3}{2},\ x_2=2,\ x_3=\frac{5}{2},\ x_4=3,\ x_5=\frac{7}{2},\ x_6=4$

**5.** $[-1,4]\quad n=5\quad h=\frac{4-(-1)}{5}=1$

$x_0=-1,\ x_1=0,\ x_2=1,\ x_3=2,\ x_4=3,\ x_5=4$

**7.** $f(x)=x^2\quad [0,3]\quad n=6\quad h=\frac{1}{2}$

**a.** $\int_0^3 x^2dx \approx \frac{h}{2}\left[f(0)+2f\left(\frac{1}{2}\right)+2f(1)+2f\left(\frac{3}{2}\right)+2f(2)+2f\left(\frac{5}{2}\right)+f(3)\right]$

$=\frac{1}{4}\left[0+\frac{1}{2}+2+\frac{9}{2}+8+\frac{25}{2}+9\right]=\frac{1}{4}\left[\frac{73}{2}\right]=9.13$

**b.** $\int_0^3 x^2dx \approx \frac{h}{3}\left[f(0)+4f\left(\frac{1}{2}\right)+2f(1)+4f\left(\frac{3}{2}\right)+2f(2)+4f\left(\frac{5}{2}\right)+f(3)\right]$

$=\frac{1}{6}[0+1+2+9+8+25+9]=9$

**c.** $\int_0^3 x^2dx=\left.\frac{x^3}{3}\right|_0^3=\frac{3^3}{3}-\frac{0^3}{3}=9$

**d.** Simpson's Rule is more accurate.

**9.** $f(x)=\frac{1}{x^2}\quad [1,2]\quad n=4\quad h=\frac{1}{4}$

**a.** $\int_1^2 \frac{1}{x^2}dx \approx \frac{h}{2}\left[f(1)+2f\left(\frac{5}{4}\right)+2f\left(\frac{3}{2}\right)+2f\left(\frac{7}{4}\right)+f(2)\right]=\frac{1}{8}\left[1+\frac{32}{25}+\frac{8}{9}+\frac{32}{49}+\frac{1}{4}\right]\approx\frac{1}{8}[4.072]=0.51$

**b.** $\int_1^2 \frac{1}{x^2}dx \approx \frac{h}{3}\left[f(1)+4f\left(\frac{5}{4}\right)+2f\left(\frac{3}{2}\right)+4f\left(\frac{7}{4}\right)+f(2)\right]=\frac{1}{12}\left[1+\frac{64}{25}+\frac{8}{9}+\frac{64}{49}+\frac{1}{4}\right]\approx\frac{1}{12}[6.01]=0.50$

**c.** $\int_1^2 x^{-2}dx=-x^{-1}\Big|_1^2=-\left(\frac{1}{2}-1\right)=\frac{1}{2}$

**d.** Simpson's Rule is more accurate.

**11.** $f(x)=x^{1/2}\quad [0,4]\quad n=8\quad h=\frac{1}{2}$

**a.** $\int_0^4 x^{1/2}dx \approx \frac{h}{2}\left[f(0)+2f\left(\tfrac{1}{2}\right)+2f(1)+2f\left(\tfrac{3}{2}\right)+2f(2)+2f\left(\tfrac{5}{2}\right)+2f(3)+2f\left(\tfrac{7}{2}\right)+f(4)\right]$

$=\frac{1}{4}[0+1.4142+2+2.4495+2.8284+3.1623+3.4641+3.7417+2]\approx\frac{1}{4}[21.0602]=5.27$

**b.**

$\int_0^4 x^{1/2}dx \approx \frac{h}{3}\left[f(0)+4f\left(\tfrac{1}{2}\right)+2f(1)+4f\left(\tfrac{3}{2}\right)+2f(2)+4f\left(\tfrac{5}{2}\right)+2f(3)+4f\left(\tfrac{7}{2}\right)+f(4)\right]$

$=\frac{1}{6}[0+2.8284+2+4.8990+2.8284+6.3246+3.4641+7.4833+2]\approx\frac{1}{6}[31.8278]=5.30$

**c.** $\int_0^4 x^{1/2}dx=\frac{2}{3}x^{3/2}\Big|_0^4=\frac{2}{3}\left(4^{3/2}-0\right)=5.33$

**d.** Simpson's Rule is more accurate.

**13.** $f(x)=\sqrt{x^3+1}\quad [0,2]\quad n=4\quad h=\frac{1}{2}$

**a.** $\int_0^2 \sqrt{x^3+1}\,dx \approx \frac{h}{2}\left[f(0)+2f\left(\tfrac{1}{2}\right)+2f(1)+2f\left(\tfrac{3}{2}\right)+f(2)\right]$

$=\frac{1}{4}[1+2.121+2.828+4.183+3]=\frac{1}{4}[13.132]=3.283$

**b.** $\int_0^2 \sqrt{x^3+1}\,dx \approx \frac{h}{3}\left[f(0)+4f\left(\tfrac{1}{2}\right)+2f(1)+4f\left(\tfrac{3}{2}\right)+f(2)\right]$

$=\frac{1}{6}[1+4.243+2.828+8.367+3]=\frac{1}{6}[19.438]=3.240$

**15.** $f(x)=e^{-x^2}\quad [0,1]\quad n=4\quad h=\frac{1}{4}$

**a.** $\int_0^1 e^{-x^2}dx \approx \frac{h}{2}\left[f(0)+2f\left(\tfrac{1}{4}\right)+2f\left(\tfrac{1}{2}\right)+2f\left(\tfrac{3}{4}\right)+f(1)\right]$

$=\frac{h}{2}[1+1.879+1.558+1.140+0.368]=\frac{1}{8}[5.945]=0.743$

**b.** $\int_0^1 e^{-x^2}dx \approx \frac{h}{3}\left[f(0)+4f\left(\tfrac{1}{4}\right)+2f\left(\tfrac{1}{2}\right)+4f\left(\tfrac{3}{4}\right)+f(1)\right]$

$=\frac{h}{3}[1+3.758+1.558+2.279+0.368]=\frac{1}{12}[8.963]=0.747$

**17.** $f(x)=\ln\left(x^2-x+1\right)\quad [1,5]\quad n=4\quad h=1$

**a.** $\int_1^5 \ln\left(x^2-x+1\right)dx \approx \frac{h}{2}[f(1)+2f(2)+2f(3)+2f(4)+f(5)]$

$=\frac{h}{2}[0+2.197+3.892+5.130+3.045]=\frac{1}{2}[14.264]=7.132$

**b.** $\int_1^5 \ln\left(x^2-x+1\right)dx \approx \frac{h}{3}[f(1)+4f(2)+2f(3)+4f(4)+f(5)]$

$=\frac{h}{3}[0+4.394+3.892+10.260+3.045]=\frac{1}{3}[21.591]=7.197$

**19.** $f(x)=f(x)\quad [1,4]\quad n=5\quad h=\frac{3}{5}$

$$\int_1^4 f(x)dx \approx \frac{h}{2}[f(1)+2f(1.6)+2f(2.2)+2f(2.8)+2f(3.4)+f(4)]$$
$$=\frac{3}{10}[1+4.4+3.6+5.8+9.2+2.1]=\frac{3}{10}[26.1]=7.8$$

**21.** $f(x)=f(x)\quad [1.2, 3.6]\quad n=6\quad h=0.4$

$$\int_{1.2}^{3.6} f(x)dx \approx \frac{h}{3}[f(1.2)+4f(1.6)+2f(2)+4f(2.4)+2f(2.8)+4f(3.2)+f(3.6)]$$
$$=\frac{2}{15}[6.1+19.2+6.2+8.0+5.6+22.4+9.7]=\frac{2}{15}[77.2]=10.3$$

**23.** $f(t)=100\frac{e^{0.1t}}{t+1}\quad [0,2]\quad n=4\quad h=\frac{1}{2}$

$$\int_0^2 f(t)dt \approx \frac{h}{3}\left[f(0)+4f\left(\tfrac{1}{2}\right)+2f(1)+4f\left(\tfrac{3}{2}\right)+f(2)\right]$$
$$=\frac{1}{6}[100+280.34+110.52+185.89+40.71]=\frac{1}{6}[717.46]=119.58$$

**25.** $C(x)=\left(x^2+1\right)^{3/2}+1000\quad [30,33]\quad n=3\quad h=1$

$$\frac{1}{33-30}\int_{30}^{33} C(x)dx \approx \frac{h}{6}[f(30)+2f(31)+2f(32)+f(33)]$$
$$=\frac{1}{6}[28{,}045+61{,}675+67{,}632+36{,}987]=\frac{1}{6}[194{,}339]=\$32{,}390$$

Note: This is the average total cost for that number of units.

**27.** $PS = p_1x_1 -$ area under supply curve

$$\text{Area} \approx \frac{h}{3}[f(0)+4f(10)+2f(20)+4f(30)+2f(40)+4f(50)+f(60)]$$
$$=\frac{10}{3}[120+1040+760+1800+1080+2520+680]=\frac{10}{3}[8000]=26{,}666.67$$

$PS = 680(60)-26{,}666.67=\$14{,}133.33$

**29.** $N\text{ of units} \approx \frac{h}{2}[f(0)+2f(1)+2f(2)+2f(3)+2f(4)+f(5)]$
$$=\frac{1}{2}[250+495.2+490.8+486.6+482.6+239.5]=\frac{1}{2}[2444.70]=1222.35$$

**31.** **a.**

| $x$ | 0.0 | 0.2 | 0.4 | 0.6 | 0.8 | 1.0 |
|---|---|---|---|---|---|---|
| $L_{1990}(x)-L_{2012}(x)$ | 0.0 | 0.003 | 0.005 | 0.013 | 0.029 | 0.0 |

**b.** Using the Trapezoid rule with $h=0.2$:

$$2\int_0^1 [L_{1990}(x)-L_{2012}(x)]dx = 0.2\left[0.0+2(0.003)+2(0.005)+2(0.013)+2(0.029)+0.0\right]=0.020$$

**c.** Positive; 1990

**33.** **a.** Yes, either can be used.

**b.** Simpson's Rule is more accurate

**c.** $\text{Area} \approx \frac{h}{3}\left[f(0)+4f(10)+2f(20)+4f(30)+2f(40)+4f(50)+2f(60)+4f(70)+f(80)\right]$

$= \frac{10}{3}\left[0+60+36+72+60+108+48+92+0\right] = \frac{10}{3}\left[476\right] = 1586.67 \text{ ft}^2$

# Chapter 13: Definite Integrals: Techniques of Integration

## *Chapter 13 Review Excercises*

1. $\sum_{k=1}^{8}(k^2+1)=\sum_{k=1}^{8}k^2+\sum_{k=1}^{8}1$
$=\frac{8(9)(17)}{6}+8\cdot 1$
$=204+8=212$

2. $\sum_{i=1}^{n}\frac{3i}{n^3}=\frac{3}{n^3}\sum_{i=1}^{n}i=\frac{3}{n^3}\cdot\frac{n(n+1)}{2}=\frac{3(n+1)}{2n^2}$

3. $f(x)=3x^2$
$\text{Area}=\frac{1}{6}\cdot f\left(\frac{1}{6}\right)+\frac{1}{6}\cdot f\left(\frac{2}{6}\right)+\cdots+\frac{1}{6}\cdot f\left(\frac{6}{6}\right)$
$=\frac{1}{6}\cdot\frac{3}{36}+\frac{1}{6}\cdot 3\cdot\frac{4}{36}+\frac{1}{6}\cdot 3\cdot\frac{9}{36}+\cdots+\frac{1}{6}\cdot 3\cdot\frac{36}{36}$
$=\frac{3}{216}(1+4+9+16+25+36)$
$=\frac{1}{72}(91)=\frac{91}{72}$

4. $A=\lim_{n\to\infty}3\left(\frac{i}{n}\right)^2\left(\frac{1}{n}\right)$
$=\lim_{n\to\infty}\frac{3}{n^3}\sum_{i=1}^{n}i^2=\lim_{n\to\infty}\frac{3}{n^3}\left[\frac{n(n+1)(2n+1)}{6}\right]$
$=\lim_{n\to\infty}\frac{(n+1)(2n+1)}{2n^2}=1$

5. $\text{Area}=\int_0^1 3x^2dx=x^3\Big|_0^1=1-0=1$

6. $\int_1^3\left(x^3-4x+5\right)dx=\left(\frac{x^4}{4}-\frac{4x^2}{2}+5x\right)\Big|_1^3$
$=\left(\frac{81}{4}-18+15\right)-\left(\frac{1}{4}-2+5\right)=14$

7. $\int_1^4 4x^{3/2}dx=\frac{8}{5}x^{5/2}\Big|_1^4=\frac{8}{5}(32-1)=\frac{8}{5}(31)=\frac{248}{5}$

8. $\int_{-3}^{2}\left(x^3-3x^2+4x+2\right)dx$
$=\left(\frac{1}{4}x^4-x^3+2x^2+2x\right)\Big|_{-3}^{2}$
$=4-8+8+4-\left(\frac{81}{4}+27+18-6\right)$
$=8-\frac{237}{4}=-\frac{205}{4}$

9. $\int_0^5\left(x^3+4x\right)dx=\left(\frac{1}{4}x^4+2x^2\right)\Big|_0^5$
$=\frac{625}{4}+50=\frac{825}{4}$

10. $\int_{-1}^{3}(3x+4)^{-2}\,dx=-\frac{1}{3(3x+4)}\Big|_{-1}^{3}$
$=-\frac{1}{3}\left(\frac{1}{13}-1\right)=\frac{4}{13}$

11. $\int_{-3}^{-1}(x+1)dx=\left(\frac{1}{2}x^2+x\right)\Big|_{-3}^{-1}$
$=\left(\frac{1}{2}-1\right)-\left(\frac{9}{2}-3\right)=-2$

12. $\int_2^3\frac{x^2}{2x^3-7}dx=\frac{1}{6}\int_2^3\frac{6x^2}{2x^3-7}dx$
$=\frac{1}{6}\ln\left(2x^3-7\right)\Big|_2^3=\frac{1}{6}\ln 47-\frac{1}{6}\ln 9$

13. $\int_{-1}^{2}\left(x^2+x\right)dx=\left(\frac{1}{3}x^3+\frac{1}{2}x^2\right)\Big|_{-1}^{2}$
$=\left(\frac{8}{3}+2\right)-\left(-\frac{1}{3}+\frac{1}{2}\right)=4\frac{1}{2}=\frac{9}{2}$

14. $\int_1^4\left(x^{-1}+x^{1/2}\right)dx=\left(\ln x+\frac{2}{3}x^{3/2}\right)\Big|_1^4=\ln 4+\frac{14}{3}$

**15.** $\int_0^2 5x^2(6x^3+1)^{1/2}\,dx = \frac{5}{18}\int_0^2 18x^2(6x^3+1)^{1/2}\,dx$

$= \frac{5}{18}\cdot\frac{(6x^3+1)^{3/2}}{3/2}\Big|_0^2 = \frac{5}{27}(6x^3+1)^{3/2}\Big|_0^2$

$= \frac{5}{27}(49)^{3/2} - \frac{5}{27}(1)^{3/2} = \frac{1715}{27} - \frac{5}{27} = \frac{190}{3}$

**16.** $\int_0^1 \frac{x}{x^2+1}\,dx = \frac{1}{2}\int_0^1 \frac{2x}{x^2+1}\,dx$

$= \frac{1}{2}\ln(x^2+1)\Big|_0^1 = \frac{1}{2}\ln 2 - 0 = \frac{1}{2}\ln 2$

**17.** $\int_0^1 e^{-2x}\,dx = -\frac{1}{2}e^{-2x}\Big|_0^1$

$= -\frac{1}{2}\left(\frac{1}{e^2} - e^0\right) = \frac{1}{2}(1-e^{-2})$

**18.** $\int_0^1 xe^{x^2}\,dx = \frac{1}{2}\int_0^1 2xe^{x^2}\,dx = \frac{1}{2}e^{x^2}\Big|_0^1 = \frac{1}{2}e - \frac{1}{2}$

**19.** $x = 1:\ f(x) = x^2 - 3x + 2 \quad g(x) = x^2 + 4$

$f(1) = 0 \quad g(1) = 5$

$A = \int_0^5 \left[(x^2+4) - (x^2 - 3x + 2)\right]dx$

$= \int_0^5 (2+3x)\,dx$

$= \left(2x + \frac{3}{2}x^2\right)\Big|_0^5 = \left(10 + \frac{75}{2}\right) - 0 = \frac{95}{2}$

**20.** $x^2 = 4x + 5$

$x^2 - 4x - 5 = 0$

$(x-5)(x+1) = 0$

$x = 5, -1$

$\int_{-1}^5 (4x + 5 - x^2)\,dx = \left(2x^2 + 5x - \frac{1}{3}x^3\right)\Big|_{-1}^5$

$= 75 - \frac{125}{3} - \left(-3 + \frac{1}{3}\right)$

$= \frac{108}{3} = 36$

**21.** $A = \int_{-1}^0 (x^3 - x)\,dx$

$= \left(\frac{1}{4}x^4 - \frac{1}{2}x^2\right)\Big|_{-1}^0 = 0 - \left(\frac{1}{4} - \frac{1}{2}\right) = \frac{1}{4}$

**22.** $x^3 - 1 = x - 1$

$x^3 - x = 0$

$x(x+1)(x-1) = 0$

$x = 0, -1, 1$

$\int_{-1}^0 \left[x^3 - 1 - (x-1)\right]dx + \int_0^1 \left[x - 1 - (x^3 - 1)\right]dx$

$= \int_{-1}^0 (x^3 - x)\,dx + \int_0^1 (x - x^3)\,dx$

$= \left(\frac{1}{4}x^4 - \frac{1}{2}x^2\right)\Big|_{-1}^0 + \left(\frac{1}{2}x^2 - \frac{1}{4}x^4\right)\Big|_0^1$

$= -\frac{1}{4} + \frac{1}{2} + \frac{1}{2} - \frac{1}{4} = \frac{1}{2}$

**23.** $\int \sqrt{x^2 - 4}\,dx$

$= \frac{1}{2}\left(x\sqrt{x^2-4} - 4\ln\left|x + \sqrt{x^2-4}\right|\right) + C$

(By formula 7)

**24.** $\int_0^1 3^x\,dx = 3^x \log_3 e\Big|_0^1 = 3\log_3 e - 1\log_3 e = 2\log_3 e$

**25.** $\frac{1}{2}\int \ln x^2\,(2x\,dx) = \frac{1}{2}x^2(\ln x^2 - 1) + C$

(By formula 14)

**26.** $\int \frac{dx}{x(3x+2)} = \frac{1}{2}\ln\left|\frac{x}{3x+2}\right| + C$

**27.** $\int x^5 \ln x\,dx$

$u = \ln x \quad dv = x^5\,dx$

$du = \frac{1}{x}dx \quad v = \frac{1}{6}x^6$

$\int x^5 \ln x\,dx = \frac{1}{6}x^6 \ln|x| - \frac{1}{6}\int x^5\,dx$

$= \frac{1}{6}x^6 \ln|x| - \frac{1}{6}\cdot\frac{x^6}{6} + C$

$= \frac{1}{6}x^6 \ln|x| - \frac{1}{36}x^6 + C$

**28.** $\int x^2 e^{-2x} dx = uv - \int v\,du$

$u = x^2 \qquad dv = e^{-2x} dx$

$du = 2x\,dx \qquad v = -\frac{1}{2}e^{-2x}$

$$\int x^2 e^{-2x} dx = -\frac{1}{2}x^2 e^{-2x} + \frac{1}{2}\int e^{-2x} 2x dx$$
$$= -\frac{1}{2}x^2 e^{-2x} + \int e^{-2x} x dx$$

Integrate by parts again.

$u = x \qquad dv = e^{-2x} dx$

$du = dx \qquad v = -\frac{1}{2}e^{-2x}$

$$\int xe^{-2x} dx = -\frac{1}{2}xe^{-2x} + \frac{1}{2}\int e^{-2x} dx$$
$$= -\frac{1}{2}xe^{-2x} - \frac{1}{4}e^{-2x} + C$$
$$\int x^2 e^{-2x} dx = -\frac{1}{2}x^2 e^{-2x} - \frac{1}{2}xe^{-2x} - \frac{1}{4}e^{-2x} + C$$
$$= -e^{-2x}\left(\frac{x^2}{2} + \frac{x}{2} + \frac{1}{4}\right) + C$$

**29.** $\int x(x+5)^{-1/2} dx$

$u = x \qquad dv = (x+5)^{-1/2} dx$

$du = dx \qquad v = 2(x+5)^{1/2}$

$$\int x(x+5)^{-1/2} dx = 2x\sqrt{x+5} - 2\int (x+5)^{1/2} dx = 2x\sqrt{x+5} - \frac{4}{3}(x+5)^{3/2} + C$$

**30.** $\int_1^e \ln x\,dx = uv - \int v\,du$

$u = \ln x \qquad dv = dx$

$du = \frac{1}{x}dx \qquad v = x$

$$\int_1^e \ln x\,dx = x\ln|x|\Big|_1^e - \int_1^e dx = (x\ln x - x)\Big|_1^e = e - e - (0-1) = 1$$

**31.** $\int_1^\infty \frac{1}{x+1} dx = \lim_{b\to\infty} \int_1^b \frac{1}{x+1} dx = \lim_{b\to\infty} \ln(x+1)\Big|_1^b = \lim_{b\to\infty}\left(\ln(b+1) - \ln 2\right) = \infty$

Thus, the integral diverges.

**32.** $\int_{-\infty}^{-1} \frac{200}{x^3} dx = \lim_{a\to\infty} \int_{-a}^{-1} \frac{200}{x^3} dx = \lim_{a\to\infty} -100x^{-2}\Big|_{-a}^{-1} = -100$

**33.** $\int_0^\infty 5e^{-3x} dx = \lim_{b\to\infty} \int_0^b 5e^{-3x} dx = -\frac{5}{3}\lim_{b\to\infty} e^{-3x}\Big|_0^b = -\frac{5}{3}\lim_{b\to\infty}\left(\frac{1}{e^{3b}} - e^0\right) = \frac{5}{3}$

**34.** $\int_{-\infty}^{0}\frac{x}{\left(x^2+1\right)^2}dx=\lim_{a\to\infty}\int_{-a}^{0}x\left(x^2+1\right)^{-2}dx=\lim_{a\to\infty}\frac{1}{2}\int_{-a}^{0}2x\left(x^2+1\right)^{-2}dx=\lim_{a\to\infty}-\frac{1}{2}\left(x^2+1\right)^{-1}\Big|_{-a}^{0}=-\frac{1}{2}$

**35.** **a.** $\int_1^3 2x^{-3}dx=-\frac{1}{x^2}\Big|_1^3=-\left(\frac{1}{9}-1\right)=\frac{8}{9}$ or 0.889

**b.** $\int_1^3\frac{2}{x^3}dx\approx\frac{h}{2}[f(1)+2f(1.5)+2f(2)+2f(2.5)+f(3)]$

$=\frac{1}{4}[2+1.185+0.500+0.256+0.074]=\frac{1}{4}[4.015]=1.004$

**c.** $\int_1^3\frac{2}{x^3}dx\approx\frac{h}{3}[f(1)+4f(1.5)+2f(2)+4f(2.5)+f(3)]$

$=\frac{1}{6}[2+2.370+0.500+0.512+0.074]=\frac{1}{6}[5.456]=0.909$

**36.** $\int_0^1\frac{4}{x^2+1}dx\approx\frac{h}{2}[f(0)+2f(0.2)+2f(0.4)+2f(0.6)+2f(0.8)+f(1)]$

$=\frac{1}{10}[4+7.692+6.897+5.882+4.878+2]=\frac{1}{10}[31.349]=3.135$

**37.** $\int_1^{2.2}f(x)dx\approx\frac{h}{3}[f(1)+4f(1.3)+2f(1.6)+4f(1.9)+f(2.2)]=\frac{1}{10}[0+11.2+10.2+16.8+0.6]=\frac{1}{10}[38.8]=3.9$

**38.** **a.** $n=5$
**b.** $n=6$

**39.** $M=\int_0^9 14{,}000(t+16)^{-1/2}dt=14{,}000\cdot 2(t+16)^{1/2}\Big|_0^9=28{,}000(5-4)=\$28{,}000$

**40.** $\int_0^2 1.4e^{-1.4x}dx=-e^{-1.4x}\Big|_0^2=-e^{-2.8}+e^0=1-e^{-2.8}=0.9392$

So the probability that it lasts 2 years is $1-0.9392\approx 0.061$.

**41.** Average $=\frac{1}{5-0}\int_0^5 1000e^{0.1t}dt=\frac{1}{5}\cdot\frac{1000}{0.1}e^{0.1t}\Big|_0^5=2000(1.6487-1)=\$1297.44$

**42.** $AI=\frac{1}{4}\int_0^4 50e^{0.2t}dt=\frac{250}{4}\int_0^4 0.2e^{0.2t}dt=\frac{250}{4}e^{0.2t}\Big|_0^4\approx\frac{250}{4}[2.2255-1]\approx\$76.60$

**43.** In 1969, the Gini coefficient was $2\int_0^1\left(x-x^{2.1936}\right)dx=2\left(\frac{1}{2}x^2-\frac{x^{3.1936}}{3.1936}\right)\Big|_0^1\approx 0.3737.$

In 2000, the Gini coefficient was $2\int_0^1\left(x-x^{2.4870}\right)dx=2\left(\frac{1}{2}x^2-\frac{x^{3.4870}}{3.4870}\right)\Big|_0^1\approx 0.4264.$

Income was more equally distributed in 1969.

**44. a.** $\sqrt{64-4x} = x-1$

$$64-4x = x^2-2x+1$$
$$0 = (x+9)(x-7)$$

So, equilibrium is at $(7,6)$.

**b.** $CS = \int_0^7 (64-4x)^{1/2}dx - 7\cdot 6 = -\frac{1}{4}\cdot\frac{(64-4x)^{3/2}}{3/2}\Bigg|_0^7 - 42 = -\frac{1}{6}\left(36^{3/2}-64^{3/2}\right)-42$

$$= -\frac{1}{6}(-296)-42 = 49.33-42 = \$7.33$$

**45.**
$$x-1 = \sqrt{64-4x}$$
$$x^2-2x+1 = 64-4x$$
$$x^2+2x-63 = 0$$
$$(x+9)(x-7) = 0$$
$$x = 7 \text{ (only positive values)}$$

$$PS = (7)(6) - \int_0^7 (x-1)\,dx = 42 - \left[\left(\frac{1}{2}x^2 - x\right)\Bigg|_0^7\right] = 42-17.5 = 24.50$$

**46.** $TI = \int_0^{10} 125e^{0.05t}dt = 2500e^{0.05t}\Big|_0^{10} = 2500(1.64872-1) = \$1621.803$ thousands $= \$1,621,803$

**47. a.** $PV = \int_0^5 150e^{-0.2t}e^{-0.08t}dt = \int_0^5 150e^{-0.28t}dt = \frac{150}{-0.28}\int_0^5 -0.28e^{-0.28t}dt$

$= -535.71\left[e^{-0.28t}\Big|_0^5\right] = -535.71(0.2466-1) \approx 403.604$ thousands or \$403,604.

**b.** $FV = e^{0.40}(PV) = 1.4918(PV) = \$602,106$

**48.** Average Cost $= \frac{1}{150-0}\int_0^{150}\left(x^2+40,000\right)^{1/2}dx$

$$= \frac{1}{150}\cdot\frac{1}{2}\left[x\sqrt{x^2+40,000} + 40,000\ln\left(x+\sqrt{x^2+40,000}\right)\right]\Bigg|_0^{150}$$
$$= \frac{1}{300}\left[150\sqrt{62,500} + 40,000\ln\left(150+\sqrt{62,500}\right) - \left(0+40,000\ln\sqrt{40,000}\right)\right]$$
$$= \frac{1}{300}[150(250) + 40,000(5.9915) - 40,000(5.2983)] = \frac{1}{300}(65,228) = \$217.43$$

**49.** $PS = 70(1000) - \int_0^{1000}\left(0.02x + 50 - \frac{10}{\sqrt{x^2+1}}\right)dx$

$$= 70,000 - \left(0.01x^2 + 50x - 10\ln\left|x+\sqrt{x^2+1}\right|\right)\Bigg|_0^{1000} = 70,000 - \left[(10000+50000-76.01)-(0)\right] \approx \$10,076$$

**50.** $PV = 9000\int_0^5 te^{-0.08t}\,dt$ Use integration by parts.

$$PV = 9000\left[\frac{-t}{0.08}e^{-0.08t}\Big|_0^5 + \frac{1}{0.08}\int_0^5 e^{-0.08t}\,dt\right] = 9000\left[\frac{-t}{0.08}e^{-0.08t}\Big|_0^5 + \frac{1}{-(0.08)^2}\int_0^5 -0.08e^{-0.08t}\,dt\right]$$

$$= 9000[-41.895] - 1,406,250\left[e^{-0.08t}\Big|_0^5\right] = 9000[-41.895] - 1,406,250[-0.32968] = \$86,557.44$$

**51.** $C(x) = \int\left[3 + 60(x+1)\ln(x+1)\right]dx$

Using integration by parts: $u = \ln(x+1)$ $dv = 60(x+1)dx$

$du = \dfrac{dx}{x+1}$ $v = 30(x+1)^2$

$$\int\left[3 + 60(x+1)\ln(x+1)\right]dx = 3x + \ln(x+1)\cdot 30(x+1)^2 - \int 30(x+1)^2\cdot\frac{dx}{x+1}$$

$$= 3x + 30(x+1)^2\ln(x+1) - 30\int(x+1)\,dx$$

$$= 3x + 30(x+1)^2\ln(x+1) - 30\left(\tfrac{1}{2}x^2 + x\right) + K$$

$$= 3x + 30(x+1)^2\ln(x+1) - 15x^2 - 30x + K$$

$$C(x) = 30(x+1)^2\ln(x+1) - 15x^2 - 27x + K$$

Using the concept that fixed cost is *K*, we have $C(x) = 30(x+1)^2\ln(x+1) - 15x^2 - 27x + 2000.$

**52.** $\int_1^\infty 1.4e^{-1.4x}dx = \lim\limits_{b\to\infty}\left[-\int_1^b -1.4e^{-1.4x}dx\right] = -\lim\limits_{b\to\infty} e^{-1.4x}\Big|_1^b = -\lim\limits_{b\to\infty}\left(e^{-1.4b} - e^{-1.4}\right) = e^{-1.4} \approx 0.247$

**53.** $CV = \int_0^\infty 120e^{.03t}\cdot e^{-0.06t}dt = \int_0^\infty 120e^{-0.03t}dt = \lim\limits_{a\to\infty}\int_0^a 120e^{-0.03t}dt$

$$= \frac{120}{-0.03}\lim_{a\to\infty} e^{-0.03t}\Big|_0^a = -4000\lim_{a\to\infty}\left(\frac{1}{e^{0.03a}} - 1\right) = \$4000 \text{ thousands}$$

**54.** $\int_0^2 100e^{-0.01t^2}dt \approx \frac{h}{3}[f(0) + 4f(0.5) + 2f(1) + 4f(1.5) + f(2)]$

$$= \frac{1}{6}[100 + 399.00 + 198.01 + 390.10 + 96.08] = \frac{1}{6}[1183.19] = \$197.198 \text{ (thousands)}$$

**55.** $\int_0^{10}\overline{MR}\,dx \approx \frac{h}{2}[f(0) + 2f(2) + 2f(4) + 2f(6) + 2f(8) + f(10)]$

$$= 1[0 + 960 + 1440 + 1440 + 960 + 0] = \$4,800 \text{ (hundreds) or } \$480,000$$

## Chapter 13 Test

**1.**

| $x_i$ | 0 | .5 | 1 | 1.5 |
|---|---|---|---|---|
| $f(x_1)$ | 2 | 1.936 | 1.732 | 1.323 |

$A \approx (0.5)\sum f(x_i) = (0.5)(6.991) = 3.496$

**2.** $f(x) = 5 - 2x \quad [0,1]$

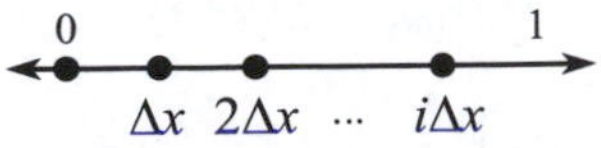


$\Delta x = \dfrac{1-0}{n} = \dfrac{1}{n}$

Rectangle 1: $\Delta x(5 - 2\Delta x)$

Rectangle 2: $\Delta x(5 - 4\Delta x)$

...

Rectangle i: $\Delta x(5 - 2i\Delta x)$

**a.** $S = \sum_{i=1}^{n}\left(5 - 2i \cdot \dfrac{1}{n}\right)\dfrac{1}{n}$

$= \dfrac{1}{n}\left(5n - \dfrac{2}{n} \cdot \dfrac{n(n+1)}{2}\right)$

$= 5 - \dfrac{n+1}{n}$ or $4 - \dfrac{1}{n}$

**b.** $\lim_{n\to\infty} S_n = \lim_{n\to\infty}\left(4 - \dfrac{1}{n}\right) = 4$

**3.** $\int_0^6 \left(12 + 4x - x^2\right)dx = \left(12x + 2x^2 - \dfrac{x^3}{3}\right)\Bigg|_0^6 = 72$

**4.** **a** $\int_0^4 (9 - 4x)\,dx = \left(9x - 2x^2\right)\Big|_0^4 = 4 - 0 = 4$

**b.** $\int_0^3 \left(8x^2 + 9\right)^{-1/2}(x\,dx) = \dfrac{1}{16}\left(\dfrac{\left(8x^2+9\right)^{1/2}}{1/2}\right)\Bigg|_0^3$

$= \dfrac{1}{8}(9 - 3) = \dfrac{3}{4}$

**c.** $\int_1^4 \dfrac{5}{4x-1} = \dfrac{5}{4}\ln(4x-1)\Big|_1^4 = \dfrac{5}{4}\ln 5$

**d.** $\int_1^{\infty} \dfrac{7}{x^2}\,dx = \lim_{a\to\infty}\left(-\dfrac{7}{x}\right)\Bigg|_1^a = \lim_{a\to\infty}\left(-\dfrac{7}{a} + 7\right) = 7$

**e.** $\int_4^4 \sqrt{x^3 + 10}\,dx = 0$

**f.** $\int_0^1 5x^2 e^{2x^3}\,dx = 5 \cdot \dfrac{1}{6}\int_0^1 e^{2x^3}\left(6x^2 dx\right)$

$= \dfrac{5}{6}e^{2x^3}\Big|_0^1 = \dfrac{5}{6}\left(e^2 - e^0\right) = \dfrac{5}{6}\left(e^2 - 1\right)$

**5.** **a.** $\int 3xe^x\,dx = 3xe^x - \int 3e^x\,dx = 3xe^x - 3e^x + C$

$u = 3x \qquad dv = e^x dx$

$du = 3dx \qquad v = e^x$

**b.** $\int x\ln(2x)\,dx = \dfrac{x^2 \ln 2x}{2} - \int \dfrac{x}{2}dx$

$= \dfrac{x^2 \ln 2x}{2} - \dfrac{x^2}{4} + C$

**6.** $\int_1^3 f(x)dx + \int_3^4 f(x)dx = \int_1^4 f(x)dx$, so

$2\int_1^3 f(x)dx = 2\left[\int_1^4 f(x)dx - \int_3^4 f(x)dx\right]$

$= 2[3 - 7] = -8$

**7.** **a.** $\int \ln(2x)\,dx = \dfrac{1}{2}\int \ln(2x)(2dx)$

$= \dfrac{1}{2} \cdot 2x\left(\ln(2x) - 1\right) = x\left(\ln(2x) - 1\right) + C$

(Formula 14)

**b.** $\int x\sqrt{3x-7}\,(dx)$

$= \left[\dfrac{2(3 \cdot 3x + 2 \cdot 7)(3x-7)^{3/2}}{15 \cdot 9}\right] + C$

(Formula 16)

**8.** $\int_1^4 \sqrt{x^3 + 10}\,dx \approx 16.089$

**9.** $80 = 120 - 0.2x$ gives an equilibrium quantity $x = 200$.

$CS = \int_0^{200} (120 - 0.2x)\,dx - 80 \cdot 200$

$= \left(120x - 0.1x^2\right)\Big|_0^{200} - 16000 = \$4000$

$PS = 80 \cdot 200 - \int_0^{200}\left(40 + 0.001x^2\right)dx$

$= 16000 - \left(40x + \dfrac{.001}{3}x^3\right)\Bigg|_0^{200} = \dfrac{\$16000}{3}$

**10.** **a.** $TI = \int_0^{12} 85e^{-0.01t}\,dt$

$= \dfrac{85}{-0.01}e^{-0.01t}\Big|_0^{12} = -8500(0.8869 - 1)$

$= \$961.18$ thousands

**b.** $PV = \int_0^{12} 85e^{-0.01t}e^{-0.07t}\,dt$

$= 85\int_0^{12} e^{-0.08t}\,dt = \dfrac{85}{-0.08}e^{-0.08t}\Big|_0^{12}$

$= \$655.68$ thousands

**c.** $CV = \int_0^{\infty} 85e^{-0.01t}e^{-0.07t}\,dt$

$= \lim_{a\to\infty} \int_0^a 85e^{-0.08t}\,dt$

$= \lim_{a\to\infty} \left[ \frac{85}{-0.08} e^{-0.08t} \Big|_0^a \right]$

$= \frac{85}{-0.08} \lim_{a\to\infty} \left( \frac{1}{e^{0.08a}} - 1 \right) = \$1062.5$ thousand

**11.** First, find the points of intersection.

$x^2 - x = 2x + 4$

$0 = x^2 - 3x - 4 = (x-4)(x+1)$

Points are –1 and 4.

$$A = \int_{-1}^{4} \left[ (2x+4) - (x^2 - x) \right] dx = \int_{-1}^{4} (3x + 4 - x^2)\,dx = \left( \frac{3}{2}x^2 + 4x - \frac{x^3}{3} \right) \Bigg|_{-1}^{4} = \frac{56}{3} - \left( -\frac{13}{6} \right) = \frac{125}{6}$$

**12.**

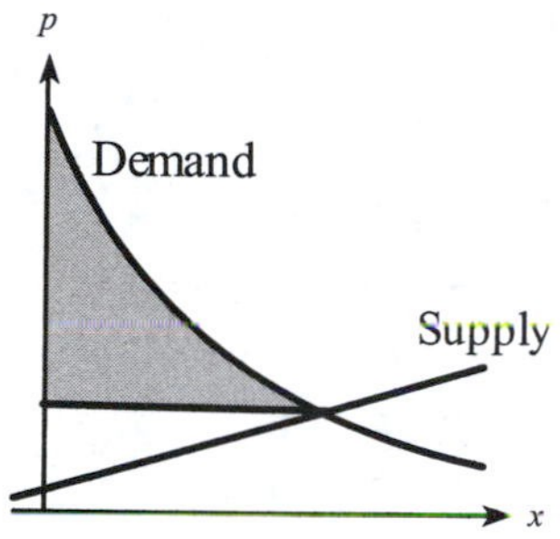


**13.** $$\text{Gini coeff} = \frac{\int_0^1 \left( x - 0.998x^{2.6} \right) dx}{1/2} = 2\left( \frac{x^2}{2} - \frac{0.998}{3.6}x^{3.6} \right) \Bigg|_0^1 = 2(0.223) = 0.446$$

$$\text{Gini coeff} = \frac{\int_0^1 \left[ x - \left( 0.57x^2 + 0.43x \right) \right] dx}{1/2} = 2\left( \frac{x^2}{2} - \frac{0.57}{3}x^3 - \frac{0.43}{2}x^2 \right) \Bigg|_0^1 = 2(0.095) = 0.19$$

The change decreases the difference in income.

**14. a.** 20.92 billion barrels

**b.** 2.067 billion barrels per year

**15.** $$\int_1^3 x \ln x\,dx \approx \frac{h}{2}\left[ f(1) + 2f(1.5) + 2f(2) + 2f(2.5) + f(3) \right] = \frac{1}{4}\left[ 0 + 1.216 + 2.773 + 4.581 + 3.296 \right]$$

$$= \frac{1}{4}\left[ 11.866 \right] = 2.967$$

**16.** $$\text{Area} \approx \frac{h}{3}\left[ f(0) + 4f(40) + 2f(80) + 4f(120) + 2f(160) + 4f(200) + f(240) \right]$$

$$= \frac{40}{3}\left[ 0 + 88 + 70 + 192 + 64 + 96 + 0 \right] = \frac{40}{3}\left[ 510 \right] = 6800 \text{ ft}^2$$

# Chapter 14: Functions of Two or More Variables

## *Exercises 14.1*

**1.** $z = x^2 + y^2$

Domain = $\{(x, y): x \text{ and } y \text{ are real}\}$

**3.** $z = \dfrac{4x-3}{y}$

Domain = $\{(x, y): x \text{ and } y \text{ are real and } y \neq 0\}$

**5.** $z = \dfrac{4x^3y - x}{2x - y}$

Domain =

$\{(x, y): x \text{ and } y \text{ are real and } 2x - y \neq 0\}$

**7.** $q = \sqrt{p_1} + 3p_2$

Domain =

$\{(p_1, p_2): p_1 \text{ and } p_2 \text{ are real and } p_1 \geq 0\}$

**9.** $z = x^3 + 4xy + y^2$

$x = 1,\ y = -1$

$z = 1 - 4 + 1 = -2$

**11.** $z = \dfrac{x-y}{x+y}$

$x = 4,\ y = -1$

$z = \dfrac{4+1}{4-1} = \dfrac{5}{3}$

**13.** $C(x_1, x_2) = 600 + 4x_1 + 6x_2$

$x_1 = 400,\ x_2 = 50$

$C(400, 50) = 600 + 1600 + 300 = 2500$

**15.** $q_1(p_1, p_2) = \dfrac{p_1 + 4p_2}{p_1 - p_2}$

$q_1(40, 35) = \dfrac{40 + 140}{40 - 35} = 36$

**17.** $z(x, y) = xe^{x+y}$

$z(3, -3) = 3e^0 = 3$

**19.** $f(x, y) = \dfrac{\ln(xy)}{x^2 + y^2}$

$f(-3, -4) = \dfrac{\ln 12}{9 + 16} = \dfrac{\ln 12}{25}$

**21.** $w = \dfrac{x^2 + 4yz}{xyz}$

At $(1, 3, 1)$ we have $w = \dfrac{1 + 12}{3} = \dfrac{13}{3}$.

**23.** $S = f(P, t) = Pe^{0.06t}$

$f(2000, 20) = 2000e^{1.2} = \$6640.23$

When \$2000 is invested for 20 years the compound amount is \$6640.23.

**25.** $Q = f(K, M, h) = \sqrt{\dfrac{2KM}{h}}$

$f(200, 625, 1) = \sqrt{\dfrac{2(200)(625)}{1}} = 500$

For the given numbers, the most economical order size is 500 units.

**27.** $S = 1.98T - 1.09(1 - H)(T - 58) - 56.9$

$A = 2.70 + 0.885T - 78.7H + 1.20TH$

Max: 97.8°F $H = 44\%$

Min: 74.7°F $H = 80\%$

Max: $S = 1.98(97.8) - 1.09(1 - 0.44)(97.8 - 58) - 56.9 = 112.5°\text{F}$

$A = 2.70 + 0.885(97.8) - 78.7(0.44) + 1.2(97.8)(0.44) = 106.3°\text{F}$

Min: $S = 1.98(74.7) - 1.09(1 - 0.8)(74.7 - 58) - 56.9 = 87.4°\text{F}$

$A = 2.70 + 0.885(74.7) - 78.7(0.8) + 1.2(74.7)(0.8) = 77.6°\text{F}$

**29. a.** $f(90, 20, 8) = \$752.80$ (Table 1, Row 5, Col 4)

When \$90,000 is borrowed for 20 years at 8%, the monthly payment needed to pay off the loan is \$752.80.

**b.** $f(160, 15, 9) = \$1622.82$ (Table 2, Row 9, Col 3)

When \$160,000 is borrowed for 15 years at 9%, the monthly payment needed to pay off the loan is \$1622.82.

**31.** $U = xy^2$

If $x = 9$, $y = 6$, then $U = 9 \cdot 6^2 = 324$.

**a.** $324 = x \cdot 9^2$ or $x = 4$ units

**b.** $324 = 81 \cdot y^2$ or $y = 2$ units

**c.**

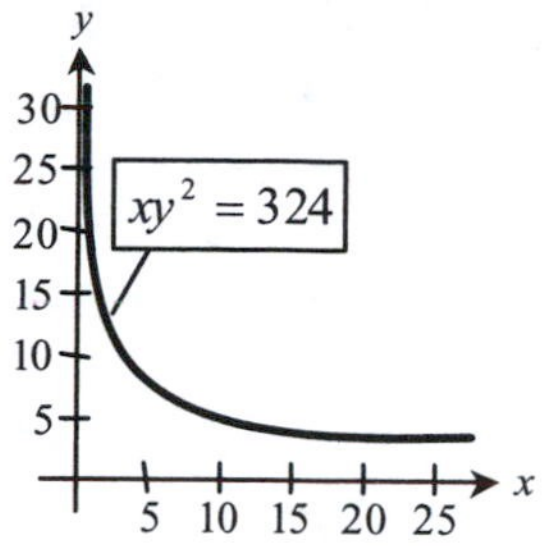


**33.** $Q = 30K^{1/4}L^{3/4}$

**a.** $Q = 30(10,000)^{1/4}(625)^{3/4}$

$= 30(10)(125) = 37,500$

**b.** $Q = 30(2K)^{1/4}(2L)^{3/4}$

$= 30(2^{1/4})K^{1/4}(2^{3/4})L^{3/4}$

$= 2\left[30K^{1/4}L^{3/4}\right]$

This is 2 times the original quantity.

**35.** $z = 20xy$

**a.** $z = 20(12)(30) = 7200$

**b.** $z = 20(10)(25) = 5000$

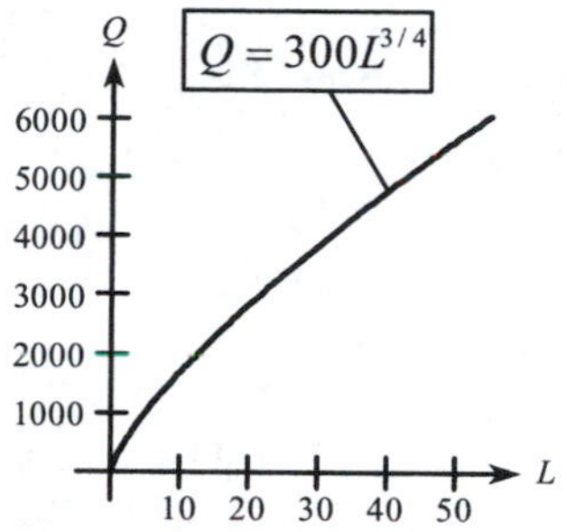


**37.** $C(x, y) = 20x + 200y$

$C(14,000,\ 20) = 20(14,000) + 200(20) = \$284,000$

## *Exercises 14.2*

**1.** $z = x^4 - 5x^2 + 6x + 3y^3 - 5y + 7$

$$\frac{\partial z}{\partial x} = 4x^3 - 10x + 6$$

$$\frac{\partial z}{\partial y} = 9y^2 - 5$$

**3.** $z = x^3 + 4x^2y + 6y^2$

$z_x = 3x^2 + 8yx$

$z_y = 4x^2 + 12y$

**5.** $f(x, y) = \left(x^3 + 2y^2\right)^3$

$$\frac{\partial f}{\partial x} = 3\left(x^3 + 2y^2\right)^2 3x^2 = 9x^2\left(x^3 + 2y^2\right)^2$$

$$\frac{\partial f}{\partial y} = 3\left(x^3 + 2y^2\right)^2 4y = 12y\left(x^3 + 2y^2\right)^2$$

**7.** $f(x, y) = \sqrt{2x^2 - 5y^2}$

$$f_x = \frac{1}{2}\left(2x^2 - 5y^2\right)^{-1/2} 4x = \frac{2x}{\sqrt{2x^2 - 5y^2}}$$

$$f_y = \frac{1}{2}\left(2x^2 - 5y^2\right)^{-1/2}(-10y) = \frac{-5y}{\sqrt{2x^2 - 5y^2}}$$

**9.** $C(x, y) = 600 - 4xy + 10x^2 y$

$\frac{\partial C}{\partial x} = -4y + 20xy$

$\frac{\partial C}{\partial y} = -4x + 10x^2$

**11.** $Q(s,t) = \frac{2s - 3t}{s^2 + t^2}$

$\frac{\partial Q}{\partial s} = \frac{(s^2 + t^2)(2) - (2s - 3t)(2s)}{(s^2 + t^2)^2}$

$= \frac{2s^2 + 2t^2 - 4s^2 + 6st}{(s^2 + t^2)^2} = \frac{2(t^2 + 3st - s^2)}{(s^2 + t^2)^2}$

$\frac{\partial Q}{\partial t} = \frac{(s^2 + t^2)(-3) - (2s - 3t)(2t)}{(s^2 + t^2)^2}$

$= \frac{-3s^2 - 3t^2 - 4st + 6t^2}{(s^2 + t^2)^2} = \frac{3t^2 - 4st - 3s^2}{(s^2 + t^2)^2}$

**13.** $z = e^{2x} + y \ln x$

$z_x = e^{2x}(2) + y \cdot \frac{1}{x} = 2e^{2x} + \frac{y}{x}$

$z_y = \ln x$

**15.** $f(x, y) = \ln(xy + 1)$

$\frac{\partial f}{\partial x} = \frac{y}{xy + 1}$

$\frac{\partial f}{\partial y} = \frac{x}{xy + 1}$

**17.** $f(x, y) = 4x^3 - 5xy + y^2$

$f_x(x, y) = 12x^2 - 5y$

$f_x(1, 2) = 12 - 5(2) = 2$

**19.** $z = 5x^3 - 4xy$

$z_x = 15x^2 - 4y$

$z_x(1, 2) = 15 - 4(2) = 7$

**21.** $z = 3x + 2y - 7e^{xy}$

$z_y = 2 - 7xe^{xy}$

$z_y(3, 0, 2) = 2 - 7(3)e^{2(0)} = 2 - 21 = -19$

**23.** $u = y^2 - x^2 z + 4x$

**a.** $\frac{\partial u}{\partial w} = 0$

**b.** $\frac{\partial u}{\partial x} = 4 - 2xz$

**c.** $\frac{\partial u}{\partial y} = 2y$

**d.** $\frac{\partial u}{\partial z} = -x^2$

**25.** $C = 4x_1{}^2 + 5x_1 x_2 + 6x_2{}^2 + x_3$

**a.** $\frac{\partial C}{\partial x_1} = 8x_1 + 5x_2$

**b.** $\frac{\partial C}{\partial x_2} = 5x_1 + 12x_2$

**c.** $\frac{\partial C}{\partial x_3} = 1$

**27.** $z = x^2 + 4x - 5y^3$

$z_x = 2x + 4, \quad z_y = -15y^2$

**a.** $z_{xx} = 2$

**b.** $z_{xy} = 0$

**c.** $z_{yx} = 0$

**d.** $z_{yy} = -30y$

**29.** $z = x^2 y - 4xy^2$

$z_x = 2xy - 4y^2, \quad z_y = x^2 - 8xy$

**a.** $z_{xx} = 2y$

**b.** $z_{xy} = 2x - 8y$

**c.** $z_{yx} = 2x - 8y$

**d.** $z_{yy} = -8x$

**31.** $f(x, y) = x^2 + e^{xy}$

$\frac{\partial f}{\partial x} = 2x + ye^{xy}, \quad \frac{\partial f}{\partial y} = xe^{xy}$

**a.** $\frac{\partial^2 f}{\partial x^2} = 2 + y^2 e^{xy}$

**b.** $\frac{\partial^2 f}{\partial y\, \partial x} = yxe^{xy} + e^{xy} \cdot 1 = e^{xy}(xy + 1)$

**c.** $\frac{\partial^2 f}{\partial x\, \partial y} = xye^{xy} + e^{xy} \cdot 1 = e^{xy}(xy + 1)$

**d.** $\frac{\partial^2 f}{\partial y^2} = x^2 e^{xy}$

**33.** $f(x,y)=y^2-\ln xy=y^2-\ln x-\ln y$

$\dfrac{\partial f}{\partial x}=-\dfrac{1}{x},\ \dfrac{\partial f}{\partial y}=2y-\dfrac{1}{y}$

**a.** $\dfrac{\partial^2 f}{\partial x^2}=\dfrac{1}{x^2}$

**b.** $\dfrac{\partial^2 f}{\partial y\,\partial x}=0$

**c.** $\dfrac{\partial^2 f}{\partial x\,\partial y}=0$

**d.** $\dfrac{\partial^2 f}{\partial y^2}=2+\dfrac{1}{y^2}$

**35.** $f(x,y)=x^3y+4xy^4$

$\dfrac{\partial f}{\partial x}=3x^2y+4y^4$

$\dfrac{\partial^2 f}{\partial x^2}=6xy$

$\dfrac{\partial^2}{\partial x^2}f(x,y)\Big|_{(1,-1)}=6(1)(-1)=-6$

**37.** $f(x,y)=\dfrac{2x}{x^2+y^2}$

**a.** $\dfrac{\partial f}{\partial x}=\dfrac{(x^2+y^2)(2)-2x(2x)}{(x^2+y^2)^2}=\dfrac{2y^2-2x^2}{(x^2+y^2)^2}$

$$\frac{\partial^2 f}{\partial x^2}=\frac{(x^2+y^2)^2(-4x)-(2y^2-2x^2)2(x^2+y^2)2x}{(x^2+y^2)^4}$$

$$=\frac{-4x(x^2+y^2)\left[(x^2+y^2)+(2y^2-2x^2)\right]}{(x^2+y^2)^4}=\frac{-4x(3y^2-x^2)}{(x^2+y^2)^3}$$

$$\frac{\partial^2 f}{\partial x^2}\Big|_{(-1,4)}=\frac{4(47)}{17^3}=\frac{188}{4913}$$

**b.** $\dfrac{\partial f}{\partial y}=2x(-1)(x^2+y^2)^{-2}(2y)=\dfrac{-4xy}{(x^2+y^2)^2}$

$$\frac{\partial^2 f}{\partial y^2}=\frac{(x^2+y^2)^2(-4x)+(4xy)2(x^2+y^2)2y}{(x^2+y^2)^4}=\frac{4x(x^2+y^2)\left[(-x^2-y^2)+4y^2\right]}{(x^2+y^2)^4}=\frac{4x(3y^2-x^2)}{(x^2+y^2)^3}$$

$$\frac{\partial^2 f}{\partial y^2}\Big|_{(-1,4)}=\frac{-4(47)}{17^3}=-\frac{188}{4913}$$

**39.** $z=x^2y+ye^{x^2}$

$z_y=x^2+e^{x^2}$

$z_{yx}=2x+e^{x^2}(2x)$

$z_{yx}\Big|_{(1,2)}=2+e^1(2)=2+2e$

**41.** $z=x^2-xy+4y^3$

$z_x=2x-y$

$z_{xy}=-1$

$z_{xyx}=0$

**43.** $w = 4x^3y + y^2z + z^3$

$w_x = 12x^2y$

**a.** $w_{xx} = 24xy, \quad w_{xxy} = 24x$

**b.** $w_{xy} = 12x^2, \quad w_{xyx} = 24x$

**c.** $w_{xyz} = 0$

**45.** $R = f(A, i)$

**a.** For a loan of \$100,000 at 8% interest, the monthly payment is \$771.82.

**b.** If the interest rate changes from 8% to 9% (on this \$100,000 loan), the monthly payment increases approximately \$66.25.

**47.** Explanation is based on business interpretation and not the mathematics of exponents.

**a.** $\frac{\partial Q}{\partial M}$ is the change in quantity ordered per unit change in sales. If the sales increase by one unit, then the order quantity should also increase.

**b.** $\frac{\partial Q}{\partial h}$ is the change in quantity ordered per unit change in holding costs. If holding costs increase, then the order quantity should decrease.

**49.** $f(x,y) = 10{,}000 - 6500e^{-0.01x} - 3500e^{-0.02y}$

$f_x = 0 - 6500e^{-0.01x}(-0.01) - 0 = 65e^{-0.01x}$

$f_x(100,100) = 65e^{-0.01(100)} \approx 23.912$

If brand 2 is held constant and brand 1 is increased from 100 to 101 liters, approximately 24,000 additional insects are killed.

**51.** $U = x^2y^2$

**a.** $\frac{\partial U}{\partial x} = 2xy^2$

**b.** $\frac{\partial U}{\partial y} = 2x^2y$

**53.** $Q = 75K^{1/3}L^{2/3}$

$\frac{\partial Q}{\partial K} = 25K^{-2/3}L^{2/3}, \quad \frac{\partial Q}{\partial L} = 50K^{1/3}L^{-1/3}$

At $K = 729$, $L = 5832$ we have $\frac{\partial Q}{\partial K} = 25 \cdot \frac{1}{81} \cdot 324 = 100$ and $\frac{\partial Q}{\partial L} = 50 \cdot 9 \cdot \frac{1}{18} = 25$.

If labor hours remain at 5832 and $K$ increases by \$1000, $Q$ will increase about 100 thousand units. If capital expenditures remain at \$729,000 and $L$ increases by one hour, $Q$ will increase about 25 thousand units.

**55.** $WC = 35.74 + 0.6215t - 35.75s^{0.16} + 0.4275ts^{0.16}$

**a.** $\frac{\partial WC}{\partial s} = 0.16(-35.75)s^{0.16-1} + 0.16(0.4275t)s^{0.16-1}$

$= -5.72s^{-0.84} + 0.0684ts^{0.84}$

**b.** At $s = 25$, $t = 10$, $\frac{\partial WC}{\partial s} = -5.72(25)^{-0.84} + 0.0684(10)(25)^{0.84}$

$= -0.337$

At this temperature and wind speed, the wind chill temperature will decrease 0.34°F if the wind speed increases 1 mph.

## *Exercises 14.3*

**1. a.** $C(x,y) = 30 + 3x + 5y$

$C(20,3) = 30 + 60 + 15 = \$105$

**b.** $C_x = 3$

**3.** $C(x,y) = 30 + 2x + 4y + \frac{xy}{50}$

**a.** $C_x = 2 + \frac{y}{50}$

**b.** $C_y = 4 + \frac{x}{50}$

# Chapter 14: Functions of Two or More Variables

**5.** $C(x,y)=20x+70y+\dfrac{x^2}{1000}+\dfrac{xy^2}{100}$

**a.** $C_x=20+\dfrac{x}{500}+\dfrac{y^2}{100}$

$C_x(10,24)=20+\dfrac{1}{50}+\dfrac{576}{100}=\$25.78$

The total cost will increase approximately \$25.78 if raw material costs increase to \$11 per pound.

**b.** $C_y=70+\dfrac{xy}{50}$

$C_y(10,24)=70+\dfrac{240}{50}=\$74.80$

The total cost will increase \$74.80 if labor costs increase to \$25 per hour.

**7.** $C(x,y)=30+x^2+3y+2xy$

**a.** $C_x=2x+2y$

$C_x(8,10)=16+20=\$36$

If $y$ remains at 10, the expected change in cost for a 9th unit of $x$ is \$36.

**b.** $C_y=3+2x$

$C_y=(8,10)=3+16=\$19$

If $x$ remains at 8, the expected change in cost for an 11th unit of $y$ is \$19.

**9.** $C(x,y)=x\sqrt{y^2+1}$

**a.** $C_x=\sqrt{y^2+1}$

**b.** $C_y=x\cdot\dfrac{1}{2}\left(y^2+1\right)^{-1/2}(2y)=\dfrac{xy}{\sqrt{y^2+1}}$

**11.** $C(x,y)=1200\ln(xy+1)+10{,}000$

**a.** $C_x=1200\cdot\dfrac{1}{xy+1}\cdot y=\dfrac{1200y}{xy+1}$

**b.** $C_y=1200\cdot\dfrac{1}{xy+1}\cdot x=\dfrac{1200x}{xy+1}$

**13.** $z=\sqrt{4xy}=2x^{1/2}y^{1/2}$

**a.** $z_x=2y^{1/2}\cdot\dfrac{1}{2}x^{-1/2}=\sqrt{\dfrac{y}{x}}$

**b.** $z_y=2x^{1/2}\cdot\dfrac{1}{2}y^{-1/2}=\sqrt{\dfrac{x}{y}}$

**15.** $z=x^{1/2}\ln(y+1)$

**a.** $z_x=\dfrac{1}{2}x^{-1/2}\ln(y+1)=\dfrac{\ln(y+1)}{2\sqrt{x}}$

**b.** $z_y=x^{1/2}\cdot\dfrac{1}{y+1}\cdot 1=\dfrac{\sqrt{x}}{y+1}$

**For 17 and 19,** $z=\dfrac{11xy-0.0002x^2-5y}{0.03x+3y}$.

**17.** At $x=300$ and $y=500$ we have $z=\dfrac{11(150{,}000)-0.0002(90{,}000)-2500}{9+1500}\approx 1092$

**19.** $z_x=\dfrac{(0.03x+3y)(11y-0.0004x)-(11xy-0.0002x^2-5y)(0.03)}{(0.03x+3y)^2}$

$z_x(300,500)=\dfrac{1509(5499.88)-(1{,}650{,}000-18-2500)(0.03)}{(1509)^2}$

$\approx 3.62$

If 500 acres are planted, the expected change in the productivity for the 301st hour of labor is 3.62 crates.

**21.** $z=400x^{3/5}y^{2/5}$

**a.** $\dfrac{\partial z}{\partial x}=\dfrac{240y^{2/5}}{x^{2/5}}$

**b.**

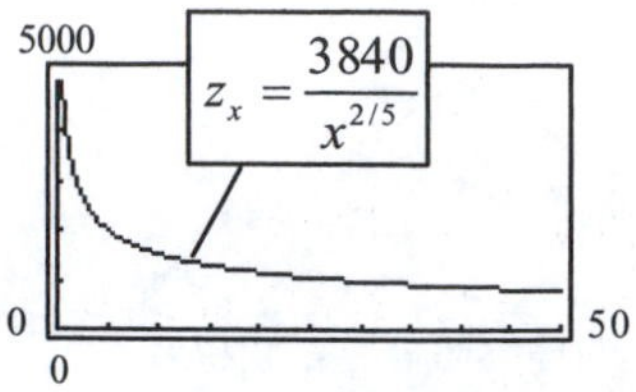


**c.** $\frac{\partial z}{\partial y} = \frac{160x^{3/5}}{y^{3/5}}$

**d.**

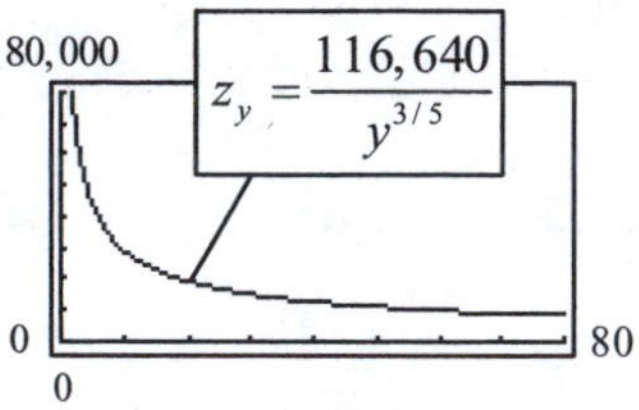


**e.** An increase in capital investment or work hours results in an increase in productivity since both partials are positive. As capital investment or work hours continue to increase, the increase in productivity is diminishing.

**23.** $q_1 = 300 - 80 - 32 = 188$

$q_2 = 400 - 50 - 80 = 270$

**25.** Setting $q_1 = q_2$, we have

$$300 - 8p_1 - 4p_2 = 400 - 5p_1 - 10p_2$$
$$-8p_1 + 5p_1 = 100 - 10p_2 + 4p_2$$
$$p_1 = -\frac{100}{3} + 2p_2 \quad (1)$$

Any values $p_1$ and $p_2$ satisfying (1) and that make $q_1$ and $q_2$ nonnegative will be a solution. Two solutions are $(6.67, 20)$ and $(20, 26.67)$.

**27.** $q_A = 400 - 3p_A - 2p_B$, $q_B = 250 - 5p_A - 6p_B$

**a.** $\frac{\partial q_A}{\partial p_A} = -3$

**b.** $\frac{\partial q_A}{\partial p_B} = -2$

**c.** $\frac{\partial q_B}{\partial p_B} = -6$

**d.** $\frac{\partial q_B}{\partial p_A} = -5$

**e.** Complementary since (b) and (d) are negative.

**29.** $q_A = 5000 - 50p_A - \frac{600}{p_B + 1}$

$q_B = 10,000 - \frac{400}{p_A + 4} + \frac{400}{p_B + 4}$

**a.** $\frac{\partial q_A}{\partial p_A} = -50$

**b.** $\frac{\partial q_A}{\partial p_B} = -600(-1) \cdot \frac{1}{(p_B + 1)^2} = \frac{600}{(p_B + 1)^2}$

**c.** $\frac{\partial q_B}{\partial p_B} = \frac{-400}{(p_B + 4)^2}$

**d.** $\frac{\partial q_B}{\partial p_A} = \frac{400}{(p_A + 4)^2}$

**e.** Competitive since (b) and (d) are positive.

**31. a.** Competitive: as the price of one type of car declines, demand for the other declines

**b. i.** $q_{\text{NEW}} = 2600 - p_{\text{NEW}}/30 + p_{\text{USED}}/15$

$q_{\text{USED}} = 750 - 0.25\, p_{\text{USED}}/30 + 0.0125\, p_{\text{NEW}}$

**ii.** Since the mixed partials are both positive (1/15 and 0.0125), the products are competitive.

# Chapter 14: Functions of Two or More Variables

## *Exercises 14.4*

**1.** $z = 9 - x^2 - y^2$

$z_x = -2x,\ z_y = -2y$

Critical point is at $x = 0$, $y = 0$.

$z_{xx} = -2,\ z_{yy} = -2,\ z_{xy} = 0$

$D = (-2)(-1) - (0)^2 > 0$ with $z_{xx} < 0$

A relative maximum at $(0, 0, 9)$.

**3.** $z = x^2 + y^2 + 4$

$z_x = 2x,\ z_y = 2y$

Critical point at $x = 0$, $y = 0$.

$z_{xx} = 2,\ z_{yy} = 2,\ z_{xy} = 0$

$D = (2)(2) - 0^2 > 0$ with $z_{xx} > 0$

A relative minimum at $(0, 0, 4)$.

**5.** $z = x^2 - y^2 + 4x - 6y + 11$

$z_x = 2x + 4,\ z_y = -2y - 6$

$z_x = 0$ at $x = -2$ and $z_y = 0$ at $y = -3$.

Critical point is at $(-2, -3)$.

$z_{xx} = 2,\ z_{yy} = -2,\ z_{xy} = 0$

$D = (2)(-2) - 0^2 < 0$

There is a saddle point at $(-2, -3, 16)$.

**7.** $z = x^2 + y^2 - 2x + 4y + 5$

$z_x = 2x - 2,\ z_y = 2y + 4$

$z_x = 0$ at $x = 1$ and $z_y = 0$ at $y = -2$. Critical point is at $(1, -2)$.

$z_{xx} = 2,\ z_{yy} = 2,\ z_{xy} = 0$

$D = (2)(2) - 0^2 > 0$ with $z_{xx} > 0$

A relative minimum at $(1, -2, 0)$.

**9.** $z = x^2 + 6xy + y^2 + 16x$

$z_x = 2x + 6y + 16,\ z_y = 6x + 2y$

$z_y = 0$ gives $y = -3x$. $z_x = 0$ and $y = -3x$ gives $2x + 6(-3x) + 16 = 0$.

Thus, $x = 1$ and $y = -3(1) = -3$. So, $(1, -3)$ is a critical point.

$z_{xx} = 2,\ z_{yy} = 2,\ z_{xy} = 6$

$D = (2)(2) - 6^2 < 0$ and thus, there is neither a relative maximum nor a relative minimum at $(1, -3, 8)$. It is a saddle point.

**11.** $z = 24 - x^2 + xy - y^2 + 36y$

$z_x = -2x + y,\ z_y = x - 2y + 36$

$z_x = 0$ yields $y = 2x$. $z_y = 0$ and $y = 2x$ yields $x - 2(2x) + 36 = 0$.

Solving, we have $x = 12$ and $y = 24$ are the critical values.

$z_{xx} = -2,\ z_{yy} = -2,\ z_{xy} = 1$

$D = (-2)(-2) - 1^2 > 0$ with $z_{xx} < 0$

There is a relative maximum at $(12, 24, 456)$.

**13.** $z = x^2 + xy + y^2 - 4y + 10x$

$z_x = 2x + y + 10,\ z_y = x + 2y - 4$

$z_x = 0$ or $2x + y = -10$ and $z_y = 0$ or $x + 2y = 4$.

Solving these two equations, we have a critical point at $(-8, 6)$.

$z_{xx} = 2,\ z_{yy} = 2,\ z_{xy} = 1$

$D = (2)(2) - 1^2 > 0$ with $z_{xx} > 0$

There is a relative minimum at $(-8, 6, -52)$.

**15.** $z = x^3 + y^3 - 6xy$

$z_x = 3x^2 - 6y,\ z_y = 3y^2 - 6x$

$z_x = 0$ or $y = \frac{1}{2}x^2$. $z_y = 0$ and $y = \frac{1}{2}x^2$ gives $\frac{3}{4}x^4 - 6x = 0$. $3x^4 - 24x = 0$ yields $3x(x^3 - 8) = 0$. At $x = 0$ we have $y = 0$ and at $x = 2$ we have $y = 2$.

Thus, the critical points are $(0, 0)$ and $(2, 2)$.

$z_{xx} = 6x,\ z_{yy} = 6y,\ z_{xy} = -6$

At $(0, 0)$: $D = (0)(0) - (-6)^2 < 0$

There is a saddle point at $(0, 0, 0)$.

At $(2, 2)$: $D = (12)(12) - (-6)^2 > 0$ with $z_{xx} > 0$.

There is a relative minimum at $(2, 2, -8)$.

**17.** $\sum x = 18,\ \sum y = 97,\ \sum xy = 465,\ \sum x^2 = 86$

$b = \frac{18(97) - 4(465)}{324 - 4(86)} = 5.7$

$a = \frac{97 - 5.7(18)}{4} = -1.4$

$\hat{y} = a + bx = 5.7x - 1.4$

**19.** $P(x, y) = 10x + 6.4y - 0.001x^2 - 0.025y^2$

$P_x = 10 - 0.002x,\ P_y = 6.4 - 0.05y$

$P_x = 0$ if $x = 5000$ and $P_y = 0$ if $y = 128$.

Critical point is $(5000, 128)$.

$P_{xx} = -0.002,\ P_{yy} = -0.05,\ P_{xy} = 0$

$D = (-0.002)(-0.05) - 0^2 > 0$

$D > 0$ with $P_{xx} < 0$ means that there is a maximum profit when 5000 pounds of Kisses and 128 pounds of Kreams are sold. The maximum profit is $P(5000, 128)$=\$25409.60.

**21.** $W = 13xy(20 - x - 2y)$

$= 260xy - 13x^2y - 26xy^2$

$W_x = 260y - 26xy - 26y^2 = 26y(10 - x - y^2)$

$W_y = 260x - 13x^2 - 52xy = 13x(20 - x - 4y)$

If $W_x = 0$, then $26y(10 - x - y) = 0$. Thus, $y = 0$ or $y = 10 - x$.

If $W_y = 0$, then $13x(20 - x - 4y) = 0$. Thus, $x = 0$ or $x = 20 - 4y$.

Critical points are $(0,0)$, $(0,10)$, and $(20,0)$ using $x = 0$ and $y = 0$.

Using $x = 20 - 4(10 - x)$, we have $x = \frac{20}{3}$. For $x = \frac{20}{3}$, we have $y = 10 - \frac{20}{3} = \frac{10}{3}$. So, $\left(\frac{20}{3}, \frac{10}{3}\right)$ is also a critical point.

$W_{xx} = -26y,\ W_{yy} = -52x,\ W_{xy} = 260 - 26x - 52y$

At $(0,0)$, $(0,10)$, and $(20,0)$, $D < 0$ and thus, there is no maximum or minimum there. At $\left(\frac{20}{3}, \frac{10}{3}\right)$, $D > 0$ and since $W_{xx} < 0$, there is a maximum weight gain with $x = \frac{20}{3}$ and $y = \frac{10}{3}$.

The maximum weight is

$W = 13\left(\frac{20}{3}\right)\left(\frac{10}{3}\right)\left(20 - \frac{20}{3} - 2\left(\frac{10}{3}\right)\right) \approx 1926$ lb.

**23.** $P = 3.78x^2 + 1.5y^2 - 0.09x^3 - 0.01y^3$

$P_x = 7.56x - 0.27x^2,\ P_y = 3y - 0.03y^2$

$P_x = 0$ if $0.27x(28 - x) = 0$ or $x = 0, 28$.

$P_y = 0$ if $3y(1 - 0.01y) = 0$ or $y = 0, 100$.

Critical points are $(0,0)$, $(0,100)$, $(28,0)$, and $(28,100)$.

$P_{xx} = 7.56 - 0.54x,\ P_{yy} = 3 - 0.06y,\ P_{xy} = 0$

At $(0,0)$ $D = (7.56)(3) - 0^2 > 0$. But $P_{xx} > 0$.

At $(0,100)$ $D = (7.56)(-3) - 0^2 < 0$.

At $(28,0)$ $D = (-7.56)(3) - 0^2 < 0$.

At $(28,100)$ $D = (-7.56)(-3) - 0^2 > 0$.

Since $P_{xx} < 0$, there is a maximum at $x = 28$ and $y = 100$. The maximum production is $P(28,100) = 5987.84 \approx 5988$.

**25.** $C = xy$

$R = p_1x + p_2y$

$= (70 - x)x + (80 - y)y$

$= 70x - x^2 + 80y - y^2$

$P = R - C = 70x - x^2 + 80y - y^2 - xy$

$P_x = 70 - 2x - y$

$P_y = 80 - 2y - x$

$P_x = 0$ if $y = 70 - 2x$. $P_y = 0$ if $x = 80 - 2y$.

Solving this system, we get $x = 20$ and $y = 30$.

Thus, the critical point is $(20,30)$.

$P_{xx} = -2,\ P_{yy} = -4,\ P_{xy} = -1$

$D = (-2)(-4) - (-1)^2 > 0$.

Since $P_{xx} < 0$, there is a maximum at $(20,30)$.

The maximum is at $P(20,30) = \$1900$ thousand.

**27.** $A = xy + 2y \cdot \dfrac{500,000}{xy} + 2x \cdot \dfrac{500,000}{xy}$

$= xy + \dfrac{1,000,000}{x} + \dfrac{1,000,000}{y}$

$A_x = y - \dfrac{10^6}{x^2}, \quad A_y = x - \dfrac{10^6}{y^2}$

Combining $A_x = 0$ and $A_y = 0$, we have

$x = \dfrac{10^6}{\frac{10^{12}}{x^4}} = \dfrac{x^4}{10^6}, \; x \neq 0.$

Thus, $x^3 = 10^6$ or $x = 100$.

The critical point is $(100,100)$.

$A_{xx} = \dfrac{2 \cdot 10^6}{x^3}, \quad A_{yy} = \dfrac{2 \cdot 10^6}{y^3}, \quad A_{xy} = 1$

$D = (2)(2) - 1^2 > 0$ with $A_{xx} > 0$.

There is a minimum for length and width equal 100 and height equal 50.

**29.** $P = R - C = 12x - 18y - 2x^2 + 2xy - y^2 - 11$

$P_x = 12 - 4x + 2y, \quad P_y = 18 + 2x - 2y$

$P_x = 0$ if $y = 2x - 6$ and $P_y = 0$ if $y = x + 9$.

Solving the system yields $(15,24)$ as the critical point.

$P_{xx} = -4, \; P_{yy} = -2, \; P_{xy} = 2$

$D = (-4)(-2) - 2^2 > 0$

Since $P_{xx} < 0$, maximum profit occurs with 15 $A$'s and 24 $B$'s. The maximum profit is \$295 thousand.

**31.** **a.** eat-in = 2400; take-out = 3800

**b.** eat-in @ \$3.60; take-out @ 3.10; max profit = \$12,480

**c.** Change pricing: more profitable

**For 33 and 35 use the linear regression feature of the graphing calculator.**

**33.** **a.** $\hat{y} = 0.81x - 2400$

**b.** $m = 0.81$ means that for every \$1 that males earn, females earn \$0.81.

**c.** The slope would probably be smaller. Women were paid much less than males in 1965.

**35.** **a.** $y = 0.06254x + 6.1914$ (where $x$ is the number of years after 2000 and $y$ is in billions)

**b.** $y(18) \approx 7.317$ billion

**c.** The world population is changing at the rate of 0.06254 billion persons per year past 2000.

## *Exercises 14.5*

**1.** The objective function is

$F(x, y, \lambda) = x^2 + y^2 + \lambda(x + y - 6)$.

$F_x = 2x + \lambda, \; F_y = 2y + \lambda, \; F_\lambda = x + y - 6$

Setting each of these equal to zero and combining, we have $x - y = 0$ and $x + y = 6$.

This yields $x = 3$ and $y = 3$. The minimum is $z = 3^2 + 3^2 = 18$.

**3.** The objective function is

$F(x, y, \lambda) = 3x^2 + 5y^2 - 2xy + \lambda(x + y - 5)$.

$F_x = 6x - 2y + \lambda, \; F_y = 10y - 2x + \lambda$

$F_\lambda = x + y - 5$

Setting $F_x = 0$ and $F_y = 0$, we have

$6x - 2y = 10y - 2x$ or $x = \dfrac{3}{2}y.$

Using $F_\lambda = 0$, we have $\dfrac{3}{2}y + y - 5 = 0$ or $y = 2$.

Now $x = \dfrac{3}{2}(2) = 3.$

Thus, the minimum occurs at $(3,2)$.

The minimum is $z = 27 + 20 - 12 = 35$.

**5.** The objective function is

$F(x, y, \lambda) = x^2 y + \lambda(x + y - 6)$.

$F_x = 2xy + \lambda, \; F_y = x^2 + \lambda, \; F_\lambda = x + y - 6$

Using $F_x = 0$ and $F_y = 0$, we have $x^2 = 2xy$ or $x(x - 2y) = 0$. Thus, $x = 0$ or $x = 2y$.

Using $F_\lambda = 0$, we have $x = 0$ and $y = 6$.

Also, $2y + y - 6 = 0$ yields $y = 2$ and so $x = 4$.

At $(0,6)$, $z = 0$ and at $(4,2)$, $z = 16 \cdot 2 = 32$.

The maximum is at $(4,2)$.

**7.** The objective function is
$F(x,y,\lambda)=2xy-2x^2-4y^2+\lambda(x+2y-8)$.
$F_x=2y-4x+\lambda,\ F_y=2x-8y+2\lambda$
$F_\lambda=x+2y-8$
$F_x=0$ means $\lambda=4x-2y$. $F_y=0$ means $\lambda=-x+4y$. Combining we have $5x-6y=0$.
$F_\lambda=0$ gives $x+2y=8$.

These two equations yield $x=3$, $y=\frac{5}{2}$.
The maximum is $F=15-18-25=-28$.

**9.** The objective function is
$F(x,y,\lambda)=xy+\lambda(9x^2+25y^2-450)$.
$\frac{\partial F}{\partial x}=y+18\lambda x,\ \frac{\partial F}{\partial y}=x+50\lambda y$
$\frac{\partial F}{\partial \lambda}=9x^2+25y^2-450$
Setting each of these equal to zero and solving the first two for $\lambda$ in terms of $x$ and $y$ yields $9x^2-25y^2=0$. The third equation gives $9x^2+25y^2=450$. Together, these two equations yield $18x^2=450$, or $x=5$. The possible maximum is 15 at $(5,3)$. Testing values near $x=5,\ y=3$, and satisfying the constraint shows that the function is maximized there.

**11.** The objective function is
$F(x,y,z,\lambda)=x^2+y^2+z^2+\lambda(x+y+z-3)$.
$F_x=2x+\lambda,\ F_y=2y+\lambda,$
$F_z=2z+\lambda,\ F_\lambda=x+y+z-3$
$F_x=F_y=F_z=0$ or $x=y=z=-\frac{\lambda}{2}$.
$F_\lambda=0$ yields $-\frac{3\lambda}{2}=3$ or $\lambda=-2.$
Then, $x=y=z=1$.
The minimum is $1+1+1=3$.

**13.** The objective function is
$F(x,y,z,\lambda)=xz+y+\lambda(x^2+y^2+z^2-1)$.
$F_x=z+2\lambda x,\ F_y=1+2\lambda y,$
$F_z=x+2\lambda z,\ F_\lambda=x^2+y^2+z^2-1$
$F_x=F_z=0$ gives $x-2\lambda x=z-2\lambda z$ or $x(1-2\lambda)-z(1-2\lambda)=0.$ So, $x=z$ or $1-2\lambda=0.$ If $x=z$, then $1+2\lambda=0.$
The solutions are $\lambda=\pm\frac{1}{2}$. If $\lambda=\frac{1}{2}$, then $F_y=0$ gives $y=-1$. Also, $z+x=0$ and $F_\lambda=0$ means that $x^2+1+z^2-1=0$ or $x=z=0$. The critical point is $(0,-1,0)$. If $\lambda=-\frac{1}{2}$, then $F_y=0$ gives $y=1$. From $F_x=F_z$ we get $z=x$. From $F_\lambda=0$ we get $x=z=0$. The critical point is $(0,1,0)$. At $(0,-1,0)$, we have $w=0-1=-1$. At $(0,1,0)$, we have $w=0+1=1$. The maximum is $w=1$ at $(0,1,0)$.

**15.** The objective function is
$F(x,y,\lambda)=xy^2+\lambda(3x+6y-18)$.
$F_x=y^2+3\lambda,$
$F_y=2xy+6\lambda,$
$F_\lambda=3x+6y-18$
$F_x=F_y=0$ or $-\frac{y^2}{3}=\frac{-xy}{3}$ or $xy-y^2=0$ or $y(y-x)=0$ thus $y=0$ or $y=x.$
$F_\lambda=0$: If $y=0$, then $x=6$. If $x=y$, then $3y-6=0$ or $y=2$. If $y=0$, then $x=6$. At $(0,6)$, we have $U=6\cdot 0=0$ and at $(2,2)$ we have $U=2\cdot 4=8.$ The maximum occurs at $(2,2)$.

**17.** The objective function is
$F(x,y,\lambda)=x^2y+\lambda(2x+3y-120)$.
$F_x=2xy+2\lambda,\ F_y=x^2+3\lambda,\ F_\lambda=2x+3y-120$
$F_x=0$ gives $\lambda=-xy$. $F_y=0$ and $F_x=0$ give $x^2-3xy=0$ so $x=0$ or $x=3y$.
$F_\lambda=0$: If $x=0$, then $y=40$.
If $x=3y$, then $y=\frac{40}{3}$ and $x=40$.
At $(0,40)$, we have $U=0\cdot 40=0$.
At $\left(40,\frac{40}{3}\right)$, we have $U=1600\cdot\frac{40}{3}=\frac{64,000}{3}$.
The maximum is at $\left(40,\frac{40}{3}\right)$.

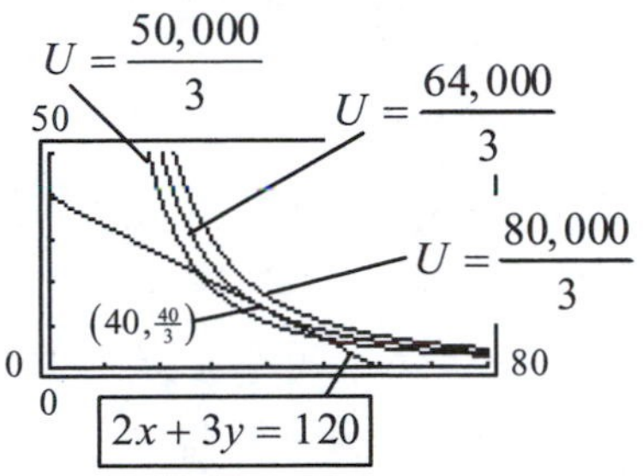


**19.** **a.** The objective function is
$F(x,y,\lambda)=400x^{0.6}y^{0.4}+\lambda(150x+100y-100,000)$
$F_x=\dfrac{240y^{0.4}}{x^{0.4}}+150\lambda,\ F_y=\dfrac{160x^{0.6}}{y^{0.6}}+100\lambda$
$F_\lambda=150x+100y-100,000$
$F_x=0$ and $F_y=0$ give $\dfrac{240y^{0.4}}{150x^{0.4}}=\dfrac{160x^{0.6}}{100y^{0.6}}$
or $x=y$.
$F_\lambda=0$: $250x-100,000=0$ or $x=400$.
$250y-100,000=0$ or $y=400$.

**b.** $-\lambda=\dfrac{240(400)^{0.4}}{150(400)^{0.4}}=\dfrac{8}{5}=1.6$
An additional 1.6 units are produced for each additional dollar spent on production.

**c.**

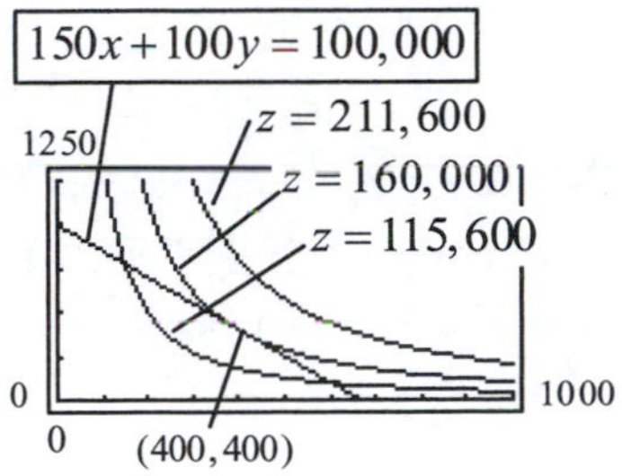


**21.** The objective function is
$F(x,y,\lambda)=x^2+1200+3y^2+800+\lambda(x+y-1200)$.
$F_x=2x+\lambda,\ F_y=6y+\lambda,\ F_\lambda=x+y-1200$
$F_x=F_y=0$ yields $x-3y=0$.
$F_\lambda=0$ yields $x+y=1200$.
Solving these 2 equations yields $(900,300)$.
To minimize cost let $x=900$ and $y=300$.

**23.** The objective function is
$F(x,y,\lambda)=20x+y^2+4xy+\lambda(x+y-30,000)$.
$F_x=20+4y+\lambda,\ F_y=2y+4x+\lambda,$
$F_\lambda=x+y-30,000$
$F_x=F_y=0$ yields $2x-y=10$.
$F_\lambda=0$ yields $x+y=30,000$.
Solving these 2 equations gives maximum revenue if the amount spent on print advertising $x=\$10,003.33$ and the amount spent on cable advertising $y=\$19,996.67$.

**25.** The objective function is
$F(x,h,\lambda)=x^2+4xh+\lambda(x^2h-500,000)$.
$F_x=2x+4h+2xh\lambda,\ F_h=4x+x^2\lambda$
$F_\lambda=x^2h-500,000$
$F_h=0$ means $x=0$ or $x\lambda=-4$. $x$ cannot be 0.
$F_x=0$ and $x\lambda=-4$ gives $2x+4h-8h=0$ or $x=2h$. $F_\lambda=0$ gives $4h^3=500,000$ or $h=50$. With $x=2h=100$, the dimensions are 100 by 100 by 50.

# Chapter 14: Functions of Two or More Variables

## *Chapter 14 Review Exercises*

**1.** $z = \dfrac{3}{2x - y}$

Domain = $\{(x, y): x, y \in \text{Reals and } y \neq 2x\}$.

**2.** $z = \dfrac{3x + 2\sqrt{y}}{x^2 + y^2}$

Domain: $\{(x, y): x$ and $y$ are real with $y \geq 0$ and $(x, y) \neq (0,0)\}$

**3.** $w(x, y, z) = x^2 - 3yz$

$w(2,3,1) = 4 - 3(3)(1) = -5$

**4.** $Q(K, L) = 70K^{2/3}L^{1/3}$

$Q(64{,}000,\ 512) = 70(1600)(8) = 896{,}000$

**5.** $z = 5x^3 + 6xy + y^2$

$z_x = 15x^2 + 6y$

**6.** $z = 12x^5 - 14x^3y^3 + 6y^4 - 1$

$\dfrac{\partial z}{\partial y} = -42x^3y^2 + 24y^3$

**7.** $z = 4x^2y^3 + \dfrac{x}{y}$

$z_x = 8xy^3 + \dfrac{1}{y},\quad z_y = 12x^2y^2 - \dfrac{x}{y^2}$

**8.** $z = \left(x^2 + 2y^2\right)^{1/2}$

$z_x = \dfrac{1}{2}\left(x^2 + 2y^2\right)^{-1/2}(2x) = \dfrac{x}{\sqrt{x^2 + 2y^2}}$

$z_y = \dfrac{1}{2}\left(x^2 + 2y^2\right)^{-1/2}(4y) = \dfrac{2y}{\sqrt{x^2 + 2y^2}}$

**9.** $z = (xy + 1)^{-2}$

$z_x = -2(xy + 1)^{-3}(y) = -\dfrac{2y}{(xy + 1)^3}$

$z_y = -2(xy + 1)^{-3}(x) = -\dfrac{2x}{(xy + 1)^3}$

**10.** $z = e^{x^2y^3}$

$z_x = e^{x^2y^3} \cdot 2xy^3 = 2xy^3e^{x^2y^3}$

$z_y = e^{x^2y^3} \cdot 3x^2y^2 = 3x^2y^2e^{x^2y^3}$

**11.** $z = e^{xy} + y\ln x$

$z_x = e^{xy}(y) + y \cdot \dfrac{1}{x} = ye^{xy} + \dfrac{y}{x}$

$z_y = e^{xy}(x) + \ln x = xe^{xy} + \ln x$

**12.** $z = e^{\ln xy} = xy$

$z_x = y,\ z_y = x$

**13.** $f(x, y) = 4x^3 - 5xy^2 + y^3$

$f_x = 12x^2 - 5y^2$

$f_x(1,2) = 12 - 20 = -8$

**14.** $z = 5x^4 - 3xy^2 + y^2$

$z_x = 20x^3 - 3y^2,\quad z_x(1,\ 2) = 20 - 12 = 8$

**15.** $z = x^2y - 3xy$

$z_x = 2xy - 3y,\quad z_y = x^2 - 3x$

**a.** $z_{xx} = 2y$

**b.** $z_{yy} = 0$

**c.** $z_{xy} = 2x - 3$

**d.** $z_{yx} = 2x - 3$

**16.** $z = 3x^3y^4 - \dfrac{x^2}{y^2}$

$z_x = 9x^2y^4 - \dfrac{2x}{y^2},\quad z_y = 12x^3y^3 + \dfrac{2x^2}{y^3}$

**a.** $z_{xx} = 18xy^4 - \dfrac{2}{y^2}$

**b.** $z_{yy} = 36x^3y^2 - \dfrac{6x^2}{y^4}$

**c.** $z_{xy} = 36x^2y^3 + \dfrac{4x}{y^3}$

**d.** $z_{yx} = 36x^2y^3 + \dfrac{4x}{y^3}$

**17.** $z = x^2e^{y^2}$

$z_x = 2xe^{y^2},\quad z_y = x^2e^{y^2} \cdot 2y$

**a.** $z_{xx} = 2e^{y^2}$

**b.** $z_{yy} = x^2e^{y^2} \cdot 2 + 2yx^2e^{y^2} \cdot 2y$

$= 2x^2e^{y^2} + 4x^2y^2e^{y^2}$

**c.** $z_{xy} = 2xe^{y^2} \cdot 2y = 4xye^{y^2}$

**d.** $z_{yx} = 4xye^{y^2}$

**18.** $z = \ln(xy+1)$

$z_x = \dfrac{y}{xy+1},\ z_y = \dfrac{x}{xy+1}$

**a.** $z_{xx} = \dfrac{-y^2}{(xy+1)^2}$

**b.** $z_{yy} = \dfrac{-x^2}{(xy+1)^2}$

**c.** $z_{xy} = \dfrac{(xy+1) - y \cdot x}{(xy+1)^2} = \dfrac{1}{(xy+1)^2}$

**d.** $z_{yx} = \dfrac{(xy+1) - x \cdot y}{(xy+1)^2} = \dfrac{1}{(xy+1)^2}$

**19.** $z = 16 - x^2 - xy - y^2 + 24y$

$z_x = -2x - y,\ z_y = -x - 2y + 24$

$z_x = 0$ gives $y = -2x$. Using $z_y = 0$ and $y = -2x$ yields $x = -8$ and thus, $y = 16$.

Critical point at $(-8,16)$.

$z_{xx} = -2,\ z_{yy} = -2,\ z_{xy} = -1$

$D = (-2)(-2) - (-1)^2 > 0$

Since $D > 0$ and $z_{xx} < 0$, there is a relative maximum of $z = 208$ at $(-8,16)$.

**20.** $z = x^3 + y^3 - 12x - 27y$

$z_x = 3x^2 - 12,\ z_y = 3y^2 - 27$

$z_x = 0$ gives $x = \pm 2$. Using $z_y = 0$ and $y = \pm 3$

The critical points are $(2,3)$ and $(-2,-3)$.

$z_{xx} = 6,\ z_{yy} = 6,\ z_{xy} = 0$

$D = (6)(6) - 0 = 36 > 0$. The point $(2,-3,38)$ and $(-2,3,-38)$ are the saddle points.

A relative maximum at $(2,3,-70)$. A relative minimum at $(-2,-3,70)$.

**21.** The objective function is

$F(x,y,\lambda) = 4x^2 + y^2 + \lambda(x+y-10)$

$F_x = 8x + \lambda,\ F_y = 2y + \lambda,\ F_\lambda = x + y - 10$

$F_x = F_y = 0$ gives $y = 4x$. $y = 4x$ and $F_\lambda = 0$ gives $5x - 10 = 0$ or $x = 2$. So, $y = 8$.

The minimum is $z = 4(4) + 64 = 80$ at $(2,8)$.

**22.** The objective function is

$F(x,y,\lambda) = x^4y^2 + \lambda(x+y-9)$

$F_x = 4x^3y^2 + \lambda,\ F_y = 2x^4y + \lambda,$

$F_\lambda = x + y - 9$

$F_x = F_y = 0$ gives $4x^3y^2 = 2x^4y$ or $2x^3y(2y - x) = 0$ so $x = 0$, $y = 0$, or $x = 2y$.

$F_\lambda = 0$: If $x = 0$, then $y = 9$. If $y = 0$, then $x = 9$. If $x = 2y$, then $y = 3$ and $x = 6$.

At $(0,9)$ $z = 0 \cdot 81 = 0$.

At $(9,0)$, $z = 6561 \cdot 0 = 0$.

At $(6,3)$, $z = 6^4 \cdot 3^2 = 11{,}664$.

The maximum value is 11,664 at $(6,3)$.

**23.** **a.** At $x = 6$ and $y = 15$, $U = 6^2 \cdot 15 = 540$.

**b.** To retain the same value of $U$ at $y = 60$, we have $540 = x^2 \cdot 60$ or $x^2 = 9$ or $x = 3$.

**24.** **a.** $A = f(100,6)$

$$= \frac{1200(100)\left[\left(1 + \frac{6}{1200}\right)^{240} - 1\right]}{6}$$

$\approx \$46{,}204.09$

**b.** A person who contributes \$250 per year at an interest rate of 7.8% will accumulate \$143,648 over 20 years.

**c.** A 1% change in the interest rate (from 7.8% to 8.8% will result in an extra \$17,770 over 20 years.

**d.** $$\frac{\partial A}{\partial R} = \frac{1200\left[\left(1 + \frac{r}{1200}\right)^{240} - 1\right]}{r}$$

$$\left.\frac{\partial A}{\partial R}\right|_{(250,7.8)} = \frac{1200\left[\left(1 + \frac{7.8}{1200}\right)^{240} - 1\right]}{7.8}$$

$\approx \$574.59$

This means that investing an extra \$1 per month for 20 years will result in an extra \$574.59 in accumulated value.

**25. a.** $B = f(1000,20) = \dfrac{3(1000)}{400 - 400e^{-0.0897(20)}}$

$\approx 8.996$ thousand, or \$8,996

**b.** $\dfrac{\partial B}{\partial V} = \dfrac{3}{400 - 400e^{-0.0897t}}$

$\left.\dfrac{\partial B}{\partial V}\right|_{(1000,20)} = \dfrac{3}{400 - 400e^{-0.0897(20)}}$

$\approx 0.009$

When the benefits are paid for 20 years, if the account value changes from 1000 to 1001 (thousand dollars), the monthly benefit increases by about \$9.

**c.** $B(V,t) = 3V\left(400 - 400e^{-0.0897t}\right)^{-1}$

$\dfrac{\partial B}{\partial t} = -3V\left(400 - 400e^{-0.0897t}\right)^{-1}\left(35.88e^{-0.0897t}\right)$

$\left.\dfrac{\partial B}{\partial t}\right|_{(1000,20)} \approx -0.161$

This means that when the account value is \$1,000,000, if the duration of benefits changes from 20 to 21 years, the monthly benefit decreases by about \$161.

**26. a.** Because raising the dollars spent for advertising, while holding the price constant, should increase the sales of the product.

**b.** Answers may vary. A higher price may cause less people to buy a product, or people may take the higher price to mean better quality (or style in the case of clothing brands) and buy more.

**27.** $C(x,y) = x^2\sqrt{y^2+13}$

**a.** $C_x = 2x\sqrt{y^2+13}$

$C_x(20,6) = 40\sqrt{36+13} = 280$

**b.** $C_y = \dfrac{x^2 y}{\sqrt{y^2+13}}$

$C_y(20,6) = \dfrac{400\cdot 6}{\sqrt{36+13}} = \dfrac{2400}{7}$

**28.** $Q = 80K^{1/4}L^{3/4}$, $K = 625$, $L = 4096$

$\dfrac{\partial Q}{\partial K} = \dfrac{20L^{3/4}}{K^{3/4}}$, $\dfrac{\partial Q}{\partial L} = \dfrac{60K^{1/4}}{L^{1/4}}$

For the given values $\dfrac{\partial Q}{\partial K} = 81.92$ and

$\dfrac{\partial Q}{\partial L} = 37.5$ (hundreds).

When work-hours are fixed at 4096, an increase of \$1000 in expenditures results in an increase of 8192 units produced. When expenditures are fixed at \$625,000 an increase of one work-hour results in an increase of 3750 units produced.

**29.** $q_A = 400 - 2p_A - 3p_B$

$q_B = 300 - 5p_A - 6p_B$

**a.** $\dfrac{\partial q_A}{\partial p_A} = -2$

**b.** $\dfrac{\partial q_B}{\partial p_B} = -6$

**c.** The two products are complementary since $\dfrac{\partial q_A}{\partial p_B}$ and $\dfrac{\partial q_B}{\partial p_A}$ are both negative.

**30.** $q_A = 800 - 40p_A - 2(p_B+1)^{-1}$

$q_B = 1000 - 10(p_A+4)^{-1} - 30p_B$

$\dfrac{\partial q_A}{\partial p_A} = -40$

$\dfrac{\partial q_A}{\partial p_B} = \dfrac{2}{(p_B+1)^2}$

$\dfrac{\partial q_B}{\partial p_A} = \dfrac{10}{(p_A+4)^2}$

$\dfrac{\partial q_B}{\partial p_B} = -30$

Competitive since $\dfrac{\partial q_A}{\partial p_B}$ and $\dfrac{\partial q_B}{\partial p_A}$ are both positive.

**31.** $P(x,y) = 40x + 80y - x^2 - y^2$

$P_x = 40 - 2x$, $P_y = 80 - 2y$

$P_x = 0$ when $x = 20$ and $P_y = 0$ when $y = 40$.

Critical point at $(20,40)$.

$P_{xx} = -2$, $P_{yy} = -2$, $P_{xy} = 0$

$D = (-2)(-2) - (0)^2 > 0$ with $P_{xx} < 0$ means profit is maximized when $x = 20$ and $y = 40$.

$P = \$2000$

**32.** $C(x,y) = 22{,}500 - 12x - 30y + 0.03x^2 + 0.01y^2$

$C_x = -12 + 0.06x$

$C_y = -30 + 0.02y$

Setting these two equations equal to zero and solving when $x, y \ge 0$ gives

$$\begin{aligned} -12 + 0.06x &= 0 & -30 + 0.02y &= 0 \\ 0.06x &= 12 & 0.02y &= 30 \\ x &= 200 & y &= 1500 \end{aligned}$$

$C_{xx} = 0.06,\ C_{yy} = 0.02,\ C_{xy} = 0$

$D = (0.06)(0.02) - (0) > 0$ with $C_{xx} > 0$, which means that $(200, 1500)$ is a relative minimum. Therefore, plant I should produce 200 units and plant II should produce 1500 units to minimize costs.

**33.** The objective function is $F(x, y, \lambda) = x^2y + \lambda(4x + 5y - 60)$

$F_x = 2xy + 4\lambda,\ F_y = x^2 + 5\lambda$

$F_\lambda = 4x + 5y - 60$

$F_x = 0$ and $F_y = 0$ give $\dfrac{-xy}{2} = \dfrac{-x^2}{5}$ or $5xy - 2x^2 = 0$ or $x(5y - 2x) = 0$. Since $x$ cannot be 0, we have

$5y - 2x = 0$ or $y = \dfrac{2}{5}x$. So, $4x + 5 \cdot \frac{2}{5}x - 60 = 0$ or $x = 10$. Then $y = \frac{2}{5} \cdot 10 = 4$.

**34. a.** The objective function is $F(x, y, \lambda) = 300x^{2/3}y^{1/3} + \lambda(50x + 50y - 75{,}000)$

$F_x = \dfrac{200y^{1/3}}{x^{1/3}},\ F_y = \dfrac{100x^{2/3}}{y^{2/3}}$

$\dfrac{\partial F}{\partial \lambda} = 50x + 50y - 75{,}000$

$F_x = F_y = 0$ gives $\dfrac{4y^{1/3}}{x^{1/3}} = \dfrac{2x^{2/3}}{y^{2/3}}$ or $x = 2y$.

$F_\lambda = 0$ and $x = 2y$ give $150y - 75{,}000 = 0$ or $y = 500$, so $x = 1000$.

Production is maximized at $x = 1000,\ y = 500$.

**b.** $-\lambda = \dfrac{4y^{1/3}}{x^{1/3}} = \dfrac{4\sqrt[3]{500}}{\sqrt[3]{1000}} \approx 3.17$ means each additional dollar spent on production results in an additional 3.17 units produced.

**c.**

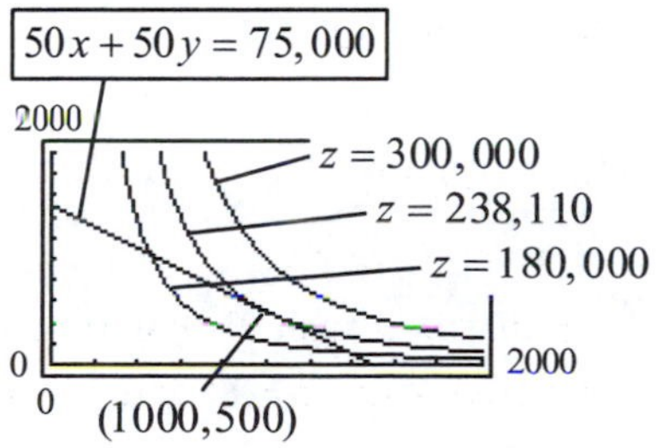

**35.** **a.** Use a graphing utility to obtain
$\hat{y} = 2375x + 39630$

**b.** If $x = 15$, $\hat{y} = \$75,255$.

**36.** **a.** Use a graphing utility to obtain
$\hat{y} = 0.736x + 28.4$

**b.** The slope $m = 0.736$ means that for every 1000 women entering the workforce, there are about 736 men entering it.

## *Chapter 14 Test*

**1.** $f(x, y) = \dfrac{2x + 3y}{\sqrt{x^2 - y}}$

**a.** Domain =
$\{(x, y) : x \text{ and } y \text{ are real and } y < x^2\}$

**b.** $f(-4, 12) = \dfrac{-8 + 36}{2} = 14$

**2.** $z = 5x - 9y^2 + 2(xy + 1)^5$

$z_x = 5 + 10(xy + 1)^4(y) = 5 + 10y(xy + 1)^4$

$z_y = -18y + 10(xy + 1)^4(x)$

$= -18y + 10x(xy + 1)^4$

$z_{xx} = 40y(xy + 1)^3(y) = 40y^2(xy + 1)^3$

$z_{yy} = -18 + 40x(xy + 1)^3(x)$

$= -18 + 40x^2(xy + 1)^3$

$z_{xy} = z_{yx} = 10y \cdot 4(xy + 1)^3(x) + (xy + 1)^4(10)$

$= 10(xy + 1)^3(5xy + 1)$

**3.** $z = 6x^2 + x^2y + y^2 - 4y + 9$

$z_x = 12x + 2xy = 2x(6 + y)$

$z_y = x^2 + 2y - 4$

$z_x = 0$ if $x = 0$ or $y = -6$

If $x = 0$ and $z_y = 0$, then $y = 2$. If $y = -6$ and $z_y = 0$, then $x^2 = 16$ or $x = \pm 4$.

Critical points are $(0, 2)$, $(4, -6)$, and $(-4, -6)$.

$z_{xx} = 12 + 2y,\ z_{yy} = 2,\ z_{xy} = 2x$

$D = z_{xx} \cdot z_{yy} - (z_{xy})^2$

At $(0, 2)$: $D = 16 \cdot 2 - 0 > 0$

At $(4, -6)$: $D = 0 \cdot 2 - 64 < 0$

At $(-4, -6)$: $D = 0 \cdot 2 - 64 < 0$

$D > 0$ and $z_{xx} > 0$ yield a relative minimum at $(0, 2)$. $D < 0$ yields saddle points at $(4, -6)$ and $(-4, -6)$.

**4.** $Q = 10K^{0.45}L^{0.55},\ K = 10,\ L = 1590$

**a.** $Q = 10(10)^{0.45}(1590)^{0.55} = \$1625$ thousands

**b.** $Q_K = \dfrac{4.5L^{0.55}}{K^{0.55}},\ Q_K = \dfrac{4.5(1590)^{0.55}}{10^{0.55}} = 73.11$

If labor hours remain constant and capital investment increases to \$11,000, the monthly production should increase \$73.11 thousands.

**c.** $Q_L = \dfrac{5.5K^{0.45}}{L^{0.45}},\ Q_L = \dfrac{5.5(10)^{0.45}}{(1590)^{0.45}} = 0.56$

If capital investment remains constant and labor hours increase to 1591, the monthly production should increase \$0.56 thousands.

**5.** **a.** $f(94.5, 25, 7) = \$667.91$

When \$94,500 is borrowed for 25 years at 7%, compounded monthly, the monthly payment is \$667.91.

**b.** $\frac{\partial f}{\partial r}(94.5, 25, 7) = \$60.28$

If the rate increases from 7% to 8%, the monthly payment will increase \$60.28. (The \$94,500 and 25 years remain fixed.)

**c.** $\frac{\partial f}{\partial n}(94.5, 25, 7)$ would be negative. If the loan amount remains at \$94,500 and the percent remains at 7%, increasing the time to pay off the loan will decrease the monthly payment, and vice versa.

**6.** $f(x,y) = 2e^{x^2y^2}$

$\frac{\partial f}{\partial y} = 2e^{x^2y^2}\left(2x^2y\right) = 4x^2ye^{x^2y^2}$

$\frac{\partial^2 f}{\partial x \partial y} = 4x^2y\left(e^{x^2y^2} \cdot 2xy^2\right) + e^{x^2y^2}(8xy)$

$= 8xye^{x^2y^2}\left(1 + x^2y^2\right)$

**7.** Find $\frac{\partial q_1}{\partial p_2}$ and $\frac{\partial q_2}{\partial p_1}$ and compare their signs. If both are positive, this means the products are competitive. If both are negative, this means the products are complementary.

$\frac{\partial q_1}{\partial p_2} = -5, \quad \frac{\partial q_2}{\partial p_1} = -4$

These products are complementary.

**8.** $P = 915x - 30x^2 - 45xy + 975y - 30y^2 - 3500$

$P_x = 915 - 60x - 45y = 15(61 - 4x - 3y)$

$P_y = -45x + 975 - 60y = 15(65 - 3x - 4y)$

Set $P_x$ and $P_y$ to 0 and solve.

$3x + 4y = 65$

$4x + 3y = 61$

The solution is $x = 7$ and $y = 11$.

$P_{xx} = -60, \quad P_{yy} = -60, \quad P_{xy} = -45$

$D = (-60)(-60) - (-45)^2 > 0$ and $P_{xx} < 0$ means that $P$ is a maximum when $x = \$7, \quad y = \$11$, and $P = \$5065$.

**9.** $U = x^3y$ with budget constraint $30x + 20y = 8000$.

The objective function is $F(x, y, \lambda) = x^3y + \lambda(30x + 20y - 8000)$

$\frac{\partial F}{\partial x} = 3x^2y + 30\lambda, \quad \frac{\partial F}{\partial y} = x^3 + 20\lambda$

$\frac{\partial F}{\partial \lambda} = 30x + 20y - 8000$

Set each partial to zero and solve for $\lambda$.

$\lambda = -\frac{x^2y}{10}$ and $\lambda = -\frac{x^3}{20}$ yield $2x^2y = x^3$ or $x^2(x - 2y) = 0$, $x \neq 0$. Substituting $x = 2y$ into $\frac{\partial F}{\partial \lambda}$ gives $80y - 8000 = 0$ or $y = 100$. Then $x = 200$. The function is maximized at 800,000,000 when $x = 200$ and $y = 100$.

**10.** **a.** $\hat{y} = 0.24x + 5.78$

**b.** The fit is excellent.

**c.** 16.58%